Developments in Plant and Soil Sciences

VOLUME 89

The titles published in this series are listed at the end of this volume.

Actin: A Dynamic Framework for Multiple Plant Cell Functions

Edited by

C. J. STAIGER

F. BALUŠKA

D. VOLKMANN

and

P.W. BARLOW

KLUWER ACADEMIC PUBLISHERS

DORDRECHT / BOSTON / LONDON

A C.I.P. Catalogue record for this book is available from the Library of Congress.

ISBN 0-7923-6412-0

Published by Kluwer Academic Publishers,
P.O. Box 17, 3300 AA Dordrecht, The Netherlands.

Sold and distributed in North, Central and South America
by Kluwer Academic Publishers,
101 Philip Drive, Norwell, MA 02061, U.S.A.

In all other countries, sold and distributed
by Kluwer Academic Publishers,
P.O. Box 322, 3300 AH Dordrecht, The Netherlands.

Printed on acid-free paper

Printed in the Netherlands.

Contents

Acknowledgements

The editors wish to extend their thanks to Martin Hauskrecht for his invaluable assistance in compiling the individual chapters into a completed volume, to Heidi Geithmann for her kind help in producing the cover, and to Thorsten Ott for technical help during the process of proof reading. DV and FB gratefully acknowledge the support of AGRAVIS (Bonn) by Deutsche Agentur für Raumfahrtangelegenheiten (Bonn) and Ministerium für Wissenschaft und Forschung (MWF, Düsseldorf). CJS acknowledges financial support from the US Department of Agriculture-NRICGP, the Department of Energy-Energy Biosciences Division, and the Showalter Fund of the Purdue Research Foundation. PB wishes to thank the BBSRC for its support.

Preface

The phenotype of any organism is defined as the result of co-ordinated reciprocal interactions between the genotype, different cell structures and the environment. In addition to nuclear and organellar DNA, there are several information-bearing cell structures which essentially contribute to growth, development, and morphogenesis of eukaryotic organisms. These structures, which comprise nuclear matrix, diverse cytoskeletal elements, membranes and extracellular matrix; are assembled into a structural and informational continuum that extends between the nucleus to the cell periphery [1]. All of the information hosted by these structures is handed over to the two daughter cells during cell division and is, moreover, itself essential for the very accomplishment of mitosis and cytokinesis. The cell body concept of the late Daniel Mazia [2], which has now been extended to plants, takes into account the information provided by this information-bearing structural continuum and has been shown to be of vital importance for plant growth, polarity, and morphogenesis. This book focuses on one component of this structural continuum, the actin cytoskeleton, and highlights the use of plant actin filaments as a dynamic framework for multiple plant cellular functions. During the 1970s, the evidence for an actin cytoskeleton was successfully introduced to the plant cell biology and it was considered mainly as an intracellular scaffold and transport device that supports motilities of membraneous structures and organelles. In the meantime, this mechanistic view of the actin cytoskeleton has been extended to touch upon its roles in signaling, in responses to extrinsic and intrinsic stimuli, and in cell-to-cell communication. Within the context of the extreme variety of isoforms of plant actin and actin-binding proteins, it appears that the dynamic plant actin cytoskeleton serves as an essential buffer [3] for a wide range of complex biological phenomena, including metabolic pathways and signal transduction chains. These exciting new developments are highlighted in this volume.

1. Ingber, FASEB J 13 (Suppl): S3-S15, 1999
2. Epel & Schatten, Trends Cell Biol 8: 416-418, 1998
3. Meagher et al., Plant Cell 11: 995-1005, 1999

Chapter 1

THE SIGNIFICANCE OF DIVERSITY IN THE PLANT ACTIN GENE FAMILY

Studies in Arabidopsis

Richard B Meagher, Elizabeth C McKinney and Muthugapatti K Kandasamy

Genetics Department, Life Science Building, University of Georgia, Athens GA 30602 USA

Key words: actin–chloroplast interactions, cytoskeleton, evolution, gene expression, isovariants, monoclonal antibodies, multigenerational studies, mutants, selective constraint

Abstract: Recent progress in understanding the plant actin gene family is reviewed, focusing on the *Arabidopsis* actins. Taking an evolutionary perspective, we have focused on the functional significance of the conserved but ancient vegetative and reproductive actin classes, which date back to the origin of vascular plants. We propose that the conservation of ancient family members is due to differential gene regulation and/or to functional differences among isovariants. The eight functional actin genes are widely dispersed on four of the five *Arabidopsis* chromosomes. Each of the five actin gene subclasses are strongly expressed at some time and place during plant development, and they are highly differentially regulated. A handful of surface epitope differences among plant and vertebrate actins enabled the isolation of general and subclass-specific anti-plant actin monoclonal antisera. These reagents give an excellent resolution to the switch from vegetative to reproductive actin protein expression during floral development. Combined with refined fixation protocols, these reagents resolve the intimate relationship between the chloroplasts and the actin cytoskeleton in leaf cells. Sequence-based screening procedures were developed for the isolation of the first mutant alleles of plant actins. These mutants have strong deleterious effects on the survival of plants and are effectively lethal mutations over several generations. Sequence differences among the co-expressed plant actin isovariants should produce complex dynamics within actin filaments and with actin-binding proteins. Future work on the significance of this ancient family will focus on the cell biology, genetics, and biochemistry of the isovariants.

C.J. Staiger et al. (eds.), Actin: A Dynamic Framework for Multiple Plant Cell Functions, 3–27.

1. INTRODUCTION

The plant actin-based cytoskeleton continues to fascinate biologists because of the myriad processes in which it is known or thought to participate. These processes include programming cell polarity, cell division plane determination, cell elongation, cell wall development, transport of vesicles and macromolecules, cytoplasmic streaming, positioning membrane receptors, and orientation and movement of organelles (Fowler & Quatrano, 1997; Meagher et al., 1999a, 1999b; Nick, 1999; Staehelin & Hepler, 1996; Staiger, 2000; Williamson, 1993). The first several of these functions are essential for directing plant development. Progress in dissecting plant actin functions has been complicated by the existence of an ancient and diverse gene family; the lack of suitable reagents for distinguishing individual genes, transcripts, and proteins; the difficulty in preserving plant microfilaments for microscopy; the lack of a simple genetic system with actin mutants; and the lack of an *in vitro* system in which to study the biochemical properties of purified plant actins. In the past five years tremendous progress has been made in understanding the plant actin cytoskeleton by focusing on the actin family in the model plant *Arabidopsis*.

The significance of actin gene and protein diversity will be addressed in the following sections. This includes a discussion of the age, origin, and relationships within the plant actin family, suggesting it arose early in land plant evolution, when the first leaves evolved from sterilized reproductive structures. The spatial and temporal expression patterns of the eight functional *Arabidopsis* actin genes revealed that the vegetative and reproductive actin classes and the five subclasses have distinct regulation. Surface epitope differences among plant and vertebrate actin isovariants may be functionally significant and allowed the isolation of strong anti-actin monoclonal antibodies. These reagents were used 1) to explore the expression of the vegetative and reproductive actin protein classes as they undergo a developmental switch during flower development, 2) to distinguish late pollen-specific actins from vegetative actins, and 3) to examine the association between chloroplasts and the actin cytoskeletal network.

The last three sections present research on the genetics of the *Arabidopsis* actins. Mapping members of the *Arabidopsis* family revealed that they are widely dispersed in the genome. Intimate knowledge of all plant actin gene sequences allowed sequence-based screening to isolate actin mutants without knowledge of phenotype. Future analysis of mutant phenotypes is supported by population studies demonstrating that the loss of the function of each actin gene is deleterious.

2. ELUCIDATING EVOLUTIONARY RELATIONSHIPS WITHIN THE PLANT ACTIN FAMILY

Initial analysis of soybean, tomato, potato, petunia, maize, and rice actin gene families revealed unexpected diversity among the encoded plant actin proteins. Based on an analysis of this diversity, we hypothesized that the ancient and divergent classes of plant actin genes have been preserved throughout vascular plant evolution because they have unique patterns of regulation and encode proteins with distinct functions (Hightower & Meagher, 1986; Meagher et al., 1999a, 1999b). We used *Arabidopsis* to explore this proposal (McDowell et al., 1996b) because its genome is much less complex than other plant species and it is easy to manipulate genetically (Baird & Meagher, 1987). The *Arabidopsis* actin gene family has only eight compact functional genes and two pseudogenes.

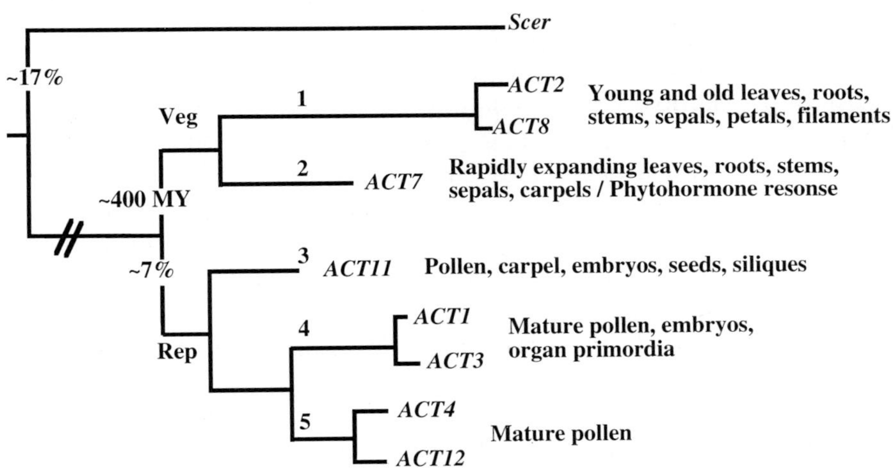

Figure 1. A gene tree resolves the eight functional *Arabidopsis* actins in two ancient classes, vegetative (Veg) and reproductive (Rep). The tree is based on Replacement Nucleotide Substitutions (RNS) that alter the amino acid sequence (*ACT2* vs *ACT1* = 6.8% RNS). Based on sequence relatedness and their distinct and dominant expression patterns the actins can be further resolved into five subclasses, as indicated. Isovariant subclasses 1 and 2 (marked on branches) encode vegetative class actins, while isovariant subclasses 3, 4, and 5 encode reproductive actins. The two plant actin classes which diverged from a common ancestral sequence about 400 million years ago differ by about 7% in amino acid sequence. All the plant actins differ from yeast actin by about 17% in amino acid sequence.

The components of our hypothesis were studied with quantitative evolutionary measurements, which demonstrated that genes in the actin family are ancient and divergent (McDowell et al., 1996b), as shown in Figure 1. The most ancient split separated vegetative and reproductive actins from a common ancestral sequence ~400 million years ago (MY), shortly after plants diverged from a green algal ancestor (Meagher et al., 1999b). These two subclasses differ by ~7% in amino acid sequence, slightly more than the amount that separates cytoplasmic actins from muscle actins in vertebrates (Meagher et al., 1999a). The two classes can be subdivided further into five ancient subclasses (1–5 in Fig. 1) that each differ from one another by more than 2.5% in amino acid sequence. The members of the most closely related pairs of genes within a subclass (e.g., *ACT2 vs 8; 1 vs 3, 4 vs 12*) encode proteins that differ by only a single amino acid change. However, analysis of silent nucleotide substitutions within codons revealed that even members of these pairs are highly divergent and have not shared common ancestry for more than 30–50 MY (McDowell et al., 1996b). Because mutations could easily have randomized DNA and encoded protein sequences during this time period, the gene for each protein isovariant must be under strong selective constraint.

Parallel studies demonstrated the potential existence of vegetative and reproductive classes of plant profilins (Huang et al., 1996b), actin depolymerizing factors, α-tubulins and ß-tubulins (Meagher et al., 1999b). The existence of separate vegetative and reproductive subclasses of cytoskeletal gene families suggests that hundreds of cytoskeletal genes may have evolved in conjunction with the actins as the first leaves evolved from sterile reproductive structures in early land plants (Meagher et al., 1999b). One possible explanation for the existence of this split of cytoskeletal gene families would be that an ancient genomic duplication and subsequent gene divergence accompanied the duplication and divergence of these organ structures.

The realization that two or more actin isovariants are commonly expressed in the same cell led us to consider other selective constraints that might be acting to preserve multiple diverse actin isovariants. For example, the co-expression of multiple actin isovariants in the same cell should result in "isovariant dynamics" that allow for more complex cytoskeletal responses. We define isovariant dynamics as the temporal and biochemical expansion and buffering of responses of a biological system resulting from the simultaneous expression and interaction of multiple isovariants of a protein (Meagher et al., 1999a). Plant actin isovariant dynamics, could for example, effect filament quality and length by altering interactions with actin-binding proteins (ABPs) and heteropolymer formation. This, in turn, could result in changes in the rate and duration of organellar subcellular movement, cell elongation, or cell division; alterations in cellular viscoelasticity; broadening the cytoskeletal response to external signals; and buffering the plant cytoskeleton to various stresses. For example, if the rate and affinity constants for binding one particular profilin are different for each actin isovariant expressed during pollen tube growth, then isovariant dynamics will play a critical role in regulating this process.

3. DISCOVERING THE SPATIAL AND TEMPORAL EXPRESSION PATTERNS OF THE EIGHT EXPRESSED *ARABIDOPSIS* ACTIN GENES

The complex expression patterns of the eight functional *Arabidopsis* actin genes were determined based on steady-state RNA levels and the expression of actin 5' region translational fusions to a reporter in transgenic plants (An et al., 1996a, 1996b; Huang et al., 1996a, 1997; McDowell et al., 1996a, 1996b). Figure 2 shows total *Arabidopsis* mRNA from a variety of organs and pollen resolved by agarose-gel electrophoresis, imprinted to a nylon membrane, and hybridized with gene-specific probes for a vegetative class actin, *ACT7*, and for a reproductive actin, *ACT1*. It is clear that *ACT7* mRNA is highly expressed in roots, stems, leaves, the floral organ complex, and siliques, but is not detected in pollen. *ACT1* mRNA expression complements this pattern, with weak expression in all organs, but exceptionally strong expression in pollen.

Confirmation and further quantification of these data came from an independent assessment of steady-state mRNA levels using reverse transcriptase mediated (RT) polymerase chain reaction (PCR) and gene-specific probes. The RT-PCR assays were quantified by controlling the input levels of cDNA for each sample in a two fold dilution series. Variation in steady-state mRNA levels among the different organs and pollen range from only eight-fold for *ACT2* (An et al., 1996b) to 5000-fold for *ACT1* (An et al., 1996a). *ACT1* mRNA is 1000-fold higher in pollen than in leaf or stem, while *ACT7* mRNA is 100-fold higher in leaf, stem, and flower than in pollen (McDowell et al., 1996a). Among the ten *Arabidopsis* actins examined, only the two pseudogenes, *ACT5* and *ACT9*, do not appear to express detectable levels of RNA in these organs or pollen.

It seemed likely that the majority of actin gene transcriptional regulation, and possibly some aspects of translational regulation, would be controlled by the 5' end of the sequence. Thus, translational fusions to a ß-glucuronidase (GUS) reporter were made for all eight, expressed actin genes.

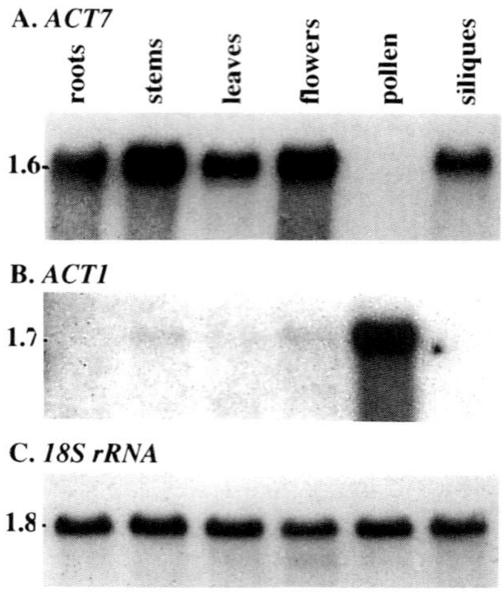

Figure 2. The expression of the various actin gene encoded mRNAs can be divided into vegetative and reproductive patterns. Representative mRNA expression patterns are shown for *ACT7* (veg) and *ACT1* (rep). **A.** Northern analysis of *ACT7* mRNA expression. Total RNA from roots, stems, leaves, flowers, pollen, and siliques was resolved in agarose gels, bloted to nylon membrane, and probed with the 3' UTR from *ACT7* mRNA. **B.** Northern analysis of *ACT1* mRNA expression. **C.** Northern analysis of 18S rRNA expression on the same blot as shown in A to demonstrate equal loading and transfer of total RNA among samples.

The actin/GUS reporter patterns articulated from over 20 independent fusion constructs were examined at three stages of development (seedlings, adolescent plants, floral tissues in adults) in more than 200 transgenic plants. Results for the five ancient subclasses of actin (*ACT2/8, ACT7, ACT11, ACT1/3,* and *ACT4/12*) are summarized in Figure 1 and they are relatively consistent within each subclass. The eight functional genes are all strongly expressed at some time and place during plant development. The expression patterns were consistent in 90% of the ten independent transgenic plant lines examined for each construct, suggesting that genomic position of the actin/reporter transgene had little impact on the quality of actin gene regulation. There is an excellent correlation between mRNA abundance and GUS activity in most organs (with a few notable exceptions) (An et al., 1996a), suggesting that expression at the organ level is strongly influenced by transcriptional and/or translational regulation encoded within the 5' end of each gene. Examples of the distinct temporal and spatial patterns of expression of *ACT7/GUS* and *ACT1/GUS* constructs are shown in Figures 3A–D.

3.1 *ACT7* (subclass 2) is expressed in young tissues and responds to phytohormones

The *ACT7/GUS* fusion was strongly and preferentially expressed in all young, developing vegetative tissues of transgenic *Arabidopsis* plants with a pattern similar to and in some cases stronger than *ACT2/GUS* in young tissues (McDowell et al., 1996a). However, unlike *ACT2/GUS* expression. which persists in tissues that are as much as 4 to 6 weeks old, *ACT7/GUS* expression did not persist in older tissues and is usually undetectable in fully expanded tissues that are more than 3 to 7 days old. *ACT7* was the only *Arabidopsis* actin gene strongly expressed in the expanding hypocotyl and seed coat, and as such it complements the lack of expression of *ACT2* in these organs. *ACT7/GUS* is the primary actin responding to most phytohormones (e.g., auxins, cytokinins, abscisic acid). The *ACT7* promoter region contains a large number of corresponding phytohormone response elements (McDowell et al., 1996a).

3.2 *ACT1* and *ACT3* (subclass 4) expression predominates in the embryo sac, pollen and organ primordia

Both *ACT1/GUS* and *ACT3/GUS* constructs were preferentially and strongly expressed in pollen, pollen tubes, and in all organ primordia, including those in roots, shoots, and the inflorescence (An et al., 1996a). Moderate expression was also detected throughout the developing embryo sac, being more strongly stained at the micropylar end. This pattern overlaps briefly in time with *ACT11/GUS*. *ACT1/GUS* expression does not continue throughout seed and silique development as does *ACT11/GUS*. While *ACT4* and *ACT12* have the same late pollen expression pattern as *ACT1* and *ACT3*, they appear to be less strongly expressed. *ACT1* is more highly expressed in pollen than any other actin gene, including *ACT3*.

4. ANALYZING ACTIN PROTEIN ISOVARIANTS

In order to make significant progress examining the differential expression of actin isovariants in organs, cells, and various cytoskeletal structures, we needed to be able to distinguish one actin protein subclass from another. Although monoclonal antibodies (MAbs) are powerful tools for characterizing specific protein isovariants, few such reagents exist even for animal actins, and no MAbs have been described that distinguish different plant actin isovariants. In our experience, most of the available actin antibodies are relatively weak against plant actins and need to be used at high protein concentrations. The major reason for the lack of high quality immune reagents is that actins are well-conserved among all eukaryotes and are notoriously poor immunogens.

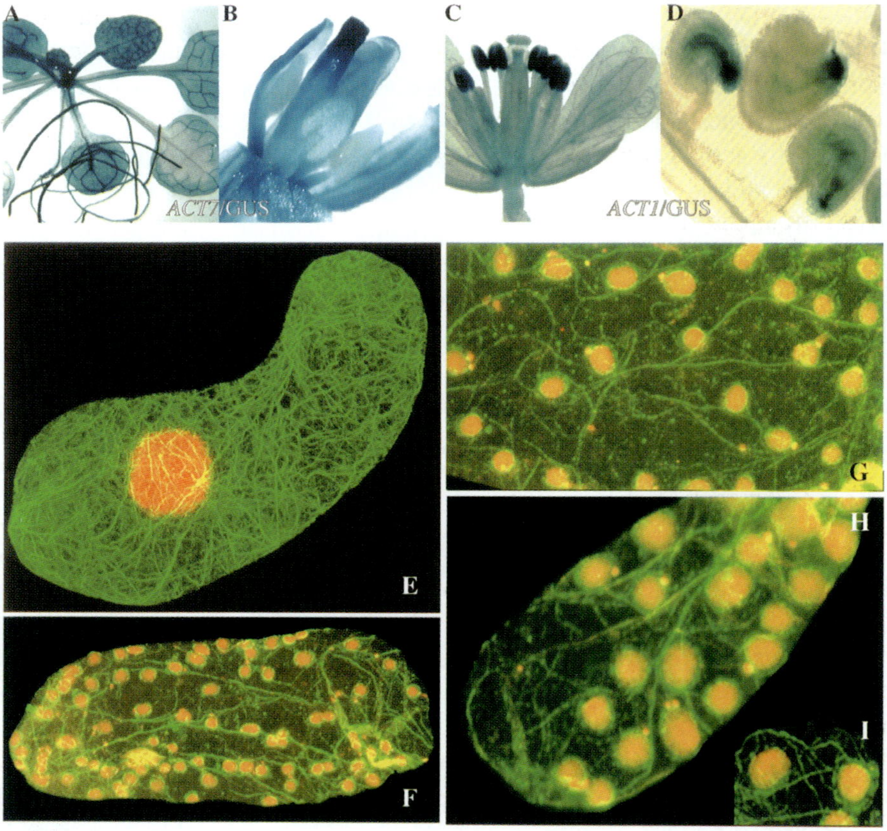

Figure 3. Expression pattern of actin gene-reporter fusions and staining of actin protein in leaf cells. **A–D**. Actin 5' regions of one vegetative actin and one reproductive actin drive reporter expression in transgenic *Arabidopsis*. **A & B**. The expression of the vegetative class *ACT7/GUS* fusion is restricted primarily to young leaves, stems, roots, sepals, stigma, and style. It is not expressed in most stages of pollen or ovule development. **C & D**. The expression of the reproductive class *ACT1/GUS* fusion is restricted primarily to pollen and ovules. **E–I**. Immunofluorescence staining of actin and visualization of actin-chloroplast association in *Arabidopsis* leaf cells. **E**. A chemically-fixed suspension cell has a complex, three-dimensional array of actin bundles and filaments. DAPI stained nucleus is indicated in yellow-orange. **F–I**. Cryofixed leaf cells have chloroplasts (yellow-orange) that are attached directly to cortical actin bundles encircling the cell cytoplasm or indirectly to these bundles through thin actin filaments. Each chloroplast is partially or completely contained within a basket of actin filaments. Actin filaments and bundles are shown in green (E–H).

The task of making actin antibodies is made worse by the fact that of the 12–15% amino acid differences that exist between plant and vertebrate actins most are conservative amino acid changes that are probably not immunogenic. However, on the surface of the 3D structure of plant actins we mapped more than a dozen non-conservative amino acid substitutions (charged residue and backbone changes) that distinguish one *Arabidopsis* actin isovariant from another (McDowell et al., 1996b). This is unlike the case for vertebrate actin families, where nearly all the amino acid changes are conservative. Even more importantly, those non-conservative differences that distinguish plant actin subclasses are a large subset (33%) of all the non-conservative amino acid differences between plant and animal actins. For at least one plant actin isovariant, the distinguishing nonconservative residue differs from that found in all animal actins. Thus, in spite of their overall high level of conservation, these diverged plant isovariants have potential immuno-epitopes for making class- or subclass-specific MAbs in mice.

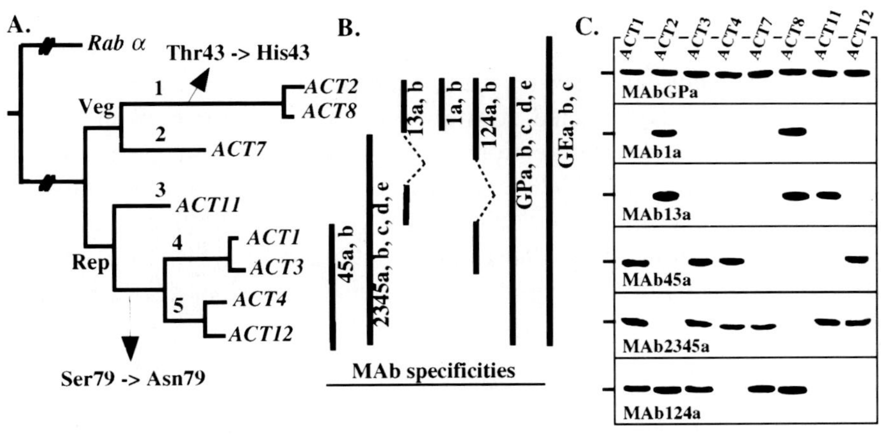

Figure 4. Isovariant specificity of MAbs to the eight expressed *Arabidopsis* actins. **A.** *Arabidopsis* actin gene tree is shown for comparison (Fig. 1). Nonconservative changes at Thr43 and Asn79 are required for the epitopes recognized by MAb2345 and MAb45a, respectively. **B.** The isovariant specificity of each anti-actin MAb is indicated relative to the gene tree in Fig. 1A. The vertical line indicates which actins in the gene tree react with each category of MAb. For example, MAb45a and MAb45b are Monoclonal Antibodies derived from two different mice (a and b) that react with two reproductive actin subclasses 4 and 5. MAbGPa reacts with all five subclasses of plant actin and has General Plant specificity. MAbGEa reacts with all *Arabidopsis* actins as well as rabbit muscle actin (Rab α) and thus has General Plant and Eukaryotic actin specificity. **C.** Differential reactivity of six categories of anti-actin MAbs with the eight *Arabidopsis* actins on Western blots. A secondary goat anti-mouse antibody conjugated to alkaline phosphatase is used to detect each bound MAb. All lanes were loaded with an equal amount of total actin protein (2 µg) made in *E. coli*. The 45-kD actin band is indicated by a line on the right side of each Western blot.

Based on this potential immunogenicity, we undertook the challenging task of producing specific MAbs against the *Arabidopsis* actin proteins. We expressed the eight different *Arabidopsis* actin isovariants in *E. coli*, purified the recombinant actin proteins, renatured them to soluble monomeric form, and used them to immunize more than 60 mice (five or more mice for each isovariant). The mice were boosted extensively (2 to 4 times) and, based on titers in whole serum, only the spleens of mice with the highest anti-actin titers were used to make hybridoma fusion lines. Hybridomas producing anti-actin antibodies were still rare, and fewer than 1 in 1000 lines produced an actin MAb that reacted strongly on ELISAs. This is much less frequent than we find MAbs for most proteins. From numerous different fusions, we isolated 26 general and 12 subclass-specific anti-actin MAbs with the specificities summarized in Figure 4B and shown relative to the gene tree for actins in Figure 4A.

Western blots presenting the specificity of a few representative MAbs are shown in Figure 4C. Thirty percent of these MAbs (12/38) distinguish different plant actin subclasses based on ELISAs and Western blots, fitting our calculation for the fraction of possible isovariant specific immuno-epitopes. This made the approach quite fruitful. Ninety percent of the MAbs examined so far have reacted strongly with actins both in plant extracts (Western blots) and in tissue sections. Three of the isovariant specific MAbs, MAb45a, MAb13a and MAb2345a, have been characterized thoroughly and we know the precise amino acid change that created the epitope (Kandasamy et al., 1999; and unpublished data). Combinations of these MAbs were used to clarify problems of differential actin protein expression, discussed in the following sections.

5. EXPRESSION OF THE VEGETATIVE AND REPRODUCTIVE ACTIN PROTEIN CLASSES UNDERGOES A DEVELOPMENTAL SWITCH DURING FLOWER DEVELOPMENT

Reproductive actin proteins are generally induced in tobacco during anther development, whereas vegetative actin synthesis declines, as shown in Figure 5. Although labeling with MAbGPa, a general plant actin specific antibody, revealed only a marginal increase in the level of total actins, the actin subclass-specific antibodies MAb13a and MAb45a clearly revealed a dramatic switch in the expression of actins from completely vegetative to predominantly reproductive during late stages of pollen maturation. Such a developmental switch in the expression of actin isovariants has also been shown in *Arabidopsis* (Kandasamy et al., 1999) and thus may occur in many angiosperms. The pollen-specific reproductive actins may be specifically required in the mature pollen for its dramatically polarized tip growth, rapid streaming and vesicle transport, movement of the sperm cells and vegetative nucleus through the pollen tube, and other unique pollen functions.

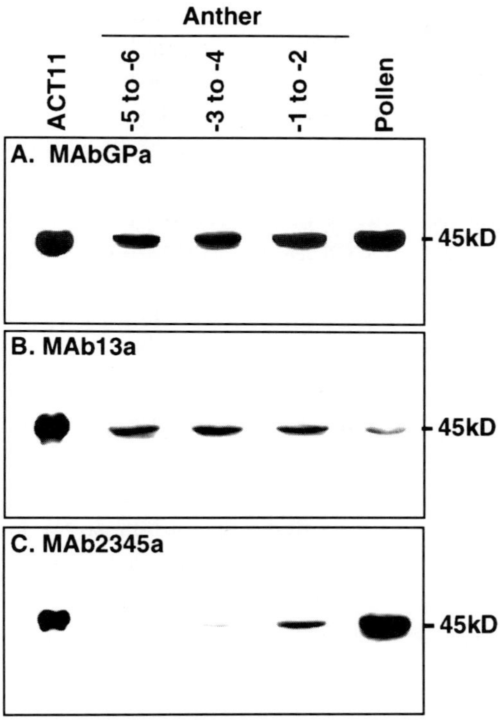

Figure 5. Subclass-specific MAbs demonstrate the induction of late pollen-specific actin proteins with the coordinate loss of vegetative actin proteins during tobacco anther development. **A**. MAbGPa is a general a plant actin antibody demonstrating that total actin increases only slightly during anther development. **B**. A slight drop in the combined expression of ACT2, ACT8, and ACT11 is demonstrated as anthers develop (beginning 6 days prior to anthesis) and their expression is relatively low in mature pollen as measured by monoclonal antibody MAb13a. The residual actin expression in mature pollen detected by this antibody is most likely due to its reaction with ACT11 protein. **C**. A 100-fold increase in reproductive actin protein expression is measured in mature pollen using MAb2345a. The two Western blots are normalized to a standard amount of ACT11 protein (0.25 μg), which reacts with both antisera. Similar results were obtained in *Arabidopsis* (Kandasamy et al., 1999).

6. LATE-POLLEN SPECIFIC ACTINS CAN BE DISTINGUISHED FROM VEGETATIVE ACTINS AT THE CELLULAR LEVEL USING AN ANCIENT EPITOPE THAT PREDATES ANGIOSPERMS

Based on extremely strong mRNA and promoter/reporter expression in mature *Arabidopsis* pollen, the actin subclasses 4 and 5 are defined as reproductive actins. However, the promoter/GUS fusion of all four genes in these subclasses showed measurable staining in young vascular tissue of leaves, stems, and roots blurring their distinction from vegetative actins. Thus, we could not help but be concerned with how well RNA and reporter expression reflect protein expression. Furthermore, we have to ask if the distinction among the three subclasses of reproductive actin isovariants is of any real importance to higher plant actin protein expression in plants distant from *Arabidopsis*?

Subclass-specific monoclonal antibodies allowed us to examine these questions. MAb45a specifically recognizes two closely-related reproductive subclasses of actins (ACT1 & 3, 4 & 12) on the basis of ELISA and immunoblot analysis, as shown in Figure 4C. MAb45a reacts strongly with actins expressed in mature pollen and pollen tubes in *Arabidopsis* and tobacco, but not significantly with actins in other tissues (Kandasamy et al., 1999). MAb45a reacts with pollen-specific actin in the mature pollen from diverse angiosperms (e.g., tobacco, rice, maize) and the recently derived gymnosperm (*Ephedra*) but not with more ancient groups of gymnosperms (e.g., pines, cycads). This suggests that MAb45a recognizes an epitope that arose in an ancestor common to angiosperms and advanced gymnosperms more than 220 million years ago, confirming our preliminary dating of these branches in the gene tree (McDowell et al., 1996b). Peptide competition studies demonstrate that an epitope containing Asn79 is essential to confer pollen-specificity to MAb45a. The 15 amino acid ACT1 peptide centered on and containing Asn79 has strong inhibition activity when pre-incubated with MAb45a. Other actin peptides or an identical peptide that is substituted with the Ser79, the residue found in the other four *Arabidopsis* actins, have no inhibitory activities, even when used at 1000-fold higher concentrations (Kandasamy et al., 1999). Another antibody, MAbGPa, which recognizes an epitope common among actins ranging from angiosperms to algae and other protists, but not found in animal actins, was used as a positive control. Used together in immunocytochemical assays, MAb45a and MAbGPa can distinguish actin filaments in pollen tubes from those in vegetative tissues in a wide variety of angiosperms. This work establishes for the first time that an ancient subset of actin isovariants are expressed late in pollen development in all higher plants (Kandasamy et al., 1999).

7. EXAMINING TEMPORAL AND SPATIAL EXPRESSION PATTERNS OF PLANT ACTIN PROTEINS: BASKETS OF ACTIN FILAMENTS LINK CHLOROPLASTS TO THE CYTOSKELETAL NETWORK

The plant actin cytoskeleton has been examined for over 25 years. The early work concentrated on large algal cells in the 1970s (Kersey et al., 1976; Palevitz et al., 1974; Palevitz & Hepler, 1975) and a second great burst of activity in the 1980s resulted from the use of fluorescently-labeled phalloidin to look at filament structures in a few angiosperms (Parthasarathy, 1987; Parthasarathy et al., 1985). Several technical barriers have limited progress beyond these earlier studies. First, among the problems, plant actin filaments are relatively unstable to paraformaldehyde fixation protocols. Second, there have been few high-quality immunological reagents that bind strongly to plant actins, and phalloidin probably does not bind all plant actins equivalently, since the binding sequence is variable among plant actins (Meagher, 1991). Progress with advanced chemical fixation and freeze substitution techniques has resulted in some elegant work (Baskin et al., 1996; Ding et al., 1991; Lancelle et al., 1987; Sonobe & Shibaoka, 1989). However, overall we are still relatively ignorant about the details of F-actin structures in plant cells and tissues compared to those in animals. For examples, little is known about the precise higher plant filament structures involved in cell polarity, cytoplasmic streaming and organelle movement, organelle orientation, development of complex cell types and organs, transport in roots, and the response of the cytoskeleton to phytohormones. We feel that these new plant fixation protocols combined with our strong new immunological reagents could lead to a renaissance in articulating plant actin structures in many plant cell and developmental laboratories. Knowledge about these structures is also essential to our characterization of the phenotypes of plant actin mutants. In addition to analyses of pollen tubes and cultured cells, we recently explored actin-organelle interactions.

Cytoplasmic streaming and associated organelle movement and positioning have been ascribed to actin- and tubulin-regulated processes (Williamson, 1993). Using combinations of rapid-freeze fixation and freeze-substitution techniques, MAbGPa that detects all eight expressed *Arabidopsis* actins and their filaments (Figs. 4C, 3E), and confocal microscopy, we have clearly identified the linkage between actin filaments and chloroplasts in *Arabidopsis* leaf cells (Kandasamy & Meagher, 1999). As shown in Figures 3F–3I, chloroplasts encapsulated in baskets of actin filaments are attached to longitudinally oriented cortical actin bundles either directly or indirectly via thin actin filaments. Latrunculin B, a drug which disrupts actin filaments even at very low concentrations (1–2 μM), results in aberrant clustering of the otherwise normally arrayed chloroplasts (not shown). Absolutely no association of microtubules with chloroplasts could be detected in these cells nor did the tubulin-disrupting drug oryzalin have any effect on chloroplast positioning (Kandasamy & Meagher, 1999).

8. GENETIC MAPPING REVEALS THAT MEMBERS OF THE *ARABIDOPSIS* ACTIN GENE FAMILY ARE WIDELY DISPERSED IN THE GENOME

Based on DNA and protein sequence divergence data, we have predicted that any pair of *Arabidopsis* actin gene family members have been diverged from each other for 30 to 400 MY. Thus, there should have been sufficient time for these genes to move around the plant genome subsequent to their duplication from ancestral sequences. The *Arabidopsis* actin genes were mapped to dispersed locations in the genome (McKinney & Meagher, 1998), as shown in Figure 6. This was done by following the segregation of cleaved amplified polymorphisms (CAPs) in recombinant inbred lines made from two *Arabidopsis* ecotypes and by hybridization to characterized yeast artificial chromosomes, YACs. The YAC mapping data confirmed and refined the positions determined by CAPs mapping. As predicted, these sequences are highly dispersed relative to each other. They are located on four of the five *Arabidopsis* chromosomes. Four actin genes are found on chromosome 3, but they are genetically dispersed. Only the two pseudogenes are closely physically linked. No significant linkage was found between reproductive or vegetative actins and their potential counterparts in the tubulin or profilin gene families. This diversity of map positions also gave us new insights into the evolution of plant genomes and the ancient origin of the plant actin gene family (McKinney & Meagher, 1998).

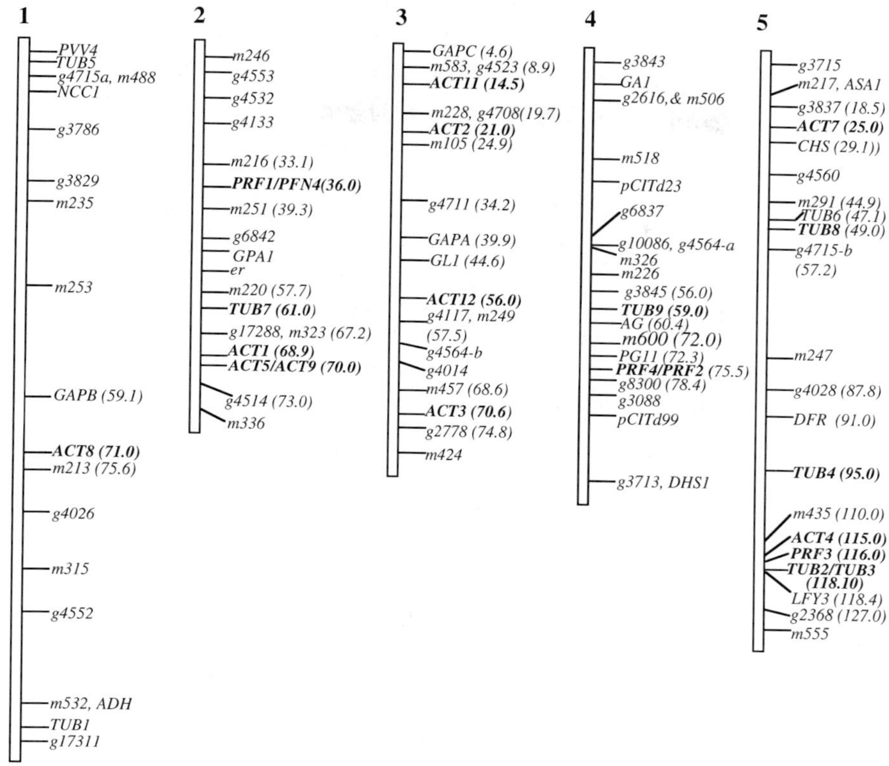

Figure 6. Chromosomal map positions of the various *Arabidopsis* actin genes. Locations of the various *Arabidopsis* actins, profilins and ß-tubulins are shown on a linkage map as determined by CAPS mapping and confirmed by YAC mapping (McKinney & Meagher, 1998). The majority of CAPs markers shown were derived from Lister et al. (1993) and enhanced by markers reported by Jarvis et al. (1994). The tubulin genes, *TUB1, TUB5* and *TUB6*, were mapped previously using RFLPs (McGrath et al., 1993), but their map positions have not been confirmed by CAPs mapping. The profilin genes *PRF1, PRF2* and *PRF4* (Huang et al., 1996b) were also identified in an independent study as *PFN1*, *PFN2* and *PFN3*, respectively (Christensen et al., 1996).

Knowing the chromosomal map positions accurately for each of the functional actin genes, we may be able to identify actin mutants among the hundreds of existing embryo lethal mutants (Patton et al., 1991). Alternatively, we should be able to construct large numbers of insertional mutants in actins using closely linked transposons (Osborne et al., 1995; Parinov et al., 1999).

9. SEQUENCE-BASED SCREENING ENABLED THE ISOLATION OF INSERTION MUTANTS IN PLANT ACTIN GENES WITHOUT KNOWLEDGE OF PHENOTYPE

The accessibility of plant mutants, when conducting investigations of physiological, developmental, and cell biological processes, has been as important as the availability of cloned versions of the affected gene. For genes like the actins, however, phenotypic screens are not available and specific actin mutations that cause lethality are difficult to detect. What has been missing is a simple and reliable strategy for using the gene or protein sequence to identify plant mutants. In collaboration with Ken Feldmann (Univ. Arizona), we developed a rapid screen for T-DNA insertions in plant genes (McKinney et al., 1995) similar to that developed in the animals *Drosophila melanogaster* and *Caenorhabiditis elegans* (Ballinger & Benzer, 1989; Zwaal et al., 1993). The general strategy is outlined in Figure 7A. A primer pair (i.e., one homologous to the end of the inserted element and with its 3' end facing outward and one primer within the target gene) is used to amplify the sequences at the junction of the insertion. Multiple combinations of element and target gene primers may be used in separate reactions to cover all possible pairs, orientations and positions of the element within the gene. Mutations that have a recessive lethal phenotype can be isolated as heterozygotes by this approach. (These are perhaps better described as hemizygotes, because there is no T-DNA insertion allele at the corresponding position on the wild-type chromosome). Furthermore, mutations with no obvious, selectable phenotype can also be identified in the homozygous or heteroygous state. The power of this approach is the ability to screen hundreds or thousands of possible insertion lines as a single DNA pool, working backward to the mutant line with the wanted mutation. This screening method is now widely viewed as the most direct route to obtaining *Arabidopsis* mutants, particularly when no assumptions can be made about phenotype or when a large number of genetically-dispersed genes are being analyzed simultaneously (Krysan et al., 1996). Using this method we have isolated several plant actin mutants. The insertions in the first three mutants we identified are mapped in Figure 7B. The *act2-1* mutant is essentially null for *ACT2* mRNA expression, while the *act7-1* mutant expresses *ACT7* mRNA with an altered 3' UTR. A similar screening method has been developed by for identifying *Mu*-element insertions in known genes in maize (Benson et al., 1996; Mena et al., 1996).

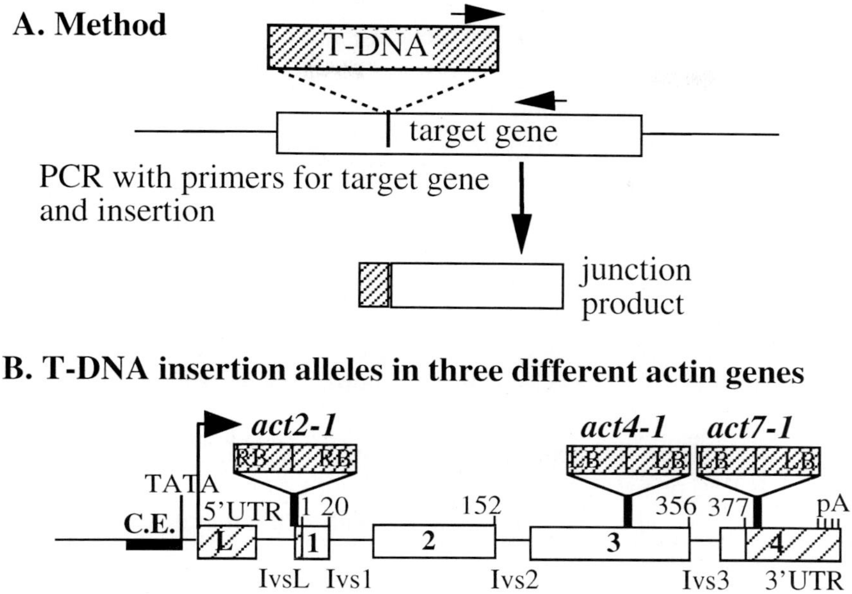

Figure 7. Sequence-based identification of insertion mutations in three *Arabidopsis* actin genes and their physical position within each gene. **A**. General model for identifying insertions in target genes by PCR with one insertion element border primer and one target gene primer. **B**. Map of a typical plant actin gene with the sites of T-DNA insertions indicated for three different mutant alleles (*act2-1*, *act4-1*, *act7-1*). The five exons present in most plant actin genes are in boxes and are labeled L, 1, 2, 3, 4. The *ACT2* gene lacks intron 1, fusing exons 1 and 2 indicated in this diagram. Numbers above the exons indicate codon positions in the 377-codon containing actin gene. Abbreviations represent: Ivs = intervening sequence; L = exon encoding the mRNA leader; LB = left border T-DNA; RB = right border T-DNA; C.E. = conserved cis elements; UTR = untranslated region of transcript; pA = polyadenylation sites.

This screening procedure was enhanced by the application of a number of technical advances, including the following: 1) The use of degenerate primers to screen for mutants in any member of a gene family allows for more efficient screening. This has worked for actins and even for the poorly-conserved profilins, where both primers were more than 10^5-fold degenerate. Ironically, the contaminating background relative to the correct product band is lower with degenerate primers than with unique primers (McKinney et al., 1995), perhaps because the concentration of each specific primer in the population is very low during the first few cycles, when most product artifacts are established in the DNA product population. 2) The identification of nested PCR products can rapidly confirm that products being examined are from a bona fide insertion in the gene of interest. Final confirmation of the disrupted gene comes from sequencing these products, which is also made more reliable by using nested primers (McKinney et al., 1995). 3) The use of AT rich ends on primers lowers the amount of spurious background products (Crameri & Stemmer, 1993, and the references therein). This is contrary to the long-accepted but erroneous dogma used by most computer primer design programs that GC-rich ends increase specificity. 4) The use of primers very near the T-DNA element borders (i.e., within 100 bp) can reduce the background by lowering the amount of substrate consumed by the PCR amplification of border-border junction products from multiple, adjacent T-DNA elements at a site. This does result in a small percentage of T-DNA insertions with heavily truncated borders being missed by the PCR screening procedure (McKinney et al., 1995). 5) The use of a PCR screening strategy to monitor the segregation of unwanted T-DNA insertions with LB/LB, RB/RB, or LB/RB fusions that may not be present in the target gene mutation (McKinney et al., 1995) aids in isolating the mutants within a clean genetic background.

10. MONITORING ALLELE FREQUENCY OVER MULTIPLE GENERATIONS REVEALS EACH ACTIN MUTATION AS HIGHLY DELETERIOUS: HOW COMMON IS TRUE REDUNDANCY WITHIN PLANT GENE FAMILIES?

Insertion mutants were first characterized in *Arabidopsis* actin genes *ACT2*, *ACT4*, and *ACT7*, representing three distinct subclasses of actin (Fig. 1). Surprisingly, adult plants homozygous for the *act2-1*, *act4-1*, and *act7-1* mutant alleles appeared robust, morphologically normal, and fully fertile when germinated and grown well separated on soil. Our molecular evolutionary studies demonstrated that these genes have accumulated very few amino acid sequence changes (~1% per 50–100 MY) over the last 30–400 million years (McDowell et al., 1996b) and all three genes are strongly expressed and exhibit distinct temporal and spatial expression patterns. This suggests that all of them are under strong selective constraint. Thus, the lack of any obvious morphological phenotype was puzzling. Now that reverse-genetic approaches are used to isolate mutants, reports of possible genetic redundancy are common in plants, animals, fungi, and protists. However, little convincing proof has been presented to demonstrate true genetic redundancy (Gilliland et al., 1998), even for gene family members with a high degree of sequence relatedness like the actins.

We examined the first three actin mutants in multigenerational studies to test if these actin genes were under selective constraint or were truly redundant (Gilliland et al., 1998). When the plants were grown as populations descending from a single heterozygous parent, all three mutant alleles were found at extremely low frequencies relative to the wild-type in the F_2 generation, as shown in Table 1.

Table 1. Frequency of genotypes segregating for mutant actin alleles in the F_2 generation starting with a F_0 heterozygote.

F_0 Genotype -->		***ACT2/ act2-1***	***ACT4/ act4-1***	***ACT7/ act7-1***
F_2 Genotype	**F_2 Expected %**	**Observed % (# plants)**	**Observed % (# plants)**	**Observed % (# plants)**
A_1A_1	37.5%	38.96% (30)	56.7% (55)	38.4% (28)
A_1A_2	25%	38.96% (30)	27.8% (27)	54.8% (40)
A_2A_2	**37.5%**	**22.1% (17)**	**15.5% (15)**	**6.8% (5)**
% (#plants)	100%	100% (77)	100% (97)	100% (73)
Allele Frequency A_1:A_2	0.5 : 0.5	0.58 : 0.42	0.71 : 0.29	0.66 : 0.34
Probab. of X^2		P=4.1 x 10^{-3}	P=1.4 x 10^{-5}	P=2.5 x 10^{-10}

The table shows genotypic frequencies found in a sampling of the F_2 generation of the heterozygous lines containing the *act2-1*, *act4-1* or *act7-1* T-DNA insertion mutations. A_1 and A_2 denote the wild-type and mutant T-DNA actin alleles, respectively. Chi square values were calculated for the observed versus the expected allele frequencies. The results were highly significant with two degrees of freedom: one would expect to get this much deviation from the expected result by chance less than one time in 100 trials (P □ 10^{-2}) for the *act2-1* allele, less than once in 10^4 trials for the *act4-1* allele (P □ 10^{-4}), and less than one in 10^9 trials for the *act7-1* allele (P □ 10^{-9}). A repetition of the experiment following the *act2-1* allele gave the similar mutant frequencies in the F_2.

The results differed significantly from those expected for inbreeding populations under normal Mendelian segregation with no selection. For example, 37.5% of the plants in a population descended from a heterozygous parent should have been homozygous mutants in the F_2 generation. In fact, homozygous mutants made up less than 6.8–22.1% of the population for each mutant. Thus, all three mutant alleles appear to be highly deleterious. These data deviated significantly from Mendelian segregation without selection, results that would only happen by chance once in 10^2 to 10^9 replicas of each segregation analysis experiment. By monitoring genotypes for four generations, the *act2-1* mutant allele was found at normal frequencies in the F_1, but significantly lower frequencies than expected in the F_2 and F_3 generations (Fig. 8A, solid lines). Obviously, these genes are under strong selective constraint under normal growth conditions and are not truly redundant in spite of the lack of obvious morphological phenotype.

In collaboration with Marjorie Asmussen (Univ. Georgia), we developed a mathematical model to interpret actin genotype and allele frequency dynamics in selfing populations such as produced with *Arabidopsis* (Asmussen et al., 1998). These were explicit time-dependent solutions for the frequencies at diallelic loci in diploid populations under any combination of fertility selection (*f*), viability selection (*v*), and gametic selection (*m*) via meiotic drive (deviation from 50:50 Mendelian ratios in gametes produced by heterozygotes). The results of our modeling showed that with no outcrossing, allelic variation is always maintained under gametic selection alone, but with any fertility or viability differences, variation will ordinarily be maintained if and only if the net fitness of heterozygotes significantly exceeds that of both homozygotes.

A case study was made of the *Arabidopsis act2-1* mutant allele over multiple generations in populations initiated from a heterozygous plant (Fig. 8A, dashed line). The observed genotypic frequency dynamics were consistent with those predicted by our model for a deleterious, incompletely recessive mutant, in either fertility or viability but not significantly subject to gametic selection. As shown in Figure 8B, allele frequency dynamics for the *act2-1* allele (dashed lines) match the theoretical curves (solid lines) when the homozygous mutants (A_2A_2) have 70% of the viability (*v*) of wild-type (A_1A_1), and the heterozygotes (A_1A_2) have 87% of the viability of wild-type. The sum of the X^2 values for all three genotypes as compared to this model totaled 1.1, giving an exceptionally good fit (Fig. 8B), where the summed X^2 values based on a segregation with no selection was 11 (Fig. 8A). A simple recessive viability mutant model did not fit the *act2-1* experimental data as well as one with directional selection (i.e., the fitness of the heterozygous genotype is intermediate between that of the wild-type and homozygous mutant). It is logical that the reduced fitness associated with this mutant allele occurs in the 2n sporophytic portion of the life cycle (viability or fertility), because *ACT2* is the most highly expressed gene in adult vegetative tissues. Based on this viability selection model, the *act2-1* allele would vanish from a relatively large population in 20 generations, which is only 20 years for a natural *Arabidopsis* population. Thus, while an initial analysis of mutant morphology in soil grown plants suggested redundancy, *ACT2* appears to be an essential gene. Over several generations the *act2-1*, *act4-1* and *act7-1* disruptions act effectively as lethal mutations.

Because these mutant alleles have a genetic phenotype influencing plant survival, they must also have physical phenotypes altering some aspect of plant growth. Close observations of the seedling root morphology of *act2-1* and *act7-1* mutants under extreme growth conditions has revealed clear physical abnormalities (Gilliland & McKinney, unpublished results) that will be described in future publications.

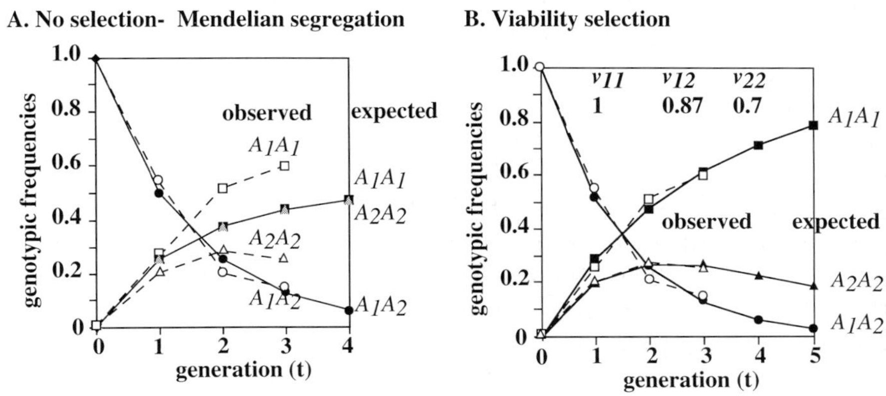

Figure 9. Dynamics of *act2-1* genotype frequencies observed in a multigenerational study (dashed lines) and compared to two predicted models (solid lines). **A**. Genotype frequencies are compared to the prediction for Mendelian segregation with no selection (Gilliland et al., 1998). **B**. Genotype frequencies are compared to viability selection as mathematically modeled to fit these experimental data (Asmussen et al., 1998). A_1 represents the wild-type *ACT2* allele, and A_2 the mutant *act2-1* allele. See text for a full explanation.

11. CONCLUSIONS AND FUTURE DIRECTIONS

By focusing on the eight functional *Arabidopsis* actins, great progress has been made in understanding the significance of complex ancient actin gene families in plants. This research has been strengthened by combining cell biological, molecular genetic, and biochemical approaches. Detailed expression studies demonstrate that the individual genes are highly regulated and that the most extreme members in the family complement each other in their vegetative and reproductive expression patterns. New immune reagents, improved fixation protocols, and microscope technology allow the plant cytoskeleton to be examined in increasingly beautiful detail and suggest a coming renaissance of cytoskeletal images. Results with some of the actin subclass specific antisera demonstrate that actin induction in response to auxin stimulus and expression of the late pollen specific actins can be extrapolated to other distant plant species. While it is clear that mutations in *Arabidopsis* actins have serious deleterious effects on plant survival, the cell, developmental, and conditional phenotypes of individual mutants need to be described to fully understand actin function. Many more mutants and mutant alleles are needed to sort out the functions of the individual isovariants. The ancient age and distinct sequences of the different actin isovariants suggest that they are functionally-distinct proteins. Their co-expression should lead to complex isovariant dynamics that both expand and buffer plant responses to their environment (Meagher et al., 1999a).

A deeper understanding of the function and dynamics of the various actin cytoskeletal processes should be aided by some additional tools that are now under development. Transcriptional profiling of the plant genomes (Schena, 1996; Schena et al., 1995) should reveal if hundreds of other cytoskeletal genes follow the well described vegetative and reproductive expression patterns common for the first characterized cytoskeletal gene families (Meagher et al., 1999b). The dynamics of these processes can be examined at the cellular level using in vivo monitoring systems such as the recently described system using the green fluorescent protein tagged actin-binding protein talin to follow cytoskeletal rearrangements during cell development (Kost et al., 1998; see also Kost et al., this volume). An assay which monitors nuclear positioning in *Tradescantia* stamen hairs following microinjection (Gibbon et al., 1997; Staiger et al., 1994; see also Gibbon & Staiger, this volume) allows the evaluation of differences in the functional properties of individual actin-binding protein isovariants. However, a fuller understanding of the physical biochemistry of actin-based processes and the distinctions between isovariants also awaits development of cell-free systems for examining plant actin polymers and binding proteins, as commonly used to examine the animal, fungal, and protist cytoskeleton. The ectopic expression of animal isovariants has been very informative in animal systems (Fyrberg et al., 1998; Hart & Cooper, 1999) and should prove a powerful tool in plants. This approach will be aided by the relative ease with which plants can be transformed with novel gene constructs.

ACKNOWLEDGMENTS

We want to thank Gay Gragson for her constructive comments on the manuscript, Karla Pinson and other members of the Meagher laboratory for their help in all stages of this project, and Yolanda Lay for her help with ELISAs during isolation of the monoclonal antibodies. This work was funded by NIH grant #GM36397.

REFERENCES

An Y-Q, Huang S, McDowell JM, McKinney EC and Meagher RB (1996a) Conserved expression of the *Arabidopsis ACT1* and *ACT3* actin subclass in organ primordia and mature pollen. Plant Cell 8: 15-30

An Y-Q, McDowell JM, Huang S, McKinney EC, Chambliss S and Meagher RB (1996b) Strong, constitutive expression of the *Arabidopsis ACT2/ACT8* actin subclass in vegetative tissues. Plant J 10: 107-121

Asmussen MA, Gilliland LU and Meagher RB (1998) Detection of deleterious genotypes in multigenerational studies. II. Theoretical and experimental dynamics with selfing and selection. Genetics 149: 727-737

Baird WV and Meagher RB (1987) A complex gene superfamily encodes actin in petunia. EMBO J 6: 3223-3231

Ballinger DG and Benzer S (1989) Targeted gene mutations in *Drosophila*. Proc Natl Acad Sci USA 86: 9402-9406

Baskin TI, Miller DD, Vos JW, Wilson JE and Hepler PK (1996) Cryofixing single cells and multicellular specimens enhances structure and immunocytochemistry for light microscopy. J Microsc 182: 149-161

Benson RJ, Johal GS, Crane VC, Tossberg JT, Schnable PS, Meeley RB and Briggs SP (1996) Cloning and characterization of the maize *An1* gene. Plant Cell 7: 75-84

Christensen HE, Ramachandran S, Tan CT, Surana U, Dong CH and Chua N-H (1996) Arabidopsis profilins are functionally similar to yeast profilins: identification of a vascular bundle-specific profilin and a pollen-specific profilin. Plant J 10: 269-279

Crameri A and Stemmer WPC (1993) 1020-fold aptamer library amplification without gel purification. Nuc Acids Res 21: 4410

Ding B, Turgeon R and Parthasarathy MV (1991) Microfilament organization and distribution in freeze substituted tobacco plant tissues. Protoplasma 165: 96-105

Fowler JE and Quatrano RS (1997) Plant cell morphogenesis: plasma membrane interactions with the cytoskeleton and cell wall. Annu Rev Cell Dev Biol 13: 697-743

Fyrberg EA, Fyrberg CC, Briggs JR, Saville D, Beall CJ and Detchum A (1998) Functional nonequivalence of *Drosophila* actin isoforms. Biochem Genet 36: 271-287

Gibbon BC, Ren H and Staiger CJ (1997) Characterization of maize (*Zea mays*) pollen profilin function *in vitro* and in live cells. Biochem J 327: 909-915

Gilliland LU, Asmussen MA, McKinney EC and Meagher RB (1998) Detection of deleterious genotypes in multi-generational studies: I. Disruptions in individual *Arabidopsis* actin genes. Genetics 149: 717-725

Hart MC and Cooper JA (1999) Vertebrate isoforms of actin capping protein ß have distinct functions in vivo. J Cell Biol 147: 1287-1298

Hightower RC and Meagher RB (1986) The molecular evolution of actin. Genetics 114: 315-332

Huang S, An Y-Q, McDowell JM, McKinney EC and Meagher RB (1996a) The *Arabidopsis ACT4/ACT12* actin gene subclass is strongly expressed in post-mitotic pollen. Plant J 10: 189-202

Huang S, An Y-Q, McDowell JM, McKinney EC and Meagher RB (1997) The *Arabidopsis ACT11* actin gene is strongly expressed in tissues of the emerging inflorescence, pollen and developing ovules. Plant Mol Biol 33: 125-139

Huang S, Chambliss C and Meagher RB (1996b) The *Arabidopsis* profilin gene family: Evidence for an ancient split between constitutive and pollen-specific profilin genes. Plant Physiol 111: 115-126

Jarvis P, Lister C, Szabo V and Dean C (1994) Integration of CAPS markers into the RFLP map generated using recombinant inbred lines of *Arabidopsis thaliana*. Plant Mol Biol 24: 685-687

Kandasamy MK, McKinney E and Meagher RB (1999) The late pollen specific actins in angiosperms. Plant J 18: 681-691

Kandasamy M and Meagher RB (1999) Actin-organelle interactions: association with chloroplast in *Arabidopsis* leaf mesophyll cells. Cell Motil Cytoskel 44: 110-118

Kersey YM, Hepler PK, Palevitz BA and Wessels NK (1976) Polarity of actin filaments in characean algae. Proc Natl Acad Sci USA 73: 165-167

Kost B, Spielhofer P and Chua N-H (1998) A GFP-mouse talin fusion protein labels plant actin filaments in vivo and visualizes the actin cytoskeleton in growing pollen tubes. Plant J 16: 393-401

Krysan PJ, Young JC, Tax F and Sussman MR (1996) Identification of transferred DNA insertions within *Arabidopsis* genes involved in signal transduction and ion transport. Proc Natl Acad Sci USA 93: 8145-8150

Lancelle SA, Cresti M and Hepler PK (1987) Ultrastructure of the cytoskeleton in freeze substituted pollen tubes of *Nicotiana alata*. Protoplasma 140: 141-150

Lister C and Dean C (1993) Recombinant inbred lines for mapping RFLP and phenotypic markers in *Arabidopsis thaliana*. Plant J 4: 745-750

McDowell J, An Y-Q, McKinney EC, Huang S and Meagher RB (1996a) The *Arabidopsis ACT7* actin gene is expressed in rapidly developing tissues and responds to several external stimuli. Plant Physiol 111: 699-711

McDowell JM, Huang S, McKinney EC, An Y-Q and Meagher RB (1996b) Structure and evolution of the actin gene family in *Arabidopsis thaliana*. Genetics 142: 587-602

McGrath JM, Jancso MM and Pichersky E (1993) Duplicate sequences with a similarity to expressed genes in the genome of *Arabidopsis thaliana*. Theor Appl Genet 86: 880-888

McKinney EC, Ali N, Traut A, Feldmann KA, Belostotsky DA, McDowell JA and Meagher RB (1995) Sequence based identification of T-DNA insertion mutations in *Arabidopsis*: Actin Mutants *act2-1* and *act4-1*. Plant J 8: 613-622

McKinney EC and Meagher RB (1998) Members of the *Arabidopsis* actin gene family are widely dispersed in the genome. Genetics 149: 663-675

Meagher RB (1991) Divergence and differential expression of actin gene families in higher plants. Int Rev Cytol 125: 139-163

Meagher RB, Kandasamy MK and McKinney EC (1999a) Isovariant dynamics expands and buffers the responses of complex systems: The diverse plant actin gene family. Plant Cell 11: 1-12

Meagher RB, Vitale AV and McKinney EC (1999b) The evolution of new structures: Clues from plant cytoskeletal genes. Trends Genet 15: 278-284

Mena M, Ambrose BA, Meeley RB, Briggs SP, Yanofsky MF and Schmidt RJ (1996) Diversification of C-function activiy in maize flower development. Science 271: 1527-1540

Nick P (1999) Signals, motors, morphogenesis—the cytoskeleton in plant development. Plant Biol 1: 169-179

Osborne BI, Wirtz U and Baker B (1995) A system of insertional mutagenesis and chromosomal rearrangement using the Ds transposon and Cre-*lox*. Plant J 7: 687-701

Palevitz BA, Ash JF and Hepler PK (1974) Actin in the green alga, *Nitella*. Proc Natl Acad Sci USA 71: 363-366

Palevitz BA and Hepler PK (1975) Identification of actin *in situ* at the ectoplasm-endoplasm interface of *Nitella*. J Cell Biol 65: 29-38

Parinov S, Sevugan M, Ye D, Yang W-C, Kumaran M and Sundaresan V (1999) Analysis of flanking sequences from Dissociation insertion lines: A database for reverse genetics in Arabidopsis. Plant Cell 11: 2263-2270

Parthasarathy MV (1987) In situ localization of actin filaments in higher plant cells using fluorescent probes. Plant Mol Biol 5: 251-259

Parthasarathy MV, Perdue TD, Witzmun A and Alvernaz J (1985) Actin network as a normal component of the cytoskeleton in many vascular plant cells. Am J Bot 72: 1318-1323

Patton DA, Franzmann LH and Meinke DW (1991) Mapping genes essential for embryo development in *Arabidopsis thaliana*. Mol Gen Genet 227: 337-347

Schena M (1996) Genome analysis with gene expression microarrays. BioEssays 18: 427-431

Schena M, Shalon D, Davis RW and Brown PO (1995) Quatitative monitoring of gene expression patterns with a complementary DNA microarray. Science 270: 467-470

Sonobe S and Shibaoka H (1989) Cortical fine actin filaments in higher plant cells visualized by rhodamine phalloidin after pretreatment with m-maleimidobenzoyl N-hydroxysuccinimide ester. Protoplasma 148: 80-86

Staehelin LA and Hepler PK (1996) Cytokinesis in higher plants. Cell 84: 821-824

Staiger CJ (2000) Signaling to the actin cytoskeleton in plants. Annu Rev Plant Physiol Plant Mol Biol 51: 257-288

Staiger CJ, Yuan M, Valenta R, Shaw PJ, Warn RM and Lloyd CW (1994) Microinjected profilin affects cytoplasmic streaming in plant cells by rapidly depolymerizing actin microfilaments. Curr Biol 4: 215-219

Williamson RE (1993) Organelle movements. Annu Rev Plant Physiol Plant Mol Biol 44: 181-202

Zwaal RR, Broeks A, van Meurs J, Groenin JTM and Plasterk RHA (1993) Target-selected gene inactivation in *Caenorhabditis elegans* by using a frozen transposon insertion mutant bank. Proc Natl Acad Sci USA 90: 7431-7435

Chapter 2

MYOSINS

Stefanie Reichelt and John Kendrick-Jones

Structural Studies Division, MRC Laboratory of Molecular Biology, Hills Road, Cambridge, CB2 2QH, UK

Key words: cell plate, motor protein, plasmodesmata, post-cytokinetic cell wall, unconventional myosin

Abstract: Myosins are a large superfamily of actin- and ATP-dependent motor proteins, which have been divided (on the basis of their conserved head or motor domain) into at least fifteen classes. Three of these classes (class VIII, XI and XIII) are exclusively expressed in plants, and there are now at least 26 plant myosin sequences available. Whereas the myosins in animal and amoeboid cells have been studied in detail, little is known about the function(s) of the myosins in plant cells. Preliminary studies indicate that class XI myosins are involved in the rapid movement of cytoplasm – cytoplasmic streaming – while class VIII myosins are associated with the plasma membrane and plasmodesmata at the post-cytokinetic cell wall. Thus, each of the myosin classes is expected to have distinct functions but within each class the myosins may have overlapping functions. It will be interesting to test this proposal and elucidate the intracellular functions of the different classes of myosins within a single species.

1. INTRODUCTION

The myosins are a large superfamily of molecular motors which create movement and mechanical force generated by ATP-dependent interactions along actin filaments. Phylogenetic analysis has identified at least fifteen distinct classes (designated I to XV) (Cope et al., 1996; Mermall et al., 1998) on the basis of their conserved head- or motor-domain sequences. Three of these myosin classes are expressed exclusively in plants (Kendrick-Jones & Reichelt, 1999).

Each myosin has a conserved N-terminal motor domain, which contains the highly conserved ATP-binding and actin-binding sequences. The motor

C.J. Staiger et al. (eds.), Actin: A Dynamic Framework for Multiple Plant Cell Functions, 29–44.

domain is followed by a light-chain binding 'neck' region containing 1-7 copies of a so-called IQ motif (Cheney & Mooseker, 1992) which serves as a binding site for calmodulin or other EF-hand calcium binding proteins. In addition, each class of myosin contains distinct tail domains which are characteristic for each myosin and are believed to be responsible for the specific subcellular localization and function of these motors.

While the involvement of the conventional two-headed, filament-forming myosins (myosin IIs) is well established in muscular contraction and cytoplasmic contractile events in unicellular systems and mammalian cells, it is only in recent years that studies on the so-called unconventional myosins have revealed a variety of actin-based and membrane-associated motor proteins which function in processes such as vesicle transport, pigment granule movement in melanocytes, cytokinesis and such specialized functions as auditory perception (for recent reviews, see Titus, 1993; Hammer, 1994; Fath & Burgess, 1994; Hasson & Mooseker, 1995; Mooseker & Cheney, 1995; Bähler, 1996; Mermall et al., 1998).

Compared to the mammalian and protozoan myosins, little is known about the functions of these motors in higher plants (for review see: Asada & Collings, 1997; Kendrick-Jones & Reichelt, 1999). Now, however, a variety of plant myosin sequences have been identified and there are preliminary biochemical and immunofluorescence studies. The first myosin gene to be cloned and sequenced from a plant was *Arabidopsis thaliana* myosin I (AtATM1) (Knight & Kendrick-Jones, 1993). Now, at least a further 16 plant myosin genes (from organisms like *Arabidopsis*, *Helianthus*, *Acetabularia, Zea* and *Chlamydomonas*) have been cloned and sequenced (Kinkema et al., 1994; Kinkema & Schiefelbein, 1994; Vugrek & Menzel, 1997, unpublished results; Liu & Pesacreta, 1999, unpublished results; Mages et al., 1998, unpublished results), and PCR screens indicate the existence of additional myosins (Kinkema et al., 1994; Moepps et al., 1993; LaClaire et al., 1995; Plazinski et al., 1997; Vugrek & Menzel, personal communication). Furthermore, an additional 11 distinct myosin sequences have been identified in the *Arabidopsis* Genome Sequencing project, which means that there are at least 16 different plant myosin genes in *Arabidopsis*.

On the basis of their conserved motor-domain sequences, the plant myosins have been placed by phylogenetic analysis into three classes (classes VIII, XI, and XIII) (Cope et al., 1996). None of the plant myosins identified so far belong to class I or II, or to any of the other myosin classes containing animal or amoeboid myosins. The plant myosins in class XI are closely related to the animal myosin Vs and to the *Dictyostelium* myosin I (Peterson et al., 1996), but they are sufficiently different to be grouped into a separate class (Cope et al., 1996). The important point to emerge from this analysis is that all the myosin genes identified from plants belong to unique classes which do not contain animal or protozoan myosins.

2. PLANT MYOSIN CLASS VIII

The plant myosin class VIII contains so far seven myosin sequences, four of which are *Arabidopsis* myosins, two are *Helianthus* myosins and one is a *Zea* myosin. The cloned and sequenced class VIII myosins, AtATM1 and (Knight & Kendrick-Jones, 1993) and AtATM2 (Kinkema *et al.*, 1994) of *Arabidopsis thaliana*, HaMYO1 and HaMYO3 of *Helianthus annus* (Vugrek & Menzel, 1997, unpublished results), and ZmZMM3 of *Zea mays* (Liu & Pesacreta, 1999, unpublished results) are between 1099 and 1166 aa in length and have a predicted heavy chain size of around 130 kDa. Their conserved motor domains (~734 residues), which are 67% identical in AtATM1 and AtATM2, contain an unusual N-terminal extension to the core motor domain (which is defined as the residues homologous to residues 89-781 of chicken skeletal muscle myosin GgFSk) of between 170-190 residues long in AtATM1, AtVIII–A, AtVIII–B and HaMYO3. In ZmZMM3 and AtATM2, the extension is only 100 residues long with both sequences missing the extreme N-terminal region. Following the motor domain, there are 3 – 4 IQ motifs, a short region (~70 residues) of predicted α-helical coiled coil, which implies that they form dimeric molecules. The tail domains of the class VIII myosins (~150 residues) show no apparent homologies or recognizable motifs to other myosin or non-myosin proteins but share homologous regions between themselves. AtATM1, AtVIII–A and AtVIII–B also contain a predicted cell attachment site (prosite motif PS00016: RGD) at amino acid (aa) 630-632 (AtATM1), aa 624 (AtVIII–A), and aa 249-252 (AtVIII–B), and AtATM1 and AtVIII–A contain a microbody C-terminal targeting signal (prosite motif PS00342: TRY) at their C-terminal ends (aa 1153).

3. PLANT MYOSIN CLASS XI

Class XI contains 18 plant myosin sequences including the two first described class XI *A. thaliana* myosins: AtMYA1 (Kinkema & Schiefelbein, 1994) and AtMYA2 (Kinkema et al., 1994). Initially, AtMYA1 and AtMYA2 were grouped with the class V myosins, but detailed comparisons of all the available myosin sequences showed that these myosins are distinct from the class V myosins (Cope et al., 1996; Kendrick-Jones & Reichelt, 1999). The myosins in class XI are similar in molecular structure to the class

V myosins (yeast MYO2 and 4, chick brain Gg p190, and mouse dilute protein Mm Dilute) (Mooseker & Cheney, 1995); they have a predicted heavy chain size of approximately 170 kDa with a conserved motor domain (73% identity in AtMYA1 and AtMYA2) followed by five or six IQ motifs. The tail contains a region with a high probability of forming an α–helical coiled coil (~200 residues) (which in some of the myosins is interrupted by a non-coiled coil region) and is followed by a large C-terminal domain (~460 residues) without any major secondary structural motifs or homology to other non-myosin proteins.

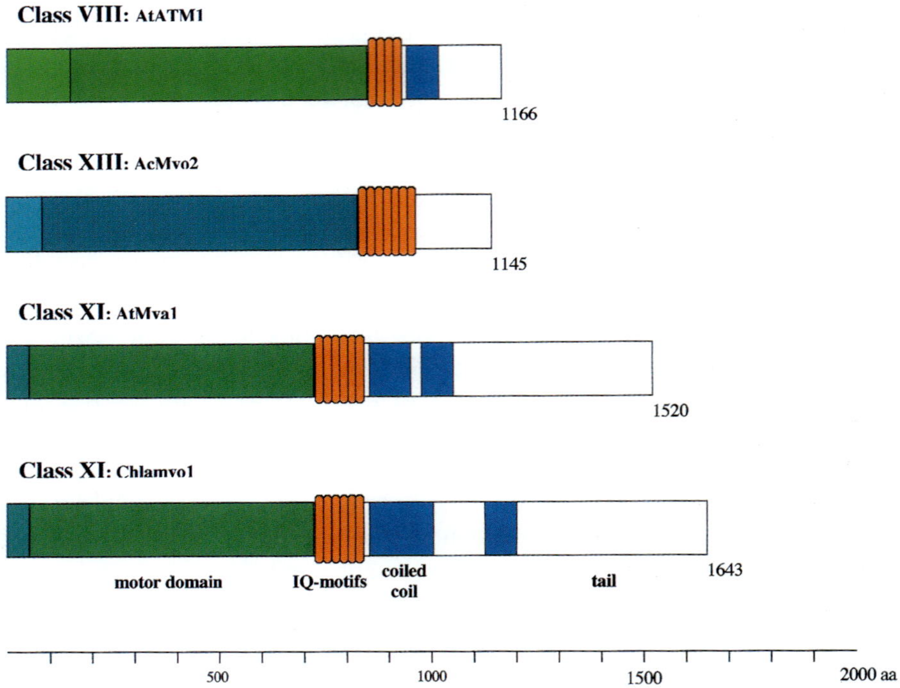

Figure 1. Schematic representations of the predicted domain structures of the plant myosins in classes VIII, XI, and XIII. Amino acid numbers are shown along the bottom. The diagram shows the predicted domain structures of the three plant myosin classes. The motor domain of around 735 residues is followed by a number of IQ motifs (3-4 in class VIII, 5-6 in class XI and 7 in class XIII) and the tail domain. The tail domains contain a coiled coil region in the myosins in classes VIII and XI, but not in class XIII.

Compared with the class VIII myosins, the N-termini of these myosins have smaller extensions of 30-70 aa. A partial sequence of another Arabidopsis myosin gene (AtMYA3) with unique neck and tail domains has also been reported (Kinkema et al., 1994). AtMYA3 contains a very similar motor-domain sequence (70% identical) to AtMYA1 and AtMYA2 followed by a neck region with five IQ motifs which is unusual because it also contains a motif for a consensus nucleotide binding site. The unique tail

Myosin VIII

NAME	LIBRARY	ACCESSION	DATE (Y-M-D)	SIZE (AA)
At ATM1	Genbank	S33812		1166 aa
At ATM2	Genbank	S51823		1101 aa
Ha MYO1	EMBL	U94781		1120 aa
Ha MYO3*	EMBL	U94783		900 aa*
Zm ZMM3	Genbank	AAD31926		1099 aa
At VIII-A	Genbank	AAD50052	99-07-07	1155 aa
At VIII-B	EMBL	CAA19731	99-09-20	1126 aa

Myosin XI

NAME	LIBRARY	ACCESSION	DATE (Y-M-D)	SIZE (AA)
At MYA1	Genbank	S46444		1520 aa
At MYA2	Genbank	S51824		1515 aa
At MYA3**	Genbank	S47105		998 aa**
Ha MYO2*	EMBL	U94782		1260 aa*
Ha MYO4	EMBL	U94784		1502 aa
Ha MYO5	EMBL	U94785		1528 aa
Zm MYO1	Genbank	AAD17931		1529 aa
Zm MYO2**	Genbank	AAD34597		1352 aa**
Cr MYO1	Genbank	AAC27525		1643 aa
At XI-A	Genbank	AAB80627	97-08-11	1736 aa
At XI-B	Genbank	AAC16753	97-08-12	1477 aa
At XI-C	Genbank	AAC14048	98-01-08	1583 aa
At XI-D	Genbank	AAC04899	98-03-04	1611 aa
At XI-E	Genbank	AAC64896	98-08-05	1556 aa
At XI-F	Genbank	AAD32282	99-02-06	1490 aa
At XI-G	Genbank	AAD21759	99-02-13	1502 aa
At XI-H*	EMBL	CAA22981	99-02-21	1446 aa*
At XI-I*	Genbank	CAB36794	99-09-20	1374 aa*

Myosin XIII

NAME	LIBRARY	ACCESSION	DATE (Y-M-D)	SIZE (AA)
Ac MYO1	Genbank	U94397		1145 aa
Ac MYO2*	Genbank	U94398		933 aa*

Figure 2. The plant myosin sequence names, accession numbers, date of submission and their sizes. This lists all the full length plant myosin sequences available so far. The first myosin sequences to be cloned and sequenced have been named by their authors (ATM1, Knight and

Kendrick-Jones, 1993; MYA1, Kinkema *et al.*, 1994). Those names have been kept in the nomenclature and have only been supplemented with the initials for the plant species they have been isolated from. All the *Arabidopsis thaliana* sequences start with the initial letters 'At', the *Helianthus annus* sequences with 'Ha', *Zea mays* sequences with 'Zm', *Chlamydomonas rheinhardii* sequences with 'Cr', *Acetabularia cliffonii* sequences with 'Ac'. The myosin sequences from the *Arabidopsis* Genome Sequencing Project have been labelled according to the class they belong to either AtVIII or AtXI followed by a letter ('A' to 'I', according to their date of submission starting with the letter A for the earliest). The accession numbers refer to the protein sequences. The myosin sequence names which are labelled with an asteriks are not full length but have been included in the phylogenetic tree because their basic core motor domains (homologous to aa 89-781 of chicken myosin II – GgFSk), which are used for the alignment, are complete. The sequences of AtMYA3 and ZmMYO2 are missing part of the N-terminal core motor domain sequence (they have been labelled with two asterisks) and have not been used in the phylogenetic tree. Preliminary analysis of the sequences of the available fragments shows that AtMYA3 and ZmMYO2 are members of class XI. AtMYA3 is highly homologous, but not identical, to AtXI–B, and ZmMYO2 is homolgue to HaMYO4.

has two predicted short coiled coil regions (29 and 36 residues) separated by a region (179 residues) containing six imperfect repeats of a novel sequence of 33 amino acids. Northern blot analyses indicate that AtMYA1, AtMYA2 and AtMYA3 exhibit different patterns of expression in flower, leaf, root, and stem (Kinkema et al., 1994). Class XI also contains three *Helianthus* myosins HaMYO2, HaMYO4 and HaMYO5 (Vugrek & Menzel, 1997, unpublished results), two *Zea* myosins, ZmMYO1 and ZmMyo2, one *Chlamydomonas* myosin CrMYO1, and further nine *Arabidopsis* myosin sequences which have been identified by the Genome Sequencing Project (AtXI–A to AtXI–B). AtMYA3 is closely related, but not identical, to AtXI–B, but because its core domain is not full length, AtMYA3 has not been included in the phylogenetic tree. Interestingly, AtXI–D, AtXI–E, and AtXI–I contain a leucine zipper motif (prosite motif PS00029) in their C-terminal tails which could be involved in dimerization.

4. PLANT MYOSIN CLASS XIII

Two myosin genes (AcMYO1 and AcMYO2) (Vugrek & Menzel, unpublished results) have been identified in the siphonous green alga, *Acetabularia cliffonii*, and have been placed in the new class (class XIII). Originally, based on their motor domain sequences, they were grouped with the class XI myosins, but analysis of their tail-domain sequences and the early divergence of algae from higher plants supports their inclusion in a new class (Cope et al., 1996). AcMYO2 (heavy chain 130 kDa) has a short unique N-terminal extension (35 residues), a conserved motor domain (720 residues), seven IQ motifs, and a tail domain (222 residues) reminiscent of some of the myosin Is, with an overall positive charge and no apparent coiled coil region. AcMYO1 is similar but has four IQ motifs and a very short tail (18 aa residues). Class VIII and XI myosins are predicted to be dimeric molecules whereas those in class XIII are predicted to be monomeric.

Further partial sequences have been obtained from the fern *Anemia phyllitidis* (Moepps et al., 1993), and, using PCR techniques, myosin-like genes from various plant species have been detected (Plazinski et al., 1997). These sequences are available under Genbank: U48781-U48791.

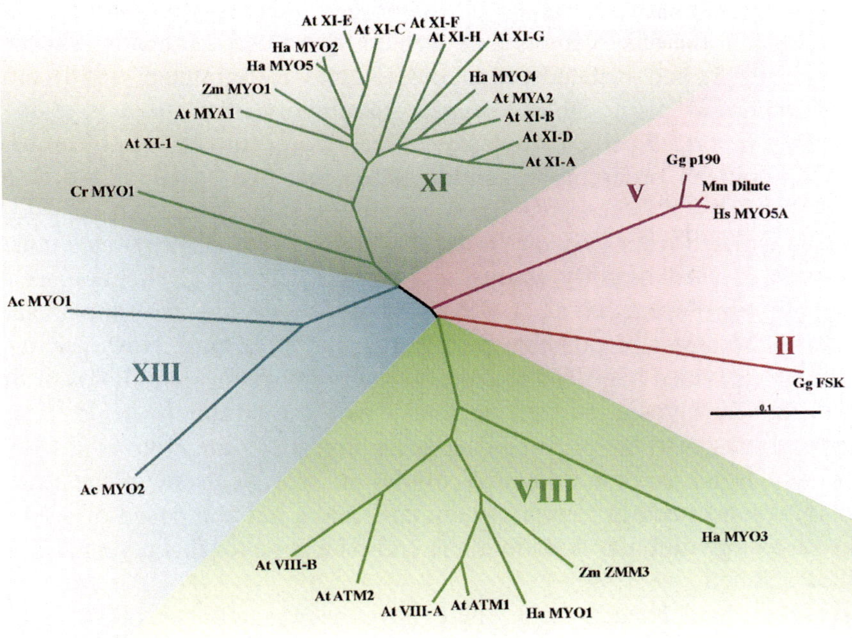

Figure 3. Unrooted phylogenetic tree of the 25 plant myosins. A preliminary (pairwise) alignment of all of the plant myosin sequences was made with a modified version of the method of Feng and Doolittle in ClustalW (Thompson *et al.*, 1994). The sequences were then trimmed to produce a core motor domain which corresponds to residues 89 to 781 of chicken pectoralis muscle myosin II (GgFSk) which was the first crystal structure of a myosin head to be solved (Rayment et al., 1993). A second guide alignment with these core motor domain sequences was produced and the tree drawn using TreeViewPPC (Page, 1996). The tree shows the three classes of myosins unique to plants, three myosin Vs and the myosin II GgFSk. The branches are coloured according to the classes, which are also indicated in Roman numerals. Percentage divergence of two sequences is proportional to the sum of the lengths of the branches connecting them. A 0.1% divergence bar is included. The tree is unrooted, given the lack of an 'outgroup' or sequence from an accepted common ancestor. For the nomenclature, see Figure 2 above.

5. BIOCHEMICAL AND IMMUNOFLUORESCENCE STUDIES

Although the functions of the myosins identified in plants have yet to be established, it is likely that one or more of them are involved in the dramatic organelle movements observed during cytoplasmic streaming which occurs in the giant cells of the alga *Chara corallina,* in lily (*Lilium*) and tobacco (*Nicotiana*) pollen tubes, in stamen hairs of *Tradescantia*, and in stalks of *Acetabularia* (for review see: Kuroda, 1990; McCurdy & Williamson, 1991; Williamson, 1993).

Cytoplasmic streaming in *Chara* is MgATP-dependent and is regulated by calcium/calmodulin and phosphorylation/dephosphorylation. This regulation and the observations that skeletal myosin coated beads (Sheetz & Spudich, 1983) and isolated organelles (Kohno & Shimmen, 1988) move along characean actin cables, suggest the involvement of a myosin in streaming. In *Chara* cells, cytoplasmic streaming occurs at rates of up to 60 $\mu m\ s^{-1}$ (about an order of magnitude faster than the speed of the fastest muscle) (Kachar & Reese, 1988). A possible candidate myosin (240 kDa) has been isolated from *Chara* (Yamamoto et al., 1994) which moves muscle F-actin in *in vitro* motility assays at a similar velocity (Yamamoto et al., 1995; Higashi-Fujime et al, 1995), strongly suggesting that the isolated protein is the motor responsible for cytoplasmic streaming. However, its in vitro actin-activated MgATPase activity is very low, considerably lower than that of muscle myosin, suggesting that it may be extremely labile. Rotary shadowed images of the purified Characean myosin (Yamamoto et al., 1995) show two heads (motor domains) similar in size to those in the muscle myosin, a short neck or regulatory domain, and a tail composed of a 50 nm rod with a large globular domain at the end. The gene for this myosin has not yet been cloned.

Besides the *Chara* system, there have been a number of studies on cytoplasmic streaming using other higher plant cell systems. For example, in *Lilium* and *Nicotiana* pollen tubes (Tang et al., 1989; Miller et al., 1995), and in various tissues from several plants (Vahey et al., 1982; Turkina et al., 1987; Qiao et al., 1989; LaClaire, 1991; Yokota et al., 1995), biochemical and immunofluorescence studies using heterologous myosin antibodies have detected myosin-like proteins. Most of these reports demonstrated spot-like labelling throughout the cytoplasm and along actin filaments, which was interpreted as showing the association of putative myosins with various vesicles and organelles (Heslop-Harrison & Heslop-Harrison, 1989; Grolig et al., 1988, 1996). An immunological study using antibodies against several non-plant myosins indicated the existence of myosin-like polypeptides at 125, 190 and 205 kDa in *Lilium* and *Nicotiana* pollen tubes (Miller et al., 1995) and showed that the antibodies were localized at different locations in the cell. A 170 kDa myosin and its CaM light chains were isolated from lily pollen tubes (Yokota & Shimmen, 1994), and an antibody generated against this 170 kDa protein showed a general distribution localized to organelles and the perinuclear region in higher plant cells (Yokota et al., 1995a, 1995b). More recent studies suggest that a 170 kDa and a 175 kDa myosins act as the motors responsible for cytoplasmic streaming in higher plant cells (Yokota et al., 1999). Two myosin-like proteins (170 kDa and 175 kDa) were biochemically isolated from *Nicotiana tabacum* Bright Yellow-2 (BY-2) cells and shown to translocate rhodamine-phalloidin-labeled F-actin *in vitro* and to participate in cytoplasmic streaming. The 175 kDa myosin was able to translocate rhodamine-phalloidin-labelled F-actin at an average velocity of about 9 $\mu m\ s^{-1}$ in the in vitro motility assay whereas the sliding velocity generated by the 170 kDa myosin was 3-4 $\mu m\ s^{-1}$. Futhermore immunoblot analysis indicated that both myosins (170 kDa and 175 kDa) are expressed in tobacco plant leaf, stem and root cells, but only the 170 kDa myosin was expressed in germinating pollen (Yokota et al., 1999). These results suggest that the two types of myosin participate in cytoplasmic streaming in the root and stem cells, but only the 170 kDa myosin is involved in cytoplasmic streaming in pollen cells. The molecular weight of the myosins so far shown to be involved in cytoplasmic streaming indicate that they belong to the myosin class XI.

Due to the multitude of myosins known today from sequence analysis, it is very likely that numerous myosins will be localized in different cell types, in different developmental stages and in different cell compartments, all probably with very specific functions.

6. MYOSIN VIII

Our recent studies on myosin VIII have shown anti-myosin localisation associated with plasmodesmata in *Zea*, *Lepidium* and *Arabidopsis* (Reichelt et al., 1997; Reichelt et al., 1999), which confirms results previously obtained with heterologous antibodies in *Allium* and *Hordeum* (Radford & White, 1998) and using an antibody raised against a mungbean myosin (Qiao et al., 1994) in *Chara* (Blackman & Overall, 1998).

In order to obtain a specific antibody against a myosin VIII, we expressed the tail domain of AtATM1 in *E. coli* and used the purified recombinant protein as an antigen. The tail region consists of a short stretch of predicted α-helical coiled coil (amino acids 950-1019) and a unique C-terminal region (residues 1020-1166) (Knight & Kendrick-Jones, 1993). These AtATM1 tail regions are similar to those in the other myosins of class VIII such as AtATM2, HaMYO1, ZmZMM3 and AtVIII–A and AtVIII–B, and are unique to the class VIII myosins since they do not show any homologies with other myosin classes or other non-myosin proteins. Therefore, the polyclonal antibody (anti-ATM1) we raised against the tail domain of ATM1 (Reichelt *et al.*, 1999) is thought to be specific for the class VIII myosins. In cell extracts, this antibody cross-reacts with polypeptides of 130 kDa, which confirms the predicted molecular weight of the class VIII myosin-1, AtATM1. Western blot analysis and immunolocalization studies have shown that the antibody cross-reacted with root tissue from *Arabidopsis thaliana, Zea mays* and *Lepidium sativum* (Reichelt et al., 1999).

6.1 Immunofluorescence localization of myosin VIII

One of the striking features of the myosin VIII localization in cress root sections is its strong localization in newly deposited cross-walls in contrast to its sparse occurrence along side walls. Figures 4a–c show the immunofluorescence localization of myosin VIII along a newly formed cell wall in roots. In the immunofluorescence image the anti-ATM1 labelling is highly concentrated along the new cell wall (Fig. 4a). In the DIC image (Fig. 4b) and the overlay of immunoflurescence and DIC (Fig. 4c) the maturing cell wall seems to consist of membranous vesicles fusing to consolidate still 'wavy' cell wall. In comparison those cell walls which are not labelled by anti-ATM appear smooth and do not show the dense concentration of vesicles.

To establish the specific stage in cell wall formation at which myosin VIII may be involved, it was necessary to compare the actin and microtubule distribution with that of the myosin VIII. Indeed, myosin VIII labelling is found most strongly when the actin and microtubules have depolymerised, which suggests that it may not be involved in the transport of cell plate precursors to the mid-line of the phragmoplast, but rather in those processes involved in cell plate maturation. In Figures 5a–c, maize cells undergoing division are double-labelled with antibodies against actin and ATM1 myosin, and in Figures 6a–c are double-labelled with antibodies against ATM1 and tubulin. In cells in prophase, actin is localised as dense irregular networks surrounding the enlarged, round nucleus (Fig. 5a). At this stage, myosin VIII labelling is concentrated along the transverse cell walls. During telophase, in the early phragmoplast region (labelled P) where the new cell wall forms between the separating daughter nuclei, actin staining can be seen as a small band in the centre of the newly forming cell plate with myosin-VIII labelling strictly confined to the opposing transverse cell walls. In late stages of cytokinesis, the phragmoplast microtubules form a ring-like structure surrounding the newly developing cell plate (Fig. 6b). Myosin VIII staining does not occur in the cell plate region in either the early or in the developing late phragmoplast nor in the spindle (Fig. 6), but is confined entirely to the opposing transverse cell walls which are strongly stained.

6.2 Association of myosin VIII with the plasma membrane and plasmodesmata

The sections through cress root tips labelled with the anti-ATM1 antibody confirmed that myosin VIII was localized along the whole transverse cell wall, but showed a distinct punctate labelling along the longitudinal walls. These sections (Figs. 6a–f) show that the staining along the newly formed cell wall reaches throughout the whole cell depth, whereas the staining along the longitudinal walls is more spot-like. So far, we can only speculate about the possible function of myosin VIII in the mature cell wall. A number of studies have shown that cell wall formation occurs in two separate stages. For instance, caffeine is known not to affect the formation of the phragmoplast cytoskeleton nor the actual deposition of the central cell plate, but affects the later stage of fusion with the mother wall and the conversion of the callose-rich plate into a less flexible wall (Hepler & Bonsignore, 1990; Valster & Hepler, 1997). Mineyuki and Gunning (1990) have reviewed these later stages of cell division and concluded that insertion and maturation factors interact with the cell plate, or pass into it from the cortical division site. Callose, a major component of the cell plate, is removed from cell plates soon after they attach to the cell wall (Northcote et al., 1989) and pectic polysaccharides are inserted (Moore & Staehlin, 1988, Samuels et al., 1995). The appearance of myosin VIII coincides, therefore, with the onset of these wall-consolidating processes. The myosin could bring the islands of membrane plate material together, or it could trigger the exocytosis of new cell-wall material. High concentrations of calcium - calmodulin are found near the phragmoplast/cell plate (Hepler, 1994) and it is possible that changes in the calcium levels at the transition between plate deposition/maturation affect the localisation and activity of myosin VIII.

Myosin VIII may have another possible role in the system involved in anchoring or directing cytoplasmic F-actin cables through or to plasmadesmata. Since F-actin goes to (or through) plasmodesmata (White et al., 1994), it has been suggested to be involved in the intercellular movement of certain viruses (McLean et al., 1995), proteins and mRNA (Lucas et al., 1996). Such particle movement is most likely to be an exaggerated form of normal directed cell transport. It is interesting therefore, that the density of plasmodesmata is far greater in the transverse end walls than in the longitudinal side walls (Juniper & Barlow, 1969).

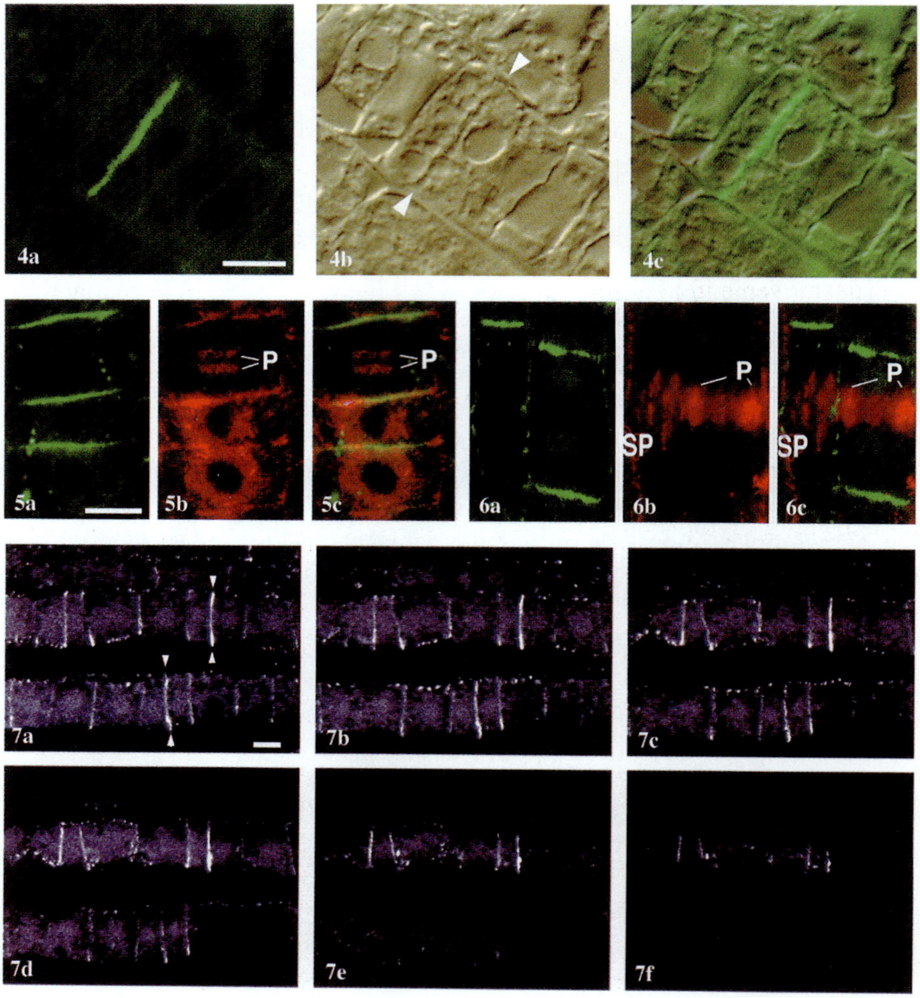

Figure 4. Immunolocalisation of myosin VIII in longitudinal cress root section. a) Myosin VIII staining is highly concentrated along a newly formed transverse cell wall; b) the corresponding DIC image shows that the cell wall is not smooth but 'wavy', and vacuoles and vesicles are densely concentrated along the new wall; c) shows an overlay of a) and b).

Figure 5. Double-labelling immunofluorescence localization of myosin VIII and actin in maize root cells. a) Myosin VIII staining in green; b) the actin staining in red; and c) is an overlay of a) and b). Myosin VIII labelling is concentrated along the transverse cell walls in prophase cells where the actin cytockeleton is a dense network surrounding the round nuclei. In telophase cells where the actin is concentrated in the early phragmoplast (labelled P) between the separating mitotic chromosomes, myosin VIII staining is not localized in the phragmoplast but along the opposing transverse cell walls.

Figure 6. Double-labelling immunofluorescence localization of myosin VIII and tubulin in maize root cells. a) Myosin VIII staining in green; b) the actin staining in red; and c) is an overlay of a) and b). In the spindle (Sp) and in the late phragmoplast stage (P), myosin VIII

staining is predominantly localized along the transverse cell walls and not concentrated in the spindle or in the mid-plane of the late phragmoplast.
Figure 7. Confocal visualisation of cress root cells labelled with anti-myosin VIII antibody. The sections were taken in 5 μm steps. The sections a-f show optical sections through the cress root revealing high concentration of myosin VIII along all transverse cell walls, especially prominent at a newly formed cell wall recognisable by the distinctly smaller cell size of the two newly separated neighbouring cells. Along the longitudinal cell walls the staining is punctate. The bar is 5 microns. Bars = 5 μm.

The movement of virus particles (McLean et al., 1995; Waigmann & Zambryski, 1995, Zambryski, 1995), and the fact that cytochalasin D enlarges the plasmodesmatal neck (White et al., 1994) indicates that there is a gating mechanism which controls the size of the pore and that this is likely to be under contractile control (Overall et al., 1982; White et al., 1994; Ding et al., 1996). Support for such a proposal is provided by two further studies. An antibody against a myosin-like mung bean protein detects a 130 kDa protein in a cytoplasmic protein extract from *Chara* and localizes at plasmodesmata in immunogold labelled sections (Blackman & Overall, 1998), and a polyclonal anti-myosin II (vertebrate smooth/skeletal muscle) antibody localises at plasmodesmata in root tissue of *Allium cepa*, *Zea mays* and *Hordeum vulgare* (Radford & White, 1998). These localization patterns and our own results (Reichelt et al., 1999) strongly suggest that myosin VIII is part of the intercellular gating complex in the plasmodesmata.

7. SUMMARY

It is evident from the myosin genes detected that plant cells contain a multitude of myosin motor proteins. In *Arabidopsis* alone, 16 different myosin genes have been identified. One of the best biochemically studied processes in plant cells is cytoplasmic streaming. This process appears to be driven in *Chara* by a 240 kDa myosin which *in vitro* translocates rhodamine-phalloidin labelled actin filaments at high speed (60 $\mu m\ s^{-1}$) (Yamamoto et al., 1994; Higashi-Fujime et al., 1995). Unfortunately, so far, this myosin has not been cloned and sequenced. However, in higher plants cytoplasmic streaming appears to be associated with at least two myosins of around 170 kDa, which show actin filament translocation *in vitro* with velocities of between 2-9 $\mu m\ s^{-1}$ (Yokota & Shimmen, 1994; Yokota et al., 1995a,b, 1999). The molecular weights of these isolated myosins would indicate they belong to the class XI myosins. Distinct from these well studied processes, our recent results show that a class VIII myosin localizes to the newly formed cell plate and may have a role in establishing the post-cytokinetic cell wall and possibly in plasmodesmata regulation (or transport to or through the plasmodesmata) (Reichelt et al., 1999). Since in animal cells myosins have been shown to be involved in processes like auditory adaptation, cytokinesis, spermatogenesis and vesicle transport, it will be very interesting to establish the diverse functions of the multitude of myosins which have now been identified in plants.

REFERENCES

Asada T and Collings D (1997) Molecular motors in higher plants. Trends Plant Sci 2: 29-37

Bähler M (1996) Myosins on the move to signal transduction. Curr Opin Cell Biol 8: 18-22

Blackman LM and Overall RL (1998) Immunolocalization of the cytoskeleton to plasmodesmata of *Chara corallina*. Plant J 14: 733-741

Cheney RE and Mooseker MS (1992) Unconventional myosins.Curr Opin Cell Biol 4: 27-35

Cope MJTV, Whisstock J, Rayment I and Kendrick-Jones J (1996) Conservation within the motor domain: Implications for structure and function. Structure 4: 969-987

Ding B, Kwon M-O and Warnberg L (1996). Evidence that actin filaments are involved in controlling the permeability of plasmodesmata in tobacco mesophyll. Plant J 10: 157-164

Fath KR and Burgess DR (1994) Membrane motility mediated by unconventional myosin. Curr Opin Cell Biol 6: 131-135

Grolig F, Williamson RE, Parke J, Miller C and Anderton BH (1988) Myosin and Ca^{2+}-sensitive streaming in the alga *Chara*: Detection of two polypeptides reacting with a monoclonal anti-myosin and their localization in the streaming cytoplasm. Eur J Cell Biol 47: 22-31

Grolig F, Schroeder J, Sawitzky H, and Lange U (1996) Partial characterization of a putative 110 kDa myosin from the green alga *Chara corallina* by in vitro binding of fluorescent F-actin. Cell Biol Int 20: 365-373

Hammer III JA (1994) The structure and function of unconventional myosins: A review. J Muscle Res Cell Motil 15: 1-10

Hasson T and Mooseker MS (1995) Molecular motors, membrane movements and physiology: Emerging roles for myosins. Curr Opin Cell Biol 7: 587-594

Hepler PK (1994) The role of calcium in cell division. Cell Calcium 16: 322-330

Hepler PK and Bonsignore CL (1990) Caffeine inhibition of cytokinesis: Ultrastructure of cell plate formation/degradation. Protoplasma 157: 182-192

Heslop-Harrison J and Heslop-Harrison Y (1989) Myosin associated with the surface of organelles, vegetative nuclei and generative cells in angiosperm pollen grains and tubes. J Cell Sci 94: 319-325

Higashi-Fujime S, Ishikawa R, Iwasawa H, Kagami O, Kurimoto E, Kohama K and Hozumi T (1995) The fastest actin-based motor protein from the green algae, *Chara*, and its distinct mode of interaction with actin. FEBS Lett 375: 151-154

Juniper BE and Barlow PW (1969) The distribution of plasmodesmata in the root tip of maize. Planta 89: 352-360

Kachar B and Reese TS (1988) The mechanism of cytoplasmic streaming in *Characean* algal cells: Sliding of endoplasmic reticulum along actin filaments. J Cell Biol 106: 1545-1552

Kendrick-Jones J and Reichelt S (1999) 'Myosin, plant' In The Guidebook of Cytoskeletal and Motor Proteins, 2nd ed., Kreiss T & Vale R (eds) Oxford: Oxford University Press

Kinkema M and Schiefelbein JW (1994) A myosin from a higher plant has structural similarities to class V myosins. J Mol Biol 239: 591-597

Kinkema M, Wang H and Schiefelbein JW (1994) Molecular analysis of the myosin gene family in *Arabidopsis thaliana*. Plant Mol Biol 26: 1139-1153

Knight AE and Kendrick-Jones J (1993) A myosin-like protein from a higher plant. J Mol Biol 231: 148-154

Kohno T and Shimmen T (1988) Accelerated sliding of pollen-tube organelles along *Characean* actin bundles regulated by Ca^{2+}. J Cell Biol 106: 1539-1543
Kuroda K (1990) Cytoplasmic streaming in plant cells. Int Rev Cytol 121: 267-307
La Claire II JW (1991) Immunolocalization of myosin in intact and wounded cells of the green alga *Ernodesmis verticillata* (Kützing) Borgesen. Planta 184: 209-217
LaClaire II JW, Chen R and Herrin DL (1995) Identification of a myosin-like protein in *Chlamydomonas rheinhardtii* (Chlorophyta) J Phycol 31: 302-306
Lucas WJ, Bouché-Pillon S, Jackson DP, Nguyen L, Baker L, Ding B and Hake S (1996) Selective trafficking of KNOTTED1 homeodomain protein and its mRNA through plasmodesmata. Science 270: 1981-1982
McCurdy DW and Williamson RE (1991) Actin and actin-associated proteins. In: Lloyd CW (ed) The Cytoskeletal Basis of Plant Growth and Form. London: Academic Press Ltd, pp 3-14
McLean BG, Zupan J and Zambryski C (1995) Tobacco mosaic virus movement protein associates with the cytoskeleton in tobacco cells. Plant Cell 7: 2101-2114
Mermall V, Post PL and Mooseker MS (1998) Unconventional myosins in cell movement, membrane traffic and signal transduction. Science 279: 527-533
Miller DD, Scordilis SP and Hepler PK (1995) Identification and localization of three classes of myosins in pollen tubes of *Lilium longiflorum* and *Nicotiana alata*. J Cell Sci 108: 2549-2653
Mineyuki Y and Gunning BES (1990) A role for preprophase bands of microtubules in maturation of new cell walls, and a general proposal of the function of preprophase band sites in cell division in higher plants. J Cell Sci 97: 527-537
Moepps B, Conrad S and Schraudolf H (1993) PCR-dependent amplification and sequence characterization of partial DNAs encoding myosin-like proteins in *Anemia phyllitidis* (L.) Sw. and *Arabidopsis thaliana* (L.) Heynh. Plant Mol Biol 21: 1077-1083
Mooseker MS and Cheney RE (1995) Unconventional myosins. Annu Rev Cell Dev Biol 11: 633-675
Northcote DH (1989) Control of plant cell wall biogenesis - An overview. ACS Symposium Series 399: 1-15
Moore PJ and Staehelin LA (1988) Immunogold localization of the cell-wall matrix polysaccharides rhamnogalacturonan-1 and xyloglucan-1 during cell expansion and cytokinesis in *Trifolium pratense* L. - implication for secretory pathways. Planta 174: 433-445
Overall RL, Wolffe J and Gunning BES (1982) Intercellular communication in Azolla roots: I. Ultrastructure of plasmodesmata. Protoplasma 111: 134-150
Page RD (1996) TreeView: an application to display phylogenetic trees on personal computers. Comput Appl Biosci 12: 357-358
Peterson DM, Urioste AS and Titus MA (1996) Dictyostelium discoideum myoJ: A member of a broadly defined myosin V class or class XI unconventional myosins. J Muscle Res Cell Motil 17: 411-424
Plazinski J, Elliott J, Hurley UA, Burch J, Arioli T and Williamson RE (1997) Myosins from Angiosperms, ferns and algae. Amplification of gene fragments with versatile PCR primers and detection of protein products with a monoclonal antibody to a conserved head epitope. Protoplasma 196: 78-86
Qiao L, Grolig F, Jablonsky PP and Williamson RE (1989) Myosin heavy chains: Detection by immunoblotting in higher plants and localization by immunofluorescence in the alga *Chara*. Cell Biol Int Rep 13: 107-117
Radford JE and White RG (1998) Localization of a myosin-like protein to plasmodesmata. Plant J 14: 743-750
Rayment I, Rypniewski WR, Schmidt-Base K, Smith R, Tomchick DR, Benning MM, Winkelmann DA, Wesenberg G and Holden HM (1993) Three-dimensional structure of myosin subfragment-1: A molecular motor. Science 261: 50-58
Reichelt S, Knight AE, Hodge TP, Baluška F, Volkmann D and Kendrick-Jones J (1997) Characterization of a plant myosin. Mol Biol Cell 7 (Suppl.) 216a

Reichelt S, Knight AE, Hodge TP, Baluška F, Šamaj J, Volkmann D and Kendrick-Jones J (1999) Characterization of the unconventional myosin VIII in plant cells and its localization at the post-cytokinetic cell wall. Plant J 19: 555-567

Samuels AL, Giddings Jr. JH and Staehelin LA (1995) Cytokinesis in tobacco BY-2 and root tip cells: A new model of cell plate formation in higher plants. J Cell Biol 130: 1345-1357

Sheetz MP and Spudich JA (1983) Myosin coated fluorescent beads move on actin cables 'in vitro'. Cell Motil Cytoskel 3: 708-709

Tang X, Hepler PK and Scordilis SP (1989) Immunochemical and immunocytochemical identification of a myosin heavy chain polypeptide in Nicotiana pollen tubes. J Cell Sci 92: 569-574

Thompson JD, Higgins DG and Gibson TJ (1994) Clustal-W - improving the sensitivity of progressive multiple alignments through sequence weighting, position-sensitive gap penalties and weight matrix choice. Nucl Acids Res 22: 4673-4680

Titus MA (1993) Myosins. Curr Opin Cell Biol 5: 77-81

Turkina MV, Kulikova AL, Sokolov OI, Bogatyrev VA and Kursanov AL (1987) Actin and myosin-filaments from the conducting tissues of *Heracleum sosnowskyi*. Plant Physiol Biochem 25: 689-696

Vahey M, Titus M, Trautwein R and Scordilis S (1982) Tomato actin and myosin: Contractile proteins from a higher land plant. Cell Motil Cytoskel 2: 131-147

Valster AH and Hepler PK (1997) Caffeine inhibition of cytokinesis: Effect on the phragmoplast cytoskeleton in living *Tradescantia* stamen hair cells. Protoplasma 196: 155-166

Waigmann E and Zambryski P (1995) Tobacco mosaic virus movement protein-mediated protein transport between trichome cells. Plant Cell 7: 2069-2079

White RG, Badelt K, Overall RL and Vesk M (1994) Actin associated with plasmodesmata. Protoplasma 180: 169-184

Williamson RE (1993) Organelle movements. Annu Rev Plant Physiol Plant Mol Biol 44: 181-202

Yamamoto K, Kikuyama M, Sutoh-Yamamoto N and Kamitsubo E (1994) Purification of actin based motor protein from *Chara chorallina*. Proc Jap Acad Ser B 70: 175-180

Yamamoto K, Kikuyama M, Sutoh-Yamamoto N, Kamitsubo E and Katayama E (1995) Myosin from alga *Chara*: Unique structure revealed by electron microscopy. J Mol Biol 254: 109-115

Yokota E and Shimmen T (1994) Isolation and characterization of plant myosin from pollen tubes of lily. Protoplasma 177: 153-162

Yokota E, McDonald AR, Liu B, Shimmen T and Palewitz BA (1995a) Localization of a 170 kDa myosin heavy chain in plant cells. Protoplasma 185: 178-187

Yokota E, Mimura T and Shimmen T (1995b) Biochemical, immunochemical, and immunohistochemical identification of myosin heavy chains in cultured cells of *Catharanthus roseus*. Plant Cell Physiol 36: 1541-1547

Yokota E, Yukawa C, Muto S, Sonobe S and Shimmen T (1999) Biochemical and immunocytochemical characterization of two types of myosins in cultured tobacco bright yellow-2 cells. Plant Physiol 121: 525-534

Zambryski P (1995) Plasmodesmata: Plant channels for molecules on the move. Science 270: 1943-1944

Chapter 3

PROFILIN

Bryan C Gibbon and Christopher J Staiger

Department of Biological Sciences, Purdue University, West Lafayette, IN 47907-1392, USA

Key words: actin-binding protein, formin, polyphosphoinositides, profilin, VASP

Abstract: Profilin has complex effects on the actin cytoskeleton; it can bind to monomeric actin and thereby inhibit polymerization, yet under other conditions it stimulates polymerization. Profilin also binds to polyphosphoinositides and proline-rich proteins which are implicated in signaling events that result in reorganization of the actin cytoskeleton. Plant profilins share many of the same biochemical properties and are structurally similar to non-plant profilins. Microinjection of profilin into plant cells inhibits cytoplasmic streaming, tip growth and cytokinesis through the depolymerization of actin microfilaments. Furthermore, measurements of profilin and actin levels in pollen are consistent with profilin sequestering the bulk of the actin monomer pool. Together these results suggest that profilin serves to sequester actin in plant cells. This simple model is complicated by the fact that there are two functionally-distinct classes of profilin, at least in maize, and in some cells both classes of profilin are expressed in the same cytoplasm. One class of profilin appears to have a high affinity for pollen actin and poly-L-proline, whereas the other class has a 2–4-fold lower affinity for actin and a 2-fold lower affinity for poly-L-proline.

1. OVERVIEW OF PROFILIN FUNCTION

Profilin is a small (12–15 kDa) actin-binding protein that has complex effects on the actin cytoskeleton. Discovered originally by Carlsson et al. (1977) as a polypeptide that copurified with actin from extracts of spleen, profilin has been found in organisms from all eukaryotic kingdoms. Profilin has since been shown to have interactions with signaling molecules such as phosphatidylinositol-4,5-bisphosphate (PtdIns(4,5)P_2; Lassing & Lindberg, 1985), proteins with proline-rich motifs (reviewed by Wasserman, 1998) and a complex of proteins that includes 2 actin-related proteins (Machesky et al.,

C.J. Staiger et al. (eds.), Actin: A Dynamic Framework for Multiple Plant Cell Functions, 45–65.

1994). These interactions are proposed to link profilin to signal transduction cascades that result in the reorganization of the actin cytoskeleton. The complex effects of profilin on actin organization undoubtedly result from the diversity of potential interactions. This chapter will review the diverse functions of profilin and will discuss what is known about the role of profilin in plant cells.

2. NON-PLANT PROFILINS

2.1 Interaction with actin

The level of filamentous actin (F-actin) in most cells is much lower than predicted by the biochemical properties of purified actin *in vitro*. Therefore, there must be mechanisms that regulate the polymerization or depolymerization of actin *in vivo*. Profilin was proposed to sequester actin monomers (G-actin) in the cytosol because it was an abundant protein and could be isolated in a high affinity 1:1 complex with G-actin (Carlsson et al., 1977). The primary effect of profilin *in vitro* is to decrease the extent of actin polymerization by binding to actin monomers and preventing their incorporation into actin filaments. Therefore, it is not surprising that when profilin is microinjected into cells, F-actin levels decrease (Cao et al., 1992; Staiger et al., 1994). This simple model has several shortcomings; in particular, the amount of profilin in many cells can not account for the size of the G-actin pool. Furthermore, when actin is overexpressed in vertebrate cells the amount of F-actin actually increases (Finkel et al., 1994), which is the opposite result one would expect if profilin sequestered G-actin. This apparent paradox was explained later in elegant biochemical experiments that demonstrate profilin can stimulate polymerization from a pool of actin complexed with another G-actin binding protein (thymosin β4). This only occurs when the fast-growing (barbed) ends of the actin filaments are free of capping proteins (Pantaloni & Carlier, 1993; Kang et al., 1999). Another activity of profilin, that can result in more actin polymer, is its ability to promote exchange of the nucleotide bound to actin (Goldschmidt-Clermont et al., 1992). Because ATP–actin polymerizes more readily than ADP–actin, this would lead to actin polymerization. There is currently a very active debate about the relative importance of these two mechanisms for promoting actin polymerization (Perelroizen et al., 1996; Korenbaum et al., 1998; Vinson et al., 1998; Selden et al., 1999).

The complexity of profilin and actin interaction is also evident from a number of studies on profilin mutants (reviewed by Ayscough, 1998). The phenotypes obtained from several profilin mutants suggest that profilin is a monomer-sequestering protein. When profilin deletion mutants are created in yeast or *Dictyostelium* there is an increase in the abundance of actin microfilaments and an inability to complete cytokinesis (Haarer et al., 1990; Haugwitz et al., 1994). These results are consistent with the loss of sequestering protein activity. Other genetic studies indicate that profilin may function to stimulate actin polymerization. When vertebrate or plant profilin is overexpressed in mammalian cells, the level of F-actin increases (Finkel et al., 1994; Rothkegel et al., 1996). In the case of vertebrate profilin, the rate of filament turnover apparently decreases (Finkel et al., 1994). Furthermore, mutations in the *chickadee* locus, which encodes profilin in *Drosophila*, result in aberrations of bristle structure and disruption of the polarity of the oocyte due to reorganization or reduction in actin microfilament abundance (Manseau et al., 1996). Interestingly, profilin appears to be required to recover from the transient depolymerization of actin that occurs in response to heat stress in yeast (Yeh & Haarer, 1996). The complex nature of the effect of profilin on actin organization is, therefore, likely to depend on the individual cell type and the particular signals that each cell receives.

2.2 Interaction with signal transduction intermediates

Remodeling of the actin cytoskeleton occurs in response to a number of intrinsic and extrinsic stimuli and profilin has been suggested to play a role in the reorganization (Schmidt & Hall, 1998). A potential link for profilin with intracellular signaling is implied by its ability to bind to membrane polyphosphoinositides, such as PtdIns(4,5)P_2 (Lassing & Lindberg, 1985; 1988) and PtdIns(3,4,5)P_3 (Lu et al., 1996). The association of profilin with PtdIns(4,5)P_2 inhibits hydrolysis by phospholipase C (PLC), unless PLC is phosphorylated (Goldschmidt-Clermont et al., 1990, 1991). This is proposed to provide a mechanism in which profilin bound to PtdIns(4,5)P_2 can be released into the cytoplasm to cause reorganization of actin upon activation of PLC by tyrosine phosphorylation (Machesky & Pollard, 1993; Sohn & Goldschmidt-Clermont, 1994). Experimental evidence that profilin is associated with PtdIns(4,5)P_2 in cells is lacking. Profilin fractionates with plasma membranes in yeast (Ostrander et al., 1995), but the experiments to demonstrate that the interaction occurs with PtdIns(4,5)P_2 *in vivo* are not convincing. Profilin II of *Acanthamoeba*, which has a high affinity for PtdIns(4,5)P_2, is localized at the periphery of migrating cells, whereas profilin I has a cytoplasmic distribution (Bubb et al., 1998). However, it remains to be demonstrated that profilin is released from interaction with PtdIns(4,5)P_2 into the cytoplasm to cause rearrangements of actin.

Profilin also interacts with a number of proteins that have specific proline-rich sequences, including vasodilator stimulated phosphoprotein

(VASP) and a number of formin-like proteins (Reinhard et al., 1995; Frazier & Field, 1997; Purich & Southwick, 1997; Wasserman, 1998). Much of what is known about the interaction between VASP and profilin comes from experiments using the pathogenic bacterium *Listeria monocytogenes* (Cossart, 1995; Carlier, 1998). *Listeria* is propelled through the cytoplasm of a host cell by polymerization of actin at one end of the bacterium. A number of host cell factors, including VASP, are involved in this process. Profilin is proposed to lower the critical concentration for actin assembly on the surface of the bacterium. Recently, the motility of *Listeria* was reconstituted *in vitro* with purified proteins (Loisel et al., 1999). These experiments demonstrate that profilin and VASP are not absolutely required for motility but rather allow a 2–3 fold increase in the rate of translocation.

The second class of proteins that interact with profilin via proline-rich sequences has a domain homologous to formin transcription factors, which are involved in limb bud initiation in vertebrates (Wasserman, 1998). The formin-homology domain 1 (FH1) proteins that interact with profilin are not transcription factors, but are cytoplasmic proteins that interact with members of the Rho family of GTPases. Monomeric Rho GTPases are molecular switches that are implicated in signal transduction cascades that lead to the reorganization of the actin cytoskeleton into specific arrays such as stress fibers, microspikes or lamellipodia (reviewed by Hall, 1994; Zigmond, 1996). The FH1 proteins function during cytokinesis in *Schizosaccharomyces* (*Cdc12*; Chang et al., 1997), *Aspergillus* (*sepA*; Harris et al., 1997) and *Drosophila* (*diaphanous*; Castrillon & Wasserman, 1994), whereas others are involved in the establishment or maintenance of polarity in *Saccharomyces* (*Bni1*; Evangelista et al., 1997) and *Drosophila* (*cappuccino*; Manseau et al., 1996). Some organisms have multiple formin-homology proteins that interact with specific Rho isoforms. For example in *Saccharomyces*, Bnr1p interacts specifically with Rho4p, whereas Bni1p interacts with Rho1p (Kohno et al., 1996; Imamura et al., 1997). Because both Bni1p and Bnr1p interact with profilin, we speculate that profilin may serve as a key transducer of multiple signals.

2.3 Association with a multiprotein complex

Profilin also interacts with a complex of 7 polypeptides that includes 2 actin-related proteins, referred to as the Arp2/3 complex. The Arp2/3 complex was originally purified by its ability to bind a profilin-agarose matrix (Machesky et al., 1994), and has subsequently been shown to bind the sides of microfilaments and nucleate actin polymerization (Mullins et al., 1998a; Welch et al., 1998). These activities are postulated to organize the meshwork of actin filaments at the leading edge of migrating cells by "dendritic nucleation" of actin polymerization (Mullins et al., 1998a). The role of the Arp2/3 association with profilin is not clear; however, it is suggested that profilin might promote nucleotide exchange on Arp2 (Mullins et al., 1998b).

Two members of the Arp2/3 complex in *Schizosaccharomyces* have been shown to interact genetically with profilin. A suppressor of the *cdc3*-124 profilin mutation, *Suppressor Of Profilin* 2 (*sop2*), restores cytokinesis at restrictive temperatures and appears to be a homologue of p40 in the *Acanthamoeba* Arp2/3 complex (Balasubramanian et al., 1996). A mutation in another component of the complex, *arp3*-c1, partially suppresses the *cdc3*-124 profilin mutation (McCollum et al., 1996). The role of these interactions in the organization of the actin cytoskeleton is not immediately apparent because the cytokinetic defect of *cdc3*-124 is rescued, yet Sop2p and Arp3p do not appear to associate with the actin ring. Understanding the role of the interaction between profilin and the Arp2/3 complex requires further investigation.

3. PLANT PROFILINS

Most of what is known about profilin biochemistry and cellular function results from the study of non-plant systems, but it has become clear that plants provide an excellent system to study the role of multiple profilin isoforms. Profilin was discovered in plants as an allergen from birch pollen by Valenta and coworkers (1991), and was later confirmed to be an actin-binding protein (Valenta et al., 1993). In this section, the focus is on the cell biology and biochemistry of profilin, rather than its impact as a plant panallergen. Most plants investigated in any detail contain multigene families of profilin that are expressed in a tissue-specific or developmentally-regulated manner (Staiger et al., 1997). The plant profilins are able to complement profilin-deletion mutants in *Dictyostelium* (Karakesisoglou et al., 1996) and *Schizosaccharomyces* (Christensen et al., 1996). This indicates that plant profilins have similar functional properties to non-plant profilins, despite sharing only ~25% amino acid sequence identity.

3.1 Multigene families

The existence of multiple profilin isoforms in plants was first demonstrated in maize by the cloning of 3 profilin isoforms from a pollen cDNA library (Staiger et al., 1993). Subsequent investigation has shown that profilin comprises a large gene family of 5–10 members in many plant species, including maize (Staiger et al., 1993; Gibbon et al., 1998; Kovar et al., 2000), *Arabidopsis* (Christensen, et al., 1996; Huang et al., 1996) and tobacco (Mittermann et al., 1995, 1996). Profilin has also been cloned and characterized from bean (Vidali et al., 1995; Guillén et al., 1999) and tomato (Yu et al., 1998). Because of the interest in profilin as a panallergen, profilins have also been cloned from timothy grass (Valenta et al., 1994; Asturias et al., 1997a), wheat (Rihs et al., 1994), bermuda grass (Asturias et al., 1997c), olive (Asturias et al., 1997b), sunflower (Asturias et al., 1998), *Mercurialis* (Vallverdu et al., 1998), *Capsicum* (Jensen-Jarolim et al., 1998), rubber tree (GenBank Acc. No. O65812 & CAA75312), cherry (#AAD29411), pear (#AAD29410), apple (#AAD29412, AAD29413 & AAD29414), soybean (#CAA11755 & CAA11756), barley (#AAA92503) and celery (#AAD29409). These plant profilins share just 25–30% amino acid identity with non-plant profilins.

Profilin isoforms are expressed in a tissue-specific and/or developmentally-regulated manner. In *Arabidopsis,* at least two profilin isoforms are expressed specifically in floral tissues or pollen and three other profilins are expressed in other tissues of the plant (Christensen et al., 1996; Huang, et al., 1996). Promoter-GUS fusions demonstrate that at least one of the latter profilin isoforms is expressed in a tissue-specific manner; profilin 2 is found specifically in vascular bundles (Christensen et al., 1996). In maize, three isoforms are expressed predominantly in pollen (Staiger et al., 1993), while two other isoforms are present throughout the plant body (Gibbon et al., 1998; Kovar et al., 2000). Comparison of available plant profilin sequences and generation of phylogenetic trees with parsimony and neighbor-joining methods show that the grass profilin isoforms form a unique clade in the phylogram. Furthermore, the pollen abundant maize profilin isoforms form a clade that is distinct from the isoforms that are expressed throughout the plant but neither share the same clade as previously identified "vegetative" or "reproductive" classes of profilin from *Arabidopsis* (Staiger, unpublished results). Therefore, the hypothesis that profilin multigene families diverged into "vegetative" and "reproductive" classes early in plant evolution (Huang et al., 1996; see Meagher et al., 1999b) should be re-evaluated with all currently available profilin sequences.

The diversity of profilin isoforms in plants raises a question: Do the different profilin isoforms have distinct functions? Several models are currently considered for the existence of cytoskeletal gene families (Meagher, 1991; Staiger et al., 1997; Meagher et al., 1999a; see also Meagher et al., this volume). First, the isoforms may be required for expression in specific tissues or at specific developmental stages. Alternatively, the profilin isoforms may have unique functional properties and therefore may be best suited for specific cell types or developmental stages. An extension of the previous possibility is that multiple profilin isoforms may be expressed in individual cells and perform distinct functions. Finally, it is possible that multiple genes are required for high levels of expression in specialized cells such as pollen. These possibilities are not mutually exclusive.

3.2 Crystal structures

One reason that highly divergent profilin isoforms from different organisms complement each other genetically is elucidated by the crystal structures of birch pollen profilin (Thorn et al., 1997) and *Arabidopsis* profilin 1 (Fedorov et al., 1997). The overall fold of the molecule is strikingly similar to other profilin molecules, with α-helices surrounding a central β-sheet (Fedorov et al., 1997; Thorn et al., 1997). However, birch profilin is somewhat unusual because the N-terminal α-helix is not parallel to the long C-terminal α-helix. Structures for profilin from vertebrates (Schutt et al., 1993; Cedergren-Zeppezauer et al., 1994; Fedorov et al., 1994b; Metzler et al., 1995) and protists (Fedorov et al., 1994a), as well as co-crystals with β-actin (Schutt, et al., 1993) and poly-L-proline (PLP) (Mahoney et al., 1997) have been published in recent years. Figure 1 shows ribbon diagrams of human platelet profilin, *Arabidopsis* profilin 1 and birch pollen profilin. Comparison of these structures reveals several interesting points. First, the most highly-conserved residues contribute sidegroups that form a hydrophobic patch for binding PLP (Fig. 1D). The co-crystal of human profilin I bound to PLP demonstrates that many of these conserved residues make direct contact with some portion of the PLP chain (Mahoney et al., 1997). The conservation of these residues further implies that they are critical for maintaining the profilin fold (Thorn et al., 1997) or that interaction with proline-rich ligands is critical for profilin function (Fedorov et al., 1997). Surprisingly, the residues predicted to contact actin, inferred from the contacts observed in the co-crystal of bovine profilin and β-actin (Schutt et al., 1993), are not well conserved among the profilins from different kingdoms (Thorn et al., 1997). The reason for this could be that each profilin may be specifically suited to interact with the actin isoforms found in that organism.

3.3 Interaction with actin

The binding of plant profilin to actin was first analyzed using vertebrate actin. Birch profilin binds to muscle α-actin and brain β/γ-actin with a K_d value of ~5 μM (Giehl et al., 1994), a value that is 10-fold higher than the K_d for vertebrate profilin I but similar to vertebrate profilin II (Table 1;

Lambrechts et al., 1995). *Arabidopsis* profilin 1 and 3 bind to rabbit skeletal muscle α-actin (RSMA) with a K_d value around 2 μM (Perelroizen et al., 1996). Interestingly, birch profilin and the *Arabidopsis* isoforms can promote polymerization of RSMA from a pool of thymosin β4–actin complexes (Perelroizen et al., 1996; Ballweber et al., 1998). The relevance of this activity has not been investigated, but may be questionable because plants do not have thymosin β4 homologs.

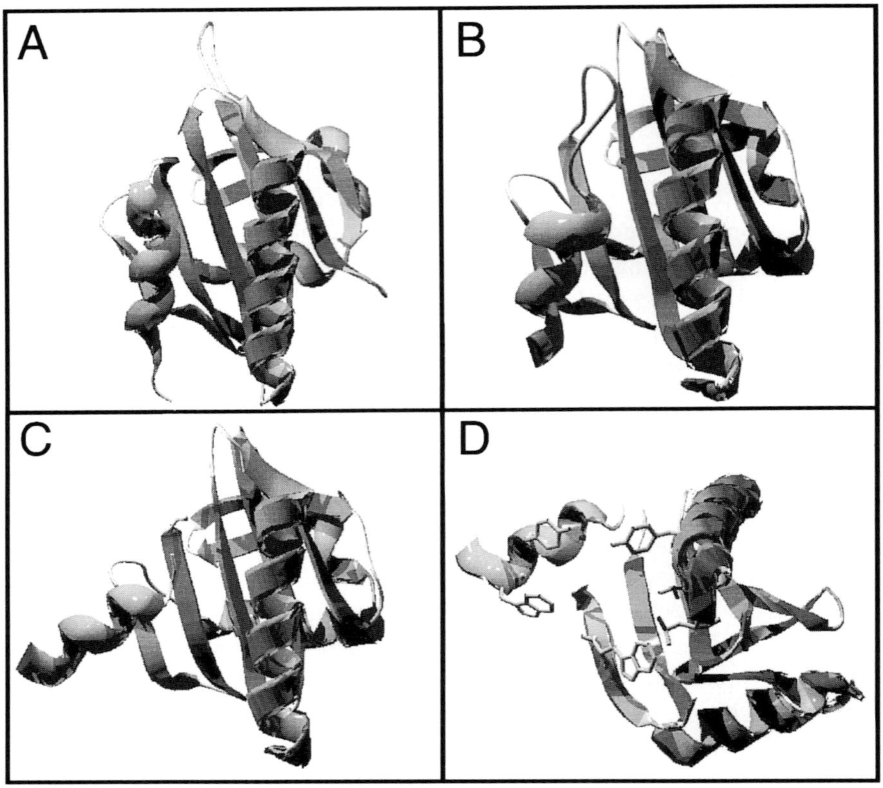

Figure 1. Ribbon diagrams of human platelet profilin (A; Fedorov et al., 1994b), *Arabidopsis* profilin 1 (B; Thorn et al., 1997) and birch pollen profilin (C; Fedorov et al., 1997). PLP binding occurs at a patch of solvent exposed hydrophobic residues (D). The hydrophobic sidegroups are diagrammed as stick figures on birch pollen profilin.

Interactions between plant profilin and plant actin have also been demonstrated. Actin copurifies with profilin from maize pollen (Staiger et al., 1993) or birch pollen (Valenta et al., 1993). Moreover, actin from the green alga *Chara* copurifies with profilin from maize coleoptiles or pea shoots that is added to the algal extracts (Ruhlandt et al., 1994). Two methods to purify polymerizable actin from maize pollen (Liu & Yen, 1992;

Ren et al., 1997) have allowed the study of the association of plant actin-binding proteins with purified plant actin *in vitro*. Because plants have many profilin isoforms it is likely that isoforms that are expressed in different tissues or at different times during development have different functional properties. The association of plant actin with multiple profilin isoforms is best characterized in maize. The recombinant and native profilin isoforms that are expressed predominantly in pollen (Class I) have an affinity for maize pollen actin that is similar to that reported for *Arabidopsis* profilin binding to skeletal muscle actin (Gibbon et al., 1997, 1998; Kovar et al., 2000; Table 1).

Table 1. **Affinity of profilin for known ligands**

	K_d values			
Profilin source[a]	Actin[b]	PLP[f]	PIP_2[b]	References
Human profilin 1	0.1–0.5[c]	359	0.2	Gieselmann et al., 1995 Hájková et al., 1997 Petrella et al., 1996 Sohn et al., 1995
Betula verrucosa	1.8–6.5[c]	150	24	Giehl, et al., 1994 Ballweber, et al., 1998 Domke et al., 1997 Drøbak et al., 1994
Zea mays				
Class I:				Gibbon et al., 1997, 1998
PRO1	1.1–1.2[d]	275–289	PRO1>PRO5[g]	Kovar et al., 2000
PRO2	ND	262		
PRO3	ND	251		
Pollen	0.7	249–250		
Class II:				
PRO4	0.3–0.4	167–173		
PRO5	0.3	164		
Endosperm	0.3	126		
Arabidopsis	1.8–2.3[c]	ND	ND	Perelroizen et al., 1996
Papaver rhoeas	ND[e]	130	ND	Clarke et al., 1998
Ricinus communis	0.07[d]	121	ND	Schobert et al., 2000

[a] Except for native maize pollen, endosperm and *P. rhoeas* pollen profilin, all proteins are recombinant.
[b] Values for K_d are reported as µM.
[c] Values for binding to RSMA at steady state
[d] Apparent K_d values for binding to maize pollen actin at steady state, in the presence of 55 µM calcium.
[e] Not determined
[f] Values are µM proline residues, except for birch which is reported as µM proline decamer
[g] Qualitative determination by inhibition of PLC activity

Interestingly, the recombinant and native maize profilin isoforms that are expressed in other tissues of the plant (Class II) have a 2–4-fold higher apparent affinity for pollen actin than the Class I profilins (Kovar et al., 2000; Table 1). These differences in affinity for pollen actin are not observed under low ionic strength conditions, indicating that physiological ionic conditions are required to obtain accurate dissociation constants. Another recombinant profilin from *Ricinus* also binds to pollen actin with very high affinity (Schobert et al., 2000). Its expression pattern and affinity for actin suggest that this dicot profilin also belongs to Class II. However, this may not be a universal property of non-pollen profilins because no difference in affinity for RSMA is observed between the *Arabidopsis* profilin isoforms (Perelroizen et al., 1996). Overall, these data suggest that profilin isoforms may serve different functions in specific tissues or at specific developmental stages.

Other potential interactions of plant profilin with actin could affect the organization of actin in cells. First, a well-established feature of many eukaryotic profilins is the ability to promote nucleotide exchange on actin (Ayscough, 1998). However, it appears that plant profilins do not share this activity. *Arabidopsis* profilins do not promote nucleotide exchange on RSMA (Perelroizen et al., 1996), birch profilin doesn't stimulate exchange on yeast actin (Eads et al., 1998), and maize profilins actually inhibit nucleotide exchange on maize pollen actin, albeit slightly (Kovar et al., 2000). Second, it has been shown recently that the interaction of maize profilin isoforms with pollen actin is dependent on Ca^{2+} (Kovar, et al., 2000). Specifically, all of the recombinant and native maize profilins tested have higher apparent K_d values for pollen actin in the presence of nM calcium concentrations. It is not clear at present whether this is due to a direct effect of Ca^{2+} on profilin or whether maize profilin has higher affinity for Ca-ATP-actin. Together, these activities could have the net effect of maintaining a pool of actin that is less readily incorporated into new polymers, in contrast to non-plant profilins which are believed to maintain a pool of readily polymerizable actin.

A quantitative analysis of profilin and actin levels in pollen suggests that the primary role of profilin is to sequester G-actin. In lily pollen, profilin and actin levels are approximately the same after germination (~25 μM) and it is argued that profilin serves to sequester G-actin because it is uniformly distributed in the cytoplasm (Vidali & Hepler, 1997). However, the concentration of F-actin was not measured. In maize, the levels of profilin and actin are also equimolar, ~125 μM. Moreover, measurement of F-actin levels reveals that the amount of total actin protein that is filamentous is very low, ~11 μM. The affinity of the endogenous profilin for pollen actin (apparent K_d = 0.9 μM; Kovar et al., 2000) is consistent with profilin serving to sequester the unpolymerized actin (Gibbon et al., 1999). This model must be tested directly by manipulating profilin levels.

Because there are many profilin and actin isoforms expressed in plants it is unclear whether in all cell types profilin will be the primary regulator of G-actin levels. During the early stages of maize microspore development, the level of profilin is insufficient to account for the measured levels of F-actin (our unpublished data). Another important possibility that has not been explored is that profilin isoforms may have different affinities or activities in the presence of particular actin isoforms (Meagher et al., 1999a). The complexity of the actin and profilin gene families in plants warrants characterization of the interaction of profilin isoforms with nonreproductive sources of actin.

3.4 Interactions with proline-rich sequences

Profilin is easily purified from plant tissues by affinity chromatography on PLP-Sepharose, and there are several detailed studies on the interaction between plant profilin and PLP. The association of birch pollen profilin was measured with titration followed by NMR analysis and is found to have a similar affinity for PLP as *Acanthamoeba* profilin, ~200 μM proline decamers (Domke et al., 1997; Table 1). Birch pollen profilin also colocalizes with endogenous profilin when overexpressed in vertebrate cells (Rothkegel et al., 1996), and it was proposed that this colocalization was due to interaction with VASP because birch profilin binds the $(GPPPPP)_3$ peptide sequence found in VASP (Reinhard et al., 1995). The relevance of these interactions remains unclear, however, because there are no reports of VASP-like proteins or proteins with formin-homology domains in plants.

The interaction of profilin isoforms with proline-rich sequences could result in the differential regulation of profilin function in individual cells. Recent evidence indicates that the binding of multiple profilin molecules to proline-rich sequences derived from VASP increases the ability of vertebrate profilin to stimulate or nucleate actin assembly (Jonckheere et al., 1999). In addition to being better actin-binding proteins, the Class II profilins bind to PLP with higher affinity than the Class I isoforms (Table 1; Gibbon et al., 1998; Kovar et al., 2000). Poppy pollen profilin, binds to PLP with an affinity that is higher than observed for the maize pollen profilin isoforms: however, actin binding under physiological conditions has not been tested (Clarke et al., 1998). *Ricinus* profilin also has a very high affinity for PLP (Schobert et al., 2000). As mentioned for actin binding, the possibility that high affinity for PLP is a general feature of Class II profilin isoforms awaits biochemical analysis of multiple isoforms from a dicot.

3.5 Links to signaling cascades

There are several indications that plant profilins are linked to signal transduction cascades. The binding of birch pollen profilin to PtdIns(4,5)P_2 was investigated by virtue of its ability to inhibit a plant-derived PLC, and the K_d value is estimated to be 24 μM (Drøbak et al., 1994). Furthermore, two maize pollen profilin isoforms show marked differences in their ability to inhibit PLC; a Class II isoform is significantly less effective than a Class I isoform (Kovar et al., 2000; Table 1). These data indicate that interaction of profilin with PtdIns(4,5)P_2 may have effects on phosphoinositide signaling in plants by differentially affecting PLC activity.

The self-incompatibility response of field poppy (*Papaver rhoeas*) is a powerful system to study signal-dependent phosphorylation in pollen. Cell-free extracts have been used to study the phosphorylation of a number of pollen polypeptides in response to S-protein challenge (Rudd et al., 1996, 1997). Interestingly, the addition of small amounts of profilin to these extracts inhibits the phosphorylation of several specific polypeptides, whereas other proteins are unaffected or show increased phosphorylation (Clarke et al., 1998). The profilin isoforms that are added make a difference in the observed effect, indicating that individual isoforms may impact signal transduction cascades in unique ways. These studies provide circumstantial evidence that profilin can exert complex control of signal transduction cascades in plant cells.

3.6 Subcellular localization

The localization of profilin in plant cells has been analyzed principally in developing or germinated pollen. Early reports on the localization of profilin indicate that profilin is localized preferentially in the tip of the growing pollen tube (Mittermann et al., 1995). However, a fluorescent analog of lily profilin has a uniform distribution in living pollen tubes (Vidali & Hepler, 1997) and immunogold labeling of *Ledbouria* pollen shows a relatively uniform distribution in the vegetative and generative cell cytoplasm (Hess et al., 1995). Therefore, in mature pollen, profilin is likely to be uniformly distributed, but the possibility that specific isoforms are differentially localized has not been addressed.

Several studies indicate that profilin is not localized uniformly in all cell types. Microinjection of fluorescently-labeled birch profilin into the green alga *Micrasterias* reveals that profilin accumulates in the nucleus (Holzinger et al., 1997). Fluorescently-labeled maize profilin microinjected into mature *Tradescantia* stamen hair cells also accumulates in the nucleus (von Witsch & Staiger, unpublished data). Moreover, profilin is found to be present in the vegetative and generative nuclei of pollen (Hess & Valenta, 1997), and birch pollen profilin expressed in mammalian cells also accumulates in the nucleus (Rothkegel et al., 1996). One possible role for profilin in the nucleus is regulation of signal transduction, perhaps by binding to nuclear polyphosphoinositides. There are also recent reports of the association of profilin with proteins or protein complexes that are localized in the nucleus (Zhao et al., 1998; Giesemann et al., 1999).

Evidence is also accumulating for a non-uniform distribution of profilin in the cytoplasm of other cell types or developmental stages. In microspores of birch pollen, profilin associates with the plasma membrane (von Witsch et al., 1998). Likewise, immunogold labeling of rapid-freeze, freeze-substituted microspores of rice also reveals that profilin at the plasma membrane (Huang, Zee & Staiger, in prep). In growing root hairs of cress and *Arabidopsis*, profilin forms an apical 'cap' that disappears when the root hairs terminate growth (Braun et al., 1999). The mechanism of profilin association with the plasma membrane is not clear; however, it is attractive to suppose that the interaction is mediated by polyphosphoinositides. The possibility that profilin interacts with membrane-associated proteins that have proline-rich sequences also can not be ignored.

3.7 Cellular responses to excess profilin

Analyses of the biochemical properties and localization of profilin are inadequate to fully assess the function of profilin. The manipulation of profilin levels in plant cells has been accomplished by microinjection of purified profilin. Initially it was shown that microinjection of birch profilin into *Tradescantia* stamen hair cells results in the rapid cessation of streaming and destruction of transvacuolar strands (Staiger et al., 1994). Profilin reduces the abundance of actin microfilaments, consistent with the model that profilin functions to sequester actin under these conditions (Staiger et al., 1994; Karakesisoglou et al., 1996). Microinjection of profilin into growing *Micrasterias* inhibits cell growth in a dose-dependent manner

(Holzinger et al., 1997). Profilin inhibits cell plate formation when injected into dividing *Tradescantia* stamen hair cells and also disrupts streaming and wound plug formation in expanding interphase cells (Valster et al., 1997). All of these effects are proposed to be due to the depolymerization of actin caused by rapidly increasing the cellular concentration of profilin, rather than perturbing signaling or binding to proline-rich sequences.

The effect of several maize profilin isoforms on the actin-dependent position of the nucleus in mature *Tradescantia* stamen hair cells has been compared (Gibbon et al., 1997, 1998; Kovar et al., 2000). These experiments suggest that the affinity of the microinjected material for actin is the primary determinant of their ability to cause displacement of the nucleus (Gibbon, et al., 1998; Kovar, et al., 2000). The effect in live cells is complicated, however, because microinjection of a mutant of maize profilin 1 that had higher affinity for PLP than the wild type had a more pronounced effect on nuclear position than the wild-type profilin (Gibbon et al., 1998). This suggests that profilin function in plants can be modified by binding to proline-rich sequences. Alternatively, the endogenous profilin may be bound to such sequences, and when dislodged into the cytoplasm causes increased depolymerization of actin. In either case, the binding of profilin to proteins with proline-rich sequences is likely to be important for regulating profilin function in plant cells.

4. CONCLUSIONS AND PERSPECTIVE

Plant profilins share many properties, such as binding with high affinity to actin, PLP and PtdIns(4,5)P_2, with their non-plant counterparts. In other respects, for example inhibition of nucleotide exchange, they are fundamentally different than non-plant profilins. These functional properties may reflect different functions for profilin in the unique milieu of the plant cell cytoplasm. The best evidence indicates that, at least in pollen, profilin serves to sequester actin. Yet, the presence of multiple profilin isoforms with distinct functional properties suggests that specific isoforms may be precisely tuned for specific functions in the cell. Although there is emerging evidence that this is the case, it remains unclear whether this hypothesis is correct.

Several types of experiment should be performed to test whether multiple profilin isoforms are important for regulating actin dynamics in plants. First, the over- and under-expression of profilin isoforms would be very useful to test the importance of the different classes. If profilin simply serves to sequester actin, the level of F-actin will be expected to change inversely with respect to the level of profilin. It was precisely this sort of experiment that led to the surprising result that overexpression of profilin could result in more F-actin (Finkel et al., 1994). Second, examination of the ability of plant profilins to sequester plant actin in the presence or absence of capping proteins will be central to understanding whether plant profilin–actin complexes can add to the barbed end of actin microfilaments. Finally, it is clear that the expression of multiple actin isoforms is tissue-specific and/or developmentally regulated. Therefore, it will be useful to investigate the association of profilins with actin purified from a tissue other than pollen to determine whether the function of profilin is indeed suited to specific actin isoforms in specific tissues (Meagher et al., 1999a; Meagher et al., this volume).

ACKNOWLEDGEMENTS

Work in our laboratory is sponsored by the USDA-NRICGP and the Showalter Fund of the Purdue Research Foundation

REFERENCES

Asturias JA, Arilla MC, Bartolomé B, Martínez J, Martínez A and Palacios R (1997a) Sequence polymorphism and structural analysis of timothy grass pollen profilin allergen (Phl p 11). Biochim Biophys Acta 1352: 253-257

Asturias JA, Arilla MC, Gómez-Bayón N, Aguirre M, Martinez A, Palacios R and Martinez J (1998) Cloning and immunological characterization of the allergen Hel a 2 (profilin) from sunflower pollen. Mol Immunol 35: 469-478

Asturias JA, Arilla MC, Gómez-Bayón N, Martinez J, Martinez A and Palacios R (1997b) Cloning and expression of the panallergen profilin and the major allergen (Ole e 1) from olive tree pollen. J Allergy Clin Immunol 100: 365-372

Asturias JA, Arilla MC, Gómez-Bayón N, Martinez J, Martinez A and Palacios R (1997c) Cloning and high level expression of *Cynodon dactylon* (Bermuda grass) pollen profilin (Cyn d 12) in *Escherichia coli*: purification and characterization of the allergen. Clin Exp Allergy 27: 1307-1313

Ayscough KR (1998) *In vivo* functions of actin-binding proteins. Curr Opin Cell Biol 10: 102-111

Balasubramanian MK, Feoktistova A, McCollum D and Gould KL (1996) Fission yeast Sop2p: a novel and evolutionarily conserved protein that interacts with Arp3p and modulates profilin function. EMBO J 15: 6426-6437

Ballweber E, Giehl K, Hannappel E, Huff T, Jockusch BM and Mannherz HG (1998) Plant profilin induces actin polymerization from actin:β–thymosin complexes and competes directly with β-thymosins and with negative co-operativity with DNase I for binding to actin. FEBS Lett 425: 251-255

Braun M, Baluska F, von Witsch M and Menzel D (1999) Redistribution of actin, profilin and phosphatidylinositol-4,5-bisphosphate in growing and maturing root hairs. Planta 209: 435-443

Bubb MR, Baines IC and Korn ED (1998) Localization of actobindin, profilin I, profilin II, and phosphatidylinositol-4,5-bisphosphate (PIP_2) in *Acanthamoeba castellanii*. Cell Motil Cytoskeleton 39: 134-146

Cao L-G, Babcock GG, Rubenstein PA and Wang Y-L (1992) Effects of profilin and profilactin on actin structure and function in living cells. J Cell Biol 117: 1023-1029

Carlier M-F (1998) Control of actin dynamics. Curr Opin Cell Biol 10: 45-51

Carlsson L, Nyström L-E, Sundkvist I, Markey F and Lindberg U (1977) Actin polymerizability is influenced by profilin, a low molecular weight protein in non-muscle cells. J Mol Biol 115: 465-483

Castrillon DH and Wasserman SA (1994) *diaphanous* is required for cytokinesis in *Drosophila* and shares domains of similarity with the products of the *limb deformity* gene. Development 120: 3367-3377

Cedergren-Zeppezauer ES, Goonesekere NCW, Rozycki MD, Myslik JC, Dauter Z, Lindberg U and Schutt CE (1994) Crystallization and structure determination of bovine profilin at 2.0 Å resolution. J Mol Biol 240: 459-475

Chang F, Drubin D and Nurse P (1997) cdc12p, a protein required for cytokinesis in fission yeast, is a component of the cell division ring and interacts with profilin. J Cell Biol 137: 169-182

Christensen HEM, Ramachandran S, Tan C-T, Surana U, Dong C-H and Chua N-H (1996) *Arabidopsis* profilins are functionally similar to yeast profilins: identification of a vascular bundle-specific profilin and a pollen-specific profilin. Plant J 10: 269-279

Clarke SR, Staiger CJ, Gibbon BC and Franklin-Tong VE (1998) A potential signaling role for profilin in pollen of *Papaver rhoeas*. Plant Cell 10: 967-979

Cossart P (1995) Actin-based bacterial motility. Curr Opin Cell Biol 7: 94-101

Domke T, Federau T, Schlüter K, Giehl K, Valenta R, Schomburg D and Jockusch BM (1997) Birch pollen profilin: structural organization and interaction with poly-(L-proline) peptides as revealed by NMR. FEBS Lett 411: 291-295

Drøbak BK, Watkins PAC, Valenta R, Dove SK, Lloyd CW and Staiger CJ (1994) Inhibition of plant plasma membrane phosphoinositide phospholipase C by the actin-binding protein, profilin. Plant J 6: 389-400

Eads JC, Mahoney NM, Vorobiev S, Bresnick AR, Wen KK, Rubenstein PA, Haarer BK and Almo SC (1998) Structure determination and characterization of *Saccharomyces cerevisiae* profilin. Biochemistry 37: 11171-11181

Evangelista M, Blundell K, Longtine MS, Chow CJ, Adames N, Pringle JR, Peter M and Boone C (1997) Bni1p, a yeast formin linking Cdc42p and the actin cytoskeleton during polarized morphogenesis. Science 276: 118-122

Fedorov AA, Ball T, Mahoney NM, Valenta R and Almo SC (1997) The molecular basis for allergen cross-reactivity: crystal structure and IgE-epitope mapping of birch pollen profilin. Structure 5: 33-45

Fedorov AA, Magnus KA, Graupe MH, Lattman EE, Pollard TD and Almo SC (1994a) X-ray structures of isoforms of the actin-binding protein profilin that differ in their affinity for phosphatidylinositol phosphates. Proc Natl Acad Sci USA 91: 8636-8640

Fedorov AA, Pollard TD and Almo SC (1994b) Purification, characterization and crystallization of human platelet profilin expressed in *Escherichia coli*. J Mol Biol 241: 480-482

Finkel T, Theriot JA, Dise KR, Tomaselli GF and Goldschmidt-Clermont PJ (1994) Dynamic actin structures stabilized by profilin. Proc Natl Acad Sci USA 91: 1510-1514

Frazier JA and Field CM (1997) Actin cytoskeleton: Are FH proteins local organizers? Curr Biol 7: R414-R417

Gibbon BC, Kovar DR and Staiger CJ (1999) Latrunculin B has different effects on pollen germination and tube growth. Plant Cell 11: 2349-2363

Gibbon BC, Ren H and Staiger CJ (1997) Characterization of maize (*Zea mays*) pollen profilin function *in vitro* and in live cells. Biochem J 327: 909-915

Gibbon BC, Zonia LE, Kovar DR, Hussey PJ and Staiger CJ (1998) Pollen profilin function depends on interaction with proline-rich motifs. Plant Cell 10: 981-993. [A printed correction appears in Plant Cell 11, 1603]

Giehl K, Valenta R, Rothkegel M, Ronsiek M, Mannherz H-G and Jockusch BM (1994) Interaction of plant profilin with mammalian actin. Eur J Biochem 226: 681-689

Gieselmann R, Kwiatkowski DJ, Janmey PA and Witke W (1995) Distinct biochemical characteristics of the two human profilin isoforms. Eur J Biochem 229: 621-628

Giesemann T, Rathke-Hartlieb S, Rothkegel M, Bartsch JW, Buchmeier S, Jockusch BM and Jockusch H (1999) A role for polyproline motifs in the spinal muscular atrophy protein SMN: profilins bind to and colocalize with SMN in nuclear gems. J Biol Chem 274: 37908-37914

Goldschmidt-Clermont PJ, Furman MI, Wachsstock D, Safer D, Nachmias VT and Pollard TD (1992) The control of actin nucleotide exchange by thymosin β4 and profilin. A potential regulatory mechanism for actin polymerization in cells. Mol Biol Cell 3: 1015-1024

Goldschmidt-Clermont PJ, Kim JW, Machesky LM, Rhee SG and Pollard TD (1991) Regulation of phospholipase C-γ1 by profilin and tyrosine phosphorylation. Science 251: 1231-1233

Goldschmidt-Clermont PJ, Machesky LM, Baldassare JJ and Pollard TD (1990) The actin-binding protein profilin binds to PIP_2 and inhibits its hydrolysis by phospholipase C. Science 247: 1575-1578

Guillén G, Valdés-López V, Noguez R, Olivares J, Rodriguez-Zapata LC, Pérez H, Vidali L, Villanueva MA and Sánchez F (1999) Profilin in *Phaseolus vulgaris* is encoded by two genes (only one expressed in root nodules) but multiple isoforms are generated *in vivo* by phosphorylation on tyrosine residues. Plant J 19: 497-508

Haarer BK, Lillie SH, Adams AE, Magdolen V, Bandlow W and Brown SS (1990) Purification of profilin from *Saccharomyces cerevisiae* and analysis of profilin-deficient cells. J Cell Biol 110: 105-114

Hájková L, Sjögren CB, Korenbaum E, Nordberg P and Karlsson R (1997) Characterization of a mutant profilin with reduced actin-binding capacity: effects *in vitro* and *in vivo*. Exp Cell Res 234: 66-77

Hall A (1994) Small GTP-binding proteins and the regulation of the actin cytoskeleton. Annu Rev Cell Biol 10: 31-54

Harris SD, Hamer L, Sharpless KE and Hamer JE (1997) The *Aspergillus nidulans sepA* gene encodes an FH1/2 protein involved in cytokinesis and the maintenance of cellular polarity. EMBO J 16: 3474-3483

Haugwitz M, Noegel AA, Karakesisoglou J and Schleicher M (1994) Dictyostelium amoebae that lack G-actin-sequestering profilins show defects in F-actin content, cytokinesis and development. Cell 79: 303-314

Hess MW, Mittermann I, Luschnig C and Valenta R (1995) Immunocytochemical localisation of actin and profilin in the generative cell of angiosperm pollen: TEM studies on high-pressure frozen and freeze-substituted *Ledebouria socialis* Roth (Hyacinthaceae). Histochemistry 104: 443-451

Hess MW and Valenta R (1997) Profilin revealed in pollen nuclei: immuno-electron microscopy of high-pressure frozen *Ledebouria socialis* Roth (Hyacinthaceae). Sex Plant Reprod 10: 283-287

Holzinger A, Mittermann I, Laffer S, Valenta R and Meindl U (1997) Microinjection of profilins from different sources into the green alga *Micrasterias* causes transient inhibition of cell growth. Protoplasma 199: 124-134

Huang S, McDowell JM, Weise MJ and Meagher RB (1996) The Arabidopsis profilin gene family: evidence for an ancient split between constitutive and pollen-specific profilin genes. Plant Physiol 111: 115-126

Imamura H, Tanaka K, Hihara T, Umikawa K, Kamei T, Takahashi K, Sasaki T and Takai Y (1997) Bni1p and Bnr1p: downstream targets of the Rho family small G-proteins which interact with profilin and regulate actin cytoskeleton in *Saccharomyces cerevisiae*. EMBO J 16: 2745-2755

Jensen-Jarolim E, Santner B, Leitner A, Grimm R, Scheiner O, Ebner C and Breiteneder H (1998) Bell peppers (*Capsicum annuum*) express allergens (profilin, pathogenesis-related protein P23 and Bet v 1) depending on the horticultural strain. Int Arch Allergy Immunol 116: 103-109

Jonckheere V, Lambrechts A, Vandekerckhove J and Ampe C (1999) Dimerization of profilin II upon binding the $(GP_5)_3$ peptide from VASP overcomes the inhibition of actin nucleation by profilin II and thymosin β4. FEBS Letters 447: 257-263

Kang F, Purich DL and Southwick FS (1999) Profilin promotes barbed-end actin filament assembly without lowering the critical concentration. J. Biol. Chem. 274: 36963-36972

Karakesisoglou I, Schleicher M, Gibbon BC and Staiger CJ (1996) Plant profilins rescue the aberrant phenotype of profilin-deficient *Dictyostelium* cells. Cell Motil Cytoskeleton 34: 36-47

Kohno H, Tanaka K, Mino A, Umikawa M, Imamura H, Fujiwara T, Fujita Y, Hotta K, Qadota H, Watanabe T, Ohya Y and Takai Y (1996) Bni1p implicated in cytoskeletal control is a putative target of Rho1p small GTP binding protein in *Saccharomyces cerevisiae*. EMBO J 15: 6060-6068

Korenbaum E, Nordberg P, Björkegren-Sjögren C, Schutt CE, Lindberg U and Karlsson R (1998) The role of profilin in actin polymerization and nucleotide exchange. Biochemistry 37: 9274-9283

Kovar DR, Drøbak BK and Staiger CJ (2000) Maize profilin isoforms are functionally distinct. Plant Cell 12: In press

Lambrechts A, Van Damme J, Goethals M, Vandekerckhove J and Ampe C (1995) Purification and characterization of bovine profilin II. Actin, poly(L-proline) and inositolphospholipid binding. Eur J Biochem 230: 281-286

Lassing I and Lindberg U (1985) Specific interaction between phosphatidylinositol 4,5-bisphosphate and profilactin. Nature 314: 472-474

Lassing I and Lindberg U (1988) Specificity of the interaction between phosphatidylinositol 4,5-bisphosphate and the profilin:actin complex. J Cell Biochem 37: 255-267

Liu X and Yen L-F (1992) Purification and characterization of actin from maize pollen. Plant Physiol 99: 1151-1155

Loisel TP, Boujemaa R, Pantaloni D and Carlier M-F (1999) Reconstitution of actin-based motility of *Listeria* and *Shigella* using pure proteins. Nature 401: 613-616

Lu PJ, Shieh WR, Rhee SG, Yin HL and Chen CS (1996) Lipid products of phosphoinositide 3-kinase bind human profilin with high affinity. Biochemistry 35: 14027-14034

Machesky LM, Atkinson SJ, Ampe C, Vandekerckhove J and Pollard TD (1994) Purification of a cortical complex containing two unconventional actins from *Acanthamoeba* by affinity chromatography on profilin-agarose. J Cell Biol 127: 107-115

Machesky LM and Pollard TD (1993) Profilin as a potential mediator of membrane-cytoskeleton communication. Trends Cell Biol 3: 381-385

Mahoney NM, Janmey PA and Almo SC (1997) Structure of the profilin-poly-L-proline complex involved in morphogenesis and cytoskeletal regulation. Nature Struc Biol 4: 953-960

Manseau L, Calley J and Phan H (1996) Profilin is required for posterior patterning of the *Drosophila* oocyte. Development 122: 2109-2116

McCollum D, Feoktistova A, Morphew M, Balasubramanian M and Gould KL (1996) The *Schizosaccharomyces pombe* actin-related protein, Arp3, is a component of the cortical actin cytoskeleton and interacts with profilin. EMBO J 15: 6438-6446

Meagher RB (1991) Divergence and differential expression of actin gene families in higher plants. Int Rev Cytol 125: 139-163

Meagher RB, McKinney EC and Kandasamy MK (1999a) Isovariant dynamics expand and buffer the responses of complex systems: The diverse plant actin gene family. Plant Cell 11: 995-1005

Meagher RB, McKinney EC and Vitale AV (1999b) The evolution of new structures: clues from plant cytoskeletal genes. Trends Genet 15: 278-284

Metzler WJ, Farmer BT, Constantine KL, Friedrichs MS, Lavoie T and Mueller L (1995) Refined solution structure of human profilin I. Prot Sci 4: 450-459

Mittermann I, Heiss S, Kraft D, Valenta R and Heberle-Bors E (1996) Molecular characterization of profilin isoforms from tobacco (*Nicotiana tabacum*) pollen. Sex Plant Reprod 9: 133-139

Mittermann I, Swoboda I, Pierson E, Eller N, Kraft D, Valenta R and Heberle-Bors E (1995) Molecular cloning and characterization of profilin from tobacco (*Nicotiana tabacum*): increased profilin expression during pollen maturation. Plant Mol Biol 27: 137-146

Mullins RD, Heuser JA and Pollard TD (1998a) The interaction of arp2/3 complex with actin: nucleation, high affinity pointed end capping, and formation of branching networks of filaments. Proc Natl Acad Sci USA 95: 6181-6186

Mullins RD, Kelleher JF, Xu J and Pollard TD (1998b) Arp2/3 complex from *Acanthamoeba* binds profilin and cross-links actin filaments. Mol Biol Cell 9: 841-852

Ostrander DB, Gorman JA and Carman GM (1995) Regulation of profilin localization in *Saccharomyces cerevisiae* by phosphoinositide metabolism. J Biol Chem 270: 27045-27050

Pantaloni D and Carlier M-F (1993) How profilin promotes actin filament assembly in the presence of thymosin β4. Cell 75: 1007-1014

Perelroizen I, Didry D, Christensen H, Chua N-H and Carlier M-F (1996) Role of nucleotide exchange and hydrolysis in the function of profilin in actin assembly. J Biol Chem 271: 12302-12309

Petrella EC, Machesky LM, Kaiser DA and Pollard TD (1996) Structural requirements and thermodynamics of the interaction of proline peptides with profilin. Biochemistry 35: 16535-16543

Purich DL and Southwick FS (1997) ABM-1 and ABM-2 homology sequences: consensus docking sites for actin-based motility defined by oligoproline regions in *Listeria* ActA surface protein and human VASP. Biochem Biophys Res Comm 231: 686-691

Reinhard M, Giehl K, Abel K, Haffner C, Jarchau T, Hoppe V, Jockusch BM and Walter U (1995) The proline-rich focal adhesion and microfilament protein VASP is a ligand for profilins. EMBO J 14: 1583-1589

Ren H, Gibbon BC, Ashworth SL, Sherman DM, Yuan M and Staiger CJ (1997) Actin purified from maize pollen functions in living plant cells. Plant Cell 9: 1445-1457

Rihs HP, Rozynek P, May TK, Welticke B and Baur X (1994) Polymerase chain reaction based cDNA cloning of wheat profilin: a potential plant allergen. Int Arch Allergy Immunol 105: 190-194

Rothkegel M, Mayboroda O, Rohde M, Wucherpfennig C, Valenta R and Jockusch BM (1996) Plant and animal profilins are functionally equivalent and stabilize microfilaments in living animal cells. J Cell Sci 109: 83-90

Rudd JJ, Franklin FCH and Franklin-Tong VE (1997) Ca^{2+}-independent phosphorylation of a 68 kDa pollen protein is stimulated by the self-incompatibility response in *Papaver rhoeas*. Plant J 12: 507-514

Rudd JJ, Franklin FCH, Lord JM and Franklin-Tong VE (1996) Increased phosphorylation of a 26-kd pollen protein is induced by the self-incompatibility response in *Papaver rhoeas*. Plant Cell 8: 713-724

Ruhlandt G, Lange U and Grolig F (1994) Profilins purified from higher plants bind to actin from cardiac muscle and to actin from a green alga. Plant Cell Physiol 35: 849-854

Schmidt A and Hall MN (1998) Signaling to the actin cytoskeleton. Annu Rev Cell Dev Biol 14: 305-338

Schobert C, Gottschalk M, Kovar DR, Staiger CJ, Yoo B-C and Lucas WJ (2000) Characterization of *Ricinus communis* phloem profilin, RcPRO1. Plant Mol Biol, In press

Schutt CE, Myslik JC, Rozycki MD, Goonesekere NCW and Lindberg U (1993) The structure of crystalline profilin-β-actin. Nature 365: 810-816

Selden LA, Kinosian HJ, Estes JE and Gershman LC (1999) Impact of profilin on actin-bound nucleotide exchange and actin polymerization dynamics. Biochemistry 38: 2769-2778

Sohn RH, Chen J, Koblan KS, Bray PF and Goldschmidt-Clermont PJ (1995) Localization of a binding site for phosphatidylinositol 4,5-bisphosphate on human profilin. J Biol Chem 270: 21114-21120

Sohn RH and Goldschmidt-Clermont PJ (1994) Profilin: at the crossroads of signal transduction and the actin cytoskeleton. BioEssays 16: 465-472

Staiger CJ, Gibbon BC, Kovar DR and Zonia LE (1997) Profilin and actin depolymerizing factor: modulators of actin organization in plants. Trends Plant Sci 2: 275-281

Staiger CJ, Goodbody KC, Hussey PJ, Valenta R, Drøbak BK and Lloyd CW (1993) The profilin multigene family of maize: differential expression of three isoforms. Plant J 4: 631-641

Staiger CJ, Yuan M, Valenta R, Shaw PJ, Warn RM and Lloyd CW (1994) Microinjected profilin affects cytoplasmic streaming in plant cells by rapidly depolymerizing actin microfilaments. Curr Biol 4: 215-219

Thorn KS, Christensen HEM, Shigeta R, Huddler D, Shalaby L, Lindberg U, Chua N-H and Schutt CE (1997) The crystal structure of a major allergen from plants. Structure 5: 19-32

Valenta R, Ball T, Vrtala S, Duchêne M, Kraft D and Scheiner O (1994) cDNA cloning and expression of timothy grass (*Phleum pratense*) pollen profilin in *Escherichia coli*: comparison with birch pollen profilin. Biochem Biophys Res Comm 199: 106-118

Valenta R, Duchêne M, Pettenburger K, Sillaber C, Valent P, Bettelheim P, Breitenbach M, Rumpold H, Kraft D and Scheiner O (1991) Identification of profilin as a novel pollen allergen; IgE autoreactivity in sensitized individuals. Science 253: 557-560

Valenta R, Ferreira F, Grote M, Swoboda I, Vrtala S, Duchêne M, Deviller P, Meagher RB, McKinney E, Heberle-Bors E, Kraft D and Scheiner O (1993) Identification of profilin as an actin-binding protein in higher plants. J Biol Chem 268: 22777-22781

Vallverdu A, Asturias JA, Arilla MC, Gómez-Bayón N, Martinez A, Martinez J and Palacios R (1998) Characterization of recombinant *Mercurialis annua* major allergen Mer a 1 (profilin). J Allergy Clin Immunol 101: 363-370

Valster AH, Pierson ES, Valenta R, Hepler PK and Emons AMC (1997) Probing the plant actin cytoskeleton during cytokinesis and interphase by profilin microinjection. Plant Cell 9: 1815-1824

Vidali L and Hepler PK (1997) Characterization and localization of profilin in pollen grains and tubes of *Lilium longiflorum*. Cell Motil Cytoskeleton 36: 323-338

Vidali L, Pérez HE, López VV, Noguez R, Zamudio F and Sánchez F (1995) Purification, characterization, and cDNA cloning of profilin from *Phaseolus vulgaris*. Plant Physiol 108: 115-123

Vinson VK, De La Cruz EM, Higgs HN and Pollard TD (1998) Interactions of *Acanthamoeba* profilin with actin and nucleotides bound to actin. Biochemistry 37: 10871-10880

von Witsch M, Baluška F, Staiger CJ and Volkmann D (1998) Profilin is associated with the plasma membrane in microspores and pollen. Eur J Cell Biol 77: 303-312

Wasserman S (1998) FH proteins as cytoskeletal organizers. Trends Cell Biol 8: 111-115

Welch MD, Rosenblatt J, Skoble J, Portnoy DA and Mitchison TJ (1998) Interaction of human Arp2/3 complex and the *Listeria monocytogenes* ActA protein in actin filament nucleation. Science 281: 105-108

Yeh J and Haarer BK (1996) Profilin is required for the normal timing of actin polymerization in response to thermal stress. FEBS Lett 398: 303-307

Yu L-X, Nasrallah J, Valenta R and Parthasarathy MV (1998) Molecular cloning and mRNA localization of tomato pollen profilin. Plant Mol Biol 36: 699-707

Zhao K, Wang WD, Rando OJ, Xue YT, Swiderek K, Kuo A and Crabtree GR (1998) Rapid and phosphoinositol-dependent binding of the SWI/SNF-like BAF complex to chromatin after T lymphocyte receptor signaling. Cell 95: 625-636

Zigmond SH (1996) Signal transduction and actin filament organization. Curr Opin Cell Biol 8: 66-73

Chapter 4

ACTIN DEPOLYMERIZING FACTOR

David R Kovar and Christopher J Staiger

Department of Biological Sciences, Purdue University, West Lafayette, IN 47907-1392 USA

Key words: actin-binding protein, actin depolymerizing factor, cofilin, pH, phosphorylation, polyphosphoinositides, profilin, treadmilling

Abstract: Unlike other actin-binding proteins, actin depolymerizing factor (ADF) binds to both filamentous and globular actin, and causes the rapid depolymerization of actin filaments. ADF localizes to sites of active actin dynamics, including growing maize root hairs. ADF activity is regulated by a variety of stimuli implicated in affecting actin organization, such as pH changes, phosphorylation and polyphosphoinositides. Our current understanding of ADF's role in regulating actin dynamics has been greatly enhanced through examination of the biochemical properties of plant ADFs. ADF depolymerizes actin filaments by a combination of severing activity and enhancing the rate of actin monomer dissociation from the pointed end. As a result, ADF-induced depolymerization creates a large pool of actin subunits. When the ends of actin filaments are uncapped, this pool is able to rapidly repolymerize. However, when filament ends are capped, repolymerization can not occur. Therefore, depending upon the presence of actin-filament capping factors, ADF either drives filament turnover or induces total depolymerization.

1. INTRODUCTION

The actin cytoskeleton redistributes during pollen germination, when pollen undergoes the self-incompatibility response, during responses of root hairs to bacterial nodulation factors, in light-mediated plastid orientation and in non-host resistance to fungal attack (reviewed in Nick, 1999; Staiger, 2000). Several signaling intermediates are implicated in these processes, including changes in cytosolic Ca^{2+} and pH, polyphosphoinositides, and monomeric G-proteins, but how these impact the plant actin cytoskeleton is not fully understood. It is quite clear that links between external stimuli and actin reorganization will involve specific actin-binding proteins.

C.J. Staiger et al. (eds.), Actin: A Dynamic Framework for Multiple Plant Cell Functions, 67–85.

Furthermore, the rate of actin-filament turnover must be extremely high to explain these dramatic changes in actin organization. In non-plant systems, actin-filament turnover within cells is 100–200-fold faster than the rate of turnover measured for purified actin *in vitro* (Theriot, 1997). Recent work demonstrates that the enhanced turnover of actin filaments in cells can be explained, in large part, by the properties of the actin depolymerizing factor (ADF) family of proteins (Carlier et al., 1997). Additionally, ADF activity is regulated by pH changes, phosphorylation status and binding to polyphosphoinositides, suggesting that it may play a pivotal role in affecting the organization of the plant actin cytoskeleton in response to extrinsic and intrinsic stimuli.

Several excellent reviews covering the ADF family have been published recently (Moon & Drubin, 1995; Theriot, 1997; Lappalainen et al., 1998; Maciver, 1998; Bamburg, 1999; Bamburg et al., 1999; Carlier et al., 1999). Here, a general review of the ADF family is presented, with special emphasis on the increasing information about plant ADFs and their role in actin organization.

2. SEQUENCE AND STRUCTURE OF ADF-FAMILY MEMBERS

2.1 Identification and primary amino acid sequences

Almost 30 members of the ADF family have been identified. These proteins contain between 118 and 168 amino acids (15–22 kDa) and have a variety of names, including ADF (Bamburg et al., 1980), cofilin (Maekawa et al., 1984), depactin (Mabuchi, 1981), destrin (Maekawa et al., 1984; Nishida et al., 1984a) and actophorin (Cooper et al., 1986). ADF was identified originally as a protein from embryonic chick brain extracts that depolymerized actin filaments and formed a 1:1 complex with monomeric actin (G-actin) (Bamburg et al., 1980). Cofilin was purified from porcine brain and characterized by the ability to cosediment with, rather than depolymerize, actin filaments (F-actin) (Nishida et al., 1984b). Subsequently, proteins with biochemical characteristics similar to both proteins were identified in many eukaryotic cells (for a detailed list see Bamburg, 1999). Sequencing of the proteins, or of the transcripts that encode these proteins, reveals a marked similarity. ADF and cofilin from a single organism share ~70% amino acid sequence identity. Despite subtle quantitative differences, their general biochemical properties are conserved (Moon & Drubin, 1995; Maciver et al., 1998; Ressad et al., 1998). This group of proteins is now referred to as the ADF/cofilin family (Moon & Drubin, 1995).

In plants, ADF is present as a small multigene family of 2–4 members (Lopez et al., 1996). Plant ADF was first identified during a search for pollen-specific transcripts in lily and subsequently from *Brassica napus* (Kim et al., 1993). Maize contains two pollen-specific ADF transcripts (*ZmADF1*, *ZmADF2*) and one transcript that is present in all tissues except pollen (*ZmADF3*) (Rozycka et al., 1995; Lopez et al., 1996). In wheat, ADF expression is cold-induced (Danyluk et al., 1996). Two *Arabidopsis* ADFs were identified from EST databases and completely sequenced (accession numbers U48938 and U48939). The deduced amino acid sequences of plant ADF (139–142 residues; ~16 kDa) share only 28–35% identity with vertebrate sequences. Interestingly, even within a single plant, ADF sequences can be quite diverse. ZmADF1 and ZmADF2 are 94% identical to each other, but only ~57% identical to ZmADF3 (Lopez et al., 1996).

2.2 Overall Structure and Actin-Binding Domains

The NMR structure of vertebrate ADF (destrin) and crystal structures of yeast cofilin and *Acanthamoeba* cofilin (actophorin) have been solved (Hatanaka et al., 1996; Fedorov et al., 1997; Leonard et al., 1997) and demonstrate a striking similarity. The overall fold of ADF-family members is comprised of a central mixed β-sheet (4 to 6 strands) with two α-helices on each side, and is termed the ADF-homology domain (Lappalainen et al., 1998). Generally, vertebrate ADFs are larger than plant and lower eukaryote family members, with the extra residues inserted into loops and at the C-terminus. Unfortunately, no plant ADF structure has been solved for direct comparison.

Mutagenesis and biochemical studies in yeast indicate that two actin-binding regions are present on the overall fold; one site contains residues necessary for binding to both G- and F-actin, and another site contains residues required for interaction with F-actin only (Lappalainen et al., 1997). These regions are probably also important for plant ADF function, since the plant sequences are 52% identical to the same region of the cofilin sequence. Additionally, targeted ZmADF3 mutants have been generated in these regions that are deficient for G- and F-actin binding, or for F-actin binding only (Jiang et al., 1997b).

Because crystallization of F-actin has not been possible, electron cryomicroscopy is used to infer how proteins bind to F-actin. Image reconstruction of F-actin decorated with human cofilin indicates that ADF locks the actin filament into a structural change that increases the twist by 5° per actin subunit (McGough et al., 1997); this probably has important functional implications (see below). It would be interesting to see if plant ADF induces similar structural changes upon binding plant F-actin.

3. *IN VITRO* BIOCHEMICAL PROPERTIES

The ability to reconstitute interactions between actin-binding proteins and actin provides exquisite insight into the molecular control of actin cytoskeletal organization and dynamics. For the ADF family, much of our current knowledge comes from biochemical studies of plant ADF. In this section, the pioneering studies of Carlier and colleagues on *Arabidopsis* ADF (AtADF1), are compared and contrasted with important work from other labs, including the analysis of maize ADFs by Hussey and coworkers

3.1 Actin-binding properties of ADF

In striking contrast with either profilin, which binds only to G-actin (Gibbon & Staiger, this volume), or fimbrin, which binds only to F-actin (McCurdy & Staiger, this volume), ADF interacts with both forms of actin. ADF family members are characterized by the ability to bind to G- and F-actin at a 1:1 stoichiometry, and induce rapid depolymerization of actin filaments through a pH-dependent mechanism. The affinity of ADF for G- or F-actin depends on the adenine nucleotide associated with actin. AtADF1 has an 80-fold higher affinity for ADP–G-actin than it does for ATP–G-actin (Carlier et al., 1997; Ressad et al., 1998). Similar properties are reported for ZmADF3 (Jiang et al., 1997a; Gungabissoon et al., 1998) as well as other members of the ADF family (Hawkins et al., 1993; Hayden et al., 1993; Maciver & Weeds, 1994; Blanchoin & Pollard, 1998; Ressad et al., 1998; Blanchoin & Pollard, 1999).

A higher affinity for ADP–actin, compared with ATP–actin, is important for ADF function, because actin filaments contain two functionally-distinct ends. Chemical differences at the two ends result in a highly dynamic (barbed) end where subunit addition is significantly faster than at the other (pointed) end (Sheterline et al., 1998). Consequently, the minimum concentration of G-actin that is needed for assembly (critical concentration), is much lower at the barbed end than at the pointed end. Therefore, actin monomers are added primarily to the barbed end and are lost from the pointed end. Polarity within the filament is also established by hydrolysis of the nucleotide associated with each actin subunit. Treadmiling, the flux of subunits through an actin filament, is maintained by the favored polymerization of ATP–G-actin at the barbed end and the subsequent depolymerization of ADP–G-actin from the pointed end. Some time after barbed-end addition of ATP–G-actin, ATP is hydrolyzed, leaving ADP-P_i–F-actin. Eventually the phosphate is released, resulting in a length of filament composed of ADP–actin subunits. ADP–actin subunits dissociate from the pointed end during depolymerization. These reactions are depicted in Figure 1. Filament treadmilling is thought to be necessary for many dynamic actin-dependent processes (Cramer, 1999). Cell motility, for example, is powered by net polymerization at the front of movement and depolymerization at the rear (reviewed in Carlier & Pantaloni, 1997).

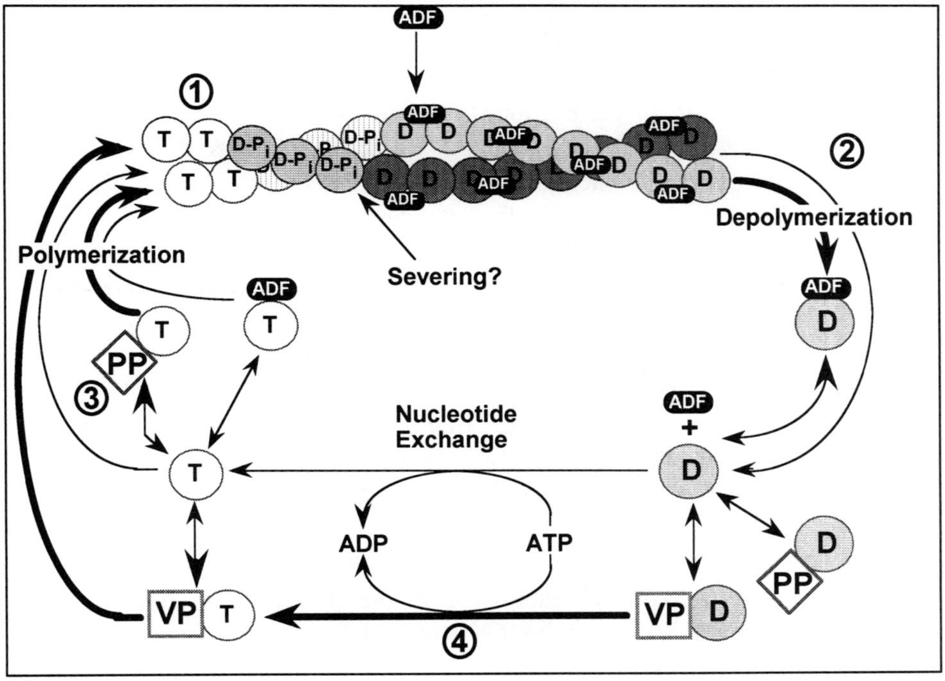

Figure 1. Model for the enhanced turnover of actin filaments by ADF [modified from Didry et al., 1998; Bamburg, 1999].

1. Actin filaments are comprised of a fast-growing barbed end with a cap of ATP–actin subunits (T), a region of ADP-P_i–actin (D-P_i) and a slow-growing pointed end with ADP–actin subunits (D).

2. ADF decorates ADP–F-actin, resulting in the rapid depolymerization of the filament from the pointed end. This increases the pool of ATP–G-actin that can repolymerize at the barbed end. When barbed ends are capped, ADF will induce total depolymerization, without repolymerization. Severing may also contribute to the increased rate of depolymerization and/or polymerization.

3. Because profilin accelerates the addition of ATP–G-actin at the barbed end, ADF and plant profilin (PP) increase the rate of filament treadmilling synergistically.

4. In addition to promoting barbed-end polymerization, vertebrate profilin (VP) also enhances the exchange of ADP for ATP on G-actin resulting in an even greater increase in the rate of filament treadmilling in the presence of ADF.

Interestingly, the treadmilling rate of F-actin in non-plant cells can be 100-fold greater than it is *in vitro* (Wang, 1985; Theriot, 1997). Treadmilling is limited by the rate of ADP–actin monomer loss from the pointed end. Early work had shown that ADF depolymerizes actin filaments (reviewed in Moon & Drubin, 1995). However, the mechanism wasn't understood until it was found that AtADF1 could accelerate the depolymerization rate at the pointed end by up to 30-fold *in vitro* (Carlier et al., 1997). Other members of the ADF family, including ZmADF3 (Jiang et al., 1997a), also greatly increase the off-rate of actin subunits at the pointed end (Du & Frieden, 1998; Maciver et al., 1998; Ressad et al., 1998; Blanchoin & Pollard, 1999).

The ability of ADF to increase the rate of pointed-end disassembly can be explained by several biochemically-determined properties: (1) cooperative binding to F-actin that results in a structural change in the filament; (2) an enhanced affinity for ADP–F- and G-actin when compared with ATP–F- and G-actin; and (3) a preference for associating with G-actin as opposed to F-actin (Carlier & Pantaloni, 1997; Blanchoin & Pollard, 1999). Because ADF does not bind to ADP-P_i–F-actin (Carlier et al., 1997), it will tend to decorate 'older' regions of actin filaments that have undergone ATP hydrolysis and P_i release (ADP–F-actin). ADF binding to F-actin is cooperative (Hawkins et al., 1993; Hayden et al., 1993; Ressad et al., 1998), implying that some actin filaments will be saturated with ADF, while other filaments will be bare. Perhaps due to the structural change, the ADF–ADP–G-actin complex dissociates rapidly from the pointed end. Because ADF inhibits nucleotide exchange on G-actin (Nishida, 1985), an increase in ATP–G-actin concentration is initially prevented. However, since ADF and ADP–G-actin are in rapid equilibrium, ADP–G-actin will be quickly released from ADF. Nucleotide exchange can then occur, resulting in an increase in the concentration of ATP–G-actin. These 'activated' subunits can then assemble onto the barbed ends of 'newer' filaments (Fig. 1). Increases in the concentrations of both ADF–ADP–G-actin and ATP–G-actin have been measured directly in the presence of AtADF1, indicating that this model is plausible (Didry et al., 1998).

Although Carlier (1997) argues that the increase in dissociation rate of actin subunits from the pointed end is sufficient to promote a faster rate of filament turnover, others have shown by light and electron microscopy that ADF severs F-actin (Cooper et al., 1986; Maciver et al., 1991, 1998). Severing would assist actin filament depolymerization, and subsequent repolymerization, by creating additional filament ends for subunit loss and addition. ADF induced 'fraying' of filament ends has also been suggested to contribute to the increased depolymerization rate (McGough & Chiu, 1999). While the significance of severing activity is still a matter of debate (Theriot, 1997; Maciver, 1998; Carlier et al., 1999), there is general agreement that ADF has a weak severing activity that, in combination with increasing pointed-end depolymerization, probably contributes to an increased actin-treadmilling rate (Du & Frieden, 1998; Maciver et al., 1998; Blanchoin & Pollard, 1999; Ressad et al., 1999).

3.2 Properties of ADF in the presence of other actin-binding proteins

It has been demonstrated directly that AtADF1 increases the steady-state treadmilling rate by at least 25-fold over the rate of actin alone (Carlier et al., 1997). This finding is supported by several observations. ADF increases the rate of actin-based motility of the pathogenic bacteria *Listeria* (Carlier et al., 1997; Rosenblatt et al., 1997). Cofilin increases actin-filament turnover in yeast (Lappalainen & Drubin, 1997). Overexpression of cofilin in *Dictyostelium* increases membrane ruffling and enhances motility (Aizawa et al., 1996). However, treadmilling rates in non-plant cells have been reported to be even greater than are predicted from the effects of ADF on actin behavior *in vitro*. Therefore, it was hypothesized that another actin-binding protein, which can drive polymerization at the barbed end, might further increase the *in vitro* treadmilling rate (Blanchoin & Pollard, 1998; Carlier, 1998).

3.2.1 Profilin

Profilin is a ubiquitous G-actin-binding protein (see Gibbon & Staiger, this volume) that can, under certain conditions, promote polymerization by participating in barbed-end specific assembly of ATP–G-actin (Pollard & Cooper, 1984; Pantaloni & Carlier, 1993; Kang et al., 1999). Profilin alone does not greatly enhance filament treadmilling (Perelroizen et al., 1996). However, in the presence of both AtADF1 and plant profilin, treadmilling of vertebrate actin increases 75-fold (Didry et al., 1998). This synergistic effect can be explained by ADF-induced depolymerization of ADP–actin at the pointed end with profilin-promoted polymerization of ATP–G-actin occurring at the barbed end (Fig. 1). Interestingly, in the presence of AtADF1 and vertebrate profilin, the treadmilling rate increases by 125-fold (Didry et al., 1998). The difference between vertebrate and plant profilin is presumably due to the ability of vertebrate, but not plant profilin (Perelroizen et al., 1996; Eads et al., 1998; Kovar et al., 2000), to enhance nucleotide exchange on G-actin. Therefore, ADP–G-actin depolymerized by ADF would be exchanged for ATP–G-actin faster in the presence of vertebrate profilin when compared with plant profilin (Fig. 1). It will be important to determine the treadmilling rate of actin filaments in different types of plant cells, and to test whether the rate of plant actin treadmilling *in vitro*, in the presence of both plant ADF and profilin, is similar to the rates measured *in vivo*.

3.2.2 Capping and nucleating factors

Because the function of ADF is based primarily on the ability to increase the rate of actin-subunit loss from filament ends, the availability of filament ends will have important effects on ADF activity. Thus, it is not surprising that F-actin end-capping factors modulate ADF function. In the presence of the barbed-end capping protein gelsolin, AtADF1 is exclusively a depolymerizing agent, inducing total filament destruction (Carlier et al., 1997). However, as a result, AtADF1 causes a greater increase in the concentration of ATP–G-actin in the presence of gelsolin compared with in the absence of gelsolin (Didry et al., 1998; Ressad et al., 1999). This large pool of ATP–G-actin would then be available for rapid association onto filaments with free barbed ends. But how are free barbed ends created in the presence of a barbed-end capping factor? It's possible that the concentration of the capping factor is not sufficient to keep all barbed ends capped, that ADF severs filaments and creates new barbed ends, or that new barbed ends are nucleated *de novo* by some other factor. These possibilities are not mutually exclusive.

The actin-related protein (Arp2/3) complex nucleates filaments that elongate at their barbed ends (Mullins et al., 1998; Welch et al., 1998; reviewed in Machesky & Gould, 1999). In the presence of gelsolin, AtADF1 significantly enhances vertebrate Arp2/3 complex-induced actin nucleation (Ressad et al., 1999). Therefore, it has been suggested that two populations of filaments are present at steady state: barbed-end capped filaments which are rapidly depolymerized by ADF, and actively growing filaments with free barbed ends (Carlier et al., 1999; Ressad et al., 1999).

This model is supported by the observation that ADF binds faster to gelsolin-capped F-actin compared with uncapped F-actin (Ressad et al., 1998). Further support comes from the recent finding that ADF, the Arp2/3 complex and a barbed-end capping protein are essential for actin-based motility of the pathogenic bacteria *Listeria* and *Shigella* (Loisel et al., 1999). Under other conditions, the Arp2/3 complex may prevent ADF-induced depolymerization by capping pointed ends (Mullins et al., 1998). ADF and the Arp2/3 complex co-localize in the actin depolymerization resistant, leading edge of *Xenopus laevis* keratocytes (Svitkina & Borisy, 1999). Several members of the Arp2/3 complex have been identified from EST databases or as part of the genome sequencing effort in *Arabidopsis* (Klahre & Chua, 1999).

3.2.3 Side-binding proteins

Inhibiting ADF from binding to F-actin, thereby preventing actin depolymerization, will have important implications for the stability of pre-existing filaments. ADF competes with tropomyosin, an actin filament-stabilizing protein, and the fungal toxin phalloidin, for binding to F-actin (Bernstein & Bamburg, 1982; Nishida et al., 1987; Ressad et al., 1998). This is probably due to the change in actin filament twist induced by ADF (McGough et al., 1997) rather than the presence of overlapping binding sites on F-actin.

If the change in twist of ADF-decorated actin filaments inhibits the binding of some F-actin binding proteins, it might be expected that other actin-binding proteins will bind to ADF-decorated filaments selectively. The actin interacting protein 1 (AIP1), originally identified in a two-hybrid screen with yeast actin as the bait (Amberg et al., 1995), was also identified from *Xenopus* oocyte extracts via the ability to bind to an ADF affinity column (Okada et al., 1999). By itself, AIP1 has little affect on actin dynamics *in vitro*, but in the presence of ADF, extremely low concentrations of AIP1 rapidly depolymerize F-actin and enhance filament severing (Okada et al., 1999; Rodal et al., 1999). How AIP1 and ADF interact and regulate each other, and what role AIP1 plays *in vivo*, is not known.

4. REGULATION OF ADF ACTIVITY

ADF activity is also regulated by changes in pH, phosphorylation, binding to polyphosphoinositides and expression of unique isoforms. Phosphorylation and binding to polyphosphoinositides disrupt the ability of ADF to bind to actin, whereas changes in pH and the expression of unique isoforms alter the effect of ADF on actin dynamics by quantitatively affecting its biochemical properties. As outlined above, an 'active' ADF will have dramatically different effects on actin dynamics, causing either total filament depolymerization or stimulating filament treadmilling, depending upon the presence of other cellular factors.

4.1 pH

Most members of the ADF family depolymerize more actin with increasing pH (Moon & Drubin, 1995; Carlier et al., 1997; Du & Frieden, 1998; Gungabissoon et al., 1998; Ressad et al., 1998). For instance, cofilin has a greater depolymerizing activity at elevated pH (>7.3–7.5) than at lower pH (Yonezawa et al., 1985). Some ADF-family members have a strong pH-dependence (human ADF), whereas others have a weak (AtADF1, chick cofilin) or no pH-dependence (*Acanthamoeba* actophorin; Maciver et al., 1998). Increased depolymerization is characterized by a reduction in the concentration of F-actin, with an increase in the concentration of G-actin. For example in the presence of human ADF, the concentration of G-actin is 3.5-fold higher at pH 8.2 than at pH 6.5 (Ressad et al., 1998). As described earlier, an increase in the pool of ADF–ADP–G-actin can increase the rate of treadmilling. Therefore, the physiological effect of changes in pH may be to regulate filament treadmilling as well as filament concentration and stability (Carlier et al., 1999). Accordingly, human ADF increases filament turnover to rates that are 3-fold higher at pH 8.0 than at pH 7.0 (unpublished data cited in Carlier et al., 1999).

Changes in cellular pH could play an important role in regulating plant ADF. The concentration of G-actin in the presence of AtADF1 is 3-fold higher at pH 8.2 compared with at pH 6.5 (Carlier et al., 1997). Similarly, ZmADF3 primarily binds to filaments at low pH without a major reduction in F-actin, but increasingly depolymerizes F-actin as pH shifts from 6 to 9 (Gungabissoon et al., 1998). Recent studies have demonstrated cytosolic pH gradients in growing pollen tubes (Messerli & Robinson, 1998; Feijó et al., 1999). An alkaline band (pH of at least 7.4, Feijó et al., 1999), is located at the base of an acidic tip region (as low as pH 6.0, Messerli & Robinson, 1998). In the presence of ADF, the alkaline band could designate a zone of rapid actin-filament turnover necessary for pollen tube growth. However, it would then be expected that the more acidic tip region should contain comparatively more F-actin than the alkaline band. This does not appear to be the case (Miller et al., 1996; Kost et al., 1998). Perhaps other actin-binding proteins, especially those that are sensitive to the tip-focused Ca^{2+}-gradient, play a predominant role in organizing actin at the extreme apex of the pollen tube (Kovar et al., 2000; Yokota & Shimmen, and Gibbon & Staiger, this volume).

4.2 Phosphorylation

Most ADF family members can be phosphorylated on serine 3 (Ser 6 in plant ADF), one of only nine amino acids that are absolutely conserved among known ADF sequences. Phosphorylated ADF, purified from vertebrate brain (Morgan et al., 1993), *Acanthamoeba* (Blanchoin et al., 2000), or phosphorylated *in vitro* (Moon & Drubin, 1995; Bamburg, 1999), does not bind to G- or F-actin. When Ser 6 of AtADF1 and ZmADF3 is replaced with Asp, to mimic phosphorylation, binding to both G- and F-actin, as well as the ability to increase pointed-end depolymerization, are greatly reduced (Ressad et al., 1998; Smertenko et al., 1998). The atomic structure of phosphorylated *Acanthamoeba* actophorin is identical to the unphosphorylated form of actophorin, suggesting that inactivation of actin-binding is due to steric hindrance rather than a conformational change (Blanchoin et al., 2000).

In non-plant cells, ADF is phosphorylated by the LIM kinases, which are activated by the Rho-family GTPases (Arber et al., 1998; Yang et al., 1998; Sumi et al., 1999). LIM kinase does not phosphorylate plant ADF (pers. commun. cited in Bamburg, 1999). Instead, ZmADF3 can be phosphorylated by a calcium-stimulated protein kinase from plant cell extracts (Smertenko et al., 1998). A simple model is that the tip-focused gradient of cytoplasmic calcium (1–10 μM) at the apex of growing pollen tubes (Rathore et al., 1991; Miller et al., 1992) deactivates ADF. It will be interesting to identify the specific calcium-sensitive kinase that is responsible for plant ADF phosphorylation as well as the phosphatase responsible for dephosphorylation.

ADF can be reactivated by dephosphorylation (Agnew et al., 1995; Moriyama et al., 1996), which occurs rapidly in many non-plant cells in response to stimuli that promote actin reorganization (reviewed in Moon & Drubin, 1995). Dephosphorylation of ADF correlates with a change from an even distribution throughout the cortical cytoplasm, to localization at areas of active actin dynamics such as the cleavage furrow, the neuronal growth cone or the leading edge of ruffled membranes (Bamburg & Bray, 1987; Suzuki et al., 1995; Abe et al., 1996; Nagaoka et al., 1996; Obinata et al., 1997).

4.3 Polyphosphoinositides

ADF is also regulated by interactions with phosphatidylinositol 4-phosphate (PIP) and 4,5-bisphosphate (PIP_2) (Yonezawa et al., 1990, 1991; Kusano et al., 1999). These phospholipids bind to ADF near the actin-binding domains and inhibit interaction with both G- and F-actin. An interaction with phospholipids has been shown for plant ADF *in vitro* (Gungabissoon et al., 1998).

4.4 Isoforms with unique functional properties

Many plant cytoskeletal proteins are encoded by multigene families that are expressed in a tissue-specific manner (Meagher, 1991; Meagher et al., 1999, and this volume). It is possible that these multiple isoforms simply ensure redundancy. However, a more intriguing possibility is that isoforms are functionally distinct. Consistent with this model, the maize profilin family contains at least 5 members which fall into two distinct functional classes (Gibbon & Staiger, this volume; Gibbon et al., 1998; Kovar et al., 2000).

Although the biochemical properties of maize or *Arabidopsis* ADF isoforms have not been compared, biochemical differences have been described for other ADF family members. The ADF–G-actin complex itself can polymerize when at sufficient concentration (Carlier et al., 1997). However, the concentration of ADF–G-actin required for assembly, is not always equal between ADFs from the same organism. For example, chick cofilin–ADP–G-actin polymerizes at a 2.5-fold lower concentration compared with chick ADF–ADP–G-actin (unpublished data cited in Bamburg, 1999). These biochemical differences explain why cofilin and ADF were originally identified via either an ability to cosediment with actin filaments (cofilin) or by depolymerization activity (ADF). *C. elegans* contains two ADF isoforms produced by alternative splicing of a single gene (McKim et al., 1994). One ADF strongly depolymerizes actin filaments, whereas the other binds to filaments but causes little depolymerization (Ono & Benian, 1998).

5. *IN VIVO* FUNCTIONS OF ADF

The participation of ADF in actin-dependent cellular processes, such as cytokinesis, endocytosis, embryo tissue development and movement, has been confirmed genetically (Moon & Drubin, 1995; Bamburg, 1999; Bamburg et al., 1999). Furthermore, the complete, or almost complete, loss-of-function is lethal in yeast (Iida et al., 1993; Moon et al., 1993), *Drosophila* (Gunsalus et al., 1995) and *C. elegans* (McKim et al., 1994). ADF is widely distributed among most tissues in higher organisms, but the expression level of specific isoforms varies (reviewed in Bamburg, 1999). Vertebrate cells contain 20 μM (~2% of soluble protein) ADF, compared with ~60 μM actin (Koffer & Daridan, 1985; Koffer et al., 1988). ADF localizes to highly dynamic actin-rich regions of cells including the leading edge and ruffling membrane of motile cells (Bamburg & Bray, 1987; Yonezawa et al., 1987), neuronal growth cones, contractile-ring of dividing cultured cells (Nagaoka et al., 1995), as well as the cleavage furrow of dividing *Xenopus* oocytes (Abe et al., 1996).

The localization of maize ADF during root hair initiation and elongation, and in pollen tubes has been examined (Jiang et al., 1997a; Hussey, pers. comm.). ADF distributes throughout the cytoplasm during the early stages of root hair development in maize roots. As the hair emerges and the actin filament bundles focus on the site of tip growth, ADF redistributes to the tip of the emerging hair and remains there as elongation proceeds (Jiang et al., 1997a). Redistribution of ADF may coincide with its activation by either dephosphorylation or pH changes. In contrast, another tip-growing cell, the pollen tube has a uniform cytoplasmic distribution of ADF, and does not specifically accumulate at the tube apex (Hussey, pers. comm.). Thus, spatially-restricted distribution may not be essential for activity. Instead, evenly-distributed ADF may sense the local environment in pollen tubes.

Increasing the cellular concentration of plant actin-binding proteins and examining the effects on actin-based function can be accomplished by microinjection into *Tradescantia* stamen hair cells (Staiger et al., 1994; Gibbon & Staiger, this volume; McCurdy & Staiger, this volume). Microinjection of pollen-specific recombinant ZmADF1 into stamen hairs causes rapid (2–3 min) cessation of cytoplasmic streaming and the disappearance of F-actin-rich, longitudinal, transvacuolar strands (Hussey et al., 1998). After 20–45 min, streaming recovers, but shows a cortical, transverse pattern. Recovery of cytoplasmic streaming coincides with the emergence of a new array of transverse actin filaments, which coalign with the cortical microtubules. Further studies might test the effects of ZmADF3 or AtADF1, for which *in vitro* biochemical data have been obtained. In addition, mutants of ZmADF3 that are defective for G- and F-actin binding (Jiang et al., 1997b), or that are constitutively phosphorylated (Smertenko et al., 1998), could be used to address the roles of specific properties of ADF and its association with other regulators.

6. CONCLUSIONS AND PERSPECTIVES

Although we assume that the turnover of actin filaments in plants cells must be extremely fast to explain the dramatic changes in actin organization observed during many cellular processess (Nick, 1999; Staiger, 2000), this has not been demonstrated directly. In yeast, the actin depolymerizing compound latrunculin was used to argue that cortical actin-patches are composed of rapidly turned-over actin monomers (Lappalainen & Drubin, 1997). Treatment with latrunculin caused rapid disassembly of the actin patches, an effect that was reduced in the presence of ADF mutants defective in actin-depolymerizing activity. This result indicates that yeast ADF enhances actin turnover *in vivo*. Treatment of mature maize pollen grains, as well as actively growing pollen tubes, with latrunculin significantly reduced F-actin levels (Gibbon et al., 1999). Based on the findings in yeast, it might be expected that over-expression of ADF in pollen would increase the sensitivity of pollen F-actin to latrunculin.

ADF is a complex protein that probably plays multiple roles in plant cells. We now know a great deal about the capabilities of ADF in regulating actin assembly and disassembly, as well as what factors regulate ADF activity *in vitro*. It will be extremely important over the next decade to work towards an understanding of how ADF regulates actin dynamics *in vivo*. The production of both gain- and loss-of-function plants should be very helpful. It is anticipated that altering ADF levels will have dramatically different effects depending upon the plant tissue or individual cell type examined. To date, no mutation in a plant actin-binding protein has been characterized. The over- and under-expression of specific ADF protein isoforms may help elucidate the requirement for a large gene family as well as shed light onto the actin-dependent processes which require their activity.

ACKNOWLEDGEMENTS

We are grateful to David Collings and Patrick Hussey for reading the manuscript and making useful suggestions. Work in our laboratory is funded by the USDA/NRICGP and the DOE Energy Biosciences Division.

REFERENCES

Abe H, Obinata T, Minamide LS and Bamburg JR (1996) *Xenopus laevis* actin-depolymerizing factor/cofilin: A phosphorylation-regulated protein essential for development. J Cell Biol 132: 871-885

Agnew BJ, Minamide LS and Bamburg JR (1995) Reactivation of phosphorylated actin depolymerizing factor and identification of the regulatory site. J Biol Chem 270: 17582-17587

Aizawa H, Sutoh K and Yahara I (1996) Overexpression of cofilin stimulates bundling of actin filaments, membrane ruffling, and cell movement in *Dictyostelium*. J Cell Biol 132: 335-344

Amberg DC, Basart E and Botstein D (1995) Defining protein interactions with yeast actin in vivo. Nature Struct Biol 2: 28-34

Arber S, Barbayannis FA, Hanser H, Schneider C, Stanyon CA, Bernard O and Caroni P (1998) Regulation of actin dynamics through phosphorylation of cofilin by LIM-kinase. Nature 393: 805-809

Bamburg JR (1999) Proteins of the ADF/cofilin family: Essential regulators of actin dynamics. Annu Rev Cell Dev Biol 15: 185-230

Bamburg JR and Bray D (1987) Distribution and cellular localization of actin depolymerizing factor. J Cell Biol 105: 2817-2825

Bamburg JR, Harris HE and Weeds AG (1980) Partial purification and characterization of an actin depolymerizing factor from brain. FEBS Lett 121: 178-182

Bamburg JR, McGough A and Ono S (1999) Putting a new twist on actin: ADF/cofilins modulate actin dynamics. Trends Cell Biol 9: 364-370

Bernstein BW and Bamburg JR (1982) Tropomyosin binding to F-actin protects the F-actin from disassembly by brain actin-depolymerizing factor (ADF). Cell Motil Cytoskel 2: 1-8

Blanchoin L and Pollard TD (1998) Interaction of actin monomers with *Acanthamoeba* actophorin (ADF/cofilin) and profilin. J Biol Chem 273: 25106-25111

Blanchoin L and Pollard TD (1999) Mechanism of interaction of *Acanthamoeba* actophorin (ADF/cofilin) with actin filaments. J Biol Chem 274: 15538-15546

Blanchoin L, Robinson RC, Choe S and Pollard TD (2000) Phosphorylation of *Acanthamoeba* actophorin (ADF/cofilin) blocks interaction with actin without a change in atomic structure. J Mol Biol 295: 203-211

Carlier M-F (1998) Control of actin dynamics. Curr Opin Cell Biol 10: 45-51

Carlier M-F, Laurent V, Santolini J, Melki R, Didry D, Xia G-X, Hong Y, Chua N-H and Pantaloni D (1997) Actin depolymerizing factor (ADF/cofilin) enhances the rate of filament turnover: Implication in actin-based motility. J Cell Biol 136: 1307-1322

Carlier M-F and Pantaloni D (1997) Control of actin dynamics in cell motility. J Mol Biol 269: 459-467

Carlier M-F, Ressad F and Pantaloni D (1999) Control of actin dynamics in cell motility. Role of ADF/cofilin. J Biol Chem 274: 33827-33830

Cooper JA, Blum JD, Williams RC and Pollard TD (1986) Purification and characterization of actophorin, a new 15,000-dalton actin-binding protein from *Acanthamoeba castellanii.* J Biol Chem 261: 477-485

Cramer LP (1999) Role of actin-filament disassembly in lamellipodium protrusion in motile cells revealed using the drug jasplakinolide. Curr Biol 9: 1095-1105

Danyluk J, Carpentier E and Sarhan F (1996) Identification and characterization of a low temperature regulated gene encoding an actin-binding protein from wheat. FEBS Lett 389: 324-327

Didry D, Carlier M-F and Pantaloni D (1998) Synergy between actin depolymerizing factor/cofilin and profilin in increasing actin filament turnover. J Biol Chem 273: 25602-25611

Du J and Frieden C (1998) Kinetic studies on the effect of yeast cofilin on yeast actin polymerization. Biochemistry 37: 13276-13284

Eads JC, Mahoney NM, Vorobiev S, Bresnick AR, Wen KK, Rubenstein PA, Haarer BK and Almo SC (1998) Structure determination and characterization of *Saccharomyces cerevisiae* profilin. Biochemistry 37: 11171-11181

Fedorov AA, Lappalainen P, Fedorov EV, Drubin DG and Almo SC (1997) Structure determination of yeast cofilin. Nature Struct Biol 4: 366-369

Feijó JA, Sainhas J, Hackett GR, Kunkel JG and Hepler PK (1999) Growing pollen tubes possess a constitutive alkaline band in the clear zone and a growth-dependent acidic tip. J Cell Biol 144: 483-496

Gibbon BC, Kovar DR and Staiger CJ (1999) Latrunculin B has different effects on maize pollen germination and tube growth. Plant Cell 11: 2349-2363

Gibbon BC, Zonia LE, Kovar DR, Hussey PJ and Staiger CJ (1998) Pollen profilin function depends on interaction with proline-rich motifs. Plant Cell 10: 981-994 [A printed correction appears in Plant Cell 11, 1603].

Gungabissoon RA, Jiang C-J, Drøbak BK, Maciver SK and Hussey PJ (1998) Interaction of maize actin-depolymerising factor with actin and phosphoinositides and its inhibition of plant phospholipase C. Plant J 16: 689-696

Gunsalus KC, Bonaccorsi S, Williams E, Verni F, Gatti M and Goldberg ML (1995) Mutations in *twinstar*, a *Drosophila* gene encoding a cofilin/ADF homologue, result in defects in centrosome migration and cytokinesis. J Cell Biol 131: 1243-1259

Hatanaka H, Ogura K, Moriyama K, Ichikawa S, Yahara I and Inagaki F (1996) Tertiary structure of destrin and structural similarity between two actin-regulating protein families. Cell 85: 1047-1055

Hawkins M, Pope B, Maciver SK and Weeds AG (1993) Human actin depolymerizing factor mediates a pH-sensitive destruction of actin filaments. Biochemistry 32: 9985-9993

Hayden SM, Miller PS, Brauweiler A and Bamburg JR (1993) Analysis of the interaction of actin depolymerizing factor with G- and F-actin. Biochemistry 32: 9994-10004

Hussey PJ, Yuan M, Calder G, Khan S and Lloyd CW (1998) Microinjection of pollen-specific actin-depolymerizing factor, ZmADF1, reorientates F-actin strands in *Tradescantia* stamen hair cells. Plant J 14: 353-357

Iida K, Moriyama K, Matsumoto S, Kawasaki H, Nishida E and Yahara I (1993) Isolation of a yeast essential gene, *COF1*, that encodes a homologue of mammalian cofilin, a low-M(r) actin-binding and depolymerizing protein. Gene 124: 115-120

Jiang C-J, Weeds AG and Hussey PJ (1997a) The maize actin-depolymerizing factor, ZmADF3, redistributes to the growing tip of elongating root hairs and can be induced to translocate into the nucleus with actin. Plant J 12: 1035-1043

Jiang C-J, Weeds AG, Khan S and Hussey PJ (1997b) F-actin and G-actin binding are uncoupled by mutation of conserved tyrosine residues in maize actin depolymerizing factor (ZmADF). Proc Natl Acad Sci USA 94: 9973-9978

Kang F, Purich DL and Southwick FS (1999) Profilin promotes barbed-end actin filament assembly without lowering the critical concentration. J Biol Chem 274: 36963-36972

Kim SR, Kim Y and An G (1993) Molecular cloning and characterization of anther-preferential cDNA encoding a putative actin-depolymerizing factor. Plant Mol Biol 21: 39-45

Klahre U and Chua N-H (1999) The *Arabidopsis* ACTIN-RELATED PROTEIN 2 (AtARP2) promoter directs expression in xylem precursor cells and pollen. Plant Mol Biol 41: 65-73

Koffer A and Daridan M (1985) Actin-regulating activities in cultured BHK cells. J Cell Sci 75: 239-257

Koffer A, Edgar AJ and Bamburg JR (1988) Identification of two species of actin depolymerizing factor in cultures of BHK cells. J Musc Res Cell Motil 9: 320-328

Kost B, Spielhofer P and Chua N-H (1998) A GFP-mouse talin fusion protein labels plant actin filaments *in vivo* and visualizes the actin cytoskeleton in growing pollen tubes. Plant J 16: 393-401

Kovar DR, Drøbak BK and Staiger CJ (2000) Maize profilin isoforms are functionally distinct. Plant Cell 12: In press

Kusano K, Abe H and Obinata T (1999) Detection of a sequence involved in actin-binding and phosphoinositide-binding in the N-terminal side of cofilin. Mol Cell Biochem 190: 133-141

Lappalainen P and Drubin DG (1997) Cofilin promotes rapid actin filament turnover *in vivo*. Nature 388: 78-82

Lappalainen P, Fedorov EV, Fedorov AA, Almo SC and Drubin DG (1997) Essential functions and actin-binding surfaces of yeast cofilin revealed by systematic mutagenesis. EMBO J 16: 5520-5530

Lappalainen P, Kessels MM, Cope MJTV and Drubin DG (1998) The ADF homology (ADF-H) domain: A highly exploited actin-binding module. Mol Biol Cell 9: 1951-1959

Leonard SA, Gittis AG, Petrella EC, Pollard TD and Lattman EE (1997) Crystal structure of the actin-binding protein actophorin from *Acanthamoeba.* Nature Struct Biol 4: 369-373

Loisel TP, Boujemaa R, Pantaloni D and Carlier M-F (1999) Reconstitution of actin-based motility of *Listeria* and *Shigella* using pure proteins. Nature 401: 613-616

Lopez I, Anthony RG, Maciver SK, Jiang C-J, Khan S, Weeds AG and Hussey PJ (1996) Pollen specific expression of maize genes encoding actin depolymerizing factor-like proteins. Proc Natl Acad Sci USA 93: 7415-7420

Mabuchi I (1981) Purification from starfish eggs of a protein that depolymerizes actin. J Biochemistry 89: 1341-1344

Machesky LM and Gould KL (1999) The Arp2/3 complex: a multifunctional actin organizer. Curr Opin Cell Biol 11: 117-121

Maciver SK (1998) How ADF/cofilin depolymerizes actin filaments. Curr Opin Cell Biol 10: 140-144

Maciver SK, Pope BJ, Whytock S and Weeds AG (1998) The effect of two actin depolymerizing factors (ADF/cofilins) on actin filament turnover: pH sensitivity of F-actin binding by human ADF, but not of *Acanthamoeba* actophorin. Eur J Biochem 256: 388-397

Maciver SK and Weeds AG (1994) Actophorin preferentially binds monomeric ADP-actin over ATP-bound actin: consequences for cell locomotion. FEBS Lett 347: 251-256

Maciver SK, Zot HG and Pollard TD (1991) Characterization of actin filament severing by actophorin from *Acanthamoeba castellanii.* J Cell Biol 115: 1611-1620

Maekawa S, Nishida E, Ohta Y and Sakai H (1984) Isolation of low molecular weight actin-binding proteins from porcine brain. J Biochem 95: 377-385

McGough A and Chiu W (1999) ADF/cofilin weakens lateral contacts in the actin filament. J Mol Biol 291: 513-519

McGough A, Pope B, Chiu W and Weeds A (1997) Cofilin changes the twist of F-actin: Implications for actin filament dynamics and cellular function. J Cell Biol 138: 771-781

McKim K, Matheson C, Marra MA, Wakarchuk MF and Baillie DJ (1994) The *Caenorhabditis elegans unc-60* gene encodes proteins homologous to a family of actin-binding proteins. Mol Gen Genet 242: 346-357

Messerli MA and Robinson KR (1998) Cytoplasmic acidification and current influx follow growth pulses of *Lilium longiflorum* pollen tubes. Plant J 16: 87-91

Miller DD, Callaham DA, Gross DJ and Hepler PK (1992) Free Ca^{2+} gradient in growing pollen tubes of *lilium.* J. Cell Sci 110: 7-12

Miller DD, Lancelle SA and Hepler PK (1996) Actin microfilaments do not form a dense meshwork in *Lilium longiflorum* pollen tube tips. Protoplasma 195: 123-132

Moon A and Drubin DG (1995) The ADF/cofilin proteins: Stimulus-responsive modulators of actin dynamics. Mol Biol Cell 6: 1423-1431

Moon AL, Janmey PA, Louie KA and Drubin DG (1993) Cofilin is an essential component of the yeast cortical cytoskeleton. J Cell Biol 120: 421-435

Morgan TE, Lockerbie RO, Minamide LS, Browning MD and Bamburg JR (1993) Isolation and characterization of a regulated form of actin depolymerizing factor. J Cell Biol 122: 623-633

Moriyama K, Iida K and Yahara I (1996) Phosphorylation of Ser-3 of cofilin regulates its essential function on actin. Genes Cells 1: 73-86

Mullins RD, Heuser JA and Pollard TD (1998) The interaction of Arp2/3 complex with actin: Nucleation, high affinity pointed end capping, and formation of branching networks of filaments. Proc Natl Acad Sci USA 95: 6181-6186

Nagaoka R, Kusano K, Abe H and Obinata T (1995) Effects of cofilin on actin filamentous structures in cultured muscle cells: Intracellular regulation of cofilin actin. J Cell Sci 108: 581-593

Nagaoka R, Minami N, Hayakawa K, Abe H and Obinata T (1996) Quantitative analysis of low molecular weight G-actin-binding proteins, cofilin, ADF and profilin, expressed in developing and degenerating chicken skeletal muscles. J Muscl Res Cell Motil 17: 463-473

Nick P (1999) Signals, motors, morphogenesis: the cytoskeleton in plant development. Plant Biol 1: 169-179

Nishida E (1985) Opposite effects of cofilin and profilin from porcine brain on rate of exchange of actin-bound adenosine 5'-triphosphate. Biochemistry 24: 1160-1164

Nishida E, Iida K, Yonezawa N, Koyasu S, Yahara I and Sakai H (1987) Cofilin is a component of intranuclear and cytoplasmic rods induced in cultured cells. Proc Natl Acad Sci USA 84: 5262-5266

Nishida E, Maekawa S, Muneyuki E and Sakai H (1984a) Action of a 19K protein from porcine brain on actin polymerization: A new functional class of actin-binding proteins. J Biochem 95: 387-398

Nishida E, Maekawa S and Sakai H (1984b) Cofilin, a protein in porcine brain that binds to actin filaments and inhibits their interactions with myosin and tropomyosin. Biochemistry 23: 5307-5313

Obinata T, Nagaokayasuda R, Ono S, Kusano K, Mohri K, Ohtaka Y, Yamashiro S, Okada K and Abe H (1997) Low molecular-weight G-actin binding proteins involved in the regulation of actin assembly during myofibrillogenesis. Cell Struct Funct 22: 181-189

Okada K, Obinata T and Abe H (1999) XAIP1: a *Xenopus* homologue of yeast actin interacting protein 1 (AIP1), which induces disassembly of actin filaments cooperatively with ADF cofilin family proteins. J Cell Sci 112: 1553-1565

Ono S and Benian GM (1998) Two *Caenorhabditis elegans* actin depolymerizing factor/cofilin proteins, encoded by the *unc-60* gene, differentially regulate actin filament dynamics. J Biol Chem 273: 3778-3783

Pantaloni D and Carlier M-F (1993) How profilin promotes actin filament assembly in the presence of thymosin ß4. Cell 75: 1007-1014

Perelroizen I, Didry D, Christensen H, Chua N-H and Carlier M-F (1996) Role of nucleotide exchange and hydrolysis in the function of profilin in actin assembly. J Biol Chem 271: 12302-12309

Pollard TD and Cooper JA (1984) Quantitative analysis of the effect of *Acanthamoeba* profilin on actin filament nucleation and elongation. Biochemistry 23: 6631-6641

Rathore KS, Cork RJ and Robinson KR (1991) A cytoplasmic gradient of Ca^{2+} is correlated with the growth of lily pollen tubes. Develop. Biol 148: 612-619

Ressad F, Didry D, Egile C, Pantaloni D and Carlier M-F (1999) Control of actin filament length and turnover by actin depolymerizing factor (ADF/cofilin) in the presence of capping proteins and ARP2/3 complex. J Biol Chem 274: 20970-20976

Ressad F, Didry D, Xia GX, Hong Y, Chua N-H, Pantaloni D and Carlier M-F (1998) Kinetic analysis of the interaction of actin-depolymerizing factor (ADF)/cofilin with G- and F-actins: Comparison of plant and human ADFs and effect of phosphorylation. J Biol Chem 273: 20894-20902

Rodal AA, Tetreault JW, Lappalainen P, Drubin DG and Amberg DC (1999) Aip1p interacts with cofilin to disassemble actin filaments. J Cell Biol 145: 1251-1264

Rosenblatt J, Agnew BJ, Abe H, Bamburg JR and Mitchison TJ (1997) *Xenopus* actin depolymerizing factor cofilin (XAC) is responsible for the turnover of actin filaments in *Listeria monocytogenes* tails. J Cell Biol 136: 1323-1332

Rozycka M, Khan S, Lopez I, Greenland AJ and Hussey PJ (1995) A *Zea mays* pollen cDNA encoding a putative actin-depolymerizing factor. Plant Physiol 107: 1011-1012

Sheterline P, Clayton J and Sparrow JC (1998) Actin. Protein Profile 4: 1-272

Smertenko AP, Jiang C-J, Simmons NJ, Weeds AG, Davies DR and Hussey PJ (1998) Ser6 in the maize actin-depolymerizing factor, ZmADF3, is phosphorylated by a calcium-stimulated protein kinase and is essential for the control of functional activity. Plant J 14: 187-194

Staiger CJ (2000) Signaling to the actin cytoskeleton in plants. Annu Rev Plant Physiol Plant Mol Biol 51: 257-288

Staiger CJ, Yuan M, Valenta R, Shaw PJ, Warn R and Lloyd CW (1994) Microinjected profilin affects cytoplasmic streaming in plant cells by rapidly depolymerizing actin microfilaments. Curr Biol 4: 215-219

Sumi T, Matsumoto K, Takai Y and Nakamura T (1999) Cofilin phosphorylation and actin cytoskeletal dynamics regulated by Rho- and Cdc42-activated LIM-kinase 2. J Cell Biol 147: 1519-1532

Suzuki K et al., (1995) Activation induces dephosphorylation of cofilin and its translocation to the plasma membrane in neutrophil-like differentiated HL-60 cells. J Biol Chem 270: 19551-19556

Svitkina TM and Borisy GG (1999) Arp2/3 complex and actin depolymerizing factor cofilin in dendritic organization and treadmilling of actin filament array in lamellipodia. J Cell Biol 145: 1009-1026

Theriot JA (1997) Accelerating on a treadmill: ADF/cofilin promotes rapid actin filament turnover in the dynamic cytoskeleton. J Cell Biol 136: 1165-1168

Wang Y-L (1985) Exchange of actin subunits at the leading edge of living fibroblasts: Possible role of treadmilling. J Cell Biol 101: 597-602

Welch MD, Rosenblatt J, Skoble J, Portnoy DA and Mitchison TJ (1998) Interaction of human Arp2/3 complex and *Listeria monocytogenes* ActA protein in actin filament nucleation. Science 281: 105-108

Yang N, Higuchi O, Ohashi K, Nagata K, Wada A, Kangawa K, Nishida E and Mizuno K (1998) Cofilin phosphorylation by LIM-kinase 1 and its role in Rac-mediated actin reorganization. Nature 393: 809-812

Yonezawa N, Homma Y, Yahara I, Sakai H and Nishida E (1991) A short sequence responsible for both phosphoinositide binding and actin binding activities of cofilin. J Biol Chem 266: 17218-17221

Yonezawa N, Nishida E, Iida K, Yahara I and Sakai H (1990) Inhibition of the interaction of cofilin, destrin, and deoxyribonuclease I with actin by phosphoinositides. J Biol Chem 265: 8382-8386

Yonezawa N, Nishida E, Koyasu S, Maekawa S, Ohta Y, Yahara I and Sakai H (1987) Distribution among tissues and intracellular localization of cofilin, a 21 kDa actin-binding protein. Cell Struct Funct 12: 443-452

Yonezawa N, Nishida E and Sakai H (1985) pH control of actin polymerization by cofilin. J Biol Chem 260: 14410-14412

Chapter 5

FIMBRIN

David W McCurdy[1] and Christopher J Staiger[2]

1 Discipline of Biological Sciences, School of Biological and Chemical Sciences, The University of Newcastle, New South Wales, Australia
2 Department of Biological Sciences, Purdue University, West Lafayette, IN 47907-1392 USA

Key words: *Arabidopsis,* actin-binding protein, cross-linking, gene family, microinjection

Abstract: Fimbrins are a class of conserved actin-binding proteins with a modular organization consisting of two tandemly repeated actin-binding domains and an N-terminal calcium-binding headpiece. The tandem arrangement of the two actin-binding domains enables fimbrins to cross-link actin filaments into tightly packed bundles that are typical of those seen in microvilli and other surface structures, such as microspikes. Fimbrin homologues have been identified in *Arabidopsis thaliana* and show the same general domain organization as non-plant fimbrins. AtFim1 binds to and cross-links pollen F-actin in a calcium-independent manner, and protects actin filaments from profilin-induced depolymerization both *in vitro* and in living stamen hair cells. The possible role of fimbrins in organizing the supramolecular architecture of the actin cytoskeleton in plant cells is discussed in light of the large size of the actin gene family and the presence of other actin-bundling proteins.

1. INTRODUCTION

A striking feature of the actin cytoskeleton in higher plants is its complex organization into morphologically-distinct arrays within the cell, and the dynamic reorganization of these arrays during the cell cycle. The actin network is organized to form nuclear baskets, large subcortical bundles, and various levels of cortical networks which are either random (Traas et al., 1987; Sonobe & Shibaoka, 1989) or more organized (McCurdy et al., 1990). These arrays consist of actin filaments formed into bundles of various complexity and extent, and thus mechanisms exist to regulate not only recruitment of filaments into bundles, but also their distinctive supramolecular morphology and presumably different physiological

C.J. Staiger et al. (eds.), Actin: A Dynamic Framework for Multiple Plant Cell Functions, 87–102.

functions. In animal cells, numerous classes of actin-binding proteins (ABPs) regulate the higher-order structure of the actin cytoskeleton and its response to local and global signalling events. To begin to unravel the molecular machinery responsible for controlling the supramolecular complexity of the actin cytoskeleton in higher plants, a homologue of the actin filament bundling protein fimbrin was cloned from *Arabidopsis thaliana* (McCurdy & Kim, 1998). This review covers the biology of fimbrins, first from the perspective of the detailed structural and functional information obtained from studies in animal and protist cells, and then provides an overview of the current understanding of fimbrins in plant cells.

2. NON-PLANT FIMBRINS

Fimbrins are a family of highly conserved actin filament cross-linking or bundling proteins identified in diverse eukaryotic cells. Fimbrin was first identified as a 68-kD structural protein from the core actin bundle of intestinal brush border microvilli, and later was shown to be abundant in actin-rich surface structures such as membrane ruffles, microspikes and cell adhesion sites (Bretscher & Weber, 1980; Bretscher, 1981; Glenney et al., 1981). Sequence analysis of chicken intestinal fimbrin (de Arruda et al., 1990) revealed that it belongs to a family including human L- and T-plastin, the former of which was identified as a protein phosphorylated in response to growth factors and phorbol esters (Goldstein et al., 1985), and now shown to be actin filament cross-linking proteins (Namba et al., 1992; Lin et al., 1994; Arpin et al., 1994). Members of the fimbrin/plastin family have a modular organization consisting of an N-terminal domain (~100 residues) that contains two Ca^{2+}-binding EF-hand motifs (Fig. 1). The C-terminal two-thirds of the protein contains two, tandemly-arranged actin-binding domains (ABDs). The two ABDs are homologous to the single ABD of actin cross-linking proteins such as α-actinin, spectrin, dystrophin and ABP120 (Matsudaira, 1991). These proteins are long, rod-shaped molecules with the ABD positioned at one end, and function as homodimers to cross-link actin filaments into loose two- or three-dimensional networks. Fimbrin is unique in this family in that it has two ABDs and thus is able to function as a monomer to cross-link adjacent actin filaments into tight bundles of uniform polarity (Bretscher, 1981).

Database searches based on statistical analyses of sequences related by secondary structure revealed that the two conserved ABDs of the fimbrin/plastin family are composed of two calponin-homology (CH)-like domains (Castresana & Saraste, 1995). Calponin, a microfilament-associated protein that regulates myosin and tropomyosin binding in smooth muscle, binds to F-actin via a single CH domain. Several proteins involved in G-protein-mediated signalling, like IQGAP and vav, also contain a single CH domain. Since this domain is sufficient to bind F-actin (Way et al., 1992; Winder et al., 1995), it has been proposed that the CH domain serves to link signalling molecules with the actin cytoskeleton (Castresana & Saraste, 1995; Hanein et al., 1998). This proposal is controversial, however, as others have argued that not all CH domains are structurally or functionally equivalent, and thus the presence of a single CH domain in a polypeptide does not confer actin-binding capabilities per se (Stradal et al., 1998).

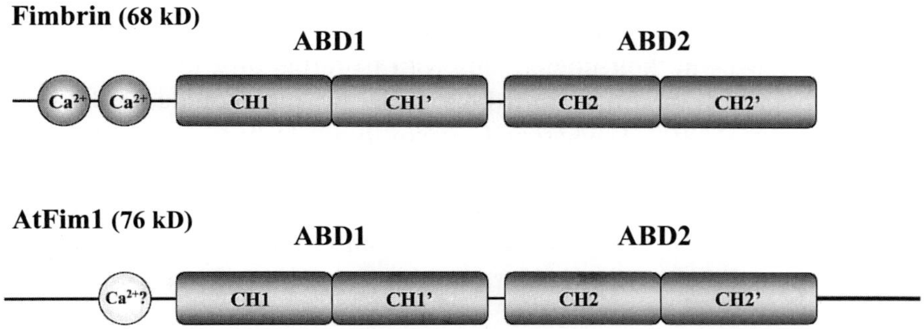

Figure 1. Modular organization of fimbrins. Non-plant fimbrins (68–70 kD) typically contain an N-terminal Ca^{2+}-binding domain composed of two EF-hand motifs (circles), followed by two tandem repeats of a conserved actin-binding domain (ABD1 and ABD2). Each ABD is composed of two calponin homology (CH)-subdomains, although as discussed in the text, the two CH subdomains of each ABD may not be structurally or functionally equivalent. AtFim1 (76 kD) from *A. thaliana* is similar in overall organization to non-plant fimbrins, except that the N-terminal domain contains only one weak EF-hand motif and the C-terminal tail is extended by about 65 residues.

Despite the overall structural conservation of fimbrins, considerable functional differences are observed between members of the fimbrin/plastin family. In vertebrates, different isoforms are expressed in a cell type-specific manner (Lin et al., 1994; Arpin et al., 1994), and this is speculated to cause functional differences in the organization of the actin cytoskeleton in these cells (Arpin et al., 1994). For example, L- and T-plastin associate with different populations of actin filaments in fibroblast-like or epithelial-like cell lines, and have different morphological effects on actin filament structures such as stress fibres and microvilli (Arpin et al., 1994). Both L- and T-plastins are also localized to cell-substratum adhesion sites via the formation of adhesion-dependent fimbrin-vimentin complexes in macrophages (Correia et al., 1999). A vimentin-binding site has been identified in the CH1 domain of fimbrin, and the complexes of fimbrin with vimentin subunits are proposed to play a role in the co-regulated assembly of the actin and intermediate filament (i.e., vimentin) cytoskeletons at sites of cell adhesion (Correia et al., 1999).

In *Saccharomyces cerevisiae*, *SAC6* encodes a fimbrin homologue which is involved in the maintenance of cell polarity via its ability to bundle actin filaments (Adams et al., 1991), and is also required for the internalization step of endocytosis (Kübler & Riezman, 1993). Of the three isoforms of human fimbrin, only T- and L-plastin can substitute for Sac6p *in vivo* and complement a *sac6* null mutant (Adams et al., 1995). Both human I-plastin and chicken fimbrin are unable to complement this mutation (Adams et al., 1995). These functional differences within the fimbrin/plastin family also reflect differences in their interactions with isoforms of actin. For example, T-plastin binds to rabbit α-actin from muscle, but also nonmuscle ß- and γ-actin (Arpin et al., 1994), whereas L-plastin binds to human ß-actin but not to rabbit muscle α-actin (Namba et al., 1992). Similarly, *Dictyostelium* plastin binds to filaments of ß/γ-actin from bovine spleen, but not muscle α-actin (Prassler et al., 1997), and yeast fimbrin (Sac6p) binds to a yeast F-actin affinity column but not one prepared from rabbit α-actin (Drubin et al., 1988).

Considerable variation amongst the fimbrin/plastin family members also occurs at the level of calcium regulation. The actin-bundling activity of human L-plastin and *Dictyostelium* plastin is inhibited by sub-millimolar levels of calcium (Namba et al., 1992; Prassler et al., 1997), whereas the bundling activity of chicken fimbrin is not greatly affected by calcium (Bretscher, 1981). Presumably, structural differences in the N-terminal 'headpiece' domain of fimbrins cause the different responses seen to calcium; this region is the most divergent among mammalian fimbrins (de Arruda et al., 1990), and critical residues necessary for calcium binding are missing in the EF-hand-like motifs in yeast fimbrin (Adams et al., 1991).

The picture emerging is that while the fimbrin/plastin family is considered to be highly conserved, and most fimbrins from diverse cell types are presumably involved in forming tightly-packed actin bundles of uniform polarity, sufficient structural and functional differences existing within the family indicate that one avenue for complex regulation may reside at the level of differential interactions between fimbrin and actin isoforms (see also Meagher et al., this volume). Regulation at this level clearly provides the potential for pleiomorphic changes in the actin cytoskeleton in response to different cellular needs.

3. TOWARD AN ATOMIC MODEL FOR FIMBRIN CROSS-LINKING F-ACTIN

The crystal structure of ABD1 from human fimbrin (Fig. 2A) shows that it is a dumbbell-shaped molecule composed of two α-helical subdomains corresponding to the two CH domains (CH1 and CH1') connected by a long central α-helix (Goldsmith et al., 1997). Each CH subdomain contains four major α-helices connected by two shorter helical segments. Goldsmith et al. (1997) suggest that the compactness of the two CH subdomains within ABD1 represent stable, independent-folding modules. This is consistent with observations that isolated N-terminal CH domains from different ABPs are capable of forming stable actin-binding modules (Way et al., 1992; Winder et al., 1995). However, as pointed out by Gimona and colleagues (Stradal et al., 1998; Gimona and Winder, 1998), not all CH domains are equivalent, and isolated C-terminal CH domains typically show little if any ability to bind to F-actin.

Recently, an atomic model of a fimbrin cross-link in an actin bundle has been determined by combining the crystal structure of ABD1 (Goldsmith et al., 1997) with helical reconstructions of three-dimensional maps derived from electron cryomicroscopy of N375, a fragment containing the N-terminal calcium domain and ABD1 of fimbrin, bound to F-actin (Hanein et al., 1998). The resulting model shows that the binding surface on N375 is predominantly localized in CH1 and in the connecting region between the two CH domains (Figs. 2B, 2C). In addition, the residues on actin that are most likely to interact with N375 are located within subdomain 2 (terminology after Lorenz et al., 1993) of the barbed-end actin subunit (n), and subdomain 1 of the adjacent pointed-end actin subunit (n + 2) (Figs. 2B, 2C). These predicted sites correspond well with actin-binding segments identified by peptide-binding and genetic studies, as well as with actin residues implicated in binding to fimbrin in yeast (Honts et al., 1994). The model further predicts that the dumbell-shaped ABD1 is aligned perpendicular to the axis of the actin filament and sits in the groove between actin subunits (n) and (n + 2) (Fig. 2B). This configuration accounts for how fimbrin can cross-link neighbouring filaments in a tightly-packed bundle with center-to-center spacing of 140Å (Matsudaira et al., 1983). The N-terminus of the two CH domains faces the actin filament while the C-terminus faces away, thus positioning the second of the two ABDs (ABD2) of fimbrin away from the surface of the filament in an orientation capable of binding a neighbouring actin filament (Hanein et al., 1998). Furthermore, the model offers a structural explanation for how the N-terminal calcium-binding domain inhibits fimbrin binding to actin in the presence of calcium. This domain is proposed to occupy the crevice between the two CH domains of ABD1 and to act as a wedge to disrupt the topology of actin-binding sites on the opposite surface of ABD1. This configuration is consistent with proteolytic patterns of fimbrin generated in the presence or absence of calcium (Hanein et al., 1998). However, as described above, not all fimbrins show calcium regulation or contain clearly recognizable EF-hand motifs in the N-terminal headpiece domain. Thus, subtle structural variations to the model proposed by Hanein and colleagues may exist amongst the family of fimbrin/plastin proteins.

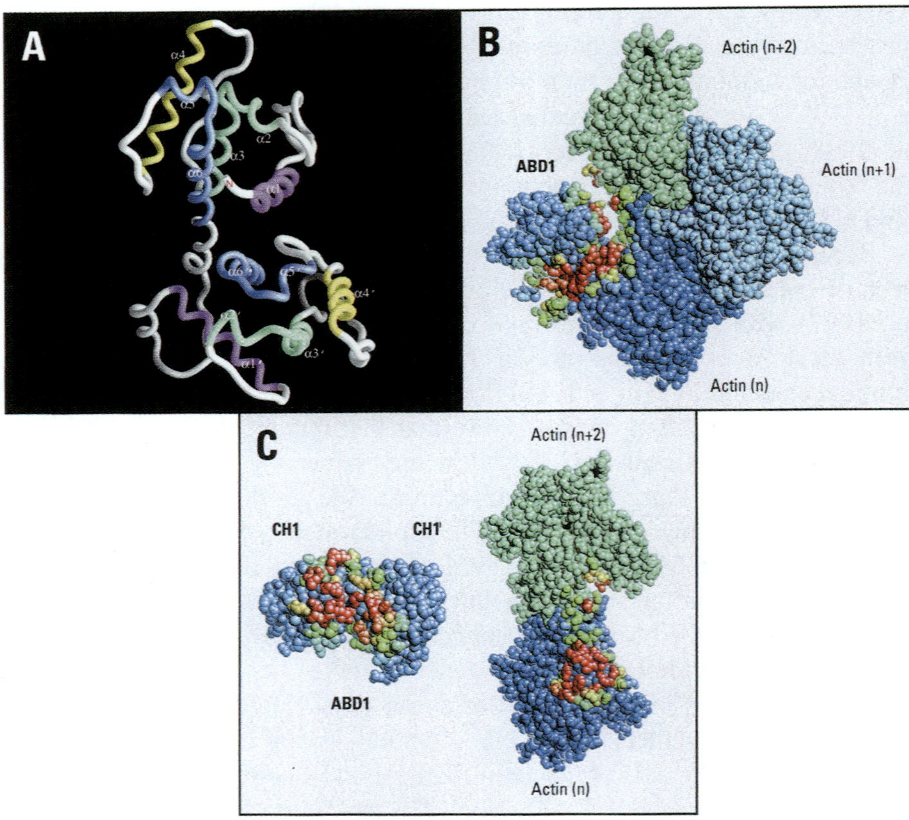

Figure 2. **A**. Structure of the N-terminal actin-binding domain (ABD1) of human fimbrin. The dumbbell-shaped structure consists of two CH-domains, the upper CH1 subdomain connected by a long central helix (α6) to the second, C-terminal CH1' subdomain. Each CH1 subdomain contains four prominent α-helices (α1, α3, α4 and α6), which are connected by two shorter segments, α2 and α5. Homologous helices in the two CH1 subdomains are color-coded to highlight their structural similarities. (Image provided by S Goldsmith & S Almo; reproduced with permission from Nature Structural Biology). **B, C**. Space filling model of ABD1 of human T-fimbrin binding to F-actin. (**B**) Three subunits of F-actin are shown (barbed end at bottom) with ABD1 sitting perpendicular to the filament axis in the groove made between actin subunits (n) and (n+2). The residues are color-coded to show interaction probabilities as

defined by Hanein et al. (1998), with red being the highest and blue the lowest. (**C**) Space-filling model of above where both the actin filament and ABD1 are rotated 90° from the view in (**B**) to highlight the residues predicted to be involved in the ABD1-actin interface. The majority of the interface residues are in CH1 and the connecting region with CH1', and those in F-actin in subunit (n). (Images provided by D Hanein & N Volkmann; reproduced with permission from Nature Structural Biology).

4. PLANT FIMBRINS

4.1 *ATFIM1* from *Arabidopsis thaliana*

Several classes of ABPs which regulate the polymerization and/or depolymerization of actin filaments in plant cells have been identified (see Gibbon & Staiger, and Kovar & Staiger, this volume), but few if any genes for factors which regulate the supramolecular architecture of the actin cytoskeleton have been discovered (see also Yokota & Shimmen, this volume). As part of a search for such genes, *ATFIM1* was cloned from an *A. thaliana* cDNA library by a degenerate PCR strategy that targetted conserved amino acid domains in fimbrin/plastin (McCurdy & Kim, 1998). The full *ATFIM1* sequence revealed that the predicted 76-kD protein, AtFim1, is structurally related to fimbrins in properties required for filament cross-linking, but also contains novel features. The overall design of AtFim1 shows that it is composed of three domains; a so-called N-terminal 'headpiece' of some 113 residues, a middle region that contains two ABDs arranged in tandem, and a novel C-terminal extension of some 65 amino acids (Fig. 1) (McCurdy & Kim, 1998). The two ABDs, encompassing residues 114–366 (ABD1) and 367–622 (ABD2), are up to 46% and 43% identical, respectively, to similar domains in non-plant fimbrins such as chicken and *Dictyostelium* plastin (Table 1). Each ABD in AtFim1 also contains a region of 27 amino acids which is up to 74% identical to a putative actin-binding motif identified in the larger family of actin cross-linking proteins represented by fimbrins, α-actinin, spectrin and dystrophin (Adams et al., 1991). The presence of the two ABDs, and the highly conserved actin-binding motifs within each domain, implied strongly that AtFim1 functions as an actin filament cross-linking protein (McCurdy & Kim, 1998).

An interesting feature of AtFim1 is the presence of some 65 amino acids that extend beyond the typical C-terminus of non-plant fimbrins (Fig. 1). This C-terminal tail contains several clusters of negatively-charged amino acids, a feature which may indicate a role in mediating interactions with other proteins in the cell. Alternatively, the extended tail may simply reflect a functionally-neutral evolutionary variant of fimbrin structure which has escaped selection pressure due to its location in the overall fold of the molecule.

TABLE 1. Actin-binding domains (ABD) 1 and 2 of AtFim1 compared to the equivalent domain in non-plant fimbrins

	AtFim1	
	ABD1	**ABD2**
Human plastin	45.4 (71.3)	36.8 (64)
Chicken fimbrin	46.2 (70.5)	36.8 (64.4)
Dictyostelium plastin	45.6 (69)	42.9 (67.9)
Yeast fimbrin	44.8 (68.6)	40.9 (65.1)

Values for percent identity (similarity) were calculated using the default settings of BESTFIT

4.2 Functional characterization of AtFim1

The ability of recombinant AtFim1 to bind to F-actin (Kovar et al., submitted), prepared from polymerization-competent maize pollen actin by the method of Ren et al. (1997), has recently been determined. The use of plant actin was considered important for these studies given the variability in the way fimbrins interact with actins from different sources. In these studies bacterially-expressed AtFim1 was shown to bind pollen F-actin in a calcium-independent manner (Table 2). Furthermore, unlike human T-plastin, a known calcium-inhibited actin cross-linking protein, the ability of AtFim1 to cross-link F-actin into supramolecular assemblies that are pelletable by low-speed centrifugation was also demonstrated to be calcium-independent (Table 2).

TABLE 2. AtFim1 binds to and cross-links pollen F-actin by a calcium-independent mechanism.

	% Sedimented AtFim1[a]	% Cross-linked actin[b]	
	- F-actin	- AtFim1	
$[Ca^{2+}]$ 4.6 nM	15 ± 8.0	11 ± 5.3	
	+ F-actin	+ AtFim1	+ T-plastin
$[Ca^{2+}]$ 4.6 nM	66 ± 2.9	64 ± 7.3	63 ± 5.6
100 nM	74 ± 5.3	60 ± 2.5	
1.0 μM	73 ± 8.7	63 ± 7.2	
12 μM	73 ± 3.7	61 ± 1.0	
1.0 mM	65 ± 3.1	64 ± 2.0	13 ± 7.6

[a]AtFim1 (1.0 μM) alone or AtFim1 plus pollen G-actin (3.0 μM) were incubated under polymerizing conditions in the presence of varying concentrations of Ca^{2+}, and after 90 min at 22°C were centrifuged at 200,000 *g* for 1 h. The supernatant and pellet fractions were subjected to SDS-PAGE and the percent of AtFim1 in the pellet was determined by scanning densitometry. Values represent an average from 3 or more experiments (± SD).
[b]Actin (3.0 μM) alone, actin plus AtFim1 (1.0 μM), or actin plus human T-plastin (1.0 μM) were incubated as described above and then centrifuged at 13,500 *g* for 30 min. The percent of actin in the pellet was determined as above.
Data from Kovar et al., submitted

The effect of AtFim1 on actin cytoskeletal architecture in live cells was also determined by microinjection into *Tradescantia virginiana* stamen hair cells. Injected AtFim1 caused cessation of cytoplasmic streaming and inhibited the dynamic movements of transvacuolar strands within 2–4 min (Kovar et al., submitted). These results are markedly different from those observed following microinjection of profilin, an actin-monomer binding protein that can promote either polymerization or depolymerization of actin *in vitro* (Staiger et al., 1997; see also Gibbon & Staiger, this volume). Recombinant ZmPRO5 (maize profilin) caused transvacuolar strands to become thinner and eventually snap, leading to collapse of the centrally-positioned nucleus to the periphery of the cell. This result is interpreted as profilin inducing the depolymerization of the actin filaments within the transvacuolar strands that position the interphase nucleus in the centre of the cell. The presence of AtFim1 appears to stabilize the actin network in microinjected cells, an interpretation that is supported by the observation that AtFim1 can reduce, in a concentration-dependent manner, the rate of nuclear displacement that occurs following injection of ZmPRO5 (Kovar et al., submitted). The ability of AtFim1-bound F-actin to resist depolymerization by profilin was also demonstrated *in vitro*. Pollen actin polymerized in the presence of ZmPRO5 alone, or ZmPRO5 plus AtFim1, was subjected to high-speed sedimentation and the amount of F-actin pelleted determined by SDS-PAGE. These experiments demonstrated that the average percent of actin depolymerized by profilin in the absence of AtFim1 was 43.3 ± 9.5, compared to 17.0 ± 7.6 in the presence of AtFim1 (Kovar et al., submitted). These results indicate that actin filaments in the presence of AtFim1 were protected from depolymerization induced by ZmPRO5, a conclusion that supports the interpretation of the *in vivo* microinjection experiments.

4.3 A fimbrin gene family in *Arabidopsis thaliana*

Genomic sequencing has revealed at least three fimbrin-like sequences distributed throughout the *Arabidopsis* genome. *ATFIM2* and *ATFIM3* encode predicted proteins of 73.6 kD and 73.4 kD, respectively, with considerable variation in calculated isoelectric points (Table 3). The overall domain organization of AtFim2 and AtFim3 is similar to AtFim1, with two tandemly-arranged ABDs up to 73.4% identical to the corresponding domains in AtFim1, plus an N-terminal 'headpiece' region which ranges from 44.2–55.7% identity between the three plant sequences. Both AtFim2 and AtFim3 have extended C-terminal domains that are not present in the non-plant fimbrins. Of these three fimbrins, AtFim1 has the longest tail and AtFim2 the shortest. The three *Arabidopsis* fimbrins contain conserved amino acid residues predicted from the T-fimbrin atomic model to be involved in binding to F-actin. Of the five T-fimbrin residues in N375 scoring the highest estimates for interaction probabilities (Hanein et al., 1998), four (G208, Q233, K236 and E270) are conserved in all three *Arabidopsis* fimbrins. Other high-scoring residues in T-fimbrin (e.g., A206, I207, V212, N213 and I237) are also either identical or conservative substitutions in the *Arabidopsis* homologues.

TABLE 3 Properties of *Arabidopsis* fimbrin polypeptides

	MW[a]	pI[a]	Chromosome Number	GenBank number
AtFim1	76,898	6.52	4	CAB36516
AtFim2	73,664	8.41	?	AAB97847
AtFim3	73,442	6.35	2	AAD22331

[a]Calculated from the predicted amino acid sequence with PEPSTAT.

Unlike the conservation of structure seen in the ABDs of fimbrin, substantial variability occurs in the N-terminal "headpiece" between non-plant fimbrins and the *Arabidopsis* proteins. In particular, variation is seen in the *Arabidopsis* sequences when compared to the residues involved in calcium co-ordination by the two EF-hand motifs in T-fimbrin. While some corresponding residues in the *Arabidopsis* fimbrins are conserved, especially in AtFim1 (McCurdy & Kim, 1998), several critical replacements in all three *Arabidopsis* sequences suggest a low capacity for calcium binding (data not shown). This conclusion is supported by the observation that AtFim1 binds to and cross-links pollen F-actin by a calcium-independent mechanism (Table 2).

Preliminary phylogenetic analysis of the three *Arabidopsis* fimbrins places *ATFIM1* and *ATFIM3* in a subgroup distinct from *ATFIM2* (data not shown). Interestingly, *ATFIM1* is expressed in all major organs examined (McCurdy & Kim, 1998). Promoter-GUS gene fusion analysis in transgenic plants has been used to demonstrate expression of *ATFIM1* in all actively dividing tissues, whereas *ATFIM2* expression is reportedly restricted to the root cap only (Christensen et al., 1997). The expression pattern of *ATFIM3* in *Arabidopsis* is presently unknown.

4.4 Other plant fimbrins

A cDNA encoding a partial fimbrin-like sequence from wheat has been reported (Cruz-Ortega et al., 1997). The truncated polypeptide contains the two tandemly-arranged ABDs characteristic of known fimbrins, and also contains the two highly-conserved 27-amino acid regions predicted to be actin-binding sites in each of the two larger ABDs. Interestingly, however, the wheat fimbrin is more like the non-plant fimbrins in that it does not contain an extended C-terminal tail region. The wheat fimbrin was isolated as part of a study to identify root proteins up-regulated in response to aluminum toxicity. The fimbrin-like sequence was up-regulated within 12 h in an aluminum-tolerant cultivar, whereas up-regulation of fimbrin-like transcripts in an aluminum-sensitive cultivar occurred after 24 h (Cruz-Ortega et al., 1997). This observation is intriguing when considered in the light of results from single cell experiments showing that tension, or rigidity, within the actin cytoskeleton network is elevated upon exposure to aluminum ions (Grabski & Schindler, 1995). One possibility is that responses to cellular stress associated with aluminum toxicity involve modulation of the cytoskeleton via an increase in fimbrin-induced cross-linking of actin filaments. Reorganization of the cytoskeleton in response to elevated aluminum has also been documented in tobacco BY-2 cells (Blancaflor et al., 1998).

Several EST clones encoding fimbrin-like sequences from tomato (Genbank accession numbers AI774774, AI489788), pine (AI813184) and a hybrid aspen (AI163828) are present in the public database. The available protein sequence data from these ESTs show features characteristic of fimbrin-like sequences, including regions of the conserved ABDs (data not shown). Clone AI489788 was isolated from tomato carpel tissue, and the second tomato clone, AI774774, isolated from a *Pseudomonas*-resistant cultivar, contains an extended C-terminal tail six residues longer than AtFim3. Phylogenetic analysis comparing these partial sequences with the *Arabidopsis* fimbrins and the wheat sequence indicates no clear ortholog relationships between the members, suggesting that the family of plant fimbrin sequences may be large.

4.5 Functions of plant fimbrins

The conserved structure of the *Arabidopsis* fimbrins, coupled with the demonstrated F-actin cross-linking activity of AtFim1 (Table 2), provides strong support for the conclusion that these proteins are responsible for forming cross-linked actin filaments in plant cells, presumably by a calcium-insensitive mechanism. The widespread expression of AtFim1 in *Arabidopsis* (McCurdy & Kim, 1998) suggests that this protein may be used as part of a common mechanism for the assembly of actin bundles, perhaps for the formation of the large subcortical bundles which are typically present in most plant cells. The contribution of other fimbrins to such a process is not known, however it may be possible that different fimbrins could then be used for the formation of actin bundles in discrete domains of the cell, or in response to different signals. The complexity of the actin cytoskeleton suggests that different levels of organization, even within the one cell, may involve different sets of actin-interacting proteins. Clearly, determining whether different cells or tissues express multiple fimbrin isoforms, and if so whether these isoforms are localized predominantly to discrete cellular domains, will be important information to obtain.

Other proteins which regulate the bundling of actin filaments in plants, such as the *Arabidopsis* homologs of the calcium-dependent bundling protein villin (Klahre et al., 2000), and the 135-kD actin-bundling protein from lily pollen tubes, also a villin-like protein, have been identified recently (Yokota et al., 1998; Vidali et al., 1999; see also Yokota & Shimmen, and Vidali & Hepler, this volume). Interestingly, both proteins are expected to bundle actin filaments *in vivo* in a calcium-dependent or calcium-calmodulin-dependent manner, unlike that predicted for the fimbrins. The actin network undergoes dramatic reorganization during the cell cycle (e.g., McCurdy & Gunning, 1990) and in response to various environmental signals such as light (Waller & Nick, 1997) or pathogen attack (Kobayashi et al., 1994). Such reorganizations presumably involve altering the actin-bundling status of the cytoskeleton, thus it will be fascinating to unravel the possible contributions of each of the classes of plant actin-bundling proteins to such processes.

5. FUTURE DIRECTIONS

Determining the size of the fimbrin gene family in *Arabidopsis*, and the detailed expression of the gene products at both the cellular and subcellular levels, will clearly be important in determining the level of potential complexity of the interactions of fimbrins with the actin cytoskeleton. A particularly important issue to address is whether different fimbrin isoforms are expressed in the same cell, and if so whether they are responsible for bundling discrete or overlapping populations of actin filaments. Coupled to this question will be whether expression of the different fimbrin isoforms is differentially regulated in response to environmental signals. Obtaining detailed biochemical information on the affinity (i.e., K_d) for plant actin of all of the three known *Arabidopsis* fimbrins, and the possible regulation of these interactions by other factors, will be important in understanding the contributions of each to the regulation of the actin cytoskeleton. Such information is relevant to the concept of isovariant dynamics, defined as the "temporal and biochemical expansion of a biological systems's responses as a result of the simultaneous expression and interaction of multiple isovariants of a protein" (Meagher et al., 1999; Meagher et al., this volume). Isovariant dynamics is argued to facilitate enormous flexibility in the dynamic behaviour of the plant actin cytoskeleton (Meagher et al., 1999). Central to this argument is that the dynamic capacity of the actin cytoskeleton would be expanded even further by the expression of multiple isoforms of ABPs, each potentially with different affinity constants for actin and each capable of responding differently to environmental signals. Given this scenario, plus the existence of other actin-bundling proteins known in plants (Yokota et al., 1998; Klahre et al, 2000), the capacity for isovariant dynamics in the plant actin cytoskeleton is extreme (Meagher et al., 1999).

Following the lead of Meagher and colleagues, who have begun to dissect plant actin function genetically (see Meagher et al., this volume), the cellular function of fimbrins should be analyzed with reverse-genetic approaches. Of particular utility is the PCR-based strategy for identification of T-DNA insertions that has uncovered mutations in several actin genes (McKinney et al., 1995; Gilliland et al., 1998). It is satisfying to note that the multiplicity of actin genes does not simply provide redundancy to plant developmental processes, such that morphological defects can be observed when individual actin genes are disrupted (Gilliland et al., 1999). Fimbrin insertional mutants could be examined cytologically for defects in the distribution and function of actin cables and bundles. Moreover, the mutant plants should be tested for sensitivity to bacterial and fungal pathogens and other environmental stimuli that impinge on actin organization and function (Staiger, 2000).

Importantly, most if not all of the experimental tools to address the questions raised above are currently available and have been successfully employed to study various aspects of the plant cytoskeleton. The combined application of these complementary approaches to the study of the fimbrin gene family will enable a better understanding of how these proteins contribute to the functioning of the actin cytoskeleton in plant cells.

ACKNOWLEDGEMENTS

We thank Steven Almo and Sharon Goldsmith (Albert Einstein College of Medicine) as well as Dorit Hanein (Brandeis University) and Niels Volkmann (The Burnham Institute) for providing figures. We are especially grateful to David Kovar (Purdue University) for sharing his unpublished data. This work was supported by an Australian Research Council grant to DWM and a DOE Energy Biosciences grant to CJS.

REFERENCES

Adams AEM, Botstein D and Drubin DG (1991) Requirement of yeast fimbrin for actin organization and morphogenesis *in vivo*. Nature 354: 404-408

Adams AEM, Shen W, Lin C-S, Leavitt J and Matsudaira P (1995) Isoform-specific complementation of the yeast *sac6* null mutation by human fimbrin. Mol Cell Biol 15: 69-75

Arpin M, Friederich E, Algrain M, Vernel F and Louvard D (1994) Functional differences between L- and T-plastin isoforms. J Cell Biol 127: 1995-2008

Blancaflor EB, Jones DL and Gilroy S (1998) Alterations in the cytoskeleton accompany aluminum-induced growth inhibition and morphological changes in primary roots of maize. Plant Physiol 118: 159-172

Bretscher A (1981) Fimbrin is a cytoskeletal protein that crosslinks F-actin *in vitro*. Proc Natl Acad Sci USA 78: 6849-6853

Bretscher A and Weber K (1980) Fimbrin, a new microfilament-associated protein present in microvilli and other cell surface structures. J Cell Biol 86: 335-340

Brower SM, Honts JE and Adams AEM (1995) Genetic analysis of the fimbrin-actin binding interaction in *Saccharomyces cerevisiae*. Genetics 140: 91-101

Castresana J and Saraste M (1995) Does vav bind to F-actin through a CH domain? FEBS Lett 374: 149-151

Christensen HEM, Mathur J and Chua N-H (1997) Cloning and characterization of the fimbrin gene family in *Arabidopsis thaliana*. 5th Int Congr Plant Mol Biol, Singapore, Abstr 278

Correia I, Chu D, Chou Y-H, Goldman RD and Matsudaira P (1999) Integrating the actin and vimentin cytoskeletons: adhesion-dependent formation of fimbrin-vimentin complexes in macrophages. J Cell Biol 146: 831-842

Cruz-Ortega, R, Cushman JC and Ownby JD (1997) cDNA clones encoding 1,3-ß-glucanase and a fimbrin-like cytoskeletal protein are induced by Al toxicity in wheat roots. Plant Physiol 114: 1453-1460

de Arruda MV, Watson S, Lin C-S, Leavitt J and Matsudaira P (1990) Fimbrin is a homologue of the cytoplasmic phosphoprotein plastin and has domains homologous with calmodulin and actin gelation proteins. J Cell Biol 111: 1069-1079

Drubin DG, Miller KG and Botstein D (1988) Yeast actin-binding proteins: evidence for a role in morphogenesis. J Cell Biol 107: 2551-2561

Gilliland LU, McKinney EC and Meagher RB (1999) Analysis of vegetative mutants in the conserved actin gene family of *Arabidopsis*. Mol Biol Cell 10 (Suppl): 22a

Gilliland LU, McKinney EC, Asmussen MA and Meagher RB (1998) Detection of deleterious genotypes in multigenerational studies. I. Disruptions in individual *Arabidopsis* actin genes. Genetics 149: 717-725

Gimona M and Winder SJ (1998) Single calponin homology domains are not actin-binding domains. Curr Biol 8: R674-R675

Glenney JR Jr, Kaulfus P, Matsudaira P and Weber K (1981) F-actin binding and bundling properties of fimbrin, a major cytoskeletal protein of microvillus core filaments. J Biol Chem 256: 9283-9288

Goldsmith SC, Pokala N, Shen W, Fedorov AA, Matsudaira P and Almo S (1997) The structure of an actin-cross-linking domain from human fimbrin. Nature Struct Biol 4: 708-712

Goldstein D, Djeu J, Latter G, Burbeck S and Leavitt J (1985) Abundant synthesis of the transformation-induced protein of neoplastic human fibroblasts, plastin, in normal lymphocytes. Cancer Res 45: 5643-5647

Grabski S and Schindler M (1995) Aluminum induces rigor within the actin network of soybean cells. Plant Physiol 108: 897-901

Hanein D, Volkmann N, Goldsmith S, Michon A-M, Lehman W, Craig R, DeRosier D, Almo S and Matsudaira P (1998) An atomic model of fimbrin binding to F-actin and its implications for filament cross-linking and regulation. Nature Struct Biol 9: 787-792

Honts JE, Sandrock TS, Brower SM, O'Dell JL and Adams AEM (1994) Actin mutations that show suppression with fimbrin mutations identify a likely fimbrin-binding site on actin. J Cell Biol 126: 413-422

Klahre U, Friederich E, Kost B, Louvard D and Chua N-H (2000) Villin-like actin-binding proteins are expressed ubiquitously in *Arabidopsis*. Plant Physiol 122: 35-47

Kobayashi I, Kobayashi Y and Hardham AR (1994) Dynamic reorganization of microtubules and microfilaments in flax cells during the resistance response to flax rust infection. Planta 195: 237-247

Kovar DR, Staiger CJ, Weaver EA and McCurdy DW (2000) Functional characterization of AtFim1, an actin filament cross-linking protein from *Arabidopsis thaliana*. Submitted

Kübler E and Riezman H (1993) Actin and fimbrin are required for the internalization step of endocytosis in yeast. EMBO J 12: 2855-2862

Lin C-S, Shen W, Chen ZP, Tu Y-H and Matsudaira P (1994) Identification of I-plastin, a human isoform expressed in intestine and kidney. Mol Cell Biol 14: 2457-2467

Lorenz M, Popp D and Holmes KC (1993) Refinement of the F-actin model against X-ray fiber diffraction data by the use of a directed mutation algorithm. J Mol Biol 234: 826-836

Matsudaira P (1991) Modular organization of actin cross-linking proteins. Trends Biochem Sci 16: 87-92

Matsudaira P, Mandelkow E, Renner W, Hesterburg LK and Weber K (1983) Role of fimbrin and villin in determining the interfilament distances of actin bundles. Nature 301: 209-214

McCurdy DW and Gunning BES (1990) Reorganization of cortical actin microfilaments and microtubules at preprophase and mitosis in wheat root tip cells: a double label immunofluorescence study. Cell Motil Cytoskel 15: 76-87

McCurdy DW and Kim M (1998) Molecular cloning of a novel fimbrin-like cDNA from *Arabidopsis thaliana*. Plant Mol Biol 36: 23-31

McKinney EC, Ali N, Traut A, Feldmann KA, Belostotsky DA, McDowell JM and Meagher RB (1995) Sequence-based identification of T-DNA insertion mutations in *Arabidopsis*: actin mutants *act2-1* and *act4-1*. Plant J 8: 613-622.

Meagher RB, McKinney EC and Kandasamy MK (1999) Isovariant dynamics expands and buffers the responses of complex systems: the diverse plant actin gene family. Plant Cell 11: 1-12

Namba Y, Ito M, Zu Y, Shigesada K and Maruyama K (1992) Human T cell L-plastin bundles actin filaments in a calcium-dependent manner. J Biochem 112: 503-507

Prassler J, Stocker S, Marriott G, Heidecker M, Kellermann and Gerisch G (1997) Interaction of a *Dictyostelium* member of the plastin/fimbrin family with actin filaments and actin-myosin complexes. Mol Biol Cell 8: 83-95

Ren H, Gibbon BC, Ashworth SL, Sherman DM, Yuan M and Staiger CJ (1997) Actin purified from maize pollen functions in living plant cells. Plant Cell 9: 1445-1457

Sonobe S and Shibaoka H (1989) Cortical fine actin filaments in higher plant cells visualized by rhodamine-phalloidin after pretreatment with m-maleimidobenzoyl N-hydroxysuccinimide ester. Protoplasma 148: 80-86

Staiger CJ (2000) Signaling to the actin cytoskeleton in plant cells. Annu Rev Plant Physiol Plant Mol Biol 51: 257-288

Staiger CJ, Gibbon BC, Kovar DR and Zonia LE (1997) Profilin and actin depolymerizing factor: modulators of actin organization in plants. Trends Plant Sci 2: 275-281

Stradal T, Kranewitter W, Winder SJ and Gimona M (1998) CH domains revisited. FEBS Lett 431: 134-137

Traas JA, Doonan JH, Rawlins DJ, Shaw PJ, Watts J and Lloyd CW (1987) An actin network is present throughout the cell cycle of carrot cells and associates with the dividing nucleus. J Cell Biol 105: 387-395

Vidali L, Yokota E, Cheung A, Shimmen T and Hepler PK (1999) The 135 kDa actin-bundling protein from *Lilium longiflorum* pollen is the plant homologue of villin. Protoplasma 209: 283-291

Waller F and Nick P (1997) Response of actin microfilaments during phytochrome-controlled growth of maize seedlings. Protoplasma 200: 154-162

Way M, Pope B and Weeds AG (1992) Evidence for functional homology in the F-actin binding domains of gelsolin and α-actinin: Implications for the requirements of severing and capping. J Cell Biol 119: 835-842

Winder SJ, Hemmings L, Maciver SK, Bolton SJ, Tinsley JM, Davies KE, Critchley DR and Kendrick-Jones J (1995) Utrophin actin binding domain: analysis of actin binding and cellular targeting. J Cell Sci 108: 63-71

Yokota E, Takahara K and Shimmen T (1998) Actin-bundling protein isolated from pollen tubes of lily. Biochemical and immunocytochemical characterization. Plant Physiol 116: 1421-1429

Chapter 6

CHARACTERIZATION OF NATIVE ACTIN-BINDING PROTEINS FROM POLLEN

Myosin and the actin-bundling proteins, 135-ABP and 115-ABP

Etsuo Yokota and Teruo Shimmen

Department of Life Science, Faculty of Science, Himeji Institute of Technology, Harima Science Park City, Hyogo 678–1297 Japan

Key words: actin-bundling proteins, calcium regulation, calmodulin, cytoplasmic streaming, myosin

Abstract: A myosin (170 kDa) and two types of actin-bundling proteins (135-ABP and 115-ABP) were isolated from pollen tubes of lily, *Lilium longiflorum*. The myosin consists of a 170-kDa heavy chain with calmodulin light chain. On the basis of immunolocalization studies using antiserum against the 170-kDa heavy chain and the analysis of the sliding velocity of F-actin *in vitro*, this myosin was suggested to be a motor responsible for cytoplasmic streaming. The motile activity of 170-kDa myosin was suppressed by Ca^{2+} through the calmodulin light chain. On the other hand, both ABPs arranged actin filaments into bundles that serve as tracks for cytoplasmic streaming. The bundling activity of 135-ABP was inhibited by Ca^{2+}-calmodulin. Physiological studies revealed that Ca^{2+} induces not only the inactivation of motor activity for cytoplasmic streaming but also causes the fragmentation and disorganization of actin bundles in pollen tubes, which should be correlated with the regulation of cytoplasmic streaming by Ca^{2+}. Hence, the Ca^{2+} sensitivity in the activities of 170-kDa myosin and 135-ABP, acting through calmodulin, is suggested to be a molecular basis for the Ca^{2+} regulation of cytoplamsic streaming in pollen tubes.

1. POLLEN AS AN IDEAL SYSTEM FOR ACTIN-BINDING PROTEIN STUDIES

Actin filaments, one of the major components of the cytoskeleton, are organized into a highly-ordered architecture and are involved in various

C.J. Staiger et al. (eds.), Actin: A Dynamic Framework for Multiple Plant Cell Functions, 103–118.

kinds of cellular activities. It is well established that the architecture and functions of filamentous actin are regulated by numerous classes of mechanochemical motor proteins, myosins, (Mooseker & Cheney, 1995; Mermall et al., 1998) and actin-binding proteins, including cross-linking proteins, severing proteins, capping proteins, side-binding proteins and monomer-sequestering proteins (Stossel et., 1985; Pollard & Cooper, 1986).

In the vegetative cells of pollen grains, it is well known that the actin cytoskeleton undergoes dynamic changes in organization during hydration, activation and germination (Pierson & Cresti, 1992; see also Vidali & Hepler, this volume). Before hydration, actin exists as fusiform or spiculate structures that are considered to be the storage forms. Upon hydration and activation, these rearrange to form a network (Heslop-Harrison et al., 1986; Tiwari & Polito, 1988a). In most parts of growing pollen tubes, actin filaments are arranged into bundles with an orientation parallel to the long axis of tubes (Perdue et al., 1985; Pierson et al., 1986; Miller et al., 1996). In the tip region of tubes, however, only a few thin bundles and short individual actin filaments in random orientation have been observed by electron microscopy of samples prepared using the rapid-freeze fixation technique (Lancelle et al., 1987; Tiwari & Polito, 1988b; Lancelle and Hepler, 1992; Miller et al., 1996). Pollen tube organelles and vesicles actively stream along the actin-filament bundles (cytoplasmic streaming), and vegetative nuclei and generative cells are transported to the growing tip (Pierson & Cresti, 1992; Li et al., 1997; Taylor & Hepler, 1997). Myosin is believed to be the motor for cytoplasmic streaming and for the transport of generative cells and vegetative nuclei (Miller et al., 1995: Li et al., 1997; Taylor & Hepler, 1997). Myosin is also implicated in cytoplasmic streaming in somatic cells of the plant body (Nagai, 1993; Williamson, 1993; Shimmen & Yokota, 1994; see also Grolig & Pierson, this volume). Because of the small number of vacuoles containing proteases, pollen and pollen tubes are ideal materials for the isolation and biochemical analyses of myosin and actin-binding proteins. We have established isolation procedures for myosin and two types of actin-bundling proteins from lily pollen tubes. In this chapter, we summarize the biochemical and immunocytochemical properties of these proteins and introduce recent molecular biological evidence for the 135-kDa actin-bundling protein (135-ABP).

2. ISOLATION PROCEDURE FOR MYOSIN AND ACTIN-BUNDLING PROTEINS

Myosins and some types of actin-binding proteins, such as cross-linking, bundling, capping and side-binding proteins, are expected to possess the ability to bind filamentous actin (F-actin) in an ATP-dependent and ATP-independent manner, respectively. When F-actin prepared from chicken breast muscle is added to a crude protein extract prepared from germinating pollen, in the presence of casein as an antiproteolysis agent, the 170-kDa polypeptide of myosin heavy chain, and the 135-kDa and 115-kDa polypeptides of actin-bundling proteins are major proteins routinely found in the pellets following high speed sedimentation. The 170-kDa polypeptide is released from F-actin in the presence of ATP (Yokota & Shimmen, 1994), whereas the 135-kDa and 115-kDa polypeptides are dissociated from F-actin under high ionic-strength conditions (Yokota et al., 1998). Hence, sedimentation with F-actin and subsequent extraction of co-precipitants under different conditions is a useful general method for separating myosin and actin-bundling proteins. Each protein is successively purified by conventional column chromatographies. In the case of myosin, its activity in each purification step is monitored by assaying for motile activity *in vitro* (Kron & Spudich, 1986) using rhodamine-phalloidin-labeled F-actin (Kohno et al., 1992; Yokota & Shimmen, 1994). Using these methods, myosin and fractions containing myosin(s) have been identified and prepared from cultured cells of tobacco (BY-2 cells; Yokota et al., 1999b) and *Catharanthus* (Yokota et al., 1995a), and from leaves of the aquatic angiosperm *Vallisneria* (Takagi et al., 1995).

2.1 Myosin

Myosin isolated from lily pollen tubes (Fig. 1A) has a 170-kDa heavy chain (Yokota & Shimmen, 1994). Eukaryotic myosins are generally composed of heavy and light chains (Mooseker & Cheney, 1995; Wolenski, 1995). In many cases, calmodulin or related proteins from the EF-hand superfamily are associated with the myosin heavy chains; these are the light chains. In the case of plants, myosin heavy chains identified by molecular methods have IQ-motif repeat domains for binding calmodulin or related proteins (Knight & Kendrick-Jones, 1993; Kinkema & Schiefelbein, 1994; Kinkema et al., 1994). An 18-kDa polypeptide that crossreacts with antibodies raised against spinach calmodulin (Muto & Miyachi, 1984) is present in isolated the 170-kDa myosin fraction (Figs. 1B, 1C). Association with the 170-kDa heavy chain under low Ca^{2+} concentrations indicates that the 18-kDa polypeptide is the light chain of 170-kDa myosin (Yokota et al., 1999a). The calmodulin light chain dissociates from the 170-kDa heavy chain in the presence of Ca^{2+} at concentrations higher than 10^{-5} M.

The ATPase activity of 170-kDa myosin is stimulated up to 60-fold by F-actin (Yokota & Shimmen, 1994). Hence, this myosin has an F-actin-activated ATPase, like other myosins isolated biochemically thus far. The

170-kDa myosin translocates F-actin *in vitro* with an average velocity between 5 to 8 μm/sec, consistent with the velocity of cytoplasmic streaming observed in living pollen tubes of lily.

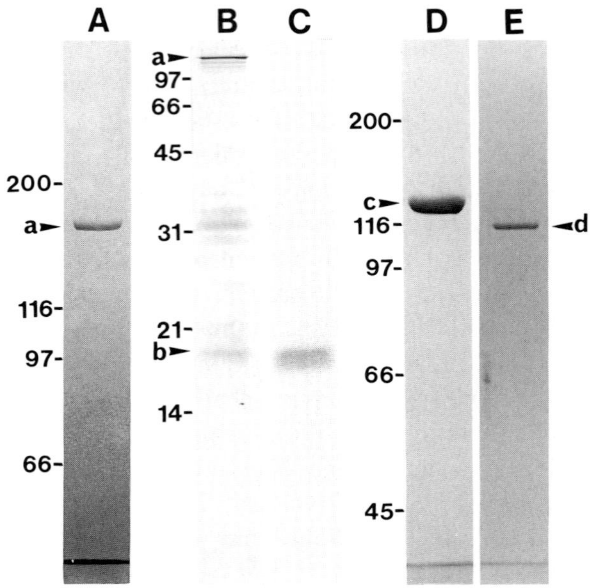

Figure 1. Purity of the 170-kDa myosin, 135-ABP and 115-ABP from lily pollen. **A**, Coomassie staining of a 6% gel showing the 170-kDa myosin fraction. Arrowhead (a) indicates the 170-kDa heavy chain. **B**, Coomassie staining of a 15% gel containing the 170-kDa myosin fraction. Arrowhead (b) indicates the 18-kDa polypeptide. C, Immunoblotting with an antiserum raised against spinach calmodulin. **D**, Coomassie staining of a 7.5% gel containing purified 135-ABP. Arrowhead (c) indicates the 135-kDa polypeptide. **E**, Coomassie staining of a 7.5% gel containing purified 115-ABP. Arrowhead (d) indicates the 115-kDa polypeptide. Molecular weight ($x10^{-3}$) of standard proteins is indicated on the left.

When casein is not included during preparation of the crude extract, 145-kDa (140-kDa) and 120-kDa polypeptides are detected in the fractions along with the 170-kDa heavy chain (Kohno et al., 1992; Yokota & Shimmen, 1994). When analyzed by polyacrylamide gel electrophoresis, the intensity of these two bands is increased; conversely, that of the 170-kDa band is decreased. Moreover, an 110-kDa polypeptide becomes apparent after prolonged storage of the myosin fraction. These results, together with the immunoreactivity of these bands with antibodies raised against 170-kDa heavy chain and 145-kDa polypeptide, indicate that the 145-kDa, 120-kDa and 110-kDa polypeptides are proteolytic fragments of the 170-kDa heavy chain. The sliding velocity of F-actin *in vitro* induced by myosin fractions that include these proteolytic fragments is lower than that observed in the presence of 'intact' myosin fractions. Furthermore, ATPase activity in this fraction is stimulated only 2–3-fold by F-actin (Kohno et al., 1992). Hence, casein is a potent and useful agent to protect myosin from proteolysis.

On the basis of immunocytochemical studies using antibodies against non-plant myosins and physiological and electron microscopic analyses using characean cells, the following model for the molecular basis of cytoplasmic streaming is proposed: Myosin associated with the membrane surface of organelles and vesicles moves and slides along actin filaments using the energy from hydrolysis of ATP (Williamson, 1993; Shimmen & Yokota, 1994). This model is supported by the reconstitution of organelle movement along actin cables in characean cells. When organelles isolated from *Chara* cells are introduced with ATP into tonoplast-free *Nitella* cells in which intrinsic myosins are inactivated by N-ethylmaleimide, they are translocated along *Nitella* actin cables (Shimmen & Tazawa, 1982). By a similar reconstitution procedure, it has been shown that organelles isolated from lily pollen tubes also move along characean actin cables (Kohno & Shimmen, 1988; Kohno et al., 1990). Immunocytochemical analysis, using antiserum raised against the 170-kDa heavy chain, reveals that myosin is localized on the surface of organelles with various sizes from lily and tobacco pollen tubes (Yokota et al., 1995b). Furthermore, this myosin is generally distributed in the somatic cells of higher plants, in cultured cells of *Arabidopsis*, tobacco BY-2 (Yokota et al., 1995b, 1999b) and *Catharanthus* cells (Yokota et al., 1995a), as well as in leaves of *Vallisneria* (Takagi et al., 1995). Myosin is also localized on organelles and the perinuclear region in these cultured cells. Together with the results showing that the 170-kDa myosin is capable of translocating F-actin *in vitro* with a velocity consistent with cytoplasmic streaming in living cells, these localization studies suggest that 170-kDa myosin acts as a motor responsible for cytoplasmic streaming in higher plant cells.

Interestingly, the velocity of F-actin sliding *in vitro* that is induced by other plant myosins, *Chara* myosin (Yamamoto et al., 1994, 1995; Higashi-Fujime et al., 1995) and BY-2 175-kDa myosin (Yokota et al., 1999b) is comparable to the velocity of cytoplasmic streaming in the same living cells. There is no significant difference in sliding velocity induced by the 170-kDa myosin when plant actin prepared from BY-2 cells or when skeletal muscle actin is used (Igarashi et al., 1999). Furthermore, myosin subfragment 1 prepared from animal skeletal muscle (S-1) translocates plant actin with velocity similar to skeletal muscle actin. Therefore, we suggest that the velocity of cytoplasmic streaming in a given cell type is determined by the properties of myosin and not of actin.

2.2 Actin-bundling proteins

Two types of actin-bundling protein have been purified from lily pollen tubes. One is composed of a 135-kDa polypeptide (135-ABP; Fig. 1D), and the other of a 115-kDa polypeptide (115-ABP; Fig. 1E). The 135-ABP binds stoichiometrically to F-actin filaments and bundles them *in vitro* (Fig. 2A), even in the presence of ATP or high concentrations of Ca^{2+} (Yokota et al., 1998). In bundles formed by 135-ABP and F-actin, cross-bridges with an axial periodicity of about 32 nm are frequently observed (Fig. 2B). The actin filaments in a bundle have a uniform polarity as demonstrated by S-1 decoration (Fig. 2C; Yokota & Shimmen, 1999). Immunocytochemical analysis, using an antibody raised against the 135-kDa polypeptide, reveals that 135-ABP is expressed not only in pollen tubes, but also in the somatic cells of higher plants, root hair cells of *Hydrocharis* (Tominaga et al., 1999, 2000) and all organs of the lily plant (Vidali et al., 1999). Moreover, 135-ABP is co-localized with actin-filament bundles in lily pollen tubes (Yokota et al., 1998; Vidali et al., 1999) and the root hair cells (Tominaga et al., 1999, 2000). In the root hair cells of *Hydrocharis*, the direction of cytoplasmic streaming is basipetal in the transvacuolar strands that penetrate central vacuoles and acropetal in the subcortical regions (Shimmen et al., 1995; Tominaga et al., 1997). Actin filaments in a bundle that serve as tracks for cytoplasmic streaming in the root hair cells exhibit uniform polarity, whereas the polarity of bundles in the transvacuolar strands is opposite to that in the subcortical regions (Tominaga et al., 2000). These observations suggest that 135-ABP arranges actin filaments into bundles of a uniform polarity and consequently determines the direction of cytoplasmic streaming in plant cells.

The 115-ABP binds sub-stoichiometrically to F-actin filaments and bundles them in an ATP- and Ca^{2+}-independent manner (Nakayasu et al., 1998). Several kinds of actin-binding proteins that possess the ability to form actin-filament bundles in non-plant cells have been identified also in plant cells (see also McCurdy & Staiger, Muday, and Stankovic et al., this volume). For example, annexins with molecular mass of 34 and 35 kDa (Boustead et al., 1989; Smallwood et al., 1990; Calvert et al., 1996), elongation factor 1α with molecular mass of 49- to 50-kDa (Yang et al., 1993; Collings et al., 1994; Clore et al., 1996), 220- to 240-kDa polypeptides crossreacting with anti-spectrin antibodies (Kaminskyj & Heath, 1995; Michaud et al., 1991; Faraday & Spanswich, 1993) and a fimbrin-like genes (Cruz-Ortega et al., 1997; McCurdy & Kim, 1998). The transcript for *Arabidopsis* fimbrin encodes a predicted protein with molecular weight of 76-kDa (McCurdy & Kim, 1998). Judging from the molecular mass data, 115-ABP is unequivocally different from those proteins. In *Acetabularia* cells, the perinuclear region stains with antibodies against chicken α-actinin (Tischendorf et al., 1988). α-Actinin is a dimeric actin cross-linking protein comprised of 90- to 110-kDa subunits (Stossel et al., 1985; Pollard & Cooper, 1986; Matsudaira, 1991; Otto, 1994). However, 115-ABP shows different antigenicity when compared with animal α-actinin (Nakayasu et al., 1998), indicating that 115-ABP is distinct from the α-actinin-like component identified immunocytochemically in *Acetabularia* cells. Furthermore, the antiserum against 135-ABP does not recognize 115-ABP (Nakayasu et al., 1998). Hence, the possibility that 115-ABP is a proteolytic product of 135-ABP can be ruled out.

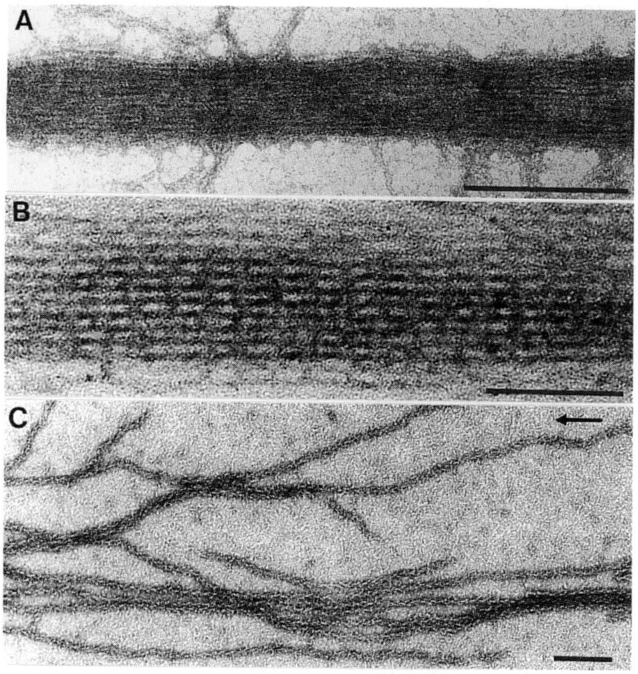

Figure 2. Electron micrographs of F-actin bundles formed in the presence of 135-ABP. **A**, Actin filament bundles following negative staining. Bar, 200 nm. **B**, A thin-section of the bundle. Bar, 100 nm. **C**, Actin filament bundle incubated with subfragment 1 of skeletal muscle myosin (S-1). Arrow in the upper right hand corner indicates the orientation of S-1 arrowheads. Bar, 200 nm.

Since the preparation of antibodies crossreacting specifically with 115-ABP has not been successful, there is no information about the localization and function of this protein in plant cells. In non-plant cells, two or three types of actin-bundling or cross-linking proteins are involved in the arrangement of an actin-filament bundle in a single cell (Furukawa & Fechheimer, 1997; Tilney et al., 1998). Therefore, it is probable that 115-ABP with 135-ABP function cooperatively in arranging actin filaments into bundles in pollen.

3. CALCIUM REGULATION OF ABP AND MYOSIN ACTIVITY

In growing pollen tubes, a tip-focused Ca^{2+} gradient is indispensable for tip growth (Rathore et al., 1991; Miller et al., 1992; Pierson et al., 1994, 1996) and tube orientation (Malhó et al., 1994, 1995; Malhó & Trewavas, 1996). In the case of lily pollen tubes, Ca^{2+} concentration is estimated to be more than 3 μM at the tip and gradually decreases to a basal level of 0.2 μM within 20 μm from the tip (Pierson et al., 1994, 1996). In the tip region, which is referred to as the clear zone, active cytoplasmic streaming is not observed (Pierson et al., 1990, 1994; Lancelle & Hepler, 1992). As described above, actin-filament bundles are fewer, finer, and in random orientation near the tip. This disorganization of actin-filament bundles is believed to be due to the tip-focused Ca^{2+} gradient, and the following evidence is provided for this notion. When growing pollen tubes are treated with caffeine, which diminishes the tip-focused Ca^{2+} gradient (Pierson et al., 1994), tip growth ceases and extensive actin-filament bundles extend into the tip (Miller et al., 1996; Lancelle et al., 1997). Concomitantly, active cytoplasmic streaming begins in this region. These observations suggest that high concentrations of Ca^{2+} in the tip region normally suppress actin-bundle formation and cytoplasmic streaming. As described above, the F-actin binding and bundling activities of 135-ABP are independent of Ca^{2+} alone. However, these activities are markedly suppressed by calmodulin when Ca^{2+} is >2 μM (Yokota et al., 1997; Yokota, Muto & Shimmen, submitted). The effective concentration of calmodulin that inactivates 135-ABP is >1 μM. Furthermore, a calmodulin antagonist, W-7 cancels the inhibitory effect of Ca^{2+}-calmodulin, indicating that the specific interaction of Ca^{2+}-calmodulin with 135-ABP results in the inhibition of 135-ABP binding to F-actin. A tip-focused distribution of calmodulin was reported in chemically-fixed pollen tubes by using an antibody against calmodulin (Tirlapur et al., 1994). However, it was recently demonstrated that the distribution of fluorescein-conjugated calmodulin microinjected into pollen tubes is uniform (Moutinho et al., 1998a). The concentration of calmodulin has been estimated to be 1.3 μM in barley aleurone cells (Schuurink et al., 1996), 4 μM in a carrot cell line (Fisher et al., 1996) and 11 μM in stamen hair cells of *Tradescantia* (Vos & Hepler, 1998). Therefore, even if calmodulin is uniformly dispersed throughout pollen tubes as reported by Moutinho et al. (1998a), it is reasonable to assume that calmodulin is present in the tip region of pollen tubes at micromolar levels, which is sufficient to induce the inhibition of 135-ABP binding to F-actin. Hence, the suppression of actin-filament binding and bundling activities of 135-ABP by Ca^{2+}-calmodulin is a potential mechanism by which cytoplasmic streaming is inhibited near the tip of pollen tubes.

Recently, the cloning of a cDNA from a lily pollen library demonstrated that 135-ABP is a plant homologue of villin (Vidali et al., 1999). Villin functions in the arrangement of actin filaments into bundles in intestinal brush border and in echinoderm and *Drosophila* oocytes (Friedrich et al., 1990; Furukawa & Fecheimer, 1997). This protein is constructed from two domains: a gelsolin/severin domain and a head piece domain in the C-terminal region (Friedrich et al., 1990). At concentrations of Ca^{2+} $<10^{-6}$ M, villin bundles F-actin filaments through the head piece domain, whereas at Ca^{2+} concentrations $>10^{-6}$ M F-actin severing and capping activities residing in the gelsolin/severin domain are stimulated. At present, the severing and capping activities have not been demonstrated for isolated 135-ABP (Yokota et al., 1998). It has been reported that actin filaments in lily pollen tubes are fragmentated when intracellular Ca^{2+} concentration is elevated above 10^{-6} M using a Ca^{2+}-ionophore (Kohno & Shimmen, 1987). This suggests the presence of a factor responsible for fragmentation of actin filaments in a Ca^{2+}-dependent manner. Assuming that 135-ABP possesses these activities, we propose that it will regulate actin assembly and disassembly and actin dynamics in a Ca^{2+}-dependent manner at the tip of pollen tubes.

Another way that Ca^{2+} may regulate pollen tube tip organization is revealed in studies in which the organelles isolated from lily pollen tubes are translocated along actin cables in characean cells. This translocation is inhibited by Ca^{2+} at concentrations higher than 10^{-6} M, indicating that the myosin responsible for cytoplasmic streaming is equipped with a Ca^{2+}-sensitive mechanism (Kohno & Shimmen, 1988). By comparison, the motile activity *in vitro* and the F-actin-activated ATPase activity of 170-kDa myosin are suppressed in the presence of $>10^{-6}$ M Ca^{2+} (Yokota et al., 1999a). The Ca^{2+}-induced inhibition of motile activity *in vitro* reaches a plateau at 2.5 μM, corresponding to Ca^{2+} concentration at the tip of elongating pollen tubes of lily (Pierson et al., 1994, 1996). In this Ca^{2+} concentration range, the inhibition of 170-kDa myosin is reversible (Yokota et al., 1999a). However, Ca^{2+} at concentrations higher than 10^{-5} M irreversibly inhibits the activity. The basis for this is likely explained by the observation that the calmodulin light chain is dissociated from 170-kDa heavy chain. In myosin I from the brush border of intestine (Collins et al., 1990; Wolenski et al., 1993) or liver (Williams & Coluccio, 1994), or in myosin V from brain (Cheney et al., 1993), Ca^{2+} modulates the motile activity *in vitro* by binding to the calmodulin light chain. Ca^{2+} inhibits the motile activity of these myosins by partial dissociation of calmodulin from the heavy chain. The impaired activity is restored upon addition of exogenous calmodulin. Also in the case of 170-kDa myosin, the inactivated motile activity is restored to some extent under the low concentrations of Ca^{2+} by supplying calmodulin that has been isolated from lily pollen tubes (Yokota et al., 1999a). These results suggest that Ca^{2+} regulates the activities of 170-kDa myosin by binding to calmodulin light chains and that this Ca^{2+} sensitivity is also one of the molecular bases for the regulation of cytoplasmic streaming near the tip of living pollen tubes.

Ca^{2+} is believed to play an essential and crucial role in the suppression of cytoplasmic streaming not only in pollen tubes, but also in somatic cells (Nagai, 1993; Shimmen & Yokota, 1994). A 175-kDa myosin isolated from BY-2 cells is composed of 175-kDa heavy chain and calmodulin light chains (Yokota et al., 1999b). Based on the sliding velocity of F-actin *in vitro*, this myosin is also suggested to be one of the motor proteins responsible for cytoplasmic streaming in BY-2 cells. This motile activity is inhibited by $>10^{-6}$ M Ca^{2+}. Immunocytochemical studies using antibodies against non-plant myosins, suggest that several types of myosins, including myosin I, II and V, are involved in cytoplasmic streaming and in the transport of vegetative nuclei and generative cells in pollen tubes (Miller et al., 1995). As described above, the activities of some types of myosin I and myosin V of animal cells are regulated by binding of Ca^{2+} to calmodulin light chains. Therefore, it is possible that calmodulin light chains generally provide the Ca^{2+} sensitivity to myosins involved in cytoplasmic streaming and in the transport of vegetative nuclei and generative cells.

Based on the pharmacological analysis of characean cells, another mechanism for Ca^{2+}-regulation of cytoplasmic streaming through the modulation of myosin activity can be suggested. The phosphorylation-dephosphorylation of myosin correlates with cessation of cytoplasmic streaming (Tominaga et al., 1987). In this model, it is hypothesized that myosin is inactivated by phosphorylation events that are catalyzed by Ca^{2+}-dependent protein kinase, and that motility is restored by dephosphorylation. Although there is no direct evidence showing that the myosin molecules responsible for cytoplasmic streaming in characean cells are phosphorylated or dephosphorylated, this model is supported by the following results. The activity of a Ca^{2+}-dependent protein kinase from soybean cells is enhanced by micromolar concentrations of Ca^{2+} (Harmon et al., 1987; Putnam-Evans et al., 1990), which is also sufficient to induce the cessation of cytoplasmic streaming in intact or permeabilized characean cells (Williamson & Ashley, 1982; Tominaga et al., 1983). This protein kinase co-localizes with organelles and along actin cables in characean cells (McCurdy & Harmon, 1992a), and it is reported that the putative light chains of myosin in characean cells are phosphorylated by this protein kinase (McCurdy & Harmon, 1992b). Surprisingly, isolated myosin from *Chara* cells does not show Ca^{2+}-sensitive motility or F-actin-activated ATPase activity (Yamamoto et al., 1994). At present, there are no studies demonstrating that Ca^{2+}-dependent phosphorylation-dephosphorylation of myosin correlates with the regulation of cytoplasmic streaming in higher plants. Indeed, the isolated 170-kDa and 175-kDa myosin show significant Ca^{2+}-sensitivity for their motile and F-actin-activated ATPase activity (Yokota et al., 1999a, 1999b), in contrast to *Chara* myosin. However, it has been demonstrated that a Ca^{2+}-dependent protein kinase resides along actin filaments in *Allium* root cells (Putnam et al., 1989). Furthermore, a similar kinase is present in an active state near the tip of elongating pollen tubes (Moutinho et al., 1998b). Therefore, the possibility that Ca^{2+}-dependent phosphorylation-dephosphorylation of myosin as a basis for the regulation of cytoplasmic streaming in higher plant cells can not be excluded. This could act in concert with Ca^{2+}-binding to calmodulin light chains.

REFERENCES

Boustead CM, Smallwood M, Small H, Bowles DJ and Walker JH (1989) Identification of calcium-dependent phospholipid-binding proteins in higher plant cells. FEBS Lett 244: 456-460

Calvert CM, Gant SJ and Bowles DJ (1996) Tomato annexins p34 and p35 bind to F-actin and display nucleotide phosphodiesterase activity inhibited by phospholipid binding. Plant Cell 8: 333-342

Cheney RE, O'Shea MK, Heuser JE, Coelho MV, Wolenski JS, Espreafico EM, Forscher P, Larson RE and Mooseker MS (1993) Brain myosin-V is a two-headed unconventional myosin with motor activity. Cell 75: 13-23

Clore AM, Dannenhoffer JM and Larkins BA (1996) EF-1α associated with a cytoskeletal network surrounding protein bodies in maize endosperm cells. Plant Cell 8: 2003-2014

Collings DA, Wasteneys GO, Miyazaki M and Williamson RE (1994) Elongation factor 1α is a component of the subcortical actin bundles of characean algae. Cell Biol Int 18: 1019-1024

Collins K, Sellers JR and Matsudaira P (1990) Calmodulin dissociation regulates brush border myosin I (110-kD-calmodulin) mechanochemical activity in vitro. J Cell Biol 110:1137-1147

Cruz-Ortega R, Cushman JC and Ownby JD (1997) cDNA clones encoding 1,3-ß-glucanase and a fimbrin-like cytoskeletal protein are induced by Al toxicity in wheat roots. Plant Physiol 114: 1453-1460

Faraday CD and Spanswick RM (1993) Evidence for a membrane skeleton in higher plants. A spectrin-like polypeptide co-isolates with rice root plasma membranes. FEBS Lett 318: 313-316

Fisher DD, Gilroy S and Cyr RJ (1996) Evidence for opposing effects of calmodulin on cortical microtubules. Plant Physiol 112: 1079-1087

Friederich E, Pringault E, Arpin M and Louvard D (1990) From the structure to the function of villin, an actin-binding protein of the brush border. BioEssays 12: 403-408

Furukawa R and Fechheimer M (1997) The structure, function, and assembly of actin filament bundles. Int Rev Cytol 175: 29-90

Harmon AC, Putnam-Evans C and Cormier MJ (1987) A calcium-dependent but calmodulin-independent protein kinase from soybean. Plant Physiol 83: 830-837

Heslop-Harrison J, Heslop-Harrison Y, Cresti M, Tiezzi A and Ciampolini F (1986) Actin during pollen germination. J Cell Sci 86: 1-8

Higashi-Fujime S, Ishikawa R, Iwasawa H, Kagami O, Kurimoto E, Kohama K and Hozumi T (1995) The fastest actin-based motor protein from the green algae, *Chara*, and its distinct mode of interaction with actin. FEBS Lett 375: 151-154

Igarashi H, Vidali L, Yokota E, Sonobe S, Hepler PK and Shimmen T (1999) Actin filaments purified from tobacco cultured BY-2 cells can be translocated by plant myosin. Plant Cell Physiol 40: 1167-1171

Kaminskyj SGW and Heath IB (1995) Integrin and spectrin homologues, and cytoplasm-wall adhesion in tip growth. J Cell Sci 108: 849-856

Kinkema M and Schiefelbein J (1994) A myosin from a higher plant has structural similarities to class V myosins. J Mol Biol 239: 591-597

Kinkema M, Wang H and Schiefelbein J (1994) Molecular analysis of the myosin gene family in *Arabidopsis thaliana*. Plant Mol Biol 26: 1139-1153

Knight AE and Kendrick-Jones J (1993) A myosin-like protein from a higher plant. J Mol Biol 231: 148-154

Kohno T, Chaen S and Shimmen T (1990) Characterization of the translocator associated with pollen tube organelles. Protoplasma 154: 179-183

Kohno T, Ishikawa R, Nagata T, Kohama K and Shimmen T (1992) Partial purification of myosin from lily pollen tubes by monitoring with in vitro motility assay. Protoplasma 170: 77-85

Kohno T and Shimmen T (1987) Ca^{2+}-induced fragmentation of actin filaments in pollen tubes. Protoplasma 141: 177-179

Kohno T and Shimmen T (1988) Accelerated sliding of pollen tube organelles along *Characeae* actin bundles regulated by Ca^{2+}. J Cell Biol 106: 1539-1543

Kron SJ and Spudich JA (1986) Fluorescent actin filaments move on myosin fixed to a glass surface. Proc Natl Acad Sci USA 83: 6272-6276.

Lancelle SA, Cresti M and Hepler PK (1987) Ultrastructure of the cytoskeleton in freeze-substituted pollen tubes of *Nicotiana alata*. Protoplasma 140: 141-150

Lancelle SA, Cresti M and Hepler PK (1997) Growth inhibition and recovery in freeze-substituted *Lilium longiflorum* pollen tubes: structural effects of caffeine. Protoplasma 196: 21-33

Lancelle SA and Hepler PK (1992) Ultrastructure of freeze-substituted pollen tubes of *Lilium longiflorum*. Protoplasma 167: 215-230

Li Y-Q, Moscatelli A, Cai G and Cresti M (1997) Functional interactions among cytoskeleton, membranes, and cell wall in the pollen tube of flowering plants. Int Rev Cytol 176: 133-199

Malhó R, Read ND, Pais MS and Trewavas AJ (1994) Role of cytosolic free calcium in the reorientation of pollen tube growth. Plant J 5: 331-341

Malhó R, Read ND, Trewavas AJ and Pais MS (1995) Calcium channel activity during pollen tube growth and reorientation. Plant Cell 7: 1173-1184

Malhó R and Trewavas AJ (1996) Localized apical increases of cytosolic free calcium control pollen tube orientation. Plant Cell 8: 1935-1949

Matsudaira P (1991) Modular organization of actin crosslinking proteins. Trends Biochem Sci 16: 87-92

McCurdy DW and Harmon AC (1992a) Calcium-dependent protein kinase in the green alga *Chara*. Planta 188: 54-61

McCurdy DW and Harmon AC (1992b) Phosphorylation of a putative myosin light chain in *Chara* by calcium-dependent protein kinase. Protoplasma 171: 85-88

McCurdy DW and Kim M (1998) Molecular cloning of a novel fimbrin-like cDNA from *Arabidopsis thaliana*. Plant Mol Biol 36: 23-31

Mermall V, Post PL and Mooseker MS (1998) Unconventional myosins in cell movement, membrane traffic, and signal transduction. Science 279: 527-533

Michaud D, Guillet G, Rogers PA and Charest PM (1991) Identification of a 220kDa membrane-associated plant cell protein immunologically related to human ß-spectrin. FEBS Lett 294: 77-80

Miller DD, Callaham DA, Gross DJ and Hepler PK (1992) Free Ca^{2+} gradient in growing pollen tubes of *Lilium*. J Cell Sci 101: 7-12

Miller DD, Lancelle SA and Hepler PK (1996) Actin microfilaments do not form a dense meshwork in *Lilium longiflorum* pollen tube tips. Protoplasma 195: 123-132

Miller DD, Scordilis SP and Hepler PK (1995) Identification and localization of three classes of myosins in pollen tubes of *Lilium longiflorum* and *Nicotiana alata*. J Cell Sci 108: 2549-2563

Mooseker MS and Cheney RE (1995) Unconventional myosins. Annu Rev Cell Dev Biol 11: 633-675

Moutinho A, Love J and Trewavas AJ (1998a) Distribution of calmodulin protein and mRNA in growing pollen tubes Sex Plant Reprod 11: 131-139

Moutinho A, Trewavas AJ and Malhó R (1998b) Relocation of a Ca^{2+}-dependent protein kinase activity during pollen tube reorientation. Plant Cell 10: 1499-1509

Muto S and Miyachi S (1984) Production of antibody against spinach calmodulin and its application to radioimmunoassay for plant calmodulin. Z Pflanzenphysiol 114: 421-431

Nagai R (1993) Regulation of intracellular movements in plant cells by environmental stimuli. Int Rev Cytol 145: 251-310

Nakayasu T, Yokota E and Shimmen T (1998) Purification of an actin-binding protein composed of 115-kDa polypeptide from pollen tubes of lily. Biochem Biophys Res Comm 249: 61-65

Otto JJ (1994) Actin-bundling proteins. Curr Opin Cell Biol 6: 105-109

Perdue TD, Mandayam V and Parthasarathy V (1985) In situ localization of F-actin in pollen tubes. Eur J Cell Biol 39: 13-20

Pierson ES and Cresti M (1992) Cytoskeleton and cytoplasmic organization of pollen and pollen tubes. Int Rev Cytol 140: 73-125

Pierson ES, Derksen J and Traas JA (1986) Organization of microfilaments and microtubules in pollen tubes grown in vitro or in vivo in various angiosperms. Eur J Cell Biol 41: 14-18

Pierson ES, Lichtscheidl IK and Derksen J (1990) Structure and behaviour of organelles in living pollen tubes of *Lilium longiflorum*. J Exp Bot 41: 1461-1468

Pierson ES, Miller DD, Callaham DA, Shipley AM, Rivers BA, Cresti M and Hepler PK (1994) Pollen tube growth is coupled to the extracellular calcium ion flux and the intracellular calcium gradient: effect of BAPTA-type buffers and hypertonic media. Plant Cell 6: 1815-1828

Pierson ES, Miller DD, Callaham DA, van Aken J, Hackett G and Hepler PK (1996) Tip-localized calcium entry fluctuates during pollen tube growth. Dev Biol 174: 160-173

Pollard TD and Cooper JA (1986) Actin and actin-binding proteins. A critical evaluation of mechanisms and functions. Annu Rev Biochem 55: 987-1035

Putnam-Evans CL, Harmon AC and Cormier MJ (1990) Purification and characterization of a novel calcium-dependent protein kinase from soybean. Biochemistry 29: 2488-2495

Putnam-Evans C, Harmon AC, Palevitz BA, Fechheimer M and Cormier MJ (1989) Calcium-dependent protein kinase is localized with F-actin in plant cells. Cell Motil Cytoskel 12: 12-22

Rathore KS, Cork RJ and Robinson KR (1991) A cytoplasmic gradient of Ca^{2+} is correlated with the growth of lily pollen tubes. Dev Biol 148: 612-619

Schuurink RC, Chan PV and Jones RL (1996) Modulation of calmodulin mRNA and protein levels in barley aleurone. Plant Physiol 111: 371-380

Shimmen T, Hamatani M, Saito S, Yokota E, Mimura T, Fusetani N and Karaki H (1995) Roles of actin filaments in cytoplasmic streaming and organization of transvacuolar strands in root hair cells of *Hydrocharis*. Protoplasma 185: 188-193

Shimmen T and Tazawa M (1982) Reconstitution of cytoplasmic streaming in *Characeae*. Protoplasma 113: 127-131

Shimmen T and Yokota E (1994) Physiological and biochemical aspects of cytoplasmic streaming. Int Rev Cytol 155: 97-139

Smallwood MF, Gurr SJ, McPherson MJ, Roberts K and Bowles DJ (1990) Purification and partial sequence analysis of plant annexins. Biochem J 281: 501-505

Stossel TP, Chaponnier C, Ezzell RM, Hartwig JH, Janmey PA, Kwiatkowski DJ, Lind SE, Smith DB, Southwick FS, Yin HL and Zaner KS (1985) Nonmuscle actin-binding proteins. Annu Rev Cell Biol 1: 353-402

Takagi S, Yokota E, Shimmen T and Nagai R (1995) Motor protein activity for cytoplasmic streaming detected in *Vallisneria* leaves. Plant Cell Physiol 36: s132

Taylor LP and Hepler PK (1997) Pollen germination and tube growth. Annu Rev Plant Physiol Plant Mol Biol 48: 461-491

Tilney LG, Connelly PS, Vranich KA, Shaw MK and Guild GM (1998) Why are two different cross-linkers necessary for actin bundle formation in vivo and what does each cross-link contribute? J Cell Biol 143: 121-133

Tirlapur UK, Scali M, Moscatelli A, Casino CD, Cai G, Tiezzi A and Cresti M (1994) Confocal image analysis of spatial variations in immunocytochemically identified calmodulin during pollen hydration, germination and pollen tube tip growth in *Nicotiana tabacum* L. Zygote 2: 63-68

Tischendorf G, Sawitzky D and Werz G (1987) Antibodies specific for vertebrate actin, myosin, actinin, or vinculin recognize epitopes in the giant nucleus of the the marine green alga *Acetabularia*. Cell Motil Cytoskel 7: 78-86

Tiwari SC and Polito VS (1988a) Spatial and temporal organization of actin during hydration, activation, and germination of pollen in *Pyrus communis* L.: a population study. Protoplasma 147: 5-15

Tiwari SC and Polito VS (1988b) Organization of the cytoskeleton in pollen tubes of *Pyrus communis*: a study employing conventional and freeze-substitution electron microscopy, immunofluorescence, and rhodamine-phalloidin. Protoplasma 147: 100-112

Tominaga M, Morita K, Sonobe S, Yokota E and Shimmen T (1997) Microtubules regulate the organization of actin filaments at the cortical region in root hair cells of *Hydrocharis*. Protoplasma 199: 83-92

Tominaga M, Yokota E, Vidali L, Sonobe S, Hepler PK and Shimmen T (1999) Dynamics of cytoskeletons in root hair cell of *Hydrocharis*: mechanism of actin filament bundling. Plant Cell Physiol 40: s49

Tominaga M, Yokota E, Vidali L, Sonobe S, Hepler PK and Shimmen T (2000) The role of plant villin in the organization of the actin cytoskeleton, cytoplasmic streaming and the architecture of the transvacuolar strand in root hair cells of *Hydrocharis*. Planta, In press

Tominaga Y, Shimmen T and Tazawa M (1983) Control of cytoplasmic streaming by extracellular Ca^{2+} in permeabilized *Nitella* cells. Protoplasma 116: 75-77

Tominaga Y, Wayne R, Tung HYL and Tazawa M (1987) Phosphorylation-dephosphorylation is involved in Ca^{2+}-controlled cytoplasmic streaming of characean cells. Protoplasma 136: 161-169

Vidali L, Yokota E, Cheung AY, Shimmen T and Hepler PK (1999) The 135 kDa actin-bundling protein from *Lilium longiflorum* pollen is the plant homologue of villin. Protoplasma 209: 283-291

Vos JW and Hepler PK (1998) Calmodulin is uniformly distributed during cell division in living stamen hair cells of *Tradescantia virginiana*. Protoplasma 201: 158-171

Williams R and Coluccio LM (1994) Novel 130-kD rat liver myosin-I will translocate actin filaments. Cell Motil Cytoskel 27: 41-48

Williamson RE (1993) Organelle movements. Annu Rev Plant Physiol Plant Mol Biol 44: 181-202

Williamson RE and Ashley CC (1982) Free Ca^{2+} and cytoplasmic streaming in the alga *Chara*. Nature 296: 647-651

Wolenski JS, Hayden SM, Forscher P and Mooseker MS (1993) Calcium-calmodulin and regulation of brush border myosin-I MgATPase and mechanochemistry. J Cell Biol 122: 613-621

Wolenski JS (1995) Regulation of calmodulin-binding myosins. Trends Cell Biol 5: 310-316

Yamamoto K, Kikuyama M, Sutoh-Yamamoto N and Kamitsubo E (1994) Purification of actin based motor protein from *Chara corallina*. Proc Japan Acad 70: 175-180

Yamamoto K, Kikuyama M, Sutoh-Yamamoto N, Kamitsubo E and Katayama E (1995) Myosin from alga *Chara*: unique structure revealed by electron microscopy. J Mol Biol 254: 109-112

Yang W, Burkhart W, Cavallius J, Merrick WC and Boss WF (1993) Purification and characterization of a phosphatidylinositol 4-kinase activator in carrot cells. J Biol Chem 268: 392-398

Yokota E, Mimura T and Shimmen T (1995a) Biochemical, immunochemical and immunohistochemical identification of myosin heavy chains in cultured cells of *Catharanthus roseus*. Plant Cell Physiol 36: 1541-1547

Yokota E, Muto S and Shimmen T (1997) Actin bundling activity of P-135-ABP isolated from lily pollen tubes is regulated by calcium-calmodulin. Cell Struct Funct 22: 685

Yokota E, Muto S and Shimmen T (1999a) Inhibitory regulation of higher-plant myosin by Ca^{2+} ions. Plant Physiol 119: 231-239

Yokota E, McDonald AR, Liu B, Shimmen T and Palevitz BA (1995b) Localization of a 170 kDa myosin heavy chain in plant cells. Protoplasma 185: 178-187

Yokota E and Shimmen T (1994) Isolation and characterization of plant myosin from pollen tubes of lily. Protoplasma 177: 153-162

Yokota E and Shimmen T (1999) The 135-kDa actin-bundling protein from lily pollen tubes arranges F-actin into bundles with uniform polarity. Planta 209: 264-266

Yokota E, Takahara K and Shimmen T (1998) Actin-bundling protein isolated from pollen tubes of lily. Biochemical and immunocytochemical characterization. Plant Physiol 116: 1421-1429

Yokota E, Yukawa C, Muto S, Sonobe S and Shimmen T (1999b) Biochemical and immunocytochemical characterization of two types of myosins in cultured tobacco bright yellow-2 cells. Plant Physiol 121: 525-534

Chapter 7

SUCROSE METABOLISM AND THE ACTIN CYTOSKELETON: SuSy AS ACTIN-BINDING PROTEIN

Heike Winter[1] and Steven C Huber[2]

1 Department of Plant Physiology, University of Osnabrück, Barbarastr. 11, D–49069 Osnabrück, Germany
2 USDA/ARS and Dept. of Crop Science, North Carolina State University, Raleigh, NC 27606, USA

Key words: actin binding, cellulose synthesis, glycolysis, protein phosphorylation, sucrose synthase (SuSy), sugar sensing

Abstract: The recent finding that sucrose synthase (SuSy) is an actin-binding protein in plant cells provides a first link between carbohydrate metabolism and the actin cytoskeleton. SuSy is expressed in heterotrophic tissue, where it is suggested to play a role in sink strength, cell wall synthesis and partitioning of sucrose into starch synthesis and glycolysis. The enzyme is found in the soluble fraction as well as in association with the plasmamembrane or the actin cytoskeleton. Reversible phosphorylation of a single serine residue on SuSy is part of the mechanism determining its partitioning between the soluble and membrane fraction. Binding to actin of the soluble and presumably phosphorylated form is metabolically regulated. Actin association of SuSy is dependent on the binding of sucrose to a regulatory sucrose-binding site that is distinct from the catalytic side. The function of SuSy interaction with filamentous actin is still unknown but several possibilities are presented.

1. SUSY IN CARBOHYDRATE METABOLISM

In most higher plants sucrose is the major transport form of assimilated carbon from the photosynthetically active "source" tissue to the heterotropic "sink" organs. In cells of sink organs, imported sucrose can be stored in the vacuole or metabolized by invertases or SuSy in the cytosol. An irreversible hydrolytic cleavage of sucrose into glucose and fructose is catalyzed by

C.J. Staiger et al. (eds.), Actin: A Dynamic Framework for Multiple Plant Cell Functions, 119–128.

invertases. SuSy on the other hand catalyzes the reversible reaction of UDP and sucrose into UDP-glucose and fructose, thereby conserving the energy of the glycosidic bond:

$$\text{Sucrose} + \text{UDP} \leftrightarrow \text{UDP-glucose} + \text{fructose}$$

SuSy has been shown to play a major role in sink strength and carbon partitioning between the synthesis of polyglucans (starch, cellulose or callose) and the glycolytic pathway (Zrenner et al., 1995; Déjardin et al., 1997; Chourey et al., 1998). Even though the underlying mechanisms by which these different functions of SuSy are regulated are still unknown, several aspects of its regulation on the transcriptional and posttranslational levels have been identified.

In most higher plants SuSy is encoded by two or three nonallelic genes, which are expressed differentially in response to sucrose and anaerobiosis and in a tissue specific manner.

2. POSTTRANSLATIONAL MODIFICATION OF SUSY

SuSy is subject to posttranslational modification by reversible protein phosphorylation (Huber et al., 1996; Zhang & Chollet, 1997) on a single serine residue (ser-15 in maize SS2). This phosphorylation site is highly conserved among most SuSy proteins of different plant species. In extracts from elongating maize leaf tissue and soybean nodules, the protein kinase activity phosphorylating SuSy was strictly Ca^{2+}-dependent with molecular weights of ~55 kD and 65 kD, respectively. It has also been reported that a phospholipid-stimulated, Ca^{2+}-dependent protein kinase co-purifies with SuSy from maize seedlings and is capable of phosphorylating SuSy (Lindblom et al., 1997). In addition to a soluble, Ca^{2+}-dependent protein kinase, a membrane-associated, Ca^{2+}-independent protein kinase activity is capable of phosphorylating SuSy (Datcheva et al., 1998). A specific protein phosphatase activity has not yet been identified.

The physiological role of SuSy phosphorylation is still not clear. The phosphorylated form has a higher affinity for sucrose in the cleavage direction with no apparent change in the synthetic reaction (Huber et al., 1995; Winter et al., 1997; Nakai et al., 1998). A more dramatic effect is the increase in surface hydrophobicity of purified, soluble and presumbly highly phoshorylated SuSy from maize elongating leaf tissue, that is caused by incubation with alkaline phosphatase *in vitro*. The enzyme is no longer soluble but precipitates out of solution. This indicates a possible involvement of the phosphorylation status of SuSy with its localization on the membrane.

3. MEMBRANE ASSOCIATION OF SUSY

SuSy is the most abundant UDP-glucose binding protein on the plasmamembrane of cotton fiber cells (Amor et al., 1995). The synthesis of cellulose and callose from [^{14}C]-sucrose in digitonin-permeabilized cotton fiber cells suggested a potential role of SuSy in channeling UDP-glucose into cellulose and callose synthesis. This is further supported by the finding that immobilized tobacco plasma membrane sheets are capable of synthezising β-glucan microfibrils from sucrose and UDP (Hirai et al., 1998). An association of SuSy with the plasmamembrane also occurs in maize endosperm and stem pulvini (Carlson & Chourey, 1996; Winter et al., 1997). The amount of membrane-associated SuSy is variable *in vivo* and correlates with the rate of cell elongation in maize pulvini and changes in the developmental state in cotton seeds (Winter et al., 1997; Ruan et al., 1997). Recently, SuSy has also been found in the Golgi-enriched fraction of maize coleoptiles (Buckeridge et al., 1999) and at the symbiosome membrane surrounding bacteroids in soybean nodules (Zhang et al., 1999).

In vivo ^{32}P-labeling of SuSy in young maize leaves shows a lower degree of ^{32}P-incorporation in membrane-associated SuSy (enzyme protein basis) when compared with the soluble form. *In vitro* phosphorylation of the membrane fraction released phosphorylated SuSy protein into the soluble phase, while incubation of a crude extract with alkaline phosphatase increased the amount of SuSy partitioned into the membrane fraction (Winter et al., 1997). Together with the decrease in hydrophobicity following incubation with alkaline phosphatase, these results indicate that reversible protein phosphorylation of SuSy may be at least part of the mechanism determining its localization on the membrane or in cytosol. Whether the membrane-association of SuSy is in fact an association of the SuSy protein with the lipid bilayer itself or caused by interaction of SuSy with a membrane protein is still unknown.

4. ASSOCIATION OF SUSY WITH ACTIN FILAMENTS

SuSy was not only found in the soluble and membrane fractions, but also in the Triton X-100 insoluble "particulate" fraction from elongating maize leaf extract. A major component of the particulate fraction is the cytoskeleton. Immunoprecipitation of SuSy with a monoclonal antibody shows that actin, but not tubulin, co-immunoprecipitates with SuSy from the soluble fraction. Since the soluble fraction contains primarily monomeric (G-)actin, whereas the particulate fraction contains filamentous (F-)actin, this suggests that SuSy might be capable of binding to either form. No co-immunoprecipitation of actin with SuSy from the membrane fraction is observed. When the tissue from the leaf elongation zone of young maize plants is infiltrated with phalloidin prior to extraction, to further reduce depolymerization of actin filaments, ~15% of the total SuSy co-sedimented with the actin filaments, whereas the other part stays in the soluble and microsomal fraction. These studies indicate that only the soluble, and presumably more highly phosphorylated, SuSy protein is capable of binding to actin (Winter et al., 1998).

In vitro studies with purified rabbit muscle actin have been performed to further characterize the association of SuSy with actin filaments. The binding of purified maize SuSy to actin is direct and saturates at a molar ratio of SuSy to actin (in filamentous form) of 1:5. Thus, one SuSy tetramer can bind to actin filament(s) consisting of ~20 subunits (Winter et al., 1998). This binding ratio could reflect steric hinderance due to the size of the SuSy protein (360 kD for the tetramer) or might indicate a possible function in crosslinking actin filaments. Most actin crosslinking proteins have a stoichiometry of binding to actin of 1:4–1:6 (Matsudaira, 1991).

5. REGULATION OF SUSY-ACTIN ASSOCIATION

The observation that actin only co-immunoprecipitates with SuSy from the soluble, but not from the membrane fraction could indicate that the phosphorylation state of SuSy is important for the association with actin. Reversible protein phosphorylation of serine residues has been shown to be involved in regulating the association of proteins with the actin cytoskeleton. Phosphorylated animal muscle phosphofructokinase has a higher affinity for actin than the dephosphorylated form, whereas in the cases of destrin, cofilin and plant actin depolymerizing factor (ZmADF) phosphorylation of an N-terminal serine inhibits the association with actin (Luther & Lee, 1986; Smertenko et al., 1998; Agnew et al., 1995; Moriyama et al., 1996). Direct evidence that phosphorylation of SuSy protein is part of the mechanism that regulates its binding to actin is still missing.

Recently it was found that the association of purified maize SuSy from the soluble fraction with rabbit muscle F-actin in vitro is subject to metabolic regulation. The association is strictly dependent on the presence of sucrose.

Raffinose and trehalose can induce binding as well, whereas other carbohydrates like hexoses or sorbitol did not show any effect and neither did the other substrates, UDP or UDP-glucose. Since raffinose and trehalose are not present in maize tissues at physiologically relevant concentrations, the effector *in vivo* is most likely sucrose itself (Winter et al., in preparation)

Interestingly raffinose is not a substrate for SuSy cleavage activity and does not compete with sucrose for binding to the catalytic sucrose binding site, even though it can bind to SuSy and induce its association with actin. This was the first indication that there must be a regulatory sucrose binding site that is distinct from the catalytic one and might be involved in the metabolic regulation of SuSy binding to actin filaments. This hypothesis was further supported by the finding that the affinities of purified maize SuSy for sucrose cleavage activity (K_m= 12 mM) and the actin-binding activity ($S_{0.5}$ ~ 114 mM) differed by an order of magnitude. Even though there are no data available on the *in vivo* concentrations of sucrose in the cytosol of maize sink tissue, by analogy with other plant tissues and species, concentrations of 80 mM sucrose in the cytosol might well be in the physiological range (Gerhardt et al., 1987; Winter et al., 1994).

In addition to differences in the affinity and specificity for sucrose, the catalytic activity and actin-binding activity display different saturation kinetics for sucrose. The sucrose concentration curve for the cleavage activity is hyperbolic. Binding of sucrose to the catalytic sucrose binding site(s) is therefore a non-cooperative process between the four subunits of the SuSy protein. In contrast, the sucrose dependence of the actin binding activity of SuSy shows sigmoidal saturation kinetics characteristic of cooperativity between the subunits in the SuSy protein in sucrose binding to a regulatory site. As expected, the Hill coefficient for sucrose binding to the catalytic site is 1, whereas sucrose-induced actin-binding activity has a Hill-coefficient of 2.4 for allosteric behavior suggesting at least 3 regulatory sucrose binding sites (Winter et al., in preparation).

These findings suggest that SuSy has two distinct sucrose binding sites; a high affinity, high specificity catalytic site and a low affinity, low specificity regulatory site. It could be hypothezised that binding of sucrose to the regulatory sites induces a conformational change, exposing the actin-binding site(s) on SuSy. SuSy is therefore not only a sucrose metabolizing enzyme, but can also 'sense' changes in the cytosolic concentration of sucrose that might effect the amout of SuSy protein bound to the actin cytoskeleton. The actual function of SuSy-actin association is still unknown. The actin-binding site on SuSy has not been identified yet. Alignments of actin-binding sites identified from other F-actin binding proteins (Tellam et al., 1989; SwissProt) did not show any significant similarities.

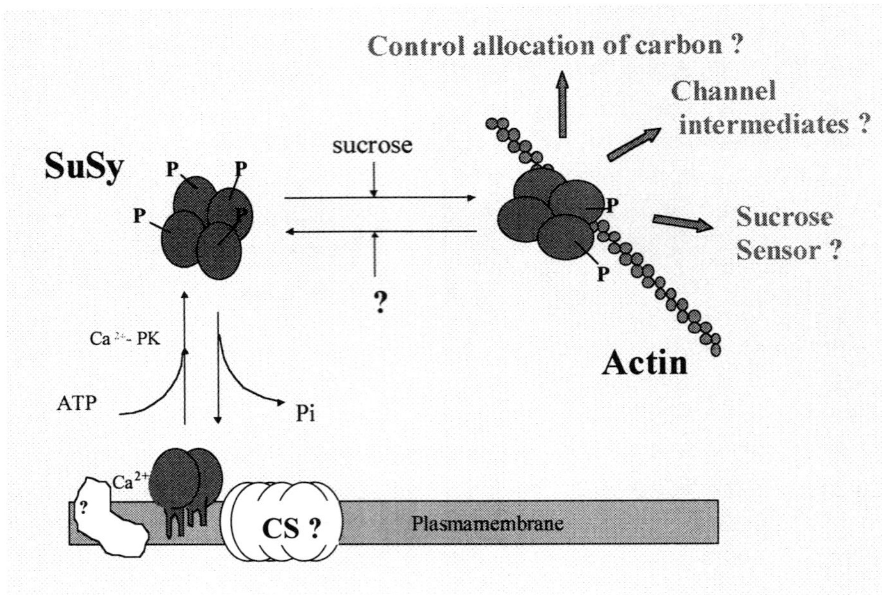

Figure 1. Hypothetical model of subcellular localization of SuSy and its regulation. CS, synthase; PK, protein kinase.

6. GLYCOLYSIS AND THE CYTOSKELETON

SuSy is the first metabolic enzyme in plants for which an association with the actin cytoskeleton has been shown. However, the finding that some glycolytic enzymes like aldolase, phosphofructokinase and phosphofructo-phoshatase show an increased partitioning into the "particulate" fraction, correlating with a stimulation of respiration in aging of carrot and sugar beet storage root slices, may indicate their association with cytoskeletal elements (Moorhead & Plaxton, 1988).

In animal cells, especially from muscle tissue, the glycolytic enzymes phosphofructokinase (PFK), aldolase (ALD), glyceraldehyde-phosphate dehydrogenase (GAPDH), phosphoglycerate kinase (PGK), pyruvate kinase (PK), and lactate dehydrogenase (LDH) have been shown to associate with the actin cytoskeleton and in some cases with microtubules as well (reviewed by Knull & Walsh, 1992). These associations are induced by anaerobiosis or insulin. The regulatory mechanisms include reversible protein phosphorylation, i.e. PFK and GAPDH, and metabolic regulation by intermediates of the glycolytic pathway (Luther & Lee, 1986; Wu et al.,

1997; Masters, 1996). Some of the enzymes are associated with the actin cytoskeleton by binding to another glycolytic enzyme.

Different physiological functions of these associations have been porposed and are still a matter of discussion. It has been shown that binding to F-actin affects the activity of some glycolytic enzymes. GAPDH, ALD, and LDH are all inhibited upon binding to either actin or tubulin, whereas F-actin activates PFK two-fold. On the other hand, it has been demonstrated that channelling of intermediates can occur throughout the entire glycolytic pathway (Clegg & Jackson, 1990). It is suggested that the cytoskeleton does not serve as a structural support for active glycolytic channeling, but rather regulates glycolytic activity with the release of enzyme-enzyme complexes from the cytomatrix.

Interestingly, ALD and GAPDH have also been shown to be structural components of the cytoskeleton in animal cells. Aldolase, for example, mediates the association of F-actin with the glucose transporter (GLUT4) and its translocation from inner membranes to the plasmamembrane by insulin-stimulated exocytosis (Kao et al., 1999).

The actin-binding site on animal aldolase has been identified and shows a high degree of sequence identity with other actin-binding proteins like gelsolin, severin, and most interestingly, with β-actin itself (O'Reilly & Clarke, 1993).

	aa res.	sequence
β-actin	363	**D E S G P S I V H R**
aldolase Dm	33	**D E S G P T M** G **K R**
aldolase Arath	29	**D E S T** E **T I** G **K R**

Table 1. Sequence alignment of the actin binding site from animal aldolase with β-actin and the cytosolic isoform of aldolase from *Arabidopsis*. Letters in bold show identical and homologous amino acids compared to the actin sequence. (Dm, *Drosophila melanogaster*; Arath, *Arabidopsis thaliana*)

As shown in Table 1, the actin-binding site of animal aldolase is a sequence that is contained in the cytosolic form of plant aldolase (*Arabidopsis thaliana*), suggesting a potential for an association with the actin cytoskeleton in plant cells.

7. CRITICAL ASPECTS AND FUTURE RESEARCH

The finding that SuSy is capable of binding to actin *in vitro* and *in situ*, opens up a new perspective about a possible interaction between structural elements and carbohydrate metabolism in cells. If the association between SuSy and actin and its regulation by the cytosolic sucrose concentration occurs *in vivo*, it suggests an involvement of the actin cytoskeleton in sucrose sensing. It is still unknown whether SuSy is enzymatically active when associated with the cytoskeleton, or whether the binding causes inhibition of its enzymatic activity as has been shown for animal phosphofructokinase. It is also possible that the membrane-associated form of SuSy can interact with actin filaments as well, and further testing of the ability of SuSy to bind to monomeric G-actin is needed.

As for aldolase in animal cells, two possible functions could be hypothesized about the physiological role of SuSy-actin interactions. First, the interaction could be necessary for channelling of substrates between glycolysis or polyglucan synthesis. Second, the binding could play a structural role in the organization of the actin cytoskeleton, e.g., by crosslinking actin filaments as shown for proteins like elongation factor-1α (see Stankovic et al., this volume) and fimbrin (see McCurdy & Staiger, this volume).

ACKNOWLEDGEMENTS

A co-operative investigation of the USDA/ARS and the North Carolina ARS, Raleigh, NC 27695-7643. The research was supported in part by grants from the US Department of Energy (Grant DE-AI05-91 ER 20031 to SCH), NASA (Grant NAGW 4984) and the Deutsche Forschungsgemeinschaft (Wi 1876/1-1 to HW). Mention of a trademark or proprietary product does not constitute a guarantee or warranty of the product by the NC ARS or the USDA and does not imply its approval to the exclusion of other products that may also be suitable.

REFERENCES

Agnew BJ, Minamide LS and Bamburg JR (1995) Reactivation of phosphorylated actin depolymerizing factor and identifiation of the regulatory site. J Biol Chem 270: 17582-17587

Amor Y, Haigler CH, Johnson S, Wainscott M and Delmer DP (1995) A membrane-associated form of sucrose synthase and its potential role in synthesis of cellulose and callose in plants. Proc Natl Acad Sci USA 92: 9353-9357

Buckeridge MS, Vergara CE and Carpita NC (1999) The mechanism of synthesis of a mixed-linkage(1→3),(1→4) β-D-glucan in maize. Plant Physiol 120: 1105-1116

Carlson SJ and Chourey PS (1996) Evidence for plasma membrane-associated forms of sucrose synthase in maize. Mol Gen Genet 252: 303-310

Clegg JS and Jackson SA (1990) Metabolism and the channeling of glycolytic intermediates in permeabilized L-929 cells. Arch Biochem Biophys 278: 452-460

Datcheva M, Buster D, Vulliet R and Delmer D (1998) Membrane-associated sucrose synthase: mechanism of membrane association and role in glucan synthesis. Plant Polysaccharide Symposium, Davis, CA

Chourey PS, Taliercio EW, Carlson SJ and Ruan YL (1998) Genetic evidence that the two isoenzymes of sucrose synthase present in developing maize endosperm are critical, one for cell wall integrity and the other for starch biosynthesis. Mol Gen Genet 259: 88-96

Déjardin A, Rochat C, Wuillèm S and Boutin JP (1997) Contribution of sucrose synthase, ADP-glucose pyrophosphorylase and starch synthase to starch synthesis in developing pea seeds. Plant Cell Environ 20: 1421-1430

Gerhardt R, Stitt M and Heldt HW (1987) Subcellular metabolite levels in spinach leaves. Plant Physiol 83: 399-403

Hirai N, Sonobe S and Hayashi T (1998) In situ synthesis of β-glucan microfibils on tobacco plasma membrane sheets. Proc Natl Acad Sci USA 95: 15102-15106

Huber SC, Huber JL, Liao PC, Gage DA, McMicheal RW Jr, Chourey PS, Hannah LC and Koch KE (1996). Phosphorylation of serine-15 of maize leaf sucrose synthase. Plant Physiol 112: 793-802

Kao AW, Noda Y, Johnson JH, Pessin JE and Saltie AR (1999) Aldolase mediates the association of F-actin with the insulin-responsive glucose transporter Glut4. J Biol Chem 274: 17742-17747

Knull HR and Walsh JL (1992) Association of glycolytic enzymes with the cytoskeleton. Curr Topic Cell Regul 33: 15-30

Lindblom S, Ek P, Muszynska G, Ek B, Szczegielniak J and Engström L (1997) Phosphorylation of sucrose synthase from maize seedlings. Acta Biochim Polonica 44:809-818

Luther MA and Lee JC (1986) The role of phosphorylation in the interaction of rabbit muscle phosphofuctokinase with f-actin. J Biol Chem 4: 1753-1759

Masters C (1996) On the role of the cytoskeleton in metabolic compartmentation. In: Hesketh JE (ed) The Cytoskeleton, Vol 2, JAI Press, pp 1-30

Matsudaira P (1991) Modular organization of actin crosslinking proteins. Trends Biol Sci 16: 87-92

Moorhead GBG and Plaxton WC (1988) Binding of glycolytic enzymes to a particulate fraction in carrot and sugar beet storage roots: dependence on metabolic state. Plant Physiol 86: 348-351

Moriyama K, Lida K and Yaharra I (1996) Phosphorylation of ser-3 of cofilin regulates its essential function on actin. Genes Cells 1: 73-86

Nakai T, et al. (1998) An increase in apparent affinity for sucrose of mung bean sucrose synthase is caused by in vitro phosphorylation or directed mutagenessis of Ser^{11}. Plant Cell Physiol 39: 1337-1341

O'Reilly G and Clarke F (1993) Identification of an actin binding region in aldolase. FEBS Lett 321: 69-72

Ruan YL, Chourey PS, Delmer DP and Perez-Grau L (1997) The differential expression of sucrose synthase in relation to diverse patterns of carbon partitioning in developing cotton seeds. Plant Physiol 115: 325-385

Smertenko AP, Jiang CJ, Simmons NJ, Weeds AG, Davies DR and Hussey PJ (1998) Ser6 in the maize actin-depolymerizing factor, ZmADF3, is phosphorylated by a calcium-stimulated protein kinase and is essential for the control of functional activity. Plant J 14: 187-193

Tellam RL, Morton DJ and Clarke FM (1989) A common theme in the amino acid sequences of actin and many actin-binding proteins? Trends Biol Sci 14: 30-33

Winter H, Huber JL and Huber SC (1997) Membrane association of sucrose synthase: changes during the graviresponse and possible control by protein phosphorylation. FEBS Lett 420: 151-155.

Winter H, Huber JL and Huber SC (1998) Identification of sucrose synthase as an actin binding protein. FEBS Lett 430: 205-208

Winter H, Robinson DG and Heldt HW (1994) Subcellular volumes and metabolite concentrations in spinach leaves. Planta 193: 532-555

Wu K, Aoki C, Elste A, Rogalski-Wilk AA and Siekevitz P (1998) The synthesis of ATP by glycolytic enzymes in the postsynaptic density and the effect of endogenously generated nitric oxide. Proc Natl Acad Sci USA 94: 13273-13278

Zhang XQ and Chollet R (1997) Seryl-phosphorylation of soybean nodule sucrose synthase (nodulin-100) by Ca^{2+}-dependent protein kinase. FEBS Lett 410: 126-130

Zhang XQ, Lund AA, Sarath G, Cerny RL, Roberts DM and Chollet R (1999) Soybean nodule sucrose synthase (Nodulin-100): further analysis of its phosphorylation using recombinant and authentic root-nodule enzymes. Arch Biochem Biophys 371: 70-82

Zrenner R, Salanoubat M, Willmitzer L and Sonnewald U (1995) Evidence of the crucial role of sucrose synthase for sink strength using transgenic potato plants (*Solanum tuberosum* L.) Plant J 7: 97-107

Chapter 8

ACTIN IN PROTEIN SYNTHESIS AND PROTEIN BODY FORMATION

Bratislav Stanković[1], Amy Clore[2], Shunnosuke Abe[3], Brian Larkins[4] and Eric Davies[5]

1 Wisconsin Center for Space Automation and Robotics, University of Wisconsin, Madison, WI 53706, USA
2 Forestry Department, North Carolina State University, Raleigh, NC 27695, USA
3 Lab of Molecular Cell Biology, Faculty of Agriculture, Ehime University, Matsuyama, Japan
4 Department of Plant Sciences, University of Arizona, Tucson, AZ 85721, USA
5 Botany Department, North Carolina State University, Raleigh, NC 27695, USA

Key words: actin, endosperm, maize, *Oryza sativa*, polyribosomes, protein body, rice, *Zea mays*

Abstract: We review the recent evidence that plant cells, like animal cells, have mRNA associated with their cytoskeleton, and that much of this mRNA is present in polyribosomes. One of these cell types is the maize endosperm cell which contains storage proteins residing inside the lumen of the endoplasmic reticulum (ER). Data from both wild type and mutant maize endosperm support the proposition that the cytoskeleton is the scaffold upon which both polyribosomes (especially those synthesizing zein) and membranes adhere. While much of the data implies a major role for the actin cytoskeleton, additional roles are suggested for the microtubule system and for elongation factor, eEF1A, which might help connect the two cytoskeleton networks. Results from rice endosperm have been especially beneficial to understanding the role of the cytoskeleton, insofar as rice has two entirely different kinds of protein bodies, the cereal-type and the legume-type. The former derive directly from the ER, store prolamines, and are closely associated with the cytoskeleton, while the latter arise from the coalescence of ER-derived vesicles, store glutelins, but are not associated with the cytoskeleton. The roles of the cytoskeleton appear to be in the segregation of specific mRNAs to specific sub-cellular locations, the enhanced translation of these mRNAs, and perhaps the accumulation of these water-insoluble storage proteins within the protein body.

C.J. Staiger et al. (eds.), Actin: A Dynamic Framework for Multiple Plant Cell Functions, 129–143.

1. INTRODUCTION

The cytoskeleton of animal cells serves as an attachment site for a sub-population of ribosomes, the cytoskeleton-bound polyribosomes (polysomes), and a burgeoning number of reports suggests that the cytoskeleton plays a fundamental role in the sub-cellular transport, partitioning and anchoring of mRNAs in animals (reviewed in Condeelis, 1995; St. Johnston, 1995; Hesketh & Pryme, 1996; Singer, 1996; Jansen, 1999). Sequences that direct localization of cytoplasmic mRNAs to the cytoskeleton have been mapped to the 3' untranslated region of mRNAs (Hesketh & Pryme, 1996; Singer, 1996; Jansen, 1999). Components of the translational machinery, such as elongation factors, directly bind F-actin, and there is evidence for co-localization of mRNA, elongation factors, and ribosomes with actin filaments (Condeelis, 1995). It is also known that common physiological regulators (e.g. pH) mediate both the interaction of elongation factor 1α with the cytoskeleton, as well as the rate of protein synthesis (Liu et al., 1996). In animal cells, it has been shown that the associations of mRNAs and components of the translational machinery with the cytoskeleton allow control over such parameters as the rate of protein synthesis, the synthesis of proteins in defined subcellular locations, and co-translational assembly of macromolecular structures. Therefore, such associations can play dramatic roles in the processes of cellular differentiation and the establishment of polarity. Adopting this paradigm from animal experimental systems, plant biologists have begun to investigate possible cytoskeleton–mRNA–ribosome interactions by looking at the subcellular localization of translational regulators in plant tissues actively involved in protein synthesis.

The study of plant ribosomes extends almost 30 years, with early experiments on the isolation of polyribosomes from plant tissues focusing on development of techniques that prevented ribonuclease (RNase) activity. The outcome of these efforts was the design of buffers that yielded both free polyribosomes and membrane-bound polyribosomes (Larkins & Davies, 1975). However, most of the agents used to inhibit RNase activity, including high levels of monovalent cations (Tris, NH_4^+ and K^+) and nuclease inhibitors such as heparin and ribonucleoside vanadyl complexes, also caused fragmentation of the actin filaments with the concomitant release of polyribosomes (Davies et al., 1991; Davies and Abe, 1995; Abe and Davies, 1995). Despite the fact that high ionic strength buffers destroy actin filaments, actin remains associated with ribosomes isolated in such buffers (Abe et al., 1994; Zak et al., 1995), suggesting that ribosome–actin binding is stronger than actin–actin binding. As a prerequisite for examining the existence of cytoskeleton–ribosome interactions in plants, a cytoskeleton-stabilizing buffer (CSB) was developed that allowed actin to be isolated as large fragments of actin filaments (F-actin), rather than small fragments or monomeric G-actin (Abe & Davies, 1991, 1995).

2. THE MAIZE ENDOSPERM AS A MODEL SYSTEM

The developing maize endosperm is a useful system to study the plant cytoskeleton, and it is likely to provide novel insights into cytoskeletal structure and function in plants. Cells of the endosperm accumulate large amounts of starch and storage proteins (zeins), which are co-translationally inserted into the lumen of the rough endoplasmic reticulum (RER) where they coalesce into spherical protein bodies (Larkins & Hurkman, 1978).

Multiple experimental approaches have been used to investigate the nature of the cytomatrix in maize endosperm. Biochemical fractionation, density gradient centrifugation, immunoblotting, and immunohistochemistry in conjunction with light, fluorescence and electron microscopy provide multiple lines of evidence for the existence of microfilament–protein body–membrane interactions. Abe et al. (1991) used subcellular fractionation, supplemented with fluorescence microscopy, to demonstrate that protein bodies isolated from developing maize endosperm by centrifugation at exceedingly low (30x to 60x) g forces are encircled by, and enmeshed in, a network of actin filaments (Fig. 1).

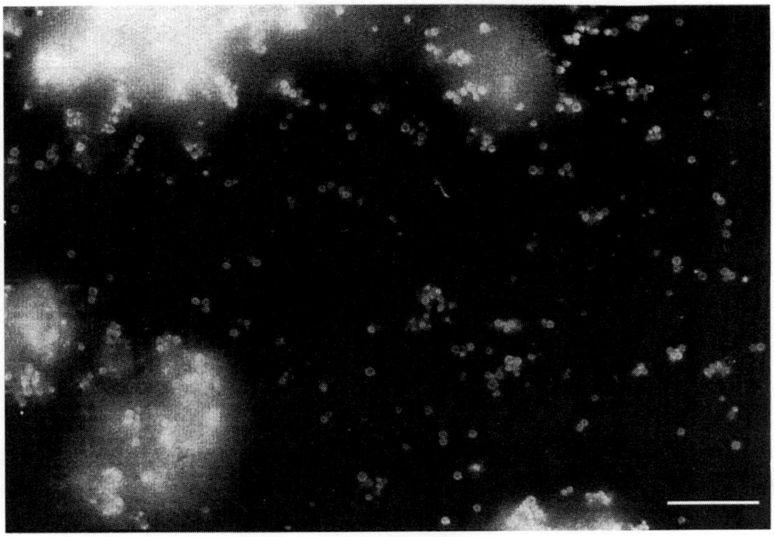

Figure 1. Protein bodies surrounded by actin sediment at exceedingly low g forces. Maize seeds were homogenized in cytoskeleton-stabilizing buffer (CSB) and centrifuged for 5 min at 30x g. The pellet was subsequently stained with rhodamine-phalloidin and examined using

fluorescence microscopy. Individual protein bodies and protein body aggregates are visible. Bar = 10 μm. Modified and reprinted with permission from Protoplasma (Abe et al., 1991).

The microfilament–protein body complex sedimented at 30x g, even in the presence of a non-ionic detergent. The protein bodies were also associated with and/or surrounded by RNA (presumably ribosomes), and treatment of the protein body pellet with non-ionic detergent, which removes virtually all of the membrane but none of the actin, had no effect on the associated polyribosomes (Abe et al., 1991).

Sucrose gradient analysis of maize endosperm homogenized in various buffers provided additional evidence for the existence of polyribosome–cytoskeleton associations in this tissue (Davies et al. 1993). Indeed, ionic reagents that are routinely used to prevent RNase activity in plant tissues, including Tris, heparin, KCl, and deoxycholate (DOC), were all found to depolymerize actin and to liberate ribosomes/polysomes from the cytoskeleton (Davies et al., 1993).

Experiments using homogenates of maize endosperm isolated in a CSB indicated that protein bodies are surrounded by filamentous actin, polyribosomes, and membranes (ER), with the polyribosomes attached via actin to the protein bodies (Stanković et al., 1993). The nature of the associations between these organelles was investigated using fluorescence microscopy in conjunction with buffers of different ionic strength (Fig. 2).

Treatment with non-ionic detergent, which removed most of the RER membranes, did not dislodge the polyribosomes from around the protein bodies nor did it disrupt the association with actin. In contrast, treatment with ionic detergent or Tris removed all of the actin and most of the ribosomes (Stanković et al., 1993). Homogenization of maize endosperm in a cytoskeleton-stabilizing buffer, followed by gradient centrifugation, produced a large peak of UV-absorbing material with a density of approximately 1.3 g ml^{-1} (Stanković et al., 1999). The nature of the peak was investigated by employing "variable-term" sucrose density centrifugation (Davies & Abe, 1995) and analysis with immunoprecipitation and/or double staining with fluorochromes. Its main constituents were protein body/cytoskeleton/polyribosome complexes, showing numerous protein bodies encircled with actin, RNA and membranes (Stanković et al., 1999).

These studies demonstrated associations between filamentous actin, polyribosomes and protein bodies following homogenization of developing endosperm in a cytoskeleton-stabilizing buffer. Although the studies did not demonstrate *in situ* associations of these components, they raised the prospect that the actin filaments may be involved in attaching zein polyribosomes to the RER surrounding protein bodies.

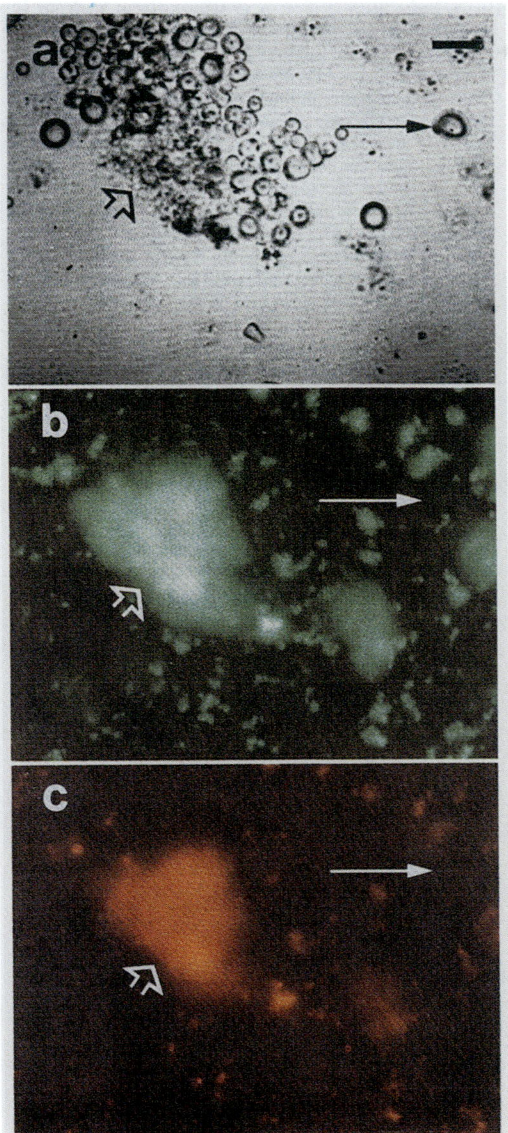

Figure 2. Co-localization of RNA and actin on and around protein bodies. Homogenates of maize endosperm were double - stained with thiazole orange and rhodamine-phalloidin, and observed under phase contrast; (a), blue filter for RNA (b), and green filter for actin (c). The open arrows indicate protein body aggregates, and the closed arrows indicate starch grains. Bar = 10 μm. Reprinted with permission from Protoplasma (Stanković et al., 1993).

To determine whether the cytoskeleton is associated with protein bodies *in situ*, and whether elongation factor 1α (eEF1A; formerly referred to as

EF-1 α; Browning, 1996) is associated with the cytoskeleton in intact maize endosperm cells, Clore et al. (1996) performed indirect immunofluorescence and confocal microscopy.

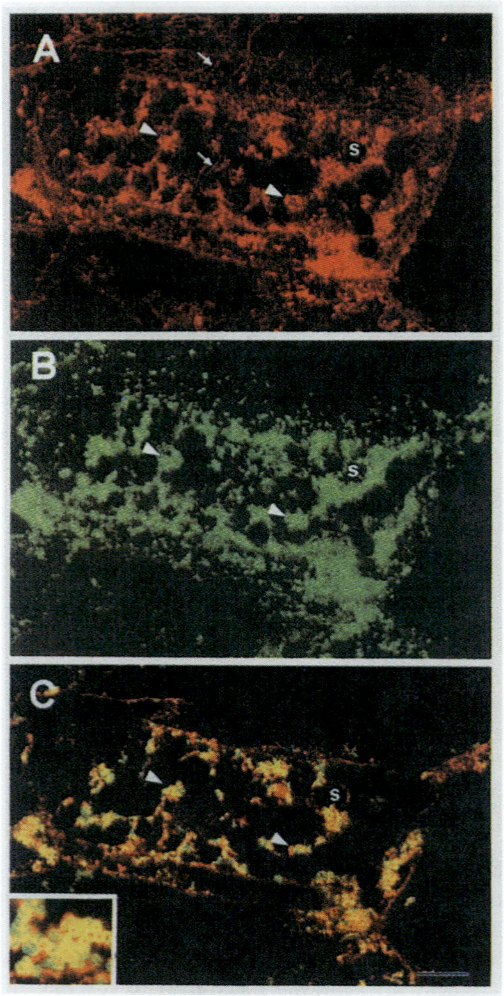

Figure 3. Immunological labeling of actin (A) and γ-zein (B) in a maize endosperm cell. Superimposition of images of actin and γ-zein from a single optical section resulted in a yellow color (C), suggesting that actin surrounds clusters of protein bodies (arrowheads).

Arrows indicate actin filaments. The inset shows a higher magnification view of the actin surrounding protein bodies. S=starch. Bar = 20 μm. Copyrighted by ASPP and reprinted with permission from The Plant Cell (Clore et al., 1996).

Specifically, the distributions of actin (both monomeric and filamentous), microtubules and eEF1A in interphase cells both before and after the onset of storage product (starch and storage protein) deposition were studied. A dramatic reorganization of the cytoskeleton was found to occur during this period. In young cells which have not yet accumulated starch or protein bodies, the actin and tubulin were found mostly in association with nuclei and/or in the cell cortex. Following the start of storage product deposition, the actin was visualized as fine filaments in the cortex and cytoplasm as well as in clusters in the cytoplasm between the starch grains (Fig. 3).

Clore et al. (1996) found that not only did the distribution of these clusters of actin closely match the location of protein bodies (as identified by immunolabeling with γ-zein) but also actin closely surrounds these bodies (Fig. 3). Finally, it was found that eEF1A is localized around protein bodies where it is complexed with actin. These studies showed that protein bodies are enmeshed in complexes of eEF1A and actin, and are found juxtaposed with microtubules. This raised the prospect that the cytoskeleton may be involved in protein body formation (Clore et al., 1996).

Electron micrographs of endosperm tissue sections (not shown) and pelleted homogenates (Fig. 4) reveal numerous associations of filaments with ribosomes and protein bodies. Higher magnification suggests the presence of a myriad of filaments approximately 8 nm in diameter emanating from (the material surrounding) individual protein bodies (Fig. 4; Abe et al., 1991).

3. USE OF MUTANTS TO INVESTIGATE THE ACTIN–PROTEIN BODY–RIBOSOME INTERACTIONS

There are several maize mutants that affect the protein content and physical properties of the endosperm. The maize *opaque-2* mutant has an altered pattern of storage protein synthesis which results in reduced protein content (Mertz et al., 1964). This is a consequence of a defective (mutant) transcription factor, and it leads to a soft, starchy endosperm. Consequently, *opaque-2* has smaller protein bodies than wild type (W64A$^+$) (Geetha et al., 1991). Until recently, it was not known whether the associations between the actin filaments, the membranes, and the polyribosomes in *opaque-2* are different from those in the wild type. Biochemical and immunoblot data suggest that the interactions of protein bodies with the actin filament/membrane complex are different in the endosperm of wild-type and *opaque-2* maize (Stanković et al., 1999).

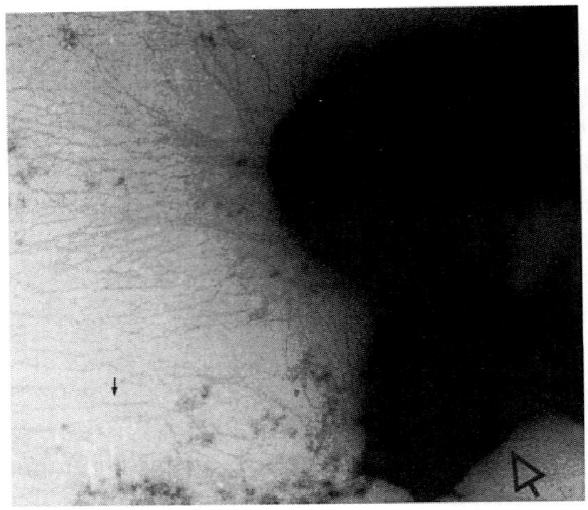

Figure 4. Electron micrograph of maize endosperm homogenate showing protein bodies decorated with numerous filaments. The material from a 30x g pellet was negatively stained and observed using electron microscopy (magnification x 35,000). The large, open arrow indicates a protein body; the small arrow indicates one of the myriad of filaments. Reprinted with permission from Protoplasma (Abe et al., 1991).

In both the wild-type and mutant endosperm, protein bodies are associated with polyribosomes which have similar cytoskeleton associations. However, the wild type has but one major type of protein body which is larger and more homogenous than in *opaque-2*; they also have a relatively lower membrane/protein ratio. In contrast, protein bodies in *opaque-2* have a greater polyribosome/protein body ratio and appear in two different forms: smaller (less dense) ones almost devoid of zein, with cytoskeleton proteins (and presumably polyribosomes) attached, and larger (denser) ones with abundant amounts of zein but lacking cytoskeleton-associated polyribosomes. Thus, the comparisons of both the UV-absorbance profiles and the immunoblot data suggest that the protein bodies from these two maize types associate differently with the membranes and the cytoskeleton.

4. THE CYTOARCHITECTURE OF ACTIN–PROTEIN BODY–RIBOSOME INTERACTIONS

It is important to understand the way in which the components of the microfilament–membrane–polyribosome complex are linked with one another. Earlier work with membrane-associated polyribosomes showed that the polyribosome could be attached to the protein body membrane via the mRNA (so-called "dangling" polyribosomes), the ribosomes, the nascent polypeptide chains, or any combination of these (Davies & Larkins, 1980). Efforts were made to determine which of these modes of attachment occur in the endosperm cytomatrix. Treatment of the polysomal pellet with RNase did not release polyribosomes, although they were severely degraded, showing that the RNase was active and implying that the polyribosomes were not anchored solely by the mRNA (Davies et al., 1993). Treatment with increasing concentrations of Tris caused release of increasingly large polyribosomes, strongly implying that ribosome-microfilament interactions were being disrupted by Tris. Treatment with protease K released polyribosomes, but also disrupted the actin filaments, thus it could not be determined whether nascent polypeptides are involved in scaffolding. Treatment of maize endosperm with RNase released some polyribosomes, but had no effect on the membrane–actin filament complex (Davies et al., 1993), implying that ribosomes are not, in fact, the scaffold upon which the other components are associated. Similarly, treatment of isolated protein bodies with non-ionic detergent removes essentially all the membrane, but leaves polyribosomes associated with the actin filaments (Stanković et al., 1993), implying that the membranes are also not the bonding element.

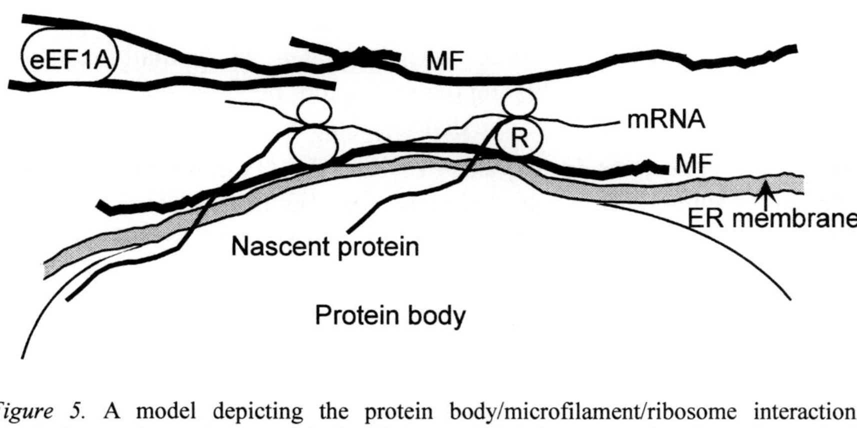

Figure 5. A model depicting the protein body/microfilament/ribosome interactions in developing maize endosperm. Actin filaments (AFs) juxtapose the sites of protein body formation. The endoplasmic reticulum surrounds the protein bodies and is enmeshed in a

network of AFs. Elongation factor 1A (eEF1A) appears to interconnect these AFs. Ribosomes are attached to the AFs and to the underlying ER membrane through the 60S subunit. Nascent proteins are co-translationally inserted into the ER lumen as the protein body is being formed.

However, treatment of protein bodies with ionic detergent or with Tris, caused fragmentation of the actin filaments and release of the majority of the polyribosomes, implying that the actin filaments attach directly to the ribosomes. In cytochalasin-D treated cells, labelling for γ-zein showed that protein bodies remained intact while part of the eEF1A antigen was removed, suggesting that eEF1A forms a complex with actin (Clore et al., 1996). Since membranes also associate with the actin cytoskeleton (Abe et al., 1992; Ito et al., 1994), the current presumption is that the F-actin is the scaffold supporting both membranes and polyribosomes (Davies et al., 1998).

The findings outlined above are summarized in the model shown in Figure 5 where polyribosomes are attached to the cytoskeleton and/or RER membrane. The polyribosomes in maize endosperm appear to be more firmly associated with the cytoskeleton than with ER membrane and might be considered cytoskeleton-membrane-bound polyribosomes or cytomatrix-bound polyribosomes (Zambetti et al., 1990; Stanković et al., 1993).

5. POSSIBLE FUNCTIONS OF THE ACTIN FILAMENT–PROTEIN BODY–POLYRIBOSOME COMPLEXES IN CEREAL ENDOSPERM

The physiological significance of the association of ribosomes with actin filaments in cereal endosperm is not known at present, but the actin filament network might provide a scaffold and a railroad on which storage protein RNA can be directed, anchored and translated in spatially-restricted subcellular compartments. This regulation could be developmental since Clore et al. (1996) observed that the distributions of actin and eEF1A change during development. The mRNA–actin interactions are most likely indirect, utilizing untranslated regions on the mRNA through protein(s) that mediate the attachment (Davies et al., 1993; Muench et al., 1998; Okita et al., 1998).

This idea is also supported in the rice endosperm system. Rice endosperm differs from that of most other cereals in having two entirely different kinds of protein body. One type is like that in maize endosperm which stores prolamines and is formed directly from the RER. The other protein body is more typical of dicotyledons, such as those in legume cotyledons. It is formed by coalescence of ER vesicles into a storage vacuole that stores glutelins (Krishan et al., 1986). The prolamine protein bodies, unlike the glutelin protein bodies, are associated primarily with actin filaments but also with some microtubules (Muench et al., 1998; Okita et al., 1998). These authors showed that prolamine mRNAs play an important role in binding the polysomes to the cytoskeleton (Muench et al., 1997; Okita et al., 1998) and they have evidence that, as in animal systems, the 3' untranslated region of prolamine mRNA has a "zip code" that targets the mRNA to the cytoskeleton (Okita, personal communication). The rice endosperm also furnishes an excellent example in plants of mRNA enrichment to different sub-cellular domains. Using blot hybridisation and *in situ* hybridisation, Okita and co-workers (Li et al., 1993) have shown that glutelin mRNA is enriched 2.5:1 over prolamine mRNA in the glutelin protein bodies, while prolamine mRNA is enriched 7–10:1 over glutelin mRNA in the prolamine protein bodies. Furthermore, very recent results from the same group (Muench et al., 2000) show that prolamine protein bodies are more abundant than glutelin protein bodies near the cell periphery while glutelin protein bodies are more abundant than prolamine protein bodies towards the cell center. In a smaller proportion of cells as many as 90% of the prolamine protein bodies appear to be found in a 5-μm strip around the cell periphery. Since these protein bodies contain 7–10 times more prolamine mRNA than glutelin mRNA on their surface, this furnishes an excellent example from the plant literature of a clear enrichment of a specific mRNA to a specific sub-cellular domain. This recent work also shows that actin (filaments and bundles), as well as microtubules, are more abundant at the cell periphery, and suggests they might be involved in anchoring the protein body–polysome–ER membrane complex to the plasma membrane (Muench et al., 2000).

Not only might the actin cytoskeleton be involved in protein synthesis through its ability to direct different mRNAs to different sub-cellular domains, but it might also function to enhance the rate of protein synthesis, as has been found with pea stem and root tissue, both *in vivo* and *in vitro* (Davies et al., 1998). These authors showed that the cytoskeleton with polyribosomes attached translated *in vitro* about 2–3 times more effectively than did the released cytoskeleton-bound polyribosomes, and about 10 times more than the free- and membrane-bound polyribosomes. *In vivo*, radioactive label was incorporated into cytoskeleton-bound polyribosomes about 10 times faster than into free- and membrane-bound polyribosomes (Davies et al., 1998). A similar large enhancement of protein synthesis by cytoskeleton-associated polyribosomes has also been noted in rice. Okita's group showed that the cytoskeleton-associated prolamine mRNA in rice was translated about 10 times faster *in vivo* than the non-cytoskeleton-associated glutelin mRNA (Li et al., 1993; Okita et al., 1994). In marked contrast, Larkins group, working with oat endosperm (Boyer et al., 1992), found that the oat globulin mRNA (presumably not associated with the cytoskeleton) was translated *in vivo* about 10 times more efficiently than avenin mRNA, a prolamine. However, it is not yet known whether avenin mRNAs are associated with the cytoskeleton.

Some studies suggest that the endosperm cytoskeleton might contribute significantly to the nutritional quality of the grain. Habben et al. (1995) found a high correlation (r^2=0.9) between the content of lysine in the endosperm and the concentration of eEF1A (EF-1α). eEF1A bundles actin (Edmonds, 1993) and also bundles (Durso & Cyr, 1994) and severs (Shiina et al., 1994) microtubules. eEF1A itself only accounts for approximately 2% of the total lysine (Sun et al., 1997). Therefore, it was hypothesized that a relationship exists between eEF1A and other lysine-rich, cytoskeleton-associated proteins, all of which are increased in high lysine genotypes (Clore et al., 1996). Indeed, it was noted that eEF1A is present along with actin in the protein body fraction following sucrose gradient separation of endosperm homogenates (J. Habben, personal communication).

It is possible that this elaborate cytoskeleton does not develop for the purpose of protein body formation. Alternative functions may include structural support of the endomembrane system, or the localization of starch-synthesizing enzymes, or other carbohydrate metabolising enzymes (Winter et al., 1998). In any case, cereal endosperm may provide an opportunity to study cytoskeletal functions beyond cell division and cell wall formation because the cytomatrix interactions described here occur in interphase cells.

6. PROSPECTS

Exciting discoveries lie ahead since many questions remain to be answered about the formation and function of the microfilament arrays in cereal endosperm. For example, are the actin–ribosome–protein body interactions present in cereal crops other than maize and rice? What is the significance of the actin–ribosome–protein body interactions? To what degree might the actin filaments be involved in endosperm morphogenesis? By what mechanisms do the changes in the actin cytoskeleton (Clore et al., 1996) occur, and what factors initiate and regulate the redistribution? How quickly do the changes in actin reorganization in maize endosperm occur? Initial observations suggest this may occur within 24–48 h during mid-stage III, 10–16 days after pollination, but the timing has yet to be systematically studied. Does eEF1A cross-link the actin around protein bodies or somehow link it to the microtubule array based on its actin-bundling and microtubule-bundling activities? What is the nature of the association between the RER surrounding protein bodies and the actin filaments? Is this interaction mediated by certain proteins, and if so, what are their identities?

It is possible that the actin cytoskeleton in cereal endosperm traffics and/or anchors storage-protein mRNAs to sites of protein body formation. Li et al. (1993) found that different classes of storage-protein mRNAs segregate to different regions of the RER in rice endosperm cells, and it has been suggested that the cytoskeleton might play a role in their localization (Okita et al., 1994). By using microinjection, cytoskeleton disruption and *in situ* hybridization it should be possible to investigate some of the above questions, as well as to determine whether or not the actin filaments colocalize with zein mRNAs.

It is also possible that other, as yet unidentified, lysine-rich proteins reside with the cytoskeleton and increase concomitantly with eEF1A in high lysine genotypes. This hypothesis could explain why eEF1A has such high predictive value for the lysine content of the grain (Moro et al., 1996). Molecular and genetic strategies are being used to identify genes that are up-regulated coordinately with eEF1A.

In summary, there is abundant evidence for the existence of associations among protein bodies, actin filaments, ribosomes, key components of the protein synthesis machinery, and ER membranes in maize endosperm. This implies a potentially significant function for the actin cytoskeleton in modulating protein synthesis in a variety of ways. Since the only known function of polyribosomes is to synthesize proteins, their attachment to the cytoskeleton surrounding protein bodies in the endosperm might regulate the process of translation and protein body formation, including the synthesis of specific types of zeins. The actin filaments in cereal endosperm might facilitate the localization of specific mRNAs to sub-cellular domains (Okita et al., 1994). It is becoming increasingly likely that a significant role of the plant cytoskeleton lies in orchestrating the spatial distribution of mRNAs and proteins within cells, as well as enhancing the rate of protein synthesis.

REFERENCES

Abe S and Davies E (1991) Isolation of F-actin from pea stems: Evidence from fluorescence microscopy. Protoplasma 163: 51-61

Abe S, You W and Davies E (1991) Protein bodies in maize endosperm are enclosed by and enmeshed in F-actin. Protoplasma 165: 139-149

Abe S, Ito Y and Davies E (1992) Cosedimentation of actin, tubulin and membranes inthe cytoskeleton fractions from peas and mouse 3T3 cells. J Exp Bot 43: 941-949

Abe S, Ito Y and Davies E (1994) Association of cytoskeletal proteins in the membrane-bound polysome fraction from peas using conventional polysome isolation buffers. Plant Physiol Biochem 3: 547-554

Abe S and Davies E (1995) Methods for isolation and analysis of the cytoskeleton. In: Galbraith DW, Bourque DP and Bohnert HJ (eds) Methods in Cell Biology, vol 50. New York: Academic Press, pp 223-236

Boyer SK, Shotwell MA and Larkins BA (1992) Evidence for the translational control of storage protein gene-expression in oat seeds. J Biol Chem 267: 17449-17457

Browning KS (1996) The plant translational apparatus. Plant Mol Biol 32: 107-144

Clore AM, Dannenhoffer JM and Larkins BA (1996) EF-1α is associated with a cytoskeletal network surrounding protein bodies in maize endosperm cells. Plant Cell 8: 2003-2014

Condeelis J (1995) Elongation factor 1 α, translation and the cytoskeleton. Trends Biochem Sci 20: 169-170

Davies E and Larkins BA (1980) Ribosomes. In: Stumpf PK and Conn EE (eds) Plant Biochemistry: A Comprehensive Treatise, vol. 1. New York: Academic Press, pp 413-435

Davies E, Fillingham BD, Ito Y and Abe S (1991) Evidence for the existence of cytoskeleton-bound polysomes in plants. Cell Biol Int 15: 973-981

Davies E, Comer EC, Lionberger JM, Stanković B and Abe S (1993) Cytoskeleton-bound polysomes in plants. III. Polysome-cytoskeleton-membrane interactions in maize endosperm. Cell Biol Int 17: 331-340

Davies E and Abe S (1995) Methods for isolation and analysis of polyribosomes. In: Galbraith DW, Bourque DP and Bohnert HJ (eds) Methods in Cell Biology, vol 50. New York: Academic Press, pp. 209-222

Davies E, Abe S, Larkins BA, Clore AM, Quatrano RS and Weidner S (1998) The role of the cytoskeleton in plant protein synthesis. In: Bailey-Serres J and Gallie DR (eds) A Look Beyond Transcription: Mechanisms Determining mRNA Stability and Translation in Plants. Amer Soc Plant Physiol, Rockville, MD, USA, pp 115-124

Durso NA and Cyr RJ (1994) Beyond translation: Elongation factor-1 α and the cytoskeleton. Protoplasma 180: 99-105

Edmonds BT (1993) ABP50 - an actin-binding elongation factor-1-alpha from *Dictyostelium discoideum*. J Cell Biochem 52: 134-139

Geetha KB, Lending CR, Lopes MA, Wallace JC and Larkins BA (1991) *Opaque-2* modifiers increase α-zein synthesis and alter its spatial distribution in maize endosperm. Plant Cell 3: 1207-1219

Habben JE, Moro GL, Hunter BG, Hamaker BR and Larkins BA (1995) Elongation factor 1 α concentration is highly correlated with the lysine content in maize endosperm. Proc Natl Acad Sci USA 92: 8640-8644

Hesketh JE and Pryme IF (1996) Association of mRNAs and polyribosomes with the cytoskeleton: Potential roles in transport and compartmentalization of mRNAs. In: The Cytoskeleton, vol 2 Role in Cell Physiology, JAI Press, Stamford, CT, USA, pp 31-58

Ito Y, Abe S and Davies E (1994) Co-localization of cytoskeleton proteins and polysomes with a membrane fraction from peas. J Exp Bot 45: 253-259

Jansen RP (1999) RNA-cytoskeleton interactions. FASEB J 13: 455-466

Krishnan HB, Franceschi VR and Okita TW (1986) Immunochemical studies on the role of the golgi complex in protein body formation in rice seeds. Planta 169: 471-480

Larkins BA and Davies E (1975) Polyribosomes from peas. V. An attempt to characterize the total free and membrane-bound polysomal population. Plant Physiol 55: 749-756

Larkins BA and Hurkman WJ (1978) Synthesis and deposition of zein in protein bodies of maize endosperm. Plant Physiol 62: 256-263

Li XX, Franceschi VR and Okita TW (1993) Segregation of storage protein mRNAs on the rough endoplasmic reticulum membranes of rice endosperm cells. Cell 72: 869-879

Liu G, Edmonds BT and Condeelis J (1996) pH, EF-1α and the cytoskeleton. Trends Cell Biol 6: 168-171

Mertz ET, Bates LS and Nelson OE (1964) Mutant gene that changes protein composition and increases lysine content of maize endosperm. Science 145: 279-280

Moro GL, Habben JE, Hamaker BR and Larkins BA (1996) Characterization of the variability in lysine content for normal and *opaque2* maize endosperm. Crop Sci 36: 1651-1659

Muench DG and Okita TW (1997) The storage proteins of rice and oat. In: Larkins BA and Vasil IK (eds) Cellular and Molecular Biology of Plant Development, Dordrecht: Kluwer Academic Press, pp 289-330

Muench DG, Wu Y, Coughlan SJ and Okita TW (1998) Evidence for a cytoskeleton-associated binding site involved in prolamine mRNA localization to the protein bodies in rice endosperm tissue. Plant Physiol 116: 559-569

Muench DG, Chuong SDX, Franceschi VR and Okita TW (2000) Developing protein bodies are primarily associated with the cortical cytoskeleton in rice endosperm cells. Planta, in press

Okita TW, Li X and Roberts MW (1994) Targeting of mRNAs to domains of the endoplasmic reticulum. Trends Cell Biol 4: 91-96

Okita TW, Choi S-B, Ito H, Muench DG, Wu Y and Zhang F (1998) Entry into the secretory system – the role of mRNA localization. J Exp Bot 49: 1081-1090

Shiina N, Gotoh Y, Kubomura N, Iwamatsu A and Nishida E (1994) Microtubule severing by elongation factor 1α. Science 266: 282-285

Singer RH (1996) RNA: Traffic report. Trends Cell Biol 6: 486-489

St Johnston D (1995) The intracellular localization of messenger RNAs. Cell 81: 161-170

Stanković B, Abe S and Davies E (1993) Co-localization of polysomes, cytoskeleton, and membranes with protein bodies from maize endosperm. Protoplasma 177: 66-72

Stanković B, Abe S, Azama K, Shibata K, Ito Y, Weidner S and Davies E (1999) Associations of maize protein bodies with cytoskeleton, membranes, and ribosomes in the endosperm of wild type and *opaque-2* mutant. Acta Physiol Plant 21: in press

Sun YJ, Carneiro N, Clore AM, Moro GL, Habben JE and Larkins BA (1997) Characterization of maize elongation factor 1A and its relationship to protein quality in the endosperm. Plant Physiol 115: 1101-1107

Winter H, Huber JL and Huber SC (1998) Identification of sucrose synthase as an actin-binding protein. FEBS Lett 430: 205-208

Zak EA, Bocharova MA, Sokolov IA and Klyachko NL (1995) Plant polysomes bound to the cytoskeleton. Doklady Bot Sci 343-345: 37-39

Zambetti G, Wilming L, Fey EG, Penman S and Stein G (1990) Differential association of membrane-bound and non-membrane bound polysomes with the cytoskeleton. Exp Cell Res 191: 246-255

Chapter 9

CORTICAL ACTIN INTERACTS WITH THE PLASMA MEMBRANE AND MICROTUBULES

David A Collings[1,2] and Nina S Allen[1]

1 Department of Botany, North Carolina State University, Raleigh NC 27695, USA
2 Department of Biological Sciences, Purdue University, West Lafayette IN 47907, USA

Key words: actin, endoplasmic reticulum, membrane interactions, microtubules, plant morphogenesis

Abstract: Cortical actin comprises a dynamic array of single actin microfilaments that interacts with the plasma membrane of most plant cells. These interactions, mediated by as yet unknown actin-binding proteins, may function to anchor the entire actin cytoskeleton, and contribute to cell signaling. In cells elongating by diffuse growth, cortical actin is typically transversely oriented, parallel to microtubules. Drug studies demonstrate that in some cases, cortical actin is necessary for the precise re-arrangements of cortical microtubules that control cell expansion and plant morphogenesis. However, in other cases, cortical actin depends on the cortical microtubules for its organization. This suggests that the cortical cytoskeleton is a closely controlled system involving feedback between actin and microtubules.

Interphase plant cells contain a range of actin arrays, including extensive bundles of actin microfilaments that serve as the tracks for cytoplasmic streaming around the nucleus, in transvacuolar strands and in the subcortex, and a dynamic array of shorter filaments in the outer cortex of the cell, most notably in cells undergoing diffuse growth. This cortical actin, first visualized by Traas et al. (1987), has proved difficult to preserve and image, but similar arrays can be seen with more effective labeling and immunofluorescence methods (Sonobe & Shibaoka, 1989), with rhodamine-phalloidin injections (Cleary, 1995), or with fusions between green fluorescent protein (GFP) and the actin-binding protein talin (Kost et al., 1998). Cortical actin arrays comprise single actin filaments or possibly bundles of several filaments, occur adjacent to the plasma membrane, and in elongating cells organize into transverse arrays parallel to the transversely-

C.J. Staiger et al. (eds.), Actin: A Dynamic Framework for Multiple Plant Cell Functions, 145–163.

aligned cortical microtubules (MTs). Our intention in this review is to investigate the possible functions of cortical actin in interphase cells, paying particular attention to actin's interactions with the plasma membrane and MTs, limiting citations of descriptive literature to initial and pertinent studies.

While cortical actin occurs adjacent to the plant plasma membrane (section 1), this does not prove that direct interactions occur between actin and the membrane. Plant cells are stationary, and use turgor pressure and the cell wall to provide mechanical support. By comparison, animal cells, notably those that change shape, use a membrane-bound actin cytoskeleton, comprising short filaments and longer stress fibers, to generate motility and support for the plasma membrane, with the maintenance of cell shape depending on the presence of membrane-binding proteins that weld actin to the membrane. So while it might seem that plant cells would not require plasma membrane-associated actin, evidence exists for direct actin interactions with the plasma membrane in plants, possibly playing important roles in signal transduction, cell development and plant morphogenesis (section 1) (Lloyd et al., 1996; Miller et al., 1997; Volkmann & Baluška, 1999).

While cortical actin can colocalize with MTs when viewed by immunofluorescence, this does not prove that interactions occur, as such small distances are not resolvable with light microscopy. However, studies with electron microscopy show that physical interactions can occur (section 2.1) as do experiments with agents directed against actin or MTs (sections 2.3, 2.4). Significantly, actin's ability to interact with cortical MTs, known to control cell expansion and plant form by orienting cellulose microfibrils (Giddings & Staehelin, 1991), suggests how responses to external stimuli that are mediated by actin might modify plant morphogenesis.

1. INTERACTIONS BETWEEN CORTICAL ACTIN AND THE PLASMA MEMBRANE

Until recently, the few reports of interactions between the cortical actin cytoskeleton and the plasma membrane were indirect, neither visualizing binding nor determining its nature. Despite immunological identification of actin associated with purified plasma membrane vesicles (Faraday & Spanswick, 1993; Sonesson & Widdell, 1993; Cox & Muday, 1994; Janßen et al., 1996; Butler et al., 1998), despite this actin representing different isoforms of actin compared to cytoplasmic fractions (Janßen et al., 1996), and despite vesicles containing a spectrin-like protein (Faraday & Spanswick, 1993), actin organization in plasma membrane vesicles has not been reported. Similarly, cytochalasin treatments demonstrate that actin can impart tension on the plasma membrane (Hahne & Hoffman, 1984), and result in increased current through a plasma membrane potassium channel (Hwang et al., 1997). Further indirect evidence for actin interactions with the plasma membrane comes from the ability of peptide sequences containing the RGD motif to modify actin-based cytoplasmic streaming (Wayne et al., 1992; Ryu et al., 1997), which, from analogy to animal systems (Clark & Brugge, 1995; Ruoslahti, 1996), suggests that actin might interact with the extracellular matrix through integrin-like proteins in the plasma membrane.

1.1 Actin is retained on membrane ghosts

Electron microscopy studies of MTs associated with the inner face of washed plasma membranes patches (membrane ghosts) may have directly visualized plasma membrane-associated actin, as the presence of occasional, unidentified thin filaments was reported (Doohan & Palevitz, 1980; van der Valk et al., 1980). Similar observations were made in whole cells by dry cleaving (Traas, 1984) and in cross sections (Ding et al., 1991). Kobayashi (1996) modified the membrane ghost method and used labeled-phalloidin to visualize membrane-associated actin of *Zinnia* mesophyll cells during tracheary element induction. Further modifications visualized randomly-organized actin and MTs associated with the plasma membranes of tobacco BY-2 cells (Fig. 1A) (Collings et al., 1998, 1999) and maize coleoptile cells (our unpublished data). Actin's interactions with the plasma membrane are independent of MTs, as actin remains on ghosts when MTs depolymerize in propyzamide or during cell division (Fig. 1B) (Collings et al., 1998, 1999).

Membrane ghosts allow testing of the membrane-associated cytoskeleton. Kobayashi (1996) demonstrated that high free-calcium does not destabilize membrane-associated actin and, following similar experiments conducted with MTs (Sonobe & Takahashi, 1994), showed that fluorescently-tagged rabbit muscle actin can bind to ghosts from which the plant actin is stripped. However, for maximizing the potential of the ghost method, protocols that avoid the protein cross-linker MBS and allow immunolocalization may prove more useful. Such methods confirm that calcium does not remove membrane-associated actin, although MTs depolymerize (Fig. 1C), and demonstrate interactions between cortical actin and MTs (section 2.4) (Collings et al., 1998).

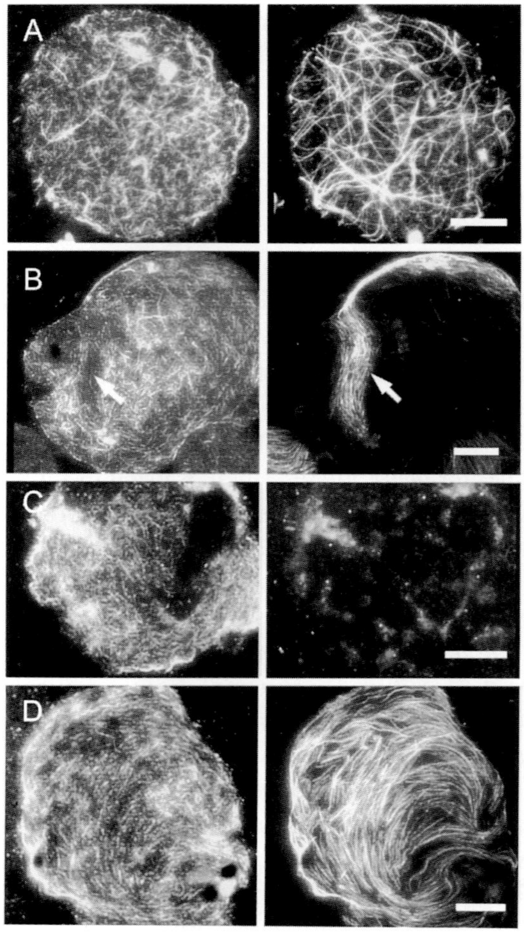

Figure 1. Ghosts retain membrane-associated actin and microtubules (MTs). Ghosts were immunolabeled with antibodies against actin (left) and tubulin (right), and viewed by fluorescence microscopy. Bars, 10 µm. (**A**) Control ghosts have randomly organized actin and MTs that exhibit only partial colocalization. (**B**) Ghosts produced from cells undergoing mitosis lack MTs, except for occasional division specific arrays such as the preprophase bands (arrow). As these ghosts retain actin, actin interactions with the plasma membrane are independent of MTs. However, actin was depleted in the preprophase band. (**C**) Ghosts from control cells were washed with a high calcium solution that depolymerized the MTs but not the cortical actin. (**D**) Ghosts produced from cells treated with 10 µM taxol prior to and during cell wall digestion retained aligned actin arrays, parallel to MTs. This demonstrates that MTs can control the organization of the cortical actin. **A**, **B** & **D** are from Collings et al., (1999) with permission from the Japanese Society of Plant Physiologists; **C** is from Collings et al., (1998) with permission from the American Society of Plant Physiologists.

While only limited use has been made of the ghosts for probing cortical actin, two experiments conducted on cortical MTs demonstrate the method's considerable power. Microtubule instability in high calcium (Fig. 1C) is caused by calmodulin. Low calcium washes prior to the high calcium remove calmodulin desensitizing the MTs to calcium, but the re-addition of exogenous calmodulin re-sensitizes the MTs (Cyr, 1991). Ghosts also demonstrate that the ATP-dependent loss of MTs from preprophase bands on ghosts can be blocked by the kinase inhibitor staurosporine (Katsuta & Shibaoka, 1992). Interestingly, while actin and MTs colocalize in preprophase bands in whole cells, actin only rarely colocalizes with MTs in preprophase bands on ghosts, although actin occurs elsewhere on such ghosts (Fig. 1B) (Collings et al., 1999). This difference has not been investigated, but it does suggest that the presence or absence of cortical MTs can modify the stability of cortical actin.

1.2 Actin's functions at the plasma membrane

Actin-binding proteins (ABPs) mediate the binding of actin to the plasma membrane. In animal cells, extensive membrane-bound actin arrays have been dissected at the molecular, genetic and biochemical levels, identifying numerous ABPs. However, these actin arrays vary from organism to organism and from cell type to cell type, with different casts of ABPs being involved in each array (Hitt & Luna, 1994). As the organization of cortical actin in plants is different to animal cells, binding of plant actin to the plasma membrane might be mediated in different ways, and by different proteins. Most known plant ABPs (profilin, fimbrin, actin-depolymerization factor, most myosins) have not been thoroughly tested for the ability to bind membranes, although myosin VIII localizes near the plasma membrane (Reichelt et al., 1999). There is, however, immunological evidence for a plant spectrin-like protein (Michaud et al., 1991; de Ruijter et al., 1998) associated with the plasma membrane (Faraday & Spanswick, 1993), although neither sequence data nor evidence for actin-binding properties yet exist. Advances have, however, been made in identifying ABPs at the plasma membrane. F-actin overlay procedures have been used to isolate a 125-kDa protein in zucchini plasma membranes that has the biochemical characteristics of an ABP (Peterman & Luna, 1999; Kaye Peterman, Wellesley College, pers. comm.).

Actin associated with the plasma membrane may play numerous roles, but it is unlikely that cytoplasmic streaming is one of these. Observation of streaming in cells microinjected with fluorescent-phalloidin confirms that longitudinal streaming patterns match the longitudinal organization of subcortical actin bundles, unlike the transverse distribution of cortical actin (Cleary, 1995). Cortical actin may, however, help anchor the subcortical actin bundles. At sites where transvacuolar strands meet the cortex of protoplasts, slight indentations in the plasma membrane reveal tension from the strands pulling on the membrane. Cytochalasin treatments release the tension, demonstrating that subcortical actin bundles are anchored to the membrane (Hahne & Hoffman, 1984). In *Vallisneria* mesophyll cells, the longitudinal pattern of cytoplasmic streaming is sensitive to proteases in the external solution, suggesting that some transmembrane protein links the extracellular matrix to the subcortical actin bundles (Ryu et al., 1995). Provocatively, application of RGD-containing peptides that inhibit integrin activity mimic protease treatments suggesting that integrins contribute to the link between the cell wall and actin (Ryu et al., 1997).

In animal cells, integrins mediate interactions between the extracellular matrix and the plasma membrane-associated actin cytoskeleton, forming a vital link in numerous signal transduction pathways. Research into integrin-mediated signaling often uses peptides containing the RGD motif that mimic extracellular matrix proteins by binding to integrins and consequently disrupting normal signaling (Clark & Brugge, 1995; Ruoslahti, 1996). Genes for integrin subunits have not been documented in plants, but evidence indicates the presence of integrin-like proteins. Antibodies raised against animal integrins detect several proteins in plant extracts (Gens et al., 1996; Canut et al., 1998; Labouré et al., 1999), while RGD-containing peptides modify the growth of callus cultures (Schindler et al., 1989; Labouré et al 1999), aggregate isolated protoplasts (Canut et al., 1998), alter cytoplasmic streaming (Wayne et al., 1992; Ryu et al., 1997), and fragment cortical MTs in protoplasts (Barthou et al., 1999). While Barthou et al. proposed a direct interaction between integrins and MTs to account for this result, evidence suggests that cortical actin interacts with and stabilizes cortical MT arrays (section 2.3) meaning that interactions between integrin-like proteins and MTs need not be direct.

Cortical actin may function in the organization of the cortical endoplasmic reticulum (ER) and Golgi apparatus, and in the delivery of secretory vesicles to the plasma membrane. Plant ER takes several forms which interact with actin in different ways. Cytochalasin inhibits the streaming of subcortical strands of ER, but only slightly modifies the organization of the cortical ER array, a stable polygonal network adjacent to the plasma membrane (Allen & Brown, 1988; Quader et al., 1989; Boevink et al., 1998). However, cytochalasin treatments disrupt the movement and organization of plant Golgi (Satiat-Jeunemaitre et al., 1996; Boevink et al., 1998; Nebenführ et al., 1999) and block vesicle secretion (Mollenhauer & Morré, 1976; Vaughan & Vaughn, 1987). Some clarification of actin's role in the secretory pathway come from studies with Golgi-targeted GFP. Microinjection of rhodamine-phalloidin into leaf epidermal cells of transformed tobacco shows that Golgi moves along actin associated with the cortical ER network (Boevink et al., 1998), although the relative roles of cortical arrays and subcortical bundles in these processes need to be determined. Similarly, the movement of Golgi in transformed tobacco BY-2 cells follows the same tracks as general cytoplasmic streaming, although the rate of Golgi movement increases slightly on MT depolymerization (Nebenführ et al., 1999).

Other ABPs may assist in functional interactions with the plasma membrane. Annexins are a family of calcium- and phospholipid-binding proteins that function in a variety of roles in animal cells, and several annexins bind actin (Gerke & Moss, 1997). Annexins have also been characterized from plants (Clark & Roux, 1995), and one of them, p35, acts with calcium to modulate exocytosis from maize root cap protoplasts (Carroll et al., 1998). Plant annexins p34 and p35 also bind specifically to F-actin (Calvert et al., 1996), and are enriched during F-actin affinity chromatography of plasma membrane vesicle extracts (Hu et al., in prep.). These observations lead to speculation that actin-binding annexins are involved in the calcium-dependent regulation of Golgi-mediated secretion of myosin-coated vesicles.

Cortical actin may play numerous other roles when bound to the plasma membrane. While some of these functions such as pathogen defense, wound healing, and signaling during root hair nodulation are dealt with elsewhere in this volume, the possible role of actin in cell polarity determination will be discussed here. Actin distribution in elongating interphase cells is polarized, with cortical actin on longitudinal walls being transverse but with bundles typically being longitudinal. However, the organization of cortical actin adjacent to the transverse walls at the ends of cells is less well understood. As any expansive growth in this wall is unpolarized, there is no requirement for transverse MTs and transverse actin. Cortical actin adjacent to the transverse walls may, however, be important in organizing bundle arrays as the anchoring of longitudinal bundles through the plasma membrane to the end wall is sensitive to proteases and RGD-containing peptides (Ryu et al., 1995; 1997). RGD-modulation of cytoplasmic streaming in *Chara* is also mediated by end walls (Wayne et al., 1992). Transverse walls of newly formed and young cells accumulate both actin, and plant myosin VIII. As this myosin occurs at only low levels in the cytoplasm and in actively expanding walls (Reichelt et al., 1999), it may function in anchoring longitudinal actin bundles, rather than in vesicle trafficking or exocytosis.

Membrane-associated actin in animal cells can delineate specific membrane domains, corralling proteins including membrane channels and pumps into precise locations (Kasumi & Sako, 1996). Thus, differences in actin between the sides and ends of a plant cell could translate into functional differences. The auxin efflux carrier distributes to the basal plasma membrane in vascular cells of *Arabidopsis* stem inflorescences (Gälweiler et al 1998). As the NPA-binding protein that regulates the auxin efflux carrier can bind actin (Cox & Muday, 1994; Butler et al., 1998; Muday, this volume), it is possible that the restriction of the efflux carrier to end walls is controlled by interactions with the cortical actin cytoskeleton. However, spatial control of the distributions of the actin-regulated potassium channel of guard cells (Hwang et al., 1997) and of other unknown actin-interacting channels, remain to be determined.

2. INTERACTIONS BETWEEN CORTICAL ACTIN AND MICROTUBULES

Studies of living animal cells have shown that interactions between actin, MTs, intermediate filaments and their associated proteins control cell dynamics and provide a structural basis for cell integrity (Gavin, 1997). The cortical actin cytoskeleton plays the major role in shape determination, with actin reorganization modulated by extracellular interactions and microtubule-dependent activities (Small et al., 1999). For example, actin polymerization controls the formation of lamellipodia and filopodia during cell motility: MT polymerization/depolymerization dynamics possibly control actin dynamics through the activation of the Rac and Rho signaling pathways (Waterman-Storer & Salmon, 1999). Similarly, the interaction of actin, MTs and intermediate filaments can provide a structural basis for cell integrity (the tensegrity principle), conveying mechanical forces through cells (Ingber, 1997).

There is now strong evidence for functional interactions between actin and MTs in plants. While this research has primarily involved drug studies and observation of fixed tissue, recent molecular studies show promise for studying these interactions in living cells.

2.1 Colocalization of actin and MTs

Actin and MTs colocalize extensively in plant cells. Electron microscopy has shown that thin filaments accompany many cortical MTs (Lancelle et al., 1987; Ding et al., 1991; Lancelle & Hepler, 1991; Tominaga et al., 1997; other references cited therein). Immunogold (Lancelle & Hepler, 1991) and heavy meromyosin labeling (Tominaga et al., 1997) demonstrate that these filaments are actin. Actin filaments lie ~10–15 nm away from the MTs, and are linked by cross-bridges (Lancelle et al., 1987; Ding et al., 1991). Similar structures form *in vitro* through the copolymerization of animal actin, tubulin and microtubule-associated proteins (MAPs) (Griffith & Pollard, 1978).

Numerous immunofluorescence studies have also demonstrated extensive colocalization of transverse actin and MTs (eg Katsuta et al., 1990).

2.2 Drug studies demonstrate interactions

The traditional approach to confirming interactions between colocalized cortical actin and MTs uses a range of pharmacological agents specific for either actin or MTs. Typically, if the disruption of MTs modifies actin organization or function, or *vice versa*, then this is evidence that the actin and MTs interact. Two different classes of drugs modify MTs properties; while taxol stabilizes MTs, various herbicides (oryzalin, propyzamide, cremart) and the alkaloid colchicine bind to tubulin dimers preventing polymerization, and drive the tubulin dimer/microtubule equilibrium in favour of depolymerization. Similarly, three classes of drugs modify actin's properties: the peptide phalloidin and the macrolide jasplakinolide stabilize filaments, other macrolides such as latrunculin bind to actin monomers preventing polymerization, forcing the G/F-actin equilibrium in favour of depolymerization, while cytochalasin caps the barbed-end of filaments, disrupting actin filaments and function. The action of cytochalasin on plant actin is, however, complicated. The cortical array is often disrupted at cytochalasin concentrations that do not inhibit cytoplasmic streaming or disrupt bundled actin, suggesting that the cortical array is a more dynamic structure (Wernicke & Jung, 1992; Collings et al., 1995). Furthermore, cytochalasin and latrunculin can exhibit subtly different effects on actin organization and function (Gibbon et al., 1999; Mathur et al., 1999).

Changes induced in the plant cytoskeleton during inhibitor experiments are often subtle and not readily detectable. For example, actin disruption with cytochalasin does not generally destabilize MT organization. As described in the following sections, however, cytochalasin often prevents the reorganization of MTs that normally occurs during cell growth. Similarly, MT disruption does not generally modify the actin cytoskeleton or cytoplasmic streaming, but will often make the actin more sensitive to disruption with cytochalasin. These results indicate that the actin and MT networks are intricately and closely connected, as they are in animal cells.

2.3 Actin can control MT organization

Numerous systems exist where actin disruption with cytochalasin inhibits MT re-organization. In developing *Zinnia* tracheary elements, re-orientation of MTs and actin from longitudinal to transverse results in MT bundles arranged into reticulate arrays adjacent to the plasma membrane, with actin patches between the MTs. Cytochalasin inhibits MT re-orientation to transverse, although the formation of MT bundles proceeds normally giving longitudinally-oriented bundles (Kobayashi et al., 1988). In *Allium*, where MTs and actin colocalize in the preprophase band, cytochalasin prevents the narrowing of the MT preprophase band that occurs as the cell approaches prophase (Mineyuki & Palevitz, 1990; Eleftheriou & Palevitz, 1992). In cotton hairs that rapidly elongate by diffuse growth, the maintenance of transverse MTs requires the constant polymerization of new MTs. Long term treatments with low cytochalasin concentrations eventually cause the MT arrays to re-orient to oblique (Seagull, 1990). During tip growth in the protonemata of the fern *Adiantum*, the circular array of transverse MTs that tracks immediately behind the growing tip is destabilized by cytochalasin (Kadota & Wada, 1992). In developing wheat mesophyll cells, where specific MT re-arrangements are necessary for the development of wall lobes, low concentrations of cytochalasin that disrupt only cortical actin and not the subcortical bundles, inhibit MT re-organization and lobe growth (Wernicke & Jung, 1992). In synchronized tobacco BY-2 cells, the high concentrations of cytochalasin required to disrupt actin prevent the re-establishment of normal MTs, including transverse cortical arrays, following cell division (Hasezawa et al., 1998). And in the outer epidermis of azuki bean hypocotyls, cytochalasin modifies the cyclic re-orientation of the cortical MTs that results in poly-lamellate wall deposition to favour longitudinally-oriented MTs and hence reduced cell elongation (Takesue & Shibaoka, 1998).

These observations suggest that cortical actin helps stabilize transverse cortical MT arrays. In all these cases, cytochalasin disrupts changes in MT configurations, rather than causing rapid changes to existing, stable arrays, although the long term, low cytochalasin treatments show that regular transverse arrays and directed cell expansion can eventually be disrupted (Seagull, 1990). Thus, these results are consistent with studies demonstrating that actin disruption with either cytochalasin (Thimann et al., 1992; Baskin & Bivens, 1995) or butanedione monoxime, a putative myosin inhibitor (Baskin & Bivens, 1995), reduce the rate of cell elongation.

2.4 Microtubules can control actin organization

Cortical actin organization may also depend on MTs. Microtubule depolymerization results in irregular changes to the actin cytoskeleton and the loss of transverse cortical actin in carrot suspension cells (Traas et al., 1987), modifies the patterning of cytoplasmic streaming in onion epidermal cells (Allen & Brown, 1988), randomizes cortical actin in developing tracheary elements of *Zinnia* (Kobayashi et al., 1988) and makes actin bundles in rye root tip cells more sensitive to disruption by cold, although

MT stabilization with taxol increases bundle stability (Chu et al., 1993). Membrane ghosts have also been used to demonstrate actin's dependence on MTs for alignment in the same tobacco BY-2 system that Hasezawa et al., (1998) used to demonstrate dependence of MTs upon actin for organization. Elongating BY-2 cells have transversely-aligned cortical MTs and actin (Katsuta et al., 1990; Collings et al., 1998), but these randomize during digestion of the cell wall so that ghosts retain random actin and MTs (Fig. 1A). However, when MTs are stabilized with taxol prior to and during wall digestion, ordered MT arrays occur on ghosts and actin randomization is prevented, so that actin on ghosts remains aligned to the MTs (Fig. 1D) (Collings et al., 1998).

Microtubule disruption can also make actin more sensitive to cytochalasin, even if MT depolymerization itself does not visibly affect actin. In the alga *Nitella*, MT depolymerization leaves cortical actin unaffected and does not slow cytoplasmic streaming, and actin disruption does not modify the cortical MTs (but see Foissner & Wasteneys, this volume). However, MT depolymerization makes cells ten-fold more sensitive to cytochalasin when both inhibitors are applied concurrently (Wasteneys & Williamson, 1991; Collings et al., 1996). Because a layer of chloroplasts separates MTs from the actin bundles that generate cytoplasmic streaming in *Nitella* (Wasteneys & Williamson, 1991), the sensitization effect may result from an increase in the concentration of MAPs that follows MT depolymerization, which could interact with and destabilize the actin bundles (Collings et al., 1996). As discussed in section 2.5, several animal MAPs can interact with actin, although similar proteins have not yet been found in plants.

A similar effect occurs in the root hairs of *Hydrocharis,* where actin and MTs do colocalize. Microtubule disruption with propyzamide affects neither actin organization nor cytoplasmic streaming, while actin disruption does not affect MT organization. However, in the presence of propyzamide, cytochalasin removal results in the recovery of aberrant, random actin arrays that do not support cytoplasmic streaming. Normal functioning of actin can only return when MTs are allowed to repolymerize (Tominaga et al., 1997). It is possible that MAP-mediated destabilization of actin might also explain these observations.

2.5 Actin-MT interactions mediated by ABPs and MAPs

Numerous proteins can link the actin of animal cells to MT either directly, or indirectly in a variety of configurations. For example, crude solutions of exogenous MAPs, actin and tubulin copolymerize *in vitro* (Griffith & Pollard, 1978; Sider et al., 1999), the microtubule actin cross-linking factor (MACF) colocalizes with actin and MT *in vivo* and stabilizes actin and MT *in vitro* (Leung et al., 1999), the MAP CLIP-190 binds to a *Drosophila* myosin VI (Lantz & Miller, 1998), and the microtubule motor kinesin can bind to rat myosin V (Huang et al., 1999). In plants, however, the best candidate for a protein that can interact with both actin and MTs is the protein synthesis component EF1α. EF1α bundles actin in *Dicytostelium* (Yang et al 1990), can bundle actin to MTs in *Physarum* (Itano & Hatano 1991), and can bind γ-tubulin (Marchesi & Ngo 1993). Biochemical evidence suggests that plant EF1α is both a MT-bundling (Moore et al 1998) and an actin-binding protein (Yang et al 1993). In tobacco BY-2 cells, EF1α colocalizes to both actin and MTs, leading to speculation about its role in mediating actin-MT interactions (Hasezawa et al 1998). Unfortunately, no further candidates for proteins mediating actin-MT interactions are known from plants, although with the rapid identification of new plant ABPs and MAPs, such proteins will undoubtedly be discovered.

2.6 An actin-MT feedback system?

To account for the observations in pharmacological studies, we propose that a system exists where transversely-aligned cortical actin and MTs interact with and stabilize each other, and that both actin and MT are necessary during cytoskeletal re-organizations that occur during cell elongation. Furthermore, we propose that this type of feedback system might be common to most plant cells showing diffuse growth, and might form a necessary component in the control of plant morphogenesis. The current literature is, however, insufficient to demonstrate such a general phenomenon, and indeed examples exist where actin disruption with cytochalasin does not appear to inhibit MT reorganization (Hush & Overall, 1992). However, because the cytoskeleton is highly dynamic, it is impractical to confirm this hypothesis through immunofluorescence microscopy and pharmacological approaches. Instead, molecular anlayses of living cells will be required to demonstrate how wide-spread actin-MT interactions are in plant cells.

2.7 Molecular approaches to actin-MT interactions

In recent years, significant progress in understanding actin-MT interactions in animals has been made by investigating living cells (Small et al., 1999; Waterman-Storer & Salmon, 1999). To determine how prevalent actin-MT interactions might be in plants, it will be necessary for cytoskeletal organization to be viewed in living cells, in conjunction with the identification and analysis of further cytoskeletal mutants, and of further ABPs and MAPs.

Actin, and thus actin-MT interactions, can be studied in living cells with microinjected fluorescent-phalloidin and GFP-talin (Kost et al., 1998; Mathur et al., 1999). Microinjected actin-depolymerization factor disrupts streaming in *Tradescantia* stamen hairs, with subsequent rhodamine-phalloidin injections showing that longitudinal actin bundles reorganize into transverse cortical bundles that are parallel to the cortical MTs, implying that MTs can control actin organization (Hussey et al., 1998). Microinjected fluorescent-tubulin, GFP-tubulin (Ueda et al., 1999) and GFP-MAP4 contructs (Marc et al., 1998) also demonstrate MT dynamics in living cells. While gravity-induced cortical MT reorganization occurs in tubulin-injected maize coleoptiles (Himmelspach et al., 1999), actin's involvement in this has not been investigated.

While mutant analysis is potentially the most powerful way to investigate cortical actin's interactions, its application so far has been limited, possibly by difficulties in isolating appropriate mutants and in interpreting data in the absence of gene sequences. For example, no reports exist of actin organization in the *ton* mutant of *Arabidopsis*, where interphase MTs are randomized and where the preprophase band is non-existent (Traas et al., 1995). However, cell elongation in the *Yin-Yang* mutant of rice, isolated because of its partial resistance to a MT-depolymerizing herbicide, is more sensitive to cytochalasin than wild-type plants (Wang & Nick, 1998), while several maize meiotic mutants that show division-specific disruptions to MT arrays show accompanying disruptions in their actin cytoskeletons (Staiger & Cande, 1991). The identities of the *Yin-Yang* and maize gene products, and how they relate to actin and MTs, remain unknown.

The *Arabidopsis* trichome provides a potentially interesting living system in which to investigate of actin-MT interactions (see also Szymanski, this volume), as live-cell imaging can be coupled to mutational analysis (Mathur et al., 1999). Application of cytochalasin to developing trichomes results in distorted growth, similar to several known mutations that have abnormal actin organization as observed by immunofluorescence (Szymanski et al., 1999) and GFP-talin (Mathur et al., 1999), while MT disruption inhibits trichome polarization but does not apparently modify actin organization (Mathur et al., 1999; Szymanski et al., 1999). Unfortunately, MT organization has not been reported in *Arabidopsis* trichomes, but because these cells undergo diffuse growth during much of their development, it might be expected that they have transverse MT arrays similar to those found in cotton (Seagull, 1990). With no interactions between actin and MT being apparent in trichomes, these observations stand in contrast to the various pharmacological studies discussed in previous sections, and the proposed feedback model. Only further careful observations in trichomes and other cell types can clarify this situation.

ACKNOWLEDGMENTS

The authors thank Ann Cleary, Gloria Muday, Kaye Peterman and Geoff Wasteneys for discussions, and acknowledge support from the Japanese Society for the Promotion of Science (DAC), NASA NSCORT grant NAGW-4984 (DAC & NSA), and NCARS research grants (NSA).

REFERENCES

Allen NS and Brown DT (1988) Dynamics of the endoplasmic reticulum in living onion epidermal cells in relation to microtubules, microfilaments and intracellular particle movement. Cell Motil Cytoskel 10: 153-163

Barthou H, Petitprez M, Brière C, Souvré A and Alibert G (1999) RGD-mediated membrane-matrix adhesion triggers agarose-induced embryoid formation in sunflower protoplasts. Protoplasma 206: 143-151

Baskin TI and Bivens NJ (1995) Stimulation of radial expansion in *Arabidopsis* roots by inhibitors of actomyosin and vesicle secretion but not by various inhibitors of metabolism. Planta 197: 514-521

Boevink P, Oparka K, Santa Cruz S, Martin B, Betteridge A and Hawes C (1998) Stacks on tracks: the plant Golgi apparatus traffics on an actin/ER network. Plant J 15: 441-447

Butler JH, Hu S, Brady S, Dixon MW and Muday GK (1998) *In vitro* and *in vivo* evidence for actin association of the naphthylphthalamic acid-binding protein from zucchini hypocotyls. Plant J 13: 291-301

Calvert CM, Gant SJ and Bowles DJ (1996) Tomato annexins p34 and p35 bind to F-actin and display nucleotide phosphodiesterase activity inhibited by phospholipid binding. Plant Cell 8: 333-342

Canut H, Carrasco A, Galaud J, Cassan C, Bouyssou H, Vita N, Ferrara P and Pont-Lezica R (1998) High affinity RGD-binding sites at the plasma membrane of *Arabidopsis thaliana* links the cell wall. Plant J 16: 63-71

Carroll AD, Moyen C, van Kesteren P, Tooke F, Battey NH and Brownlee C (1998) Ca^{2+}, annexins, and GTP modulate exocytosis from maize root cap protoplasts. Plant Cell 10: 1267-1276

Chu B, Kerr P and Carter JV (1993) Stabilizing microtubules with taxol increases microfilament stability during freezing of rye root tips. Plant Cell Environ 16: 883-889

Clark EA and Brugge JS (1995) Integrins and signal transduction pathways: the road taken. Science 268: 233-239

Clark GB and Roux SJ (1995) Annexins of plant cells. Plant Physiol 109: 1133-1139

Cleary AL (1995) F-actin redistributions at the dividing site in living *Tradescantia* stomatal complexes as revealed by microinjection of rhodamine-phalloidin. Protoplasma 185: 152-165

Collings DA, Asada T, Allen NS and Shibaoka H (1998) Plasma membrane-associated actin in Bright Yellow 2 tobacco cells: Evidence for interaction with microtubules. Plant Physiol 118: 917-928

Collings DA, Asada T and Shibaoka H (1999) Plasma membrane ghosts form differently when produced from microtubule-free tobacco BY-2 cells. Plant Cell Physiol 40: 36-46

Collings DA, Wasteneys GO and Williamson RE (1995) Cytochalasin rearranges cortical actin of the alga *Nitella* into short stable rods. Plant Cell Physiol 36: 765-772

Collings DA, Wasteneys GO and Williamson RE (1996) Actin-microtubule interactions in the alga *Nitella:* analysis of the mechanism by which microtubule depolymerization potentiates cytochalasin's effects on streaming. Protoplasma 191: 178-190

Cox DN and Muday GK (1994) NPA binding activity is peripheral to the plasma membrane and is associated with the cytoskeleton. Plant Cell 6: 1941-1953

Cyr RJ (1991) Calcium/calmodulin affects microtubule stability in lysed protoplasts. J Cell Sci 100: 311-317

de Ruijter NCA, Rook MB, Bisseling T and Emons AMC (1998) Lipochito-oligosaccharides re-initiate root hair tip growth in *Vicia sativa* with high calcium and spectrin-like antigen at the tip. Plant J 13: 341-350

Ding B, Turgeon R and Parthasarathy MV (1991) Microfilament organization and distribution in freeze substituted tobacco plant tissues. Protoplasma 165: 96-105

Doohan ME and Palevitz BA (1980) Microtubules and coated vesicles in guard-cell protoplasts of *Allium cepa* L. Planta 149: 389-401

Eleftheriou EP and Palevitz BA (1992) The effect of cytochalasin D on preprophase band organisation in root tip cells of *Allium.* J Cell Sci 103: 989-998

Faraday CD and Spanswick RM (1993) Evidence for a membrane skeleton in higher plants. FEBS Lett 318: 313-316

Gavin RH (1997) Microtubule-microfilament synergy in the cytoskeleton. Int Rev Cytol 173: 207-242

Gälweiler L, Guan C, Müller A, Wisman E, Mendgen K, Yephremov A and Palme K (1998) Regulation of polar auxin transport by AtPIN1 in *Arabidopsis* vascular tissue. Science 282: 2226-2230

Gens JS, Reuzeau C, Doolittle KW, McNally JG and Pickard BG (1996) Covisualization by computational optical-sectioning microscopy of integrin and associated proteins at the cell membrane of living onion protoplasts. Protoplasma 194: 215-230

Giddings Jr TH and Staehelin LA (1991) Microtubule-mediated control of microfibril deposition: a re-examination of the hypothesis. In: Lloyd CW (ed) The Cytoskeletal Basis of Plant Growth and Form. London: Academic Press, pp 85-99

Gerke V and Moss SE (1997) Annexins and membrane dynamics. Biochim Biophys Acta 1357: 129-154

Gibbon BC, Kovar DR and Staiger CJ (1999) Latrunculin B has different effects on pollen germination and tube growth. Plant Cell 11: 2349-2363

Griffith LM and Pollard TD (1978) Evidence for actin filament-microtubule interaction mediated by microtubule-associated proteins. J Cell Biol 78: 958-965

Hahne G and Hoffmann F (1984) The effect of laser microsurgery on cytoplasmic strands and cytoplasmic streaming in isolated plant protoplasts. Eur J Cell Biol 33: 175-179

Hasezawa S, Sano T and Nagata T (1998) The role of microfilaments in the reoganization and orientation of microtubules during the cell cycle transition from M phase to G_1 phase in tobacco BY-2 cells. Protoplasma 202: 105-114

Himmelspach R, Wymer CL, Lloyd CW and Nick P (1999) Gravity-induced reorientation of cortical microtubules observed *in vivo.* Plant J 18: 449-453

Hitt AL and Luna EJ (1994) Membrane interactions with the actin cytoskeleton. Curr Opin Cell Biol 6: 120-130

Huang J, Brady ST, Richards BW, Stenoien D, Resau JH, Copeland NG and Jenkins NA (1999) Direct interaction of microtubule- and actin-based transport motors. Nature 397: 267-270

Hush JM and Overall RL (1992) Re-oreintation of cortical F-actin is not necessary for wound-induced microtubule re-orientation and cell polarity establishment. Protoplasma 169: 97-106

Hussey PJ, Yuan M, Calder G, Khan S and Lloyd CW (1998) Microinjection of pollen-specific actin-depolymerizing factor, ZmADF1, reorientates F-actin strands in *Tradescantia* stamen hair cells. Plant J 14: 353-357

Hwang J, Suh S, Yi H, Kim J and Lee Y (1997) Actin filaments modulate both stomatal opening and inward K^+-channel activities in guard cells of *Vicia faba* L. Plant Physiol 115: 335-342

Ingber DE (1997) Tensegrity: the architectural basis of cellular mechanotransduction. Annu Rev Physiol 59: 575-599

Itano N and Hatano S (1991) F-actin bundling protein from *Physarum polycephalum*: purification and its capacity for co-bundling of actin filaments and microtubules. Cell Motil Cytoskel 19: 244-254

Janßen M, Hunte C, Schulz M and Schnabl H (1996) Tissue specification and intracellular distribution of actin isoforms in *Vicia faba* L. Protoplasma 191: 158-163

Kadota A and Wada M (1992) The circular arrangement of cortical microtubules around the subapex of tip-growing fern protonemata is sensitive to cytochalasin B. Plant Cell Physiol 33: 99-102

Katsuta J, Hashiguchi Y and Shibaoka H (1992) The role of the cytoskeleton in positioning of the nucleus is premitotic tobacco BY-2 cells. J Cell Sci 95: 413-422

Katsuta J and Shibaoka H (1992) Inhibition by kinase inhibitors of the development and the disappearance of the preprophase band of microtubules in tobacco BY-2 cells. J Cell Sci 103: 397-405

Kobayashi H (1996) Changes in the relationship between actin filaments and the plasma membrane in cultured *Zinnia* cells during tracheary element differentiation investigated by using plasma membrane ghosts. J Plant Res 109: 61-65

Kobayashi H, Fukuda H and Shibaoka H (1988) Interrelation between the spatial disposition of actin filaments and microtubules during the differentiation of tracheary elements in cultured *Zinnia* cells. Protoplasma 143: 29-37

Kost B, Spielhofer P and Chua N-H (1998) A GFP-mouse talin fusion protein labels plant actin filaments *in vivo* and visualizes the actin cytoskeleton in growing pollen tubes. Plant J 16: 393-401

Kusumi A and Sako Y (1996) Cell surface organization by the membrane skeleton. Curr Opin Cell Biol 8: 566-574

Labouré A, Faik A, Mandaron P and Falconet D (1999) RGD-dependent growth of maize calluses and immunodetection of an integrin-like protein. FEBS Lett 442: 123-128

Lancelle SA, Cresti M and Hepler PK (1987) Ultrastructure of the cytoskeleton in freeze-substituted pollen tubes of *Nicotiana alata*. Protoplasma 140: 141-150

Lancelle SA and Hepler PK (1991) Association of actin with cortical microtubules revealed by immunogold localization in *Nicotiana* pollen tubes. Protoplasma 165: 167-172

Lantz VA and Miller KG (1998) A class VI unconventional myosin is associated with a homologue of a microtubule-binding protein, cytoplasmic linker protein-170, in neurons and at the posterior pole of *Drosophila* embryos. J Cell Biol 140: 897-910

Lloyd CW, Drøbak BK, Dove SK and Staiger CJ (1996) Interactions between the plasma membrane and the cytoskeleton in plants. In: Smallwood M, Knox JP, Bowles DJ (eds) Membranes: Specialized Functions in Plants. Oxford: BIOS Scientific, pp 1-20

Leung CL, Sun D, Zheng M, Knowles DR and Liem RKH (1999) Microtubule actin cross-linking factor (MACF): a hybrid of dystonin and dystrophin that can interact with the actin and microtubule cytoskeletons. J Cell Biol 147: 1275-1285

Marc J, Granger CL, Brincat J, Fisher DD, Kao T, McGrubbin AG and Cyr RJ (1998) A GFP-MAP4 reporter gene for visualizing cortical microtubule rearrangements in living epidermal cells. Plant Cell 10: 1927-1939

Marchesi VT and Ngo N (1993) In vitro assembly of multiprotein complexes containing alpha, beta and gamma tubulins, heat shock protein HSP70, and elongation factor 1α. Proc Natl Acad Sci USA 90: 3028-3032

Mathur J, Spielhofer P, Kost B and Chua N-H (1999) The actin cytoskeleton is required to elaborate and maintain spatial patterning during trichome cell morphogenesis in *Arabidopsis thaliana*. Development 126: 5559-5568

Michaud D, Guillet G, Rogers PA and Charest PM (1991) Identification of a 220 kDa membrane-associated plant cell protein immunologically related to human β-spectrin. FEBS Lett 294: 77-80

Miller DD, de Ruijter NCA and Emons AMC (1997) From signal to form: aspects of the cytoskeleton-plasma membrane-cell wall continuum in root hair tips. J Exp Bot 48: 1881-1886

Mineyuki Y and Palevitz BA (1990) Relationship between preprophase band organization, F-actin and the division site in *Allium*. J Cell Sci 97: 283-295

Mollenhauer HH and Morré DJ (1976) Cytochalasin B, but not colchicine, inhibits migration of secretory vesicles in root tips of maize. Protoplasma 87: 39-48

Moore RC, Durso NA and Cyr RJ (1998) Elongation factor-1α stabilizes microtubules in a calcium/calmodulin-dependent manner. Cell Motil Cytoskel 41: 168-180

Nebenführ A, Gallagher LA, Dunahy TG, Frohlick JA, Mazurkiewicz AM, Meehl JB and Staehelin LA (1999) Stop-and-go movements of plant Golgi stacks are mediated by the acto-myosin system. Plant Physiol 121: 1127-1141

Niggli V (1996) Membrane-cytoskeleton. In: Hesketh JE, Pryme IF (eds) The Cytoskeleton. Structure and Assembly. Greenwich: JAI Press, pp 123-168

Peterman K and Luna EJ (1999) p125, a plasma membrane-associated F-actin-binding protein. Plant Physiol 120: s71

Quader H, Hofmann A and Schnepf E (1989) Reorganization of the endoplasmic reticulum in epidermal cells of onion bulb scales after cold stress: involvement of cytoskeletal elements. Planta 177: 273-280

Reichelt S, Knight AE, Hodge TP, Baluška F, Šamaj J, Volkmann D and Kendrick-Jones J (1999) Characterization of the unconventional myosin VIII in plant cells and its localization at the post-cytokinetic cell wall. Plant J 19: 555-567

Ruoslahti E (1996) RGD and other recognition sequences for integrins. Annu Rev Cell Dev Biol 12: 697-715

Ryu J-H, Mizuno K, Takagi S and Nagai R (1997) Extracellular components implicated in the stationary organization of the actin cytoskeleton in mesophyll cells of *Vallisneria*. Plant Cell Physiol 38: 420-432

Ryu J, Takagi S and Nagai R (1995) Stationary organization of the actin cytoskeleton in *Vallisneria*: the role of stable microfilaments at the end walls. J Cell Sci 108: 1531-1539

Satiat-Jeunemaitre B, Steele C and Hawes C (1996) Golgi-membrane dynamics are cytoskeleton dependent: a study on golgi stack movement induced by brefeldin A. Protoplasma 191: 21-33

Schindler M, Meiners M and Cheresh DA (1989) RGD-dependent linkage between cell wall and plasma membrane: consequences for growth. J Cell Biol 108: 1955-1965

Seagull RW (1990) The effects of microtubule and microfilament disrupting agents on cytoskeletal arrays and wall deposition in developing cotton fibres. Protoplasma 159: 44-59

Sider JR, Mandato CA, Weber KL, Zandy AJ, Beach D, Finst RJ, Skoble J and Bement WM (1999) Direct observation of microtubule-f-actin interaction in cell free lysates. J Cell Sci 112: 1947-1956

Small JV, Clemens R and Kaverina I (1999) Functional design in the actin cytoskeleton, Curr Opin Cell Biol 11: 54-60

Sonesson A and Widell S (1993) Cytoskeleton components of inside-out and right-side-out plasma membrane vesicles from plants. Protoplasma 177: 45-52

Sonobe S and Shibaoka H (1989) Cortical fine actin filaments in higher plant cells visualized by rhodamine-phalloidin after pretreatment with m-maleimidobenzoyl N-hydroxysuccinimide ester. Protoplasma 148: 80-86

Sonobe S and Takahashi S (1994) Association of microtubules with the plasma membrane of tobacco BY-2 cells in vitro. Plant Cell Physiol 35: 451-460

Staiger CJ (2000) Signaling to the actin cytoskeleton in plants. Annu Rev Plant Physiol Plant Mol Biol 51: 257-288

Staiger CJ and Cande WZ (1991) Microfilament distribution in maize meiotic mutants correlates with microtubule organization. Plant Cell 3: 637-644

Szymanski DB, Marks DM and Wick SM (1999) Organized F-actin is essential for normal trichome morphogenesis in *Arabidopsis*. Plant Cell 11: 2331-2347

Takesue K and Shibaoka H (1998) The cyclic reorientation of cortical microtubules in epidermal cells of azuki bean epicotyls: the role of actin filaments in the progression of the cycle. Planta 205: 539-546

Thimann KV, Reese K and Nachmias VT (1992) Actin and the elongation of plant cells. Protoplasma 171: 153-166

Tominaga M, Morita K, Sonobe S, Yokota E and Shimmen T (1997) Microtubules regulate the organization of actin filaments at the cortical region in root hair cells of *Hydrocharis*. Protoplasma 199: 83-92

Traas J (1984) Visualization of the membrane bound cytoskeleton and coated pits of plant cells by means of dry cleaving. Protoplasma 119: 212-218

Traas JA, Doonan JH, Rawlins DJ, Shaw PJ, Watts J and Lloyd CW (1987) An actin network is present in the cytoplasm throughout the cell cycle of carrot cells and associates with the dividing nucleus. J Cell Biol 105: 387-395

Traas J, Bellini C, Nacry P, Kronenberger J, Bouchez D and Caboche M (1995) Normal differentiation patterns in plants lacking microtubular preprophase bands. Nature 375: 676-677

Ueda K, Matsuyama T and Hashimoto T (1999) Visualization of microtubules in living cells of transgenic *Arabidopsis thaliana*. Protoplasma 206: 201-206

van der Valk P, Rennie PJ, Connolly JA and Fowke LC (1980) Distribution of cortical microtubules in tobacco protoplasts. An immunofluorescence microscopic and ultrastructural study. Protoplasma 105: 27-43

Vaughan MA and Vaughn KC (1987) Effects of microfilament disrupters on microfilament distribution and morphology in maize root cells. Histochemistry 87: 129-137

Volkmann D and Baluška F (1999) Actin cytoskeleton in plants: from transport networks to signaling networks. Microsc Res Tech 47: 135-154

Wang Q and Nick P (1998) The auxin response of actin is altered in the rice mutant *Yin-Yang*. Protoplasma 204: 22-33

Wasteneys GO and Williamson RE (1991) Endoplasmic microtubules and nucleus-associated actin rings in *Nitella* internodal cells. Protoplasma 162: 86-98

Waterman-Storer CM and Salmon ED (1999) Positive feedback interactions between microtubule and actin dynamics during cell motility. Curr Opin Cell Biol 11: 61-67

Wayne R, Staves MP and Leopold AC (1992) The contribution of the extracellular matrix to gravisensing in characean cells. J Cell Sci 101: 611-623

Wernicke W and Jung G (1992) Role of cytoskeleton in cell shaping of developing mesophyll of wheat (*Triticum aestivum* L.). Eur J Cell Biol 57: 88-94

Yang F, Demma M, Warren V, Dharmawardhane S and Condeelis J (1990) Identification of an actin-binding protein from *Dictyostelium* as elongation factor 1α. Nature 347: 494-496

Yang W, Burkhart W, Cavallius WC and Boss WF (1993) Purification and characterization of a phosphatidylinositol 4-kinase activator in carrot cells. J Biol Chem 268: 392-398

Chapter 10

CYTOPLASMIC STREAMING: FROM FLOW TO TRACK

Franz Grolig[1] and Elisabeth S Pierson[2,3]

1 Fachbereich Biologie/Botanik, Philipps-Universität, Lahnberge, D–35032 Marburg, Germany; 2 Department of Experimental Botany, Graduate School for Experimental Plant Sciences; 3 Present address: General Instrumentation, Katholieke Universiteit Nijmegen, Toernooiveld 1, NL–6525 ED Nijmegen, The Netherlands

Key words: actin filaments, *Chara*, cytoplasmic streaming, living cell imaging, microscopy, motility, myosin, organelle track, quantitative analysis

Abstract: This chapter combines classical and recent views on cytoplasmic streaming in plants, an efficient mechanism to achieve convection in the plant cytosol. Bundles of actin filaments with an unidirectional polarity and organelle-associated myosin motors provide the molecular base for bulk flow, typical of characean internodal cells, and for independent movements of organelles. Many physiological and environmental factors influence cytoplasmic streaming, often via modulation of the free cytosolic Ca^{2+} level which in turn regulates myosin activity. Quantitative analysis of single organelle tracks recorded by high-performance digital microscopy, for example in *Spirogyra* and pollen tubes, is an effective approach to characterize organelle movements by meaningful parameters. The combination of advanced imaging techniques and force measurements will contribute to trace down the molecular mechanisms underlying cytoplasmic streaming.

1. MOTILITY AND THE ACTOMYOSIN SYSTEM IN PLANT CELLS

Motility is a pertinent feature of life and can be found at all levels of organization. The intracellular distribution and movement of macromolecules and organelles is mainly non-random, but depends on interaction between actin polymers (actin filaments = AFs) or tubulin polymers (microtubules = MTs), and motor proteins (myosins, dyneins and kinesins).

C.J. Staiger et al. (eds.), Actin: A Dynamic Framework for Multiple Plant Cell Functions, 165–190.

These proteins convert chemical energy into mechanical work in form of a step-wise, unidirectional walk (Howard, 1997; Rice et al., 1999). The polarity of MTs and AFs determines the direction of movement of the associated motor protein. With the exception of a single myosin (Wells et al., 1999), all known myosins move towards the AF plus-end (the fast-growing or 'barbed' end; for polarity of actin-based plant organelle translocation: Kersey et al., 1976; Kohno et al., 1990). Both the instantaneous and the net translocation velocities of cellular constituents (regarded as 'cargo') show high variability. Translocation and positioning may involve the (inter)action of both MT- and AF-based forces ('tensegrity' : Ingber, 1993). One of the most intriguing examples of precise positioning (where both cargo and destination are highly specified) is the premitotic relocation of the nucleus in plant cells (reviews: Wick, 1991; Grolig, 1998). Conversely, in the case of cytoplasmic streaming, the cargo is abundant and the destination little confined. Cytoplasmic streaming is supposed to have evolved primarily as a means of intracellular convection to compensate for the limited efficiency of mere diffusion to support metabolism in cells larger than 10 μm (Hocharka, 1999). Accordingly, the covered distances and intensity of cytoplasmic movements tend to be most prominent in large vacuolate and/or polarized cells. Various patterns of cytoplasmic streaming have been defined in plant and other cells: agitation, circulation, rotation, reverse fountain-like streaming and multistriate streaming (Kuroda, 1990; Kiermayer et al., 1979). Qualitative and quantitative characteristics of the cytoplasmic streaming (e.g. pattern, velocity, shape of tracks and other intricate parameters detailed below) reflect physiological and environmental conditions of the cell.

Cytoplasmic streaming in plant cells is essentially achieved by the actomyosin system, with velocities in the range from 10^{-7} to 10^{-4} m s^{-1}. In general, the movement is most vigorous at interphase, but declines markedly during mitosis (e.g. *Adiantum* protonemata: Wada et al., 1982; *Spirogyra*: Sawitzky & Grolig, 1995) in concert with reduction and reorganization of the actin cytoskeleton (Staiger & Schliwa, 1987). During interphase, in most plant cells three AF-arrays can be discerned: 1) a cortical meshwork of mostly short, probably non-bundled AFs, 2) a subcortical array of long, mostly unipolar AF-bundles, extending also within transvacuolar strands and 3) a perinuclear network (overviews in: Parthasarathy et al., 1985; Staiger & Schliwa, 1987; Sonobe & Shibaoka, 1989; other chapters of this volume). These arrays are contiguous and support translocation of particles throughout the cell. Long range translocation of organelles occurs along the subcortical and transvacuolar AF-bundles (*in vivo* observations of AF-associated motility: Hepler et al., 1993; Fig. 1). Cytoplasmic streaming in plant cells is affected by a variety of inhibitors of the actin cytoskeleton: cytochalasins (CB, CD; reviews on plant cells: Kamiya, 1986; Emons et al., 1991; Pierson & Cresti, 1992), phalloidin and phallacidins (Barak et al., 1980), mycalolide

B (Shimmen et al., 1995), the cyclodepsipeptide jasplakinolide (Holzinger & Meindl, 1997; Sawitzky et al., 1999), chondramides (Kunze et al., 1995), myrmicacin (Shimmen & Tazawa, 1985) and latrunculin (Kandasamy & Meagher, 1999; Gibbon et al., 1999). A few plant cells lack (subcortical) AF-bundles and consequently cytoplasmic streaming: e.g. root cap cells involved in graviperception (Baluška et al., 1997), the alga *Mougeotia* bearing light-steered chloroplast positioning (Grolig, 1992; Mineyuki et al., 1995) and angiosperm male reproductive cells (Palevitz & Liu, 1992).

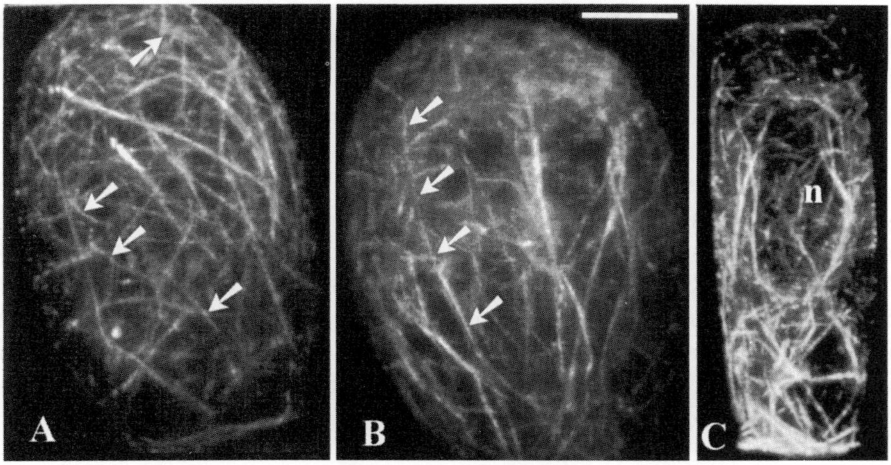

Figure 1. Actin filaments in living staminal hair cells of *Tradescantia virginiana*. (**A** and **B**) show two projections of images of a maturing cell taken at different focus levels. (**C**) shows a young cell. These staminal hair cells were loaded with rhodamine-phalloidin by microinjection and imaged with a confocal laser scanning microscope, while still exhibiting cytoplasmic streaming (method described in Valster et al., 1997). (**A**) shows the criss-cross arrays (arrows) located in the peripheral cytoplasm, (**B**) the long bundles of actin filaments in tran-svacuolar strands (arrows) and (**C**) the F-actin network around the nucleus (n). (Pierson & Emons, unpublished data). Bar: 10 µm.

The architecture and the dynamics of the actin cytoskeleton are orchestrated by a number of actin-binding (ABPs) and actin-associated proteins (non-plant systems: van Troys et al., 1999; in plants, profilins, actin-depolymerizing factors, α-actinin, spectrin, fimbrin, actin-bundling proteins of 115 KDa and 135 kDa, annexins, PIK-A49, and Rho-type GTPases: de Ruijter & Emons, 1999; other chapters in this volume). In spite of the divergence and differential expression of actin gene families in angiosperms (Meagher et al., 1999; Meagher et al., this volume) the sites of putative interaction of various actin isoforms with ABPs seems to be well conserved (Reece et al., 1992). Although AFs are obviously anchored in the cell cortex, their putative binding sites, which can be probed with heterolog F-actin

(Kobayashi, 1996), are not well known so far. Immunological and physiological evidence suggests that plant cells contain adhesion molecules (i.e. vitronectin, N-cadherin, α- and β-catenin and integrins showing the typical RGD / RYD motif). These molecules may primarily interact with intracellular membranes (Baluška et al., 1999). Postulated transmembrane adhesion proteins and focal adhesion points (review: Jauh & Lord, 1995) have not been demonstrated as yet. Myosins have been determined in plants by sequence analysis (Reichelt & Kendrick-Jones, this volume), immunodetection and -localization with antibodies raised against non-plant myosins (references in Miller et al., 1995), and purification and characterization *in vitro* (Yokota & Shimmen, this volume). AFs and MTs may corroborate intimately in organizing the cytoplasm and (indirectly) supporting organelle movement (Salitz & Schmitz, 1989; Tominaga et al., 1997; reviews: Emons et al., 1991 and references herein; Cai et al.,1996).

Extensive reviews exist on generalities (Kamiya, 1960, 1962, 1981, 1986; Kuroda, 1990; Williamson, 1992, 1993; Shimmen, 1992; Shimmen & Yokota, 1994) and specific aspects (Wagner & Grolig, 1992; Nagai, 1993; Wada et al., 1993) of cytoplasmic streaming and organelle movements in plants. In this chapter we aim to conciliate the classical and update view on bulk flow streaming in characean internodal cells together with recent insights in the movement of organelles gained from quantitative analyses of single particle tracks.

2. VIEWS ON CYTOPLASMIC STREAMING

2.1 The primer: experiments on characean internodal cells

The work of Corti (1774) on the massive translocation of particles within the giant internodal cells of the green alga *Chara* is the first known report on cytoplasmic streaming. In the early 1950s, Noburo Kamiya and Kiyoko Kuroda chose the same object for their pioneering work on cytoplasmic streaming. Their studies laid the foundations for the nowadays rapidly radiating field of research on the (plant) cytoskeleton. Because a variety of manipulations can be easily performed in these large cylindrical cells (several cm long, diameter up to 1 mm; cellular organization: Fig. 2), they still provide stimulating tools to investigate intracellular motility (not only) in plants. The *Charophyceae* are the taxonomic group closest to the green algal predecessor of the higher plants (Mattox & Stewart, 1984; Kranz et al., 1995). In view of the vital role and the high conservation of the molecular components, experimental findings on the motility machineries in these algal cells are of fundamental relevance for motility research in higher plants (Grolig, 1992; McIntosh et al., 1995; Sawitzky & Grolig, 1995).

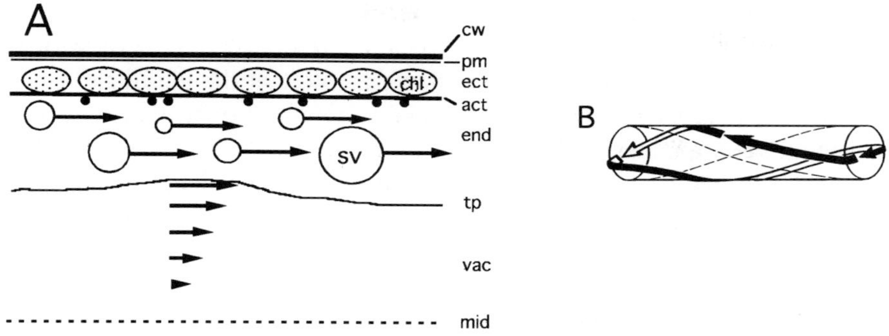

Figure 2. Organization of rotational streaming in characean internodal cells. (**A**) depicts the gross internal organization of an internodal cell. The diagram represents the upper half of a longitudinal midplane view. The cytoplasmic layer (10 to 20 μm thick) surrounding the large central vacuole (vac, drawn undersized here) can be functionally divided in endoplasm (end) and ectoplasm (ect) or cortex. The endoplasm contains many small vacuoles (sv), endoplasmic reticulum and other organelles and moves as bulk flow in a more or less coherent manner. The streaming (direction and velocity indicated by black arrows) is transmitted to the vacuole where it decreases rapidly towards zero in the middle of the cell (mid = dashed line) due to viscous coupling of an opposing stream in the lower half of the cell (not shown) (Nothnagel & Webb, 1982). The stationary ectoplasm contains a densely packed, single layer of immobile chloroplasts (chl), arranged in slightly helically oriented files parallel to the streaming. At the boundary between ecto- and endoplasm, small organelles (black dots) can be seen as moving intermittently and independently from each other along massive bundles of F-actin (act). Further details: cell wall (cw), plasma membrane (pm) and tonoplast (tp). (**B**) Streaming of the bulk endoplasm occurs as unidirectional rotation of a closed ribbon throughout the cell. Only where the edges of the ribbon meet (with endoplasm necessarily streaming in opposite direction) narrow zones without chloroplasts and without streaming are found (the so-called 'neutral lines' or 'indifferent zones'; dashed lines). In reality, the ratio of cell diameter and length is much smaller than drawn here.

The velocity profile of streaming endoplasm (up to 100 μm s^{-1}) indicates that the force for streaming is generated at the subcortical boundary between ecto- and endoplasm (Kamiya & Kuroda, 1956). Ablation of chloroplasts with a focused light beam (Kamitsubo, 1966), creates a window through which the movement of organelles can be viewed along straight fibers located at the interface between ecto- and endoplasm (Fig. 2). Experiments based on targeted damage (Hayashi, 1964; Kamitsubo, 1966) revealed that these fibers are the structural prerequisite for endoplasmic streaming. Each fiber contains more than fifty 6-nm filaments (Nagai & Rebhun, 1966; Pickett-Heaps, 1967; Kamiya & Nagai, 1982) of uniform polarity in accordance with the direction of streaming (meromyosin decoration: Palevitz et al., 1974; Williamson, 1974; Kersey et al, 1976). Immunofluorescence detection through the chloroplast layer (Williamson & Toh, 1979) and, more clearly, from the vacuolar side (Grolig et al., 1988; Fig. 3), as well as labeling with fluorescent phallotoxin after perfusion (Barak et al., 1980) or microinjection (Wasteneys et al., 1996) and effects of treatment with CB (Nagai & Kamiya, 1977) converge to demonstrate that the guiding filaments are made of F-actin. Biochemical analysis has shown that characean internodal cells express a single isoform of actin (M_r: 43 kDa; isoelectric point: 5.5; Williamson et al., 1987). The large, highly vacuolate internodal cell develops from a subapical uninucleate cell. Elongation of this primary cell is accompanied by a profound reorganization of the actin cytoskeleton, to begin with the formation of a reticulate network of F-actin closely associated with the nucleus. Gradually, AF bundles arrange to parallel fibers of identical polarity, which first become evenly spaced and then organize as discrete sets of bundles along each cortical choloroplast file. A cortical array of fine AF bundles interconnects by strands moving through the layer of chloroplasts with the subcortical F-actin bundles. Overall, characean internodal cells and cells of higher plants show a similar organization of their actin cytoskeleton (Wasteneys et al., 1996).

CB and N-ethylmaleimid (NEM, a non-specific sulfhydryl-poisoning agent that inhibits also myosin) were together elegantly used to show by combined centrifugal displacement of the endoplasm and differential treatment of the endoplasm-rich and -poor half-cell that the putative myosin motor resides in the streaming endoplasm and the corresponding F-actin in the stationary ectoplasm (Chen & Kamiya 1975). By use of counteracting centrifugal (Kamiya & Kuroda, 1958) or hydrodynamic forces (Tazawa, 1968) the motive force of cytoplasmic streaming was estimated to be ~ 1 dyne cm^{-2} (= 0.1 pN 10 μm^{-2}) for the bulk endoplasm, but about a 10-fold higher for the most peripheral layer of organelles (Kamitsubo & Kikuyama, 1994). Polystyrene beads bind endogenous myosin in live *Chara* cells and move at velocities equal to those of endoplasmic organelles (Chaen et al., 1995). The maximal velocity was severalfold larger than that of beads previously coated with muscle myosin (Oiwa et al., 1990), suggesting that native *Chara* myosin has a considerably higher and probably continuously maintained cycling rate in comparison to muscle myosin (Oiwa et al., 1990). Centrifugation of beads opposite to the streaming (positive load) revealed a nearly straight steady force – bead velocity curve, in contrast to muscle myosin (Chaen et al., 1995). Therefore, the authors supposed that *Chara* myosin might detach only for a short time from the actin during its working cycle. Further centrifugation experiments showed that the rigor linkage in the absence of ATP can be broken much easier than the (also reversible) Ca^{2+}-elicited linkage.

Severe perturbation of the cell evokes an action potential (Biskup et al., 1999) which propagates along the elongate cell, elicits a rise of the cytosolic free Ca^{2+}-concentration $[Ca^{2+}]_{cyt}$ presumably by release from internal stores (Plieth et al., 1988b) and causes a transient cessation of cytoplasmic streaming. While streaming ceases readily upon rise of $[Ca^{2+}]_{cyt}$ from a resting value of ~ 0.2 μM up to ~7 μM (peak after about 0.4 s), subsequent recovery of cytoplasmic streaming takes several minutes, even after $[Ca^{2+}]_{cyt}$ has returned to its resting value (Williamson & Ashley, 1982; Tazawa et al., 1987).

Regulatory aspects of the interaction between myosin and the F-actin bundles for the streaming were explored by ingenious use of cell models. Perfusion of well-defined solutions through the elongate characean internodal cells either removes the whole vacuole and some endoplasm (Williamson, 1975), or the vacuolar content only, with the tonoplast being permeabilized (Tominaga & Tazawa, 1981; Tominaga et al., 1983; Shimmen & Tazawa, 1982). The procedure introduced by Kamiya and Kuroda already in 1955 (review: Kamiya, 1986) makes use of the particular advantage of a stabile cortical cytoplasm with massive subcortical F-actin bundles. Experimental manipulation of the physico-chemical parameters in cell models demonstrated, among others, that cytoplasmic streaming ceases in the absence of Mg-ATP, but that this effect is fully reversible (review: Tazawa & Shimmen, 1987). In plasmalemma-permeabilized cells, cytoplasmic streaming is (increasingly and reversibly) impeded if $[Ca^{2+}]_{cyt}$ exceeds the critical level of about 0.2 μM as found in intact cells. Down-regulation of streaming appears to involve inhibitory, Ca^{2+} -dependent phosphorylation: antagonists of protein phosphatases inhibit streaming in plasmalemma-permeabilized cells even at low $[Ca^{2+}]_{cyt}$, while application of ATP-γ-S prevents reversion of Ca^{2+}-induced inhibition of streaming. Calmodulin antagonists do not affect inhibition of cytoplasmic streaming by increased $[Ca^{2+}]_{cyt}$, but prevent the subsequent reversion at low $[Ca^{2+}]_{cyt}$. Calmodulin therefore may trigger dephosphorylation to reset inhibition (Tominaga et al., 1987).

After intra- and inter-specific transplantation, organelles retain the capability to move on F-actin bundles in perfused, NEM-pretreated characean cells (Shimmen & Tazawa, 1982; Kohno & Shimmen 1988b). The intrinsic Ca^{2+}-sensitivity and other parameters of various purified myosins, coupled to organelle-sized microbeads, could be tested on the characean F-actin tracks as 'standard' substratum (Sheetz & Spudich, 1983, Sheetz et al., 1986). Some of the myosins from non-characean sources moved considerably faster on the massive AF-bundles than in the descent cell, but none of the myosins reached the velocity of the endogenous characean streaming (reviewed by Shimmen, 1988).

A monoclonal pan-antimyosin (Parke et al., 1986) detects two polypeptides of 205 and 110 kDa and localizes the epitope along the AF-bundles and on the surface of endoplasmic, motile organelles. Cells perfused with ATP-free solution retain many of the endoplasmic organelles, including aggregates of endoplasmic strands that likely represent endoplasmic reticulum (ER; Fig. 3; compare with: Bradley, 1973; Kachar & Reese, 1988), and apparently much of the myosin, in a rigor-like association with the AF-bundles (Grolig et al., 1988). Accordingly, ATP removes the AF-associated motile organelles (and obviously most of the myosin). The ER apparently provides the coupling means for coherent movement of the bulk endoplasm, as previously claimed on the basis of hydrodynamic considerations (Nothnagel & Webb, 1982).

As detailed above, downregulation of myosin in characean endoplasm appears to involve inhibitory, Ca^{2+}-dependent phosphorylation. An antibody against a Ca^{2+}-dependent, AF-associated protein kinase purified from soybean (Putnam-Evans et al., 1989) detects in *Chara* a 50 kDa protein which shows an intracellular distribution similar to that of myosin (McCurdy & Harmon, 1992a). The protein phosphorylates an endogenous polypeptide immunodetected by an antibody against the regulatory light chain of a mammalian myosin (McCurdy & Harmon, 1992b).

The dynamics of the massive subcortical AF-bundles in characean internodal cells appear to be low, as indicated by their high stability during perfusion and their resistance to CB (Nagai & Kamiya, 1977) and ADP-ribosylation of actin (Grolig et al., 1996a). Upon wounding, however, the cortical and subcortical actin cytoskeleton readily reorganizes into an intricate AF meshwork which is a prerequisite for allocation of the wound wall material (Foissner & Wasteneys, 1997; Foissner & Wasteneys, this volume). The cortical actin in *Nitella* rearranges into short stabile rods in the presence of CB (Collings et al., 1995). Intriguingly, inhibition of cytoplasmic streaming by CB is potentiated by simultaneous depolymerization of the cortical MTs (Collings et al., 1996); possibly released tubulin or MT-associated proteins sensitize the AF-bundles for the action of CB. The regeneration of AF-bundles at sites where chloroplasts and the overlying AFs have been experimentally removed depends on endoplasmic flow supported by the surviving upstream AF-bundles, i.e. polymerization of new AFs appears to occur only in direction of streaming. Extension of such bundles with exposed plus ends is completely inhibited by low concentrations of CB which do not affect the streaming.

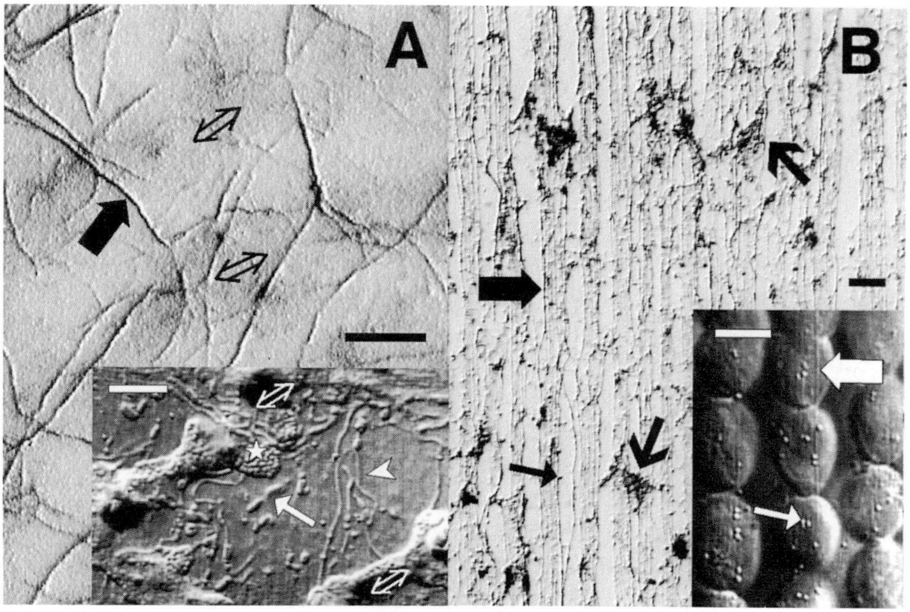

Figure 3. Arrangement of AF-bundles underlying agitational and rotational streaming in respectively *Spirogyra crassa* and *Chara corallina* internodal cells. (**A**) and (**B**) are both derived from fluorescence micrographs in which contrast has been enhanced by inversion and shadowcast. (**A**) *S. crassa*: rhodamine-phalloidin labeling, performed according to Grolig (1990), visualizes thick F-actin bundles (thick black arrows) and thinner bundles in between. Biheaded arrows indicate the axes of the out-of-focus chloroplast bands; the vertical axis of the image corresponds with the cell long axis. Insert: video-enhanced DIC view of the peripheral cytoplasm of *S. crassa,* showing two ellipsoidal mitochondria (small arrow) in line and elongating tubules (arrowhead) along which vesicles migrate, all moving along (unresolved) tracks of F-actin; chloroplast grana (asterisk). (**B**) *C. corallina*: immunofluorescence of myosin associated with F-actin, viewed from the vacuolar side of a perfused, longitudinally cut cell (Grolig et al., 1988). Indirectly visualized are sets of parallel F-actin bundles (black thick arrows; running parallel to the cell long axis here), and, attached to these bundles, numerous small organelles (small arrow) and entangled membrane tubules (probably ER; middle-sized arrow). Insert: video-enhanced DIC view of a similarly treated cell. The image shows a set of parallel thick AF-bundles (white thick arrow) running on top of chloroplast files. In the absence of ATP, numerous small organelles (small arrow) stick to the bundles. Bars: 10 μm.

Actin appears to have a tendency to self-assemble into bundles which can be anchored to the cortex in the presence or absence of plastids. The alignment of regenerated F-actin bundles parallels that of the flow generated by the surviving actin, suggesting flow-induced self-alignment (without template) as an organization principle. In long-term experiments the average number of bundles associated with the chloroplast files decreased in the presence of CB, however, no fragmentation of bundles was observed and the remaining bundles grew in length (like the cell) (Williamson & Hurley, 1986; Williamson et al, 1986).

Cytoplasmic drops from characean cells are readily enclosed by a reforming tonoplast-like membrane. In such drops, the addition of both Mg-ATP and muscle heavy meromyosin (Kuroda & Kamiya, 1975) prolongs the rotational movement of enclosed circular chloroplast chains (Jarosch, 1956a). The AF-cables associated with the rotating chloroplasts, visible in dark field microscopy, appear very tense, but only when moving (Higashi-Fujime, 1980). In endoplasm isolated from characean internodal cells a bimodal organelle velocity distribution (mean velocities of 11 and 62 $\mu m\ s^{-1}$, respectively) was observed (Kachar, 1985). Upon addition of rabbit skeletal F-actin ER tubules elongate – most likely due to tubule-associated myosin (Higashi-Fujime, 1988). Assays on cytoplasmic fragments pioneered the in vitro reconstitution of the basic actomyosin motility in form of ATP-fueled movement of fluorescently labeled heterologous F-actin over immobilized (characean) motor protein (Yamamoto et al., 1994 and 1995; Grolig et al., 1996b; Higashi-Fujime et al., 1995; Rivolta et al., 1995). *In vitro* motility revealed F-actin translocation velocities similar to those of cytoplasmic streaming.

2.2 Ca^{2+} as mediator of physiological challenges

The inhibitory effects of Ca^{2+} on cytoplasmic streaming have been elucidated in detail for characean internodal cells (see section 2.1) and for lily pollen tubes (Kohno & Shimmen, 1988a,b), but have been reported also for other cells (Doree & Picard, 1980; Woods et al., 1984a,b; Takagi et al., 1990; Wacker & Schnepf, 1990). In higher plant cells $[Ca^{2+}]_{cyt}$ above 10^{-5} M likely causes dissociation of a calmodulin light chain from myosin (Yokota et al., 1999); above 10^{-6} M $[Ca^{2+}]_{cyt}$ the Ca^{2+}-calmodulin complex possibly blocks the AF-bundling activity of a putative villin equivalent (135-ABP) and thereby destabilizes AFs (Yokota & Shimmen, this volume).

A number of physiological factors affect cytoplasmic streaming, most of them probably via modulation of $[Ca^{2+}]_{cyt}$ (Shimmen & Yoshida, 1993). The onset of light elicits a decrease of the $[Ca^{2+}]_{cyt}$ (Miller & Sanders, 1987; Johnson et al., 1995) which promotes the circulation of organelles (characean cells: Plieth et al., 1988a,b; *Vallisneria gigantea* mesophyll cells: Takagi, 1997; Takagi, this volume). Although chilling effects are very variable among species (Lewis, 1956), in most cases streaming is reduced below 10°C (for exact relationship in characean cells: Takamatsu et al., 1993), and blocked completely between 5°C and 0°C, probably not only due to impaired energetics, but also because of perturbed Ca^{2+} homeostasis (Woods et al., 1984a,b; Minorsky, 1985). A massive restructuring of the ER upon cold treatment of cold-tolerant onion epidermal cells without major changes of the actin cytoskeleton suggests a prime ($[Ca^{2+}]_{cyt}$ -mediated) effect on the force generating component (Quader et al., 1989). Similar conclusions were drawn from extracellular treatments with membrane-permeable weak acids (Felle, 1988; Tominaga et al., 1998). In tip-growing cells, oscillations in $[Ca^{2+}]_{cyt,}$ coupled to natural fluctuations, artificial treatments or in response to external signaling (Mahló & Trewavas, 1996; Pierson et al., 1996; Emons et al., this volume; Snowman et al., this volume), or variations in cytosolic pH (Tominaga et al., 1998; Feijó et al., 1999) not only inflect the growth rate of the cell, but also modify the pattern and velocity of streaming. Wounding (Foissner & Wasteneys, this volume), osmotic challenge (*Nitella*: Tazawa et al., 1994, 1995; *Lamprothamnium papulosum*: Shepherd et al., 1999), and, to a certain extent, gravity (Wayne et al., 1990; Staves, 1997; discussed in Braun & Wasteneys, this volume) infer on the streaming. The inhibitory sensitivity of root hairs to high concentrations of auxins (~ 10^{-5} M) is increased by raising the external Ca^{2+} level (applying IAA to tomato: Ayling et al., 1996), and by lowering the pH in the culture medium (applying NAA to *Hydrocharis*, Tominaga et al., 1998). Thus, cumulative stress may have aggravating effects on streaming.

3. QUANTITATIVE ANALYSES OF STREAMING

Quantification of organelle movement begins with high-performance light microscopy (Inoué, 1986; Shotton, 1993a; Lichtscheidl & Foissner, 1996; i.e. differential interference contrast, ultraviolet light, or phase contrast imaging, respectively Figs. 3, 4 and 5). Nowadays, subcellular structures are seldomly recorded live or with time-lapse on films (reviews: Kamiya, 1986; Pierson & Cresti, 1992), but preferably visualized with video or digital cameras (Inoué, 1986; red or ultraviolet light imaging: Wada et al., 1982; Lichtscheidl & Foissner, 1996; Fig. 4) that produce images with high spatial and temporal resolution which can be digitally enhanced (Shotton, 1993b; examples in Figs. 3, 4 and 5). With these improvements the impact of errors in measurements has been shifted to new frontiers (discussions on error analysis: Weiss et al., 1990; de Win et al., 1998).

When organelles in a microscopic view clearly contrast in brightness or color with respect to their surrounding, their trajectories can be automatically tracked (example in Fig. 5) e.g. by subtracting the input image from a reference image in a time-series (e.g. Gualtieri & Coltelli, 1991a,b), or by applying thresholding (Weiss et al., 1986) within a range of pixel values. So-called erosion functions are useful to determine the center of gravity of organelles (the baricenter). Contrast can be derived from scattering or absorbance properties of the organelle and from labeling with specific fluorochromes (Haughland, 1996) or organelle-tagged (G)FPs (Tsien, 1998; Ludin & Matus, 1999; in plants: Kost et al., 1998; Kost et al., this volume). In cases in which automatic functions are inadequate to track organelles, interactive approaches sometimes give an outcome (example of procedure, programming and application: de Win et al., 1998, de Win, 1997; Fig. 4). A more global approach to evaluate the movement of untraceable organelles, for instance Golgi vesicles (Pierson et al., 1990; Foissner et al., 1996), is to perform bulk flow velocimetry with Laser Doppler spectroscopy (Mustacich & Ware, 1976; Steer et al., 1984; Ackers et al., 1994) or to measure the pixel-by-pixel changes in brightness of sequential frames (Mineyuki et al, 1983).

Ideally, data on organelle displacements should be collected in three-dimensions (review: Stelzer, 1999). Progress has come from direct stereo viewing in double-focal video microscopy of split images or double recording using anaxial illumination (reviewed in de Win, 1997), and more recently from depth fluorescence microscopy (Mermall et al., 1994) or the application of optical sectioning and 3D-reconstruction in ultrathin wide-field (Inoué, 1994; Farkas et al., 1993; Frymier et al., 1995), conventional confocal (Fig. 1) or multiphoton microscopy (Xu et al., 1996; Centone & White, 1998). Advances in computation technology allow projection and deconvolution of series of images, stereoscopic display and pseudocolor animation (Shotton, 1993b).

3.1 Rational analyses on single tracks

The simplest and often most elucidating parameter to quantify movement is to determine the transit time taken by various organelles to cross the straight distance between two arbitrary points (Tsuchiya et al., 1991). Weiss et al. (1990) describe three approaches to measure the average velocity in the major direction of movement: 1) the slope of the line fitting the positional data over time, 2) the average velocity measured frame by frame (coming close to the so-called instantaneous velocity), and 3) the ratio of distance and time for the tracked path. In Brownian-like motion, for example, path lengths per time unit fit a Gaussian-like distribution, whereas saltatory, active movements show a discontinuous profile over time (Rebhun, 1972; application to *Nitella* and *Chara*: Kamiya, 1981).

Intracellular displacements of individual particles comprise spatio-temporal variations in velocity (decelerations, accelerations, pauses and turns), which partially reflect the forces involved in particle-cytoskeleton interaction. Morevoer, particular organelles (e.g. polygons of cortical ER and stacks of Golgi) display specific patterns of movement ('behavior'; Quader et al., 1987; Grolig, 1990; Pierson et al., 1990; Lichtscheidl, 1995; Boevink et al., 1998). Observations and measurements of streaming in staminal hair cells of *Setcreasea purpurea* (Tucker & Allen, 1986) and other plant cells (Grolig, 1990; Lichtscheidl, 1995; de Win et al., 1999) indicate that organelles move 1) along well-defined tracks, 2) in unequal steps, 3) at different rates, and 4) sometimes in opposite direction, in both transvacuolar strands and the peripheral cytoplasm. From a study by Bartnik et al. (1990) it appears that the radial translocation of Golgi vesicles within the tip of growing *Chara* rhizoids does not occur with the same frequency as the tangential movement, suggesting that ER aggregates in the tip region contribute to adjust the organelle migration.

To gain better insight in the AF-myosin based cytoplasmic dynamics in tobacco pollen tubes, a variety of quantitative motion analyses have been applied to reconstructed tracks (de Win et al., 1997). The so-called regularity quotient (Q_r; de Win et al., 1999; an adaptation of the mean square diffusional displacement method; Jarosch, 1956b; Qian, et al., 1991), gives a measure for the directionality of movement, being 2.0 for pure Brownian-like movement and 4.0 for uniform, linear movement.

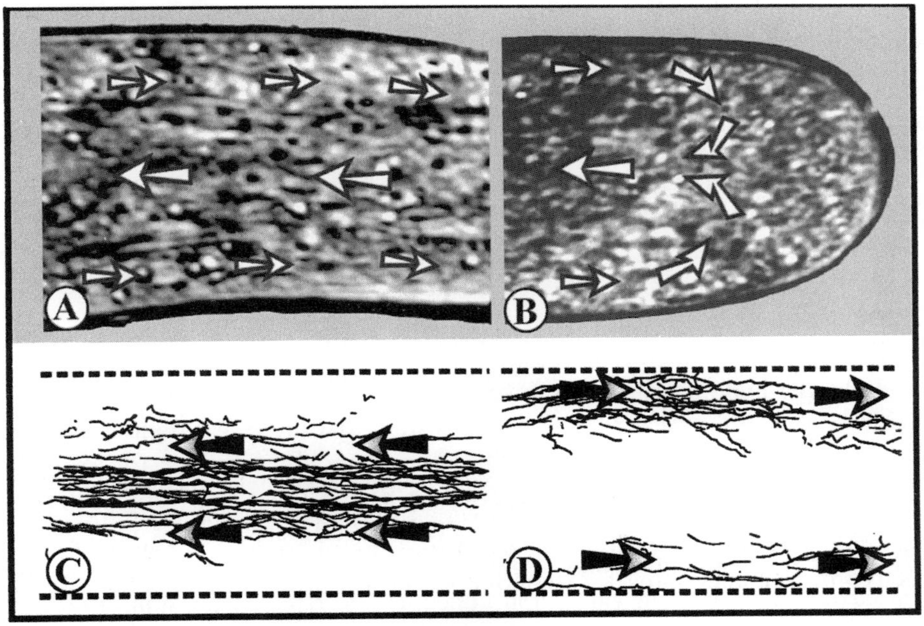

Figure 4. Cytoplasmic streaming in a tobacco pollen tube based on single tracks. (**A**) and (**B**) show individual organelles visualized by ultraviolet light (360 nm) video-enhanced transmission microscopy. (**C**) and (**D**) represent parts of reconstructed tracks recorded (de Win et al. (1998); time interval 0.40 s; region between 19 and 37 μm from the tip; the cell boundary is indicated by dashed lines) and analyzed with the arithmetical dissection method (de Win et al., 1999). (**C**) depicts parts of tracks that have a vectorial component directed toward the base of the tube (arrows to the left). These sections are nearly exclusively located

in the center of the tube. (**D**) depicts parts of the track with a tip-directed component (arrows to the right), located all in the flanks of the tube. The total of all displacements gives the impression of cytoplasmic streaming according to a reverse fountain-like pattern, but single organelles appear to move as individual elements. The tip region of the tube is occupied by Golgi vesicles which show rapid Brownian-like movements. Pollen tube diameter: ~10 µm. (Lichtscheidl, de Win, Derksen, Pierson, unpublished data).

The progressiveness ratio (the net distance between the last and first position versus the total distance covered over all displacements) is designed to evaluate the degree of straightness of trajectories, being in 1.0 for ideally rectilinear trajectories and gradually decreasing toward 0 for coiled trajectories. Calculations of the Qr and progressiveness ratio on ~15.000 displacements in untreated tobacco pollen tubes have demonstrated how the movement becomes less organized and less vectorial closer to the tube tip and that the behavior of organelles in the tip resembles that of organelles extracted from the cell. (de Win et al., 1999). The arithmetical dissection method (de Win et al., 1999), discerns ten movement types on account of trends in the persistence of movement in a custom-defined (e.g. cell axes) coordinate system. The main motion characteristics (main direction and velocity of movements and rate of directionality in consecutive displacements) of ~900 tobacco pollen tube organelles have been determined to assess the impact of the location within the tube and the effect of cytochalasin treatment in a statistically reliable manner (Fig. 4). Three main outcomes have emerged from this analysis: 1) pollen tube organelles move primarily as individual elements, that is asynchronously and according to heterogeneous patterns, but with global similarities in vectorial behavior within each region of the tube; 2) the level of directionality increases with the distance from the tip; 3) friction may impede motility in lanes with bidirectional movement, as deduced from the lower velocity in such lanes. Clearly, the main direction of movement is opposite in the flanks of the cell with respect to the center (Fig. 4), suggesting that these two regions contain arrays of AFs of antiparallel polarity. Actin-bundling proteins, for example the 135Kda protein recently identified in pollen tubes and root hair cells (Yokota et al., 1998; Tominaga et al., 1999, 2000), are perhaps involved in the polar arrangement of these AFs. Region-based curve analysis, a method based on track dilation and graph concepts (Worring & de Win, 1996; de Win et al., 1997), allows to classify the level of complexity of (any series of) organelle trajectories and helps to assess the consistency in the polarity of movement and the probability of contact between an organelle and its guiding element.

3.2 pN forces, nm steps and perspectives

High resolution video-enhanced microscopy allows the determination of baricenters with a nanometer precision, thus, beyond the limits of resolution of the light microscope and in the range of the step sizes of the organelle-associated motor molecules. From the Poisson-distributed distances of stochastic vesicle displacements during initial resumption of the cytoplasmic streaming after an experimentally elicited stop, the step size of a vesicle-associated myosin in internodal cells of *Nitella* was calculated to be about 10^{-7} m (Nemoto et al. 1995) which is about tenfold more than the step-size of muscle myosin (Finer et al., 1994). Time series analysis of *in vivo* organelle movements along the characean actin bundles (Uchida et al., 1999) revealed a power spectrum with two peaks near 4.0 and 7.5 Hz, suggesting that the myosin molecules attached to the surfaces of organelles work cooperatively and not independently. Less than 1% of the observed organelles moved for a few 10^{-1} s in opposite direction of the streaming. This movement had almost the same power spectrum, indicating a similar dynamics. In *Spirogyra crassa*, high-resolution analysis of seemingly steady movements of mitochondria has revealed stochastic variations of displacement distances within the 40 ms sampling interval (Fig. 5). Variable alignment of the mitochondrial long axis with the track axis largely contributes to the impression of passive translocation ('streaming'). Improved alignment correlates with increased mitochondrial velocity (Fig. 5). Although viscous drag (Berg, 1983) may enforce alignment, the observed increases in velocity largely exceed the expectable gain and therefore most likely result from an increased number of motor molecules interacting with the underlying AF-bundle.

The application of optical traps (Simmons et al., 1996) or magnetic fields (Uchida et al., 1998) in conjunction with high-performance microscopy permits the appraisal of picoNewton forces and nanometer-scale displacements involved in chemomechanical intracellular motility processes, and the assessment of the tensile and flexural properties even of single AFs and MTs (Schindler, 1995; Arai et al., 1999; Fisher & Komomeisky, 1999; Kitamura et al., 1999). Application of these approaches to *in vitro* motility assays (e.g. with beads coated with myosins resulting from targeted mutations) as well as to the more complex live cell (Felgner et al., 1997) will contribute to trace down functional aspects and dynamic components of the supramolecular aggregates underlying the 'streaming of fluid' reported for *Chara* by Corti in 1774.

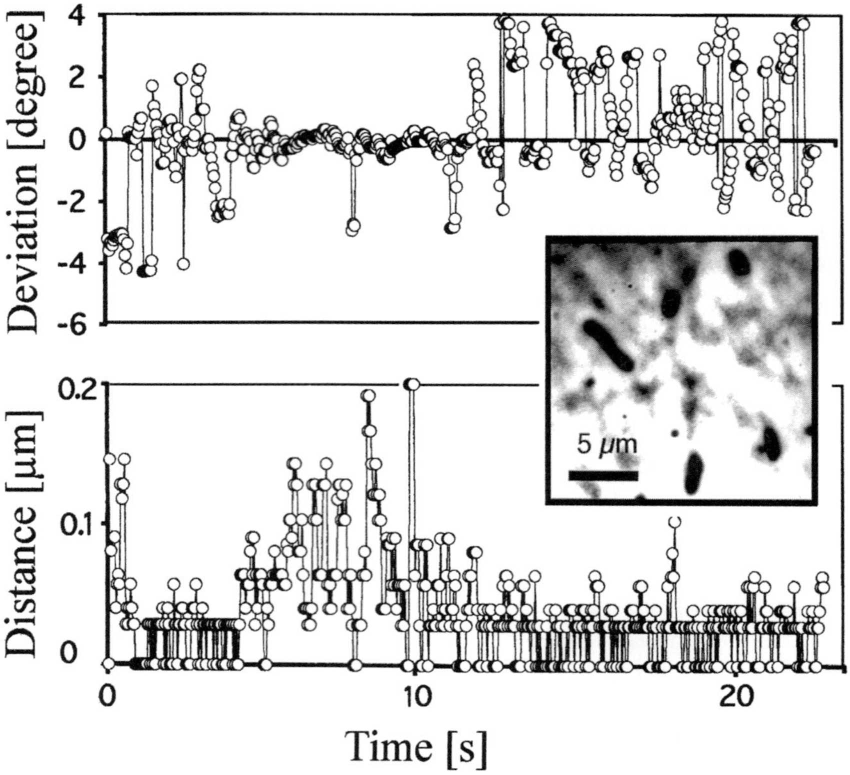

Figure 5. High resolution analysis of organelle tracks in *Spirogyra.* Fast software (MedeaLab®, Erlangen, Germany) in combination with double buffering frame grabber technology, developed for tracking of free motile organisms (Häder & Vogel, 1991), was adapted for automatic real time tracking of plant organelles. The position, main axis, outline and area of objects extracted by a gray level threshold band are determined from their chain code (Freeman, 1980; Jiang & Buhnke, 1991). Eeach object is tracked by searching the nearest neighborhood and checking the object´s form parameters. Continual, actomyosin-based movements of mitochondria (viewed by video-enhanced phase contrast; length up to 5 µm, diameter ~ 1 µm; see insert) in the peripheral cytoplasm (depth ~ 5 µm) of *S. crassa* were tracked in a virtually two-dimensional field of observation. The x-axis denotes the progress of time in the upper and lower graph. The top y-axis denotes the deviation (in degree) of the mitochondrial long axis from the track axis. The bottom y-axis denotes the displacement distance covered within the 40 ms sampling interval. The track shows stochastic variations of the displacement distances. Transiently improved coalignment of the mitochondrial long axis with the track axis (upper diagram) correlates with increased velocity of the mitochondrium (lower diagram), indicating that an increased number of mitochondrium-associated motor molecules interacts with the underlying AF bundle (Grolig, Weintraut & Hanneforth, unpublished data).

Acknowledgements

The authors thank Drs IK Lichtscheidl (Univ of Vienna), J Derksen (University of Nijmegen), AMC Emons (University of Wageningen) and AHN de Win (University of Nijmegen) for the use of unpublished data and/or hospitality, J Derksen in particular for his valuable comments on the manuscript. F.G. acknowledges support by the Deutsche Forschungsgemeinschaft.

References

Ackers D, Hejnowicz Z and Sievers A (1994) Cytoplasmic streaming and gravity effect in characean internodal cells measured by laser-Doppler-velocimetry. Protoplasma 179: 61-71

Arai Y, Yasuda R, Akashi KI, Harada Y, Miyata H, Kinosita K Jr and Itoh H (1999) Tying a molecular knot with optical tweezers. Nature 399: 446-448

Ayling SM and Clarkson DT (1996) The cytoplasmic streaming response to tomato root hairs to auxin: The role of calcium. Austr J Plant Physiol 23: 699-708

Baluška F, Kreibaum A, Vitha S, Parker JS, Barlow PW and Sievers A (1997) Central root cap cells are depleted of endoplasmic microtubules and actin microfilament bundles: Implications for their role as gravity-sensing statocytes. Protoplasma 196: 212 223

Baluška F, Šamaj J and Volkmann D (1999) Proteins reacting with cadherin and catenin antibodies are present in maize showing tissue-, domain-, and development-specific associations with endoplasmic-reticulum membranes and actin microfilaments in root cells. Protoplasma 206: 174-187

Barak LS, Yocum RR, Nothnagel EA and Webb WW (1980) Fluorescence staining of the actin cytoskeleton in living cells with 7-nitrobenz-2-oxa-1,3-diazole-phallacidin. Proc Natl Acad Sci USA 77: 980-984

Bartnik E, Hejnowicz Z and Sievers A (1990) Shuttle-like movements of Golgi vesicles in the tip of growing *Chara* rhizoids. Protoplasma 159: 1-8

Berg HC (1983) Random Walks in Biology. Princeton: Princeton University Press

Biskup B, Gradmann D and Thiel G (1999) Calcium release from $InsP_3$-sensitive internal stores initiates action potential in *Chara*. FEBS Lett 453: 72-76

Boevink P, Oparka K, Santa Cruz S, Martin B, Betteridge A and Hawes C (1998) Stacks on tracks: The plant Golgi apparatus traffics on an actin/ER network. Plant J 15: 441-447

Bradley MO (1973) Microfilaments and cytoplasmic streaming: Inhibition of streaming with cytochalasin. J Cell Sci 12: 327-343

Cai G, Moscatelli A, Del Casino C and Cresti M (1996) Cytoplasmic motors and pollen tube growth. Sex Plant Reprod 9: 59-64

Centone VE and White JG (1998) Multiphoton excitation provides optical sections from deeper within scattering specimens than confocal imaging. Biophys J 75: 2015-2024

Chaen S, Inoue J and Sugi H (1995) The force-velocity relationship of the ATP-dependent actin-myosin sliding causing cytoplasmic streaming in algal cells, studied using a centrifuge microscope. J Exp Biol 198: 1021-1027

Chen JCW and Kamiya N (1975) Localization of myosin in the internodal cell of *Nitella* as suggested by differential treatment with N-ethylmaleimide. Cell Struct Funct 1: 1-9

Collings DA, Wasteneys GO and Williamson RE (1995) Cytochalasin rearranges cortical actin of the alga *Nitella* into short, stable rods. Plant Cell Physiol 36: 765-772

Collings DA, Wasteneys GO and Williamson RE (1996) Actin-microtubule interactions in the alga *Nitella*: Analysis of the mechanism by which microtubule depolymerization potentiates cytochalasin's effects on streaming. Protoplasma 191: 178-190

Corti B (1774) Osservazioni Microscopiche sulla Tremella e sulla Circolazione del Fluido in una Pianta Acquajuola. University of Lucca.

de Ruijter NCA and Emons AMC (1999) Actin-binding proteins in plant cells. Plant Biol 1: 26-35

de Win AHN (1997). Quantitative analysis of organelle movements in pollen tubes. PhD thesis. Catholic University of Nijmegen. Appendix: computer programs

de Win AHN, Worring M, Derksen J and Pierson ES (1997) Classification of organelle trajectories using region-based curve analysis. Cytometry 29: 136-146

de Win AHN, Pierson ES, Timmer C, Lichtscheidl IK and Derksen J (1998) An interactive computer-assisted position acquisition procedure designed for the analysis of organelle movement in pollen tubes. Cytometry 32: 263-267

de Win AHN, Pierson ES and Derksen J (1999) Rational analyses of organelle trajectories in tobacco pollen tubes reveal characteristics of the actomyosin cytoskeleton. Biophys J 76: 1648-1658

Doree M and Picard A (1980) Release of Ca^{2+} from intracellular pools stops cytoplasmic streaming in *Tradescantia* staminal hairs. Experientia 36: 1291-1292

Emons AMC, Pierson ES and Derksen J (1991) Cytoskeleton and intracellular movement in plant cells. In: Cheremisinoff PN and Ferrante LM (eds) Biotechnology: Current Progress. Lancaster: Technomic Publ Comp, pp 311-335

Farkas DL, Baxter G, DeBiasio R, Gough A, Nederlof MA, Pane D, Pane J, Patek DR, Ryan KW and Taylor DL (1993) Multimode light microscopy and the dynamics of molecules, cells, and tissues. Annu Rev Physiol 55: 785-817

Feijó JA, Sainhas J, Hackett GR, Kunkel JG and Hepler PK (1999) Growing pollen tubes possess a constitutive alkaline band in the clear zone and a growth-dependent acidic tip. J Cell Biol 144: 483-496

Felgner H, Grolig F, Müller O and Schliwa M (1997) *In vivo* manipulation of internal cell organelles. In: Sheetz MP (ed) Laser Tweezers in Cell Biology, Meth Cell Biol 55: 195-203

Felle H (1988) Cytoplasmic free calcium in *Riccia fluitans* and *Zea mays*: Interaction of Ca^{2+} and pH? Planta 176: 248-255

Finer JT, Simmons RM and Spudich JA (1994) Single myosin molecule mechanics: Piconewton forces and nanometre steps. Nature 368: 113-119

Fisher ME and Kolomeisky AB (1999) The force exerted by a molecular motor. Proc Natl Acad Sci USA 96: 6597-6602

Foissner I and Wasteneys GO (1997) A cytochalasin-sensitive actin filament network is a prerequisite for local wound wall deposition in Nitella internodal cells. Protoplasma 200: 17-30

Foissner I, Lichtscheidl IK and Wasteney GO (1996) Actin-based vesicle dynamics and exocytosis during wound wall formation in *Characean* internodal cells. Cell Motil Cytoskel 35: 35-48

Freeman H (1980) Analysis and manipulation of lineal map data. In: Freeman H and Pieroni GG (eds) Map Data Processing. New York: Academic Press, pp 151-168

Frymier P, Ford RM, Berg HC and Cummings PT (1995) Three-dimensional tracking of motile bacteria near a solid planar surface. Proc Natl Acad Sci USA 92: 6195-6199

Gibbon BC, Kovar DR and Staiger CJ (1999) Latrunculin B has different effects on pollen germination and tube growth. Plant Cell 11: 2349-2364

Grolig F (1990) Actin-based organelle movements in interphase *Spirogyra*. Protoplasma 155: 29-42

Grolig F (1992) The cytoskeleton of the Zygnemataceae. In: Menzel D (ed) The Cytoskeleton of the Algae. Boca Raton: CRC Press, pp 165-193

Grolig F (1998) Nuclear centering in *Spirogyra*: Force integration by microfilaments along microtubules. Planta 204: 54-63

Grolig F, Just I and Aktories K (1996a) ADP-ribosylation of actin from the green alga *Chara corallina* by *Clostridium botulinum* C2 toxin and *Clostridium perfringens* iota toxin. Protoplasma 193: 77-81

Grolig F, Schröder J, Sawitzky H and Lange U (1996b) Partial characterization of a putative 110 kDa myosin from the green alga *Chara corallina* by *in vitro* binding of fluorescent F-actin. Cell Biol Int 20: 365-373

Grolig F, Williamson RE, Parke J, Miller C and Anderton BH (1988) Myosin and Ca^{2+}-sensitive streaming in the alga *Chara*: Detection of two polypeptides reacting with a monoclonal anti-myosin and their localization in the streaming endoplasm Eur J Cell Biol 47: 22-31

Gualtierri P and Coltelli P (1991a) A real-time automated system for the analysis of moving images. J Comput Assist Microscop 3: 15-21

Gualtierri P and Coltelli P (1991b) An image-processing system, motion analysis oriented (IPS-100), applied to microscopy. Comp. Method Progr Biomed 36: 15-25

Häder DP and Vogel K (1991) Simultaneous tracking of flagellates in real time by image analysis. J Meth Biol 30: 63-72

Haughland R (1996) Handbook of Fluorescent Probes and Research Chemicals. Eugene: Molecular Probes

Hayashi, T (1964) The role of the cortical gel layer in cytoplasmic streaming. In: Allen RD and Kamiya N (eds) Primitive Motile Systems in Cell Biology. New York: Academic Press, pp 19-29

Hepler PK, Cleary AL, Gunning BES, Wadworh P, Wasteney GO and Zhang DH (1993) Cytoskeletal dynamics in living plant cells. Cell Biol Int 17: 127-142

Higashi-Fujime S (1980) Active movement *in vitro* of bundles of microfilaments isolated from *Nitella* cell. J Biol Chem 87: 569-578

Higashi-Fujime S (1988) Actin-induced elongation of fibers composed of cytoplasmic membrane from *Nitella*. Protoplasma Suppl 2: 27-36

Higashi-Fujime S, Ishikawa R, Iwasawa H, Kagami O, Kurimoto E, Kohama K and Hozumi T (1995) The fastest actin-based motor protein from the green alga, *Chara*, and its distinct mode of interaction with actin. FEBS Lett 375: 151-154

Hocharka PW (1999) The metabolic implications of intracellular circulation. Proc Natl Acad Sci USA 96: 12233-12239

Holzinger A and Meindl U (1997) Jasplakinolide, a novel actin targeting peptide, inhibits cell growth and induces actin filament polymerization in the green alga *Micrasteria*. Cell Motil Cytoskel 38: 365-372

Howard J (1997) Molecular motors: Structural adaptations to cellular functions. Nature 389: 561-567

Ingber DE (1993) Cellular tensegrity: Defining new rules of design that govern the cytoskeleton. J Cell Sci 104: 613-627

Inoué S (1986) Video Microscopy. New York: Plenum Press

Inoué S (1994) Ultrathin optical sectioning and dynamic volume investigation with conventional light microscopy. In: Stevens JK, Mills LR and Trogadis JE (eds) Three Dimensional Confocal Microscopy: Volume Investigation of Biological Specimens. San Diego: Academic Press, pp 397-419

Jarosch R (1956a) Plasmaströmung und Chloroplastenrotation bei Characeen. Phyton 6: 87-108

Jarosch R (1956b) Die Impulsrichtungsänderungen bei der Induktion der Protoplasma-strömung. Protoplasma 67: 478-486

Jiang XY and Bunke H (1991) Ein konturbasierter Ansatz zur Berechnung von Momenten. In: Radig B (ed) Mustererkennung 1991, 13. DAGM-Symposium, Heidelberg: Springer-Verlag, pp 143-150

Jauh GY and Lord E (1995) Movement of the tube cell in the lily style in the absence of the pollen grain and the spent pollen tube. Sex Plant Reprod 8: 168-172

Johnson CH, Knight MT, Kondo T, Masson P, Sedbrook J, Haley A and Trewavas A (1995) Circadian oscillations of cytosolic and chloroplastic free calcium in plants. Science 269: 1863-1865

Kachar B (1985) Direct visualization of organelle movement along actin filaments dissociated from characean algae. Science 227: 1355-1357

Kachar B and Reese TS (1988) The mechanism of cytoplasmic streaming in characean algal cells: Sliding of endoplasmic reticulum along actin filaments. J Cell Biol 106: 1545-1552

Kamitsubo E (1966) Motile protoplasmic fibrils in cell of the *Characeae*. II. Linear fibrillar structure and its bearing on protoplasmic streaming. Proc Jap Acad 42: 640-643

Kamitsubo E and Kikuyama M (1994) Measurement of motive force for cytoplasmic streaming in characean internodes. Protoplasma 180: 153-157

Kamiya N (1960) Physics and chemistry of protoplasmic streaming. Annu Rev Plant Physiol 11: 323-340

Kamiya N (1962) Protoplasmic streaming. In: Ruhland W (ed) Encyclopedia of Plant Physiology (Vol 17/2) Berlin: Springer-Verlag, pp 979-1035

Kamiya N (1981) Physical and chemical basis of cytoplasmic streaming. Annu Rev Plant Physiol 32: 205-236

Kamiya N (1986) Cytoplasmic streaming in giant algal cells: A historical survey of experimental approaches. Bot Mag Tokyo 99: 441-467

Kamiya N and Kuroda K (1956) Velocity distribution of the cytoplasmic streaming in *Nitella* cells. Bot Mag Tokyo 69: 544-554

Kamiya N and Kuroda K (1958) Measurement of the motive force of the protoplasmic rotation in *Nitella*. Protoplasma 50: 144-148

Kamiya N and Nagai R (1982) Structural similarity between actin bundles from characean algal cells and sea urchin oocytes. J Mol Biol 155: 168-172

Kandasamy MK and Meagher RB (1999) Actin-organelle interaction: Association with chloroplast in *Arabidopsis* leaf mesophyll cells. Cell Motil Cytoskel 44: 110-118

Kersey YM, Hepler PK, Palevitz BA and Wessels NK (1976) Polarity of actin filaments in characean algae. Proc Natl Acad Sci USA 73: 165-167

Kiermayer O, Url WG and Galle HK (1979) Protoplasmaströmung. Scientific Film C 1295. Göttingen: IWF

Kitamura K, Tokunaga M, Iwane AH and Yanagida T (1999) A single myosin head moves along an actin filament with regular steps of 5.3 nanometres. Nature 397: 129-134

Kobayashi H (1996) Changes in the relationship between actin filaments and the plasma membrane in cultured *Zinnia* cells during tracheary element differentiation investigated by using plasma membrane ghosts. J Plant Res 109: 61-65

Kohno T and Shimmen T (1988a) Mechanism of Ca^{2+} inhibition of cytoplasmic streaming in lily pollen tubes. J Cell Sci 91: 501-509

Kohno T and Shimmen T (1988b) Accelerated sliding of pollen tube organelles along *Characeae* actin bundles regulated by Ca^{2+}. J Cell Biol 106: 1539-1543

Kohno T, Chaen S and Shimmen T (1990) Characterization of the translocator associated with pollen tube organelles. Protoplasma 154: 179-183

Kost B, Spielhofer P and Chua NH (1998) A GFP-mouse talin fusion protein labels plant actin filaments *in vivo* and visualizes the actin cytoskeleton in growing pollen tubes. Plant J 16: 393-401

Kranz HD, Miks D, Siegler ML, Capesius I, Sensen CW and Huss VAR (1995) The origin of land plants: Phylogenetic relationships among charophytes, bryophytes, and vascular plants inferred from complete small-subunit ribosomal RNA gene sequences. J Mol Evol 41: 74-84

Kunze B, Jansen R, Sasse F, Höfle G and Reichenbach H (1995) Chondramides A-D, new antifungal depsipeptides from *Chondromyces crocatus* (Myxobacteria). J Antibiotics 48: 1262-1266

Kuroda K (1990) Cytoplasmic streaming in plant cells. Int Rev Cytol 121: 267-307

Kuroda K and Kamiya N (1975) Active movement of *Nitella* chloroplast *in vitro*. Proc Jap Acad 51: 774-777

Lewis DA (1956) Protoplasmic streaming in plants sensitive and insensitive to chilling temperatures. Science 124: 75-77

Lichtscheidl IK (1995) Inner epidermal cells of *Allium cepa* bulb scales investigated by modern light microscopical techniques. 149 pp, Nr 47. Wien: Österreichisches Bundesinstitut für den Wissenschaftlichen Film

Lichtscheidl IK and Foissner I (1996) Video microscopy of dynamic plant cell organelles: Principles of the technique and practical application. J Microsc 181: 117-128

Ludin B and Matus A (eds) (1999) GFP in Motion: Live Fluorescence Imaging using Green Fluorescent Protein. CD-ROM, Elsevier Sciences.

Mahló R and Trewavas AJ (1996) Localized apical increases of cytosolic free calcium control pollen tube orientation. Plant Cell 8: 1935-1949

Mattox KR and Stewart KD (1984) Classification of the green algae: a concept based on comparative cytology. In: Irvine DEG and John DM (eds) Systematics of the Green Algae. London: Academic Press, pp 29-72

McCurdy DM and Harmon AC (1992a) Calcium-dependent protein kinase in the green alga Chara. Planta 188: 54-61

McCurdy DM and Harmon AC (1992b) Phosphorylation of a putative myosin light chain in *Chara* by calcium-dependent protein kinase. Protoplasma 171: 85-88

McIntosh K, Pickett-Heaps JD and Gunning BES (1995) Cytokinesis in *Spirogyra*: Integration of cleavage and cell-plate formation. Int J Plant Sci 156: 1-8

Meagher RB, McKinney EC and Kandasamy MK (1999) Isovariant dynamics expand and buffer the responses of complex systems: the diverse plant actin gene family. Plant Cell 11: 995-1005

Mermall V, McNelly JG and Miller KG (1994) Transport of cytoplasmic particles catalysed by an unconventional myosin in living *Drosophila* embryos. Nature 369: 560-562

Miller AJ and Sanders D (1987) Depletion of cytosolic free calcium by photosynthesis. Nature 326: 397-400

Miller DD, Scordilis SP and Hepler PK (1995) Identification and localization of three classes of myosins in pollen tubes of *Lilium longiflorum* and *Nicotiana alata*. J Cell Sci 108: 2549-2563

Mineyuki Y, Yamada M, Takagi M, Wada M and Furuya M (1983) A digital image processing technique for the analysis of particle movements: Its application to organelle movements during mitosis in *Adiantum* protonemata. Plant Cell Physiol 24: 225-234

Mineyuki Y, Kataoka H, Masuda Y and Nagai R (1995) Dynamic changes in the actin cytoskeleton during the high-fluence rate response of the *Mougeotia* chloroplast. Protoplasma 185: 222-229

Minorsky PV (1985) An heuristic hypothesis of a chilling injury in plants: A role for calcium as the primary physiological transducer of injury. Plant Cell Environ 8: 75-94

Mustacich RV and Ware BR (1976) A study of protoplasmic streaming in *Nitella* by laser doppler spectroscopy. Biophys J 16: 373-388

Nagai R (1993) Regulation of intracellular movements in plant cells by environmental stimuli. Int Rev Cytol 145: 251-309

Nagai R and Kamiya N (1977) Differential treatment of *Chara* cells with cytochalasin B with special reference to its effect on cytoplasmic streaming. Exp Cell Res 108: 231-237

Nagai R and Rebhuhn LI (1966) Cytoplasmic microfilaments in streaming *Nitella* cells. J Ultrastruct Res 14: 571-589

Nemoto T, Uchida G, Takamatsu A and Tsuchiya Y (1995) Stochastic behavior of organelle motion in Nitella internodal cells. Biochem Biophys Res Comm 214: 1102-1107

Nothnagel EA and Webb WW (1982) Hydrodynamic models of viscous coupling between motile myosin and endoplasm in characean algae. J Cell Biol 94: 444-454

Oiwa K, Chaen S, Kamitsubo E, Shimmen T and Sugi H (1990) Steady-state force-velocity relation in the ATP-dependent sliding movement of myosin-coated beads on actin cables *in vitro* studied using a centrifuge microscope. Proc Natl Acad Sci USA 87: 7893-7897

Palevitz BA, Ash JF and Hepler PK (1974) Actin in the green alga *Nitella*. Proc Natl Acad Sci USA 71: 363-366

Palevitz BA and Liu B (1992) Microfilaments (F-actin) in generative cells and sperm: An evaluation. Sex Plant Reprod 5: 89-100

Parke J, Miller C and Anderton BH (1986) Higher plant myosin heavy-chain identified using a monoclonal antibody. Eur J Cell Biol 41: 9-13

Parthasarathy MV, Perdue TD, Witztum A and Alvernaz J (1985) Actin network as a normal component of the cytoskeleton in many vascular plant cells. Am J Bot 72: 1318-1323

Pickett-Heaps JD (1967) Ultrastructure in *Chara* sp. I. vegetative cells. Aust J Biol Sci 20: 539-551

Pierson ES and Cresti M (1992) Cytoskeleton and cytoplasmic organization of pollen and pollen tubes. Int Rev Cytol 140: 73-125

Pierson ES, Lichtscheidl IK and Derksen J (1990) Structure and behaviour of organelles in living pollen tubes of *Lilium longiflorum*. J Exp Bot 41:1461-1468

Pierson ES, Miller DD, Callaham DA, van Aken J, Hackett G, and Hepler PK (1996) Tip-localized calcium entry fluctuates during pollen tube growth. Dev Biol 174: 160-173

Plieth C, Sattelmacher B and Hansen UP (1998a) Light-induced cytosolic calcium transients in green plant cells. II. The effect on a K^+ channel as studied by a kinetic analysis in *Chara corallina*. Planta 207: 52-59

Plieth C, Sattelmacher B, Hansen UP and Thiel G (1998b) The action potential in Chara: Ca^{2+} release from internal stores visualized by Mn^{2+}-induced quenching of fura-dextran. Plant J 13: 167-175

Putnam-Evans C, Harmon AC, Palevitz BA, Fechheimer M and Cormier MJ (1989) Calcium-dependent protein kinase is localized with F-actin in plant cells. Cell Motil Cytoskel 12: 12-22

Qian H, Sheetz MP and Elson EL (1991) Single particle tracking: Analysis of diffusion and flow in two-dimension systems. Biophys J 60: 910-921

Quader H, Hofmann A and Schnepf E (1987) Shape and movement of the endoplasmic reticulum in onion bulb epidermis cells: Possible involvement of actin. Eur J Cell Biol 47: 17-26

Quader H, Hofmann A and Schnepf E (1989) Reorganization of the endoplasmic reticulum in epidermal cells of onion bulb scales after cold stress: Involvement of cytoskeletal elements. Planta 177: 273-280

Rebhun LI (1972) Polarized intracellular particle transport: Saltatory movements and cytoplasmic streaming. Int Rev Cytol 32: 93-137

Reece KS, McElroy D and Wu R (1992) Function and evolution of actins. Evol Biol 26: 1-34

Rice S, Lin AW, Safer D, Hart CL, Naber N, Carragher BO, Clan SM, Pechatnikova E, Wilson-Kubalek EM, Whittaker M, Pate E, Cooke R, Taylor EW, Milligan RA and Vale RD (1999) A structural change in the kinesin motor protein that drives motility. Nature 402: 778-784

Rivolta MN, Urrutia R and Kachar B (1995) A soluble motor from the alga *Nitella* supports fast movement of actin filaments *in vitro*. Biochim Biophys Acta 1232: 1-4

Salitz A and Schmitz K (1989) Influence of microfilament and microtubule inhibitors applied by immersion and microinjection on circulation streaming in staminal hairs of *Tradescantia blossfeldiana*. Protoplasma 153: 37-45

Sawitzky H and Grolig F (1995) Phragmoplast of the green alga *Spirogyra* is functionally distinct from the higher plant phragmoplast. J Cell Biol 130: 1359-1371

Sawitzky H, Liebe S, Willingale-Theune J and Menzel D (1999) The anti-proliferative agent jasplakinolide rearranges the actin cytoskeleton of plant cells. Eur J Cell Biol 78: 424-433

Schindler M (1995) The optical displacement assay (CODA): Measurements of cytoskeletal tension in living plant cells with a laser optical trap. Methods Cell Biol 49: 69-82

Sheetz MP and Spudich JA (1983) Movement of myosin-coated fluorescent beads on actin cables *in vitro*. Nature 303: 31-35

Sheetz MP, Block SM and Spudich JA (1986) Myosin movement *in vitro*: A quantitative assay using oriented actin cables from *Nitella*. Meth Enzymol 134: 531-544

Shepherd VA, Beilby MJ and Heslop D J (1999) Ecophysiology of the hypotonic response in the salt-tolerant charophyte alga *Lamprothamnium papulosum*. Plant Cell Environ 22: 333-346

Shimmen T (1988) Characean actin bundles as a tool for studying actomyosin-based motility. Bot Mag Tokyo 101: 533-544

Shimmen T (1992) The characean cytoskeleton: dissecting the streaming mechanism. In Menzel D (ed) The Cytoskeleton of the Algae. Boca Raton: CRC Press, pp 297-314

Shimmen T and Tazawa M (1982) Cytoplasmic streaming in the cell model *Nitella*. Protoplasma 112: 101-106

Shimmen T and Tazawa M (1985) Mechanism of inhibition of cytoplasmic streaming by myrmicacin (ß-Hydroxydecanoic acid) in *Chara* and *Spirogyra*. Protoplasma 127: 93-100

Shimmen T and Yokota E (1994) Physiological and biochemical aspects of cytoplasmic streaming. Int Rev Cytol 155: 97-139

Shimmen T and Yoshida S (1993) Analysis of temperature dependence of cytoplasmic streaming using tonoplast-free cells of *Characeae*. Protoplasma 176: 174-177

Shimmen T, Hanatani M, Saito S, Yokota E, Mimura T, Fusetani N and Karaki H (1995) Role of the actin filaments in cytoplasmic streaming and organization of transvacuolar strands in root hair cells of *Hydrocharis*. Protoplasma 185: 188-193

Shotton D (1993a) An introduction to the electronic acquisition of light microscopes images. In: Shotton DM (ed) Electronic Light Microscopy: The Principles and Practice of Video-Enhanced Contrast, Digital Intensified Fluorescence and Confocal Laser Scanning Microscopy. New York: Wiley-Liss, pp 1-38

Shotton D (1993b) An introduction to digital image processing and image display in electronic light microscopy. In: Shotton DM (ed) Electronic Light Microscopy: The Principles and Practice of Video-Enhanced Contrast, Digital Intensified Fluorescence and Confocal Laser Scanning Microscopy. New York: Wiley-Liss, pp 39-70

Simmons RM, Finer JT, Chu S and Spudich JA (1996) Quantitative measurements of force and displacement using an optical trap. Biophys J 70: 1813-1822

Sonobe S and Shibaoka H (1989) Cortical fine actin filaments in higher plant cells visualized by rhodamine-phalloidin after pretreatment with m-maleimidobenzoyl N-hydroxysuccinimide ester. Protoplasma 148: 80-86

Staiger CJ and Schliwa M (1987) Actin localization and function in higher plants. Protoplasma 141: 1-12

Staves MP (1997) Cytoplasmic streaming and gravity sensing in *Chara* internodal cells. Planta 203: S79-S84

Steer MV, Picton JM and Earnshaw JC (1984) Diffusive motions in living cytoplasm probed by Laser Doppler microscopy. J Microsc 134: 143-149

Stelzer EHK (1999) Three-dimensional microscopy. In Handbook of Computer Vision and Applications. Vol 1. New York: Academic Press, pp 541-562

Takagi S (1997) Photoregulation of cytoplasmic streaming: cell biological dissection of signal transduction pathway. J Plant Res 110: 299-303

Takagi S, Yamamoto KT, Furuya M and Nagai R (1990) Cooperative regulation of cytoplasmic streaming and Ca^{2+}-fluxes by P_{fr} and photosynthesis in Vallisneria-mesophyll cells. Plant Physiol 94: 1702-1708

Takamatsu A., Aoki T and Tsuchiya Y (1993) Ca^{2+} effect on protoplasmic streaming in Nitella internodal cell. Biophys J 64: 182-186

Tazawa M (1968) Motive force for the cytoplasmic streaming in *Nitella*. Protoplasma 65: 207-222

Tazawa M and Shimmen T (1987) Cell motility and ionic relations in characean cells as revealed by internal perfusion and cell models. Int Rev Cytol 109: 259-312

Tazawa M, Kikuyama M and Shimmen T (1976) Electric characteristics and cytoplasmic streaming of *Characeae* cells lacking tonoplast. Cell Struct Funct 1: 165-176

Tazawa M, Shimmen T and Mimura T (1987) Membrane control in the *Characeae*. Annu Rev Plant Physiol 38: 95-117

Tazawa M., Yoko-o T, Mimura T and Kikuyama M (1994) Intracellular mobilization of Ca^{2+} and inhibition of cytoplasmic streaming induced by transcellular osmosis in internodal cells of *Nitella flexilis*. Plant Cell Physiol 35: 63-72

Tazawa M, Shimada K and Kikuyama M (1995) Cytoplasmic hydration triggers a transient increase in cytoplasmic Ca^{2+} concentration in *Nitella flexilis*. Plant Cell Physiol 36: 335-340

Tominaga M, Morita K, Sonobe S, Yokota E and Shimmen T (1997) Microtubules regulate the organization of actin filaments at the cortical region in root hair cells of *Hydrocharis*. Protoplasma 199: 83-92

Tominaga M, Sonobe S and Shimmen T (1998) Mechanism of inhibition of cytoplasmic streaming by auxin in root hair cells of *Hydrocharis*. Plant Cell Physiol 39: 1342-1349

Tominaga M, Yokota E, Vidali L, Sonobe S, Hepler PK and Shimmen T (1999) Dynamics of cytoskeletons in root hair cell of *Hydrocharis*: Mechanism of actin filament bundling. Plant Cell Physiol 40: s49

Tominaga M, Yokota E, Vidali L, Sonobe S, Hepler PK and Shimmen T (2000) The role of plant villin in the organization of the actin cytoskeleton, cytoplasmic streaming and the architecture of the transvacuolar strand in root hair cells of *Hydrocharis*. Planta, in press

Tominaga Y and Tazawa M (1981) Reversible inhibition of cytoplasmic streaming by intracellular Ca^{2+} in tonoplast-free cells of *Chara australis*. Protoplasma 109: 103-111

Tominaga Y, Shimmen T and Tazawa M (1983) Control of cytoplasmic streaming by extracellular Ca^{2+} in permeabilized *Nitella* cells. Protoplasma 116: 75-77

Tominaga Y, Wayne R, Tung HYL and Tazawa M (1987) Phosphorylation-dephosphorylation is involved in Ca^{2+} -controlled cytoplasmic streaming of characean cells. Protoplasma 136: 161-169

Tsien RY (1998) The green fluorescent protein. Annu Rev Biochem 67:509-544

Tsuchiya Y, Yamazaki H and Aoki T (1991) Steady and transient behaviors of protoplasmic streaming in *Nitella* internodal cell. Biophys J 59: 249-251

Tucker EB and Allen Strömgren N (1986) Intracellular particle motions (cytoplasmic streaming) in staminal hairs of *Setcreasea purpurea*: Effect of azide and low temperature. Cell Motil Cytoskel 6: 305-313

Uchida G, Mizukami Y, Nemoto T and Tsuchiya Y (1998) Sliding motion of magnetizable beads coated with *Chara* motor protein in a magnetic field. J Phys Soc Jap 67: 345-350

Valster AH, Pierson ES, Valenta R, Hepler PK and Emons AMC (1997) Probing the plant actin cytoskeleton during cytokinesis and interphase by profilin microinjection. Plant Cell 9: 1815-1824

Van Troys M, Vanderkerckhove J and Ampe C (1999) Structural modules in actin-binding proteins: Towards a new classification. Biochem Biophys Acta 1448: 323-348

Vidali L, Yokota E, Cheung AY, Shimmen T and Hepler PK (2000) ABP-135 from *Lilium longiflorum* pollen is the plant homologue of villin. Protoplasma, in press
Wacker I and Schnepf E (1990) Effects of nifedipine, verapamil, and diltiazem on tip growth in *Funaria hygrometrica.* Planta 180: 492-501
Wada M, Mineyuki Y and Furuya M (1982) Change in the rate of organelle movement during progression of the cell cycle in *Adiantum* protonemata. Protoplasma 113: 132-136
Wada M, Grolig F and Haupt, W (1993) Light-oriented chloroplast positioning. Contribution to progress in photobiology. J Photochem Photobiol B: 17: 3-25
Wagner G and Grolig F (1992) Algal chloroplast movements. In: Melkonian M (ed) Algal Cell Motility. New York: Chapman & Hall, pp 39-72
Wasteneys GO, Collings DA, Gunning BES, Hepler PK and Menzel D (1996) Actin in living and fixed characeaen internodal cells: identification of a cortical array of fine actin strands and chloroplast actin rings. Protoplasma 190: 25-38
Wayne R, Staves MP and Leopold AC (1990) Gravity-dependent polarity of cytoplasmic streaming in *Nitellopsis*. Protoplasma 155: 43-57
Weiss DG, Keller F, Gulden J and Maile W (1986) Towards new classification of intracellular particle movements based on quantitative analyses. Cell Motil Cytoskel 6: 128-135
Weiss DG, Galfe G, Gulden J, Seitz-Tutter D, Langford GM, Struppler A and Weindl A (1990) Motion analysis of intracellular objects: Trajectories with and without visible tracks. In: Alt W and Hoffmann G (eds) Biological Motion: Lecture Notes in Biomatematics. Vol. 89 Berlin: Springer Verlag, pp 95-116
Wells AL, Lin AW, Chen LQ, Safer D, Cain SM, Hasson T, Carragher BO, Milligan RA and Sweeney HL (1999) Myosin VI is an actin-based motor that moves backwards. Nature 401: 505-508
Wick SM (1991) Spatial aspects of cytokinesis in plant cells. Curr Opin Cell Biol 3: 253-260
Williamson RD (1974) Actin in the alga, *Chara corallina.* Nature 248: 801-802
Williamson RD (1975) Cytoplasmic streaming in *Chara*: A cell model activated by ATP and inhibited by cytochalasin B. J Cell Sci 17: 655-668
Williamson RE (1992) Cytoplasmic streaming in characean algae: Mechanism, regulation by Ca^{2+} and organization. In: M Melkonian (ed) Algal Cell Motility. New York: Chapmann & Hall, pp 73-98
Williamson RE (1993) Organelle movements. Annu Rev Plant Physiol Plant Mol Biol 44: 181-202
Williamson RD and Toh BH (1979) Motile models of plant cells and the immunofluorescent localization of actin in a motile *Chara* cell model. In: Hatano S, Ishikawa H and Sato H (eds) Cell Motility: Molecules and Organization. Tokyo: University of Tokyo Press, pp 339-346
Williamson RE and Ashley CC (1982) Free Ca^{2+} and cytoplasmic streaming in the alga *Chara.* Nature 296: 647-651
Williamson RE and Hurley UA (1986) Growth and regrowth of actin bundles in *Chara*: bundle assembly by mechanisms differing in sensitivity to cytochalasin. J Cell Sci 85: 21-32
Williamson RE, Hurley UA and Perkin JL (1986) Regeneration of actin bundles in *Chara*: Polarized growth and orientation by endoplasmic flow. Eur J Cell Biol 34: 221-228
Williamson RE, McCurdy DW, Hurley UA and Perkin JL (1987) Actin of *Chara* giant internodal cells. Plant Physiol 85: 268-272
Woods CM, Reid MS and Patterson BD (1984a) Response to chilling stress in plant cells. I. Changes in cyclosis and cytoplasmic structure. Protoplasma 121: 8-16
Woods CM, Polito VS and Reid MS (1984b) Response to chilling stress in plant cells. II. Redistribution of intracellular calcium. Protoplasma 121: 17-24
Worring M and de Win AHN (1996). Analysis of complex motion patterns using region based curve analysis. Proc Int Conf Pattern Recognition, Vienna, pp 388-392

Xu C, Zipfel W, Shear JB, Williams RM and Webb WW (1996) Multiphoton fluorescence excitation: New spectral windows for biological nonlinear microscopy. Proc Natl Acad Sci 93: 10763-10768

Yamamoto K, Kikuyama M, Sutoh-Yamamoto N and Kamitsubo E (1994) Purification of actin based motor protein from *Chara corallina*. Proc Jap Acad 70: 175-180

Yamamoto K, Kikuyama M, Sutoh-Yamamoto N, Kamitsubo E and Katayama E (1995) Myosin from alga *Chara*: Unique structure revealed by electron microscopy. J Mol Biol 254: 109-112

Yokota E, Takahara KI and Shimmen T (1998) Actin-bundling protein isolated from pollen tubes of lily: Biochemical and immunocytochemical characterization. Plant Physiol 116: 1421-1429

Yokota E, Muto S and Shimmen T (1999) Inhibitory regulation of higher-plant myosin by Ca^{2+} ions. Plant Physiol 119: 231-239

Chapter 11

MOTILITY OF ENDOPLASMIC RETICULUM IN PLANT CELLS

Irene K Lichtscheidl[1] and František Baluška[2]

1 Institute of Plant Physiology, University of Vienna, Althanstrasse 14, A–1091 Vienna, Austria
2 Botanisches Institut der Universität Bonn, Kirschallee 1, D–53115 Bonn, Germany

Key words: actin, endoplasmic reticulum, motility, myosin, organelle movement

Abstract: The endoplasmic reticulum (ER) can be observed in living plant cells by fluorescence and video microscopy. Both its distribution and its dynamic properties seem largely to depend on actin and myosin. Actin filaments (AFs) provide a framework for the spacing of the cortical, lace-like ER elements and of the ER strands. Moreover, AFs are held responsible for the extension of ER tubules and lamellae as well as for their motility. Synchronized movements of ER and other organelles of plant cells seem to depend on co-localizations of ER and AFs. We summarize the growing body of evidence that the motility of plant ER is based preferentially on the actin cytoskeleton.

1. INTRODUCTION

The endoplasmic reticulum (ER) traverses the cytoplasm of mature plant cells within the cortical areas subjacent to the cell wall as well as in transvacuolar strands. It forms a lace-like net whose quantity and shape are highly variable and correspond to the developmental, physiological as well as the ecological situation of the cell. ER membranes occur in the form of thin tubules with dimensions (80 nm diameter) below the limit of resolution of the light microscope. These tubules either can become inflated into small elongated or spherical vacuoles or extend as thin lamellar sheets. The structure and function of this multipurpose organelle have been well investigated, and results were summarized in some recent reviews (Hepler et al., 1990; Vitale et al., 1993; Staehelin, 1997). One reason why the ER is able to fulfill its many tasks, often simultanously, is that it is highly mobile

C.J. Staiger et al. (eds.), Actin: A Dynamic Framework for Multiple Plant Cell Functions, 191–201.

and dynamic. This dynamic behaviour, however, has so far been investigated only in few cases in animal and plant cells, which is due to the size of the membranes. These latter become visible mainly in the static images from the electron microscope (EM). With the introduction of membrane specific fluorescent dyes, the development of electronic light microscopy, and the possibility to tag specific proteins with green fluorescent protein (GFP), it is now possible to observe the membranous elements of ER in living cells and, hence, to analyze their motile behaviour and interactions with other organelles.

2. TECHNIQUES FOR IN VIVO VISUALIZATION OF ER

Aside from the most important analysis of the structure and function of ER by different EM techniques (reviewed by Hepler et al., 1990; Staehelin, 1997), knowledge about its dynamic behaviour has long been sought through analysis of living cells in the light microscope. In an early paper, Drawert and Rueffer-Bock (1964) used chlorotetracycline in onion inner epidermal cells and showed the fluorescence of ER tubules and lamellae which is due to the high content of calcium in the ER; this technique still proves useful (Allen & Brown, 1988). Alternatively, fluorescence of ER can be obtained by treating cells with carbocyanine dyes, such as DiOC6(3) (Terasaki et al., 1984; Terasaki, 1990). Again, onion inner epidermal cells served as a model system (Quader & Schnepf, 1986; Quader et al., 1987, 1989), but other cells were also tested, as for instance caulonemal cells of *Funaria* during their development (McCauley & Hepler, 1992). A shadow was cast on the elegance of this technique, however, when it was found out that the dye, or a combination of such dyes with anti-cytoskeletal drugs, might be toxic to the cells (Quader & Schnepf, 1986; Lee & Chen, 1988).

An alternative means of study, therefore, was sought in the application of video microscopy (Inoué, 1986; Shotton, 1988; Lichtscheidl & Foissner, 1995). By using ultraviolet microscopy (Lichtscheidl & Url, 1987) and video-enhanced contrast light microscopy (Allen & Brown, 1988; Lichtscheidl & Weiss, 1988), the distribution and the dynamics of ER tubules and lamellae in living cells could be observed without any cytochemical treatment (Lichtscheidl & Url, 1990).

With the introduction of molecular tools, like GFP-constructs, we are now able to observe fluorescing ER and/or associated organelles in living cells (Boevink et al., 1998; Hawes et al., 1999; Marc et al., 1999; Nebenführ et al., 1999; Ridge et al., 1999). Overexpression of the fusion proteins seems not to disturb either the structure or the function of the organelles (e.g. Nebenführ et al., 1999; Ridge et al., 1999).

3. MORPHOLOGY OF ER

The image of ER in mature plant cells that was developed with the above techniques is that of a continuum of tubular and lamellar elements within both the parietal cytoplasm subjacent to the cell wall and the transvacuolar cytoplasmic strands (Hepler et al., 1990; Lichtscheidl & Hepler, 1996; Staehelin, 1997). Using light microscopy, its morpholgy and dynamics are difficult to observe within the transvacuolar strands, because here the ER tubules are tightly packed. We can see, however, that they are parallel and interconnected and seem to form a polygonal net which is stretched in two directions. Knowledge about the organization and dynamics of the ER, therefore, comes mainly from observations of the cortical cytoplasm in which the membrane tubules combine at three-way junctions into polygonal meshes. The mesh width may differ greatly within and between cells (Lichtscheidl & Url, 1990), and depends on developmental, physiological and ecological conditions (e.g. Quader et al., 1989; Quader & Fast, 1990; McCauley & Hepler, 1992; Hoftberger & Lütz-Meindl, 1999). In addition, tubules sometimes radiate from focal arrays thus forming asters (Hepler, 1981; Allen & Brown, 1988). They can coalign and form small bundles, and may also become flattened into thin lamellar sheets (Drawert & Rüffer-Bock, 1964; Lichtscheidl & Url, 1990).

The shape and distribution of the lace-like reticulum seems to depend on actin, although there is also one report which describes the disintegration of ER by high doses of colchicine (Allen & Brown, 1988). When actin filaments (AFs) are disrupted by treatment with cytochalasin D, tubules accumulate into patches and fuse to form lamellar sheets (Hensel, 1987; Allen & Brown, 1988; Quader et al., 1987, 1989; Knebel et al., 1990), or meshes increase in size and form large rectangular windows which seem to run parallel with cortical microtubules (MTs). In contrast to actin drugs, anti-MT agents have no effect on the endoplasmic mainstream but distort cortical streaming (Marc et al., 1999). Additional weight is given to this assumption by the colocalization of phalloidin-stained AFs with ER in ERD2::GFP fusion constructs (Boevink et al., 1998) and by pictures from the EM where ER is often associated with AFs (Lichtscheidl et al., 1990; Staehelin, 1997). MTs do occur but they are usually not co-aligned. This is in striking contrast to the ER in animal cells; although the ER is of very similar shape (reviewed by Terasaki, 1990), its distribution depends preferentially on MTs in living animal cells (Terasaki et al., 1986; Dailey & Bridgman, 1988; Lee & Chen, 1988; Lee et al., 1989; Allan & Vale, 1994; Waterman-Storer & Salmon,

1998; reviewed by Terasaki, 1990). In accordance with this, anti-actin drugs cause no disruption of ER in extracts from cultured animal cells (Dabora & Sheetz, 1988; Vale & Hotani, 1988). Nevertheless, at least some forms of ER motility are supported via actomyosin-based forces also in animal cell (for unconventional myosin V see Tabb et al., 1998).

4. ANCHORAGE AND MOTILITY OF ER

Movements of individual ER elements were described by using fluorescent dyes and video-enhanced contrast microscopy (Allen & Brown, 1988; Knebel et al., 1990; Lichtscheidl & Url, 1990; Quader, 1990), and by the use of ER::GFP constructs (Boevink et al., 1998; Nebenführ et al., 1999; Marc et al., 1999; Ridge et al., 1999).

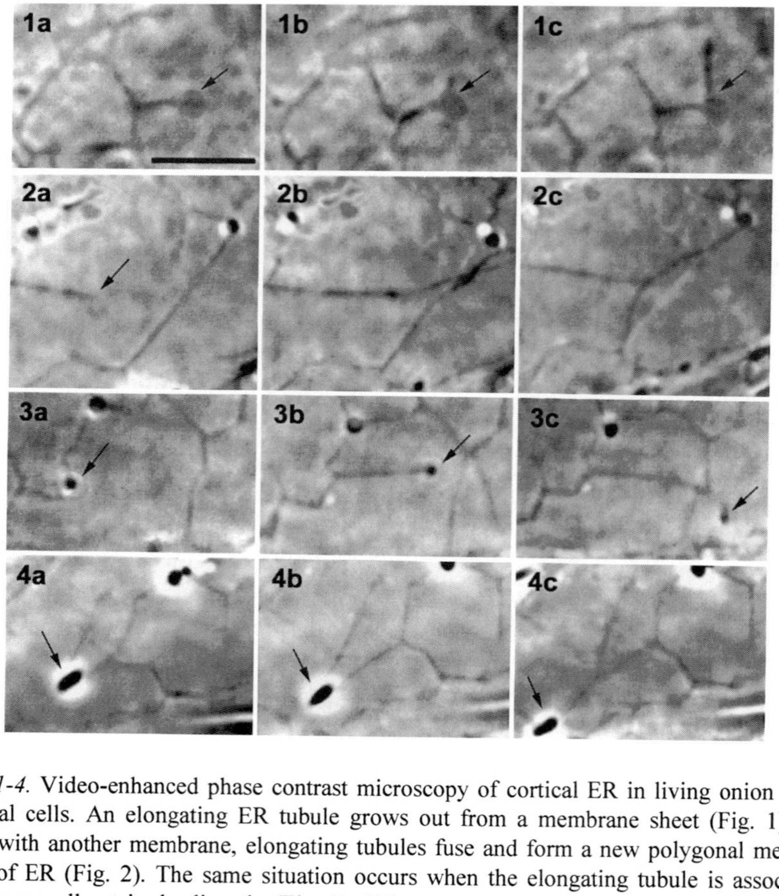

Figure 1-4. Video-enhanced phase contrast microscopy of cortical ER in living onion inner epidermal cells. An elongating ER tubule grows out from a membrane sheet (Fig. 1). On contact with another membrane, elongating tubules fuse and form a new polygonal mesh in the net of ER (Fig. 2). The same situation occurs when the elongating tubule is associated with an organelle at its leading tip (Fig. 3). When an organelle is hindered in its way by a membrane element, this can be stretched to some extent (Fig. 4a, b). Afterwards; the organelle

continues its way and it may pull out a thread of membrane from the impeding ER (Fig. 4c). Reproduced from Lichtscheidl and Url (1990) with permission from Springer-Verlag. Bar: 5μm.

In the very periphery of the cell, closely adjacent to the cell wall, the polygonal ER net appears is rather stable and there are many sites where ER tubules and lamellae seem fixed and anchored. When tubules and lamellae dislocate, because they are pulled away by associated organelles, they return to these anchorage sites after some time. Close binding of ER to the PM was also convincingly proven by centrifugation experiments which failed to dislocate the cortical ER (Quader et al., 1987). The mechanism of this immobilization is not clear yet. It could depend on close binding of the ER to the plasma membrane, as it was observed in *Drosera* in the EM (Lichtscheidl et al., 1990; Staehelin, 1997). Support is given to this assumption by Reuzeau et al. (1997) who correlated binding sites at the plasma membrane to actin and to transmembrane proteins. They might represent so-called 'plasmalemmal control centers' which were hypothesized to govern the activities of ER within the cell and also anchor ER elements at distinct cell periphery domains. The participation of the cell wall in the stabilization of ER and associated actin and organelles has to be considered, because this feature seems characteristic of plant cells. In animal cells, to our knowledge, such fixation of peripheral ER was not described. Studies of plant ER using plasmolysis showed that some elements of the cortical ER remained tightly attached to the plasma membrane and cell wall (e.g. Oparka et al., 1996). Moreover, the cell-to-cell continuity of ER elements through plasmodesmata (Baron-Epel et al., 1988; Grabski et al., 1993; Cantrill et al., 2000) also suggests its anchorage to certain domains at the cell periphery. Interestingly, cortical ER population which associates with plasmodesmata and pit-fields is enriched with calreticulin (Baluška et al., 1999) and plant unconventional myosin VIII (Baluška & Volkmann, in preparation). Recent reports stress possible involvements of both actin and myosin in anchoring ER at domains near and at plasmodesmata (Blackman & Overall, 1998; Radford & White, 1998). Interestingly in this respect, plant unconventional myosin VIII localizes to plasmodesmata (Reichelt et al., 1999; Baluška et al., this volume) and AFs are often organized abundantly around plasmodesmata and pit-fields in root cells (Baluška et al., 2000, this volume). As plant myosin VIII associates presumably also with the PM (Reichelt et al., 1999), it might participate in anchoring cortical ER to PM at pit-fields. In support of this attractive notion, the general inhibitor of myosin ATPases 2,3-butanedione 2-monoxime (e.g., Cramer & Mitchison, 1995; May et al., 1998) induces alterations to the morphology of cortical ER elements associated with plasmodesmata (Šamaj et al., 2000).

Despite this relative stability of cortical ER, there is a constant changing of the spacing and distribution of its reticulum. Stable ER membranes serve as tracks for elongating tubules which can coalign somewhat deeper within the cytoplasm and form small bundles. From lamellar sheets as well as from tubules of the existing ER, additional membranes emerge and grow into new tubules and lamellae (Fig. 1). They course through the cytoplasm, and on contact with other membranes may fuse and form a new junction and thus add new meshes to the net (Fig. 2). Due to these events, the net of ER is getting to be denser. The junctions are not always fixed, and they may glide along the tubules. This leads to distortions of lamellar sheets and to changes of the polygons. Tubules may glide away from a certain part of the net, and rings may become constricted and finally disappear. ER thus becomes eliminated from parts of the cell and the net appears less dense (Lichtscheidl & Url, 1990). These forms of ER motility are very similar to those described in animal cells, where 'tubule branching', 'ring closure' and 'sliding' of junctions have been described (Lee & Chen, 1988). In animal cells, membrane movements depend on MTs and associated kinesins (Allan & Vale, 1994; Waterman-Storer & Salmon, 1998; reviewed by Terasaki et al., 1990). In plant cells, there is also evidence for a structural association of ER with MTs (Franke, 1971; Lancelle & Hepler, 1992; McCauley & Hepler, 1992). In addition, however, there is convincing evidence for the coalignment of ER with AFs (Lichtscheidl et al., 1990; Boevink et al., 1998). Moreover, all physiological studies point to actomyosin-driven motilities in plant cells. In extruded cytoplasm from the giant internodal cells of the alga *Nitella*, for instance, tubules of ER move along AF cables (Kachar & Reese, 1988). This process is driven by myosins and depends on a sufficient supply of ATP. The same is true for onion cells, where experiments with anti-cytoskeletal drugs revealed that actomyosin-based forces drive the movement of ER membranes (Allen & Brown, 1988; Liebe & Quader, 1994). Similar observations were made in transformed leaf epidermal cells of *Nicotiana* (Boevink et al., 1998).

5. SYNCHRONIZED MOVEMENTS OF ER AND ORGANELLES

Connections between the ER and organelles have been known for several years from static images of the EM (e.g. Harris & Oparka, 1983). When living ER was investigated, a mutual dependence of ER arrangements and organelle movements could be found (Drawert & Rüffer-Bock, 1964; Knebel et al., 1990; Lichtscheidl & Url, 1990). These findings revealed that the distribution of ER elements can be governed by adjacent organelles and this could, at the same time, influence the pattern of organelle movements:

(1) Organelles affect the distribution of ER elements; by connecting to ER tubules and lamellae, they can drag along a tail of membranes on their way through the cytoplasm. The newly formed membranes may fuse with the existing net. Organelles thus weave new meshes and strands into more complex ER arrangements (Fig. 3).

(2) When organelles course through the cytoplasm, they often follow the tracks provided by ER tubules (Lichtscheidl & Weiss, 1988; Pauluzzi et al., 1996; Boevink et al., 1998; Hawes et al., 1999), which are coaligned by AFs (Lichtscheidl et al., 1990; Boevink et al., 1998). In addition, they can be hindered in their motion by membranes which block their way (Lichtscheidl & Url, 1990): organelles then can stretch the obstructing membranes to some extent and either become repulsed or bind to these membranes. Alternatively, they continue their way whereby they often draw a membrane element out from the impeding ER element (Fig. 4). In addition, organelles become suspended in the net by binding to the membranes and adapt their shape to the geometry of the polygons (Lichtscheidl & Url, 1990).

While Drawert and Rueffer-Bock (1964) described mainly mitochondria, Allen and Schumm (1990) reported about ER-based motilities of tiny vesicles described as calciosomes, and Boevink et al. (1998), Hawes et al. (1999) and Nebenführ et al. (1999) analysed the trafficking of the whole Golgi apparatus moving along an actin/ER network. By video microscopy, we can see all sorts of organelles including mitochondria, Golgi bodies, spherosomes and microbodies at the tips of elongating membranes and to follow the tracks of the existing ER (Lichtscheidl & Url, 1990).

The synchronized movements of ER and organelles are very similar to those observed in animal cells (Lee & Chen, 1988; Lee et al., 1989; Fuhrmann et al., 1990). Again, the molecular motors involved in this task appear to be MT-based in animal cells while all evidence available point to actomyosin-based motors in higher plant cells (e.g. Allen & Schumm, 1990; Lichtscheidl & Url, 1990; Liebe & Quader, 1994; Boevink et al., 1998; Hawes et al., 1999; Nebenführ et al., 1999). It is suggested that plant ER is coaligned by AFs that serve as tracks for organelles which either travel alone or drag a tail of membrane in their wake.

6. CONCLUSIONS AND FUTURE ASPECTS

Observations of the distribution and motility of the ER in living plant cells reveal the image of a highly dynamic system of interconnected membranes, that, however, become stabilized at certain anchorage sites close to the plasma membrane and cell wall. For both anchorage and motility, as well as for the correlated movement of organelles, all points to the active involvement of the actin-myosin system. Myosins might participate in the stabilization of ER, and AFs serve as tracks for elongating new membranes with and without an organelle at their leading tip. This is in striking contrast to the ER in animal cells, which has very similar dynamic properties that depend, however, on microtubule-based motors.

Regarding the many fundamental processes that are governed or influenced by ER, including the synthesis and the transport of proteins and lipids, the regulation of ionic conditions, especially of calcium and protons, and the participation in signal perception and transduction, further studies about its motile behaviour should be performed to deepen the understanding of the physiological functions of the ER. In addition, they should lead to a better understanding of the dynamic properties of other organelles which are so often correlated with the ER.

A video showing the motility of ER in living onion inner epidermal cells is provided at http://www.univie.ac.at/pph/cell.htm

ACKNOWLEDGEMENT

We thank Drs Chris Hawes, Peter Hepler, Jan Marc and Andrew Staehelin for invaluable comments, and we are indebted to Dr. Peter Barlow for help with the preparation of the manuscript.

REFERENCES

Allan V and Vale R (1994) Movement of membrane tubules along microtubules in vitro: Evidence for specialised sites for motor attachment. J Cell Sci 107: 1885-1897

Allen NS and Brown DT (1988) Dynamics of the endoplasmic reticulum in living onion epidermal cells in relation to microtubules, microfilaments, and intracellular particle movement. Cell Motil Cytoskel 10: 153-163

Allen NS and Schumm JH (1990) Endoplasmic reticulum, calciosomes and their possible roles in signal transduction. Protoplasma 154: 172-178

Blackman LM and Overall RL (1998) Immunolocalisation of the cytoskeleton to plasmodesmata of *Chara corallina*. Plant J 14: 733-741

Baluška F, Šamaj J, Napier, R and Volkmann D (1999) Maize calreticulin localizes to plasmodesmata in root apex. Plant J 19: 481-487

Baluška F, Volkmann D and Barlow PW (2000) Actin-based domains of 'cell periphery complex' and their associations with polarized 'cell bodies' in higher plants. Plant Biol, in press

Baron-Epel O, Hernandez D, Jiang L-W, Meiners S and Schindler M (1988) Dynamic continuity of cytoplasmic and membrane compartments between plant cells. J Cell Biol 106: 715-721

Boevink P, Oparka K, Santa Cruz S, Martin B, Betteridge A and Hawes C (1998) Stacks on tracks: The plant Golgi apparatus traffics on an actin/ER network. Plant J 15: 441-447

Cantrill LC, Overall RL and Goodwin PB (1999) Cell-to-cell communication via plant endomembranes. Cell Biol Int 23: in press

Cramer LP and Mitchison TJ (1995) Myosin is involved in postmitotic cell spreading. J Cell Biol 131: 179-189

Dabora SL and Sheetz MP (1988) The microtubule-dependent formation of a tubulovesicular network with characteristics of ER from cultured cell extracts. Cell 54: 27-35

Dailey ME and Bridgman PC (1988) Relationship between the endoplasmic reticulum and microtubules in cultured nerve growth cones. J Cell Biol 107: 39a

Drawert H and Rüffer-Bock U (1964) Fluorochromierung von Endoplasmatischem Reticulum, Dictyosomen und Chondriosomen mit Tetracyclin. Ber Dtsch Bot Ges 77: 440-449

Franke WW (1971) Cytoplasmic microtubules linked to endoplasmic reticulum with cross-bridges. Exp Cell Res 66: 486-489

Fuhrmann C, Bereiter-Hahn J and Brändle K (1990) Influence of the cytoskeleton, energy supply, and protein synthesis on the structure of the endoplasmic reticulum. Protoplasma 158: 53-65

Grabski F, de Feijter AW and Schindler M (1993) Endoplasmic reticulum forms a dynamic continuum for lipid diffusion between contiguous soybean root cells. Plant Cell 5: 25-38

Harris N and Oparka J (1983) Connections between dictyosomes, ER and GERL in cotyledons of Mung bean (*Vigna radiata* L.). Protoplasma 114: 93-102

Hawes C, Brandizzi F and Andreeva AV (1999) Endomembranes and vesicle trafficking. Curr Opin Plant Biol 2: 454-461

Hensel W (1987) Cytodifferentiation of polar plant cells: Formation and turnover of endoplasmic reticulum in root statocytes. Exp Cell Res 172: 377-384

Hepler PK (1981) The structure of the endoplasmic reticulum revealed by osmium tetroxide-potassium ferricyanide staining. Eur J Cell Biol 26: 102-110

Hepler PK, Palevitz BA, Lancelle SA, McCauley MM and Lichtscheidl IK (1990) Cortical endoplasmic reticulum in plants. J Cell Sci 96: 355-373

Hoftberger M and Lütz-Meindl U (1999) Septum formation in the desmid *Xanthidium* (Chlorophyta): Effects of cytochalasin D and latrunculin B suggest the involvement of actin microfilaments. J Phycol 35: 768-777

Inoué S (1986) Video Microscopy. New York/London: Plenum Press

Kachar B and Reese TS (1988) The mechanism of cytoplasmic streaming in Characean algal cells: Sliding of endoplasmic reticulum along actin filaments. J Cell Biol 106: 1545-1552

Knebel W, Quader H and Schnepf E (1990) Mobile and immobile endoplasmic reticulum in onion bulb epidermis cells: Short- and long-term observations with a confocal laser scanning microscope. Eur J Cell Biol 52: 328-340

Lancelle SA and Hepler PK (1992) Ultrastructure of freeze-substituted pollen tubes of *Lilium longiflorum*. Protoplasma 167: 215-230

Lee C and Chen LB (1988) Behavior of endoplasmic reticulum in living cells. Cell 54: 37-46

Lee C, Ferguson M and Chen LB (1989) Construction of the endoplasmic reticulum. J Cell Biol 109: 2045-2055

Lichtscheidl IK and Foissner I (1995) Video microscopy of dynamic plant cell organelles: Principles of the technique and practical application. J Microsc 181: 117-128

Lichtscheidl IK and Hepler PK (1996) Endoplasmic reticulum in the cortex of plant cells. In: Membranes: Specialized Functions in Plants. Smallwood M, Knox JP, Bowles DJ (eds), Oxford: BIOS Scientific Publishers Ltd, pp 383-402

Lichtscheidl IK and Url WG (1987) Investigation of the protoplasm of Allium cepa inner epidermal cells using ultraviolet microscopy. Eur J Cell Biol 43: 93-97

Lichtscheidl IK and Url WG (1990) Organization and dynamics of endoplasmic reticulum in inner epidermal cells of onion bulb scales. Protoplasma 157: 203-215

Lichtscheidl IK and Weiss DG (1988) Visualization of submicroscopic structures in the cytoplasm of Allium cepa inner epidermal cells by video-enhanced contrast light microscopy. Eur J Cell Biol 46: 376-382

Lichtscheidl IK, Lancelle SA and Hepler PK (1990) Actin-endoplasmic reticulum complexes in *Drosera*: Their structural relationship with the plasmalemma, nucleus and organelles in cells prepared by high-pressure freezing. Protoplasma 155: 116-126

Liebe S and Quader H (1994) Myosin in onion (*Allium cepa*) bulb scale epidermal cells: Involvement in dynamics of organelles and endoplasmic reticulum. Physiol Plant 90: 114-124

Marc J, Granger CL and Cyr RJ (1999) Dynamic reorganization of endoplasmic reticulum along actin filaments and microtubules. ASPP Annual Meeting, Abstract 41003

May KM, Wheatley SP, Amin V and Hyams JS (1998) The myosin ATPase inhibitor 2,3-butanedione-2-monoxime (BDM) inhibits tip growth and cytokinesis in the fission yeast, *Schizosaccharomyces pombe*. Cell Motil Cytoskel 41: 117-125

McCauley MM and Hepler PK (1992) Cortical ultrastructure of freeze-substituted protonemata of the moss *Funaria hygrometrica*. Protoplasma 169: 168-178

Nebenführ A, Gallagher LA, Dunahay TG, Frohlick JA, Mazurkiewicz AM, Meehl JB and Staehelin LA (1999) Stop-and-go movements of plant Golgi stacks are mediated by the acto-myosin system. Plant Physiol 121: 1127-1141

Oparka K, Roberts AG, Roberts IM, Prior DAM and Santa Cruz S (1996) Viral coat protein is targeted to, but does not gate, plasmodesmata during cell-to-cell movement of potato virus X. Plant J 10: 805-813

Pauluzzi I, Lichtscheidl IK and Url WG (1996) Ultra-violet microscopy. Proc Royal Microsc Soc 31: 155-160

Quader H (1990) Formation and disintegration of cisternae of the endoplasmic reticulum visualized in live cells by conventional fluorescence and confocal laser scanning microscopy: Evidence for the involvement of calcium and the cytoskeleton. Protoplasma 155: 166-175

Quader H and Fast H (1990) Influence of cytosolic pH changes on the organisation of the endoplasmic reticulum in epidermal cells of onion bulb scales. Protoplasma 157: 216-224

Quader H and Schnepf E (1986) Endoplasmic reticulum and cytoplasmic streaming: Fluorescence microscopical observations in adaxial epidermis cells of onion bulb scales. Protoplasma 131: 250-252

Quader H, Hofmann A and Schnepf E (1987) Shape and movement of the endoplasmic reticulum in onion bulb epidermis cells: Possible involvement of actin. Eur J Cell Biol 44: 17-26

Quader H, Hofmann A and Schnepf E (1989) Reorganization of the endoplasmic reticulum in epidermal cells of onion bulb scales after cold stress: Involvement of cytoskeletal elements. Planta 177: 273-280

Radford JE and White RG (1998) Localization of a myosin-like protein to plasmodesmata. Plant J 14: 743-750

Reichelt S, Knight AE, Hodge TP, Baluška F, Šamaj J, Volkmann D and Kendrick-Jones J. (1999) Characterization of the unconventional myosin VIII in plant cells and its localization at the post-cytokinetic cell wall. Plant J 19: 555-569

Reuzeau C, Doolittle KW, McNally JG and Pickard B (1997) Covisualization in living onion cells of putative integrin, putative spectrin, actin, putative intermediate filaments, and other proteins at the cell membrane and in the endomembrane sheath. Protoplasma 199: 173-197

Ridge RW, Uozumi Y, Plazinski J, Hurley UA and Williamson RE (1999) Developmental transitions and dynamics of the cortical ER of Arabidopsis cells seen with green fluorescent protein. Plant Cell Physiol 40: 1253-1261

Šamaj J, Peters M, Volkmann D and Baluška F (2000) Effects of myosin ATPase inhibitor 2,3-butanedione 2-monoxime on distributions of myosins, F-actin, microtubules, and cortical endoplasmic reticulum in maize root apices. Plant Cell Physiol, in press

Shotton D (1988) Video enhanced light microscopy and its application in cell biology. J Cell Sci 89: 129-150

Staehelin LA (1997) The plant ER: A dynamic organelle composed of a large number of discrete functional domains. Plant J 11: 1151-1165

Tabb JS, Molyneaux BJ, Cohen DL, Kuznetsov SA and Langford GM (1998) Transport of ER vesicles on actin filaments in neurons by myosin V. J Cell Sci 111: 3221-3234

Terasaki M (1990) Recent progress on structural interaction of the endoplasmic reticulum. Cell Motil Cytoskel 15: 71-75

Terasaki M, Song J, Wong JR, Weiss MJ and Chen LB (1984) Localization of endoplasmic reticulum in living and glutaraldehyde-fixed cells with fluorescent dyes. Cell 38: 101-108

Terasaki M, Chen LB and Fujiwara K (1986) Microtubules and the endoplasmic reticulum are highly interdependent structures. J Cell Biol 103: 1557-1568

Vale RD and Hotani H (1988) Formation of membrane networks in vitro by kinesin-driven microtubule movement. J Cell Biol 107: 2233-2242

Vitale A, Ceriotti A and Denecke J (1993) The role of the endoplasmic reticulum in protein synthesis, modification and intracellular transport. J Exp Bot 44: 1417-1444

Waterman-Storer CM and Salmon ED (1998) Endoplasmic reticulum membrane tubules are distributed by microtubules in living cells using three distinct mechanisms. Curr Biol 8: 798-806

Chapter 12

ROLES FOR ACTIN FILAMENTS IN CHLOROPLAST MOTILITY AND ANCHORING

Shingo Takagi

Department of Biology, Graduate School of Science, Osaka University, Machikaneyama 1–16, Toyonaka, Osaka, 560–0043 Japan

Key words: actin microfilament, chloroplast, orientation movement (of chloroplasts), photosynthesis, phytochrome, *Vallisneria gigantea* (epidermal cell)

Abstract: In a variety of plant cells, chloroplasts change their intracellular arrangement and/or distribution in response to light signals, such as the direction, fluence rate and wavelength. Rapid improvements in cell biological techniques, which enable visualization of cytoskeletal organization in plant cells, has made it increasingly evident that actin microfilaments play a critical role in regulating the intracellular movements and anchoring of chloroplasts. Recent progress in this field is briefly summarized and a possible mode of action of the actin cytoskeleton in the photocontrol of motility of chloroplasts is discussed.

1. INTRODUCTION

Intracellular movements of chloroplasts have long attracted many plant scientists in cell biology as well as in ecophysiology. Among a number of environmental stimuli, in response to irradiation with light, chloroplasts exhibit conspicuously unique modes of movement (Haupt & Scheuerlein, 1990; Wada et al., 1993; Haupt, 1999). As a result of such movements, each chloroplast settles into a region where the exposure to light is maximal under light at low fluence rates and where exposure is minimal under light at high fluence rates. These light-dependent movements are consequently designated as orientation movements of chloroplasts and these movements are considered as one of the adaptive responses of plant cells which lead to efficient photosynthesis under fluctuating light conditions (Zurzycki, 1955).

C.J. Staiger et al. (eds.), Actin: A Dynamic Framework for Multiple Plant Cell Functions, 203–212.

It has been known that, in most cases, light in the blue region effectively induces the orientation movements of chloroplasts (Senger, 1980, 1984; Haupt, 1999). Light-dependent redistribution of chloroplasts was also observed in *Arabidopsis*. However, it was suggested that none of the blue light photoreceptors identified to date (Batschauer, 1998) is involved in the movements of chloroplasts (Trojan & Gabryš, 1996). The identification and characterization of the relevant blue-light photoreceptor molecules are one of the most exciting topics in photobiology in this area. On the other hand, in several cases, phytochrome has been shown to function in the regulation of chloroplast movements under low fluence rates (Haupt, 1999). Using linearly polarized actinic light, it has been assumed that the photoreceptor molecules are orderly arranged in the vicinity of the plasma membrane (Haupt & Scheuerlein, 1990; Yatsuhashi, 1996).

Cytoskeletal components, which mainly include actin microfilaments, microtubules, and their associated proteins, construct a well organized motile apparatus for the movements of chloroplasts (Nagai, 1993). It has generally been assumed that each chloroplast participates only passively in the movements of a mobile cytoplasmic matrix (Haupt & Schönbohm, 1970). By contrast, there have been some reports that suggest a direct association of putative motor proteins with the chloroplasts (La Claire, 1991; La Claire et al., 1995). In fact, some chloroplasts can apparently exhibit motile behavior, as has been recorded on several occasions by video microscopy (Izutani et al., 1990; Kadota & Wada, 1992b; Takagi et al., 1992; Kagawa & Wada, 1996). The exact localization of motor proteins involved in the generation of motive force for the movements of chloroplasts is still a critical open question.

In the siphonaceous alga *Vaucheria* (Blatt & Briggs, 1980; Blatt et al., 1980), the coenocytic green alga *Caulerpa* (Menzel & Elsner-Menzel, 1989), protonemal cells of the fern *Adiantum* (Kadota & Wada, 1992a), and the green alga *Mougeotia* (Mineyuki et al., 1995), the intracellular organization of actin filaments changes concomitantly with the light-induced movements of chloroplasts. These studies strongly suggest that the actin cytoskeleton plays an important role in the dynamic regulation of light-dependent movements of chloroplasts in plant cells.

We have obtained a substantial body of knowledge on the photoreceptor molecules as well as the cytoskeletal proteins that are implicated in these precisely regulated stimulus-response systems of plant cells. However, the numerous transduction processes intervening between the receptor systems and the effector systems remains obscure. One of the main reasons for this obscurity might reside in a shortage of analytical studies that aim at a thorough cell biological understanding of each process, from the perception of light to the regulation of chloroplast movements in a single experimental system. Leaf cells of aquatic monocotyledonous plants, such as *Elodea* and *Vallisneria*, have traditionally contributed much to the understanding of mechanisms of cell motility in higher plants (Kamiya, 1959; Staiger & Schliwa, 1987). These cells sensitively respond to light and, moreover, provide good experimental systems that could be investigated by light microscopy under excellent optical conditions. In this chapter, I will summarize our studies on the mode of photoregulation of intracellular movements of chloroplasts in epidermal cells of the leaves of *Vallisneria gigantea* Graebner. I also propose a role for the actin cytoskeleton as the molecular machinery which anchors chloroplasts to the plasma membrane.

2. ORIENTATION MOVEMENTS OF CHLOROPLASTS IN EPIDERMAL CELLS OF *VALLISNERIA GIGANTEA*

Epidermal cells of the leaves of *V. gigantea* have a large central vacuole surrounded by a thin layer of cytoplasm that contains a number of chloroplasts. These chloroplasts accumulate in the cytoplasmic layer that faces the outer periclinal wall (P side) under light at low fluence rates and in the layers that face the anticlinal walls (A sides) under light at high fluence rates. Light in the red region most effectively induces the accumulation of chloroplasts on the P side, whereas light in the blue region is exclusively effective to induce accumulation on the A sides (Izutani et al., 1990).

To dissect the cellular processes preceding the accumulation of chloroplasts on the P side, which was induced by irradiation with red light at low fluence rates, we used time-lapse video microscopy (Izutani et al., 1990; Dong et al., 1995). The behavior of chloroplasts on the P side was recorded with a silicon intensifier target camera under light of 796 nm, which has little effect on the orientation movements of chloroplasts. Even under dark conditions, the chloroplasts slowly migrate from the P side to the A sides and in the opposite direction. The chloroplasts migrate in the both directions to a similar extent at 0–0.5 chloroplasts in 1 min. The rate of migration of chloroplasts in the both directions rapidly increased several-fold within 5 min of saturating irradiation with red light (650 nm, 0.41 W m^{-2}). Thereafter, the elevated rate of migration of chloroplasts from the P side to the A sides decreased more rapidly than the elevated rate of migration of chloroplasts in the opposite direction, with the resultant progressive accumulation of chloroplasts on the P side over several tens of minutes. The light-induced, rapid increase in the rate of migration of chloroplasts between the P side and the A sides seems to be an indispensable process in fulfilling a rapid accumulation of chloroplasts on the P side in response to light at low fluence rates.

3. PHOTOCONTROL OF THE MOTILITY OF CHLOROPLASTS

Next, we attempted to examine, in a quantitative manner, light-dependent changes in the motility of individual chloroplasts on the P side (Dong et al., 1996). After recordings obtained by the time-lapse video microscopy were fed into a personal computer, the ratio of the total area over which one chloroplast moved during a defined period of time on the P side to the area of the chloroplast itself was calculated. These motion analyses demonstrated that the motility of individual chloroplasts directly determines the rate of migration of chloroplasts between the P side and the A sides. Under dark conditions, each chloroplast moves over an area of about 1.2–1.7× its own area in 2 min. A brief irradiation with red light increased the motility of chloroplasts by 10–50% in a few minutes. Far-red light (730 nm, 0.14 W m^{-2}) applied subsequently completely antagonized the effect of red light. Moreover, these reversible effects of red and far-red light were obvious even in the presence of inhibitors of photosynthesis. These observations strongly suggest that the motility of chloroplasts in the epidermal cells of *V. gigantea* is under control of type II phytochrome in terms of the fluence required and the photoreversible regulation (Furuya, 1993). The motility of chloroplasts seems to increase in the presence of the far-red light absorbing form of phytochrome (Pfr) and decrease in the presence of the red light absorbing form of phytochrome (Pr). The involvement of an acceleration of the activity of motor proteins (Takagi, 1997) and/or a decrease in the rigidity of the cytoplasmic matrix (Virgin, 1949; Takagi et al., 1992) in the Pfr-dependent rapid increase in the motility of chloroplasts is under investigation.

In the absence of inhibitors of photosynthesis, after a rapid acceleration produced by red light, the motility of chloroplasts declined drastically. Eventually, the chloroplasts became almost immobile on the P side 30–60 min after the start of irradiation with red light, either under continuous red light or under dark conditions. In contrast, in the presence of inhibitors of photosynthesis, the accelerated level of motility of chloroplasts, which had been produced by red light, was maintained for over an hour and, hence, no rapid decline in the rate of migration of chloroplasts from the P side to the A sides was observed. Consequently, both the presence of Pfr and the operation of intact photosynthesis are necessary to induce the accumulation of chloroplasts on the P side, and that the photoreceptor systems exert their effects cooperatively through modulation of the motility of each chloroplast.

4. ACTIN CYTOSKELETON IN THE EPIDERMAL CELLS

Actin microfilaments in the epidermal cells of *V. gigantea* were visualized for the first time by electron microscopy (Yamaguchi & Nagai, 1981). A filamentous structure, of 6–7 nm in diameter, runs in a parallel array as thin bundles in the vicinity of the plasma membrane. In the cytoplasmic suspension prepared from epidermal protoplasts, bundle structures with a transverse periodicity of about 40 nm were frequently observed. The optical diffraction pattern of a paracrystalline array in these bundles exhibited a close similarity to that reported in filamentous muscle actin. Moreover, these bundles could be decorated with muscle heavy meromyosin to form the characteristic arrowhead complexes.

Recently, we have succeeded in visualizing the intracellular organization of actin filaments in the epidermal cells by staining with fluorescein isothiocyanate-conjugated (FITC-conjugated) phalloidin (Dong et al., 1996). In the dark-adapted cells, bundles of actin filaments on the P side run in an almost straight course and form a network array (Fig. 1A). Actin filaments are apparently not in contact with any chloroplasts. Treatment of the epidermal cells with cytochalasin B caused a fragmentation of the bundles of actin filaments (Dong et al., 1998). Concomitantly, the chloroplasts completely lost their motility and were frozen in the cytoplasmic matrix (Izutani et al., 1990). The motility of chloroplasts was also substantially suppressed in the presence of butanedione monoxime, which is known as an non-competitive inhibitor of myosin ATPases (Cramer & Mitchison, 1995). Consequently, the movement of chloroplasts in the epidermal cells of *V. gigantea* is most probably driven by an actomyosin system, as has been shown in other higher plants (Shimmen & Yokota, 1994).

5. LIGHT-INDUCED DYNAMIC CHANGES IN THE CONFIGURATION OF ACTIN FILAMENTS

Actin filaments on the P sides of the epidermal cells change the configuration simultaneously with the red-light-induced accumulation of chloroplasts on the P side (Dong et al., 1996). A considerable number of chloroplasts on the P side seems to become surrounded by actin filaments distributed in a striking honeycomb array (Fig. 1B). To determine whether these dynamic changes in the configuration of actin filaments were the cause or effect of the loss of motility of chloroplasts – that is, whether the actin filaments rendered the chloroplasts immobile through the changes in the configuration or the loss of motility of chloroplasts induced the changes in the configuration of the actin filaments – we examined the precise time course of changes in the configuration of actin filaments upon brief irradiation with red light, as well as the time course of changes in the motility of chloroplasts on the P side (Dong et al., 1998). We revealed that there seems to be a critical period, around 10–20 min after the termination of irradiation with red light for 5 min, during which rapid changes in the configuration of actin filaments proceed in synchrony in the irradiated cells. A rapid decline in the motility of chloroplasts seems to occur during this critical period. This temporal coincidence of the two events suggests that chloroplasts might lose their motility on the P side as a result of changes in the configuration of the actin filaments.

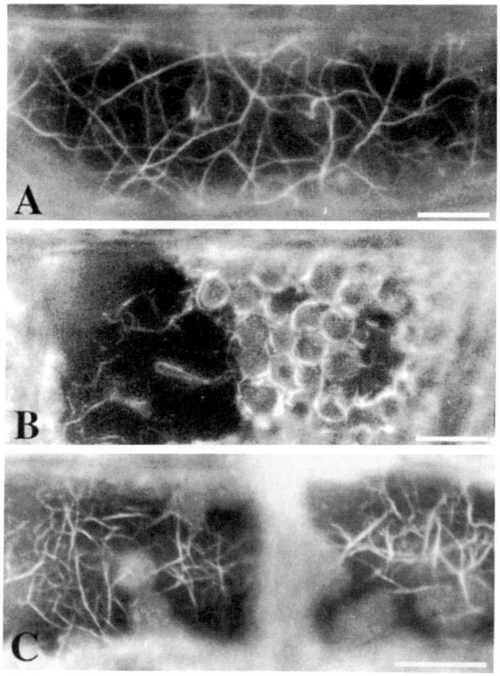

Figure 1. Typical arrays of actin filaments on the P sides of epidermal cells of *V. gigantea*. The actin microfilaments on the P sides were visualized by staining with FITC-conjugated phalloidin in a dark-adapted specimen (A), in a specimen that had been incubated in darkness for 40 min after irradiation with red light (650 nm, 0.41 W m^{-2}) for 5 min (B), and in a specimen after irradiation with blue light (450 nm, 3.0 W m^{-2}) for 5 min (C). A network (A), a honeycomb (B), and a stretched network array (C) are obvious. Bars, 10 μm.

In the presence of inhibitors of photosynthesis, red light produced no detectable change in the configuration of actin filaments on the P side. This was also true of the epidermal cells sequentially irradiated with red and far-red light (Dong et al., 1996). Both the presence of Pfr and the operation of intact photosynthesis are indispensable for the change in the configuration of actin filaments from a network to a honeycomb array. Importantly, these inductive light conditions coincide with, on the P side, an accumulation of chloroplasts and a loss of their motility.

In the case of irradiation with blue light at high fluence rates, the bundles of actin filaments on the P side seem to become much thicker and straighter (Fig. 1C). Under this condition, most of the chloroplasts on the P side moved to the A sides. This movement is induced within several minutes of irradiation with blue light (Izutani et al., 1990). The mode of movement of chloroplasts seems to rapidly change from a randomly oriented, short-range movement, which is observed under dark conditions, to a unidirectional, smooth movement. If we assume that actin filaments provide tracks for the movement of the cytoplasmic matrix, and that the chloroplasts participate passively in the movement of cytoplasmic matrix, the network array (Fig. 1A) and the blue-light induced more stretched array (Fig. 1C), respectively, might be suitable for the randomly oriented, short-range movement and the unidirectional, smoother movement of the chloroplasts.

6. ANCHORING OF CHLOROPLASTS BY ACTIN FILAMENTS

From the temporal analyses of the motility of chloroplasts and the configuration of actin filaments obtained under red light at low fluence rates, we assumed that the actin filaments might contribute to the deceleration of the motility of chloroplasts by anchoring the chloroplasts through the changing of their configuration into a honeycomb array. To ascertain this possibility, we examined the effects of centrifugal forces on the distribution of chloroplasts on the P sides in living epidermal cells (Takagi et al., 1991; Dong et al., 1998). After the inductive light treatment, the chloroplasts on the P sides became extremely resistant to centrifugal forces. Again, the effect of red light was negated either by far-red light or by the presence of inhibitors of photosynthesis. In addition, the increased resistance of chloroplasts to centrifugal forces disappeared after treatment with cytochalasin B. The red light-induced anchorage of chloroplasts on the P side might have been attributable solely to the actin filaments, which surrounded the chloroplasts as a consequence of the change in their configuration from a network to a honeycomb array.

Our tentative conclusion is that, in epidermal cells of the leaves of *V. gigantea*, actin filaments not only drive the movement of chloroplasts but also play a crucial role in anchorage of the chloroplasts along the outer periclinal wall through dynamic changes in their configuration under cooperative regulation by Pfr and photosynthesis.

7. CONCLUDING REMARKS

A general role for actin filaments in intracellular anchorage of chloroplasts has been proposed also from an ecological point of view (Terashima et al., 1995), for example, in terms of mesophyll-cell resistance to CO_2 diffusion (Evans & von Caemmerer, 1996). The following cell biological findings obtained in several plant species may support the proposal. Kobayashi et al. (1987) were probably the first to demonstrate unambiguously the association of actin filaments in reticulated arrays with chloroplasts by the staining of isolated mesophyll cells of *Zinnia* with FITC-phalloidin. Fukuda and Kobayashi (1989) proposed that these arrays of actin filaments play an important role in anchoring the chloroplasts to the plasma membrane. In protonemal cells of the fern *Adiantum*, Kadota and Wada (1992a) proposed a similar mechanism for the light-dependent anchorage of chloroplasts through changes in the configuration of actin filaments to circular arrays. And very recently, in mesophyll cells of *Arabidopsis*, the association of actin filaments with chloroplasts was clearly demonstrated by cryofixation and freeze-substitution methods (Kandasamy & Meagher, 1999).

One of the intriguing points in our findings is an involvement of photosynthesis in the light-dependent changes in the configuration of actin filaments. We at first thought that the dynamic behavior of the actin cytoskeleton should be positioned upstream in the regulation of the intracellular orientation of chloroplasts, thus bringing about an efficient operation of photosynthesis. The total inhibition of the changes in the configuration of actin filaments and the loss of motility of chloroplasts by the inhibitors of photosynthesis definitely denied the idea. A possible involvement of photosynthesis in the orientation movements of chloroplasts was also suggested in *V. spiralis* (Seitz, 1979). Undoubtedly, photosynthesis itself plays a crucial role in the regulation of actin organization to settle the chloroplasts into the appropriate intracellular destination.

Finally, we would like to propose that the actomyosin system is a molecular mechanism that enables the chloroplasts to make intimate contact with the plasma membrane. The resultant highly efficient interaction between photosynthesis and functions of the plasma membrane, such as active transport systems, might be one of the essential requirements for plants to survive under adverse environmental conditions.

REFERENCES

Batschauer A (1998) Photoreceptors of higher plants. Planta 206: 479-492

Blatt MR and Briggs WR (1980) Blue-light-induced cortical fiber reticulation concomitant with chloroplast aggregation in the alga *Vaucheria sessilis*. Planta 147: 355-362

Blatt MR, Wessells NK and Briggs WR (1980) Actin and cortical fiber reticulation in the siphonaceous alga *Vaucheria sessilis*. Planta 147: 363-375

Cramer LP and Mitchison TJ (1995) Myosin is involved in postmitotic cell spreading. J Cell Biol 131: 179-189

Dong XJ, Nagai R and Takagi S (1998) Microfilaments anchor chloroplasts along the outer periclinal wall in *Vallisneria* epidermal cells through cooperation of Pfr and photosynthesis. Plant Cell Physiol 39: 1299-1306

Dong XJ, Ryu JH, Takagi S and Nagai R (1996) Dynamic changes in the organization of microfilaments associated with the photocontrolled motility of chloroplasts in epidermal cells of *Vallisneria*. Protoplasma 195: 18-24

Dong XJ, Takagi S and Nagai R (1995) Regulation of the orientation movement of chloroplasts in epidermal cells of *Vallisneria*: Cooperation of phytochrome with photosynthetic pigment under low-fluence-rate light. Planta 197: 257-263

Evans JR and von Caemmerer S (1996) Carbon dioxide diffusion inside leaves. Plant Physiol 110: 339-346

Fukuda H and Kobayashi H (1989) Dynamic organization of the cytoskeleton during tracheary-element differentiation. Develop Growth Differ 31: 9-16

Furuya M (1993) Phytochromes: Their molecular species, gene families, and functions. Annu Rev Plant Physiol Plant Mol Biol 44: 617-645

Haupt W (1999) Chloroplast movement: From phenomenology to molecular biology. Progress Bot 60: 3-36

Haupt W and Scheuerlein R (1990) Chloroplast movement. Plant Cell Environ 13: 595-614

Haupt W and Schönbohm E (1970) Light-oriented chloroplast movements. In: Halldall P (ed) Photobiology of Microorganisms. London: Wiley-Interscience, pp 283-306

Izutani Y, Takagi S and Nagai R (1990) Orientation movements of chloroplasts in *Vallisneria* epidermal cells: Different effects of light at low- and high-fluence rate. Photochem Photobiol 51: 105-111

Kadota A and Wada M (1992a) Photoinduction of formation of circular structures by microfilaments of chloroplasts during intracellular orientation in protonemal cells of the fern *Adiantum capillus-veneris*. Protoplasma 167: 97-107

Kadota A and Wada M (1992b) Photoorientation of chloroplasts in protonemal cells of the fern *Adiantum* as analyzed by use of a video-tracking system. Bot Mag Tokyo 105: 265-279

Kagawa T and Wada M (1996) Phytochrome- and blue-light-absorbing pigment-mediated directional movement of chloroplasts in dark-adapted prothallial cells of fern *Adiantum* as analyzed by microbeam irradiation. Planta 198: 488-493

Kamiya N (1959) Protoplasmic streaming. In: Heilbrunn LV, Weber F (eds) Protoplasmatologia vol VIII 3a. Vienna: Springer-Verlag, pp 1-199

Kandasamy MK and Meagher RB (1999) Actin-organelle interaction: Association with chloroplast in *Arabidopsis* leaf mesophyll cells. Cell Motil Cytoskel 44: 110-118

Kobayashi H, Fukuda H and Shibaoka H (1987) Reorganization of actin filaments associated with the differentiation of tracheary elements in *Zinnia* mesophyll cells. Protoplasma 138: 69-71

La Claire JW II (1991) Immunolocalization of myosin in intact and wounded cells of the green alga *Ernodesmis verticillata* (Kutzing) Borgesen. Planta 184: 209-217

La Claire JW II, Chen R and Herrin DL (1995) Identification of a myosin-like protein in *Chlamydomonas reinhardtii* (Chlorophyta). J Phycol 31: 302-306

Menzel D and Elsner-Menzel C (1989) Actin-based chloroplast rearrangements in the cortex of the giant coenocytic green alga *Caulerpa*. Protoplasma 150: 1-8

Mineyuki Y, Kataoka H, Masuda Y and Nagai R (1995) Dynamic changes in the actin cytoskeleton during the high-fluence rate response of the *Mougeotia* chloroplast. Protoplasma 185: 222-229

Nagai R (1993) Regulation of intracellular movements in plant cells by environmental stimuli. Int Rev Cytol 145: 251-310

Seitz K (1979) Light induced change in the centrifugability of chloroplasts: Different action spectra and different influence of inhibitors in the low and high intensity range. Z Pflanzenphysiol 95: 1-12

Senger H (1980) The Blue Light Syndrome. Berlin Heidelberg New York: Springer-Verlag

Senger H (1984) Blue Light Effects in Biological Systems. Berlin Heidelberg New York Tokyo: Springer-Verlag

Shimmen T and Yokota E (1994) Physiological and biochemical aspects of cytoplasmic streaming. Int Rev Cytol 155: 97-139

Staiger CJ and Schliwa M (1987) Actin localization and function in higher plants. Protoplasma 141: 1-12

Takagi S (1997) Photoregulation of cytoplasmic streaming: Cell biological dissection of signal transduction pathway. J Plant Res 110: 299-303

Takagi S, Kamitsubo E and Nagai R (1991) Light-induced changes in the behavior of chloroplasts under centrifugation in *Vallisneria* epidermal cells. J Plant Physiol 138: 257-262

Takagi S, Kamitsubo E and Nagai R (1992) Visualization of a rapid, red/far-red light-dependent reaction by centrifuge microscopy. Protoplasma 168: 153-158

Terashima I, Ishibashi M, Ono K and Hikosaka K (1995) Three resistances to CO_2 diffusion: Leaf-surface water, intercellular spaces and mesophyll cells. In: Mathis P (ed) Photosynthesis: From Light to Biosphere, vol V. Dordrecht, Netherlands: Kluwer Academic Publishers, pp 537-542

Trojan A and Gabryš H (1996) Chloroplast distribution in *Arabidopsis thaliana* (L.) depends on light conditions during growth. Plant Physiol 111: 419-425

Virgin HI (1949) The relation between the viscosity of the cytoplasm, the plasma flow, and the motive force. An experimental study. Physiol Plant 2: 157-163

Wada M, Grolig F and Haupt W (1993) Light-oriented chloroplast positioning. Contribution to progress in photobiology. J Photochem Photobiol B: Biol 17: 3-25

Yamaguchi Y and Nagai R (1981) Motile apparatus in *Vallisneria* leaf cells. I. Organization of microfilaments. J Cell Sci 48: 193-205

Yatsuhashi H (1996) Photoregulation systems for light-oriented chloroplast movement. J Plant Res 109: 139-146

Zurzycki J (1955) Chloroplasts arrangement as a factor in photosynthesis. Acta Soc Bot Pol 24: 27-63

Chapter 13

ACTIN AND CYTOMORPHOGENESIS IN THE GIANT, SINGLE-CELLED GREEN ALGAE *ACETABULARIA* AND *MICRASTERIAS*

Ursula Lütz-Meindl[1] and Diedrik Menzel[2]

1 Institut für Pflanzenphysiologie, Universität Salzburg, Hellbrunner Str. 34, A–5020 Salzburg, Austria
2 Botanisches Institut der Universität Bonn, Kirschallee 1, D–53115 Bonn, Germany

Key words: Actin, alga, cell wall, cytomorphogenesis, Dasycladales, Desmidiaceae, intracellular motility, microtubules, spectrin, tip growth

Abstract: Cytomorphogenesis in the giant celled, uninucleate green algae *Acetabularia* and *Micrasterias* involves the actin cytoskeleton in different ways. *Acetabularia* in its sporophytic stage is a tip growing system which undergoes changes in the geometry of the apex in order to initiate the morphogenesis of lateral organs. The subapical wall becomes specifically modified to allow local expansion. These changes are predicted by changes in the intracellular actin pattern. *Micrasterias*, on the other hand, is a symmetrically expanding system. Modification of the cell shape is caused by spatially restricted increase in cell wall growth after exocytosis of specific vesicles. The actin cytoskeleton is required for exocytosis and vesicle transport in both growing and non-growing areas.

1. INTRODUCTION

Studies on cellular morphogenesis in plants have traditionally made use of experimental model organisms that have large cell sizes and elaborate patterns in their shape, wall sculpturing or intracellular architecture. The dasyclad green alga *Acetabularia* and the desmid green alga *Micrasterias*, have both been among the most fascinating cell types to study. They have been introduced as experimental models by Hämmerling (see Hämmerling, 1953 for summary) and by Waris (see Waris & Kallio, 1964 for summary) and Kiermayer (see Kiermayer, 1981 for summary), respectively, and have

C.J. Staiger et al. (eds.), Actin: A Dynamic Framework for Multiple Plant Cell Functions, 213–236.

since been the subject of numerous studies addressing intracellular architecture, spatial pattern formation, and organelle motility during growth, cell shape changes and cytokinesis. The current review summarizes our knowledge on the phenomenon of cellular morphogenesis in these two cell types with particular focus on the role of the actin cytoskeleton.

2. STAGES OF CYTOMORPHOGENESIS

Acetabularia: Members in this genus of siphonous green algae (order Dasycladales, Berger & Kaever, 1992) are unique in several ways. They are of giant dimensions (on the order of 0.5 to several cm in length and 50 to 500 μm in width) but still remain uninuclear for most of their life cycle. They go through distinct stages of growth and morphogenesis comparable to those seen in higher multicellular organisms, i.e., germling, juvenile, adult (with the capacity to reproduce), senescence, and death (Fig. 1, Menzel, 1994; Mandoli, 1998). As the main axis elongates in the juvenile stage, dasyclad green algae undergo repeating cycles of identical apical morphogenesis several times, before acquiring the adult morphology. The juvenile stage is very much alike in all members of the Dasycladales, even though the adult morphologies reveal differences due to permutations in size and shape of lateral organs that have occurred in the course of evolution (Fig. 2). All these features make *Acetabularia* and other members of this group very special experimental model organisms for the study of cellular morphogenesis.

From the moment when the zygote is formed by isogamete fusion *Acetabularia* is polarly constructed and remains that way through its entire life. As the zygote germinates, the cell begins to grow in length by tip growth, which during the juvenile stage is interrupted in regular intervals by the initiation of a whorl of ephemeral dicho-/trichotomously branching hairs. At maturation, the morphogenetic program, that has generated several storeys of hair whorls, is replaced by a new, somewhat modified program that governs the formation of cap rays and with that, tip growth is terminated. The cap rays are much larger lateral appendices than the hairs, but like them they are built from a subapically inserted ring of buds.

As observed by Günther Werz (1965, 1974), the morphogenetic key event in side branch formation – hairs as well as cap rays – is the abrupt change in the direction of the growth axis. This process encompasses the interruption of tip growth of the primary axis, flattening and widening of the tip, positioning of initial growth buds in a ring placed at the margins of the flattened apex, and initiation of secondary tip growth in each of the buds.

Figure 1. Developmental stages of the *Acetabularia* Sporophyte.

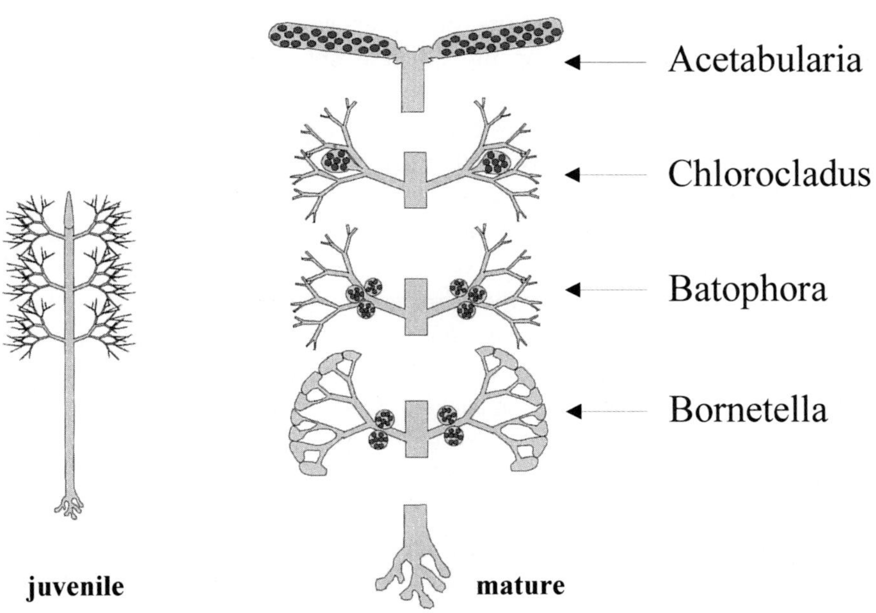

Figure 2. Permutation in size and shape of lateral organs in dasyclad species

Werz (1965) also discovered, that bud formation is, unlike the situation in most other branching cell types, preceded by the formation of lytic cavities in the inner layer of the subapical part of the wall and that these cavities are the founder structures for the new secondary apices. Branch initiation by lytic cavities in the wall is a prerequisite for the formation of the aperture that is inserted between the main axis and each side branch.

From studies of the dasyclad life cycle it is well established, that these stages circumscribe the morphogenesis of the diploid sporophyte (summary in Menzel, 1994), which is followed by morphogenesis of the haploid aplanospores inside the confinement of the sporophyte. Aplanospores represent the gametophytic stage referred to as cysts. Sporulation in the dasyclads comes with a whole new set of morphogenetic programs. Onset of cyst formation is heralded by the pale stalk stage, named after the rapid paling of the stalk due to mass exodus of chloroplasts out of the stalk into the cap ray chambers (Schulze, 1939). This mass transport of chloroplasts sets in right after the giant, primary nucleus in the rhizoid has meiotically divided. The haploid daughters get amplified in number by several rounds of mitosis and along with the chloroplasts become transported into the cap rays. They arrange in an orderly pattern in the cap ray cytoplasm, thus creating domains which then split off from one another and the parent cytoplasm in a highly orchestrated fashion. Each domain rounds up and with the moment of its complete separation becomes an independent entity, the cyst (see Menzel, 1994).

The Gametophytes are morphologically very simple round to ovoid structures with a tough cellulosic cell wall. The only morphogenetic event that requires spatial patterning is the insertion of a lid in the cellulosic cell wall above and around the position of the parietal primary cyst nucleus. During gametophytic maturation which proceeds while cysts are still confined within the walls of the cap rays, nuclei multiply by several consecutive rounds of mitosis, after which they enter a resting stage and only become determined for gamete formation once the appropriate stimuli arise.

With the opening of the lid and release of the biflagellate isogametes, the life cycle is completed.

Micrasterias: Haploid desmid cells (class Conjugatophyceae) have a completely different cell architecture. Like the model system *Micrasterias denticulata* (Fig. 3) which grows to a diameter of about 250 μm, they consist of two semicells connected by a narrow central constriction, the isthmus. The approximately 6000 different species within the family Desmidiaceae (Brook, 1981) develop a variety of different highly ornamented shapes with lobes, spines, papillary tubercles or cell wall ridges distributed over the cell surface in species specific patterns. In the genus *Micrasterias* the biradiate, symmetric morphology of one semicell with an alternating sequence of lobes and indentations mirrors the shape of the other semicell exactly and develops within 5 hours after mitosis. Development starts with the formation of a septum wall growing inward from the parental cell wall of the isthmus region towards the cell center thus separating the two semicells like a diaphragm.

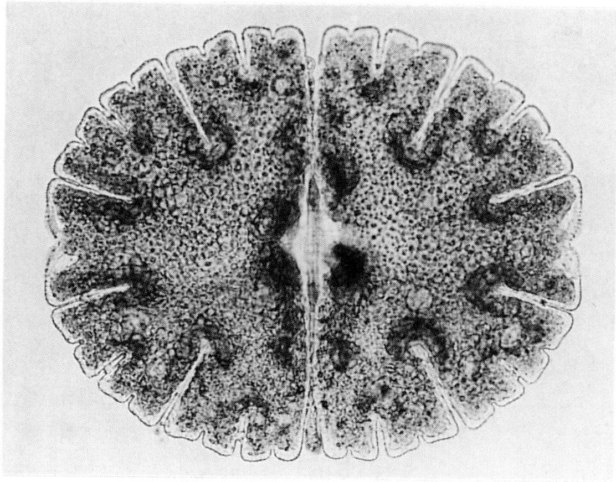

Figure 3. Morphology of *Micrasterias denticulata*

A particular population of Golgi vesicles (Kiermayer, 1981; Lütz-Meindl & Brosch-Salomon, 2000) accounts for septum formation by fusing with the plasma membrane at the rim of the parental cell wall and the edges of the growing septum wall, respectively. The following stage of primary wall formation is characterized by localized deposition of wall material according to a stage specific pattern. During the first 75 min after the onset of mitosis, a bulb develops from the non growing old semicell and expands uniformly in the main plane of the cell (Fig. 4). In *Micrasterias denticulata* pattern formation of the new semicell starts with cessation of growth at two exactly defined areas of the cell perimeter which become the deepest indentations in the final cell shape. Together with the next pair of indentations forming about 15 min later their location defines the symmetry of the species specific cell pattern. All subsequent episodes of indentation formation are achieved by dichotomic bifurcation of growth zones, i.e. the lobe tips. Incorporation of vesicles containing cell wall matrix material (Brosch-Salomon et al., 1998; Lütz-Meindl & Brosch-Salomon, 2000) during primary wall growth takes place preferentially in the areas of the growing lobes and only marginally at the indentations (Meindl et al., 1992). How confinement of fusion processes to defined but changing areas is achieved throughout the developmental course, is one of the central questions in *Micrasterias* morphogenesis. Local influxes of calcium ions leading to a stage specific calcium pattern, determine the growth zones of *Micrasterias* (Meindl, 1982; Troxell & Scheffey, 1991) but do not establish intracellular ion gradients (Holzinger et al., 1995).

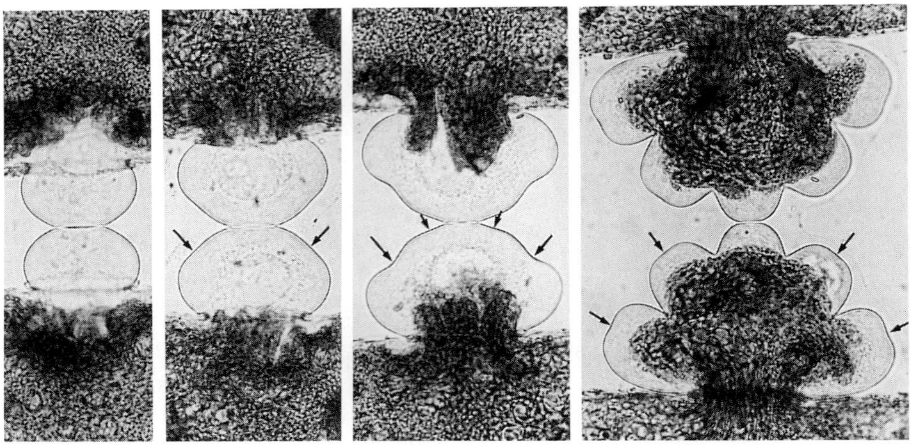

Figure 4. Stages in *Micrasterias* semicell morphogenesis. a: young undifferentiated bulb. b: formation of the first pair of indentations (arows). c: formation of the second pair of indentations (arrows). d: five-lobe stage at the moment of lobe bifurcation (arrows). From Meindl (1993) with permission.

Cell pattern formation is completed, when the growing semicell corresponds to the parental non-growing semicell in shape and size. This shape is consolidated by a cellulosic secondary wall deposited underneath the primary cell wall as a uniform layer. It lines the shape of the primary wall and contains pores distributed in a regular, mostly hexagonal pattern. The pores are inserted into the secondary wall by strictly patterned, localized exocytosis of distinct 'pore vesicles'. Shedding of the primary cell wall by a sudden synchronous onset of slime secretion through all cell wall pores terminates cell development in *Micrasterias*.

Primary cell wall growth in desmids is accompanied or followed by prominent nuclear migrations (for summary see Meindl 1992). In *Micrasterias*, the nucleus located in the central isthmus region in non-growing cells, moves into the growing semicell on a lineal path in synchrony with cell expansion and returns to its original position when growth is finished. Particularly the backward oriented migration occurs independently from cytoplasmic streaming and against the large, still expanding chloroplast (Meindl, 1983).

Expansion of the giant chloroplast of *Micrasterias* from the non-growing into the growing semicell is achieved by a gradual differentiation of an initially thick and undifferentiated chloroplast mass into a flat, plate-like organelle with lobes and indentations remodelling exactly the outer cell shape (for details see Meindl & Kiermayer, 1982). Its cleavage in the isthmus when secondary wall formation is completed, results in two independent chloroplasts typical for the adult *Micrasterias* cell.

3. ROLE OF MICROTUBULES

Often microtubules and actin filaments occur side by side in the same cytoplasmic domain which makes it necessary to probe the role of both cytoskeletal elements in further detail in order to understand the underlying mechanism of morphogenetic pattern formation. Even though the focus in this review is on actin, we should briefly review what is known about microtubules in both cell systems:

Acetabularia: Microtubules are absent from the juvenile stage altogether, they appear for the first time in the cell cycle on the surface of the premeiotic primary nucleus (Menzel, 1994; Liddle et al., 1997) and remain associated with the nuclear surfaces of the secondary nuclei that are formed after meiosis and throughout all the subsequent stages of morphogenesis (Menzel et al., 1996). The major roles of microtubules is seen in (1) the anchoring of secondary nuclei but not their movement in the cap rays prior to cyst formation (Menzel et al., 1996), (2) the maintenance of equal spacing between adjacent nuclei, (3) the positioning of the cleavage mechanism and (4) the subsequent process of lid formation (Menzel & Elsner-Menzel, 1990a). Long distance organelle transport (see below) as well as sporophytic (vegetative) morphogenesis are independent of microtubules (Menzel, 1994).

Micrasterias: As in *Acetabularia* the main function of microtubules in *Micrasterias* is associated with nuclear positioning. A microtubule complex consisting of a basket-like microtubule arrangement and connected to a rail-like microtubule band by a microtubule center ('post-telophase system of microtubules', PTS) accounts for generation of the movement and is established only during postmitotic nuclear migration (Meindl, 1983). It converts into a ring shaped microtubule band referred to as 'isthmus system of microtubules' (IS) which anchors the large nucleus in its central position when nuclear migration is finished. Microtubules, though present during growth in cortical cytoplasmic regions, do not participate in pattern formation in *Micrasterias* (Schmid & Meindl, 1992). In contrast to higher plants and to closely related desmids like *Closterium* (Hogetsu, 1992) artificial destruction of microtubules does neither impair the final cell shape nor the patterned deposition of cellulose microfibrils in both the primary and the secondary cell wall. From numerous experiments (see below) it seems likely that the role of guiding cellulose complexes in the plasma membrane has devolved to the actin filament system in *Micrasterias*. Microtubules participate at least in part in the expansion of the chloroplast from the non-growing into the growing semicell (Meindl & Kiermayer, 1982).

4. ROLE OF ACTIN FILAMENTS

4.1 Actin filament distribution during cell development

Acetabularia: The structure of the actin cytoskeleton has been determined by two specific staining methods, microdissection followed by indirect immunofluorescence (Menzel, 1986) and microwave accelerated phalloidin staining (Sawitzky et al., 1996). A system of densely packed, axially oriented actin bundles runs through the parietal cytoplasm of the stalk, into all side branches and up into the extreme apex, where they splay apart and transform into a finer meshwork. A clear zone devoid of actin filaments was not observed.

As soon as elongation growth subsides, apical swelling sets in and the tip begins to flatten in preparation for side branch initiation. The subapical actin bundle system undergoes progressive transformation into a random network configuration proceeding downwards for a long stretch of several 100 micrometers away from the apex back to the insertion point of the previously formed last whorl (Sawitzky & Menzel, unpublished). Next, a seam of more strongly condensed actin foci appears at the subapical location coinciding with the location of the lytic cell wall cavities (Fig. 5; Sawitzky & Menzel, unpublished). This indicates that the onset of expansion growth (flattening of the tip plus apical swelling) is favored by a random actin filament organization probably by creating the opportunity to form a greater number of free filament ends that could maintain intimate contacts with the submembrane region and facilitate short distance travelling of cell wall precursors to the apical cell membrane. It also indicates that positions of the cavities are spatially controlled by foci that form within the fine apical actin network. It is interesting to note that within the expanding buds of the young cap ray chambers the actin cytoskeleton remains in a random configuration, whereas in the main axis, the actin bundle system becomes quickly re-established leading directly up to the orifice into the young cap ray (Figs. 6 and 7). This suggests that long distance transport towards the entrance of the bud is an important prerequisite for growth of the cap ray, whereas organelle and vesicle transport within the growing bud must proceed in a more random fashion.

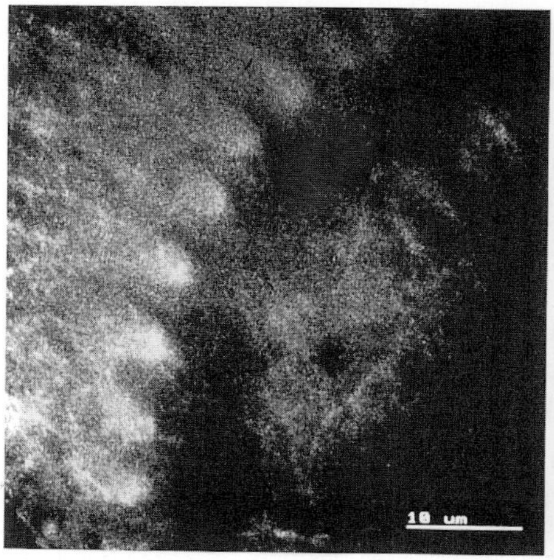

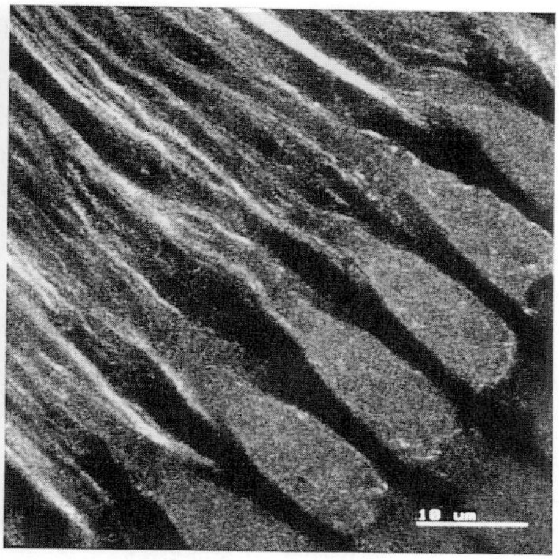

Figure 5. Condensed actin foci in *Acetabularia* along a subapical annulus at the positions where lytic cavities occur in the inner layer of the wall.

Figure 6. At a later stage in the formation of cap rays in *Acetabularia*, a subapical ring of buds has formed. Actin bundles in the stalk lead up directly to the orifice of the cap ray buds. Inside the buds, the actin cytoskeleton consists of very fine scattered filaments.

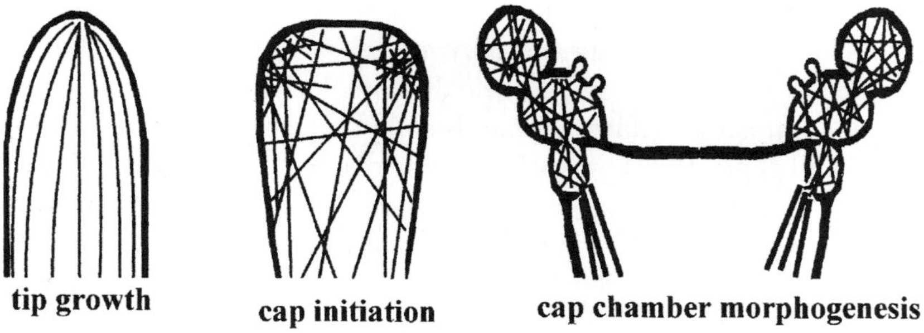

Figure 7. Principal organization of the actin cytoskeleton in the *Acetabularia* apex in the course of cap initiation. After the emergence of buds tripartite chambers develop (right)

The process of sporulation, i.e., cyst formation in the cytoplasm of the cap ray chambers relies on a spatially and temporally orchestrated interaction between the perinuclear microtubule systems and the actin cytoskeleton (for details see Menzel, 1994). First, the axial, parallel actin bundle system breaks down, giving rise to a random, dense network. Next, the distal ends of the perinuclear microtubule systems begin to break into numerous fragments, which reorient and subsequently elongate to establish a delineation in the perimeter of each cyst domain (cytoplasmic domain that transforms into a cyst). Along this perimeter more and more actin filaments appear until each of the domains is encircled by an actin filament band, which subsequently begins to contract as the encircled cytoplasmic portion swells up. The resultant protoplasmic compartments adhere along the contracted rings producing a grape-like arrangement of basally connected protoplasts within the confinement of the cap rays.

With the complete separation of the protoplasts from one another, cell wall formation sets in around the cysts starting from the apical membrane patch around the primary cyst nucleus, i.e. nuclei of all the cysts in a given cap ray chamber face outwards thereby determining cyst polarity. At the same time the future lid in the cyst wall is positioned above and around the nucleus. This process is accompanied by intricate spatial rearrangements in the perinuclear microtubule system of the cyst which positions the actual 'lid forming apparatus' (Neuhaus-Url & Schweiger, 1984; Menzel & Elsner-Menzel, 1990a). After maturation, cysts contain hundreds of small haploid nuclei randomly distributed through the parietal cytoplasm each with its own radially symmetrical perinuclear microtubule system. The actin cytoskeleton at this stage consists of a large number of short bundles scattered in the cytoplasm without any preferential orientation or any type of organelle association.

Micrasterias: For the visualization of actin filaments in *Micrasterias* specific techniques had to be employed. Both microinjection of fluorescently labeled phalloidin into living and growing cells (Meindl et al., 1994) as well as phalloidin staining of glutaraldehyde/formaldehyde fixed non-embedded cells or anti-actin-incubated butyl-methyl-metacrylate embedded cells (Pflügl-Haill et al., 1999) gave comparable results. Within the first two hours of growth until the cell has developed five lobes (2 pairs of indentations) single actin filaments are randomly distributed all over the growing semicell. From this stage on, until the end of primary cell wall growth a distinct and compact actin filament network is developed between the plasma membrane and the surface of the chloroplast (Fig. 8). It corresponds to the actin filament arrangement of the non-growing semicell and reaches far into cortical regions of the growing lobes. In living cells continuous dynamic changes in the arrangement of F-actin cables are visible after microinjection of fluorescently labeled phalloidin, yet the net-like distribution is maintained. The long actin cables establishing this network apparently do not have any preferential orientation, nor does the actin filament distribution perceptibly correlate with the developing cell pattern. The latter finding is inconsistent with earlier results on only glutaraldehyde fixed cells of another *Micrasterias* species (Ueda & Noguchi, 1988) where the authors describe thick actin bundles traversing the single lobes as central axes and terminating at the lobe tips.

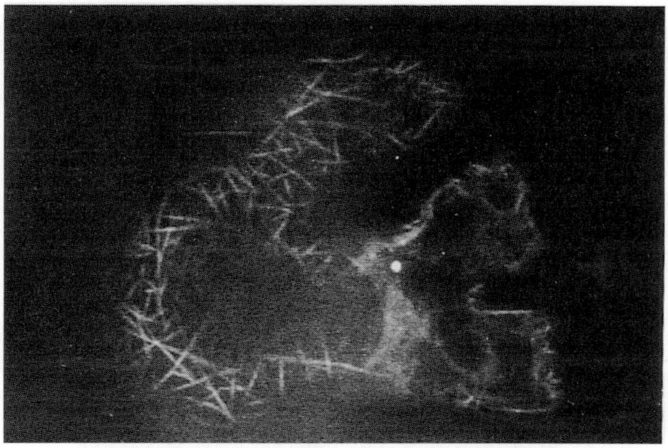

Figure 8. Compact actin filament network in *Micrasterias* between the plasma membrane and the surface of the chloroplast corresponding to the actin filament arrangement of the non-growing semicell. From Meindl et al. (1994) with permission.

In addition to the cortical actin filament network, recent findings (Meindl et al., 1994; Pflügl-Haill et al., 1999) have revealed a second F-actin system connected to the migrating nucleus. As soon as the nucleus leaves its central position in the isthmus, bundles of actin filaments, converging from a knob-like structure become visible at its isthmus-facing halve. When the nucleus has moved away from the isthmus the actin cables merge into one thick bundle running towards the isthmus and terminating exactly in its center (Fig. 9). Optical serial sections reveal that the spatial arrangement of the actin filament system surrounding the nucleus corresponds to a microtubule aggregation visualized in earlier electron microscopic investigations in the same area of the cell. The knob-like actin filament-converging center co-localizes with a microtubule center which gives rise to a basket-like microtubule system surrounding the nucleus at its isthmus-facing side at the one hand, and to a rail-like microtubule band terminating exactly in the center of the isthmus area at the other hand (Meindl, 1983; Meindl, 1992).

When the nucleus returns to its central position in the isthmus, the actin filament system is shifted into the non-growing semicell together with the microtubule basket before both gradually disappear (summarized from Meindl, 1983, 1992; Meindl et al., 1994). In both microinjection experiments and studies on fixed cells phalloidin label was not only found in the actin filament system surrounding the nucleus but also in the nuclear membrane itself.

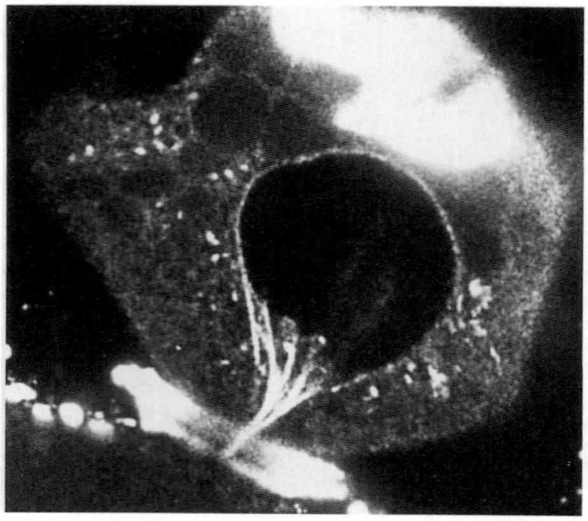

Figure 9. Actin filament system around the migrating post-telophase nucleus in *Micrasterias*. From Meindl et al. (1994) with permission

4.2 Role of the actin cytoskeleton in organelle translocation and cytomorphogenesis

Acetabularia: Evidence for a role of the actin cytoskeleton in organelle transport and morphogenesis comes from a combination of methods that enable us to visualize the actin cytoskeleton and at the same time to record abnormal behavior after the application of inhibitors that interfere with cytoskeletal elements.

Organelle motility has been studied since the 1960s by several groups (see Kamiya, 1986) and later in greater detail by Koop and Kiermayer (1980a). Particularly unusual is the so called 'headed streaming band (HSB)' motility discovered by Koop and Kiermayer (1980b), which operates by a speed 2-3 times faster than chloroplast transport and has a preference for cargo organelles such as small particles, polyphosphate granules and secondary nuclei. The mechanism that drives headed streaming bands and transport of nuclei within the bands in *Acetabularia* is not entirely understood.

Structural details of the actin cytoskeleton have been studied in cells treated with actin inhibitors such as cytochalasin D and the cyclodepsipeptide jasplakinolide (Sawitzky et al., 1999). These results suggest that both types of long distance transport, chloroplasts as well as HSBs, depend on the longitudinally oriented actin bundle system.

From careful video observations of nuclear transport in the presence of various cytoskeletal inhibitors combined with double immunofluorescence, we concluded that in contrast to the situation in *Micrasterias* microtubules where not involved in the transport mechanism even though, at this stage, a comet-tail microtubule system is attached to each secondary nucleus. However, in the subsequent stage of nuclear positioning this microtubule system is required for anchoring the nuclei in the surrounding cytoplasm (Menzel et al., 1996). After comparison of the distribution of filamentous and diffuse actin staining by immunofluorescence with the staining pattern obtained by phalloidin treatment, we postulated that actin polymerization may be the driving force for headed streaming band formation and propagation and that nuclei are cotransported with the cargo material in the head structure (Menzel, 1994; Menzel et al., 1996).

Inhibitor studies have been exhaustively performed on the process of cyst formation in *Acetabularia* (i.e. Menzel, 1988; Menzel & Elsner-Menzel, 1990b). From these studies we know that actin is important for the transformation of the coenocytic cytoplasm in the cap rays into unicellular and uninuclear cysts. The data suggest that the role of actin is associated with the cytokinetic process in two ways, first by producing tension over the area of the cyst domain to aid the bulging process and second by contracting the cytokinetic ring, which cleaves off the cyst from the parent cytoplasm.

Micrasterias: Indications for an involvement of actin microfilaments in organelle transport and cytomorphogenesis of *Micrasterias* come from observations on both the distribution of F-actin during cell growth and from numerous results obtained by the use of inhibitors. As described above formation of a distinct actin filament network coincides with the onset of pattern formation. Destruction of this network by drugs interacting with F-actin polymerization like cytochalasin D or latrunculin B (Pflügl-Haill et al., 1999) leads to retardation or inhibition of growth combined with marked aberrations in cell symmetry and pattern formation (Noguchi & Ueda, 1981; Lehtonen, 1983). Changes of the actin filament system induced by inhibitors reveal themselves as breakdown of the F-actin network visible in the confocal scanning laser microscope by the presence of bright actin clusters and short actin filament fragments (Pflügl-Haill et al., 1999).

A strong influence of actin targeting agents, particularly of cytochalasins known to specifically bind to actin and to disorganize its dynamics (Cooper, 1987) have been observed on cell shaping, cytoplasmic streaming and distribution of organelles of different *Micrasterias* species and other closely related desmid cells in earlier investigations (Tippit & Pickett-Heaps, 1974; Lehtonen, 1983; Url et al., 1993). For instance, it has been shown that low concentrated cytochalasin B targets the pattern of two different cytoplasmic streaming systems in *Micrasterias torreyi* differently (Lehtonen, 1983). Undirected oscillatory streaming in young bulbs prior to pattern formation is less affected than a second more cortical streaming system which develops with the onset of lobe formation and preforms the subsequent cell pattern.

Morphogenesis of the daughter generation is impaired when cells are treated with cytochalasin during interphase. Septum formation is inhibited and the isthmus regions lengthens to an undifferentiated middle section instead of developing normal daughter semicells (Lehtonen, 1983). Incomplete and malformed septum growth under cytochalasin D influence occurs as consequences of secretory vesicles and ER displacement in the closely related desmid *Xanthidium armatum* (Höftberger & Lütz-Meindl, 1999). Cytoplasmic vesicle-ER 'preformation bands' obviously established by actin filaments at the growing edges of the septum wall determine proper growth and orientation of the septum in untreated cells. After cytochalasin exposure the 'preformation bands' loose their orientation and instead of running in the division plane of the cell the bands mainly branch towards the postmitotic daughter nuclei. This leads to distorted, thickened and sometimes even branched septum walls which are not capable anymore of separating the daughter cell completely. In summary these results reveal that actin filaments are indispensable elements in the basic morphogenetic step in desmid cells, namely formation of the septum wall which corresponds to cell plate formation in higher plant cells. It has to be considered that during this early developmental stage a prepattern for the later morphogenesis of *Micrasterias* gets established in the septum membrane determining the basic symmetry of the cell (Kiermayer, 1981).

The effects of anti-actin agents like cytochalasins and latrunculins on cell development and primary wall growth in *Micrasterias* range from retardation of growth velocity up to a complete inhibition of differentiation in a concentration dependent manner (Tippit & Pickett-Heaps, 1974; Lehtonen, 1983; Pflügl-Haill et al., 1999). Eventually, marked malformations of the cell shape occur when the cells are able to continue growth during drug treatment. Particularly when treatment starts at young developmental stages the young bulbs either remain undifferentiated without any lobes and indentations or the sites of indentation formation are shifted to abnormal places.

At an ultrastructural level it becomes clear that transport of secretory vesicles to the plasma membrane is inhibited or retarded by cytochalasins in both *Micrasterias* (Lütz-Meindl, unpublished results) as well as in the closely related species *Euastrum* (Url et al., 1993). Large amounts of secretory vesicles are accumulated around the dictyosomes whereas their number at the growth zones is reduced. The lack of cell wall material available for growth explains inhibition or retardation of cell development by cytochalasins and demonstrates that actin filaments represent the transport system for secretory vesicles. The occurrence of cell shape aberrations, however, apparently indicates an additional role for actin filaments in morphogenesis, as a simple reduction in wall material can never account for asymmetries in the cell morphology. An additional role for actin filaments is seen in positioning and anchoring of dictyosomes which become shifted from their normal central location to more cortical areas under cytochalasin influence.

Jasplakinolide, supposed to exert opposite effects on actin filaments to those described for cytochalasins (Bubb et al., 1994) has indeed similar effects on cell growth and morphogenesis in *Micrasterias* (Holzinger & Meindl, 1997). In concentrations higher than 1.5 µM it leads to a transient retardation of growth and to marked malformations of the cell shape of *Micrasterias* (Holzinger & Meindl, 1997). The drug arrests septum formation but allows development of a secondary cell wall yet in an abnormal way. Dense abnormal accumulations of actin filaments occur in the cytoplasm of *Micrasterias* as a consequence of jasplakinolide action when investigated by means of electron microscopy. As F-actin bundles have never been observed at an ultrastructural level in untreated *Micrasterias* cells this finding indicates actin polymerizing capacities of jasplakinolide in situ. Moreover, jasplakinolide induces displacement of dictyosomes and reduces the number of primary wall material containing secretory vesicles in a similar way as cytochalasins. These results demonstrate that regardless of the detailed mode of action, any disturbance of actin filament dynamics affects cell growth and morphogenesis in *Micrasterias* and thus emphasizes the importance of a functioning actin filament system in cell differentiation.

Close spatial association of actin filament systems with the migrating nucleus and the chloroplast surface (see above) also indicate an involvement of the actin cytoskeleton in translocation of both organelles. In fact any experimental disturbance of actin microfilaments causes dislocation of the nucleus when applied during postmititoc nuclear migration and prevents spreading of the chloroplast inside the young semicell when employed at the respective stage (e.g. Pflügl-Haill et al., 1999). In contrast to microtubule inhibitors which lead to abnormal nuclear positions inside the growing semicell far away from the isthmus area (Meindl, 1992) cytochalasins or latrunculins induce an only slight displacement of the nucleus out of the center of the isthmus region. This may arise from a completely different role of microtubules and actin filaments in the migration process: It is conceivable that microtubules act only as guiding elements during the inward migration of the nucleus into the growing semicell, but are involved in generating the motive force for the backward migration into the isthmus region. Their destruction causes deviation of the nucleus from its normal lineal path into the growing semicell and prevents remigration of the nucleus to the isthmus thus leading to abnormal nuclear positions frequently close to one of the cell indentations. Actin filaments most likely account for pushing the nucleus into the growing semicell together with the inward streaming cytoplasm. Their function in nuclear remigration to the isthmus is not yet completely clear, however, it appears to be subordinate to that of the microtubule system. The nature of microtubule or actin-depending motors responsible for generating the motive force in *Micrasterias* and their involvement in the individual stages of nuclear translocation still remain to be elucidated.

4.3 Molecular motors and other molecular components of the actin cytoskeleton

In both model systems, work on the molecular components of the actin cytoskeleton is still in its infancy. However, recent progress either with methods of protein extraction and purification in combination with immunoblotting and immunofluorescence as in both model systems or with methods of molecular genetics as in *Acetabularia* have produced very promising results.

Acetabularia: Morphogenesis at the sporophyte apex, i.e., tip growth and side branch formation most likely require the formation and dynamic changes of prepatterns in the cytoplasm. While we are able to describe aspects of these changes by determining the changes occurring in the actin cytoskeleton such as bundling versus network formation and the appearance of localized actin foci, as well as the concomitant changes in the cell wall, we are still a long way from understanding the molecular mechanisms that underlie these changes.

Localization of the actin foci in the subapical ring may be a process of spatially controlled self assembly requiring several actin-binding and membrane-associated factors and is likely to be dependent upon other cues such as the geometry of the intracellular calcium gradients in the apex as suggested by Goodwin and Briere (1992). Mathematical modeling of the distribution of submembrane calcium ions by these authors led to patterns matching the subapical ring-like positioning of the actin foci. Rearrangements of the actin cytoskeleton as a consequence of a rise in intracellular calcium ions has been addressed by studying the behavior of the actin cytoskeleton in the vicinity of mechanical wounds in *Acetabularia* (Menzel & Elsner-Menzel, 1989) in the presence or absence of Ca-channel blockers. What happens in the cytoplasm at the wound edge when Ca-channels are unblocked can be described as network formation from a preexisting system of axial bundles and, subsequently, the spontaneous formation of focal points which develop into contractile centers. Coalescence of the centers leads to the formation of a contractile ring which eventually closes the wound. This process is not unlike the purse-string mechanism in fibroblast monolayers, except that in the monolayer the contractile units are separated by cell membranes (Bement et al., 1993) whereas in *Acetabularia* they are not. If actin-dependent contraction is considered an integral part of morphogenetic pattern formation in the cytoplasm of *Acetabularia*, i.e., in apical morphogenesis, the following scenario may be envisaged. Actin foci may act as local centers of increased mechanical tension, which could have an effect on the adjacent patch of plasma membrane with its associated ion channels, leading to further influx of calcium ions. The foci would also transmit tension deeper into the surrounding actin network which could be instrumental for a directional reorganization such as alignment of actin filaments and eventual establishment of new transportation pathways from greater distances towards the focal points.

In agreement with the results obtained in *Micrasterias*, it may be further speculated that the actin foci act as congregation points for dictyosomes which are known to translocate along the actin bundle system in higher plant cells (Hawes et al., 1998) as well as dictyosome-derived vesicles loaded with lytic enzymes and other modifying agents. Their localized exocytosis may lead to the formation of cavities, which in turn enlarge by turgor driven expansion so that they soon become visible on the cell surface as buds.

Comparing apical morphogenesis in *Acetabularia* with the intricate process of pattern formation in the cap ray cytoplasm that precedes cyst formation, similar processes of rearrangements in the actin cytoskeleton emerge. Involvement of actin-binding and -sequestering proteins should be expected in order to help transform the system of axial bundles into a random network, albeit on a much larger scale than at the growing apex. Restructuring of the actin cytoskeleton from axial bundles oriented in parallel to one another to evenly spaced rings, not only requires stage specific activation of new molecular partners that execute these changes (Menzel & Elsner-Menzel, 1990), but also spatial control over the distribution of these factors. A clear indication for a role of the microtubule system as spatial integrator comes from pharmacological studies. If the microtubule cytoskeleton is prevented from forming (Menzel, 1988; and unpublished data) actin rings do not form either. On the other hand, if the microtubule system becomes hypertrophied so that distal ends do not break off and reorient properly, as is the case after treatment with cyclohexemide (Menzel & Elsner-Menzel, 1990), only partial segments of the actin rings form and ultimately cyst morphogenesis fails.

Notwithstanding the role of the microtubule system as spatial controller for proper placement of actin bundle structures, we have ample evidence for the remarkable potential of the actin cytoskeleton to self-organize into ring-like configurations (Fig. 10), which in the absence of further spatial cues are placed randomly in the cytoplasm. This process has been observed in cysts which had been treated with the protein phosphatase 2A (PP2A) inhibitors okadaic acid (OKA) and calyculin A. As mentioned above, the actin cytoskeleton in the cyst consists of scattered bundles and neither organelle motility nor interactions of these bundles with other cell structures have been observed. OKA forces these actin bundles to laterally combine into wider bands and causes these bands to loop back and slide into each other head-to-tail thus forming ring- or even disk-like structures (Menzel et al., 1995). One of the most astounding observations is, that these rings rotate either by themselves or by propelling chloroplasts and other organelles around them. So it is quite obvious that intracellular motility in the cyst is inducible at a stage that naturally lacks it.

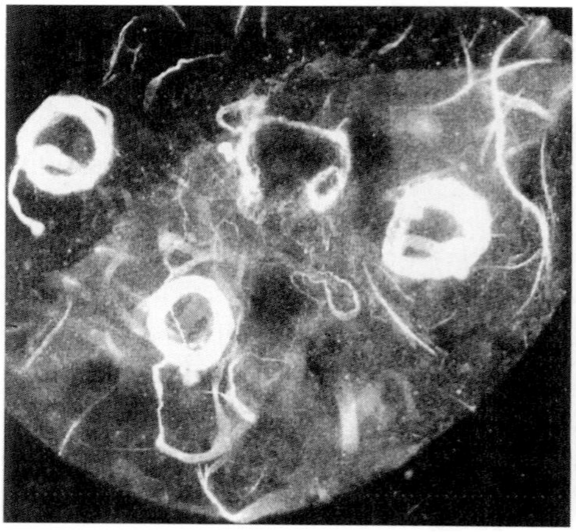

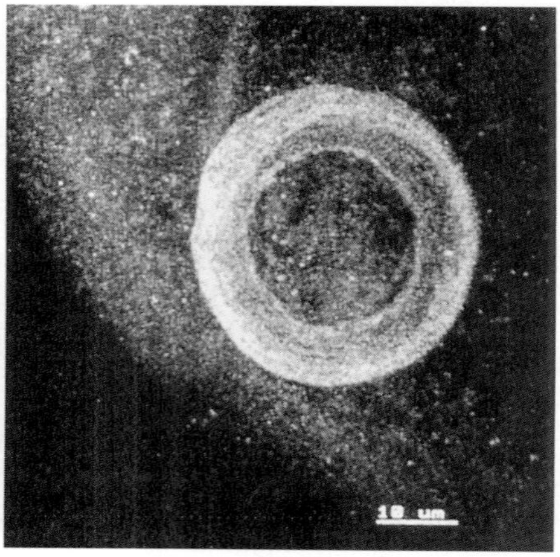

Figure 10. Ring-like configuration of actin bundles in the *Acetabularia* cyst cytoplasm have formed from scattered bundles after treatment with 0.5 μM okadaic acid for 8 h. Between 5 and 10 rings can form simultaneously within a single cyst.
Figure 11. One of these circular configurations of actin stained with an antibody against animal spectrin (Sigma S1390).

A similar conclusion can be drawn from a previous study using the methylation inhibitor EHNA (erythro-9-[3-(2-hydroxynonyl)] adenine) in *Acetabularia.* The mode of action of EHNA can be explained by the fact, that all known catalytic subunits of PP2A including the *Acetabularia* PP2A (Menzel et al., 1995) posses a C-terminal Leucine residue. Methylation of this residue is known to activate PP2A. Our data, therefore, are consistent with the prediction that EHNA inactivates PP2A and hence should have an effect similar to that of OKA.

It is very likely that actin-binding proteins are involved in OKA-induced ring formation induced naturally as well as artificially. Towards that end, we have been able to localize antigens similar to spectrin (Fig. 11), alpha-actinin and tropomyosin on OKA-induced rings in the cyst by immunofluorescence using a spectrum of heterologous antibodies (Menzel, unpublished).

Recently, progress has been made in the isolation and characterization of genes and cDNAs from *Acetabularia.* In the course of this work full length sequences of actin, myosin, PP2A and PP1 have been determined. Most interesting are the two myosin sequences retrieved from nuclear DNA and cDNA, which are probably both single headed organelle motors (see sequence comparison in Mermall et al., 1998), because one of them has an extremely short tail and the other lacks any alpha-helical coiled-coil conformations in its C-terminal tail region (Menzel & Vugrek, 1997). Which of these myosins may be responsible for actin ring formation is unknown. We should either expect a bifunctional myosin with an actin binding domain opposite to the motor domain or we may anticipate the existence of a third myosin in *Acetabularia* capable of forming dimers or filaments.

Micrasterias: Regulation of morphogenesis in *Micrasterias* is most likely a membrane-bound process based on interactions between the actin cytoskeleton and the plasma membrane. This is supported by the observation that firstly microtubules are not involved in morphogenesis, secondly that in contrast to *Acetabularia* no cytoplasmic prepattern is visible in the spatial arrangement of actin filaments during growth and that thirdly confinement of vesicle fusions to particular stage specific areas is achieved by a strictly localized rise in the calcium concentration at the plasma membrane but not by establishment of an intracellular free calcium gradient. This is also reflected by the random non-oriented distribution of secretory vesicles throughout the cytoplasm. Actin binding proteins, particularly those known to mediate membrane-actin interactions, like proteins from the spectrin superfamily, are thus expected to assume regulatory functions in multipolar growth of *Micrasterias*. In fact spectrin-like proteins have been immuno-detected on blots and localized by means of immuno-electron microscopy during cell development of *Micrasterias* using different heterologous antibodies (Holzinger et al., 1999a). The antigens were found at the plasma membrane of developing stages after high pressure freeze fixation and freeze substitution, yet without perceivable correlation to the developing cell pattern. On the other hand the presence of spectrin like proteins at membranes of vesicles containing secretory wall material and at dictyosomal membranes themselves indicate a linkage function between organelles and the actin cytoskeleton during organelle transport and suggest a role for spectrin in microdomain organization.

Actin filament dynamics and membrane-cytoskeleton interactions may be also regulated by the small G-actin binding protein profilin (Staiger et al., 1997), which may cause relative disassembly of actin filaments but may also interact with phosphoinositides and have a signaling role in cell development. In *Micrasterias* denticulata this protein was detected by means of immuno-blotting using antibodies against celery root profilin, recombinant tobacco profilin and recombinant birch profilin (Holzinger et al., 1999b). It is evenly distributed throughout the cytoplasm of developing cells when investigated by confocal microscopy after fluorescent labeling or by immunogold detection in the transmission electron microscope. However, interestingly, profilin is accumulated in the microtubule center (Fig. 12; Holzinger et al., 1999b) which connects two parts of the post-telophase microtubule system around the migrating nucleus (Meindl, 1983) and again co-localizes with an F-actin converging center (Meindl et al., 1994). This suggests a role of profilin in the process of nuclear migration. In addition, microinjection of birch profilin into living and growing *Micrasterias* cells cause transient inhibition of cell growth and differentiation and demonstrates the capability of the protein to regulate the actin cytoskeleton in *Micrasterias* (Holzinger et al., 1997). In a similar way microinjected gelsolin changes the pattern of cytoplasmic streaming and retards cell development.

Profilin is also naturally accumulated in several areas of the nucleus including nuclear pores, nucleolus and heterochromatin where it has been also found after artificially increasing the intracellular profilin level by microinjecting the FITC-labeled protein (Holzinger et al., 1997, 1999b).

5. CONCLUSIONS

Taking the apparent discrepancies that have emerged in the course of this comparison between *Acetabularia* and *Micrasterias*, one might come to the conclusion that morphogenesis in the two model systems relies on largely disparate subcellular and molecular mechanisms. It seems that determination of the spatial cell pattern in *Micrasterias* involves localized restriction of cell wall expansion, whereas cytomorphogenesis in *Acetabularia* involves localized induction and promotion of cell wall expansion (apical morphogenesis) or even shape determination in the absence of a cell wall (cyst morphogenesis). On the other hand, in each case the actin cytoskeleton governs the process of pattern formation in the cytoplasm leading up to the creation of spatially defined membrane domains which exert the ultimate control over exocytosis. Conceptually, it is less important whether the control is by restriction or by enhancement of vesicle fusion. It is, however, quite clear that the actin cytoskeleton must be physically associated with these membrane domains and spectrin is the only one molecular candidate for a multifunctional mediator of that interaction.

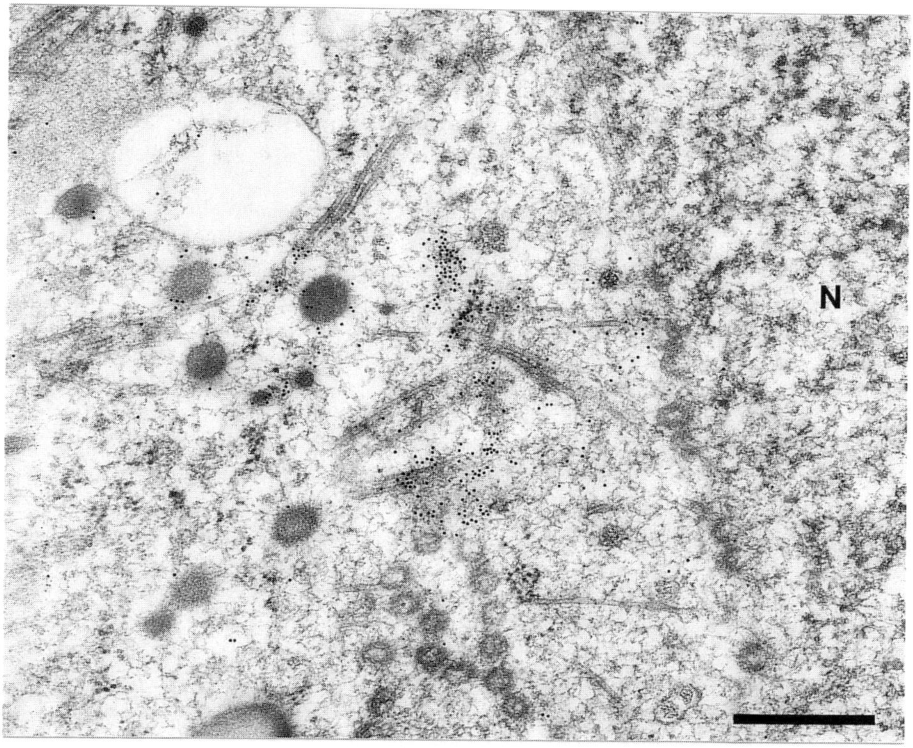

Figure 12. Immunogold localization of profilin in the microtubule center close to the nucleus (N) in *Micrasterias*. Bar 0.5 µm. From Holzinger et al. (2000)

It is also quite clear that microtubules do not have a function either in lobe formation or in apical morphogenesis in these two cell types. They do, however, have a crucial function when it comes to complex three-dimensional positioning of organelles most obviously of nuclei. This is the case for the movement of the nucleus in *Micrasterias* into the growing semicell and back again into the isthmus and likewise in *Acetabularia* when secondary nuclei are brought into position for cyst morphogenesis. And in both cases, microtubules and the actin cytoskeleton work together, prompting the hypothesis that physical and functional links must exist between the two cytoskeletal elements. Studying the molecular basis of these interactions will help to understand pattern formation in plant cells.

Also, in order to understand the mechanisms of plant cytomorphogenesis, future research will have to focus on those cytoskeleton associated proteins that operate as multifunctional linker elements in the immediate vicinity of submembrane domains of the growing and non-growing regions.

REFERENCES

Bement WM, Forscher P and Mooseker MS (1993) A novel cytoskeletal structure involved in purse string wound closure and cell polarity maintenance. J Cell Biol 121: 565-578

Berger S and Kaever MJ (1992) Dasycladales, An Illustrated Monograph of a Fascinating Algal Order. Stuttgart, New York: Georg Thieme Verlag

Brook AJ (1981) The Biology of Desmids. Oxford: Blackwell Scientific Publications

Brosch-Salomon S, Höftberger M, Holzinger A and Lütz-Meindl U (1998) Ultrastructural localization of polysaccharides and N-acetyl-D-galactosamine in the secretory pathway of green algae (Desmidiaceae). J Exp Bot 49: 145-153

Bubb MR, Senderowicz AMJ, Sausville EA, Duncan KLK and Korn ED (1994) Jasplakinolide, a cytotoxic natural product, induces actin polymerization and competitively inhibits the binding of phalloidin to F-actin. J Biol Chem 269: 14869-14871

Cooper JA (1987) Effects of cytochalasin and phalloidin on actin. J Cell Biol 105: 1473-1478

Goodwin BC and Brière C (1992) A mathematical model of cytoskeletal dynamics and morphogenesis in *Acetabularia*. In: The Cytoskeleton of the Algae. Menzel D (ed), Boca Raton, Florida: CRC Press, pp 219-238

Hämmerling J (1953) Nucleocytoplasmic relationships in the development of *Acetabularia*. Int Rev Cytol 2: 475-498

Hawes C, Oparka K and Santa Cruz S (1998) Green fluorescent protein to study endomembrane dynamics in plant cells. Biol Cell 90: 261 - 264

Höftberger M and Lütz-Meindl U (1999) Septum formation in the desmid *Xanthidium* (Chlorophyta): Effect of cytochalasin D and latrunculin B suggest the involvement of actin microfilaments. J Phycol 35: 768-777

Hogetsu T (1992) The cytoskeleton in the Desmidiales – cell morphogenesis. In: The Cytoskeleton of the Algae. Menzel D (ed), Boca Raton, Florida: CRC Press, pp 133-147

Holzinger A and Meindl U (1997) Jasplakinolide, a novel actin targeting peptide, inhibits cell growth and induces actin filament polymerization in the green alga *Micrasterias*. Cell Motil Cytoskel 38: 365-372

Holzinger A, Callaham DA, Hepler PK and Meindl U (1995) Free calcium in *Micrasterias*: Local gradients are not detected in growing lobes. Eur J Cell Biol 67: 363-371

Holzinger A, Mittermann I, Laffer S, Valenta R and Meindl U (1997) Microinjection of profilins from different sources into the green alga *Micrasterias* causes transient inhibition of cell growth. Protoplasma 199: 124-134

Holzinger A, De Ruijter N, Emons AM and Lütz-Meindl U (1999) Spectrin-like proteins in green algae (Desmidaceae). Cell Biol Int 23: 335-344

Holzinger A, Valenta R and Lütz-Meindl U (2000) Profilin localized in the nucleus associated microtubule/actin-system and the cytoplasm of the green alga *Micrasterias*. Protoplasma in press

Kamiya N (1986) Cytoplasmic streaming in giant algal cells: A historical survey of experimental approaches. Bot Mag 99: 444-467

Kiermayer O (1981) Cytoplasmic basis of morphogenesis in *Micrasterias*. In: Cytomorphogenesis in Plants. Kiermayer, O (ed), Cell Biol Monographs Vol. 8, Wien, New York: Springer Verlag, pp 147-189

Koop HU and Kiermayer O (1980a) Protoplasmic streaming in the giant unicellular green alga *Acetabularia mediterranea*. I. Formation of intracellular transport systems in the course of cell differentiation. Protoplasma 102: 147-166

Koop HU and Kiermayer O (1980b) Protoplasmic streaming in the giant unicellular green alga *Acetabularia mediterranea*: II. Differential sensitivity of movement systems to substances acting on microfilaments and microtubules. Protoplasma 102: 295-306

Lehtonen J (1983) Action of cytochalasin B on cytoplasmic streaming and morphogenesis in *Micrasterias torreyi* (Conjugatophyceae). Nord J Bot 3: 521-531

Liddle LB, Gleissberg S, Giese G and Berger S (1997) Microtubule formation on the nuclear surface during meiosis in *Acetabularia acetabulum*. Protoplasma 200: 10-16

Lütz-Meindl U and Brosch-Salomon S (1999) Cell wall secretion in the green alga *Micrasterias.* J Microsc in press

Mandoli D (1998) Elaboration of body plan and phase change during development of *Acetabularia*: How is the complex architecture of a giant unicell built? Annu Rev Plant Physiol Plant Mol Biol 49: 173-198

Meindl U (1982) Patterned distribution of membrane-associated Ca^{2+} during pore formation in *Micrasterias*. Protoplasma 112: 138-141

Meindl U (1983) Cytoskeletal control of nuclear migration and anchoring in developing cells of *Micrasterias denticulata* and the change caused by the anti microtubular herbicide amiprophos-methyl (APM). Protoplasma 118: 75-90

Meindl U (1992) Cytoskeleton-based nuclear translocation in desmids. In: Menzel D (ed) The Cytoskeleton of the Algae. Boca Raton, Florida: CRC Press, pp 133-147

Meindl U (1993) *Micrasterias* cells as a model system for research on morphogenesis. Microbiol Rev 57: 415-433

Meindl U and Kiermayer O (1982) Über die Kern- und Chloroplastenmigration von *Micasterias denticulata* Bréb. II. Die Chloroplastenmigration und ihre Veränderung durch verschiedene Stoffe. Phyton 22: 213-231

Meindl U, Lancelle S and Hepler PK (1992) Vesicle production and fusion during lobe formation in *Micrasterias* visualized by high-pressure freeze fixation. Protoplasma 170: 104-114

Meindl U, Zhang D and Hepler PK (1994) Actin microfilaments are associated with the migrating nucleus and the cell cortex in the green alga *Micrasterias.* J Cell Sci 107: 1929-1934

Menzel D (1986) Visualization of cytoskeletal changes through the life cycle in *Acetabularia.* Protoplasma 134: 30-42

Menzel D (1988) Perturbation of cytoskeletal assemblies in cyst domain morphogenesis in the green alga *Acetabularia.* Eur J Cell Biol 46: 217-226

Menzel D (1994) Cell differentiation and the cytoskeleton in *Acetabularia.* New Phytol 128: 369-393

Menzel D and Elsner-Menzel C (1989) Induction of actin-based cytoplasmic contraction in the siphonous green alga *Acetabularia* (Chlorophyceae) by locally restricted calcium influx. Bot Acta 102: 164-171

Menzel D and Elsner-Menzel C (1990a) The microtubule cytoskeleton in developing cysts of the green alga *Acetabularia*: Involvement in cell wall formation. Protoplasma 157: 52-63

Menzel D and Elsner-Menzel C (1990b) Cytoskeletal dynamics in the reproductive cycle of *Acetabularia* (Dasycladaceae, Chlorophyceae). Cyst morphogenesis and delayed cytokinesis. In: Experimental Phycology. Wiessner W, Robinson DG, Starr R (eds), Heidelberg: Springer-Verlag, pp 105-123

Menzel D and Vugrek O (1997) Muskelproteine in Pflanzen. Biologie in unserer Zeit 27: 195-203

Menzel D, Jonitz H and Elsner-Menzel, C (1996) The perinuclear microtubule system in the green alga *Acetabularia*: anchor or motility device? Protoplasma 193: 63-76

Menzel D, Vugrek O, Frank S and Elsner-Menzel C (1995) Protein phosphatase 2A, a potential regulator of actin dynamics and organelle motility in the green alga *Acetabularia.* Eur J Cell Biol 67: 179-187

Mermall V, Post P and Mooseker M (1998) Unconventional myosins in cell movement, membrane traffic, and signal transduction. Science 279: 527-533

Neuhaus-Url G and Schweiger H-G (1984) The lid forming apparatus in cysts of the green alga *Acetabularia mediterranea.* Protoplasma 122: 120-124

Noguchi T and Ueda K (1981) Effect of metabolic inhibitors on the formation of cell walls in a green alga, *Micrasterias crux-melitensis.* Plant Cell Physiol 22: 1437-1445

Pflügl-Haill M, Vidali L, Vos JW, Hepler PK and Lütz-Meindl U (1999) Changes of the actin filament system in the green alga *Micrasterias* induced by different cytoskeleton inhibitors. Protoplasma in press

Sawitzky H, Willingale-Theune J and Menzel D (1996) Improved visualization of F-actin in the green alga *Acetabularia* by microwave-accelerated fixation and simultaneous FITC-phalloidin staining. Histochem J 28: 353-360

Sawitzky H, Liebe S, Willingale-Theune J and Menzel D (1999) The anti-proliferative agent jasplakinolide rearranges the actin cytoskeleton of plant cells. Eur J Cell Biol 78: 424-433

Schmid VHR and Meindl U (1992) Microtubules do not control orientation of secondary cell wall microfibril deposition in *Micrasterias.* Protoplasma 169: 148-154

Schulze KL (1939) Cytologische Untersuchungen an *Acetabularia mediterranea* und *Acetabularia wettsteinii.* Arch Protist 92: 179-225

Staiger CJ, Gibbon BC, Kovar DR and Zonia LE (1997) Profilin and actin-depolymerizing factor: Modulators of actin organization in plants. Trends Plant Sci 2: 275-281

Tippit DH and Pickett-Heaps JD (1974) Experimental investigations into morphogenesis in *Micrasterias.* Protoplasma 81: 271-296

Troxell CL and Scheffey C (1991) Ionic currents flow through *Micrasterias* and *Closterium* cells during expansion of the primary cell wall. Planta 184: 218-225

Ueda K and Noguchi T (1988) Microfilament bundles of F-actin and cytomorphogenesis in the green alga *Micrasterias crux-melitensis.* Eur J Cell Biol 46: 61-67

Url T, Höftberger M and Meindl U (1993) Cytochalasin B influences dictyosomal vesicle production and morphogenesis in the desmid *Euastrum.* J Phycol 29: 667-674

Waris H and Kallio P (1964) Morphogenesis in *Micrasterias.* Adv Morphog 4: 45-80

Werz G (1965) Determination and realization of morphogenesis in *Acetabularia.* Brookhaven Symp Biol 18: 185-203

Werz G (1974) Fine-structural aspects of morphogenesis in *Acetabularia.* Int Rev Cytol 38: 319-367

Chapter 14

ACTIN IN CHARACEAN RHIZOIDS AND PROTONEMATA

Tip growth, gravity sensing and photomorphogenesis

Markus Braun[1] and Geoffrey O Wasteneys[2]

1 Botanisches Institut der Universität Bonn, D–53115 Bonn, Germany
2 Plant Cell Biology Group, Research School of Biological Sciences, The Australian National University, Canberra Australia

Key words: actin, chara, gravitropism, photomorphogenesis, protonemata, rhizoids, tip growth

Abstract: Unicellular tip-growing cells are frequently used for investigating structural and functional aspects of the plant cytoskeleton as well as the mechanisms of oriented tip growth. Cell extension, sensing of and the response to external stimuli are all confined to a small region at the tip, the apical dome. Rhizoids and protonemata of the characean green algae are almost identical in form and both use unique $BaSO_4$-crystal-filled statoliths for their gravity-oriented tip growth. Despite these similarities, the direction of graviresponses of rhizoids and protonemata are opposite. The actin cytoskeleton of both cell types interacts differently with the statoliths in the mediation of their positioning and movement. This cytoskeleton is also central to the mechanisms of the opposite gravitropic tip growth as well as to several other motile processes, including cytoplasmic streaming and vesicle trafficking. In this chapter, we summarize the current knowledge of the complex architecture of the actin cytoskeleton in characean rhizoids and protonemata. We present the existing models for differential gravitropic tip growth that have resulted from multiple approaches to characterize actin's various functions in the different zones of these tip-growing cells.

1. INTRODUCTION

The unicellular, gravity-sensing rhizoids and protonemata of characean algae are model systems for investigating gravitropic tip growth and cytological, molecular and functional aspects of the plant cytoskeleton. The

C.J. Staiger et al. (eds.), Actin: A Dynamic Framework for Multiple Plant Cell Functions, 237–258.

remarkably rapid, gravity-directed reorientation of characean rhizoids has attracted the attention of scientists for over a century (Zacharias, 1890; Giesenhagen, 1896, 1901; Schröder, 1904). They are tube-like cells with a highly polarized cytoplasmic zonation and their growth is limited strictly to the apical cell pole (Fig. 1). The basal zone contains a single large vacuole that is surrounded by two opposing layers of rotational endoplasmic streaming and stationary cortical cytoplasm. The 300-μm long tip region can be subdivided into a subapical and an apical zone, both characterized by a relatively stationary cytoplasm devoid of vacuoles and in which cytoplasmic streaming does not occur. The subapical zone contains randomly distributed organelles including mitochondria, proplastids, dictyosomes and the large nucleus. Extensive endoplasmic-reticulum (ER) cisternae are mainly axially oriented. The apical zone extends from the outermost tip to 35–40 μm into the cell and incorporates the exocytosis control center, the Spitzenkörper, which consists of an accumulation of secretory vesicles surrounding a unique ER aggregate. Another conspicuous feature of the apical zone is the presence of up to 50 statoliths, which cluster at a fairly constant distance from the apical cell wall. The statoliths of rhizoids and protonemata are prominent membrane-bound vesicles of up to 2 μm in diameter. They do not contain starch, as is the case in higher plant amyloplast-statoliths, but embody a matrix of carbohydrates and proteins (Wang-Cahill & Kiss, 1995) in which $BaSO_4$-crystals are arranged in a radial pattern (Schröter et al., 1975). Traces of strontium were also detected by X-ray analysis (Sievers & Schmitz, 1982). Centrifugation experiments unambiguously proved their function as statoliths (Buder, 1961) and stimulated Sievers and coworkers to explore the characean rhizoid system using various light- and electron-microscopical methods. These cytological studies were backed up with mathematical analysis (Hejnowicz & Sievers, 1971) to produce a model for the gravitropic response mechanism of rhizoids in which statolith sedimentation causes a local impediment to exocytosis, resulting in differential flank growth (Sievers 1971; Sievers & Schröter, 1971; Sievers et al., 1979; Sievers & Schnepf, 1981).

This was the first conclusive model for a complete gravitropic response chain. In contrast, the integration of various components of the gravitropic signal and response pathways in higher plant systems remains incomplete (reviewed by Sack, 1991; Sievers & Braun, 1996; Sack, 1997; Chen et al., 1999; Rosen et al., 1999). When negatively gravitropic (upward growing) characean protonemata, originally described by Fritsch (1935), were first used in photomorphogenetic and gravitropic investigations by Hodick (1993), it immediately became obvious that, together, rhizoids and protonemata would provide an ideal experimental system for comparative studies on gravity-oriented tip growth. Rhizoids and protonemata are indistinguishable in terms of cell shape and polar cytoplasmic organization but show opposite gravitropic responses (Fig. 1). Which features determine why rhizoids bend downwards whereas protonemata bend upwards in response to the same gravistimulation? Both cell types originate from nodal cells of the multicellular green thallus and commitment to positive or negative gravitropism is dependent on the environmental conditions at the time of their initiation. The positively gravitropic (downward growing) rhizoids develop when light is available in response to environmental stress or fragmentation of thalli; they penetrate mud and soil and anchor the plant segment in the sediment. Negatively gravitropic (upward growing) protonemata, however, are generated when blue light is prevented from reaching the cells, which occurs, for example, when the thallus becomes covered by sand or mud. By bending against the gravity vector, protonemata grow back into the light, an effective survival response. Once a protonema emerges from the sediment, blue light stimulates a remarkable regeneration process involving the termination of tip growth, and the formation of a relatively complex, multicellular green thallus (Pickett-Heaps, 1975; Hodick, 1993).

There is increasing evidence from studies of the last decade that the actin cytoskeleton plays an essential role in many of the specialized features of characean rhizoids and protonemata. These characteristic features include rotational cytoplasmic streaming, tip growth, statolith positioning, gravitropic orientation and photomorphogenetic differentiation. This chapter focuses on the organization of the actin cytoskeleton during these processes and how this organization may help explain the opposite gravitropic responses of rhizoids and protonemata.

2. ARRANGEMENT OF THE ACTIN CYTOSKELETON IN RHIZOIDS AND PROTONEMATA

The arrangement of the actin cytoskeleton is very similar in rhizoids and protonemata and reflects the cells' polar cytoplasmic organization. The conspicuous and rapid cytoplasmic streaming in the basal zone relies on thick, interconnected actin cables that are easily visualized with rhodamine phalloidin (Tewinkel et al., 1989; Sievers et al., 1991a; Braun, 1997) and immunolabelling (Braun & Wasteneys, 1998a,b). Just as in the diffusely expanding internodal cells (Wasteneys et al., 1996; see also other chapters in this volume), the actin cables form two hemicylindrical, slightly helical arrays, separated from each other by actin-depleted zones, at the interface between stationary cortical and streaming endoplasm.

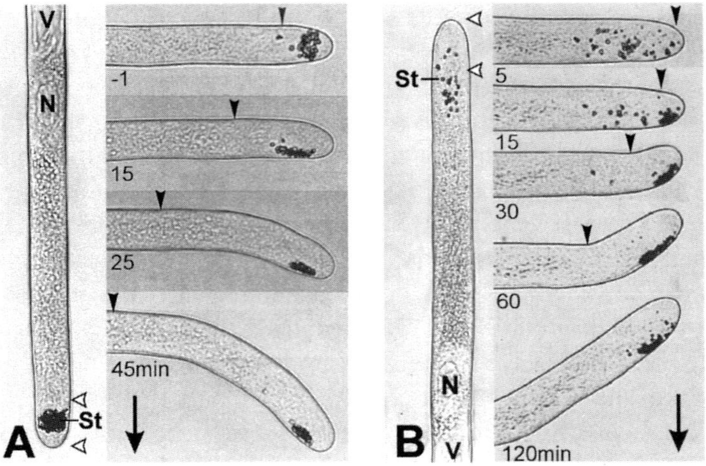

Figure 1. Gravitropic responses of rhizoids (**A**) and protonemata (**B**) of *Chara globularis*. Only the apical zones (between the white arrowheads), the subapical zones (containing the nucleus (**N**) and most of the stationary cytoplasm) and a small part of the basal zones (characterized by the large vacuole (**V**) surrounded by rotational endoplasmic streaming) is shown. **A** The positive gravitropic response of rhizoids is initiated by gravity-induced sedimentation of the statoliths (**St**) onto the lower cell flank that results in differential flank growth and a continuous downward bending of the tip (bending by bowing). **B** The negative gravitropic response of protonemata is initiated by the statoliths (**St**) which invade the apical dome and sediment asymmetrically near the growth center at the tip. This is followed by a drastic upward shift of the cell tip (bending by bulging). Arrows = direction of gravity; black arrowheads indicate the same point in each micrograph. Diameter of the cells: 30 μm. **A** modified after Braun (1996b); **B** modified after Hodick (1994).

Microinjecting small amounts of fluorescently tagged phallotoxins labelled actin cables without affecting cytoplasmic streaming (Braun & Wasteneys, 1998a). The *in vivo* phalloidin labelling pattern was very similar to that observed after immunolabelling or rhodamine-phalloidin staining fixed cells, indicating that chemical fixation reliably preserves actin's *in vivo* configuration. In addition to the subcortical actin cables, microinjection and rhodamine-phalloidin staining, but not immunolabelling, revealed ring-like actin structures associated with a specific type of organelle, possibly proplastids, in the streaming endoplasm. Similar actin rings were proposed to mediate rotational movements of numerous endopolyploid nuclei in the streaming endoplasm of *Nitella* internodes (Wasteneys & Williamson, 1991; see also Jarosch, 1956; Foissner & Wasteneys, 2000), and chloroplast actin rings have been identified in internodes of *Chara* and *Nitella* (Wasteneys et al., 1996). The streaming endoplasm of rhizoids and protonemata, however, does not contain nuclei and, to date, there is no evidence for additional rotational movements of proplastids. In protonemata, but not rhizoids, some proplastids become anchored in the cortical cytoplasm after initiation of morphogenetic differentiation (Braun & Wasteneys, 1998b). Actin rings could be involved in proplastid anchoring.

At the apical end of the basal zone, the actin cables fan out to form a dense meshwork of fine, mainly axially oriented microfilament (MF) bundles. This extensive actin meshwork penetrates the subapical and apical zones and envelopes the large nucleus and numerous organelles. In the apex, more extensively bundled actin MFs form a less dense meshwork and focus in a central actin array, which is symmetrically positioned close to the cell tip (Fig. 2B). This spherical actin array colocalizes with the ER aggregate (Sievers et al., 1991a; Braun & Wasteneys, 1998a) first described by Bartnik and Sievers (1988). Thus, the organization of the actin cytoskeleton appears more complex in characean rhizoids and protonemata than it does in many other tip-growing cells such as root hairs (Miller et al., 1999; Braun et al., 1999a), pollen tubes (Miller et al., 1996; Cai et al., 1997), moss protonemata (Walker & Sack, 1995; Meske et al., 1996), fern protonemata (Kadota et al., 1999) and fungal hyphae (Jackson & Heath, 1990; Levina et al., 1994). In particular, the presence of actin in the growing tips of pollen tubes is still a matter of debate. Several reports show a dense accumulation of actin in the pollen tube tip, suggesting that actin may play a role in vesicle delivery and docking. More recent studies, however, failed to demonstrate actin in the apical clear zone of pollen tubes prepared by rapid freeze fixation (Cai G. et al., 1997; Taylor & Hepler, 1997) or microinjected with fluorescent phalloidin (Miller et al., 1996).

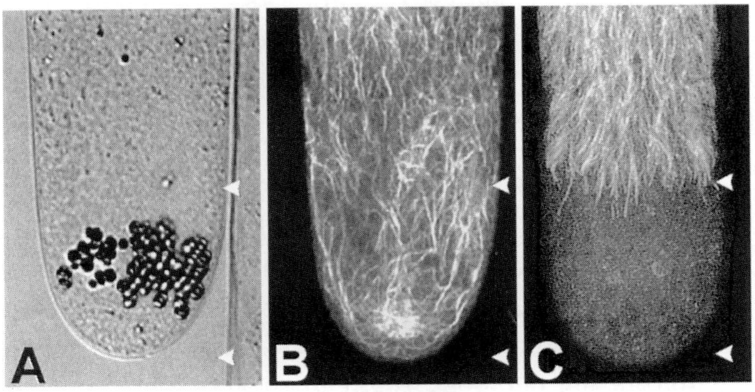

Figure 2. Apical and part of the subapical region of a Chara rhizoid. **A** Brightfield image showing the complex of statoliths located 10–30 µm above the apical cell wall. **B** The rhodamine-phalloidin labelling of the cell (shown in **A**) documents a fine meshwork of mainly

axially oriented actin MFs which appear more loosely arranged in the apical region (region between the two arrowheads) and form a dense spherical actin array close to the growth center at the tip. **C** Anti-tubulin immunofluorescence image showing an extensive meshwork of MTs in the subapical region and the absence of MTs from the apical region. Diameter of the cells: 30 µm. From Braun & Wasteneys (1998a) with permission from Springer Verlag.

In root hairs, axially oriented actin bundles in the root hair tube flare out into finer bundles in the subapical zone, but the vesicle-rich apex is apparently devoid of actin filament bundles (Miller et al., 1999). Instead, a cap-shaped pool of profilin-actin complexes has been reported at the growing root hair tips (Braun et al., 1999a). A fine meshwork of actin MFs is described for protonemata of the moss, *Ceratodon* (Walker & Sack, 1995), extending uniformly into the apical dome. In contrast, Meske et al. (1995) reported a collar-like arrangement of apical MFs and a MF-depleted tip region in caulonema cells of *Ceratodon*.

3. ACTIN-BASED CYTOPLASMIC STREAMING

The rotational cytoplasmic streaming in the basal zone of rhizoids and protonemata resembles the endoplasmic streaming in characean internodal cells (Ackers et al., 2000; see also Grolig & Pierson, this volume). Streaming is driven by the myosin-coated endoplasmic network of ER and other organelles sliding along subcortical files of actin bundles, thus mobilizing the whole endoplasm including organelles and vesicles (Kachar & Reese, 1988; for review see Kuroda, 1990). The crucial function of the two hemicylindrical arrays of actin cables for cytoplasmic streaming in characean rhizoids was demonstrated in studies using the fungal toxins cytochalasins B and D (Hejnowicz et al., 1985; Braun & Sievers, 1994). Basipetal streaming stops earlier than the acropetal streaming after application of cytochalasins. Myosin-like proteins in the basal zone have been immunofluorescently localized in a diffusely reticulate pattern, probably indicating myosin-associated ER membranes in the endoplasm of the basal zone (Braun, 1996a). In contrast to internodal cells, a constitutive difference exists in the rates of acropetal and basipetal streaming velocities, even though this difference is not reflected in the arrangement of the actin cytoskeleton in this zone. In vivo video-microscopy of rhizoids in different orientations and exposed to different accelerations in centrifuges and to microgravity during the parabolic flight of rockets revealed that a gravity-dependent component exists that affects the velocity differences merely physically and without a physiological amplification at 1*g* (Ackers et al., 2000). Under microgravity conditions both acropetal and basipetal streaming velocities increased but the velocity differences decreased (Buchen et al., 1991; Sievers et al., 1991b). What causes the endogenous difference between the actin-mediated acropetal and basipetal streaming in rhizoids and protonemata is still a matter of debate (Ackers et al., 2000).

4. ACTIN-MICROTUBULE INTERACTIONS AND POLAR CYTOPLASMIC ORGANIZATION

The interplay between microtubules (MTs) and the actin cytoskeleton is an important feature of rhizoid and protonema structure and function. Immunfluorescence and in vivo observations of MTs in living cells by injection of fluorescent tubulin have documented a clear distribution pattern for MTs under steady-state, gravistimulated, chemically and otherwise perturbed conditions (Braun & Sievers, 1994; Braun & Wasteneys, 1998a). The arrangement of MTs is different from that of the actin cytoskeleton, but MTs are also distributed in accordance with the polarized cytoplasmic organization of these cells.

In the basal zone, a peripheral network of cortical MTs is distributed into two hemicylindrical arrays that correspond with the underlying subcortical actin arrays whose confinement to the two hemicylindrical zones in turn corresponds with the separation of basipetal and acropetal routes of cytoplasmic streaming. The net axial alignment of MTs in the basal zone is also similar to that of the underlying actin bundles. MT axial alignment could reflect passive alignment by hydrodynamic forces generated at the actin-endoplasm interface. Such coalignment has been experimentally induced by removal of the chloroplast barrier normally separating cortical MTs and subcortical actin bundles in characean internodal cells (Foissner & Wasteneys, 1999), and plastids are absent or sparse in rhizoids and protonemata. MTs in the basal zone form a discontinuous array, with MTs frequently arranged in small branching x- and y-shaped clusters. Carboxyfluorescein-tubulin microinjection experiments suggest that MT assembly is dispersed throughout the region and is not confined to a defined organizing center (Braun & Wasteneys, 1998a).

In the subapical region, MTs and actin MFs colocalize in a dense network of scattered but predominantly axially oriented elements (Braun & Sievers, 1994; Braun & Wasteneys, 1998a). Microinjection of fluorescently tagged tubulin showed that the subapical MTs originate from the basal side of the nuclear envelope, indicating the nuclear envelope's MT-organizing function. MTs appear to flare out into the subapical zone but, unlike actin filaments, end abruptly 35–40 μm from the tip (Fig. 2C). MTs thus appear to be completely excluded from the apex of growing cells and this exclusion seems to be important for both gravisensing and the gravitropic response. Terminating tip growth by mechanically manipulating the cells or by illuminating protonemata causes MTs to extend into the apical zone where they impair the actin-mediated movements and gravity-directed statolith sedimentation (Braun & Wasteneys, 1998a,b). Conversely, depolymerization of MTs by oryzalin treatment dramatically alters the cell's polar organization and disturbs the arrangement of actin MFs in the apical and subapical, but not the basal, regions. The vacuole moves closer to the tip and streaming takes place in what was once stationary cytoplasm, coincident with the extension of thick actin bundles into this region. Gravitropic tip growth does not continue once the apical cytoplasm is converted into streaming endoplasm (Braun & Sievers, 1994). In contrast, disruption of the actin cytoskeleton with cytochalasin D inhibits cytoplasmic streaming (and effectively arrests tip growth) without changing the polar cytoplasmic organization or the arrangement of MTs in the basal and subapical zone (Braun & Sievers, 1994). Based on these results, it is concluded that MTs play a crucial role in the maintenance of the polar cytoplasmic zonation, a precondition for tip growth, and intimately interact with the functional arrangement of the subapical actin cytoskeleton. Similar conclusions are emerging from recent studies on tip-growing higher plant cells (Bibikova et al., 1997). The motile processes, however (i.e. cytoplasmic streaming, transport of organelles, positioning of statoliths and delivery of vesicles), are generated by the acto-myosin system.

5. ROLE OF ACTIN IN TIP GROWTH

The process of polar growth has been demonstrated to be dependent on the complexly organized apical actin cytoskeleton. Cytochalasin-induced disruption of actin inhibits vesicle trafficking and exocytosis and causes the disappearance of the prominent ER aggregate near the apex (Bartnik & Sievers, 1988; Bartnik et al., 1990), resulting in termination of tip growth (Hejnowicz & Sievers, 1981; Braun & Sievers, 1993). The spherical aggregate of ER membranes in the apex represents the structural center of the vesicle-rich Spitzenkörper whose putative function is to regulate vesicle guidance and exocytosis. The ER aggregate co-localizes with the dense apical actin array (Sievers et al., 1991a; Braun & Wasteneys, 1998a,b) and recent immunological evidence (Braun, submitted) suggests a protein with similar function and some homology to animal spectrin is specifically involved in organizing this ER-actin aggregate. This actin – spectrin interaction may play a role in anchorage and maintenance of the structural organization by providing mechanical stability to the distinct apical ER subdomain. These proteins may also provide a mechanism for the recruitment of specific membrane proteins required for the characteristic functions of the ER aggregate, i.e., regulation of vesicle trafficking (Bartnik & Sievers, 1988; Bartnik et al., 1990), and for control of the highly specific physiological environment at the tip.

A tip-high gradient of cytoplasmic free calcium was shown to be essential for tip growth in characean rhizoids and protonemata (Braun & Richter, 1999) in agreement with reports for many other tip-growing plant cell types (see Sanders et al., 1999 and references therein). The calcium gradient visualized with the single-wavelength indicator, Calcium Crimson, strongly colocalizes at the tip with dihydropyridine receptors. These are putative calcium channels (Braun & Richter, 1999). Both gradients dissipate when tip growth is inhibited. Interfering with the calcium gradient of rhizoids and protonemata by applying the calcium ionophore A23187 or the calcium-channel blocker gadolinium chloride, causes disintegration of the ER aggregate, prevents spectrin antibody labelling and causes a major reorganization and bundling of the actin MFs. Removal of these agents is followed by a restoration of the complex actin arrangement, the reformation of the ER aggregate, the reappearance of spectrin-like epitopes and resumption of tip growth.

6. ACTIN-MEDIATED POSITIONING AND MOVEMENT OF STATOLITHS

Several observations demonstrate convincingly that statolith positioning in characean rhizoids and protonemata is primarily controlled by the acto-myosin cytoskeleton. Myosin antibodies label the surface of statoliths (Braun, 1996a) and, in the steady-state, statoliths are positioned in an actin MF-rich, but MT-free, zone (Braun & Wasteneys, 1998a). Cytochalasin treatments that disrupt the actin cytoskeleton impair normal responses of statoliths to gravistimulation (Hejnowicz & Sievers, 1981; Braun & Sievers, 1993). In this section, the regulation of statolith positioning and motility by the acto-myosin cytoskeleton is discussed. Despite the very similar arrangement of actin in rhizoids and protonemata, the motile properties of statoliths in these two cell types turn out to be remarkably different. These differences help explain how positive and negative gravitropic responses are mediated.

6.1 Actin-mediated positioning of statoliths is a prerequisite for gravisensing

Acropetal transport of statoliths occurs in both protonemata and rhizoids. It is more obvious and constantly active, however, in upward tip-growing protonemata in which statoliths drop in a gravity-dependent manner up to 100 µm back into the subapical zone and are sporadically re-transported towards the tip. Cytochalasin prevents re-transport, demonstrating actin's critical involvement in generating the net-acropetal forces on statoliths that allows them to keep pace with the growing tip (Hodick, 1994; Hodick & Sievers, 1998). In rhizoids, however, the positioning of statoliths is mediated by net-basipetal forces; gravitational forces preclude the need for active retransport as long as cells are downward-pointing, but if cells are repositioned in a horizontal or upward-pointing direction, acropetal transport of statoliths is required to prevent slippage of statoliths into the basal region of the cell where they are useless for graviresponses. Indeed, removing statoliths from their original position by basipetal centrifugation abolishes graviresponsiveness (Buder, 1961; Sievers et al., 1991a), and graviresponsiveness is only restored once the displaced statoliths are re-transported to their original position. Cytochalasin experiments demonstrated that this re-transport, which can overcome basipetal accelerations of up to 70*g* (Braun & Sievers, 1993), is mediated by actin (Sievers et al., 1991a). Statoliths re-transported against basipetal centrifugal forces take longer to sediment along the gravity vector, implying that statolith–actin interactions become stronger under such conditions (Braun & Sievers, 1993). The actin cytoskeleton is thus able to adapt to altered gravitational environments in order to maintain statoliths in a strategic position.

6.2 Displacement of the Spitzenkörper by statoliths is involved in bending of protonemata

The constitutive activity of the acropetal transport mechanism in protonemata is demonstrated when cells are gravistimulated (Sievers et al., 1996; Braun, 1997). Upon horizontal positioning of protonemata, statoliths not only sediment in the direction of gravity, but are also simultaneously transported into the apical dome where they then settle asymmetrically against the apical plasma membrane (Fig. 1; Hodick, 1994; Hodick et al., 1998). This apical positioning of statoliths close to the growth center at the outermost tip is followed by upward-directed reorientation of the protonema tip. Inverting protonemata in a tip-downward orientation results in statoliths settling into the apical dome where they leave only a small region at the outermost tip free. Random asymmetric distribution of statoliths eventually initiates reorientation of the protonema tip back to the normal upright position (Fig. 1). The apical intrusion of sedimenting statoliths in protonemata is suggested to induce the negative gravitropic reorientation of the growth direction by displacing the Spitzenkörper and, thus, bringing about a repositioning of the growth center from the outermost tip to the upper flank (Sievers et al., 1996; Braun, 1997). This mode of bending was termed 'bending by bulging' (Green et al., 1970).

6.3 A basipetally directed transport mechanism excludes statoliths from the growing tip of rhizoids

In contrast to protonemata, rhizoids have an additional mechanism that prevents statoliths from sedimenting against the plasma membrane of the apical dome. Statoliths are held in a dynamically stable position, 10–30 μm above the tip where they continuously perform saltatory and rotational movements (for review see Sievers et al., 1996; Braun, 1997). By controlling this position, actin MFs prevent statolith sedimentation into the physically lowest cell apex, an outcome that would impede exocytosis. Cytochalasin treatments provide indirect evidence for actin's role in statolith positioning: disruption of the actin cytoskeleton causes the statoliths to drop onto the apical cell wall and thus inhibit tip growth (Hejnowicz & Sievers, 1981; Braun & Sievers, 1993). After removing cytochalasin, the statoliths are lifted back to their original position and tip-growth resumes (Hejnowicz & Sievers, 1981; Braun & Sievers, 1993). Thus, in normal, tip downward-oriented rhizoids, the gravity force, which pulls the statoliths towards the tip, is compensated by an internal counteracting force mediated by the acto-myosin system (Sievers et al., 1991b; Volkmann et al., 1991).

Further evidence for the existence of this counter-gravity mechanism in rhizoids has come literally from outer space. When rhizoids were subjected to the 6-min microgravity phase ($< 10^{-4}g$) of the parabolic flight of sounding rockets (TEXUS), statoliths moved basipetally and doubled their original distance from the cell tip (Fig. 3A). Treating rhizoids with cytochalasin D prior to launch caused statoliths to settle onto the apical cell wall from which they were not displaced during the microgravity phase (Fig. 3B). Removing the inhibitor resulted in a repolymerization of the actin cytoskeleton and the cells quickly recovered. This proved that the statolith lifting in cytochalasin-free rhizoids could not be caused by a simple physical effect such as random positioning of weightless particles (Buchen et al., 1993). When rhizoids were positioned horizontally prior to the launch of the TEXUS rocket, the complex of sedimented statoliths was displaced basipetally rather than towards the former upper flank during microgravity (Buchen et al., 1997).

Together, these experiments indicate that the position of statoliths in rhizoids is highly controlled and regulated in both axial directions, but only weakly controlled in the lateral direction (Braun, 1997). Interestingly, earlier optical tweezer experiments came to the same conclusion by measuring the laser output power required to move statoliths (Leitz et al., 1995); the force needed to move statoliths toward the apex is greater than the force to move the statoliths towards the flank. Thus, the actin cytoskeleton actively controls the position of the statoliths to accomplish graviresponsiveness and allows the rapid, gravity-directed sedimentation of statoliths in lateral directions, which is crucial for the initiation of the graviresponse.

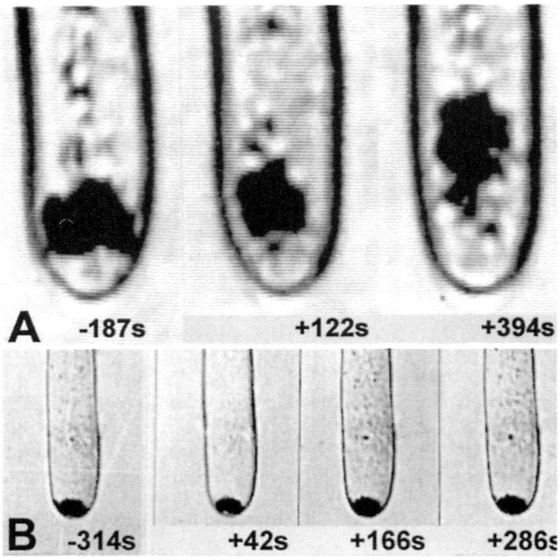

Figure 3. Videomicroscopic images showing the behaviour of statoliths of characean rhizoids during the 6-min microgravity phase of a TEXUS-rocket flight. **A** In untreated rhizoids, the statoliths nearly doubled their original distance from the tip (-187s) during the microgravity phase (+122s, +394s). **B** In a rhizoid in which the statoliths were settled into the physically lowest cell tip due to cytochalasin D treatment prior to launch (-314s), the statoliths were not displaced during the microgravity phase (+42s, +166s, +286s) but were lifted back to their original position after removing the inhibitor. Time given in seconds with respect to launch. Diameter of the cells: 30 μm. **A** modified from Volkmann et al. (1991); **B** modified from Buchen et al. (1993).

In an attempt to simulate the effect of weightlessness on statolith positioning, rhizoids were rotated on the horizontal axis of slow- and fast-rotating clinostats (Cai W. et al., 1997). Rotational velocities of 60–90 rpm resulted in a basipetal displacement of statoliths similar to but slower than the movement of statoliths during the microgravity phase of TEXUS flights. During clinostatting, however, the cluster of statoliths became more and more dispersed and eventually moved closer towards the tip again. This apical approach was interpreted as a cytoskeleton-mediated adaptational effect, something that was also observed in rhizoids grown for hours in real microgravity aboard the Space Shuttle (IML-2, S/MM05; Braun, 1997). In contrast, slower rotational velocities of the classical clinostat (1–10 rpm) failed to simulate the effect of microgravity. Statoliths were displaced from the periphery towards the central cell axis and, thereby, the complex of statoliths elongated axially, extending more basally and more apically as compared to the original position. These results suggest mechanical perturbations as the major effect of the classical clinostat, but mainly endogenous actomyosin-driven movements of the statoliths induced by fast rotations, thus properly simulating microgravity effects.

6.3.1 Gravisensing mechanisms develop even in the absence of gravity

In vivo videomicroscopy and ultrastructural analysis of rhizoids that were initiated at 1*g* (1*g* = 981 cm s^{-2}: Earth acceleration) on Earth and grown for several days in microgravity (IML-2), versus rhizoids that developed and grew exclusively in microgravity aboard the SpaceLab (IML-2) and SpaceHab (S/MM05), provided evidence that the gravisensing cells follow their endogenous program for development and morphogenesis even in the absence of net-accelerations in a microgravity environment. Deprived of directional accelerations, rhizoids radiated in all directions from the nodal complexes, and developed normal cell shape, and organization, as well as a polar distribution of organelles, including the ER-aggregate in the center of the vesicle-rich Spitzenkörper at the cell tip (Braun, 1997; Braun et al., 1999b). Furthermore, by using a centrifuge microscope (Friedrich et al., 1996), it was confirmed that the microgravity-grown rhizoids are still able to respond to acceleration stimuli, even when statoliths are positioned further away from the tip, so long as they remain in the MT-free zone (Braun et al., 1999b). Based on experiments performed aboard the Space Shuttle during the IML-2 mission and during the 13-min microgravity phase of the parabolic flight of a MAXUS-rocket (MAXUS-3), the threshold value for gravisensitivity of rhizoids can be expected to be $> 0.1g$ and clearly $< 0.3g$. These results and preliminary immunolabelling studies indicate that the organization and function of the cytoskeleton is not subjected to drastic changes in microgravity (Braun et al., 1999b). However, it also becomes evident that the actin-mediated statolith positioning is perfectly adapted to the Earth's 1*g*-environment. Therefore, gravisensitivity in microgravity is unlikely to be exactly the same as in 1*g*-controls (Braun et al., 1999b).

6.3.2 Protonemata bend by bulging, whereas rhizoids bend by bowing

The experiments described so far, whether performed in microgravity, with laser tweezers, clinostats, centrifuges, or incorporating actin-specific inhibitors, all point to the critical nature of the growth center in defining the properties that support positive, versus negative, gravitropism. In protonemata, displacement of the growth center from the apex results in 'bending by bulging' (as defined by Green et al., 1970). Actin-mediated acropetal transport of statoliths ensures that protonemata bend upward when moved out of their vertical position. In rhizoids, the transport of statoliths is net-basipetal and protects the growth center from statolith intrusion. This results in statoliths sedimenting along on the subapical flank, suppressing the growth rate along that flank (Sievers et al., 1979; Sievers & Schnepf, 1981). Differential flank growth generates 'bending by bowing' (Green et al., 1970). Because the growth rate of the statolith-free upper flank is higher than that of the lower flank, rhizoids bend downward.

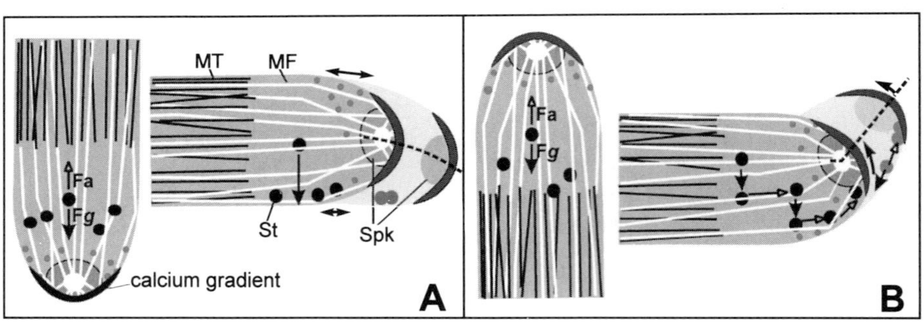

Figure 4. Schematic drawing of the apical and part of the subapical region of a characean rhizoid (**A**) and a protonema (**B**) in normal vertical and horizontal position (gravistimulated). **A** In tip-downward growing rhizoids, the statolith (**St**) position results from net-basipetally acting actin-mediated forces (**Fa**) compensating the gravity force (**Fg**). Upon tilting, the statoliths simply follow the direction of gravity and sediment onto the lower cell flank. The subsequent curvature is induced by differential extension of the opposite cell flanks. The Spitzenkörper (**Spk**) and the calcium gadient remain symmetrically located at the tip and are not directly involved in gravitropic bending. **B** In upward growing protonemata, net-acropetally acting forces mediated by the actin microfilaments (MFs) compensate the basally directed gravity force. Placed in a horizontal position, statoliths settle near the growth center at the tip by following the gravity vector and by the additional forces of the acropetally acting actin MFs. This causes a drastic shift of both the calcium gradient and the Spitzenkörper towards the upper flank and the new outgrowth occurs at that site. Until the protonema

reaches the upright position, the statoliths are continuously pushed upward towards the growth center and fall back again. MT = microtubules.

Centrifugation experiments provide further insights into bending mechanisms in protonemata and rhizoids. When centrifuged at an angle of 90° normal to the cell axis, protonemata showed a gradual reduction in statolith sedimentation-mediated growth curvature as accelerations increased from 1*g* to 8*g* (Hodick & Sievers, 1998). These strong centrifugal forces press the statoliths onto the lateral cell flank, so this probably inhibits their actin-mediated acropetal movement. Acropetal movement therefore appears to be essential for initiating bending by bulging, which requires lateral displacement of the growth center. This is clearly in contrast to the results of rhizoid centrifugation experiments using similar lateral accelerations (Braun, 1996b). Positive gravitropic bending of horizontally positioned rhizoids started abruptly when lateral centrifugal forces were introduced, but bending was only transient. Under continuous centrifugation, further bending by bowing was inhibited and rhizoids continued growth at a uniform angle to the acceleration vector. Lack of further bending appears to occur because the sedimented statoliths displace the Spitzenkörper and, thereby, the growth center is displaced towards the centripetal flank. The balance between bowing towards the acceleration vector (positive gravitropism) by differential flank growth and bulging away from the acceleration vector (negative gravitropism) may explain how the modified growth axis is maintained. Indeed, vertically positioned rhizoids can be induced to deviate from their axis if centrifugal forces are applied at an angle greater than 5° and less than 90° to the original axis. If the acceleration angle is less than 5°, statoliths sediment symmetrically into the apex and radial swelling of the tip, but without bending, occurs. At greater centrifugal angles, statoliths sediment asymmetrically, allowing both bowing and bulging to occur.

6.3.3 Spitzenkörper anchorage is more stable in rhizoids than in protonemata

The fact that apically sedimented statoliths impair bending of rhizoids but promote negative gravitropism in protonemata suggests that one major difference between protonemata and rhizoids is the relative ease with which the growth center can be displaced by sedimenting statoliths. There is evidence from centrifugation experiments (Braun, 1996b; Hodick & Sievers, 1998) and from attaching particles to the surface of bending cells (Sievers et al., 1979) that the position of the growth center at the cell tip is relatively stable and that in rhizoids the Spitzenkörper is more tightly anchored than in protonemata. Asymmetric statolith sedimentation against the apical cell wall and displacement of the Spitzenkörper occurs naturally in protonemata but requires considerable centrifugal force to happen in rhizoids. Rhizoids can be forced to respond to some extent like protonemata but only when centrifuged or by pushing statoliths aymmetrically into the apical dome with optical tweezers (Braun, unpublished results).

6.3.4 How does the actin cytoskeleton modulate gravitropic mechanisms?

The idea that the degree of Spitzenkörper anchorage is mediated by properties of the actin cytoskeleton and the activity of calcium-dependent, actin-binding proteins (Braun & Richter, 1999) is supported by recent experimental data. Immunofluorescence of spectrin in graviresponding cells labels the ER aggregate, the structural center of the Spitzenkörper. The symmetrical position of the spherical labelling pattern was drastically displaced towards the upper flank, the site of future outgrowth, during initiation of the graviresponse in protonemata, clearly before curvature was recognizable by the formation of a bulge on the upper flank (Braun, submitted). In contrast, the anti-spectrin-fluorescence labelling in rhizoids remained symmetrically positioned in the apical dome throughout the graviresponse, confirming that a repositioning of the ER aggregate is involved in the negative graviresponse of protonemata but not in the positive graviresponse of rhizoids. Further evidence comes from calcium imaging, which demonstrates a drastic shift of the calcium gradient from the tip towards the upper flank during initiation of the graviresponse in protonemata but not in rhizoids. In accordance with this observation, dihydropyridine-fluorescence, which indicates a symmetrical distribution of putative calcium channels at the tip of normal vertically growing cells, was also found to be displaced towards the upper flank only in graviresponding protonemata. These results suggest that the early asymmetric repositioning of the calcium gradient in protonemata may result from statolith-induced displacement or, more likely, from statolith-induced differential activation or inhibition of apical calcium channels. The asymmetric influx of Ca^{2+}, in turn, may mediate the repositioning of the Spitzenkörper and the growth center by differentially regulating the actin-anchorage or the activity of actin-associated proteins along the shifting calcium gradient. In *Fucus* zygotes and rhizoids, an asymmetric Ca^{2+} gradient establishes a labile new growth axis and the actin cytoskeleton is required to fix the new growth direction (Quatrano et al., 1991; Shaw & Quatrano, 1996). Similarly, the tendency for protonemata to reorient towards the former growth axis after a short gravistimulation indicates that the new growth axis induced by the upward shift of the Ca^{2+} gradient is rather labile and may also require actin cytoskeletal anchorage to stabilize the new growth direction (Braun & Richter, 1999).

7. ACTIN REORGANIZATION DURING PHOTOMORPHOGENETIC DIFFERENTIATION

Rhizoids develop in the light and continue positive gravitropic tip growth in either light or darkness. They may undergo cell division but only to produce a secondary rhizoid. In contrast, protonemata develop and perform negatively gravitropic tip growth exclusively in the absence of blue light (Hodick, 1993). Illumination is followed by cessation of tip growth and tip-ward migration of the nucleus. Eventually, the unicellular protonema divides, giving rise to a multicellular filament of relatively uniform cells which subsequently regenerates the green thallus (Braun & Wasteneys, 1998b) and is typically organized in alternating nodal cell complexes and internodal cells. The morphogenetic differentiation is proposed to be initiated by a blue-light receptor whose identity and localization is still unknown (Hodick, 1993).

The irreversible light-dependent transition from the polarly organized and tip-growing protonema into a relatively unpolarized cell is reflected by a dramatic reorganization of the cytoskeletal elements (Braun & Wasteneys, 1998b). The disappearance of the ER aggregate at the tip and the breakdown of the polar distribution of organelles is strongly correlated with the rearrangement of the polarly organized and tip-focussed actin MFs into a uniform dense meshwork of randomly oriented actin MFs. MTs inhibit statolith sedimentation and, thus, gravity perception by invading the apical cytoplasm. MTs also mediate the repositioning of the plastids from their former random location in the subapical cytoplasm towards the cortex. After termination of tip growth, the apical cytoplasmic region is no longer divided into an apical and subapical zone and subsequent cell growth is by diffuse expansion. During metaphase of the cell division in the single-celled protonema, MTs form a highly asymmetrical extranuclear spindle. The apical cell pole is characterized by abundant, aster-like cytoplasmic MTs, possibly required to stabilize the apical cytoplasm with the mitotic apparatus. Only short MTs connect with the broader posterior pole which is not associated with the cortical population of basal MTs. During mitosis, the extensive apical actin network encircles the mitotic apparatus, but is not part of it. After breakdown of the nuclear envelope, MT bundles extend from the pole to the kinetochores. However, even kinetochore MT bundles seem to be excluded from the inner compartment of the mitotic apparatus suggesting that the chromosomes may be aligned with their kinetochores close to the peripheral spindle. Actin MFs, which are not associated with the spindle MTs, are clearly coaligned along the MTs of the early phragmoplast. Interestingly, the later belt-like phragmoplast is completely depleted of actin MFs except for some actin bundles that extend from the cytoplasm towards the margins of the growing cell plate. The cell plate itself and young anticlinal cell walls show a diffuse or reticulate actin immunofluorescence pattern. A single file of relatively uniform cells is produced by several of these anticlinal cell divisions.

In contrast to mitotic spindles in the polarized unicellular protonema, the relatively non-polarized apical dome cell of the no longer tip-growing multicellular protonema has symmetric spindle poles, showing aster-like arrangements of branching MT clusters. In this cell type, only a small number of cytoplasmic actin MFs and MTs is present during mitosis. Later, only the basal cells of the multicellular filament produce nodal cell complexes by multiple periclinal divisions. The apical dome cell of the new shoot which originates from one of these nodal cells becomes the meristem initial that regularly divides to produce segment cells. During mitosis in this apical dome cell, the actin and MT arrangement resembles that of typical higher plant cells, except that preprophase bands are absent. The MT spindle is now barrel-shaped and actin MFs are mainly absent during mitosis. The segment cells frequently divide to produce the typical structure of alternating internodal cells and multicellular nodes of the characean thallus. Nodal cells produce rhizoids, lateral branchlets, or protonemata in the absence of light. The actin and MT distribution of nodal cells resembles that of higher plant meristematic cells, whereas the mature, highly vacuolated internodal cells exhibit a highly specialized cortical system of MTs and streaming-generating actin bundles that typify plant cells which elongate by diffuse rather than tip growth (see Foissner & Wasteneys, this volume). In conclusion, the cytoskeletal studies reveal some interesting and unusual features on the role of the actin and MT cytoskeleton during morphogenetic cell differentiation. The unique diversity of spindle forms and the varying cytoskeletal arrangements of actin MFs and MTs represent specific functions and different cell organization levels throughout photomorphogenesis of protonemata, thus, recapitulating aspects of the evolutionary transition from more primitive organisms to higher plants.

8. CONCLUDING REMARKS

Tip growth is certainly not simply a case of limited cell wall expansion, a relict of primitive single-celled organisms. In this chapter, it has been demonstrated that highly specialized, actin-mediated mechanisms have evolved in charophytes in response to the need for effective adaptations to specific environmental conditions. Actin's involvement in tip growth is less clearly defined in other plant cell types, including higher plant root hairs and pollen tubes. The full extent of the actin cytoskeleton or its exact organization in these cells remains the subject of considerable debate (Taylor & Hepler, 1997; Cai G. et al., 1997a). It has been argued that the absence of actin from the outermost cell tip may be essential for the process of vesicle docking and exocytosis (Miller et al., 1999) but, as the studies described in this chapter demonstrate, this certainly does not apply in the case of characean rhizoids and protonemata (and for root hairs see Braun et al., 1999a). The direction of growth in most tropistic-oriented cells, however, is determined by external gradients of chemicals or ions, attractants or repellents. Such exogenous growth cues may allow or necessitate quite a differently organized actin MF system. In root hairs, for instance, a cap-like accumulation of actin and profilin has been visualized at the tip and this may mediate their specific mode of polar growth (Braun et al., 1999a). In contrast, the protonemal and caulonemal tip-growing cells of mosses exhibit an extensive, but uniform, meshwork of actin MFs. These cells can orient phototropically and gravitropically according to external physical stimuli but their tropic responses are relatively slow. Therefore, the highly polarized cytoplasmic and cytoskeletal organization, including the unique ER aggregate of characean rhizoids and protonemata, may reflect their highly sensitive and responsive gravity-oriented tip growth.

REFERENCES

Ackers D, Buchen B, Hejnowicz Z and Sievers A (2000) The pattern of acropetal and basipetal streaming velocities in *Chara* rhizoids and protonemata, and gravity effect on the pattern as measured by laser-Doppler-velocimetry. Planta, in press

Bartnik E and Sievers A (1988) *In-vivo* observation of a spherical aggregate of endoplasmic reticulum and of Golgi vesicles in the tip of fast-growing Chara rhizoids. Planta 176: 1-9

Bartnik E, Hejnowicz Z and Sievers A (1990) Shuttle-like movements of Golgi vesicles in the tip of growing Chara rhizoids. Protoplasma 159: 1-8

Bibikova TN, Zhigilei A and Gilroy S (1997) Root hair growth in *Arabidopsis thaliana* is directed by calcium and an endogenous polarity. Planta 203: 495-505

Braun M (1996a) Immunolocalization of myosin in rhizoids of *Chara globularis* Thuill. Protoplasma 191: 1-8

Braun M (1996b) Anomalous gravitropic response of Chara rhizoids during enhanced accelerations. Planta 199: 443-455

Braun M (1997) Gravitropism in tip-growing cells. Planta 203: S11-S19

Braun M and Sievers A (1993) Centrifugation causes adaptation of microfilaments; studies on the transport of statoliths in gravity sensing Chara rhizoids. Protoplasma 174: 50-61

Braun M and Sievers A (1994) Role of the microtubule cytoskeleton in gravisensing *Chara* rhizoids. Eur J Cell Biol 63: 289-298

Braun M and Wasteneys GO (1998a) Distribution and dynamics of the cytoskeleton in graviresponding protonemata and rhizoids of characean algae: Exclusion of microtubules and a convergence of actin filaments in the apex suggest an actin-mediated gravitropism. Planta 205: 39-50

Braun M and Wasteneys GO (1998b) Reorganization of the actin and microtubule cytoskeleton throughout blue-light-induced differentiation of characean protonemata into multicellular thalli. Protoplasma 202: 38-53

Braun M and Richter P (1999) Relocalization of the calcium gradient and a dihydropyridine receptor is involved in upward bending by bulging of Chara protonemata, but not in downward bending by bowing of Chara rhizoids. Planta 209: 414-423

Braun M, Baluška F, von Witsch M and Menzel D (1999a) Redistribution of actin, profilin and phosphatidylinositol-4,5-bisphosphate (PIP_2) in growing and maturing root hairs. Planta 209: 435-443

Braun M, Buchen B and Sievers A (1999b) Electron microscopical analysis of gravisensing Chara rhizoids developed under microgravity conditions. FASEB J 13: 113-120

Buchen B, Hejnowicz Z, Braun M and Sievers A (1991) Cytoplasmic streaming in Chara rhizoids: Studies in a reduced gravitational field during parabolic flights of rockets. Protoplasma 165: 121-126

Buchen B, Braun M, Hejnowicz Z and Sievers A (1993) Statoliths pull on microfilaments. Experiments under microgravity. Protoplasma 172: 38-42

Buchen B, Braun M and Sievers A (1997) Statoliths, cytoskeletal elements and cytoplasmic streaming of Chara rhizoids under reduced gravity during TEXUS flights. In: Life Sciences Experiments Performed on Sounding Rockets (1985-1994). Nordwijk, ESA Publications Division, ESA-SP 1206, pp 71-75

Buder J (1961) Der Geotropismus der Characeenrhizoide. Ber Dtsch Bot Ges 74: (14)-(23)

Cai G, Moscatelli A and Cresti M (1997) Cytoskeletal organization and pollen tube growth. Trends Plant Sci 2: 86-91

Cai W, Braun M and Sievers A (1997) Displacement of statoliths in Chara rhizoids during horizontal rotation on clinostats. Acta Bot Exp Sinica 30: 147-155

Chen R, Rosen E and Masson PH (1999) Gravitropism in higher plants. Plant Physiol 120: 343-350

Foissner I and Wasteneys GO (1999) Microtubules at wound sites of Nitella internodal cells passively coalign with actin bundles when exposed to hydrodynamic forces generated by cytoplasmic streaming. Planta 208: 480-490

Foissner I and Wasteneys GO (2000) Nuclear crystals, lampbrush-chromosome-like structures and perinuclear cytoskeletal elements associated with nuclear fragmentation in characean internodal cells. Protoplasma, in press

Franklin-Tong VE (1999) Signaling and the modulation of pollen tube growth. Plant Cell 11: 727-738

Friedrich ULD, Joop O, Pütz C and Willich G (1996) The slow rotating centrifuge microscope NIZEMI: A versatile instrument for terrestrial hypergravity and space microgravity research in biology and material science. J Biotech 47: 225-238

Fritsch FE (1935) The Structure and Reproduction of the Algae. vol 1, London: Cambridge University Press, pp 447-469

Giesenhagen K (1896) Untersuchungen über die Characeen. Flora 82: 381-433

Giesenhagen K (1901) Über innere Vorgänge bei der gravitropischen Krümmung der Wurzeln von Chara. Ber Dtsch Bot Ges 19: 277-285

Green PB, Erickson RO and Richmond PA (1970) On the physical basis of cell morphogenesis. Ann NY Acad Sci 175: 712-731

Hejnowicz Z and Sievers (1971) Mathematical model of geotropically bending Chara rhizoids. Z Pflanzenphysiol. 66: 34-48

Hejnowicz Z and Sievers A (1981) Regulation of the position of statoliths in Chara rhizoids. Protoplasma 108: 117-137

Hejnowicz Z, Buchen B and Sievers A (1985) The endogenous difference in the rate of acropetal and basipetal cytoplasmic streaming in Chara rhizoids is enhanced by gravity. Protoplasma 125: 219-229

Hodick D (1993) The protonema of *Chara fragilis* Desv.: Regenerative formation, photomorphogenesis, and gravitropism. Bot Acta 106: 388-393

Hodick D (1994) Negative gravitropism in Chara protonemata and rhizoids: A model integrating the opposite gravitropic responses of protonemata and rhizoids. Planta 195: 43-49

Hodick D and Sievers A (1998) Hypergravity can reduce but not enhance the gravitropic response of Chara globularis protonemata. Protoplasma 204: 145-154

Hodick D, Buchen B and Sievers A (1998) Statolith positioning by microfilaments in *Chara* rhizoids and protonemata. Adv Space Res 21: 1183-1189

Jackson SL and Heath IB (1990) Visualization of actin arrays in growing hyphae of the fungus *Saprolegnia ferax*. Protoplasma 154: 66-70

Jarosch R (1956) Aktiv bewegungsfähige Plasmaelemente und Chloroplastenrotation bei Characeen. Anz Österr Akad Wiss 6: 58-60

Kadota A., Yoshizaki N and Wada M (1999) Cytoskeletal changes during resumption of tip growth in nongrowing protonemal cells of the fern *Adiantum capillus-veneris* L. Protoplasma 207: 195-202

Kachar B and Reese TS (1988) The mechanism of cytoplasmic streaming in characean algal cells: sliding of endoplasmic reticulum along actin filaments. J Cell Biol 106: 1545-1552

Kuroda K (1990) Cytoplasmic streaming in plant cells. Int Rev Cytol 121: 267-307

Leitz G, Schnepf E and Greulich KO (1995) Micromanipulation of statoliths in gravity-sensing Chara rhizoids by optical tweezers. Planta 197: 278-288

Levina NN, Lew RR and Heath IB (1994) Cytoskeletal regulation of ion channel distribution in the tip-growing organism *Saprolegnia ferax*. J Cell Sci 107: 127-134

Meske V and Hartmann E (1995) Reorganization of microfilaments in protonemal tip cells of the moss *Ceratodon purpureus* during the phototropic response. Protoplasma 188: 59-69

Meske V, Ruppert V. and Hartmann E (1996) Structural basis of the red light induced repolarization of tip-growth in caulonema cells of *Ceratodon purpureus*. Protoplasma 192: 189-198

Miller DD, Lancelle SA and Hepler PK (1996) Actin microfilaments do not form a dense meshwork in Lilium longiflorum pollen tube tips. Protoplasma 195: 123-132

Miller DD, de Ruijter NCA, Bisseling T and Emons AMC (1999) The role of actin in root hair morphogenesis: Studies with lipochito-oligosaccharide as a growth stimulator and cytochalasin as an actin perturbing drug. Plant Cell 17: 141-154

Pickett-Heaps JD (1975) Green Algae: Structure, Reproduction, and Evolution in Selected Genera. Sunderland: MA, Sinauer

Quatrano RS, Brian L, Aldridge J and Schulz T (1991) Polar axis fixation in Fucus zygotes: components of the cytoskeleton and extracellular matrix. Development (S) 1: 11-16

Rosen E, Chen R and Masson PH (1999) Root gravitropism: A complex response to a simple stimulus? Trends Plant Sci 4: 407-412

Sack FD (1991) Plant Gravity Sensing. Int Rev Cytol 127: 193-252.

Sack FD (1997) Plastids and gravitropic sensing. Planta 203: S63-S68

Sanders D, Brownlee C and Harper JF (1999) Communicating with calcium. Plant Cell 11: 691-706

Schröder (1904) Zur Statolithentheorie des Geotropismus. Beih Bot Centralbl 16: 269-288

Schröter K, Läuchli A and Sievers A (1975) Mikroanalytische Identifikation von Bariumsulfat-Kristallen in den Statolithen von *Chara fragilis* Desv. Planta 122: 213-225

Shaw SL and Quatrano RS (1996) Polar localization of a dihydropyridine receptor on living Fucus zygotes. J Cell Sci 109: 335-342

Sievers A (1971) Gravity receptors in lower plants. In: Gordon SA, Cohen MJ (eds) Gravity and the organism. Chicago: Univ Chicago Press, pp 51-63

Sievers A and Schröter K (1971) Versuch einer Kausalanalyse der geotropischen Reaktionskette im Chara-Rhizoid. Planta 96: 339-353

Sievers A and Schnepf E (1981) Morphogenesis and polarity of tubular cells with tip growth. In: Kiermayer (ed) Cytomorphogenesis in Plants, Cell Biology Monographs, vol 8 Wien, New York: Springer Verlag, pp 265-299

Sievers A and Braun M (1996) Root cap: structure and function. In: Waisel Y, Eshel A and Kafkafi U (eds) Plant Roots - The Hidden Half (2nd edn). New York: Marcel Dekker, pp 31-49

Sievers A, Heinemann B and Rodriguez-Garcia MI (1979) Nachweis des subapikalen differentiellen Flankenwachstums im Chara-Rhizoid während der Graviresponse. Z Pflanzenphysiol 91: 435-442

Sievers A and Schmitz M (1982) Röntgen-Mikroanalyse von Barium, Schwefel und Strontium in Statolithen-Kompartimenten von Chara-Rhizoiden. Ber Dtsch Bot Ges 95: 353-360

Sievers A, Kramer-Fischer M, Braun M and Buchen B (1991a) The polar organization of the growing Chara rhizoid and the transport of statoliths are actin-dependent. Bot Acta 104: 103-109

Sievers A, Buchen B, Volkmann D and Hejnowicz Z (1991b) Role of the cytoskeleton in gravity perception. In: Lloyd CW (ed) The Cytoskeletal Basis for Plant Growth and Form. London: Academic Press, pp 169-182

Sievers A, Buchen B and Hodick D (1996) Gravity sensing in tip-growing cells. Trends Plant Sci 1: 273-279

Steer MW (1990) Role of actin in tip growth. In: Heath IB (ed) Tip Growth in Plant and Fungal Cells. San Diego: Academic Press, pp 110-145

Taylor LT and Hepler PK (1997) Pollen germination and tube growth. Annu Rev Plant Physiol Plant Mol Biol. 48: 461-491

Tewinkel M, Kruse S, Quader H, Volkmann D and Sievers A (1989) Visualization of actin filament pattern in plant cells without prefixation. A comparison of differently modified phallotoxins. Protoplasma 149: 178-182

Volkmann D, Buchen B, Hejnowicz Z, Tewinkel M and Sievers A (1991) Oriented movement of statoliths studied in a reduced gravitational field during parabolic flights of rockets. Planta 85: 153-161

Walker LM and Sack FD (1995) Microfilament distribution in protonemata of the moss *Ceratodon*. Protoplasma 189: 229-237

Wang-Cahill F and Kiss JZ (1995) The statolith compartment in Chara rhizoids contains carbohydrate and protein. Amer J Bot 83: 220-229

Wasteneys GO and Williamson RE (1991) Endoplasmic microtubules and nucleus-associated actin rings in Nitella internodal cells. Protoplasma 162: 86-98

Wasteneys GO, Collings DA, Gunning BES, Hepler PK and Menzel D (1996) Actin in living and fixed characean internodal cells: Identification of a cortical array of fine actin strands and chloroplast actin rings. Protoplasma 190: 25-38

Zacharias E (1890) Über Bildung und Wachstum der Zellhaut bei *Chara foetida*. Ber Dtsch Bot Ges 8: (56)-(59)

Chapter 15

ACTIN IN CHARACEAN INTERNODAL CELLS

Morphogenesis, wounding, actin-microtubule interactions

Ilse Foissner[1] and Geoffrey O Wasteneys[2]

1 Institut für Pflanzenphysiologie, Universität Salzburg, Hellbrunnerstr. 34, A–5020 Salzburg, Austria
2 Plant Cell Biology Group, Research School of Biological Sciences, The Australian National University, G.P.O. Box 475, Canberra, ACT, Australia

Key words: actin reorganization, actin-microtubule interactions, characean internodes, morphogenesis, wounding

Abstract: This chapter describes how the actin cytoskeleton is organized throughout the development of characean internodal cells. Conspicuous cytoplasmic streaming makes these cells ideal for studying the properties of actin bundle assembly. A fine network of peripheral actin is also described and its role in vesicle secretion discussed. Wounding studies and treatments with actin agonists have been useful for elucidating further properties of both the robust subcortical actin bundles that generate streaming and the finer strands in the cortex. These studies support the idea that a range of actin binding proteins modulate the diverse properties and functions of the actin cytoskeleton. Finally, the interaction between cortical microtubules and the actin cytoskeleton, via both hydrodynamic forces and associated proteins, is explored.

1. THE CHARACEAN THALLUS AND THE CYTOPLASMIC ORGANIZATION OF THE MATURE INTERNODAL CELL

The Characeae are macroscopic green algae that inhabit fresh and brackish water (Wood & Imahori, 1965). Their thallus consists of a regular alternation of groups of small roundish or discoidal uninucleate cells (nodes) and huge cylindrical coenocytic internodal cells that may attain a length of up to 20 cm by diffuse expansion. The thallus is anchored in the substratum by delicate, tip-downward-growing rhizoids, which resemble the upward

C.J. Staiger et al. (eds.), Actin: A Dynamic Framework for Multiple Plant Cell Functions, 259–274.

growing protonemata formed during zygote germination or regeneration of thallus fragments (Braun & Wasteneys, 1998; this volume).

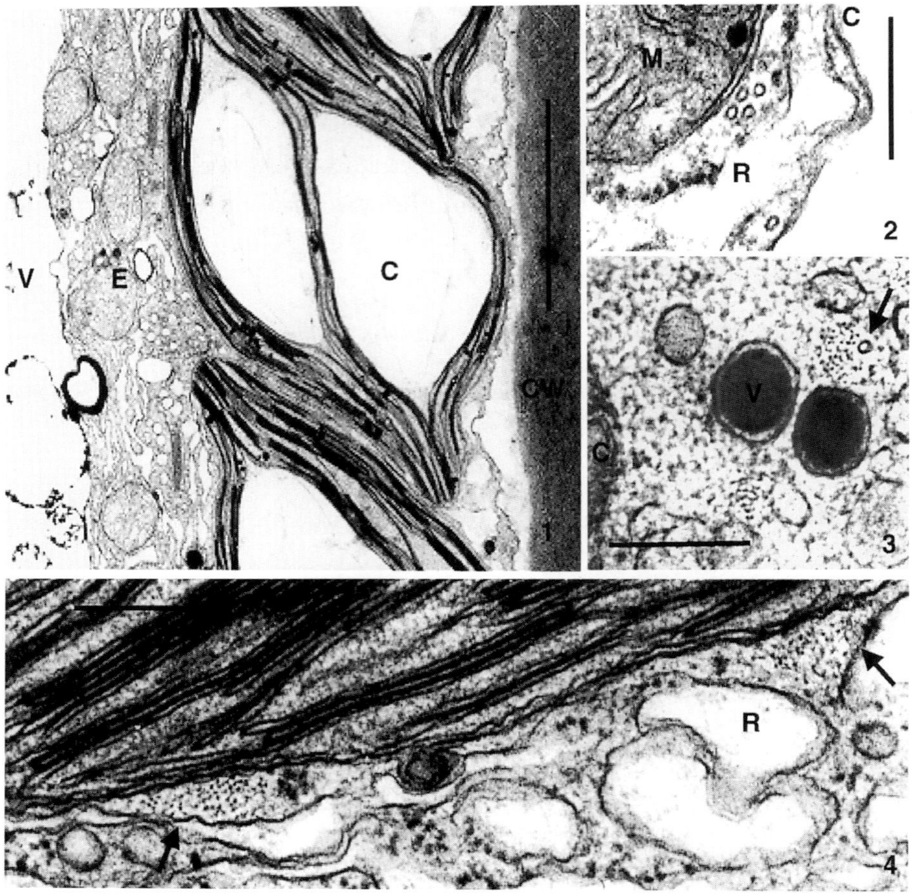

Fine structure of characean internodal cells.
Figure 1. Cross-section showing the cell wall (CW), cortical cytoplasm dominated by chloroplasts (C), motile endoplasm (E) and vacuole (V).
Figure 2. Longitudinal section through elongating internode. A single microtubule and a group of four microtubules are located along the outer and inner surface of cortical endoplasmic reticulum cisternae (R). M = mitochondrion.
Figure 3. Actin bundles and chloroplasts are widely spaced in very young internodes. Note endoplasmic microtubule (arrow) along actin bundle. V = vesicle.
Figure 4. In older cells, subcortical actin bundles (arrows) are attached to the chloroplast surface and contact endoplasmic reticulum cisternae. Bars = 3.5 μm (Fig. 1) and 300 nm (Figs. 2 – 4). Figs. 1, 2 are from Foissner & Wasteneys (1999) by permission of Springer.

The cytoplasm of mature internodal cells is divided into a stationary cortex and a motile endoplasm (Fig. 1). The cortex contains helically arranged, immobile chloroplast files. Cortical microtubules and fine actin strands are located near the plasma membrane (Figs. 2, 6a). Other organelles, including several thousand nuclei, are found in the endoplasm and glide *en masse* along the chloroplast files at velocities as high as 100 μm s^{-1}. Rotational streaming is generated by interaction of subcortical actin bundles attached to the inner side of the chloroplasts (Figs. 4, 6b; Nagai & Rebhun, 1966) with myosin attached to the membranes of endoplasmic reticulum cisternae and other organelles (reviewed by Higashi-Fujime, 1991; Kuroda, 1990; Shimmen & Yokota, 1994; Williamson, 1992; and by Grolig & Pierson, this volume).

2. REORGANIZATION OF THE ACTIN CYTOSKELETON DURING FORMATION AND GROWTH OF INTERNODAL CELLS

2.1 Uninucleate cells contain a meshwork of randomly arranged actin filaments

The huge internodal cells develop from uninucleate segment or nodal cells (Cook et al., 1998; Pickett-Heaps, 1975). The actin cytoskeleton of the uninucleate internode primordium (Wasteneys et al., 1996) is similar to that of non-differentiated higher plant cells (reviewed by Staiger & Lloyd, 1991). A dense, reticulate actin network consisting of fine, randomly arranged filaments is present throughout the cytoplasm and enmeshes the central nucleus and the cortical chloroplasts (Fig. 5). Organelles occasionally perform saltatory movements, but cytoplasmic streaming is not discernible.

2.2 The actin cytoskeleton becomes organized into a cortical array of delicate strands and thick subcortical bundles during elongation

During elongation of young internodal cells, the single nucleus enlarges, migrates to a peripheral location and divides amitotically (Wasteneys et al., 1996; Foissner & Wasteneys, 2000). The central portion of the cell becomes vacuolated but for some time remains traversed by cytoplasmic strands containing actin filaments. These, and the fine actin meshworks around nuclei and chloroplasts, gradually disappear and actin filaments become increasingly organized into bundles approximately parallel to the longitudinal axis of the cell. The bundles frequently cross over, or merge, but always form a continuous array without breaks or bundles terminating at the cell ends where they undergo a 180° turn. Chloroplasts also become arranged as loose, axial files. Initially they are not closely associated with the actin bundles (Fig. 3). Orientation of the cortical actin strands changes from random to predominantly transverse, in parallel with the cortical microtubules, as elongation proceeds (Fig. 6a; Wasteneys et al., 1996). In later stages of expansion, the chloroplast files gradually become aligned at a small angle relative to the longitudinal axis. Actin bundles become increasingly organized parallel to each other and parallel to the helical chloroplast files. Endoplasmic organelles therefore follow a helical path as they circumnavigate the internode. Endoplasmic microtubules are often found in parallel alignment to the actin bundles (Fig. 3).

The actin cytoskeleton of the thin peripheral cytoplasm in mature cells is separated into a subcortical and a cortical domain. The subcortical domain consists of discrete sets of 3 - 6 bundles along the inner side of chloroplasts (Figs. 4, 6b). The uniform polarity of actin filaments in bundles (Kersey et al., 1976; Williamson, 1975; Palevitz & Hepler, 1975) explains the strictly unidirectional streaming in internodal cells. These bundles may extend through the chloroplast layer towards the cell periphery (Wasteneys et al., 1996; Williamson et al., 1986). Delicate actin filaments between the chloroplasts and cell membrane remain in the cortex of mature cells, but they lose the transverse orientation characteristic of actin strands in growing cells (Fig. 6a). No delicate cortical actin in close contact with the cell membrane has been demonstrated at the chloroplast-free end walls of axial internodes or at chloroplast-free wound areas (see below). However, actin strands (Blackman & Overall, 1998) as well as myosin (Radford & White, 1998) are present in the plasmodesmata between nodal and internodal cells.

Chloroplasts bordering the neutral line between opposite flows of cytoplasm usually lack parallel subcortical bundles but are associated with an actin ring along their midplane, parallel to the cell surface (Wasteneys et al., 1996; compare Fig. 12). In many species, especially in those of the genus *Nitella*, actin rings, spirals and straight bundles are associated with the nuclear surface (Fig. 7; Foissner & Wasteneys, 2000). All of these actin-associated organelles spin as they move with the streaming endoplasm or support vesicle movement along their surface when immobilized (Wasteneys & Williamson, 1991, and references therein).

2.3 Self-organization in cytoplasmic droplets and strain alignment during actin-morphogenesis

Observations in cytoplasmic droplets and at wound regions (section 3) suggest that self-organization contributes to the formation of continuous bundles in elongating cells. Cytoplasm squeezed from internodal cells initially contains fine, actively moving actin filaments that fray out from subcortical bundles of detached chloroplasts (Jarosch, 1956). Within a period of several minutes to hours, the fine filaments coalesce into thick bundles. These bundles support cytoplasmic streaming that may eventually be organized in such a way that it involves the whole droplet, the inner portion of which undergoes rotational movement (Kuroda, 1968). These observations suggest that bundling proteins play a key role during morphogenesis of the actin cytoskeleton and the development of rotational streaming. Possible candidates include a 37 kDa unidentified protein resembling tropomyosin in size but not other properties (Williamson et al., 1985) and a 51 kDa EF1α homologue (Collings et al., 1994). Recent identification of fimbrin-like proteins in *Arabidopsis thaliana* (McCurdy & Kim, 1998; see also McCurdy & Staiger, this volume) should provide new probes for the plant-specific homologues of actin bundling proteins which could be present in characean actin bundles.

The helical structure of the characean internode cortex and cell wall probably results from strain deformation (Green, 1964). Despite the tight alignment of subcortical actin bundles along chloroplast files, several observations suggest that actin bundles do not depend on the prepositioning of the chloroplasts for their alignment, and it is also unlikely that actin bundles determine the orientation of chloroplast files (see discussion in Wasteneys et al., 1996). The question of chloroplast and actin bundle positioning is closely related to the nature of the anchoring protein(s). Filamentous connections between actin filaments of a bundle and the chloroplast envelope have been reported in rapidly frozen and deep-etched cells (McLean & Juniper, 1988). A fibrous layer, assumed to be responsible for the immobilization of both actin bundles and chloroplasts, was described from Triton-extracted internodes (Williamson, 1985). The molecular basis of these structures, however, remains to be elucidated. To date, the only progress towards a molecular link between the extracellular matrix and the actin cytoskeleton via the plasma membrane is from indirect evidence of integrin homologues in gravisensing internodes (Wayne et al., 1992).

3. WOUNDING-INDUCED REORGANIZATION OF THE ACTIN CYTOSKELETON

Characean internodes are a valuable experimental system for investigating local wound responses (e.g. Foissner et al., 1996, 1998). These experiments reactivate, and therefore help to identify, fundamental processes involved in exocytosis, polarity establishment and organization of the actin cytoskeleton.

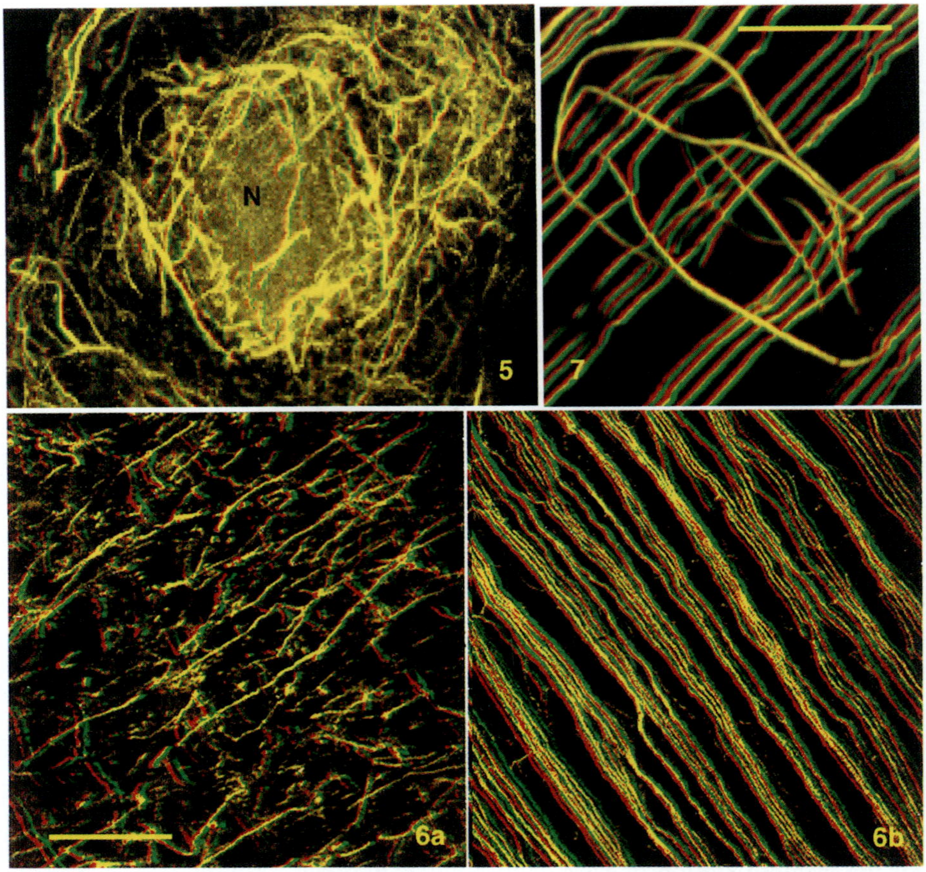

Actin immunofluorescence of elongating and mature characean internodes (stereo anaglyphs).
Figure 5. A network of actin strands traverses the cytoplasm in a uninucleate internode precursor in which bundles are beginning to assemble.
Figure 6. In elongating internodes, cortical actin strands (Fig. 6a) are predominantly oriented perpendicular to the orientation of the subcortical bundles (Fig. 6b) and the cell's long axis. Separate calculations of the same series.
Figure 7. Straight and spiral actin bundles along the surface of a *Nitella* nucleus located close to the subcortical bundles.
Fig. 6 is from Wasteneys et al. (1996) by permission of Springer. Bars = 10 μm.

3.1 Formation of an actin meshwork is required for wound wall deposition

When severe injury damages the cell wall and plasma membrane, replacement of both involves massive exocytosis and requires *de novo* formation of a peripheral actin meshwork consisting of fine, randomly arranged actin filaments (Fig. 8). This meshwork forms within minutes of injury and is followed by vesicle fusion at the plasma membrane, as visualized by high resolution video microscopy (Foissner et al., 1996, 1998). The meshwork structure most likely promotes the saltatory movements of exocytotic vesicles, which originate in the streaming endoplasm, towards and away from the wound surface (Foissner et al., 1996, 1998; Foissner & Wasteneys, 1997).

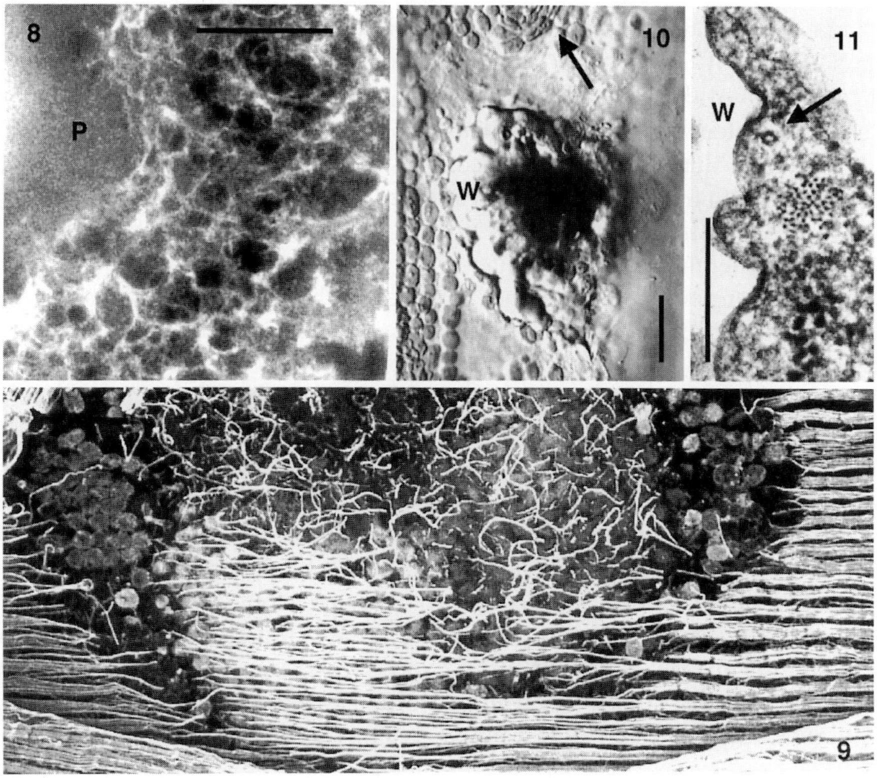

Punctured cells.
Figure 8. Plaque-like actin meshwork present 15 minutes after wounding cell. P = wound plug.
Figure 9. Subcortical AF bundles are severed and a more open meshwork is present at regions where a wound wall is being secreted (upper part of figure). Subcortical bundles have regenerated in the lower region of the window.
Figure 10. Light micrograph of puncture wound. Note thick wound wall (W) and absence of chloroplasts beneath the wound. Arrangement of chloroplast rows adjacent to the wound (arrow) indicates U-like regeneration of actin bundles across a neutral line.
Figure 11. Electron micrograph of cortical region near wound wall. Regenerated actin bundle and cortical microtubule are not cross-linked but have a similar orientation.
Fig. 8 is from Foissner & Wasteneys (1997) by permission of Springer. Bars = 20 μm (Figs. 8 – 10) and 200 nm (Fig. 11).

3.2 Subcortical actin bundle regeneration begins after wound wall secretion is complete

Once the exocytotic phase of wounding is complete, the actin meshwork disappears and subcortical actin bundles start regenerating across the wound site, eventually restoring cytoplasmic streaming. When injury is mild and the plasma membrane remains intact, the exocytosis phase is by-passed and regeneration of subcortical bundles occurs more or less immediately. Regeneration of the actin cytoskeleton following such moderate damage was first investigated in detail by Williamson et al. (1984). In the area irradiated with intense blue light, chloroplasts swell and detach from the cortex together with their associated actin bundles, leaving behind a 'window' that allows observation of the recovery process (Kamitsubo, 1972). Severed actin bundles regenerate mainly by polymerization from their barbed ends located at the upstream end of the wound and eventually reconnect with bundles at the downstream end (Figs. 9, 11). They are aligned mainly by the passive endoplasmic flow in the wound regions and, if endoplasmic flow deviates across the neutral line, regenerated bundles may even form U-turns connecting oppositely polarized subcortical bundles (compare Fig. 10). Recent investigations confirmed this work but showed that regeneration of actin bundles may also proceed from the pointed end of filaments at the downstream end of wounds as long as meshworks linger and deposition of wound wall continues at the upstream end (Foissner & Wasteneys, 1997).

Cortical actin strands are absent from healed wounds although they are prominent in nearby unwounded regions, which indicates that the delicate actin strands are easily realigned by exposure to streaming, or can even be removed from the cortex when the chloroplast barrier between the plasma membrane and endoplasm is removed (Foissner & Wasteneys, 1997). It likewise suggests that the cortex itself does not induce spreading of bundles into fine strands.

3.3 Flow-aligned actin bundles versus flow-resistant meshworks imply recruitment of diverse actin binding proteins

In contrast to the regenerated bundles, the actin filaments within the transient actin meshworks found beneath damaged plasma membrane/cell wall areas are resistant to orientation by passive endoplasmic flow. Actin meshworks require binding of cross-linking proteins that will differ from the proteins required to cross-link actin filaments into subcortical actin bundles (e.g. Ayscough, 1998). Thus, the fairly predictable series of events taking place during wounding could be useful for identifying the activity of diverse actin binding proteins. Inhibitors of RNA or protein synthesis do not prevent meshwork formation (own unpublished results), so attempts to identify such actin binding proteins by subtractive biochemical or gene expression strategies are unlikely to be successful. This result, however, demonstrates that the special conditions of the wound cytoplasm must govern binding of cross-linking proteins already present in the cytoplasm of internodal cells.

4. ACTIN REORGANIZATION FOLLOWING TREATMENT WITH CYTOCHALASIN

Cytochalasins readily disrupt the dynamic actin cytoskeletons of most animal and plant cells (e.g. McCurdy et al., 1991). In characean internodes, commonly used cytochalasins inhibit cytoplasmic streaming at concentrations from about 10^{-5} M to 10^{-4} M but the force-generating actin bundles remain intact after short exposure (Bradley, 1973; Williamson, 1975; Williamson & Hurley, 1986). Cytochalasins, however, have a marked effect on the structure of the delicate cortical actin filaments and on the regeneration of bundles following wounding, even at concentrations that inhibit but do not completely arrest cytoplasmic streaming.

4.1 Cytochalasins limit subcortical actin bundle generation and remodel the fine cortical strands into short, thick rods

Cells treated with cytochalasin B at concentrations too low to inhibit cytoplasmic streaming during their rapid growth phase have a reduced number of subcortical bundles but neither bundle continuity nor orientation is disrupted (Williamson & Hurley, 1986). In mature cells, short exposure to cytochalasin D does not disrupt subcortical bundles but generates additional endoplasmic actin, which often extends from nuclear actin rings (Collings et al., 1995).

The cortical actin filaments of characean internodes become reversibly transformed into short, thick rods by cytochalasin D, even at concentrations that do not affect streaming velocity (Fig. 12; Collings et al., 1995). Unlike the delicate cortical actin of untreated cells, actin rods survive aldehyde fixation without MBS pre-treatment (Collings et al., 1995). A stabilizing effect was also reported from cytochalasin B which, in perfused cells, makes subcortical actin bundles insoluble in low salt solutions (Williamson, 1978).

4.2 Cytochalasins modify actin regrowth at wound sites

Concentrations of cytochalasin D that arrest streaming completely abolish actin network formation, wound wall deposition and bundle regrowth when applied to cells with plasma membrane and cell wall damage (Foissner & Wasteneys, 1997). Cytochalasin concentrations that only reduce streaming velocity delay meshwork formation but the amount of F-actin present at wounds seems to be enhanced. In late stages of recovery, the actin bundles of the meshwork become very thick and intervening spaces become wider, thereby forming a conspicuous, three-dimensional lattice continuous with interwebbing subcortical bundles and cortical actin around the periphery of the wound (Fig. 13). In windowed cells, cytochalasins B (Williamson & Hurley, 1986) and D (own unpublished results) likewise block the elongation from barbed ends of actin filaments necessary for recovery of continuous bundles. Straight, discontinuous and thick bundles are formed in the windows at concentrations that do not arrest streaming (compare Fig. 15). Their orientation is random in spite of continuing passive endoplasmic flow.

4.3 Cytochalasins at low concentrations mimic the action of bundling proteins

Cytochalasins obviously have multiple effects on the actin cytoskeleton of characean internodes and other cells (compare discussion in Foissner & Wasteneys, 1997; Williamson & Hurley, 1986). High concentrations arrest cytoplasmic streaming without disassembly of subcortical actin bundles, suggesting that force generation can be inhibited without depolymerizing actin bundles. The delicate actin strands in the cortex and within the wound-induced meshworks are probably more labile because they are subjected to remodelling even at concentrations that do not inhibit streaming. The inhibition of bundle extension necessary for wound repair corresponds to the in vitro binding to the barbed end which blocks monomer addition and polymerization of filaments (reviewed by Cooper, 1987).

At lower concentrations, cytochalasins might act as nucleating centres and the number of polymerizing filaments is expected to increase (Cooper, 1987). This is consistent with the denser actin networks and thicker bundles observed at such low concentrations. Binding to the barbed end alone, however, cannot account for the formation of isolated actin rods and bundles or for the thickening of meshworks. Such extensive bundling of actin filaments suggests that cytochalasin D either itself simulates the action of a bundling protein or competes with binding sites for actin-severing proteins. From the observation that cytochalasin does not disrupt actin bundles in growing cells, it was suggested that assembly of filaments during elongation requires a mechanism other than assembly at the barbed end (Williamson & Hurley, 1986).

5. ACTIN-MICROTUBULE INTERACTIONS

In characean internodal cells, cortical microtubules and subcortical actin bundles are normally separated from each other by the chloroplast layer, making physical interactions unlikely. In the cortex, the fine actin strands are in much closer proximity to microtubules and share a similar orientation during cell elongation. Nevertheless they are not intimately associated with each other (Collings et al., 1995). Inhibitor experiments and structural data from wounded cells, however, support the existence of indirect microtubule-actin interactions.

5.1 Cortical microtubules can be flow-aligned when exposed to the streaming endoplasm

The absence of chloroplasts at wounds exposes the normally sheltered cortical microtubules to the streaming endoplasm and permits physical contact between regenerated actin bundles and reassembled cortical microtubules. Studying interactions between cortical microtubules and subcortical actin bundles in chloroplast-free windows has provided insights into the orientation properties of cortical microtubules. Microtubule orientation varies according to the size of the window (Foissner & Wasteneys, 1994, 1999). Microtubules are randomized in small windows up to 30 μm in diameter, but parallel to regenerated actin bundles in larger windows. Electron microscopy demonstrates that parallel microtubules and actin bundles in larger windows are too widely spaced to allow interaction via cross-linking proteins, even though they are no longer separated from each other by chloroplasts (Foissner & Wasteneys, 1999). Treatments that inhibit or reduce cytoplasmic streaming randomize microtubules previously oriented in the streaming direction, even when taxol is present (Foissner & Wasteneys, 1998). Orientation of regenerated microtubules beneath wound walls is also affected by wound-wall morphology.

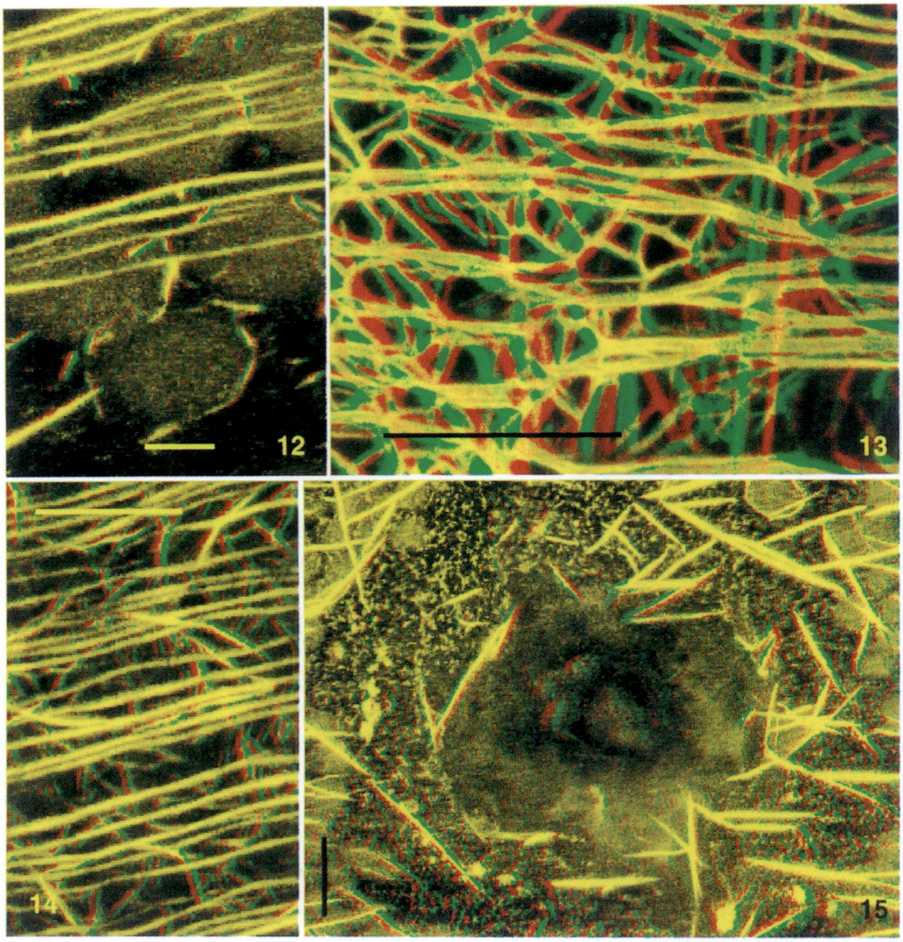

Modifications of the actin cytoskeleton following inhibitor treatments (stereo anaglyphs, actin immunofluorescence).

Figure 12. Control, unwounded cell. Note cortical actin rods and remnants of actin ring at chloroplast near neutral line in lower part of figure.

Figure 13. Regenerated actin bundles at puncture wound form a three-dimensional net.

Figure 14. Control, unwounded cell. Note greater number and length of cortical rods.

Figure 15. Regenerated actin bundles at puncture wounds are randomly oriented and discontinuous.

Figs. 12, 13 are sub-inhibitory concentrations of cytochalasin; Figs 14, 15 are sub-inhibitory concentrations of cytochalasin and microtubule-disassembling concentrations of oryzalin. Fig. 13 is from Foissner & Wasteneys (1997) by permission of Springer. Bars = 10 μm (Figs. 12, 14), 5 μm (Fig. 13) and 20 μm (Fig. 15).

Co-alignment of microtubules and actin bundles is observed beneath evenly flat wound walls (Fig. 11) but microtubules are randomized in the narrow clefts between uneven wound-wall protuberances (Foissner & Wasteneys, 1999). Together, these observations demonstrate that the normally randomized microtubules of mature internodal cells are passively aligned when exposed to hydrodynamic forces created by cytoplasmic flow (Foissner & Wasteneys, 1999). This property could be of importance in understanding microtubule orientation mechanisms that are relevant to anisotropic growth.

5.2 Microtubule disassembly potentiates the effects of cytochalasins, whereas cytochalasins enhance microtubule polymerization

Depolymerization of microtubules using a variety of unrelated drugs potentiates the inhibitory effects of cytochalasins on characean cells (Wasteneys & Williamson, 1991; Collings et al., 1996). Disassembly of microtubules lowers the cytochalasin concentrations required to inhibit streaming and to transform labile cortical actin strands into short, stable rods, reduces the time necessary for streaming to stop, and delays recovery of streaming and the conversion of cortical actin rods back into fine strands. Actin rods formed in the presence of both cytochalasin and oryzalin are significantly longer and more numerous than those formed at the same concentration of cytochalasin alone (Fig. 14). Intracellular cytochalasin concentrations are the same in microtubule-depleted and microtubule-containing cells, so uptake/efflux properties cannot explain the potentiation phenomenon (Collings et al., 1996).

Microtubule depolymerization also potentiates the effects of cytochalasin at wound sites. The absence of microtubules alone has no visible effect on the delicate actin meshwork formed during wound wall deposition (Foissner & Wasteneys, 1994) but networks formed in the presence of low cytochalasin concentrations become extensively modified. They consist of extremely thick bundles, linked to each other by thinner, often sharply bent actin bundles (unpublished results). Later stages of recovery are characterized by discontinuous, spear-like bundles (Fig. 15) that are randomly oriented despite the unidirectional endoplasmic flow.

Microtubule properties also appear to be altered when the actin cytoskeleton is challenged with cytochalasins. Although cytochalasin does not alter microtubule orientation mechanisms, it does appear to affect microtubule assembly properties. Cortical microtubules become significantly longer not only when streaming is arrested by cytochalasin treatment but also when the streaming velocity is reduced by more than 50% by incubation at 8°C (Foissner & Wasteneys, 1998). Taxol stimulates further increases in microtubule length, suggesting that streaming inhibition and taxol stimulate microtubule polymerization by distinct mechanisms.

In summary, our observations suggest a coordinated communication between actin filaments and microtubules, even though they rarely colocalize. A number of associated proteins is known to bind both to actin filaments and microtubules (for references see Collings et al., 1996). If such proteins exist in characean internodal cells, their increased ability to saturate binding sites on actin molecules might confer increased sensitivity or avidity of filamentous actin to cytochalasins. Competitive dissociation of the same proteins from cytochalasin-bound actin filaments might also increase their availability to bind and possibly stabilize microtubules or, alternatively, to stimulate tubulin polymerization, thus changing the dynamic equilibrium between polymer and dimer. This sort of ambivalence would enable an actin- and microtubule-binding protein to function as a feedback controller between the actin and microtubules, regardless of physical separation between the two elements.

ACKNOWLEDGEMENTS

This work was supported by the Österreichische Nationalbank, Austria; and the Australian Research Council.

REFERENCES

Ayscough KR (1998) *In vivo* functions of actin-binding proteins. Curr Opin Cell Biol 10: 102-111

Blackman LM and Overall RL (1998) Immunolocalisation of the cytoskeleton to plasmodesmata of *Chara corallina*. Plant J 14: 733-741

Bradley MO (1973) Microfilaments and cytoplasmic streaming: Inhibition of streaming with cytochalasin. J Cell Sci 12: 327-343

Braun M and Wasteneys GO (1998) Distribution and dynamics of the cytoskeleton in graviresponding protonemata and rhizoids of characean algae: Exclusion of microtubules and a convergence of actin filaments in the apex suggest an actin-mediated gravitropism. Planta 205: 39-50

Collings DA, Wasteneys GO, Miyazaki M and Williamson RE (1994) Elongation factor 1α is a component of the subcortical actin bundles of characean algae. Cell Biol Int 18: 1019-1024

Collings DA, Wasteneys GO and Williamson RE (1995) Cytochalasin rearranges cortical actin of the alga *Nitella* into short, stable rods. Plant Cell Physiol 36: 765-772

Collings DA, Wasteneys GO and Williamson RE (1996) Actin-microtubule interactions in the alga *Nitella*: Analysis of the mechanism by which microtubule depolymerization potentiates cytochalasin's effects on streaming. Protoplasma 191: 178-190

Cook ME, Graham LE and Lavin CA (1998) Cytokinesis and nodal anatomy in the charophycean green alga *Chara zeylandica*. Protoplasma 203: 65-74

Cooper JA (1987) Effects of cytochalasin and phalloidin on actin. J Cell Biol 105: 1473-1478

Foissner I and Wasteneys GO (1994) Injury to *Nitella* internodal cells alters microtubule organization but microtubules are not involved in the wound response. Protoplasma 182: 102-114

Foissner I and Wasteneys GO (1997) A cytochalasin-sensitive actin filament meshwork is a prerequisite for local wound wall deposition in *Nitella* internodal cells. Protoplasma 200: 17-30

Foissner I and Wasteneys GO (1998) Taxol stabilizes microtubules in characean internodal cells but does not prevent their disassembly at wound sites. Cell Biol Int 21: 866-868 (1997)

Foissner I and Wasteneys GO (1999) Microtubules at wound sites of *Nitella* internodal cells passively co-align with actin bundles when exposed to hydrodynamic forces generated by cytoplasmic streaming. Planta 208: 480-490

Foissner I and Wasteneys GO (2000) Nuclear crystals, lampbrush-chromosome-like structures and perinuclear cytoskeletal elements associated with nuclear fragmentation in characean internodal cells. Protoplasma in press

Foissner I, Lichtscheidl IK and Wasteneys GO (1996) Actin-based vesicle dynamics and exocytosis during wound wall formation in characean internodal cells. Cell Motil Cytoskel 35: 35-48

Foissner I, Lichtscheidl IK and Wasteneys GO (1998) Actin-based vesicle dynamics and exocytosis during wound wall formation in characean internodal cells. Cell Motil Cytoskel 39: Video Suppl 5

Green PB (1964) Cinematic observations on the growth and division of chloroplasts in *Nitella*. Amer J Bot 51: 334-342

Higashi-Fujime S (1991) Reconstitution of active movement in vitro based on the actin-myosin interaction. Int Rev Cytol 125: 95-138

Jarosch R (1956) Plasmaströmung und Chloroplastenrotation bei Characeen. Phyton (Arg) 6: 87-107

Kamitsubo E (1972) A 'window technique' for detailed observation of characean cytoplasmic streaming. Exp Cell Res 74: 613-616

Kersey YM, Hepler PK, Palevitz BA and Wessells NK (1976) Polarity of actin filaments in characean algae. Proc Natl Acad Sci USA 73: 165-167

Kuroda K (1968) Protoplasmic streaming in a giant plant cell. Symp Cell Chem 19: 37-43

Kuroda K (1990) Cytoplasmic streaming in plant cells. Int Rev Cytol 121: 267-307

McCurdy DW and Kim M (1998) Molecular cloning of novel fimbrin-like cDNA from the higher plant *Arabidopsis thaliana*. Plant Mol Biol 36: 23-31

McCurdy, Palevitz BA and Gunning BES (1991) Effect of cytochalasins on actin in dividing root tip cells of *Allium* and *Triticum*: A comparative immunocytochemical study. Cell Motil Cytoskel 18: 107-112

McLean B and Juniper BE (1988) Fine structure of *Chara* actin bundles, using rapid-freezing and deep-etching. Cell Biol Int Rep 12: 509-17

Nagai R and Rebhun LI (1966) Cytoplasmic microfilaments in streaming *Nitella* cells. J Ultrastruct Res 14: 571-589

Palevitz BA and Hepler PK (1975) Identification of actin *in situ* at the ectoplasm-endoplasm interface of *Nitella*. Microfilament-chloroplast association. J Cell Biol 65: 29-38

Pickett-Heaps JD (1975) Green Algae. Sunderland, Mass. Sinauer Ass. Inc.

Radford JE and White RG (1998) Localization of a myosin-like protein to plasmodesmata. Plant J 14: 743-750

Shimmen T and Yokata E (1994) Physiological and biochemical aspects of cytoplasmic streaming. Int Rev Cytol 155: 97-139

Staiger CJ and Lloyd CW (1991) The plant cytoskeleton. Curr Opin Cell Biol 3: 33-42

Wasteneys GO, Collings DA, Gunning BES, Hepler PK and Menzel D (1996) Actin in living and fixed characean internodal cells: Identification of a cortical array of fine actin strands and chloroplast actin rings. Protoplasma 190: 25-38

Wasteneys GO and Williamson RW (1991) Endoplasmic microtubules and nucleus-associated actin rings in *Nitella* internodal cells. Protoplasma 162: 86-98

Wayne R, Staves MP and Leopold AC (1992) The contribution of the extracellular matrix to gravisensing in characean cells. J Cell Sci 101: 611-623

Williamson RE (1975) Cytoplasmic streaming in *Chara*: A cell model activated by ATP and inhibited by cytochalasin B. J Cell Sci 17: 655-668

Williamson RE (1978) Cytochalasin B stabilises the sub-cortical actin bundles of *Chara* against a solution of low ionic strength. Cytobiologie 18: 107-113

Williamson RE (1985) Immobilisation of organelles and actin bundles in the cortical cytoplasm of the alga *Chara corallina* Klein ex Wild. Planta 163: 1-8

Williamson RE (1992) Cytoplasmic streaming in characean algae: Mechanism, regulation by Ca^{2+}, and organization. In: Melkonian M (ed) Algal Cell Motility. New York, London: Chapman and Hall, pp 73-98

Williamson RE and Hurley UA (1986) Growth and regrowth of actin bundles in *Chara*: Bundle assembly by mechanisms differing in sensitivity to cytochalasin. J Cell Sci 85: 21-32

Williamson RE, Hurley UA and Perkin JL (1984) Regeneration of actin bundles in *Chara*: Polarized growth and orientation by endoplasmic flow. Eur J Cell Biol 34: 221-228

Williamson RE, Perkin JL and Hurley UA (1985) Selective extraction of *Chara* actin bundles: Identification of actin and two coextracting proteins. Cell Biol Int Rep 9: 547-554

Williamson RE, Perkin JL and McCurdy DW (1986) Production and use of monoclonal antibodies to study the cytoskeleton and other components of the cortical cytoplasm of *Chara*. Eur J Cell Biol 41: 1-8

Wood RD and Imahori K (1965) Monograph of the Characeae. Weinheim, Cramer

Chapter 16

ORGANIZATION AND FUNCTIONS OF ACTIN IN HYPHAL TIP GROWTH

I Brent Heath

Biology Department, York University, 4700 Keele Street, Toronto, Ontario, M3J 1P3, Canada

Key words: actin, cell wall synthesis, hyphae, organelle motility, tip growth

Abstract: The organization of hyphal actin is not fully understood due to technical limitations, but it is labile, and most concentrated adjacent to the apical plasma membrane. Actin organization varies between groups of organisms. Oomycetes contain a 'cap' of fibrils which transforms to plaques and cables sub-apically. True fungi lack membrane-associated fibrils but contain peripheral apical plaques, cytoplasmic fibrils and often, but not always, a Spitzenkörper. While still poorly described, all species apparently contain diffuse actin permeating the cytoplasm and, enriching the growing tips, a membrane skeleton of both actin and spectrin-like proteins. Evidence is reviewed which indicates that these diverse populations of actin are directly involved in the control of hyphal tip expansion, localization of vesicle exocytosis, movement of organelles, cytoplasmic migration and the distribution of cell wall fibril synthases and ion channels.

1. INTRODUCTION

Tip growth is the dominant growth form of the fungal kingdom and the oomycetes. The latter are now known to be protists, more closely related to plants than fungi. Nevertheless, they continue to be widely studied by mycologists and they produce hyphae which are similar to fungal hyphae, hence their inclusion in this chapter. Budding yeasts, such as *Saccharomyces*, can be considered to show a very determinate type of tip growth and will also be included. Fungi and oomycetes present developmental features which make them well suited for analysis of tip growth. Sub-apical to the continuously extending main tips, all hyphal species produce branches. Branching entails assembly of the entire tip-

C.J. Staiger et al. (eds.), Actin: A Dynamic Framework for Multiple Plant Cell Functions, 275–300.

growth apparatus in a region of the cell which has long been vacated by the main tip-growth ensemble. Thus, branch formation provides an unique opportunity to analyze the normal assembly of the apparatus. It is also possible to observe formation of new tips in germinating spores and regenerating protoplasts, thus providing another opportunity to examine establishment of polarity and the tip-growth apparatus. In addition, fission yeasts, such as *Schizosaccharomyces,* show predictable determinate tip growth, thus offering the possibility of analyzing changes in the tip-growth apparatus during normal transitions from growing to non-growing.

While the focus of this review is on the roles of actin in tip growth, it is important to remember that the life cycles of fungi and oomycetes contain a number of stages with various properties and cellular processes which also involve actin. These include sporogenesis (Heath & Harold, 1992; Harold & Harold, 1992; Li & Heath, 1994; Thompson-Coffe & Zickler, 1993; Fischer & Timberlake, 1995; Jackson & Hardham, 1998), plant pathogen penetration structures (Hoch & Staples, 1983a; Bourett & Howard, 1992; Corrêa & Hoch, 1993) and cytokinesis via septa (Butt & Heath, 1988; Momany & Hamer, 1997; Bi et al., 1998). These functions of actin will only be mentioned if the observations contribute to the main topics of the chapter.

As with other walled cells that show tip growth, there are many processes which culminate in the highly localized and dynamic protrusion of the tip. Dominant among these are localized synthesis and highly regulated expansion of the apical cell wall and plasma membrane (PM), migration of cytoplasm to keep up with the advancing tip, movement of both organelles and exocytotic vesicles through the apical cytoplasm, and producing and maintaining polarized distributions of membrane proteins such as cell-wall fibril synthases and ion channels. There is evidence for a role of actin in all of these processes, but precise mechanisms are still unknown. One prerequisite for such understanding is knowledge of the organization of the apical actin. While there are discernible generalized patterns, it is likely that we have only a partial image of the total complement, and that may well be distorted! I shall start by focusing on the critical issue of organization prior to discussing the evidence for the participation of actin in each of the major processes.

2. ORGANIZATION OF TIP ACTIN

There have been two dominant methods of visualizing hyphal actin organization. One is based on fluorochrome-labeled phalloidin (most commonly rhodamine-labeled phalloidin, RP) and the other on anti-actin antibodies visualized with either a flurochrome or colloidal gold. The only other approaches have been the *in vivo* use of fluorescently-labeled actin-binding proteins (ABPs) and observation of appropriately sized filaments in transmission electron microscopy. There are four potentially serious problems with each of these approaches: 1) with phalloidin, there is evidence of false negative results (section 2.3), although no false positives have been reported, 2) with any study involving fixation prior to visualization, there are potentially serious problems of destruction and relocation. Such are likely with the highly labile actin participating in the dynamic, and thus labile, tip-growth process (sections 2.2 & 2.3). 3) With ABPs, there is the problem of differential localization, not all actin may be associated with the selected ABP. 4) With all methods, there is the problem of sensitivity. There may not be enough material to give a detectable signal, but also, in the presence of a strong signal, adjacent weaker signals may be overlooked or dismissed as 'background', yet be indicators of functionally significant quantities of actin. I shall examine aspects of each of these problems in more detail.

2.1 What we see

There is a basic dichotomy between the reported observations of hyphal tip actin. Among the oomycetes there is a prominent cap of peripheral actin fibrils which lines the apical PM and transforms subapically to more widely spaced fibrils interspersed with plaques (Heath, 1987, 1990; Jackson & Heath, 1990b; Temperli et al., 1990). The plaques appear to be undifferentiated sites of attachment of actin to the PM (Kaminskyj & Heath, 1996), possibly artefactually more concentrated by fixation (Harold & Harold, 1992). There are normally no defined actin-containing structures deeper in the cytoplasm, although the cytoplasm is permeated by a dynamic and diffuse population of unresolved actin filaments (McKerracher & Heath, 1987; Jackson & Heath, 1990a; 1993a; Heath, 1990).

Among the true fungi, the tips contain a high concentration of peripheral plaques which are most concentrated either at the extreme tip (Adams & Pringle, 1984; Anderson & Soll, 1986; Runeberg et al., 1986; Butt & Heath, 1988; Salo et al., 1989; Heath, 1990; Akashi et al., 1994; Fischer & Timberlake, 1995; Tsukamoto et al., 1996; Tinsley et al., 1996; May et al., 1998) or about 5 µm behind the tip (Hoch & Staples, 1983a; Roberson, 1992; Corrêa & Hoch, 1993; Srinivasan et al., 1996; Torralba et al., 1998a). The identity of these plaques is unclear and probably differs between species. In *Saccharomyces*, they appear to coincide with finger-like invaginations of the PM which are encircled by actin (Mulholland et al., 1994; Kaminskyj & Heath, 1996). The significance of these invaginations is obscure, they have been suggested to be involved in endocytosis (Mulholland et al., 1994), but there is no direct evidence for this. They are highly mobile along the PM (Doyle & Botstein, 1996; Waddle et al., 1996), the function of which is totally obscure. These invaginations have not been found in any of the many studies of tip ultrastructure of hyphal or fission yeast species, with the possible single exception of *Endothia* (Newhouse et al., 1983). Among these species, the plaques appear to coincide with filasomes (Kanbe et al., 1989; Bourett & Howard, 1991; Roberson, 1992; Corrêa & Hoch, 1993; Srinivasan et al., 1996), small structures of unknown function with a vesicle-like core surrounded by filamentous material (Howard, 1981) which lie predominantly adjacent to the PM. Finally, since the sub-apical plaques of the oomycetes appear to be patches of actin attached to the inside of the PM, at sites showing no obvious differentiation (Kaminskyj & Heath, 1996; Bachewich & Heath, 1997a), it is entirely possible that some or all of the plaques in the hyphal fungi are also similarly morphologically undifferentiated attachments.

In addition to the plaques, an apical structure unique to some fungi, the Spitzenkörper, is also actin-rich (Bourett & Howard, 1991; Czymmek et al., 1995; Srinivasan et al., 1996). The functions of the Spitzenkörper are unclear. It is a highly labile (Lopez-Franco & Bracker, 1996) accumulation of vesicles lying at the centre of the growing tips of many species. It has been postulated to be a structure which determines the distribution and subsequent sites of exocytosis of vesicles in the tips (Bartnicki-Garcia et al, 1989), although direct demonstration of this function is lacking and it is absent in some fungi and all oomycetes.

Saccharomyces contains clear cytoplasmic fibrils which are attached to the plaques (e.g. Karpova et al., 1998). Comparable fibrils, apparently not associated with the plaques, are also seen in *Uromyces* (Hoch & Staples, 1983a). The final structural type of apical actin among the fungi seems to be a diffuse population permeating the general cytoplasm, sometimes also associated with the nuclear envelope. Evidence for the existence of such diffuse actin, as opposed to non-specific 'background staining', is not abundant, but it has been reported in many studies (Hoch & Staples, 1983a; 1985; Marks & Hyams, 1985; Runeberg et al., 1986; McKerracher & Heath, 1987; Wittenberg et al., 1987; Butt & Heath, 1988; Salo et al., 1989; Heath, 1990; Roberson, 1992; Corrêa & Hoch, 1993; Kanbe et al., 1994; Li & Heath, 1994; Srinivasan et al., 1996).

2.2 Is what we see distorted?

An unequivocal answer to this question requires imaging all actin at high resolution and sensitivity in living hyphae. This can be achieved in principle with GFP-actin, but the only successful use of this probe is in *Saccharomyces*, where it was found to label only the actin plaques (Doyle &

Botstein, 1996). This work showed that the plaques appeared the same as in fixed cells, indicating that for these rather simple structures, fixation reveals an accurate image. The absence of staining of the cytoplasmic fibrils seen in fixed material was attributed to lack of sensitivity, indicating the current inability of the technique to address distortion issues in other arrays. Future improvements in GFP efficiency may overcome this limitation.

The related approach of using GFP-labeled ABPs has also been employed only in *Saccharomyces*, where it was again found to reveal only the peripheral plaques (Doyle & Botstein, 1996; Waddle et al., 1996). It is unclear whether this selectivity is due to lack of sensitivity, or absence of the ABPs on the cytoplasmic fibrils, but it precludes addressing the distortion issue on anything other than the plaques.

An alternative procedure has been introduction of RP into living cells via either microinjection (Corrêa & Hoch, 1993; Jackson & Hardham, 1998) or electroporation (Jackson & Heath, 1990a, 1993a). This technique is limited because not all populations of actin are accessible to RP, as discussed in section 2.3. However, these studies indicate that images of fixed material stained with either RP or antibodies can be both accurate and inaccurate, depending upon the species. In *Uromyces* germ tubes, limited comparisons indicate that all populations of actin detected in fixed material can also be detected with similar organization in living cells (Corrêa & Hoch, 1993). Conversely, in *Phytophthora* sporangia, the fixation employed failed to preserve or reveal a major population of actin, thus giving an incomplete image, although what was revealed in the fixed material was not distorted (Jackson & Hardham, 1998). In *Saprolegnia* hyphae, living cells failed to reveal most of the actin seen after fixation and thus could not address the distortion issue (Jackson & Heath, 1990a, 1993a).

Beyond these few studies, the potential distortion in the abundant fixed images can only be examined by comparing different protocols. There are reports of differences in actin patterns following different preparation and visualization protocols in a single species. These are *a priori* evidence for processing-induced distortion of the *in vivo* organization. As discussed previously (Heath, 1990), variations in seemingly minor fixative components such as buffers substantially alter the preserved patterns of RP staining (Heath, 1987). Major changes in fixative components cause major changes in actin patterns (Harold & Harold, 1992; Kaminskyj & Heath, 1994). Pre-fixation treatments with mild detergents, which do not change some forms of organelle motility, do dramatically change the actin patterns (Heath, 1988). More surprisingly, post-fixation detergent treatments can also change the patterns (Heath, 1990), thus indicating caution in interpreting immunocytochemical staining protocols which use extensive post-fixation treatments. Indeed, even when immunocytochemical patterns are comparable to those shown by simpler RP staining, the images are less detailed (Kaminskyj & Heath, 1994), possibly due to the larger antibody probes, but alterations in actin organization by the later stages in processing may also be a factor. Post-fixation changes indicate that, even with optimal fixation such as freeze substitution which avoids problems with initial ion fluxes and other potentially disruptive 'death throes' changes, the final image may not be that of the living cell.

Currently available evidence indicates that the organization of actin may, or may not, be faithfully revealed by fixation and staining protocols, depending on both species and actin populations. Therefore, great care is required in interpretating the observed patterns. However, even distorted patterns may be informative. Consistent patterns indicate underlying organization and properties, even when the observed pattern is not the same as in the living cell.

2.3 Is there more than what we see?

The basic messages from the previous section seem to be, the closer we look, the more we see (May & Hyams, 1998), and we have yet to reach the plateau. There is extensive evidence which indicates that current images have not yet revealed the entire cellular actin complement.

Fixed oomycete hyphae contain one of the most complex and extensive arrays of actin found in any hyphal species (Heath, 1987; 1988; 1990), but it took use of electroporated RP on living hyphae to reveal the less prominent, diffuse actin permeating the bulk cytoplasm (Jackson & Heath, 1990a, 1993a). Similarly diffuse cytoplasmic staining is seen in *Uromyces* germ tubes microinjected with RP (Corrêa & Hoch, 1993), but comparable studies are lacking in other species. The cytoplasmic actin is less intensely stained than the more concentrated fibrils and plaques and thus easily dismissed as "background" or simply overlooked, yet it may be functionally very important. Even in these two species we may still lack a complete picture of the actin since, in *Saprolegnia*, abnormal strong perinuclear staining can be induced (Heath, 1990), indicating the existence of nuclear envelope-attached actin which is not normally visualized as such either by microinjected RP or by staining of fixed material.

There have been three studies employing anti-actin immunolocalizations at the electron microscope level in fungi (Bourett & Howard, 1991; Roberson, 1992; Mulholland et al., 1994) and in each case, while the prominent peripheral plaques and (in *Saccharomyces*) cytoplasmic fibrils were clearly the predominant stained structures, there was significant label elsewhere in the cytoplasm. The latter may be non-specific "background" or indicative of poorly preserved diffuse actin. Interestingly, in one of these studies (Bourett & Howard, 1991), the well known septum-associated actin inexplicably did not stain, indicating that not all populations of actin may be visualized with any specific antibody. This further supports the possible existence of undetected actin populations in hyphae.

Another way of indicating the existence of hypothetically missing actin is to see what actin-like material is present in electron microscope images but is unaccounted for in material stained specifically for actin. In the well studied *Uromyces* germ tubes mentioned previously, there are actin-like filaments laterally associated with microtubules (MTs) in electron microscope preparations but these are not described in light microscope images (Heath & Heath, 1978; Hoch & Staples, 1983b). In *Fusarium* (Howard, 1981) there are actin-like filaments closely appressed to the PM and in the cytoplasm which are not accounted for by the only well known filasome-, septa- and Spitzenkörper-associated actin in these hyphae. Similarly unaccountable PM-associated fibrils have been seen in *Endothia* (Newhouse et al., 1983) and *Basidiobolus* (McKerracher & Heath, 1986b). Furthermore, the cytoplasm can be preserved as a filamentous matrix (McKerracher & Heath, 1986b, 1987; Wittenberg et al., 1987), which is actin-rich in the latter case. Even though in all of these examples there is no positive identification of the filaments as being composed of actin, the existence of a filamentous cytoplasmic matrix is interesting.

Further evidence that hyphae contain undetected actin is the known inability of current fixation protocols to preserve actin at the electron microscope level. All fixation protocols fall into two categories, chemical fixation and freeze-substitution. In the former, membrane permeability barriers are broken prior to cross-linking and immobilization of proteins. The ensuing fluxes of ions and molecules (e.g. proteases) may have devastating impacts on cellular organization and protein preservation prior to cytoplasmic stabilization. Such effects have been directly demonstrated at the light microscope level in hyphae (Kaminskyj et al., 1992b), although their likely attribution to ion fluxes is speculative. Freeze-substitution potentially avoids this problem because the initial freezing locks organization prior to fixation in the frozen state.

The overall quality of organizational preservation is superior with freeze-substitution, but loss of components is entirely possible during the thawing and subsequent processing steps. Destruction of actin at the electron microscope level has not been fully explored in hyphae, but related observations were discussed in section 2.2 and have been long known in other cells (e.g. Maupin-Szamier & Pollard, 1978). Direct evidence that freeze-substitution fails to preserve hyphal actin is the absence of actin in freeze-substituted hyphae of *Saprolegnia* (Heath et al., 1985; Heath & Kaminskyj, 1989) and the generally poorly preserved (or at least poorly visualized) filasome-associated actin in other freeze-substituted hyphal tips (e.g. Howard & Aist, 1980; Roberson & Fuller, 1988; Bourett & Howard, 1991; Roberson, 1992; Srinivasan et al., 1996; Torralba et al., 1998a). These examples of inadequate preservation of known concentrated actin in freeze-substituted hyphae show the possibility of failure to preserve or reveal less concentrated or less stable organizations.

An example of the latter hypothetical population may be the conspicuous 'zone of exclusion' adjacent to the inside of the PM in regions of growth seen in many fungal cells. This zone is typically thin, <0.1 μm, and is characterized by the exclusion of ribosomes and larger organelles and the presence of a finely fibrillar or amorphous matrix. It is, predictably, often penetrated by the secretory vesicles which undergo exocytosis to liberate cell wall components. It is well seen, although not always identified, in many studies of fungal ultrastructure (e.g. Hoch & Howard, 1980; Howard & Aist, 1980; Newhouse et al., 1983; Heath et al., 1985; Mims et al., 1988; Baba et al., 1989; Kurtz et al., 1994; Heath, 1994; Kanbe et al., 1993; Kaminskyj & Heath, 1996; Torralba et al., 1998a). It appears to be more prominent when osmium tetroxide is omitted from the substitution fluid (Heath et al., 1985), consistent with the known destructive effects of this fixative (Maupin-Szamier & Pollard, 1978).

The final indication of the potentially incomplete imaging of hyphal actin is the failure of the widely used RP to reveal all actin. This has already been mentioned for living *Saprolegnia* hyphae and is also shown in other fixed cells (Wang, 1987; Wilson et al., 1987). It has been suggested to occur in pollen tubes (Tang et al., 1989), although the failure of RP to stain actin in cytochalasin-treated tubes containing antibody-stainable filaments could have been due to cytochalasin-induced diminution of cell wall permeability to RP, a factor which would not have interfered with the antibody protocol.

Overall, there are ample reasons to believe that fungal hyphae contain more populations of actin than have been described. The potential significance of this will be discussed in the following sections. However, I shall focus on the roles of actin deduced largely from the demonstrated actin ensembles and the inhibitions of those ensembles by inhibitors or genetic manipulations.

3. FUNCTIONS OF TIP ACTIN

It is important to understand that in the highly dynamic and concentrated region of the hyphal tip, all of the diverse activities that contribute to tip growth are very intimately interrelated. Thus it is difficult to dissect cause-and-effect relationships of actin in each separable process because actin is probably involved in all and hence its effects on one process affects others. Some observations can be explained by actin functions in more than one process. However, for convenience, I shall separate the processes and deal with each in turn, using the observations interpreted in support of that process only in each section.

3.1 Morphogenesis

Tip morphogenesis involves control of tip expansion and localization of exocytosis of the vesicles which contribute both PM and cell wall precursors. However, both of these processes in turn depend upon an appropriate supply of vesicles, which in turn requires both production and transportation activity. The latter will be covered in section 3.2, the former is beyond the scope of this work and is probably largely independent of actin. Expansion also typically involves regulation of turgor pressure. While there is increasing evidence that ion channel activity, and thus ionic contributions to turgor regulation, may involve actin (partly discussed further in section 3.4) and actin may be involved in turgor pressure sensing systems, such do not yet make a clear picture and will be excluded.

3.1.1 Control of surface expansion/protrusion

Conventional wisdom indicates that surface expansion is propelled by turgor pressure and a gradient of regulated yield of the cell wall generates the hyphal shape. That the cell wall is important in maintaining the shape and integrity of hyphae is shown by loss of shape and osmotic fragility of protoplasts. However, it is unclear whether it is either the sole, or even the primary, determinant of tip extensibility, especially since it is outside of intracellular regulatory systems.

Evidence against turgor-driven tip expansion is seen in the oomycetes. They are unusual in not fully regulating their turgor pressure in response to external osmotica, permitting observation of a poor correlation between growth rates and turgor pressure (Kaminskyj et al., 1992a) and normal growth rates in the absence of measurable turgor pressure (Money & Harold, 1992, 1993). These observations indicate a protrusive force apart from turgor pressure. This force has been suggested to be similar to that of amoeboid pseudopodial protrusion, and thus based on actin (Heath, 1995b; Harold et al., 1995; Heath & Steinberg, 1999). Under normal turgor, the actin would restrain tip expansion and under zero turgor it would protrude the tip. This model removes the need to postulate evolution of a novel protrusive system for the highly unlikely natural zero turgor condition and places regulation of tip expansion inside the PM, where it would be subject to intracellular regulatory systems. Very fine control is essential to generate the typically uniform diameter of the hyphal tube shape and is unlikely to exist in the extracellular environment.

The membrane skeleton model proposes that PM-associated actin, linked via integrin-like molecules through the PM to cell wall polymers for additional strength, regulates tip extensibility (Heath, 1995b) and is based on an original hypothesis for pollen tube growth (Picton & Steer, 1982). In this model the individual cell wall polymers add strength to the membrane skeleton, but the cell wall is assembled on the shape of the apical PM (Wessels, 1988; Heath, 1995b; Kaminskyj & Heath, 1996). Detailed analysis of this model, which can be applied to any fungal system irrespective of the currently known patterns of apical actin (discussed in more detail below) has been presented previously (Heath, 1990, 1995b; Jackson & Heath, 1990b; Kaminskyj & Heath, 1996) and thus the primary evidence for it, citing only new references unavailable or omitted in the previous reviews, will be summarized here.

a) Hyphal tips are always enriched in PM-associated actin. This is true of both oomycetes, where the apical cap of PM-associated actin has a predictable form for shape generation, and the true fungi where the apical actin plaques are less obviously suited. The apical PM-associated zone of exclusion (section 2.3) present in both oomycetes and true fungi may represent the necessary additional actin both linking the plaques and forming the postulated membrane skeleton. Such could also provide the necessary system responsible for the observed motility of the *Saccharomyces* plaques, although none of the known yeast myosins (the most likely motors) are individually essential for the movement (Waddle et al., 1996).

b) In the oomycetes, the organization of the apical actin correlates with apical extensibility, being most concentrated at the extreme tip where the wall is weakest and most extensible and changing its pattern to that of the sub-apical inextensible regions at the end of the apical zone of tip extensibility.

c) The organization of the apical actin correlates with growth. Growing and non-growing tips have different actin patterns, and new tip formation in branches and spore germination is preceded by the formation of the characteristic growing apical actin patterns (Bachewich & Heath, 1998).

d) Weakening or disruption of actin with cytochalasins or latrunculins leads to ballooning of tips in diverse oomycetes and true fungi (Herr & Heath, 1982; Grove & Sweigard, 1996; Gupta & Heath, 1997; Ayscough et al., 1997). The ballooning is predictably absent under low turgor and is preceded by growth acceleration under high, but not low, turgor (Gupta & Heath, 1997). These observations are most consistent with actin restraining tip expansion under normal turgor and protruding tips under low turgor.

Fungal cytoplasm can form actin-rich protrusions (Li & Heath, 1994; Heath & Steinberg, 1999) and latrunculin-sensitive amoeboid-like cytoplasmic protrusions occur in hyphae which are repopulating their tips with cytoplasm after withdrawal of apical cytoplasm (Bachewich & Heath, 1999).

e) One of the phenotypes of gene-regulated actin disruption is increased osmotic sensitivity.

f) Morphological abnormalities are found in some ABP disruptants and a myosin 1 is essential for growth and localizes to growing tips (McGoldrick et al., 1995).

g) Actin may be reinforced or replaced by other structural components of the apical membrane skeleton. Spectrin reinforces animal cell PMs and a related protein enriches the PM in hyphal tips of oomycetes (Kaminskyj & Heath, 1995) and true fungi (Degousée et al., 2000). In the latter, it is highly tip-restricted and forms a cap corresponding very precisely with the zone of apical extension. In the true fungi, where actin is less cap-like than in the oomycetes, a spectrin-like protein may be more important to tip morphogenesis than actin. Since spectrins contain actin-binding domains, cooperative interactions that determine total membrane skeleton properties are likely. Spectrins participate in amoebal pseudopodium protrusion (Choi & Jeon, 1992), consistent with the spectrin-containing membrane skeleton both restraining and protruding hyphal tips.

h) The postulated trans-PM interactions between the membrane skeleton and cell wall polymers is supported by integrin-like proteins being concentrated in hyphal tips (Kaminskyj & Heath, 1995; Degousée et al., 2000) and having morphogenic consequences when manipulated (Corrêa et al., 1996; Gale et al., 1996). Tips also contain the predicted actin- and integrin-enriched cytoplasm-PM-cell wall adhesions which undergo changes related to growth (Kaminskyj & Heath, 1995; Bachewich & Heath, 1997a).

The above observations provide substantial support for the hypothesis that the apical PM-associated actin (with other interacting molecules) plays a major, possibly primary, role in tip morphogenesis. However, a number of these observations (a–f) could also be attributable (with appropriate assumptions) to actin functioning in vesicle exocytosis. Additional observations more directly and exclusively related to exocytosis control follow.

3.1.2 Control of exocytosis

Since the apical PM-associated actin, together with the zone of exclusion, which we have seen may also be actin-containing, lies between the source of the vesicles and the PM with which they must fuse, it must minimally part to permit exocytosis. The vesicles lying within the zone of exclusion (section

2.3) are evidence that this parting occurs, but there is no structural information in these images which explains the process.

Direct evidence for actin involvement in regulation of exocytosis is seen in the results of latrunculin disruption of actin. During the previously mentioned tip ballooning, the balloons extend sub-apically, so that previously inextensible cell surface extends (Gupta & Heath, 1997). This requires cell wall softening and probable additional wall synthesis. Wall-softening enzymes are a likely content of the vesicles, which also release the wall precursors at exocytosis (Gooday & Gow, 1990). Delocalized softening is thus likely to equate with delocalized exocytosis and is a consequence of actin disruption. Furthermore, long-term actin disruption results in massive abnormal sub-apical cell wall deposition (Bachewich & Heath, 1998), again indicating a role for actin in controlling exocytosis. During new tip formation in branching and spore germination, the tip-related actin ensembles form as radial arrays focusing on the future sites of outgrowth (Bachewich & Heath, 1998). Tip outgrowth requires softening of previously inextensible cell wall, new wall synthesis and thus localized exocytosis, a process for which the radial arrays of PM-associated actin are well suited. That they do function in this localization is shown by the absence of new tip formation and diffuse cell expansion when actin is disrupted (Bachewich & Heath, 1998; Heath et al., 2000).

Further evidence for actin's role in regulating exocytosis is seen in actin and myosin mutants where secretory vesicle accumulations, reduced secretion and loss of polarized growth are common (Novick & Botstein, 1985; Johnston et al., 1991; McGoldrick et al., 1995; Steinberg, 1998). A similar reduction in secretion also follows presumed (Thomas et al., 1974) or demonstrated (Torralba et al., 1998b) actin disruption with cytochalasin. None of these studies have elucidated the detailed mechanisms by which actin controls exocytosis, but they have incorporated myosin and supported the principle of actin involvement.

The mechanism by which actin regulates exocytosis is unclear. Vesicles may have filaments on their surfaces (Howard & Aist, 1979; Hoch & Howard, 1980; Hoch & Staples, 1983b; Heath et al., 1985), and these filaments may be actin (Mulholland et al., 1994). Reported negative actin immunolocalizations (Bourett & Howard, 1991; Roberson, 1992) may not truly indicate the absence of actin (section 2.3). If the filamentous coats are actin, actin-actin interactions mediated by myosins bringing about vesicle docking and fusion is an attractive concept, but speculative and probably simplistic.

An unusual aspect of actin's role in exocytosis regulation, apparently unique in detail to many true fungi, relates to the Spitzenkörper. This structure is a distinct globular-to-hemispherical aggregate of vesicles and actin filaments (Grove & Bracker, 1970; Bourett & Howard, 1991; Czymmek et al., 1995). It is highly labile (Lopez-Franco & Bracker, 1996), moving around coincident with changes in direction of growth (Riquelme et al., 1998) and dispersing at the cessation of growth. One can model and predict rates and morphologies of tip growth if one assumes that the Spitzenkörper is a vesicle supply centre regulating the distribution, and thus the gradient, of exocytosis of the apical vesicles (Bartnicki-Garcia et al., 1989). The Spitzenkörper is not present in all species of true fungi (e.g. Hoch & Staples, 1983b), is never seen in oomycetes, and is inessential for generation of the postulated gradients of exocytosis (Heath & Janse van Rensburg, 1996). Nevertheless, it is a major part of the tip-growth apparatus in many species, and is composed of secretory vesicles and actin. The way in which the Spitzenkörper functions as a vesicle distribution system is unclear, but if considered as part of the apical actin ensemble interacting with secretory vesicles late in their exocytotic pathway, it fits into the general concept of actin's role in exocytosis. Thus, it becomes no longer a unique structure in cell biology and elucidation of the factors which cause its unusual formation will help explain general mechanisms of actin regulation of exocytosis.

3.2 Vesicle and organelle motility

The distributions of all cytoplasmic inclusions, nuclei, mitochondria, vacuoles and secretory vesicles, are regulated in a dynamic way so that they maintain characteristic species-specific patterns in growing hyphae. Perhaps the most spectacular example of this behaviour, unique to basidiomycetes, is the amazingly regulated choreography of the nuclei in dikaryotic hyphae. Pairs of nuclei maintain position in the centre of the continuously elongating growing tips, move through clamp connections and septa at each mitosis, and rapidly migrate to take up predictable positions after mitosis (e.g. Kaminskyj et al., 1989).

There is extensive evidence that MTs and their associated motors function in some vesicle and most organelle movements (Heath, 1994; 1995a; Steinberg, 1998), but there is also evidence of a role for actin. It is the latter on which I shall focus. It is very difficult to dissect out the relative contributions of actin and MTs in these movements because they cohabit the same cytoplasm and they apparently interact. However, such interactions are beyond the scope of the present review and will be mentioned only briefly.

3.2.1 Vesicles

Most of the evidence in section 3.1.2, that actin is involved in vesicle exocytosis, can also be explained in terms of actin functioning in vesicle transport. Failure of correct vesicle transport will likely result in failure, or mislocalization, of exocytosis and secretion and abnormal vesicle accumulations. However, the long distance transport of vesicles from their sub-apical sources demands both tracks and motors. Evidence indicating that MTs, the primary alternative track system, are not directly responsible is part of the evidence for actin being the track. Thus normal bud or tip growth, which presumably requires normal vesicle transport, occurs in MT-depleted cells (Howard & Aist, 1977; Howard & Aist, 1980; Herr & Heath, 1982; Oakley & Rinehart, 1985; Hoch et al., 1987; Jacobs et al., 1988; Yokoyama et al., 1990; Steinberg, 1998). Furthermore, none of the extensive published ultrastructural studies of fungal hyphae have shown any illustrations of vesicles associated with MTs, as seen so abundantly in axons where MTs are so clearly the tracks for vesicle transport. Positive support for the track-like role for actin comes from the observed transport of vesicles towards the hyphal apex in apposition to, and aligned along, the peripheral actin filaments in oomycete hyphae (Heath & Kaminskyj, 1989). In *Schizosaccharomyces*, actin disruption blocked vesicle transport to the tips of the cells (Kanbe et al., 1993) and one of the *Saccharomyces* myosins, and thus actin also, has been directly implicated in vesicle transport (Schott et al., 1999).

Collectively, the data indicate that MTs do not play a track-like role in vesicle transport, and thus both by default and by appropriate interpretation of inhibitor and mutational studies, actin probably has this function. However, the necessary immunolocalizations of motors on the vesicles and demonstration of the actin ‘tracks’ must be accomplished before this model can be accepted.

3.2.2 Nuclei

The preponderance of evidence implicates MTs and their associated motor molecules as the prime movers of nuclei, but there are also indications of actin's participation. In *Pleurotus*, fast post-mitotic nuclear migration is independent of MTs since such are absent during migration, thus indicating an alternative system (Kaminskyj et al., 1989). In *Saccharomyces*, disruption of actin (Palmer et al., 1992) profilin (Haarer et al., 1990) and myosin (Watts et al., 1987) genes disrupts normal nuclear positioning. MTs associated with nuclear migration are laterally associated with actin (Hoch & Staples, 1985) or actin-like filaments (Heath & Heath, 1978; Hoch &

Staples, 1983b; McKerracher & Heath, 1985). In a number of species, the nuclear envelope is associated with abundant actin (Butt & Heath, 1988; Marks et al., 1986; Heath, 1990), at least at some stages of the cell cycle. Presumably this actin interacts with other parts of the cytoskeleton and is likely to have some role in nuclear migration.

It is not clear how the nucleus-associated actin participates in nuclear migration. In some cases it appears to interact with MTs, in others there are no MTs present, indicating diversity of mechanisms and alternative systems, either in different species or at different stages in the cell cycle. However, such speculations require far more detailed analysis to become clear.

3.2.3 Mitochondria

As with the nuclei, with mitochondrial movement there is evidence for both MT and actin involvement (Steinberg, 1998). With respect to actin-dependency, a) mitochondrial migration can be independent of MTs (Oakley & Rinehart, 1985), thus indicating an alternative system; b) actin disruption changes mitochondrial distribution (Drubin et al., 1993; Smith et al., 1995; Simon et al., 1995; Gupta & Heath, 1997); c) mitochondria move along actin cables (Simon et al., 1997), with which they can also interact *in vitro* (Lazzarino et al., 1994); d) there are gene products which specifically and concurrently inhibit mitochondrial movements and actin organization (Hermann et al., 1997); and e) mitochondria-associated microtubules are laterally associated with actin-like filaments (Heath & Heath, 1978; Heath et al., 1982). While these data indicate that both microtubules and actin are intimately involved in mitochondrial movements, there are no data to explain the factors which determine how these components interact, or whether there are diverse mechanisms operating in different species.

3.2.4 Vacuoles

Vacuoles are both motile and positioned. There is evidence for both MTs and actin being involved, with MTs probably playing the dominant role (Steinberg, 1998; Hyde et al., 1999). Although actin-like filaments are associated with motile vacuoles in *Pisolithus* (Shepherd et al., 1993), motility can be insensitive to actin disruption (Hyde et al., 1999). The most prominent PM-associated actin in *Saprolegnia* hyphae is unimportant to motility of one specialized type of vacuole, although a role for the more dispersed central actin is certainly possible (Heath, 1988). The shape of the larger and more prominent vacuoles in *Saprolegnia* appears to be imposed by actin (Gupta & Heath, 1997; Bachewich & Heath, 1999), thus presumably such actin is involved in vacuole motility. Furthermore, actin is known to be important to the positioning of cytokinetic vacuoles during sporulation (Heath & Harold, 1992; Li & Heath, 1994; Jackson & Hardham, 1998). Actin, a myosin and profilin are also important for correct vacuole distribution in *Saccharomyces* (Hill et al., 1996). As with the other organelles, the differences in MT versus actin functions in vacuole motility are not yet explained.

3.2.5 Summary

We have seen that motility of organelles in general involves contributions from both MTs and actin. This conclusion may reflect the fact that different systems operate in different species at different stages of development, but a more general view may be that consistent interactions must occur between the two dominant cytoskeletal components. Even when MTs are most prominently involved and their motor molecules identified (Steinberg, 1998), there is still the question of with what the motors interact in order to generate the force needed to move the organelles. In an investigation of nuclear motility in *Basidiobolus* hyphae it was concluded that the MTs interact with unknown elements of the cytoplasm, very possibly the diffuse actin discussed in section 2, and function as reinforcing elements to impart tensile strength to the cytoplasm (McKerracher & Heath, 1986a). Such interactions could then be regulated to generate motility relative to the cytoplasm or to hold station in that cytoplasm. Integrating this concept into the entire cytoskeletal/organelle motility package will be complicated, but it helps explain the participation of both components.

3.3 Cytoplasmic migration

In many developmental situations, including hyphal growth, cytoplasm with included organelles migrates through the hyphae to maintain an apical mass of cytoplasm, leaving behind highly vacuolate regions with only a thin peripheral layer of cytoplasm. This situation has been likened to a tube-dwelling amoeba (Reinhardt, 1892), a concept that is still viable and for which there is extensive supporting evidence (Heath & Steinberg, 1999). The analogy indicates apical substrate adhesions and actin-based cytoplasmic protrusive and contractile systems.

In the hyphal situation, the 'substrate' would be the inside of the cell wall. The apical actin plaques of *Saccharomyces* described in section 2 were suggested to be the equivalent of animal focal adhesions (Adams & Pringle, 1984), a concept which remains possible, but unproven for hyphae, even though they appear to be highly motile (section 2.1). In at least some hyphae, there are abundant and complex apical PM-cell wall adhesions (Kaminskyj

& Heath, 1995; Bachewich & Heath, 1997a), which are appropriately placed for cytoplasmic migration, further details of which have been discussed elsewhere (Heath, 2000).

Cytoplasmic protrusion has been discussed in section 3.1. Cytoplasmic contractions and migrations are inducible in a calcium-dependent and or latrunculin-sensitive (indicative of actin involvement) manner in hyphae (Jackson & Heath, 1992; Kaminskyj et al., 1992b; McKerracher & Heath, 1986c; Bachewich & Heath, 1999) and the contracted cytoplasm can be actin-rich (McKerracher & Heath, 1987; Heath, 1990). Presumably, the diffuse cytoplasmic actin (section 2) is responsible for these behaviours.

While both the presence of cytoplasmic migration and predicted components of its likely mechanisms are demonstrated for a limited number of hyphae, the way in which it functions and is controlled in growing hyphae remains elusive and will most surely involve very complex regulation.

3.4 Plasma membrane protein distribution

There are minimally two types of PM proteins whose distribution in hyphae is likely to be non-random; ion channels and cell-wall fibril (e.g. chitin) synthases. Growing hyphae maintain gradients of cytoplasmic Ca^{2+} (Jackson & Heath, 1993b; Hyde & Heath, 1997) and possibly H^+ (McGillviray & Gow, 1987; Roncal et al., 1993; Robson et al., 1996), although H^+ gradients are also reportedly absent (Parton et al., 1997; Bachewich & Heath, 1997b). Hyphae also generate and maintain ion currents around their tips (Harold & Caldwell, 1990; Lew, 1999). These observations indicate likely tip-concentrated appropriate ion channels, and such have been directly demonstrated (Garrill et al., 1992, 1993). The distribution of these channels is determined by PM-associated actin (Levina et al., 1994). The previously mentioned formation of the radial arrays of actin which precede branch formation (sections 3.1 & 3.2) are also well placed to recruit the necessary ion channels.

Hyphae also produce cell-wall fibrils (chitin and glucans) in a strong, tip-high gradient with the aid of PM-located synthetic enzymes (Gooday & Gow, 1990), which presumably are also localized in a similar gradient. The coincidence of enriched apical PM-associated actin and the synthases is in itself evidence that the latter gradient is generated or maintained by actin. More compelling is that, in *Schizosaccharomyces*, a wall fibril synthase colocalizes with peripheral actin, and its distribution is dependent upon that actin (Katayama et al., 1999). In regenerating protoplasts of the same species, there is a strong correlation between the timing and position of PM-associated actin and wall fibril synthesis, and both are disrupted by actin disruption (Osumi et al., 1998). In yeasts in general there is a clear trend for chitin fibril synthesis to be linked with PM-associated actin concentrations, and disruption of the latter delocalizes the former (May & Hyams, 1998). Similarly, in *Saprolegnia* hyphae, disruption of the PM-associated actin delocalizes cell wall fibril deposition (Bachewich & Heath, 1998).

Together, the above observations indicate that PM-associated actin plays a role in localizing both ion channels and cell wall fibril synthases. While the nature of the role is as yet unexplained, a number of models are possible. One is a direct interaction between the synthases and the actin filaments. This concept is similar to the situation in plant cells undergoing secondary cell wall synthesis where the orientation of the cellulose fibrils mirrors the orientation of MT-associated actin that is located adjacent to the PM (Seagull & Heath, 1980; Heath & Seagull, 1982).

4. CONCLUSIONS

We have seen that actin is, not surprisingly, involved in many, perhaps all, critical aspects of the hyphal tip-growth process. Many of the observations can be explained on different assumptions about which specific process it is involved with, because so many of the processes are both colocalized in time and space and thus presumably coregulated. Further clarification of actin's function in the tip-growth process will ultimately require very high resolution localization of both actin and its associated regulatory and motor ABPs. Such is a long way into the future. However, significant progress will be made when we are able to fully describe the location and organization of the main player – actin itself. Hopefully, the present discussion of our current shortcomings with respect to the latter, coupled with indications of the very wide importance of actin to the growth process of the fungal kingdom will stimulate the necessary future work. While this wish may seem out of place in a book dedicated primarily to the plant kingdom, it should not be forgotten that the fungi are responsible for much of the nutrient recycling, mobilization and uptake that permit plant growth, the bulk of plant degradation and recycling, and the dominant disease losses of crop plants. Perhaps a few cross-overs from the plant research community might be appropriate!

REFERENCES

Adams AEM and Pringle JR (1984) Relationship of actin and tubulin distribution to bud growth in wild-type and morphogenetic mutant *Saccharomyces cerevisiae.* J Cell Biol 98: 934-945

Akashi T, Kanbe T and Tanaka K (1994) The role of the cytoskeleton in polarized growth of the germ tube in *Candida albicans*. Microbiology 140: 271-280

Anderson JM and Soll DR (1986) Differences in actin localization during bud and hypha formation in the yeast *Candida albicans*. J Gen Microbiol 132: 2035-2047

Ayscough KR, Stryker J, Pokala N, Sanders M, Crews P and Drubin DG (1997) High rates of actin filament turnover in budding yeast and roles for actin in establishment and maintenance of cell polarity revealed using the actin inhibitor latrunculin-A. J Cell Biol 137: 399-416

Baba M, Baba N, Ohsumi Y, Kanaya K and Osumi M (1989) Three-dimensional analysis of morphogenesis induced by mating pheromone α factor in *Saccharomyces cerevisiae*. J Cell Sci 94: 207-216

Bachewich CL and Heath IB (1997a) Differential cytoplasm-plasma membrane-cell wall adhesion patterns and their relationships to hyphal tip growth and organelle motility. Protoplasma 200: 71-86

Bachewich CL and Heath IB (1997b) The cytoplasmic pH influences hyphal tip growth and the cytoskeleton-related organization. Fung Genet Biol 21: 76-91

Bachewich CL and Heath IB (1998) Radial F-actin arrays precede new hypha formation in *Saprolegnia*: implications for establishing polar growth and regulating tip morphogenesis. J Cell Sci 111: 2005-2016

Bachewich CL and Heath IB (1999) Cytoplasmic migrations and vacuolation are associated with growth recovery in hyphae of *Saprolegnia*, and are dependent on the cytoskeleton. Mycol Res 103: 849-858

Bartnicki-Garcia S, Hergert F and Gierz G (1989) Computer simulation of fungal morphogenesis and the mathematical basis for hyphal (tip) growth. Protoplasma 153: 46-57

Bi E, Maddox P, Lew D, Salmon E, McMillan J, Yeh E and Pringle J (1998) Involvement of an actomyosin contractile ring in *Saccharomyces cerevisiae* cytokinesis. J Cell Biol 142: 1301-1312

Bourett TM and Howard RJ (1991) Ultrastructural immunolocalization of actin in a fungus. Protoplasma 163: 199-202

Bourett TM and Howard RJ (1992) Actin in penetration pegs of the fungal rice blast pathogen, *Magnaporthe grisea* . Protoplasma 168: 20-26

Butt TM and Heath IB (1988) The changing distribution of actin and nuclear behavior during the cell cycle of the mite-pathogenic fungus *Neozygites* sp. Eur J Cell Biol 46: 499-505

Choi EY and Jeon KW (1992) Role of spectrin in *Amoeba proteus*, as studied by microinjection of anti-spectrin monoclonal antibodies. Exp Cell Res 199: 174-178

Corrêa A and Hoch HC (1993) Microinjection of urediospore germlings of *Uromyces appendiculatus*. Exp Mycol 17: 253-273

Corrêa A, Staples RC and Hoch HC (1996) Inhibition of thigmostimulated cell differentiation with RGD-peptides in *Uromyces* germlings. Protoplasma 194: 91-102

Czymmek KJ, Bourett TM and Howard RJ (1995) Immunolocalization of tubulin and actin in thick-sectioned fungal hyphae after freeze-substitution fixation and methacrylate de-embedment. J Microsc 181: 153-161

Degousée N, Gupta GD, Lew RR and Heath IB (2000) A putative spectrin-containing membrane skeleton in hyphal tips of *Neurospora crassa*. Fung Genet Biol 29: in press

Doyle T and Botstein D (1996) Movement of yeast cortical actin cytoskeleton visualized *in vivo*. Proc Nat Acad Sci USA 93: 3886-3891

Drubin DG, Jones HD and Wertman KF (1993) Actin structure and function: roles in mitochondrial organization and morphogenesis in budding yeast and identification of the phalloidin-binding site. Mol Biol Cell 4: 1277-1294

Fischer R and Timberlake WE (1995) *Aspergillus nidulans apsA* (anucleate primary sterigmata) encodes a coiled-coil protein required for nuclear positioning and completion of asexual development. J Cell Biol 128: 485-498

Gale C, Finkel D, Tao N, Meinke M, McClellan M, Olson J, Kendrick K and Hostetter M (1996) Cloning and expression of a gene encoding an integrin-like protein in *Candida albicans*. Proc Natl Acad Sci USA 93: 357-361

Garrill A, Jackson SL, Lew RR and Heath IB (1993) Ion channel activity and tip growth: tip-localized stretch-activated channels generate an essential Ca^{2+} gradient in the oomycete *Saprolegnia ferax*. Eur J Cell Biol 60: 358-365

Garrill A, Lew RR and Heath IB (1992) Stretch-activated Ca^{2+} and Ca^{2+}-activated K^+ channels in the hyphal tip plasma membrane of the oomycete *Saprolegnia ferax*. J Cell Sci 101: 721-730

Gooday GW and Gow NAR (1990) Enzymology of tip growth in fungi. In: Heath IB (ed) Tip Growth in Plant and Fungal Cells. San Diego: Academic Press, pp 31-58

Grove SN and Bracker CE (1970) Protoplasmic organization of hyphal tips among fungi: vesicles and Spitzenkörper. J Bacteriol 104: 989-1009

Grove SN and Sweigard JA (1996) Apical cytoplasmic organization and polar growth in *Rhizoctonia solani* are modified by cytochalasins. WWW J Biol 2: Available: http://epress.com/w3jbio/vol2/grove/grove.htm

Gupta G and Heath IB (1997) Actin disruption by latrunculin B causes turgor-related changes in tip growth of *Saprolegnia* hyphae. Fung Genet Biol 21: 64-75

Haarer BK, Lillie SH, Adams AEM, Magdolen V, Bandlow W and Brown SS (1990) Purification of profilin from *Saccharomyces cerevisiae* and analysis of profilin-deficient cells. J Cell Biol 110: 105-114

Harold FM and Caldwell JH (1990) Tips and currents: electrobiology of apical growth. In: Heath IB (ed) Tip Growth in Plant and Fungal Cells. San Diego: Academic Press, pp 59-90

Harold FM, Harold RL and Money NP (1995) What forces drive cell wall expansion? Can J Bot 73, Suppl. 1: S379-S383

Harold RL and Harold FM (1992) Configuration of actin microfilaments during sporangium development in *Achlya bisexualis*: comparison of two staining prototcols. Protoplasma 171: 110-116

Heath IB (1987) Preservation of a labile cortical array of actin filaments in growing hyphal tips of the fungus *Saprolegnia ferax*. Eur J Cell Biol 44: 10-16

Heath IB (1988) Evidence against a direct role for cortical actin arrays in saltatory organelle motility in hyphae of the fungus *Saprolegnia ferax*. J Cell Sci 91: 41-47

Heath IB (1990) The roles of actin in tip growth of fungi. Int Rev Cytol 123: 95-127

Heath IB (1994) The cytoskeleton in hyphal growth, organelle movements, and mitosis. In: Wessels JGH, Meinhardt F (eds) The Mycota. I: Growth, Differentiation and Sexuality. Berlin: Springer Verlag, pp 43-65

Heath IB (1995a) The cytoskeleton. In: Gow NAR, Gadd GM (eds) The Growing Fungus. London: Chapman Hall, pp 99-134

Heath IB (1995b) Integration and regulation of hyphal tip growth. Can J Bot 73, Suppl. 1: S131-S139

Heath IB (2000) Bridging the divide, cytoskeleton-plasma membrane-cell wall interactions in growth and development. In: Howard R J, Gow N A R (eds) The Mycota VIII: Biology of the Fungal Cell. Heidelberg: Springer Verlag, in press

Heath IB, Gupta G and Bai S (2000) Plasma membrane-associated actin filaments, but not microtubules, are essential for both polarization and hyphal morphogenesis in *Saprolegnia ferax* and *Neurospora crassa.* Fung Genet Biol 29: in press

Heath IB and Harold RL (1992) Actin has multiple roles in the formation and architecture of zoospores of the oomycetes, *Saprolegnia ferax* and *Achlya bisexualis.* J Cell Sci 102: 611-627

Heath IB and Heath MC (1978) Microtubules and organelle movements in the rust fungus *Uromyces phaseoli* var. vignae. Cytobiologie 16: 393-411

Heath IB, Heath MC and Herr F (1982) Motile systems in fungi. In: Amos WB, Duckett JG (eds) Prokaryotic and Eukaryotic Flagella. Society for Experimental Biology Symposium 35. Cambridge: Cambridge University Press, pp 563-588

Heath IB and Janse van Rensburg EJ (1996) Critical evaluation of the VSC model for tip growth. Mycoscience 37: 1-10

Heath IB and Kaminskyj SGW (1989) The organization of tip-growth related organelles and microtubules revealed by quantitative analysis of freeze-substituted oomycete hyphae. J Cell Sci 93: 41-52

Heath IB, Rethoret K, Arsenault AL and Ottensmeyer FP (1985) Improved preservation of the form and contents of wall vesicles and the Golgi apparatus in freeze substituted hyphae of *Saprolegnia.* Protoplasma 128: 81-93

Heath IB and Seagull RW (1982) Oriented cellulose microfibrils and the cytoskeleton: a critical comparison of models. In: Lloyd C W (eds) The Cytoskeleton in Plant Growth and Development. London: Academic Press, pp. 163-182

Heath IB and Steinberg G (1999) Mechanisms of hyphal tip growth: tube dwelling amoebae revisited. Fung Genet Biol 28: 79-93

Hermann GJ, King EJ and Shaw JM (1997) The yeast gene, *MDM20*, is necessary for mitochondrial inheritance and organization of the actin cytoskeleton. J Cell Biol 137: 141-153

Herr FB and Heath MC (1982) The effects of antimicrotubule agents on organelle positioning in the cowpea rust fungus, *Uromyces phaseoli* var. vignae. Exp Mycol 6: 15-24

Hill KL, Catlett NL and Weisman LS (1996) Actin and myosin function in directed vacuole movement during cell division in *Saccharomyces cerevisiae.* J Cell Biol 135: 1535-1549

Hoch HC and Howard RJ (1980) Ultrastructure of freeze-substituted hyphae of the basidiomycete *Laetisaria arvalis.* Protoplasma 103: 281-297

Hoch HC and Staples RC (1983a) Visualization of actin *in situ* by rhodamine-conjugated phalloidin in the fungus *Uromyces phaseoli.* Eur J Cell Biol 32: 52-58

Hoch HC and Staples RC (1983b) Ultrastructural organization of the non-differentiated uredospore germling of *Uromyces phaseoli* variety *typica.* Mycologia 75: 795-824

Hoch HC and Staples RC (1985) The microtubule cytoskeleton in hyphae of *Uromyces phaseoli* germlings: its relationship to the region of nucleation and to the F-actin cytoskeleton. Protoplasma 124: 112-122

Hoch HC, Tucker BE and Staples RC (1987) An intact microtubule cytoskeleton is necessary for mediation of the signal for cell differentiation in *Uromyces.* Eur J Cell Biol 45: 209-218

Howard RJ (1981) Ultrastructural analysis of hyphal tip cell growth in fungi: Spitzenkörper, cytoskeleton and endomembranes after freeze-substitution. J Cell Sci 48: 89-103

Howard RJ and Aist JR (1977) Effects of MBC on hyphal tip organization, growth, and mitosis of *Fusarium acuminatum*, and their antagonism by D_2O. Protoplasma 92: 195-210

Howard RJ and Aist JR (1979) Hyphal tip cell ultrastructure of the fungus *Fusarium:* improved preservation by freeze-substitution. J Ultrastr Res 66: 224-234

Howard RJ and Aist JR (1980) Cytoplasmic microtubules and fungal morphogenesis: ultrastructural effects of methyl benzimidazole-2-yl carbamate determined by freeze substitution of hyphal tip cells. J Cell Biol 87: 55-64

Hyde GJ, Davies D, Perasso L, Cole L and Ashford AE (1999) Microtubules, but not actin microfilaments, regulate vacuole motility and morphology in hyphae of *Pisolithus tinctorius*. Cell Motil Cytoskel 42: 114-124

Hyde GJ and Heath IB (1997) Ca^{2+} gradients in hyphae and branches of *Saprolegnia ferax*. Fung Genet Biol 21: 238-251

Jackson SL and Hardham AR (1998) Dynamic rearrangement of the filamentous actin network occurs during zoosporogenesis and encystment in the oomycete *Phytophthora cinnamomi*. Fung Genet Biol 24: 24-33

Jackson SL and Heath IB (1990a) Visualization of actin arrays in growing hyphae of the fungus *Saprolegnia ferax*. Protoplasma 154: 66-70

Jackson SL and Heath IB (1990b) Evidence that actin reinforces the extensible hyphal apex of the oomycete *Saprolegnia ferax*. Protoplasma 157: 144-153

Jackson SL and Heath IB (1992) UV microirradiations elicit Ca^{2+} dependent apex-directed cytoplasmic contractions in hyphae. Protoplasma 170: 46-52

Jackson SL and Heath IB (1993a) The dynamic behavior of cytoplasmic F-actin in growing hyphae. Protoplasma 173: 23-34

Jackson SL and Heath IB (1993b) The roles of calcium ions in hyphal tip growth. Microbiol Rev 57: 367-382

Jacobs W, Adams AEM, Szaniszlo PJ and Pringle JR (1988) Functions of microtubules in the *Saccharomyces cerevisiae* cell cycle. J Cell Biol 107: 1409-1426

Johnston GC, Prydz K, Ryd M and van Deurs B (1991) The *Saccharomyces cerevisiae* MYO2 gene encodes an essential myosin for vectorial transport of vesicles. J Cell Biol 113: 539-552

Kaminskyj SGW, Garrill A and Heath IB (1992a) The relation between turgor and tip growth in *Saprolegnia ferax*: turgor is necessary but not sufficient to determine apical extension rates. Exp Mycol 16: 64-75

Kaminskyj SGW and Heath IB (1994) A comparison of techniques for localizing actin and tubulin in hyphae of *Saprolegnia ferax*. J Histochem Cytochem 42: 523-530

Kaminskyj SGW and Heath IB (1995) Integrin and spectrin homologues, and cytoplasm-wall adhesion in tip growth. J Cell Sci 108: 849-856

Kaminskyj SGW and Heath IB (1996) Studies on *Saprolegnia ferax* suggest the general importance of the cytoplasm in determining hyphal morphology. Mycologia 88: 20-37

Kaminskyj SGW, Jackson SL and Heath IB (1992b) Fixation induces differential polarized translocations of organelles in hyphae of *Saprolegnia ferax*. J Microsc 167: 153-168

Kaminskyj SGW, Yoon KS and Heath IB (1989) Cytoskeletal interactions with post-mitotic migrating nuclei in the oyster muchroom fungus, *Pleurotus ostreatus*: evidence against a force-generating role for astral microtubules. J Cell Sci 94: 663-674

Kanbe T, Akashi T and Tanaka K (1993) Effect of cytochalasin A on actin distribution in the fission yeast *Schizosaccharomyces pombe* studied by fluorescent and electron microscopy. Protoplasma 176: 24-32

Kanbe T, Akashi T and Tanaka K (1994) Changes in the distribution of F-actin in the fission yeast *Schizosaccharomyces pombe* by arresting growth in distilled water: correlative studies with fluorescence and electron microscopy. J Elect Microsc 43: 20-24

Kanbe T, Kobayashi I and Tanaka K (1989) Dynamics of cytoplasmic organelles in the cell cycle of the fission yeast *Schizosaccharomyces pombe.* J Cell Sci 94: 647-656

Karpova T, McNally J, Moltz S and Cooper J (1998) Assembly and function of the actin cytoskeleton of yeast: relationships between cables and patches. J Cell Biol 142: 1501-1517

Katayama S, Hirata D, Arellano M, Perez P and Toda T (1999) Fission yeast α-glucan synthase Mok1 requires the actin cytoskeleton to localize the sites of growth and plays an essential role in cell morphogenesis downstream of protein kinase C function. J Cell Biol 144: 1173-1186

Kurtz MB, Heath IB, Marrinan J, Dreikorn S, Onoshi J and Douglas C (1994) Morphological effects of lipopeptides against *Aspergillus fumigatus* correlate with activities against (1,3)-β-D-glucan synthase. Antimicrob Agents Chemotherap 38: 1480-1489

Lazzarino DA, Boldogh I, Smith MG, Rosand J and Pon LA (1994) Yeast mitochondria contain ATP-sensitive reversible actin-binding activity. Mol Biol Cell 5: 807-818

Levina NN, Lew RR and Heath IB (1994) Cytoskeletal regulation of ion channel distribution in the tip-growing organism *Saprolegnia ferax*. J Cell Sci 107: 127-134

Lew RR (1999) Comparative analysis of Ca^{2+} and H^{+} flux magnitude and location along growing hyphae of *Saprolegnia ferax* and *Neurospora crassa.* Eur J Cell Biol 78: 892-902

Li J and Heath IB (1994) The behaviour of F-actin during the zoosporic phases of the chytridiomycete gut fungi *Neocallimastix* and *Orpinomyces*. Exp Mycol 18: 57-69

Lopez-Franco R and Bracker CE (1996) Diversity and dynamics of the Spitzenkörper in growing hyphal tips of higher fungi. Protoplasma 195: 90-111

Marks J, Hagan IM and Hyams JS (1986) Growth polarity and cytokinesis in fission yeast: the role of the cytoskeleton. J Cell Sci Suppl 5: 229-241

Marks J and Hyams JS (1985) Localization of F-actin through the cell division cycle of *Schizosaccharomyces pombe.* Eur J Cell Biol 39: 27-32

Maupin-Szamier P and Pollard TD (1978) Actin filament destruction by osmium tetroxide. J Cell Biol 77: 837-852

May K, Wheatly S, Amin V and Hyams J (1998) The myosin ATPase inhibitor 2,3-butanedione-2-monoxime (BDM) inhibits tip growth and cytokinesis in the fission yeast, *Schizosaccharomyces pombe*. Cell Motil Cytoskel 41: 117-125

May KM and Hyams JS (1998) The yeast cytoskeleton: the closer we look, the more we see. Fung Genet Biol 24: 110-122

McGillviray AM and Gow NAR (1987) The transhyphal electrical current of *Neurospora crassa* is carried principally by protons. J Gen Microbiol 133: 2875-2881

McGoldrick CA, Gruver C and May GS (1995) *myoA* of *Aspergillus nidulans* encodes an essential myosin I required for secretion and polarized growth. J Cell Biol 128: 577-587

McKerracher LJ and Heath IB (1985) Microtubules around migrating nuclei in conventionally-fixed and freeze-substituted cells. Protoplasma 125: 162-172

McKerracher LJ and Heath IB (1986a) Fungal nuclear behaviour analysed by ultraviolet microbeam irradiation. Cell Motil Cytoskel 6: 35-47

McKerracher LJ and Heath IB (1986b) Comparison of polyethylene glycol and diethylene glycol distearate embedding methods for the preservation of fungal cytoskeletons. J Electr Microsc Tech 4: 347-360

McKerracher LJ and Heath IB (1986c) Polarized cytoplasmic movement and inhibition of saltations induced by calcium-mediated effects of microbeams in fungal hyphae. Cell Motil Cytoskel 6: 136-145

McKerracher LJ and Heath IB (1987) Cytoplasmic migration and intracellular organelle movements during tip growth of fungal hyphae. Exp Mycol 11: 79-100

Mims CW, Richardson EA and Timberlake WE (1988) Ultrastructural analysis of conidiophore development in the fungus *Aspergillus nidulans* using freeze-substtiution. Protoplasma 144: 132-141

Momany M and Hamer JE (1997) Relationship of actin, microtubules, and crosswall synthesis during septation in *Aspergillus nidulans*. Cell Motil Cytoskel 38: 373-384

Money NP and Harold FM (1992) Extension growth of the water mold *Achlya* : interplay of turgor and wall strength. Proc Natl Acad Sci USA 89: 4245-4249

Money NP and Harold FM (1993) Two water molds can grow without measurable turgor pressure. Planta 190: 426-430

Mulholland J, Preuss D, Moon A, Wong A, Drubin D and Botstein D (1994) Ultrastructure of the yeast actin cytoskeleton and its association with the plasma membrane. J Cell Biol 125: 381-391

Newhouse JR, Hoch HC and MacDonald WL (1983) The ultrastructure of *Endothia parasitica.* Comparison of a virulent with a hypovirulent isolate. Can J Bot 61: 389-411

Novick P and Botstein D (1985) Phenotypic analysis of temperature-sensitive yeast actin mutants. Cell 40: 405-416

Oakley BR and Rinehart JE (1985) Mitochondria and nuclei move by different mechanisms in *Aspergillus nidulans.* J Cell Biol 101: 2392-2397

Osumi M, Sato M, Ishijima SA, Konomi M, Takagi T and Yaguchi H (1998) Dynamics of cell wall formation in fission yeast, *Schizosaccharomyces pombe*. Fung Genet Biol 24: 178-206

Palmer RE, Sullivan DS, Huffaker T and Koshland D (1992) Role of astral microtubules and actin in spindle orientation and migration in the budding yeast, *Saccharomyces cerevisiae.* J Cell Biol 119: 583-594

Parton RM, Fischer S, Malho R, Papasouliotis O, Jelitto TC, Leonard T and Read ND (1997) Pronounced cytoplasmic pH gradients are not required for tip growth in plant and fungal cells. J Cell Sci 110: 1187-1198

Picton JM and Steer MW (1982) A model for the mechanism of tip extension in pollen tubes. J Theor Biol 98: 15-20

Reinhardt MO (1892) Das Wachsthum der Pilzhyphen. Jahrb wiss Botanik 23: 479-566

Riquelme M, Reynaga-Peña CG, Gierz G and Bartnicki-Garcia S (1998) What determines growth direction in fungal hyphae? Fung Genet Biol 24: 101-109

Roberson RW (1992) The actin cytoskeleton in hyphal cells of *Sclerotium rolfsii.* Mycologia 84: 41-51

Roberson RW and Fuller MS (1988) Ultrastructural aspects of the hyphal tip of *Sclerotium rolfsii* preserved by freeze substitution. Protoplasma 146: 143-149

Robson GD, Prebble E, Rickers A, Hosking S, Denning DW, Trinci APJ and Robertson W (1996) Polarized growth of fungal hyphae is defined by an alkaline pH gradient. Fung Genet Biol 20: 289-298

Roncal T, Ugalde UO and Irastorza A (1993) Calcium-induced conidiation in *Penicillium cyclopium*: calcium triggers cytosolic alkalinization at the hyphal tip. J Bact 175: 879-886

Runeberg P, Raudaskoski M and Virtanen I (1986) Cytoskeletal elements in the hyphae of the homobasidiomycete *Schizophyllum commune* visualized with indirect immunofluorescence and NBD-phallacidin. Eur J Cell Biol 41: 25-32

Salo V, Niini SS, Virtanen I and Raudaskoski M (1989) Comparative immunocytochemistry of the cytoskeleton of filamentous fungi with dikaryotic and multinucleate hyphae. J Cell Sci 94: 11-24

Schott D, Ho J, Pruyne D and Bretscher A (1999) The COOH-terminal domain of Myo2p, a yeast myosin V, has a direct role in secretory vesicle targeting. J Cell Biol 147: 791-807

Seagull RW and Heath IB (1980) The organization of cortical microtubule arrays in the radish root hair. Protoplasma 103: 205-229

Shepherd VA, Orlovich DA and Ashford AE (1993) A dynamic continuum of pleomorphic tubules and vacuoles in growing hyphae of a fungus. J Cell Sci 104: 495-507

Simon VR, Karmon SL and Pon LA (1997) Mitochondria inheritance: cell cycle and actin cable dependence of polarized mitochondrial movements in *Saccharomyces cerevisiae.* Cell Motil Cytoskel 37: 199-210

Simon VR, Swayne TC and Pon LA (1995) Actin-dependent mitochondrial motility in mitotic yeast and cell-free systems: identification of a motor activity on the mitochondrial surface. J Cell Biol 130: 345-354

Smith MG, Simon VR, O'Sullivan H and Pon LA (1995) Organelle-cytoskeletal interactions: Actin mutations inhibit meiosis-dependent mitochondrial rearrangement in the budding yeast *Saccharomyces cerevisiae*. Mol Biol Cell 6: 1381-1396

Srinivasan S, Vargas MM and Roberson RW (1996) Functional, organizational, and biochemical analysis of actin in hyphal tip cells of *Allomyces macrogynus*. Mycologia 88: 57-70

Steinberg G (1998) Organelle transport and molecular motors in fungi. Fung Genet Biol 24: 161-177

Tang X, Lancelle SA and Hepler PK (1989) Fluorescence microscopic localization of actin in pollen tubes: comparison of actin antibody and phalloidin staining. Cell Motil Cytoskel 12: 216-224

Temperli E, Roos U-P and Hohl HR (1990) Actin and tubulin cytoskeletons in germlings of the oomycete fungus *Phytophthora infestans*. Eur J Cell Biol 53: 75-88

Thomas DS, Lutzac M and Manavathu E (1974) Cytochalasin selectively inhibits synthesis of a secretory protein, cellulase, in *Achlya*. Nature 249: 140-142

Thompson-Coffe C and Zickler D (1993) Cytoskeletal interactions in the ascus development and sporulation of *Sordaria macrospora.* J Cell Sci 104: 883-898

Tinsley JH, Minke PF, Bruno KS and Plamann M (1996) $p150^{glued}$, the largest subunit of the dynactin complex, is nonessential in *Neurospora* but required for nucler distribution. Mol Biol Cell 7: 731-742

Torralba S, Raudaskoski M and Pedregosa A (1998a) Effects of methyl benzimidazole-2-yl carbamate on microtubulue and actin cytoskeleton in *Aspergillus nidulans*. Protoplasma 202: 54-64

Torralba S, Raudaskoski M, Pedregosa AM and Laborda F (1998b) Effect of cytochalasin A on apical growth, actin cytoskeleton organization and enzyme secretion in *Aspergillus nidulans*. Microbiology 144: 45-53

Tsukamoto M, Yamamoto M, Tanabe S and Kamada T (1996) Cytochalasin-E-resistant mutants of *Coprinus cinereus*: isolation and genetic, biochemical, and cytological analyses. Fung Genet Biol 20: 52-58

Waddle JA, Karpova TS, Waterston RH and Cooper JA (1996) Movement of cortical actin patches in yeasts. J Cell Biol 132: 861-870

Wang Y-L (1987) Mobility of filamentous actin in living cytoplasm. J Cell Biol 105: 2811-2816

Watts FZ, Shiels G and Orr E (1987) The yeast MYO1 gene encoding a myosin-like protein required for cell division. EMBO J 6: 3499-3505

Wessels JGH (1988) A steady-state model for apical wall growth in fungi. Acta Bot Neerl 37: 3-16

Wilson P, Fuller E and Forer A (1987) Irradiations of rabbit myofibrils with an ultraviolet microbeam. II. Phalloidin protects actin in solution but not in myofibrils from depolymerization by ultraviolet light. Biochem Cell Biol 65: 376-385

Wittenberg C, Richardson SL and Reed SI (1987) Subcellular localization of a protein kinase required for cell cycle initiation in *Saccharomyeces cerevisiae*: evidence of an asociation between the CDC28 gene product and the insoluble cytoplasmic matrix. J Cell Biol 105: 1527-1548

Yokoyama K, Kaji H, Nishimura K and Miyaji M (1990) The role of microfilaments and microtubules in apical growth and dimorphism of *Candida albicans*. J Gen Microbiol 136: 1067-1075

Chapter 17

RHO GTPASES AND THE ACTIN CYTOSKELETON

Hai Li and Zhenbiao Yang

Department of Botany and Plant Sciences, University of California, Riverside, CA 92521 USA

Key words: actin-binding proteins, actin dynamics, cell morphogenesis, polar growth, Rop GTPase, signal transduction

Abstract: RHO-family GTPases act as molecular switches to modulate actin organization and dynamics. In response to a signal, RHO turns on an effector protein that directly regulates the activity of actin-binding proteins. Different RHO members function to control distinct forms of F-actin. Plants possess a unique subfamily of RHO GTPase termed Rop. Emerging evidence suggests that Rop also regulates actin organization and thus plays a crucial role in the control of cell morphogenesis in plants.

1. INTRODUCTION

In response to developmental cues, plant cells organize and remodel the interphase cortical actin, the preprophase actin, and the phragmoplast actin to modulate cell morphogenesis as well as the position and orientation of cell division. Actin regorganization in plant cells can be induced by many external signals including pathogen invasion, light, and wounding (reviewed in Staiger et al., 1997). Signal-mediated organization and dynamics of the actin cytoskeleton are also important for the development of other eukaryotic cells. In yeast, cortical actin patches are polarized in response to signals that control the site of budding. In animals, the formation of various F-actin arrays associated with filopodia, lamelopodia, and focal adhesion is also mediated by specific extracellular stimuli (see below).

How does the cell interpret various signals to control the organization of actin cytoskeleton? In animal and yeast cells, actin organization and dynamics are directly regulated by actin-binding proteins (ABPs). Several

C.J. Staiger et al. (eds.), Actin: A Dynamic Framework for Multiple Plant Cell Functions, 301–321.

classes of functionally distinct ABPs have been characterized in eukaryotic cells: 1) those that regulate polymerization and depolymerization of actin, 2) those that promotes bundling of actin filaments, 3) those that initiate actin polymerization or site branches of actin filaments, and those that anchor actin to membranes. Homologs for some of these ABPs, e.g., profilins, ADFs, fimbrins, have been identified in plants and implicated in the regulation of actin organization in plant cells (see also Yokota & Shimmen, Gibbon & Staiger, Kovar & Staiger, McCurdy & Staiger, this volume). Evidence indicates that many ABPs are the targets of signal transduction mediated by extracellular signals (Carlier, 1998; Machesky & Gould, 1999). Importantly, the localization or activity of ABPs are directly or indirectly regulated by the RHO family of small GTPases (Rohatgi et al., 1999; Hall, 1998). By integrating various extracellular and intracellular signals, RHO GTPases act as molecular switches that control the organization and dynamics of various forms of the actin cytoskeleton. RHO-related proteins are present in plants and evidence is emerging that these proteins also play a critical role in the regulation of plant actin organization. This chapter will highlight recent studies on RHO-mediated actin cytoskeleton signaling in yeast, animals, and plants with an emphasis on how this conserved signaling mechanism is used to regulate specific forms of actin cytokeleton in diverse systems.

2. THE RHO FAMILY OF GTPASES

In mid/late 1980, RHO GTPases were first identified from mammals and yeast as Ras-related monomeric small GTP-binding proteins with a molecular weight of 21–25 kD (Hall, 1990). Subsequently various RHO-like GTPases, including Rac, Cdc42, Rop, and their close relatives have been found in different eukaryotic cells (Chardin, 1993; Yang & Watson, 1993). To date, sequences for more than 90 RHO and RHO-like GTPases are available from the GenBank database. Based on their sequence similarity, these GTPases are grouped into a distinct GTPase family termed RHO (Fig. 1). Members of the RHO family share more than 45% identity at the amino acid sequence level but less than 30% identity with other families of the RAS superfamily including RAS, RAB/YPT, RAN and ARF. According to phylogenetic analyses, the RHO family is further divided into four major subfamilies (Rho, Rop, Cdc42 & Rac) and a suite of orphan members from different organisms (Li et al., 1998). Members within a subfamily generally share 70% or greater amino acid sequence identity (Chardin, 1993; Li et al., 1998). Rho is present in yeast and animals but has not been identified in plants. Whereas Cdc42 is found in both yeast and mammals, Rac and Rop seem to be specific to animals and plants, respectively (Li et al., 1998).

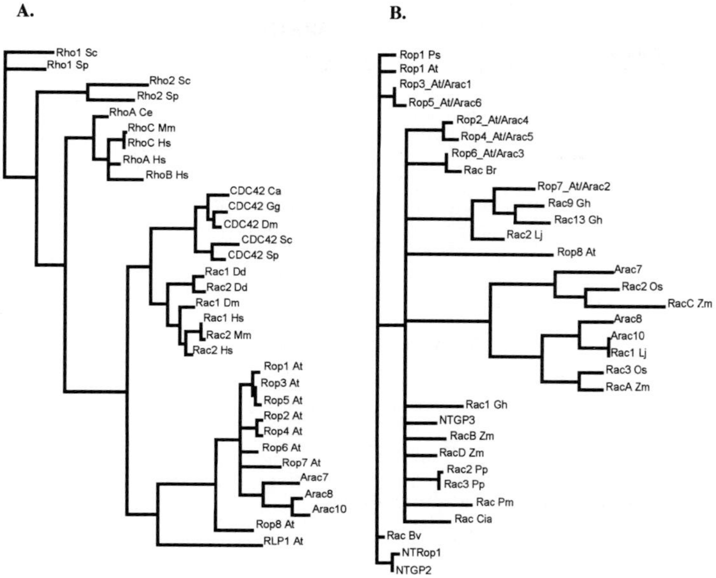

Figure 1. Phylogenetic trees for the RHO GTPase family and the Rop subfamily. Sequences for RHO GTPases were obtained from GenBank. RHO members from yeast and animals and all *Arabidopsis* RHO sequences (including 11 Rop members and RLP1) were used to constructed an unrooted phylogenetic tree (**A**) using parsimony analysis (PAUL 3.1 software). The Rop tree (**B**) was constructed from all 32 Rops available from GenBank. Abbreviations: At, *Arabidopsis thaliana*; Br, *Brassica rapa*; Bv, *Beta vulgaris*; Ca, *Candida albicans;* Ce, *Caenorhabditis elegans;* Cia, *Cicer arietinumal*; Dd, *Dictyostelium discoideum*; Dm, *Drosophila melanogaster*; Gg, *Gallus gallus*; Gh, *Gossypium hirsutum*; Hs, *Homo sapien*; Lj, *Lotus japonicus*; Mm, *Mus musculus*; Nt, *Nicotiana tabacum*; Os, *Oryza sativa*; Pm, *Picea mariana;* Pp, *Physcomitrella patens*; Ps, *Pisum sativum*; Sc, *Saccharomyces cerevisiae*; Sp *Schizosaccharomyces pombe*; Zm, *Zea mays.*

Athough RHO GTPases are involved in signaling of diverse cellular processes, e.g., cell cycle progression, gene expression, production of active oxygen species, organelle dynamics, their first known and most conserved function is to act as a molecular switch in the signaling of actin organization and dynamics (Hall, 1998). Like other G-proteins, the nature of RHO GTPase signaling, i.e., the switch between the GTP-bound "on" state and the GDP-bound "off" state, has been vital in elucidating the function of these GTPases. The switch is controlled by several regulators. Upon stimulation by a signal, a RHO-specific GDP/GTP exchange factor (RHO GEF) interacts with the GDP-bound form and replaces GDP with GTP, leading to the activation of RHO. The GTP-bound/active form then activates its effector and turns on RHO-dependent pathways. When bound GTP is hydrolyzed to GDP by an intrinsic GTPase activity, which is promoted by a GTPase-activating protein (RHO GAP), RHO becomes inactive and turns off the signaling pathway. In addition, the GDP-bound form can form a complex with a guanine nucleotide dissociation inhibitor (Rho GDI) preventing it from activation by GEF.

Consensus	FPXXYVPTVFDNYXAXVXV
Rho1_Sc	..EV.V....E.YVADVEV
Rho1_Sp	..EV.V....E.YVADVEV
Rho2_Sc	..EQ.H....E.YVTDSRV
Rho2_Sp	..TE.V....E.YVSDCRV
RhoA_Ce	..DV.V....E.YVADIEV
RhoC_Mm	..EV.V....E.YIADIEV
RhoA_Hs	..EV.V....E.YVADIEV
RhoB_Hs	..EV.V....E.YVADIEV
RhoC_Hs	..EV.V....E.YIADIEV
CDC42_Ca	..SE.V....D.YAVTVMI
CDC42_Gg	..SE.V....D.YAVTVMI
CDC42_Dm	..SE.V....D.YAVTVMI
CDC42_Sc	..AD.V....D.YAVTVMI
CDC42_Sp	..SD.V....D.YAVTVMI
Rac1_Dd	..GE.I....D.YSANVMV
Rac1_Dm	..GE.I....D.YSANVMV
Rac1_Hs	..GE.I....D.YSANVMV
Rac2_Dd	..GE.I....D.YSANVMV
Rac2_Hs	..GE.I....D.YSANVMV
Rac2_Mm	..GE.I....D.YSANVMV
Rop1At	..TD.V....D.FSANVVV
Rop2At	..TD.V....D.FSANVVV
Rop3At	..TD.V....D.FSANVVV
Rop4At	..TD.V....D.FSANVVV
Rop5At	..TD.V....D.FSANVVV
Rop6At	..TD.V....D.FSANVIV
Rop7At	..TD.V....D.FSANVVV
Rop8At	..TD.V....D.FNANVLV
Arac7	..TD.I....D.FSANVAV
Arac8	..TD.I....D.FSVNVVV
Arac10	..TD.I....D.FSANVVV

Figure 2. Comparison of RHO effector domains among different RHO subfamilies. The conserved effector domain (corresponding to residues 31–49 in Rop1At) from various RHO GTPases were manually aligned and the consensus sequence was derived. Invariant amino acid residues are marked with dots (bold face in the consensus sequence); X indicates residues that are variable or unique to each subfamily.

RHO GTPases contain 4 conserved nucleotide-binding domains (Yang & Watson, 1993). Point mutations on specific residues in these domains result in permanent binding to GDP or GTP. Replacement of a glycine residue (e.g., glycine15 in Rop1At; Li et al., 1998) in domain I with a valine residue causes a deficiency in the intrinsic GTPase activity, leading to the permanently GTP-bound mutants termed constitutively active or dominant positive mutants. These mutations are analogous to those on oncogenic RAS proteins. Introduction of such mutants into cells result in constitutive activation of a RHO-mediated pathway regardless of the existence of a signal. In contrast, replacement of threonine 20 in domain I with an asparagine residue creates GDP-bound dominant negative (or inhibitory) mutants. When overexpressed in cells, dominant negative mutants block RHO-dependent pathways by titrating out RHO GEF. In addition to the nucleotide binding domains, the effector domain of RHO GTPases, which binds RHO effectors, is also highly conserved but contains distinct amino acid residues among different subfamilies of RHO GTPases (Fig. 2). The C3 toxin from *Clostridium botulinum* is known to preferentially ADP-ribosylate the asparagine 41 residue of Rho subfamily GTPases but not Cdc42 and Rac (Aktories et al., 1989). The C3-mediated ADP ribosylation inactivates Rho by blocking the binding of Rho effectors. Because of the structural conservation involved in the dominant mutations and C3 action, they have been instrumental tools in investigating the function of Rho GTPases in various systems.

The first evidence for the role of RHO in signaling to the actin cytoskeleton was revealed by using the C3 toxin and constitutively active rhoA mutants in mammalian cells. Microinjection of C3 toxin into Swiss 3T3 fibroblast cells caused disassembly of actin microfilaments, a phenomenon normally induced by serum starvation, whereas constitutively active rho protein induces the formation of actin stress fibers in quiescent cells that is normally induced by serum (Chardin et al., 1989; Paterson et al., 1990). Subsequent studies using dominant negative and constitutively active mutants indicate that each of the three subfamilies controls a specific actin-dependent process (Ridley & Hall, 1992; Ridley et al., 1992; Luo et al., 1994; Chant & Stowers, 1995; Nobes & Hall, 1995). Rho is activated by extracellular ligands, LPA (lysophosphatidic acid), leading to the formation of stress fibers and associated focal adhesions (Ridley & Hall, 1992). Rac can be activated by growth factors, such as PDGF, insulin, and bombesin, stimulating the formation of localized actin networks at the plasma membrane to induce membrane ruffling and lamellipodia formation (Knaus et al., 1991; Ridley et al., 1992; Ridley, 1995; Van Aelst & D'Souza-Schorey, 1997). Activation of Cdc42 by bradykinin results in the formation of a fine actin meshwork at the leading edge and filopodia formation (Hall, 1994; Kozma et al., 1995; Nobes & Hall, 1995). These three subfamilies Cdc42, Rac and Rho have been placed in a hierarchical cascade wherein Cdc42 activates Rac, which in turn activates Rho (Fig. 3) (Chant & Stowers, 1995; Van Aelst & D'Souza-Schorey, 1997; Mackay & Hall, 1998).

Distinct functions for different RHO subfamilies in the regulation of specific forms of actin filaments are also found in other systems. In budding yeast, Cdc42 controls the establishment of cell polarity by regulating the dynamics of cortical actin patches that define the site of cell growth; however, Cdc42 is not required for cell growth *per se*. In contrast, Rho3 and Rho4 or Rho1 and Rho2 are required for bud growth (Tanaka & Takai, 1998). Rho has multiple functions in yeast, one of which is to regulate glucan synthase activity both directly and through a MAP kinase cascade (Drgonová et al., 1996).

Rho also interacts with the profilin-binding protein Bni1 and appears to be required for the formation of the cortical actin patches (Kohno et al., 1996; Evangelista et al., 1997; Imamura et al., 1997). In *Drosophila* epithelial cells, Cdc42 regulates polarized cell shape changes through effecting basal actin organization, whereas Rac1 modulates actin assembly at adherent junctions (Murphy & Montell, 1996).

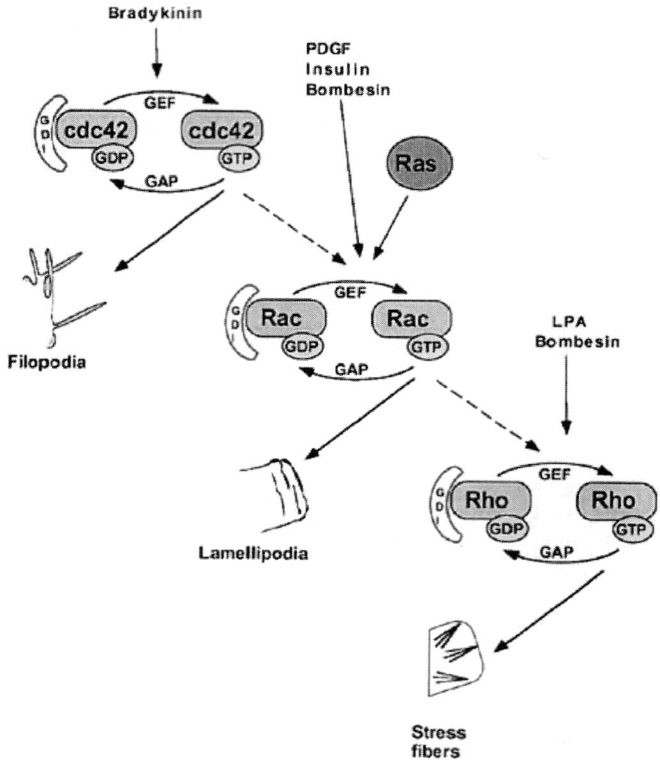

Figure 3. A RHO GTPase cascade involved in actin cytoskeletal organization in mammalian fibroblasts. Various extracellular stimuli independently trigger the activation of Cdc42, Rac, and Rho GTPases and elicit specific short-term responses such as the formation of dense cortical actin meshworks in filopodia or microspikes, cortical actin polymerization in lamelipodia, and the assembly of actin-myosin stess fibers associated with focal adhesion, respectively. Furthermore, Cdc42 activates Rac, which in turn activates RHO. The mechanism by which these agonists activate RHO GTPases may involve specific GEFs, GAPs, or GDIs.

Rac1 or Cdc42 and Rho also have distinct functions in the regulation of neuronal outgrowth (Gallo & Letourneau, 1998). Orphan members of the Rho family have also been demonstrated to have specific functions. For example, mig2, a *C. elegans* Rac-like GTPase is involved in axonal guidance but not neuronal outgrowth, whereas RacE, a *Dictyostelium* Rac-like GTPase, controls cytokinesis (Larochelle et al., 1997; Zipkin et al., 1997; Gerald et al., 1998).

3. HOW DO RHO FAMILY GTPASES CONTROL ACTIN ORGANIZATION AND REMODELING?

To understand how RHO GTPases regulate actin reorganization and how each RHO GTPase can modulate different cellular processes, extensive efforts have been devoted to identify RHO downstream targets (or effectors). Because the active form of RHO is expected to bind effectors, GTP-bound RHO or constitutively active mutants have been used for affinity chromatography or yeast two-hybrid screens. As expected, each RHO has been shown to interact with a plethora of targets, which carry out specific RHO-dependent functions (Aspenström, 1999). Known RHO effectors demonstrated to be involved in actin organization are summarized in Figure 4 (reviewed in Aspenström, 1999). These RHO effectors act directly on ABPs or are linked to ABPs through one or two intermediate components. The tight linkage between RHO and ABPs conform with the conservation of the function for RHO GTPases in the regulation of actin cytoskeleton.

Three known Rho effectors are linked to the regulation of actin organization. Mammalian p160 Rho-kinase promotes the phosphorylation of myosin light chain directly as well as indirectly through inactivation of myosin light chain phosphatase, leading to actomyosin assembly and actin bundle formation (Mackay & Hall, 1998). The Rho-kinase also phosphorylates LIM-kinase, which in turn phosphorylates cofilin (an ADF), leading to the inhibition of actin depolymerization (Maekawa et al., 1999). Another class of Rho effectors is a family of proteins containing FH (formin-homology) domains including mammalian p140mDia, *Aspergillus* SepA, *Drosophila* diaphanous protein, and yeast Bni1 and Bnr1 proteins. These effectors interact with profilin through the proline-rich FH domain, leading to increased actin polymerization (Watanabe et al., 1997; Lynch, 1997). Rho is also known to activate phosphoinositidyl phosphate kinase and thus the production of phosphatidylinositol 4,5-bisphosphate (PIP_2) (Chong et al., 1994), which in turn regulates ERMs (ABPs involved in focal adhesion) as well as profilins and the actin-capping protein gelsolin (Toker, 1998). The role of several other Rho targets (e.g., Citron, PKN, Rhotekin, and

Rhophilin) in actin organization is not clear, although Citron (also a protein kinase) appears to regulate actomyosin in the contractile ring involved in cytokinesis in animal cells (Tapon & Hall, 1997; Mackay & Hall, 1998; Aspenström, 1999).

Cdc42 and Rac GTPases share several common targets, at least based on *in vitro* interaction assays, reflecting the fact that Cdc42 and Rac are highly related (70% amino acid identity) and have nearly identical effector-binding domains (Tapon & Hall, 1997; Aspenström, 1999). However, Cdc42- or Rac-specific targets are also known (Tapon & Hall, 1997; Aspenström, 1999). Many Cdc42/Rac effectors contain the conserved Cdc42/Rac-interactive binding (CRIB) motif of 18 amino acid residues that are required and sufficient for the binding of these effectors to Cdc42/Rac (Burbelo et al., 1995). Two families of Cdc42/Rac targets, Pak (p21-activated kinase) and N-WASP (Wiskott-Aldrich syndrome protein), are among the best characterized CRIB-containing proteins. Paks are multi-functional and control the remodeling of cortical F-actins (Frost et al., 1998; Holly & Blumer, 1999). In yeast, two functionally redundant Pak homologs STE20 and CLA4 act as Cdc42 effectors to regulate the polarization of cortical actin patches to the site of bud growth, but are not required for the formation of cortical actin *per se* (Holly & Blumer, 1999). How Paks regulate the remodeling of cortical F-actin is not clear.

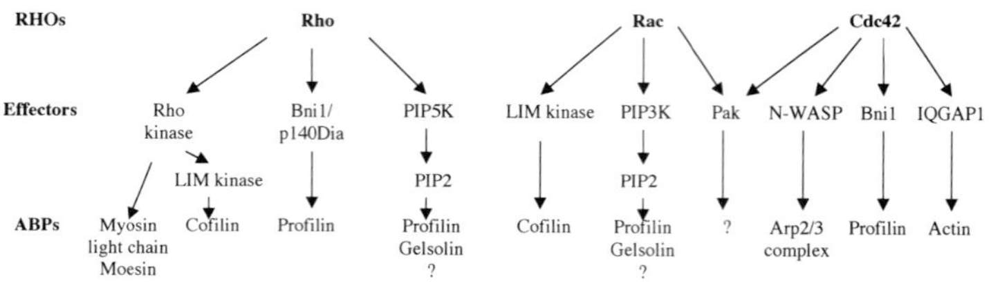

Figure 4. Roles of RHO effectors in the regulation of actin organization. Effectors listed for each RHO subfamily that have been implicated in the regulation of actin organization in yeast and/or animals (reviewed in: Hall, 1998; Mackay & Hall, 1998; Aspenström, 1999) and ABPs known to be regulated by these effectors are also indicated. Question marks indicate potential involvement that lacks direct experimental support.

N-WASP has been demonstrated to link to the actin cytoskeleton through the Arp2/3 complex that promotes actin nucleation (Ma et al., 1998; Miki et al., 1998; Rohatgi et al., 1999). N-WASP contains a PIP_2-binding site and a CRIB motif at the N-terminus and a C-terminal VCA region composed of a verprolin (a G-actin-binding protein) homology domain, a cofilin homology domain, and an acidic domain (Rohatgi et al., 1999). The VCA region binds to and activates the Arp2/3 complex. The N-terminal region of N-WASP appears to negatively regulate the VCA region probably by sterically masking the VCA region. This inhibition is released by binding of Cdc42 to the CRIB motif and PIP_2 to the PIP_2-binding site of the N-WASP N-terminal region. Homologs of WASPs exist in *Drosophila* and yeast and are likely to have a similar role in actin cytoskeleton signaling (Tapon & Hall, 1997; Aspenström, 1999). However, homologs for WASPs have not been found in plants, although an Arp2 homolog has been identified in *Arabidopsis* (Klahre & Chua, 1999).

Several other CRIB-containing Cdc42/Rac targets have also been identified but their roles in the signaling of actin organization have not been established (Tapon & Hall, 1997; Aspenström, 1999). A number of Cdc42/Rac targets lacking the CRIB motif have also been found in yeast and animals, among which IQGAP and PIP-kinase appears to have a role in the regulation of actin organization (Epp & Chant, 1997).

4. THE ROP SUBFAMILY OF RHO GTPASES

4.1 The Rop subfamily is unique to plants

Because of the structural and functional conservation of RHO GTPases in animals and fungi, the search for RHO-like GTPases from plants was pursued (Yang & Watson, 1993). A polymerase chain reaction-based approach led to the identification of a pea RHO-like GTPase, Rho1Ps (renamed Rop1Ps) (Yang & Watson, 1993). Many genes highly related to Rop1Ps were subsequently cloned from different plant species (Delmer et al., 1995; Winge et al., 1997; Li et al., 1998). Rop1Ps and its relatives share more than 75% amino acid sequence identity with each other, less than 65% with Rac, about 55% with Cdc42, and 45–50% with Rho. Because they share greatest overall sequence similarity to Rac, plant Rop1Ps relatives have often been named Racs in the literature. However, phylogenetic analyses indicate that these GTPases do not belong to any of the three major subfamilies of RHO GTPases (Rho, Rac & Cdc42) from animals and fungi. Instead, they form a distinct subfamily of Rho GTPases, termed Rop (Rho GTPases from plants) (Fig. 1). The Rop subfamily appears to be plant-specific, because

Rop homologs are not present in the yeast genome or found in animals. The distinction of Rop from other subfamilies of RHO GTPases is also supported by the presence of Rop-specific sequences in the effector domain (Fig. 3) as well as in the RHO insert region that has also been implicated in the interaction of certain RHO members with their effectors (Li et al., 1998). Furthermore, phylogenetic analyses using various methods consistently suggest that Rop evolved prior to the divergence of Cdc42 and Rac (Li et al., 1998). Thus, it was speculated that Rop may have some conserved functions similar to those of Cdc42 and Rac as well as have evolved to perform specific functions that are unique to plant cells. This speculation is clearly supported by functional studies of Rop GTPases in plants (see below). Lastly, Rop is regulated by a distinct mechanism and may have acquired novel effector proteins (see below).

Apart from the plant-specific Rop subfamily, plants also contain a Rac-related orphan member RLP1 which shares greatest similarity to the *Dictyostelium* RacE known to be involved in the regulation of cytokinesis (Collins & Johnson, 1997; Larochelle et al., 1997). Inhibition of cytoplasmic streaming in pollen tubes by injected C3 toxin, an effect that is distinct from blocking of Rop signaling, is consistent with the notion that the Rho subfamily of RHO GTPases is also present in plants (Lin & Yang, 1997). However, no Rho-related sequences from plants have been revealed to date, even though more than 70% of the *Arabidopsis* genome sequence is completed.

4.2 Why so many Rops?

Rop is the largest subfamily of RHO GTPases with 32 Rop members reported in the Genbank database. *Arabidopsis*, despite having one of the smallest genome in plants, contains at least 11 Rops. Why is there such a large gene family? One or a combination of the following explanations may account for the existence of the Rop gene family:

1) Rop members within the Rop subfamily or within a subgroup of Rops have redundant function. Functional redundancy seems to be very common in plants, as knockout of individual members of a gene family often shows no visible phenotypes.

2) Each or a subset of Rop is differentially expressed in certain cells or tissues and thus may have distinct developmental function, although its cellular function may be the same. This is supported by differential expression of Rops in various plants (Yang & Watson, 1993; Delmer et al., 1995; Winge et al., 1997; Li et al., 1998). In *Arabidopsis*, three closely related Rops, Rop1, Rop3 and Rop5, are all expressed in pollen, but Rop3 and Rop5 also are expressed in vegetative organs; whereas Rop2, Rop4, and

Rop6 are constitutively expressed in all vegetative organs. Rop7At/Arac2 is specifically expressed in *Arabidopsis* hypocotyls and roots. A cotton Rop (Rac13) is preferentially expressed in cotton fiber cells (Delmer et al., 1995).

3) Each or a subset of Rops has a distinct function in regulating one or more specific processes. This is supported by further subdivision of the Rop subfamily into distinct subgroups or orphan members, each possessing a unique C-terminal region that is presumably involved in the specific subcellular localization (Li et al., 1998). GFP-tagging experiments suggest that Rop1 with C-terminal CSIL is localized to the plasma membrane (Kost et al., 1999; Li et al., 1999), whereas Rop2 with C-terminal CAFL appears to be localized both in the plasma membrane and endosome-like organelles in tobacco BY-2 cells (our unpublished data). In pea tapetal cells, Rop has been shown to be specifically localized to the tonoplast of developing vacuoles, although which Rop is responsible for this localization is not known (Lin & Yang, unpublished data).

Phenotypic analyses of transgenic plants expressing constitutively active or dominant negative rop1 and rop2 mutants under the pollen-specific LAT52 promoter or the constitutive 35S CaMV promoter suggest that Rops are involved in multiple cellular processes, e.g., tip growth, cell expansion and morphogenesis, cell polarity control, seedling development, and vascular differentiation (Li et al., 1999; our unpublished data). Rops have also been shown to regulate the production of reactive oxygen species in rice and cotton (Kawasaki et al., 1999; Potikha et al., 1999); this role of Rops appears to be analogous to that of mammalian Racs that regulate NADPH oxidase (Heyworth et al., 1994). Together with differential expression and subcellular localization of Rops, these observations support the notion that each Rop or a subset of Rops has different roles in the regulation of specific pathways, owing to differential developmental expression or differential cellular function. Identification of rop knockout mutants should help clarify this issue.

4.3 Rops in the signaling of polar growth–acting through actin or calcium?

The function of Rop GTPases in plant cells is best characterized in pollen tubes. In *Arabidopsis*, three Rop genes, Rop1At, Rop3At and Rop5At, are expressed in mature pollen and pollen tubes, while Rop1At is pollen-specific and Rop3At and Rop5At are also expressed in vegetative parts of the plant (Li et al., 1998). Rop1Ps, an apparent pea ortholog of Rop1At, is also preferentially expressed in pea pollen tubes (Lin et al., 1996). Studies involving immunolocalization, microinjection of anti-Rop1Ps antibodies, and pollen-specific overexpression of constitutively active and dominant negative rop mutants, and sense and antisense RNAs have established a critical role for Rops in the control of tip growth in pollen tubes (Fig. 5; Lin et al., 1996; Lin & Yang, 1997; Kost et al., 1999; Li et al., 1999). The following conclusions can be drawn from these studies regarding the function of Rops in pollen tubes:

1) Rop controls the establishment of cell polarity for tip growth. Transgenic overexpression of constitutively active rop1At mutants and wild-type Rop1At in *Arabidopsis* causes delocalized growth (i.e., bulbous or swollen tubes; Li et al., 1999). Similarly, transient overexpression of similar constructs with Rop5At (also called AtRac2) in tobacco pollen also caused pollen tube tips to swell (Kost et al., 1999).

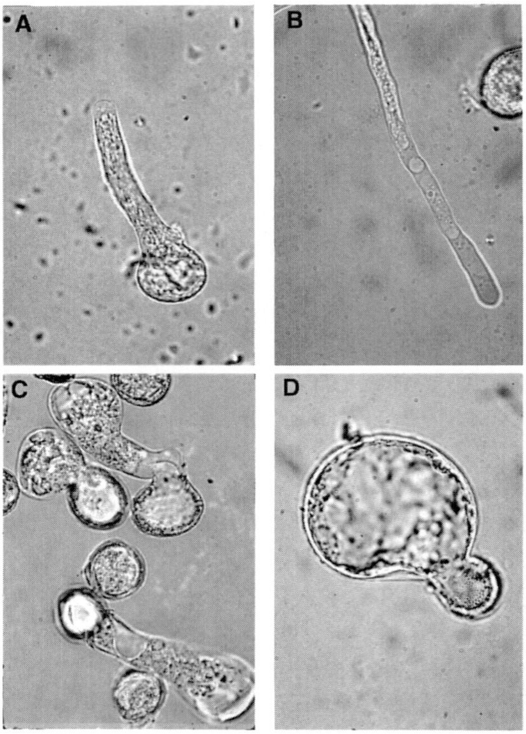

Figure 5. Functions of Rop in polar growth. Shown are typical pollen tube phenotypes in transgenic *Arabidopsis* plants expressing various forms of Rop1At under the control of LAT52 promoter (Li et al., 1999). **A**, dominant-negative rop1At mutants. **B**, untransformed plants. **C**, wild-type Rop1At overexpression. **D**, constitutively-active rop1At mutants.

2) Both localized activation and membrane targeting of Rop are implicated in the control of tip growth sites in pollen tubes. Constitutively-active rop1 mutants caused more severe delocalized growth of pollen tubes than wild-type Rop1 when overexpressed at the same level. This results suggest that Rop signaling is important for the control of cell polarity.

Furthermore, several observations suggest a role for Rop recruitment to the tip of pollen tubes in specifying the site for tip growth. First, Rop is localized to the apical region of the pollen tube plasma membrane. Second, mutations that abolish membrane localization of Rop cause Rop to lose biological activity. Last, Rop overexpression causes ectopic localization of Rop to the apical region of the plasma membrane (Li et al., 1999). Similarly, subcellular localization of Cdc42 is also implicated in the regulation of cell polarity in yeast (Ziman et al., 1993).

3) Rop signaling is essential for pollen tube growth and probably pollen germination as well. Blocking Rop signaling by microinjection of Rop1 antibodies into pea pollen tubes and transgenic overexpression of dominant negative rop1at mutants in *Arabidopsis* leads to the inhibition of pollen tube elongation but does not significantly alter pollen tube morphology (Lin & Yang, 1997; Li et al., 1999). Transient expression of dominant negative rop5At mutants in tobacco pollen also inhibits pollen tube growth (Kost et al., 1999). In transgenic lines expressing high levels of dominant negative rop1At mutants, pollen tube germination is also reduced (unpublished data). Likewise, partial suppression of Rop expression by antisense rop1At inhibited *Arabidopsis* pollen tube elongation, only when extracellular calcium levels are lower than the optimal level. Thus, Rop signaling seems to be required for localized exocytosis.

4) Rop acts upstream of the tip-localized calcium activity. A tip-focused calcium gradient and a tip-localized calcium influx are both essential for pollen tube growth and orient the direction of pollen tube growth (Garrill et al., 1993; Pierson et al., 1994; Wymer et al., 1997). It was shown that microinjected Rop1 antibodies eliminate the tip-high accumulation of intracellular calcium, while effects caused by Rop inactivation or suppression can be rescued by raising extracellular calcium (Lin & Yang, 1997; Li et al., 1999).

That Rop inactivation does not alter growth polarity and that Rop acts upstream of calcium suggest that Rop couples the temporal regulation of tip growth (i.e., the rate of growth) with the spatial control (i.e., the specification of sites of growth) (Li et al., 1999). This coupling role for Rop in growth control seems to be unique for tip-growing cells in plants. In budding yeast, Cdc42 is involved in the establishment of growth sites but not the growth *per se*, as loss-of-function cdc42 mutations cause isodiametric cell growth, whereas overexpression leads to random formation of bud sites (Johnson & Pringle, 1990). In fission yeast, loss-of-function cdc42 mutations cause growth arrest and loss of cell polarity, suggesting that Cdc42 has two distinct roles in fission yeast, i.e., growth control and polarity control (Miller & Johnson, 1994). As discussed above, yeast Cdc42 controls the establishment of cell polarity by regulating the polarization of cortical actin patches at the site of growth. The cortical actin is essential for the establishment of cell polarity in yeast. Does Rop have a similar role in cell polarity control through the organization of the actin cytoskeleton in tip-growing cells?

The answer to this question is complicated by the recent observations that little or no cortical actin is present in the extreme tip of both types of plant tip-growing cells–pollen tubes and root hairs (Lancelle & Hepler, 1992; Miller et al., 1996, 1999). However, it was reported that helical cortical actin cables were found in swollen tobacco pollen tubes induced by transient overexpression of constitutively active rop5At mutants (Kost et al., 1999). Similar actin patterns were found in transgenic *Arabidopsis* pollen tubes expressing constitutively active Rop1At (Lin & Yang, unpublished data). In wild-type tubes, cortical actin cables are oriented in parallel with the axis of tube elongation. Furthermore, a decrease or increase in actin bundling was observed in tubes expressing dominant-negative or constitutively-active rop5At mutants, respectively (Kost et al., 1999). These results appear to suggest a role for Rop in the organization of actin cables, which are thought to be involved in cytoplasmic streaming. However, these observations need to be interpreted with caution. No evidence is available to support a role for these actin cables in the control of cell polarity in pollen tubes. In addition, it is possible that the observed helical actin cables could be the secondary effect (resulting from the change in cell morphology) but not the primary effect of constitutively-active rop mutants. A kinetic analysis of changes in the actin organization versus the loss of cell polarity induced by constitutively-active rops may provide a better view of the involvement of Rops in the regulation of actin organization in pollen tubes.

Interestingly, it was shown that recombinant Rop is physically associated with a pollen tube PIP kinase activity (Kost et al., 1999), which synthesizes PIP_2. A potential role for PIPK as a Rop effector is supported by the tip-localization of GFP fused to a PIP_2-specific PH (plekstrin-homology) domain (Kost et al., 1999). In mammalian cells, Rac activates the phosphatidylinositol-4-phosphate 5–kinase (PIP5-kinase) leading to the acummulation of PIP_2 which promotes actin polymerization probably through profilin or N-WASP (Joneson et al., 1996; Lamarche et al., 1996; Van Aelst & D'Souza-Schorey, 1997; Hall, 1998; Mackay & Hall, 1998). Thus, Rop could potentially regulate pollen tube actin organization through PIP kinase activity. However, PIP_2 may also serve as a substrate for inositol triphosphate (IP_3) production, which may play a role in establishing the tip-focused Ca^{2+} gradient (Malhó et al., 1998). It is also possible that Rop and PIP_2 are involved in both the regulation of actin organization and the tip-localized calcium activity.

4.4 Rops in cell morphogenesis

Because pollen tube tip growth is considered to be a highly specialized form of cell growth and does not seem to involve tip-localized cortical actin, other cell types may serve a better system to study the potential role of Rop in the regulation of cell morphogenesis through actin organization and dynamics. In addition to the pollen-specific Rop1At, there are at least 10 other Rop genes that are differentially expressed in various *Arabidopsis* tissues (see above). Using a polyclonal antibody that reacts with different Rops, we showed that Rop proteins are present in all cells and tissues in *Arabidopsis* plants (Lin & Yang, unpublished data).

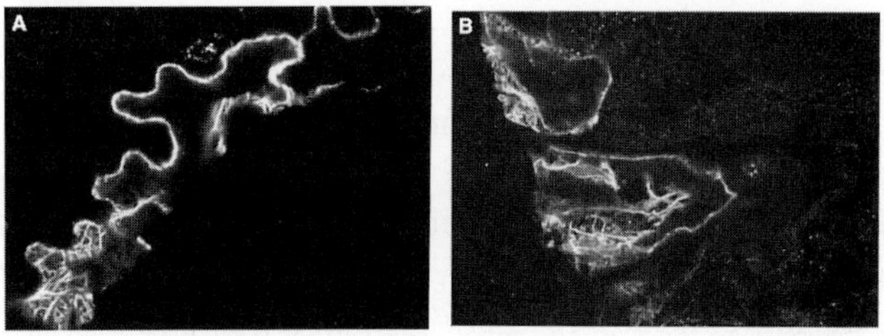

Figure 6. Alteration of actin filaments in *Arabidopsis* leaf epidermal cells expressing constitutively-active rop2At. Leaves were fixed and stained with maize actin antibody and fluorescein-labeled second antibody. Stained cells were visualized using a confocal microscope. Images shown are laser sections between the upper-side cortex and the mid-plane of epidermal pavement cells. In wild-type plants, pavement cells are puzzle-shaped and only contain cortical F-actin (**A**), whereas pavement cells in rop2 transgenic plants are irregularly-shaped and contain little cortical actin, but cytoplasmic actin cables are apparent (**B**).

To understand whether Rop has a general role in the control of cell morphogenesis and actin organization, we introduced into *Arabidopsis* plants constitutively-active rop2At mutants under the control of the constitutive 35S CaMV promoter (unpublished work). Indeed, leaf cells in the transgenic plants have altered cell morphology. Both pallisade and epidermal pavement cells become irregularly-shaped. This alteration is particularly dramatic in pavement cells. In wild-type plants, pavement cells have well defined lobes and sinuses forming intercalary puzzle-shaped cells. Recent studies suggest that the lobes and sinuses in cell walls are the result of a strictly controlled localized cell wall expansion associated with the cortical cytoskeleton (Panteris et al., 1994). In wild-type pavement cells, there is extensive cortical actin, but little cytoplasmic F-actin is found (Fig. 6). In contrast, in rop2 transgenic plants lobes and sinuses are completely lost, accompanied by a dramatic reduction in cortical F-actin. Instead, thick cytoplasmic actin cables are found in these pavement cells. These results strongly support a crucial role for Rop in the regulation of cell morphogenesis through the organization of cortical actin cytoskeleton. Analyses of rop transgenic plants involving live actin visualization should provide further insight into the relationship between Rop and actin organization.

4.5 Rop regulators and targets

To establish a role for Rop in the regulation of actin organization in plant cells, an important step is to identify components that link Rop to the actin cytoskeleton. As described above, PIP kinase may be a Rop effector, but whether it is indeed a direct target of Rop is unknown and its role in Rop signaling remains elusive. Homologs of other RHO GTPase targets have not appeared in any plant DNA databases. In an effort to identify Rop targets, we used the yeast two-hybrid method to isolate Rop-interacting proteins. Three classes of Rop-interacting proteins have been identified: RopGAPs, proteins related to receptor-like protein kinases (RLK), and a novel family of proteins containing a CRIB motif (Wu et al., 2000; Trotochaud et al., 1999; Wu & Yang, unpublished results). RopGAPs are a novel class of Rho GTPase-activating proteins, because they contain a CRIB motif to enhance the interaction of RopGAPs with Rop and to positively regulate the GTPase-stimulating activity of RopGAPs (Wu et al., 2000). Homologs of RopGAPs have also been found in other plant species (Borg et al., 1999). However, such CRIB-mediated GAPs have not been identified in non-plant systems. Overexpression of RopGAPs in pollen tubes inhibits pollen tube growth in a calcium-dependent manner, indicating that RopGAPs act as negative regulators in Rop signaling (Wu et al., 2000). We have also shown that a Rop-like protein associates with a complex containing activated CLAVATA1, an RLK involved in meristem maintenance (Trotochaud et al., 1999). RLKs may function as a direct upstream regulator to control the activity of Rops. This finding is exciting, because it represents the first example where a Rho GTPase is directly associated with a receptor kinase. Therefore, these studies have revealed surprising novel means for the regulation of Rho GTPase signaling and support the notion that the Rop subfamily represents a unique class of Rho GTPases.

It appears that plants have also evolved a novel class of CRIB-containing RHO targets. We have shown that the CRIB-containing novel Rop-interacting proteins specifically bind the active form of Rop, typical of RHO effector proteins, and do not have GAP activity (Wu & Yang, unpublished data). Thus these proteins are likely Rop effectors. However, their biological function and roles in Rop signaling remain to be determined. The absence of homologs for most of the animal and yeast RHO targets in the *Arabidopsis* database also raises an interesting question: Do plants have novel mechanisms for linking RHO GTPases to the actin cytoskeleton? The answer to this question should be forthcoming as the *Arabidopsis* genome is completely sequenced and additional Rop-interacting prorteins are identified through the yeast two-hybrid screen and other methodologies.

5. CONCLUDING REMARKS

In summary, the RHO family GTPases are conserved molecular switches that link extracellular signals to the dynamic organization and remodeling of the actin cytoskeleton in eukaryotes. Different eukaryotic phyla appear to have evolved distinct subfamilies of RHO GTPases that are involved in signaling to specific actin structures. Significant progress has been made towards understanding the role of the plant-specific Rop subfamily in the control of tip growth and cell morphogenesis. Evidence for the involvement of Rop in the regulation of actin organization is also emerging. To link Rop GTPases to the actin cytoskeleton, future work should focus on the identification and characterization of Rop targets. With Rop and ABPs as starting points and combined biochemical, genetic and cytological approaches, an understanding of how a signal modulates the organization and remodeling of the actin cytoskeleton in plants should be in sight.

REFERENCES

Aktories K, Braun U, Rosener S, Just I and Hall A (1989) The rho gene product expressed in *E. coli* is a substrate of botulinum ADP-ribosyltransferase C3. Biochem Biophys Res Comm 158: 209-213

Aspenström P (1999) Effectors for the Rho GTPases. Curr Opin Cell Biol 11: 95-102

Borg S, Pødenphant L, Jensen TJ and Poulsen C (1999) Plant cell growth and differentiation may involve GAP regulation of Rac activity. FEBS Lett 453: 341-345

Burbelo PD, Drechsel D and Hall A (1995) A conserved binding motif defines numerous candidate target proteins for both Cdc42 and Rac GTPases. J Biol Chem 270: 29071-29074

Carlier M-F (1998) Control of actin dynamics. Curr Opin Cell Biol 10: 45-51

Chant J and Stowers L (1995) GTPase cascades choreographing cellular behavior: movement, morphogenesis, and more. Cell 81: 1-4

Chardin P (1993). ras homologs: a comparison of primary structures. In: Lacal JC, McCormick F (eds) The Ras Superfamily of GTases. Boca Raton, FL: CRC Press, pp 203-229

Chardin P, Boquet P, Maduale P, Popoff MR, Rubin EJ and Gill DM (1989) The mammalian G protein rhoC is ADP-ribosylated by Clostridium botulinum exoenzyme C3 and affects actin microfilaments in vero cells. EMBO J 8: 1087-1092

Chong LD, Trynor-Kaplan A, Bokoch GM and Schwartz MA (1994) The small GTP-binding protein rho regulates a phosphatidylinositol 4-phosphate 5-kinase in mammalian cells. Cell 79: 507-513

Collins CC and Johnson DI (1997) An *Arabidopsis* expressed sequence tag cDNA encodes a Rac-like protein. Plant Physiol 113: 1463-1464

Delmer DP, Pear JR, Andrawis A and Stalker DM (1995) Genes for small GTP-binding proteins analogous to mammalian Rac are preferentially expressed in developing cotton fibers. Mol Gen Genet 248: 43-51

Drgonová J, Drgon T, Tanaka K, Kollar R, Chen GC, Ford RA, Chan CS, Takai Y and Cabib E (1996) Rho1p, a yeast protein at the interface between cell polarization and morphogenesis. Science 272: 277-279

Epp JA and Chant J (1997) An IQGAP-related protein controls actin-ring formation and cytokinesis in yeast. Curr Biol 7: 921-929

Evangelista M, Blundell K, Longtine MS, Chow CJ, Adames N, Pringle JR, Peter M and Boone C (1997) Bni1p, a yeast formin linking cdc42p and the actin cytoskeleton during polarized morphogenesis. Science 276: 118-122

Frost JA, Khokhlatchev A, Stippec S, White MA and Cobb MH (1998) Differential effects of PAK1-activating mutations reveal activity-dependent and -independent effects on cytoskeletal regulation. J Biol Chem 273: 28191-28198

Gallo G and Letourneau PC (1998) Axon guidance: GTPases help axons reach their targets. Curr Biol 8: R80-R82

Garrill A, Jackson SL, Lew RR and Heath IB (1993) Ion channel activity and tip growth: tip localized stretch-activated channels generate an essential Ca^{2+} gradient in the oomycete, *Saprolegnia ferax*. Eur J Cell Biol 60: 358-365

Gerald N, Dai J, Ting-Beall HP and De Lozanne A (1998) A role for *Dictyostelium* RacE in cortical tension and cleavage furrow progression. J Cell Biol 141: 483-492

Hall A (1990) The cellular function of small GTP-binding proteins. Science 249: 635-640

Hall A (1994) Small GTP-binding proteins and the regulation of the actin cytoskeleton. Annu Rev Cell Biol 10: 31-54

Hall A (1998) Rho GTPases and the actin cytoskeleton. Science 279: 509-514

Heyworth PG, Bohl BP, Bokoch GM and Curnutte JT (1994) Rac translocates independently of the neutrophil NADPH oxidase components $p47^{phox}$ and $p67^{phox}$. J Biol Chem 269: 30749-30752

Holly SP and Blumer KJ (1999) PAK-family kinases regulate cell and actin polarization throughout the cell cycle of *Saccharomyces cerevisiae*. J Cell Biol 147: 845-856

Imamura H, Tanaka K, Hihara T, Umikawa M, Kamei T, Takahashi K, Sasaki T and Takai Y (1997) Bni1p and Bnr1p: downstream targets of the Rho family small G-proteins which interact with profilin and regulate actin cytoskeleton in *Saccharomyces cerevisiae*. EMBO J 16: 2745-2755

Johnson DI and Pringle JR (1990) Molecular characterization of CDC42, a *Saccharomyces cerevisiae* gene involved in the development of cell polarity. J Cell Biol 111: 143-152.

Joneson T, McDonough M, Bar-Sagi D and Van Aelst L (1996) RAC regulation of actin polymerization and proliferation by a pathway distinct from Jun kinase Science 274: 1374-1376. [published erratum appears in Science 1997, 276:185]

Kawasaki T, Henmi K, Ono E, Hatakeyama S, Iwano M, Satoh H and Shimamoto K (1999) The small GTP-binding protein rac is a regulator of cell death in plants. Proc Natl Acad Sci USA 96: 10922-10926

Klahre U and Chua N-H (1999) The Arabidopsis actin-related protein2 AtARP2 promoter directs expression in xylem precursor cells and pollen. Plant Mol Biol 41: 65-73

Knaus UG, Heyworth PG, Evans T, Curnutte JT and Bokoch GM (1991) Regulation of phagocyte oxygen radical production by the GTP-binding protein Rac 2. Science 254: 1512-1515

Kohno H, Tanaka K, Mino A, Umikawa M, Imamura H, Fujiwara T, Fujita Y, Hotta K, Qadota H, Watanabe T, Ohya Y and Takai Y (1996) Bni1p implicated in cytoskeletal control is a putative target of Rho1p small GTP binding protein in Saccharomyces cerevisiae. EMBO J 15: 6060-6068

Kost B, Lemichez E, Spielhofer P, Hong Y, Tolias K, Carpenter C and Chua N-H (1999) Rac homologues and compartmentalized phosphatidylinositol 4,5-bisphosphate act in a common pathway to regulate polar pollen tube growth. J Cell Biol 145: 317-330

Kozma R, Ahmed S, Best A and Lim L (1995) The Ras-related protein Cdc42Hs and bradykinin promote formation of peripheral actin microspikes and filopodia in Swiss 3T3 fibroblasts. Mol Cell Biol 15: 1942-1952

Lamarche N, Tapon N, Stowers L, Burbelo PD, Aspenström P, Bridges T, Chant J and Hall A (1996) Rac and Cdc42 induce actin polymerization and G1 cell cycle progression independently of p65PAK and the JNK/SAPK MAP kinase cascade. Cell 87: 519-529

Lancelle SA and Hepler PK (1992) Ultrastructure of freeze-substituted pollen tubes of *Lilium longiflorum*. Protoplasma 167: 215-230

Larochelle DA, Vithalani KK and De Lozanne A (1997) Role of *Dictyostelium* racE in cytokinesis: mutational analysis and localization studies by use of green fluorescent protein. Mol Biol Cell 8: 935-944

Li H, Lin Y, Heath RM, Zhu MX and Yang Z (1999) Control of pollen tube tip growth by a Rop GTPase -dependent pathway that leads to tip-localized calcium influx. Plant Cell 11: 1731-1742

Li H, Wu G, Ware D, Davis KR and Yang Z (1998) Arabidopsis Rho-related GTPases: Differential gene expression in pollen and polar localization in fission yeast. Plant Physiol 118: 407-417

Lin Y, Wang Y, Zhu J and Yang Z (1996) Localization of a rho GTPase implies a role in tip growth and movement of the generative cell in pollen tubes. Plant Cell 8: 293-303

Lin Y and Yang Z (1997) Inhibition of pollen tube elongation by microinjected anti-Rop1Ps antibodies suggests a crucial role for Rho-type GTPases in the contol of tip growth. Plant Cell 9: 1647-1659

Lynch ED, Lee MK, Morrow JE, Welcsh PL, Leon PE and King MC (1997) Nonsyndromic deafness DFNA1 associated with mutation of a human homolog of the *Drosophila* gene *diaphanous.* Science 278: 1315-1318

Luo L, Liao YJ, Jan LY and Jan YN (1994) Distinct morphogenetic functions of similar small GTPases: *Drosophila* Drac1 is involved in axonal outgrowth and myoblast fusion. Genes Develop 8: 1787-1802

Ma L, Rohatgi R and Kirschner MW (1998) The Arp2/3 complex mediates actin polymerization induced by the small GTP-binding protein Cdc42. Proc Natl Acad Sci USA 95: 15362-15367

Machesky LM and Gould KL (1999) The Arp2/3 complex: a multi-functional actin organizer. Curr Opin Cell Biol 11: 117-121

Mackay DJG and Hall A (1998) Rho GTPases. J Biol Chem 273: 20685-20688

Maekawa M, Ishizaki T, Boku S, Watanabe N, Fujita A, Iwamatsu A, Obinata T, Ohashi K, Mizuno K and Narumiya S (1999) Signaling from Rho to the actin cytoskeleton through protein kinases ROCK and LIM-kinase. Science 285: 895-898

Malhó R, Moutinho A, van der Luit A and Trewavas AJ (1998) Spatial characteristics of calcium signaling: the calcium wave as a basic unit in plant cell calcium signaling. Phil Trans R Soc Lond B 353: 1463-1473

Miki H, Sasaki T, Takai Y and Takenawa T (1998) Induction of filopodium formation by a WASP-related actin- depolymerizing protein N-WASP. Nature 391: 93-96

Miller DD, de Ruijter NCA, Bisseling T and Emons AMC (1999) The role of actin in root hair morphogenesis: studies with lipochito-oligosaccharide as a growth stimulator and cytochalasin as an actin perturbing drug. Plant J 17: 141-154

Miller DD, Lancelle SA and Hepler PK (1996) Actin microfilaments do not form a dense meshwork in *Lilium longiflorum* pollen tube tips. Protoplasma 195: 123-132

Miller PJ and Johnson DI (1994) Cdc42p GTPase is involved in controlling polarized cell growth in *Schizosaccharomyces pombe*. Mol Cell Biol 14: 1075-1083

Murphy AM and Montell DJ (1996) Cell type-specific roles for Cdc42, Rac, RhoL in *Drosophila* oogenesis. J Cell Biol 133: 617-630

Nobes CD and Hall A (1995) Rho, Rac, and Cdc42 GTPases regulate the assembly of multimolecular focal complexes associated with actin stress fiber, lamellipodia, and filopodia. Cell 81: 53-62

Panteris E, Apostolakos P and Galatis B (1994) Sinuous ordinary epidermal cells: behind several patterns of waviness, a common morphogenetic mechanism. New Phytol 127: 771-780

Paterson HF, Self AJ, Garrett MD, Just I, Aktories K and Hall A (1990) Microinjection of recombinant p21rho induces rapid changes in cell morphology. J Cell Biol 111: 1001-1007

Pierson ES, Miller DD, Callaham DA, Shipley AM, Rivers BA, Cresti M and Hepler PK (1994) Pollen tube growth is coupled to the extracellular calcium ion flux and the intracellular calcium gradient: effect of BAPTA-type buffers and hypertonic media. Plant Cell 6: 1815-1828

Potikha TS, Collins CC, Johnson DI, Delmer DP and Levine A (1999) The involvement of hydrogen peroxide in the differentiation of secondary walls in cotton fibers. Plant Physiol 119: 849-858

Ridley AJ (1995) Rho-related proteins: actin cytoskeleton and cell cycle. Curr Opin Genet Dev 5: 24-30

Ridley AJ and Hall A (1992) The small GTP-binding protein rho regulates the assembly of focal adhesions and actin stress fibers in response to growth factors. Cell 70: 389-399

Ridley AJ, Paterson HF, Johnston CL, Diekmann D and Hall A (1992) The small GTP-binding protein rac regulates growth factor-induced membrane ruffling. Cell 70: 401-410

Rohatgi R, Ma L, Miki H, Lopez M, Kirchhausen T, Takenawa T and Kirschner MW (1999) The interaction between N-WASP and the Arp2/3 complex links Cdc42-dependent signals to actin assembly. Cell 97: 221-231

Staiger CJ, Gibbon BC, Kovar DR and Zonia LE (1997) Profilin and actin depolymerizing factor: modulators of actin organization in plants. Trends Plant Sci 2: 275-281

Tanaka K and Takai Y (1998) Control of reorganization of the actin cytoskeleton by Rho family small GTP-binding proteins in yeast. Curr Opin Cell Biol 10: 112-116

Tapon N and Hall A (1997) Rho, Rac and Cdc42 GTPases regulate the organization of the actin cytoskeleton. Curr Opin Cell Biol 9: 86-92

Toker A (1998) The synthesis and cellular roles of phosphatidylinositol 4,5-bisphosphate. Curr Opin Cell Biol 10: 254-261

Trotochaud AE, Hao T, Wu G, Yang Z and Clark SE (1999) The CLAVATA1 receptor-like kinase requires CLAVATA3 for its assembly into a signaling complex that includes KAPP and a Rho-related protein. Plant Cell 11: 393-406

Van Aelst L and D'Souza-Schorey C (1997) Rho GTPases and signaling networks. Genes Dev 11: 2295-322

Watanabe N, Madaule P, Reid T, Ishizaki T, Watanabe G, Kakizuka A, Saito Y, Nakao K, Jockusch BM and Narumiya S (1997) p140mDia, a mammalian homolog of *Drosophila diaphanous*, is a target protein for Rho small GTPase and is a ligand for profilin. EMBO J 16: 3044-3056

Winge P, Brembu T and Bones AM (1997) Cloning and characterization of rac-like cDNAs from *Arabidopsis thaliana*. Plant Mol Biol 35: 483-495

Wu G, Li H and Yang Z (2000) A novel family of *Arabidopsis* Rho GTPase-activating proteins requires a Cdc42/Rac-interactive binding motif for GTPase activation and functions as negative regulators of Rop GTPase signaling. Submitted

Wymer CL, Bibikova TN and Gilroy S (1997) Cytoplasmic free calcium distributions during the development of root hairs of *Arabidopsis thaliana*. Plant J 203: 427-439

Yang Z and Watson JC (1993) Molecular cloning and characterization of rho, a ras-related small GTP-binding protein from the garden pea. Proc Natl Acad Sci USA 90: 8732-8736

Ziman M, Preuss D, Mulholland J, O'Brien JM, Botstein D and Johnson DI (1993) Subcellular localization of Cdc42p, a *Saccharomyces cerevisiae* GTP-binding protein involved in the control of cell polarity. Mol Biol Cell 4: 1307-1316

Zipkin ID, Kindt RM and Kenyon CJ (1997) Role of a new Rho family member in cell migration and axon guidance in C. elegans. Cell 90: 883-894

Chapter 18

ACTIN IN POLLEN AND POLLEN TUBES

Luis Vidali and Peter K Hepler

Biology Department, University of Massachusetts, Amherst, MA 01003 USA

Key words: actin dynamics, ADF, alkaline band, ARP2/3 complex, calcium gradient, cofilin, cytochalasin, cytoplasmic streaming, gametophyte, gelsolin, latrunculin, myosin, pollen tube, profilin, tip growth, villin

Abstract: Flowering plants rely on the highly polarized process of pollen tube growth for delivery of the sperm cells and thus for sexual reproduction. Although it has long been known that the actin cytoskeleton is necessary for pollen grain germination and tube growth, the precise structure of the actin arrays and their regulation by actin-associated proteins is only starting to be uncovered. Long, axially-oriented bundles of actin are abundant in the shank of pollen tubes but absent in their apices. These bundles function as tracks for the myosin-dependent cytoplasmic streaming that transports organelles and secretory vesicles in a bi-directional, reverse-fountain pattern. Exocytosis of the vesicles, which is required for cell growth, is restricted to the apex of the pollen tube. To understand these growth-dependent motile processes, actin as well as many actin-binding proteins, are being isolated from pollen and characterized both *in vitro* and *in vivo*. Spatial modulation of the actin cytoskeleton may be governed in part by calcium, protons and small GTPases, which are asymmetrically distributed. Understanding how the actin cytoskeleton interacts with these different factors helps us explain the molecular basis of polarized pollen tube growth.

1. INTRODUCTION

Pollen tube growth, which transports the generative cell to the ovule for sexual reproduction in higher plants, is a complex phenomenon that involves the concerted activity of many cellular components, and among these the actin cytoskeleton is particularly prominent. It has long been known that pollen tube germination and elongation are highly sensitive to the actin-perturbing action of cytochalasins (Franke et al., 1972; Mascarenhas &

C.J. Staiger et al. (eds.), Actin: A Dynamic Framework for Multiple Plant Cell Functions, 323–345.

Lafountain, 1972), and that F-actin is present in pollen grains and tubes (Condeelis, 1974; Franke et al., 1972; Lancelle & Hepler, 1989; Perdue & Parthasarathy, 1985). It is also well established that the vigorous cytoplasmic streaming present in the pollen tube (Iwanami, 1956) is dependent on myosin and actin (Heslop-Harrison & Heslop-Harrison, 1989c; Kohno & Shimmen, 1988; Mascarenhas & Lafountain, 1972; Tang et al., 1989), and that this streaming is responsible for the transport of secretory vesicles, necessary for growth (Picton & Steer, 1981). Over the last years, many excellent reviews have covered aspects of the actin cytoskeleton in pollen grains and pollen tubes (Cai et al., 1997; Derksen et al., 1995b; Heslop-Harrison, 1987; Heslop-Harrison & Heslop-Harrison, 1989a; Mascarenhas, 1993; Pierson & Cresti, 1992; Staiger et al., 1997; Staiger & Schliwa, 1987; Steer & Steer, 1989; Taylor & Hepler, 1997). In the following chapter, we will review the recent cellular, molecular, and biochemical advances that are generating the necessary information to understand the mechanisms for actin participation in pollen tube growth. We will firstly outline the current knowledge about the structures and molecules that constitute the actin cytoskeleton in pollen grains and tubes, and secondly evaluate the models that have been developed to explain the participation of actin in pollen tube growth.

2. ACTIN STRUCTURES

The structure of actin in the pollen grain, at the light and electron microscope level, has many forms, notably fusiform and spiculate aggregates (Heslop-Harrison et al., 1986; Heslop-Harrison & Heslop-Harrison, 1992b; Tiwari & Polito, 1988b) that can appear almost crystalline (Cresti et al., 1986). These structures gradually disperse into a filament network of finer bundles (Heslop-Harrison & Heslop-Harrison, 1992a; Tiwari & Polito, 1988b), but it is not known if this process involves the disassembly and re-assembly of new filaments. Moreover, the function and composition of these structures are still not known, but they have been postulated to be storage forms of F-actin (Cresti et al., 1986; Heslop-Harrison et al., 1986). Interestingly, they can be reversibly formed by cycles of hydration-dehydration (Heslop-Harrison & Heslop-Harrison, 1992a), which suggests that they participate in stabilizing the actin for the dehydrated dormant state of the pollen grain. The fact that they can be stained with phalloidin (Heslop-Harrison & Heslop-Harrison, 1992a, 1992b; Heslop-Harrison et al., 1986; Tiwari & Polito, 1988b), indicates that they are composed of F-actin; it is likely that actin-binding proteins will also stabilize and regulate these structures.

When preparing for germination, the fine meshwork of F-actin is organized into parallel bundles oriented towards the pollen aperture, anticipating the site of pollen tube emergence (Heslop-Harrison et al., 1986; Heslop-Harrison & Heslop-Harrison, 1992b; Tiwari and Polito, 1988b). How this polarity is generated is not know. Probably spatial clues that polarize the actin cytoskeleton are present at the time of pollen grain formation, but they could also develop spontaneously, since protoplasts or subprotoplasts that have a randomly-oriented microfilament system can polarize it to generate a pollen tube-like emergence (Rutten & Derksen, 1990; Tanaka & Wakabayashi, 1992).

Once the pollen tube emerges, a constant F-actin structure is maintained that is necessary for pollen tube elongation. In both gymnosperms and angiosperms the main organization includes a system of parallel bundles of F-actin, which are longitudinally oriented along the shank of the pollen tube (de Win et al., 1996; Franke et al., 1972; Lancelle et al., 1987; Lancelle & Hepler, 1992; Lazzaro, 1996; Perdue & Parthasarathy, 1985; Tiwari & Polito, 1988a). Actin filaments can also be observed associated with cortical microtubles (Franke et al., 1972; Lancelle & Hepler, 1991; Pierson et al., 1989), in which a single microfilament aligns parallel with a microtubule. Despite its early observation, the function and nature of this microfilament/microtubule association remains unknown (see also Collings & Allen, this volume). Unlike root hairs, where the organization of the actin cytoskeleton is dependent on microtubles (Tominaga et al., 1997), in pollen tubes it seems to be independent of these elements (Heslop-Harrison & Heslop-Harrison, 1991b).

The precise structure of F-actin at the tip of the pollen tube has been a matter of debate, due to the difficulties in preserving the detailed morphology of the 'clear zone'. This area has a lower refractivity due to the accumulation of small organelles. The clear zone can be further subdivided into the apical region, composed of secretory vesicles and tubular ER, and the subapical region which in addition contains Golgi bodies and mitochondria (Heslop-Harrison & Heslop-Harrison, 1990; Lancelle et al., 1987; Lancelle & Hepler, 1992; Tiwari & Polito, 1988a). The larger organelles, which include the vacuole, amyloplasts and lipid droplets, are excluded from the clear zone. It is well known that chemical fixation drastically alters the fine structure of pollen tube tip (Doris & Steer, 1996; Heslop-Harrison & Heslop-Harrison, 1991a). However, alternatives to conventional chemical fixation that included dimethyl-sulfoxide (DMSO) permeabilization (Pierson, 1988), microwave stabilization (Heslop-Harrison & Heslop-Harrison, 1991a), or m-maleimidobenzoyl N-hydroxysuccinimide ester (MBS) (Doris & Steer, 1996), a bi-functional cross-linker known to stabilize actin filaments in plant cells (Sonobe & Shibaoka, 1989), have been explored with limited success. Since all these methodologies are prone to artifacts and variations, it is important to evaluate carefully the observations, since they could lead to erroneous conclusions. For example, using DMSO permeabilization and fluorescent-phalloidin, a dense meshwork of F-actin was reported in the tip of the pollen tube (Pierson, 1988), however, when the tubes where stabilized by microwave irradiation only a loose meshwork was visible, which extended to the very tip (Heslop-Harrison & Heslop-Harrison, 1991a).

Other researchers have challenged the presence of a dense meshwork, and even the existence of F-actin at the apex, with more reliable physical methods of fixation. With the use of rapid freezing and freeze substitution, a loose meshwork was observed from the serial sections of pollen tubes (Lancelle et al., 1987; Tiwari & Polito, 1988a). Also, fluorescent-phalloidin injections and immuno-EM, have shown that F-actin is absent from the apical region of the lily pollen tube (Miller et al., 1996). In this region, no F-actin was detectable, while fine actin bundles were detected in the subapical region. Recently, the use of a GFP-mouse talin fusion protein has also shown that in tobacco pollen tubes there is no F-actin at the extreme apex (Kost et al., 1998). Interestingly, a 'collar' of F-actin was detected at the subapical region. The streaming patterns at the apical and subapical region further strengthen the observations from rapidly frozen and living cells. Organelles show bi-directional and mainly axially-oriented translocations in the shank of the pollen tube, while they exhibit more random motions at the apical and subapical regions (de Win et al., 1999; Heslop-Harrison & Heslop-Harrison, 1991a).

Although generally organized in a manner similar to that in *in vitro* grown pollen tubes, the F-actin of pollen tubes growing *in vivo* in the style has been observed, after chemical fixation, to form star-like structures that resemble focal adhesion sites (Jauh & Lord, 1995; Pierson et al., 1986), the origin and composition of these structures is still unknown. Actin has been reported to be associated with the vegetative nucleus and the generative cell (Heslop-Harrison & Heslop-Harrison, 1989b; Tiwari & Polito, 1988b), and contribute to their shape and movement (Heslop-Harrison & Heslop-Harrison, 1991b). However, the recent studies by Kost and coworkers (Kost et al., 1998) with live tobacco pollen tubes have questioned the existence of F-actin associated with these organelles.

3. ACTIN AND ITS ASSOCIATED PROTEINS

Pollen is an excellent source from which actin and its associated proteins can be biochemically isolated. Large quantities of pollen can be harvested from anthers of some species (Liu & Yen, 1992; Vidali & Hepler, 1997; Yokota & Shimmen, 1994). Pollen proteins are very concentrated, and actin and its binding proteins are some of the most abundant proteins (Gibbon et al., 1999; Liu & Yen, 1992; Ren et al., 1997; Vidali & Hepler, 1997; Yokota & Shimmen, 1994; Yokota et al., 1998). Also, pollen can be stored frozen for long periods, without loss of protein activity, due to stabilization by a natural dehydration process (Taylor & Hepler, 1997).

3.1 Actin

Although the existence of actin in pollen was firmly established in early studies (Condeelis, 1974; Franke et al., 1972; Mascarenhas & Lafountain, 1972), its purification and initial characterization came several years later. A protocol for actin purification from pollen was reported by Yen and collaborators (Liu & Yen, 1992), who utilized acetone extraction of maize pollen grains followed by precipitation and chromatography. The isolated actin could polymerize and depolymerize (Yen et al., 1995), the polymerized filaments could be decorated with animal myosin sub-fragments, and activate animal myosin Mg-ATPase (Liu & Yen, 1995). An alternative approach to isolate maize pollen actin was developed by Staiger and coworkers (Ren et al., 1997), in which actin was isolated by adding mammalian profilin to a pollen crude extract and binding the profilin-actin complex to a poly-L-proline (PLP) column. Actin was then eluted with 1 M KCl, leaving the profilin attached to the column. This protocol produced a higher yield, and purity, and was much faster than the previously reported method. The actin has a critical concentration of 0.4–0.6 μM, kinetics of polymerization very similar to those of muscle actin, binds to animal myosin, and incorporates into the endogenous actin arrays in living plant cells (Ren et al., 1997). DNase I can also be used to isolate actin from pollen (Andersland et al., 1992; Vidali & Hepler, 1996). We isolated actin from lily pollen with a yield (1 mg/g of dry pollen) comparable to that isolated by profilin affinity (Vidali & Hepler, 1996). Unfortunately, the purified protein polymerizes poorly, probably due to partial denaturation during the harsh treatments used for its elution from DNase I–Sepharose. Another possibility is that the final preparation was contaminated with trace amounts of an actin-depolymerizing protein.

The concentration of actin has been determined only in pollen grains and tubes from lily and maize. Although in both cases polyclonal antibodies against pollen actin were used, important differences are observed. We will focus on the percentage of the total protein, because the values can be compared without the major assumptions that are necessary to calculate molarity. In lily, the total actin pool increases from 1.4 to 2.3% of the total protein after 2 h of germination (Vidali & Hepler, 1997), while in maize the concentration of actin is 5% of the total protein and does not change during pollen germination (Gibbon et al., 1999). To clarify if the reported differences are due to the methodology used or if they are of physiological relevance, it becomes necessary to know what the concentration and patterns of expression are in other plant species. Also, the concentration of F-actin has been estimated in maize pollen grains and tubes, with only 10% of the total found in the filamentous form (Gibbon et al., 1999). These values are important since they represent the only estimates to date, but they should be taken as a minimum and not absolute values, since they were obtained using fluorescent-phalloidin, which does not bind F-actin in the presence of ADF/cofilin (Ressad et al., 1998), a known pollen constituent (Kim et al., 1993; Lopez et al., 1996).

Several actin genes have been identified in plants, but only a few have been shown to be specifically expressed in pollen. In *Arabidopsis* there are eight expressed actin genes, which can be divided in two classes: reproductive and vegetative (McDowell et al., 1996). The reproductive class can be further subdivided into three subclasses, all of which are highly expressed in pollen (McDowell et al., 1996). The evolutionary selection of multiple actin genes has been attributed to either their expression patterns or true functional differences (Meagher et al., 1999; see also Meagher et al., this volume). Since actin expression is developmentally switched from vegetative to reproductive form during pollen maturation, it is likely that functional rather than regulatory reasons have generated the two classes of actin (Kandasamy et al., 1999). Monoclonal antibodies (Kandasamy et al., 1999) and deletion mutants (Gilliland et al., 1998) are being generated to further understand the existence of pollen specific isoforms.

3.2 Myosin

Myosin is the motor protein responsible for cytoplasmic streaming in the pollen tube, as in virtually all other plant cells (see Yokota & Shimmen, this volume). Two classes of myosins exist in plants, comprised of 130 kDa (class VIII) and a 170 kDa (class XI) heavy chains, respectively (Kinkema et al., 1994; Mooseker & Cheney, 1995). The biochemical properties of lily pollen myosin have been extensively studied. Using the *Nitella* perfusion assay, Kohno and coworkers demonstrated that an actin-dependent motor associates with the pollen tube organelles, and it could be inhibited by low calcium concentrations (Kohno & Shimmen, 1988). By using F-actin co-precipitation and *in vitro* motility assays, Yokota and coworkers isolated the intact motor, which is composed by a 170-kDa heavy chain dimer (Yokota & Shimmen, 1994). They showed that the motor has F-actin activated ATPase activity (Yokota & Shimmen, 1994), and calmodulin light chains (Yokota et al., 1999), which confer reversible inhibition by Ca^{2+} at concentrations between 1 and 10 μM. This myosin translocates animal and plant actin at identical rates (Igarashi et al., 1999). A cDNA corresponding to this myosin has been isolated and identified as a class XI, closely related to the class V myosins (Yokota, Vidali, Cheung, Shimmen and Hepler, unpublished data). A polyclonal antibody was generated against this protein, which by immunofluorescence, localizes it to organelles and the tip of some cells in whole mounts of chemically-fixed pollen tubes of both lily and tobacco (Yokota et al., 1995). Using heterologous antibodies, others have also identified myosin at the tip of the cells, on various punctuate structures, and on the surface of the nucleus and the generative cell (Heslop-Harrison & Heslop-Harrison, 1989c; Miller et al., 1995; Tang et al., 1989; Tirlapur et al., 1995). Interestingly, using antibodies against myosin I, II and V, distinct myosin populations have been localized to different organelles (Miller et al., 1995). Only in one report has the localization of myosin been studied at the ultrastructural level, where it was detected on vesicle-like structures by a combination of chemical fixation and heterologous antibodies (Tirlapur et al., 1995). To uncover the precise and unequivocal distribution of myosin at the ultrastructural level, it will be necessary to use rapid-freeze fixation and freeze substitution followed by immunocytology with homologous antibodies.

3.3 Profilin

Profilins are small G-actin-, phosphoinositide- and poly-L-proline-binding proteins (Machesky & Pollard, 1993; Staiger et al., 1997). They have been shown to regulate actin dynamics in conjunction with other actin-binding proteins (Theriot & Mitchison, 1993). Depending on the conditions they can inhibit filament nucleation (Carlsson et al., 1977), promote polymerization (Pantaloni & Carlier, 1993), and enhance nucleotide exchange on the actin monomer (Mockrin & Korn, 1980). In pollen, profilin was identified originally as an allergen (Valenta et al., 1991), and has been studied extensively, partly by immunologists, but also by cell biologists who have focused on its actin regulatory role (see also Gibbon & Staiger, this volume). Profilin, as actin, has isoforms that are reproductive organ specific (Huang et al., 1996). The three-dimensional structure of two profilins from plants has been solved (Fedorov et al., 1997; Thorn et al., 1997), and despite only 30% identity with other profilins, they have a highly conserved overall fold in which the N- and C-termini, forming a PLP-binding domain, are particularly well conserved. This domain contains a patch of aromatic residues conserved throughout evolution, and probably reflects the importance of this interaction for profilin activity. Pollen profilins rescue yeast (Christensen et al., 1996) and *Dictyostelium* (Karakesisoglou et al., 1996) null mutants and also function in animal cells (Rothkegel et al., 1996). They bind to actin (Gibbon et al., 1997; Giehl et al., 1994), PIP_2 (Drøbak et al., 1994) and PLP (Valenta et al., 1991), which are the classical ligands of profilin. However, plant profilins do not increase the nucleotide exchange rate of rabbit (Perelroizen et al., 1996) or yeast actin (Eads et al., 1998), a function postulated to be important for their animal counterparts (Goldschmidt-Clermont et al., 1991). Pollen from maize (Staiger et al., 1993) and lily (Vidali & Hepler, 1997) has been shown to contain multiple profilin isoforms. The differences in affinity for PLP among the maize isoforms have been correlated with their capacity to disrupt the cell architecture of *Tradescantia* stamen hair cells (Gibbon et al., 1998). Also, by an unknown mechanism, pollen profilins have been shown to affect the phosphorylating activity of poppy pollen extracts (Clarke et al., 1998). Unlike some vegetative forms, which are tyrosine-phosphorylated (Guillén et al., 1999), pollen profilin does not seem to be phosphorylated (Clarke et al., 1998). Profilins can be easily purified from pollen in large quantities by PLP-affinity chromatography; several types have been over-expressed in bacteria and their characteristics analyzed *in vitro* and *in vivo* (Gibbon et al., 1997, 1998; Perelroizen et al., 1996; Vrtala et al., 1996).

Most reports agree that profilin localizes homogeneously in the pollen grain and tube cytoplasm (Grote et al., 1995; Hess et al., 1995; Mittermann et al., 1995; Vidali & Hepler, 1997); some of these results have been obtained using immunolocalization after rapid freezing and freeze substitution (Hess et al., 1995; Vidali & Hepler, 1997) or fluorescent analog cytochemistry of live cells (Vidali & Hepler, 1997). However, only in one report was profilin localized to the cortical region of pollen grains (von Witsch et al., 1998). The validity of this report should be questioned since the results were obtained using chemical fixation, which is known to generate redistribution of soluble proteins (Melan & Sluder, 1992; Vos & Hepler, 1998). Furthermore, when observed with the EM the signal is closely, if not completely, associated with the pollen cell wall (von Witsch et al., 1998).

As in the case of actin, profilin concentration has only been estimated in the pollen from lily and maize, but for profilin, the differences are greater than for actin. In lily and maize pollen, profilin concentration does not change significantly after germination, but the estimated concentration is five times higher in maize (Gibbon et al., 1999) than in lily, where profilin is 0.5% of the total protein (Vidali & Hepler, 1997). Interestingly, both reports agree that in pollen grains the profilin concentration is equimolar with that of actin, but it is only in pollen tubes from lily that the concentration of actin increases. A possible explanation for this difference is that actin is regulated differently in the two species. Nevertheless, as in the case of actin, measurement of the concentration of profilin in pollen from other species is necessary. In any case, due to the high abundance of profilin it is very likely that in the pollen tube, nucleation and polymerization of actin will be modulated by this monomer-binding protein.

3.4 ADF/cofilin

ADF/cofilin has been shown to play a major role in the regulation of actin dynamics (Carlier et al., 1999; Kovar & Staiger, this volume). These proteins enhance the pointed-end depolymerization of actin filaments, which is the rate-limiting step for treadmilling (Carlier et al., 1997). This generates an increase of 25-fold in the turnover rate of actin filaments. ADF/cofilin has been cloned from the pollen of maize (Lopez et al., 1996), and lily (Kim et al., 1993), where it is expressed as a pollen-specific protein; it has also been cloned from *Arabidopsis* (Carlier et al., 1997). Detailed biochemical studies, performed with the maize recombinant ADF, indicate that it has F-actin and G-actin binding activity (Lopez et al., 1996). It also depolymerizes actin *in vitro* (Lopez et al., 1996) and, when injected into living stamen hair cells, it rearranges the structure of the actin cytoskeleton (Hussey et al., 1998). Its activity *in vitro* can be regulated both by phosphorylation and pH (Smertenko et al., 1998). Lily ADF has been localized at the EM level, where it was found to be mainly cytosolic in the pollen tubes, and associated with amorphous structures in the pollen grains (Chung et al., 1995). However, the lack of ultrastructural preservation, due to the use of chemical fixation, makes their interpretation difficult. More detailed work with rapidly-frozen samples, where actin is well preserved, should give a better understanding of ADF sub-cellular location.

3.5 Villin

Villins are F-actin bundling proteins composed of a gelsolin-homology domain and a second F-actin binding region called the headpiece, which allows bundling to take place (Glenney et al., 1981). Like gelsolin, they can nucleate, sever and cap actin filaments in a calcium-dependent fashion; they are also inhibited by PIP_2 (Andre et al., 1988; Glenney et al., 1981; Yin et al., 1990). Villin forms parallel bundles in which filaments have the same orientation (Coluccio & Bretscher, 1989). The recently isolated bundling protein from lily pollen, P-135-ABP (Yokota et al., 1998; see also Yokota & Shimmen, this volume) corresponds to a plant member of the villin/gelsolin family of proteins (Vidali et al., 1999). This villin localizes to actin filaments

(Vidali et al., 1999; Yokota et al., 1998), is comprised of at least 5 isoforms, and related peptides are expressed in other tissues of lily (Vidali et al., 1999). Recent evidence suggests that another bundling protein, P-115-ABP isolated from lily pollen (Nakayasu et al., 1998), is another form of villin (Yokota, Vidali, Cheung, Shimmen and Hepler, unpublished data). The filaments in the bundles formed by P-135-ABP *in vitro* have the same polarity (Yokota & Shimmen, 1999). Although its bundling activity has not been found to be Ca^{2+} regulated *in vitro* and no severing activity of the purified protein has been detected (Yokota et al., 1998), a Ca^{2+}-dependent severing activity has been inferred in lily pollen tubes (Kohno & Shimmen, 1987). More experimentation will be necessary to clarify if this villin is a Ca^{2+}-regulated severing protein *in vivo*. Sequences from 4 different villins have been identified in *Arabidopsis* (Klahre et al., 2000). As in the example of lily villin, the *Arabidopsis* villins share high homology with other members of the villin/gelsolin family of proteins. Characteristic of plant villins is a longer and highly variable region that links the gelsolin-like domain with the headpiece (Vidali et al., 1999). The *Arabidopsis* villin 2 and 3 share the highest sequence homology with the lily pollen villin, with villin 1 having the most distant relation (Vidali et al., 1999).

3.6 ARP2/3 complex

The ARP2/3 complex has been identified recently as a central regulator of the actin cytoskeleton. It is composed of 7 proteins, two of which are actin-related (Machesky et al., 1994; Machesky & Gould, 1999). This complex has been implicated in the movement of actin-containing cortical patches in yeast (Winter et al., 1997), propulsion of intracellular pathogens (Loisel et al., 1999; May et al., 1999) and leading edge motility of some animal cell types (Machesky et al., 1997; Svitkina & Borisy, 1999). One of the actin-related members of the complex has been identified in *Arabidopsis* and has been shown to be highly expressed in pollen (Klahre & Chua, 1999). Other members of the complex have also been identified as part of the *Arabidopsis*-genome sequencing project (Klahre & Chua, 1999). Since these complexes are essential for the regulation of actin in animals and yeast, it is very likely they will be important for pollen tube actin regulation as well.

3.7 Other proteins

Fimbrin is another actin-bundling protein; plant fimbrin homologues have been described in *Arabidopsis* and these are expressed in all organs analyzed (McCurdy & Kim, 1998; see also McCurdy & Staiger, this volume). It is likely that fimbrin will be expressed in pollen where, by analogy with other systems, it could play a role in the formation and regulation of actin bundles in combination with other proteins (Holtzman et al., 1993; Shibayama et al., 1987).

A caldesmon-related protein has been identified in pollen by its cross-reactivity with three different polyclonal and one monoclonal antibody (Krauze et al., 1998). The cross-reacting polypeptide binds F-actin and calmodulin, and shows a molecular weight intermediate between the muscle and non-muscle proteins from animals. The confirmation of this protein as true caldesmon awaits sequence information.

The picture emerging of the actin-associated proteins suggests that most of the groups that regulate the actin cytoskeleton in other systems will be present in pollen, but due to the remarkable characteristics and specific function of the growing pollen tube, it is likely they will be utilized and regulated in unique ways.

4. ACTIN CYTOSKELETON REGULATION

4.1 Calcium

Pollen tubes possess a steep, tip-focused Ca^{2+} gradient. Due to its multiple roles as a second messenger and by analogy with other systems, Ca^{2+} has long been suspected to regulate the actin cytoskeleton in the pollen tube (Kohno & Shimmen, 1987; Picton & Steer, 1982); this idea acquires added pertinence with the discovery of villin, a putative calcium-sensitive protein present in lily pollen tubes (Vidali et al., 1999). Myosin can also be regulated by calcium via its calmodulin light chains (Yokota et al., 1999). A calcium-dependent, calmodulin-independent kinase has been identified in pollen and was shown to localize with the actin cytoskeleton (Putnam-Evans et al., 1989). Furthermore, this protein has been suggested as the kinase responsible of ADF/cofilin phosphorylation (Smertenko et al., 1998). The finding that pollen tube growth rates and the $[Ca^{2+}]$ at the tip are coupled and oscillate, clearly identifies Ca^{2+} as a central regulator of tip growth (Holdaway-Clarke et al., 1997; Pierson et al., 1996), and it is probable that part of this regulation is due to its effects on actin and its associated proteins (Taylor & Hepler, 1997).

4.2 Protons

The recently described alkaline band, present in the subapical region of lily pollen tubes, could also be a locus for actin regulation (Feijó et al.,

1999). The principal actin-associated protein known to be regulated by pH is ADF/cofilin, which shows increased depolymerization activity at high pH (Yonezawa et al., 1985). To elucidate the precise activity of this protein in the alkaline band, it will be necessary to dissect the integrated effects of pH, phosphorylation, and the presence of other actin-binding proteins. It is likely that ADF/cofilin will participate in the remodeling of newly formed filaments and in their incorporation into the preexisting actin bundles. Another actin-bundling protein, elongation factor-1α (Edmonds et al., 1995), is also regulated by pH and could participate in the formation of actin bundles in the clear zone. However, its presence in pollen has not been described and its actin-bundling activity has not been demonstrated in plants (Durso & Cyr, 1994).

4.3 Rops

In animal cells and yeast, the small GTPases rho, rac and CDC42 have been identified as upstream regulators of the actin cytoskeleton (Hall, 1998). In plants, a family of related proteins exists, which shows more similarity to rac than to the others, and has been called 'Rop' (Yang, 1998; see also Li & Yang, this volume). In pollen tubes, Rop has been localized at the tip plasma membrane by immunocytochemistry (Lin et al., 1996) and GFP-fusions (Kost et al., 1999). By microinjecting antibodies (Lin & Yang, 1997) and transient expression of dominant negative protein (Kost et al., 1999), Rop appears to be important for pollen tube elongation. Impressively, its transient over-expression or the transient expression of dominant positive forms (Kost et al., 1999), generated balloon-shaped pollen tips that expanded iso-diametrically, losing their regular polarized growth. Pollen Rop also binds a phospholipid kinase that generates PIP_2 (Kost et al., 1999), which is a ligand and modulator of many actin-associated proteins.

5. PARTICIPATION OF ACTIN IN POLLEN TUBE GROWTH

Numerous studies have been performed to understand how pollen tubes grow, and as a result several factors have been implicated in the process including secretion, Ca^{2+}, electrical currents, wall components, small GTPases, and the actin cytoskeleton (Feijó et al., 1995; Steer & Steer, 1989; Taylor & Hepler, 1997; Yang, 1998). In the following section, we will focus on the possible role that the actin cytoskeleton plays in tube growth and how it could be regulated.

Over the years it has been well established that actin participates in pollen tube growth, nevertheless, the detailed mechanisms for its participation are poorly understood. The initial observations obtained with the electron microscope (Franke et al., 1972), together with the simultaneous inhibition of cytoplasmic streaming and growth by cytochalasin (Franke et al., 1972; Mascarenhas & Lafountain, 1972), prompted investigators to suggest that the main function of actin was the vectorial transport of secretory vesicles to the tip. Although this simple interpretation is probably correct, it fails to explain important aspects of actin regulation, such as the controlled polymerization of new microfilaments, and the polarized delivery of secretory vesicles in a cell that is vigorously and bi-directionally streaming.

In the early 1980s, in an attempt to integrate the need for Ca^{2+} and actin stability for pollen tube elongation, Picton and Steer (1982) developed a model to explain the mechanism of pollen tube growth. By analogy with amoeboid motion, this model proposed that F-actin was present at the tip of the cell in a high enough concentration to oppose the internal osmotic pressure, and suggested that Ca^{2+} stabilized this network. However, in marked contrast with amoeboid motion, they incorporated the obvious need for secretion in their model. This model failed to predict the observations that followed, namely that Ca^{2+} destabilizes actin networks in pollen tubes (Kohno & Shimmen, 1987; see also Yokota & Shimmen, and Snowman et al., this volume), and that a dense network of F-actin is not present at the tip of the cell (Kost et al., 1998; Miller et al., 1996).

Based on video-enhanced DIC observations of cytoplasmic streaming patterns in growing pollen tubes, and the discovery of myosin on the surface of pollen tube organelles (Heslop-Harrison & Heslop-Harrison, 1989c; Tang et al., 1989), another model was proposed by Heslop-Harrison and Heslop-Harrison (1990). They suggested that the F-actin bundles function as a filter, which sorts-out the smaller organelles from the larger ones, generating the observed apical zonation and helping to direct secretion to one end of the cell. This model suggested that myosin-coated organelles moving on randomly-polarized microfilaments generate the force balance responsible for the proposed sorting-out action. Confusingly, the authors invoke randomly-oriented fibrils in the subapical area of the cell, but acknowledge an organized and well-defined pattern of streaming in a more apical area. The existence of these actin zones in relation with each other, and how they are interconverted was not addressed. A similar interpretation of the function of actin emerged again from the live-cell work of Kost and coworkers (Kost et al., 1998), where a 'collar' of actin was described at the apical region of the cell; they suggested this could be a filtering mechanism that would only allow secretory vesicles to permeate to the tip of the cell.

Incorporating the most recent observations and relevant parts of previous models, we propose a new model, which explains the possible participation and regulations of actin in the process of pollen tube elongation. The main facts to consider are:

(A) F-actin is not present at the extreme apex of the cell, where mainly secretory vesicles accumulate.
(B) F-actin is present as short filaments, in a loose meshwork, or sometimes as a 'collar' immediately after the F-actin depleted zone.
(C) Cytoplasmic streaming is a bi-directional process, which follows a reverse-fountain pattern that starts deep within the clear zone, and is absent from the most apical region.
(D) The cortical region of the shank of the pollen tube has helically-oriented, parallel actin bundles.
(E) The central region has straight and longitudinal actin bundles.
(F) Organelles of all classes are present in the different parts of the pollen tube, including secretory vesicles and dictyosomes.
(G) Smaller organelles accumulate at the tip, generating the clear zone.

We will assume that cytoplasmic streaming follows the polarity of the actin bundles, and is driven by a barbed-end directed motor, an assumption supported by recent work on root hairs of *Hydrocharis,* which exhibit reverse-fountain streaming, and which reveal that the streaming direction is always towards the barbed end (Tominaga et al., 2000). Therefore, the cortical bundles that drive the acropetal streaming will have their barbed ends pointing towards the tip of the cell, while the basipetal streaming will be along bundles of opposite polarity. The smaller organelles will be able to travel at faster rates than the larger ones, since they can permeate between the larger ones and eventually accumulate at the tip, where the smallest ones, namely the secretory vesicles, will be highly enriched. This will be all driven by the acropetal cortical flow. The accumulation of smaller organelles at the tip region will in turn hinder the access of larger organelles, creating the clear zone. At the extreme apex where the Ca^{2+} concentration is very high, myosins will be inhibited, F-actin will be fragmented, and the secretory vesicles will further accumulate. One of the predictions of this model is that there will be a gradient of organelle sizes in the clear zone, which has been observed (Derksen et al., 1995a; Heslop-Harrison & Heslop-Harrison, 1990). This explains the establishment and need of apical zonation. The reverse streaming will constantly retrieve small organelles from the clear zone and maintain a balance of new and old organelles in that region, which is particularly important due the need of endocytosis for membrane recycling in that part of the cell (Derksen et al., 1995a; Picton & Steer, 1983).

To keep up with cell growth, new actin filaments need to be generated. Based on the estimated concentrations (Vidali & Hepler, 1997; Gibbon et al., 1999) and by analogy with other systems (Kaiser et al., 1999), it is likely that most of the actin will be complexed with profilin, which will inhibit the spontaneous nucleation but will maintain actin ready for polymerization at activation sites. New filaments will be nucleated from the profilin-actin pool, at the high Ca^{2+} region. These could be started, as in animal cells, via small GTPase activation of members of the ARP2/3 complex (Egile et al., 1999); alternatively Ca^{2+}-dependent nucleating proteins such as villins could play a role. Members of all three groups of proteins are known to be present in the pollen tube (Klahre & Chua, 1999; Lin et al., 1996; Vidali et al., 1999). At the very tip, F-actin will not be able to accumulate due to the Ca^{2+}-dependent severing activity of villins or similar proteins predicted to be active in the pollen tube (Kohno & Shimmen, 1987). This lack of F-actin together with the high $[Ca^{2+}]$ will facilitate the secretory process (Vitale et al., 1991). A zone of disorganized small filaments should be generated between the high calcium site and where the organized streaming starts; these filaments will be rapidly and probably cooperatively incorporated into pre-existing bundles. The polarity of incorporation will be pre-determined, since the bundles will be formed by villins, which bundle filaments in the same orientation (Coluccio & Bretscher, 1989; Yokota & Shimmen, 1999). The acropetal and the basipetal lanes in this way can be perpetuated.

This model successfully incorporates the most recent observations and does not need to invoke a sieving function for actin besides that provided by the difference in size of the streaming organelles. The model relies heavily on the need of highly-controlled actin polymerization, and the dependence of growth on stable streaming patterns, which maintain the clear zone. Supporting this are our unpublished observations that one of the first signs of latrunculin inhibition is a reduction in the length of the clear zone, concomitantly with growth inhibition. Other sources of regulation, such as the effects on ADF/cofilin of the recently described subapical alkaline band (Feijó et al., 1999), and membrane sub-compartments at the tip of the cell (Kost et al., 1999), will also prove important for the fine tuning and steering of the growth process.

6. CONCLUDING REMARKS

It is clear that more research is necessary to understand actin regulation and its participation in pollen tube growth. The biochemical characterization as well as the generation of fluorescent analogs of actin and its binding proteins is of great importance. The production of mutations, affecting different parts of actin regulation, for the dissection of genetic pathways are also necessary. Cell-free systems that complement the *in vivo* observations are fundamental to understand how actin polymerization is regulated. By understanding the actin cytoskeleton in the pollen tube we are opening a gate to advance the knowledge of plant cells in general.

REFERENCES

Andersland JM, Jagendorf AT and Parthasarathy MV (1992) The isolation of actin from pea roots by DNase I affinity chromotography. Plant Physiol 100: 1716-1723

Andre E, Lottspeich F, Schleicher M and Noegel A (1988) Severin, gelsolin, and villin share a homologous sequence in regions presumed to contain F-actin severing domains. J Biol Chem 263: 722-727

Cai G, Moscatelli A and Cresti M (1997) Cytoskeletal organization and pollen tube growth. Trends Plant Sci 2: 86-91

Carlier M-F, Laurent V, Santolini J, Melki R, Didry D, Xia GX, Hong Y, Chua N-H and Pantaloni D (1997) Actin depolymerizing factor (ADF/cofilin) enhances the rate of filament turnover: implication in actin-based motility. J Cell Biol 136: 1307-1322

Carlier M-F, Ressad F and Pantaloni D (1999) Control of actin dynamics in cell motility. Role of ADF/cofilin. J Biol Chem 274: 33827-33830

Carlsson L, Nystrom LE, Sundkvist I, Markey F and Lindberg U (1977) Actin polymerizability is influenced by profilin, a low molecular weight protein in non-muscle cells. J Mol Biol 115: 465-483

Christensen HE, Ramachandran S, Tan CT, Surana U, Dong CH and Chua NH (1996) *Arabidopsis* profilins are functionally similar to yeast profilins: identification of a vascular bundle-specific profilin and a pollen-specific profilin. Plant J 10: 269-279

Chung YY, Magnuson NS and An GH (1995) Subcellular-localization of actin depolymerizing factor in mature and germinating pollen. Mol Cells 5: 224-229

Clarke SR, Staiger CJ, Gibbon BC and Franklin-Tong VE (1998) A potential signaling role for profilin in pollen of *Papaver rhoeas*. Plant Cell 10: 967-979

Coluccio LM and Bretscher A (1989) Reassociation of microvillar core proteins: making a microvillar core *in vitro*. J Cell Biol 108: 495-502

Condeelis JS (1974) The identification of F actin of the pollen tube and protoplast of *Amaryllis belladonna*. Exp Cell Res 88: 435-439

Cresti M, Hepler PK, Tiezzi A and Ciampolini F (1986) Fibrilar strucutres in *Nicotinana* pollen: Changes in ultrastrucure during pollen activation and tube emission. In: Mulcahy DL, Bergamini-Mulcahy G, Ottaviano E (eds) Biotechnology and Ecology of Pollen. New York: Springer-Verlag, pp 283-288

de Win AHN, Knuiman B, Pierson ES, Geurts H, Kengen HMP and Derksen J (1996) Development and cellular organization of *Pinus sylvestris* pollen tubes. Sex Plant Reprod 9: 93-101

de Win AH, Pierson ES and Derksen J (1999) Rational analyses of organelle trajectories in tobacco pollen tubes reveal characteristics of the actomyosin cytoskeleton. Biophys J 76: 1648-1658

Derksen J, Rutten T, Lichtscheidl IK, de Win AHN, Pierson ES and Rongen G (1995a) Quantitative analysis of the distribution of organelles in tobacco pollen tubes: implications for exocytosis and endocytosis. Protoplasma 188: 267-276

Derksen J, Rutten T, van Amstel T, de Win A, Doris F and Steer M (1995b) Regulation of pollen tube growth. Acta Bot Neerl 44: 93-119

Doris FP and Steer MW (1996) Effects of fixatives and permeabilisation buffers on pollen tubes: implications for localisation of actin microfilaments using phalloidin staining. Protoplasma 195: 25-36

Drøbak BK, Watkins PAC, Valenta R, Dove SK, Lloyd CW and Staiger CJ (1994) Inhibition of plant plasma membrane phosphoinositide phospholipase C by the actin-binding protein, profilin. Plant J 6: 389-400

Durso NA and Cyr RJ (1994) Beyond translation: elongation factor-1α and the cytoskeleton. Protoplasma 180: 99-105

Eads JC, Mahoney NM, Vorobiev S, Bresnick AR, Wen KK, Rubenstein PA, Haarer BK and Almo SC (1998) Structure determination and characterization of *Saccharomyces cerevisiae* profilin. Biochemistry 37: 11171-11181

Edmonds BT, Murray J and Condeelis J (1995) pH regulation of the F-actin binding properties of *Dictyostelium* elongation factor 1α. J Biol Chem 270: 15222-15230

Egile C, Loisel TP, Laurent V, Li R, Pantaloni D, Sansonetti PJ and Carlier M-F (1999) Activation of the CDC42 effector N-WASP by the *Shigella flexneri* IcsA protein promotes actin nucleation by Arp2/3 complex and bacterial actin- based motility. J Cell Biol 146: 1319-1332

Fedorov AA, Ball T, Mahoney NM, Valenta R and Almo SC (1997) The molecular basis for allergen cross-reactivity: crystal structure and IgE-epitope mapping of birch pollen profilin. Structure 5: 33-45

Feijó JA, Malhó R and Obermeyer G (1995) Ion dynamics and its possible role during *in vitro* pollen germination and tube growth. Protoplasma 187: 155-167

Feijó JA, Sainhas J, Hackett GR, Kunkel JG and Hepler PK (1999) Growing pollen tubes possess a constitutive alkaline band in the clear zone and a growth-dependent acidic tip. J Cell Biol 144: 483-496

Franke WW, Herth W, Van der Woude WJ and Morré DJ (1972) Tubular and filamentous structures in pollen tubes: Possible involvement as guide elements in protoplasmic streaming and vectorial migration of secretory vesicles. Planta 105: 317-341

Gibbon BC, Ren H and Staiger CJ (1997) Characterization of maize (*Zea mays*) pollen profilin function *in vitro* and in live cells. Biochem J 327: 909-915

Gibbon BC, Zonia LE, Kovar DR, Hussey PJ and Staiger CJ (1998) Pollen profilin function depends on interaction with proline-rich motifs. Plant Cell 10: 981-93

Gibbon BC, Kovar DR and Staiger CJ (1999) Latrunculin B has different effects on pollen germination and tube growth. Plant Cell 11: 2349-2363

Giehl K, Valenta R, Rothkegel M, Ronsiek M, Mannherz HG and Jockusch BM (1994) Interaction of plant profilin with mammalian actin. Eur J Biochem 226: 681-689

Gilliland LU, McKinney EC, Asmussen MA and Meagher RB (1998) Detection of deleterious genotypes in multigenerational studies. I. Disruptions in individual *Arabidopsis* actin genes. Genetics 149: 717-725

Glenney JR, Jr., Geisler N, Kaulfus P and Weber K (1981) Demonstration of at least two different actin-binding sites in villin, a calcium-regulated modulator of F-actin organization. J Biol Chem 256: 8156-8161

Goldschmidt-Clermont PJ, Machesky LM, Doberstein SK and Pollard TD (1991) Mechanism of the interaction of human platelet profilin with actin. J Cell Biol 113: 1081-1089

Grote M, Swoboda I, Meagher RB and Valenta R (1995) Localization of profilin- and actin-like immunoreactivity in *in vitro*-germinated tobacco pollen tubes by electron microscopy after special water-free fixation techniques. Sex Plant Reprod 8: 180-186

Guillén G, Valdés-López V, Noguez R, Olivares J, Rodríguez-Zapata LC, Pérez H, Vidali L, Villanueva MA and Sánchez F (1999) Profilin in *Phaseolus vulgaris* is encoded by two genes (only one expressed in root nodules) but multiple isoforms are generated *in vivo* by phosphorylation on tyrosine residues. Plant J 19: 497-508

Hall A (1998) Rho GTPases and the actin cytoskeleton. Science 279: 509-514

Heslop-Harrison J, Heslop-Harrison Y, Cresti M, Tiezzi A and Ciampolini F (1986) Actin during pollen germination. J Cell Sci 86: 1-8

Heslop-Harrison J (1987) Pollen germination and pollen-tube growth. Int Rev Cytol 107: 1-78

Heslop-Harrison J and Heslop-Harrison Y (1989a) Actomyosin and movement in the angiosperm pollen tube: an interpretation of some recent results. Sex Plant Reprod 2: 199-207

Heslop-Harrison J and Heslop-Harrison Y (1989b) Conformation and movement of the vegetative nucleus of the angiosperm pollen tube: association with the actin cytoskeleton. J Cell Sci 93: 299-308

Heslop-Harrison J and Heslop-Harrison Y (1989c) Myosin associated with the surfaces of organelles, vegetative nuclei and generative cells in angiosperm pollen grains and tubes. J Cell Sci 94: 319-325

Heslop-Harrison J and Heslop-Harrison Y (1990) Dynamic aspects of apical zonation in the angiosperm pollen tube. Sex Plant Reprod 3: 187-194

Heslop-Harrison J and Heslop-Harrison Y (1991a) The actin cytoskeleton in unfixed pollen tubes following microwave-accelerated DMSO-permeabilisation and TRITC-phalloidin staining. Sex Plant Reprod 4: 6-11

Heslop-Harrison J and Heslop-Harrison Y (1991b) Restoration of movement and apical growth in the angiosperm pollen tube following cytochalasin-induced paralysis. Phil Trans R Soc Lond 331: 225-235

Heslop-Harrison J and Heslop-Harrison Y (1992a) Cyclical transformations of the actin cytoskeleton of hyacinth pollen subjected to recurrent vapour-phase hydration and dehydration. Biol Cell 75: 245-252

Heslop-Harrison Y and Heslop-Harrison J (1992b) Germination of monocolpate angiosperm pollen: Evolution of the actin cytoskeleton and wall during hydration, activation and tube emergence. Annals Bot 69: 385-394

Hess MW, Mittermann I, Luschnig C and Valenta R (1995) Immunocytochemical localisation of actin and profilin in the generative cell of angiosperm pollen: TEM studies on high-pressure frozen and freeze-substituted *Ledebouria socialis* Roth (Hyacinthaceae). Histochem Cell Biol 104: 443-451

Holdaway-Clarke TL, Feijó JA, Hackett GR, Kunkel JG and Hepler PK (1997) Pollen tube growth and the intracellular cytosolic calcium gradient oscillate in phase while extracellular calcium influx is delayed. Plant Cell 9: 1999-2010

Holtzman DA, Yang S and Drubin DG (1993) Synthetic-lethal interactions identify two novel genes, SLA1 and SLA2, that control membrane cytoskeleton assembly in *Saccharomyces cerevisiae*. J Cell Biol 122: 635-644

Huang S, McDowell JM, Weise MJ and Meagher RB (1996) The *Arabidopsis* profilin gene family. Evidence for an ancient split between constitutive and pollen-specific profilin genes. Plant Physiol 111: 115-126

Hussey PJ, Yuan M, Calder G, Khan S and Lloyd CW (1998) Microinjection of pollen-specific actin-depolymerizing factor, ZmADF1, reorientates F-actin strands in *Tradescantia* stamen hair cells. Plant J 14: 353-357

Igarashi H, Vidali L, Yokota E, Sonobe S, Hepler PK and Shimmen T (1999) Actin filaments purified from tobacco cultured BY-2 cells can be translocated by plant myosin. Plant Cell Physiol 40: 1167-1171

Iwanami Y (1956) Protoplasmic movement in pollen grains and tubes. Phytomorphology 6: 288-295

Jauh GY and Lord EM (1995) Movement of the tube cell in the lily style in the absence of the pollen grain and the spent pollen tube. Sex Plant Reprod 8: 168-172

Kaiser DA, Vinson VK, Murphy DB and Pollard TD (1999) Profilin is predominantly associated with monomeric actin in *Acanthamoeba*. J Cell Sci 112: 3779-3790

Kandasamy MK, McKinney EC and Meagher RB (1999) The late pollen-specific actins in angiosperms. Plant J 18: 681-691

Karakesisoglou I, Schleicher M, Gibbon BC and Staiger CJ (1996) Plant profilins rescue the aberrant phenotype of profilin-deficient *Dictyostelium* cells. Cell Motil Cytoskeleton 34: 36-47

Kim SR, Kim Y and An G (1993) Molecular cloning and characterization of anther-preferential cDNA encoding a putative actin-depolymerizing factor. Plant Mol Biol 21: 39-45

Kinkema M, Wang H and Schiefelbein J (1994) Molecular analysis of the myosin gene family in *Arabidopsis thaliana*. Plant Mol Biol 26: 1139-1153

Klahre U and Chua N-H (1999) The *Arabidopsis* actin-related protein 2 (AtARP2) promoter directs expression in xylem precursor cells and pollen. Plant Mol Biol 41: 65-73

Klahre U, Friederich E, Kost B, Louvard D and Chua N-H (2000) Villin-like actin-binding proteins are expressed ubiquitously in *Arabidopsis*. Plant Physiol 122: 35-48

Kohno T and Shimmen T (1987) Ca^{2+}-induced fragmentation of actin filaments in pollen tubes. Protoplasma 141: 177-179

Kohno T and Shimmen T (1988) Accelerated sliding of pollen tube organelles along *Characeae* actin bundles regulated by Ca^{2+}. J Cell Biol 106: 1539-1543

Kost B, Spielhofer P and Chua N-H (1998) A GFP-mouse talin fusion protein labels plant actin filaments *in vivo* and visualizes the actin cytoskeleton in growing pollen tubes. Plant J 16: 393-401

Kost B, Lemichez E, Spielhofer P, Hong Y, Tolias K, Carpenter C and Chua N-H (1999) Rac homologues and compartmentalized phosphatidylinositol 4,5-bisphosphate act in a common pathway to regulate polar pollen tube growth. J Cell Biol 145: 317-330

Krauze K, Makuch R, Stepka M and Dabrowska R (1998) The first caldesmon-like protein in higher plants. Biochem Biophys Res Comm 247: 576-579

Lancelle SA, Cresti M and Hepler PK (1987) Ultrastructure of the cytoskeleton in freeze-substituted pollen tubes of *Nicotiana alata*. Protoplasma 140: 141-150

Lancelle SA and Hepler PK (1989) Immunogold labelling of actin on sections of freeze-substituted plant cells. Protoplasma 150: 72-74

Lancelle SA and Hepler PK (1991) Association of actin with cortical microtubules revealed by immunogold localization in *Nicotiana* pollen tubes. Protoplasma 165: 167-172

Lancelle SA and Hepler PK (1992) Ultrastructure of freeze-substituted pollen tubes of *Lilium longiflorum*. Protoplasma 167: 215-230

Lazzaro MD (1996) The actin microfilaments network within elongation pollen tubes of the gymnosperm *Picea abies* (Norway spruce). Protoplasma 194: 186-194

Lin Y, Wang Y, Zhu JK and Yang Z (1996) Localization of a Rho GTPase implies a role in tip growth and movement of the generative cell in pollen tubes. Plant Cell 8: 293-303

Lin Y and Yang Z (1997) Inhibition of pollen tube elongation by microinjected anti-Rop1Ps antibodies suggests a crucial role for Rho-type GTPases in the control of tip growth. Plant Cell 9: 1647-1659

Liu X and Yen L-F (1992) Purification and characterization of actin from maize pollen. Plant Physiol 99: 1151-1155

Liu X and Yen LF (1995) Interaction of pollen F-actin with rabbit muscle myosin and its subfragments (HMM, S1). Protoplasma 186: 87-92

Loisel TP, Boujemaa R, Pantaloni D and Carlier M-F (1999) Reconstitution of actin-based motility of *Listeria* and *Shigella* using pure proteins. Nature 401: 613-616

Lopez I, Anthony RG, Maciver SK, Jiang C-J, Khan S, Weeds AG and Hussey PJ (1996) Pollen specific expression of maize genes encoding actin depolymerizing factor-like proteins. Proc Natl Acad Sci USA 93: 7415-7420

Machesky LM and Pollard TD (1993) Profilins. In: Kreis T, Vale R (eds) Guidebook to the Cytoskeletal and Motor Proteins. New York: Oxford University Press Inc. pp 66-68

Machesky LM, Atkinson SJ, Ampe C, Vandekerckhove J and Pollard TD (1994) Purification of a cortical complex containing two unconventional actins from *Acanthamoeba* by affinity chromatography on profilin-agarose. J Cell Biol 127: 107-115

Machesky LM, Reeves E, Wientjes F, Mattheyse FJ, Grogan A, Totty NF, Burlingame AL, Hsuan JJ and Segal AW (1997) Mammalian actin-related protein 2/3 complex localizes to regions of lamellipodial protrusion and is composed of evolutionarily conserved proteins. Biochem J 328: 105-112

Machesky LM and Gould KL (1999) The Arp2/3 complex: a multifunctional actin organizer. Curr Opin Cell Biol 11: 117-121

Mascarenhas JP and Lafountain J (1972) Protoplasmic streaming, cytochalasin B, and growth of the pollen tube. Tissue Cell 4: 11-14

Mascarenhas JP (1993) Molecular mechanisms of pollen tube growth and differentiation. Plant Cell 5: 1303-1314

May RC, Hall ME, Higgs HN, Pollard TD, Chakraborty T, Wehland J, Machesky LM and Sechi AS (1999) The Arp2/3 complex is essential for the actin-based motility of *Listeria monocytogenes*. Curr Biol 9: 759-762

McCurdy DW and Kim M (1998) Molecular cloning of a novel fimbrin-like cDNA from *Arabidopsis thaliana*. Plant Mol Biol 36: 23-31

McDowell JM, Huang S, McKinney EC, An YQ and Meagher RB (1996) Structure and evolution of the actin gene family in *Arabidopsis thaliana*. Genetics 142: 587-602

Meagher RB, McKinney EC and Kandasamy MK (1999) Isovariant dynamics expand and buffer the responses of complex systems. The diverse plant actin gene family. Plant Cell 11: 995-1006

Melan MA and Sluder G (1992) Redistribution and differential extraction of soluble proteins in permeabilized cultured cells: implications for immunofluorescence microscopy. J Cell Sci 101: 731-743

Miller DD, Scordilis SP and Hepler PK (1995) Identification and localization of three classes of myosins in pollen tubes of *Lilium longiflorum* and *Nicotiana alata*. J Cell Sci 108: 2549-2563

Miller DD, Lancelle SA and Hepler PK (1996) Actin microfilaments do not form a dense meshwork in *Lilium longiflorum* pollen tube tips. Protoplasma 195: 123-132

Mittermann I, Swoboda I, Pierson E, Eller N, Kraft D, Valenta R and Heberle-Bors E (1995) Molecular cloning and characterization of profilin from tobacco (*Nicotiana tabacum*): increased profilin expression during pollen maturation. Plant Mol Biol 27: 137-146

Mockrin SC and Korn ED (1980) Acanthamoeba profilin interacts with G-actin to increase the rate of exchange of actin-bound adenosine 5'-triphosphate. Biochemistry 19: 5359-5362

Mooseker MS and Cheney RE (1995) Unconventional myosins. Annu Rev Cell Dev Biol 11: 633-675

Nakayasu T, Yokota E and Shimmen T (1998) Purification of an actin-binding protein composed of 115-kDa polypeptide from pollen tubes of lily. Biochem Biophys Res Comm 249: 61-65

Pantaloni D and Carlier M-F (1993) How profilin promotes actin filament assembly in the presence of thymosin ß4. Cell 75: 1007-1014

Perdue TD and Parthasarathy MV (1985) In situ localization of F-actin in pollen tubes. Eur J Cell Biol 39: 13-20

Perelroizen I, Didry D, Christensen H, Chua N-H and Carlier M-F (1996) Role of nucleotide exchange and hydrolysis in the function of profilin in action assembly. J Biol Chem 271: 12302-12309

Picton JM and Steer MW (1981) Determination of secretory vesicle production rates by dictyosomes in pollen tubes of *Tradescantia* using cytochalasin D. J Cell Sci 49: 261-272

Picton JM and Steer MW (1982) A model for the mechanism of tip extension in pollen tubes. J theor Biol 98: 15-20

Picton JM and Steer MW (1983) Membrane recycling and the control of secretory activity in pollen tubes. J Cell Sci 63: 303-310

Pierson ES, Derksen J and Traas JA (1986) Organization of microfilaments and microtubules in pollen tubes grown *in vitro* or *in vivo* in various angiosperms. Eur J Cell Biol 41: 14-18

Pierson ES (1988) Rhodamine-phalloidin staining of F-actin in pollen after dimethyl sulphoxide permeabilisation. A comparison with the conventional formaldehyde preparation. Sex Plant Reprod 1: 83-87

Pierson ES, Kengen HMP and Derksen J (1989) Microtubules and actin filaments co-localize in pollen tubes of *Nicotiana tabacum* L. and *Lilium longiflorum* thumb. Protoplasma 150: 75-77

Pierson ES and Cresti M (1992) Cytoskeleton and cytoplasmic organization of pollen and pollen tubes. Int Rev Cytol 140: 73-125

Pierson ES, Miller DD, Callaham DA, van Aken J, Hackett G and Hepler PK (1996) Tip-localized calcium entry fluctuates during pollen tube growth. Devel Biol 174: 160-173

Putnam-Evans C, Harmon A, Palevitz BA, Fechheimer M and Cormier MJ (1989) Calcium-dependent protein kinase is localized with F-actin in plant cells. Cell Motil Cytoskeleton 12: 12-22

Ren HY, Gibbon BC, Ashworth SL, Sherman DM, Yuan M and Staiger CJ (1997) Actin purified from maize pollen functions in living plant cells. Plant Cell 9: 1445-1457

Ressad F, Didry D, Xia GX, Hong Y, Chua NH, Pantaloni D and Carlier MF (1998) Kinetic analysis of the interaction of actin-depolymerizing factor (ADF)/cofilin with G- and F-actins. Comparison of plant and human ADFs and effect of phosphorylation. J Biol Chem 273: 20894-20902

Rothkegel M, Mayboroda O, Rohde M, Wucherpfennig C, Valenta R and Jockusch BM (1996) Plant and animal profilins are functionally equivalent and stabilize microfilaments in living animal cells. J Cell Sci 109: 83-90

Rutten TLM and Derksen J (1990) Organization of actin filaments in regenerating and outgrowing subprotoplasts from pollen tubes of *Nicotiana tabacum* L. Planta 180: 471-479

Shibayama T, Carboni JM and Mooseker MS (1987) Assembly of the intestinal brush border: appearance and redistribution of microvillar core proteins in developing chick enterocytes. J Cell Biol 105: 335-344

Smertenko AP, Jiang CJ, Simmons NJ, Weeds AG, Davies DR and Hussey PJ (1998) Ser6 in the maize actin-depolymerizing factor, ZmADF3, is phosphorylated by a calcium-stimulated protein kinase and is essential for the control of functional activity. Plant J 14: 187-193

Sonobe S and Shibaoka H (1989) Cortical fine actin filaments in higher plant cells visualized by rhodamine-phalloidin after pretreatment with m-maleimidobenzoyl N-hydroxysuccinimide ester. Protoplasma 148: 80-86

Staiger CJ and Schliwa M (1987) Actin localization and function in higher plants. Protoplasma 141: 1-12

Staiger CJ, Goodbody KC, Hussey PJ, Valenta R, Drøbak BK and Lloyd CW (1993) The profilin multigene family of maize: differential expression of three isoforms. Plant J 4: 631-641

Staiger CJ, Gibbon BC, Kovar DR and Zonia LE (1997) Profilin and actin depolymerizing factor: modulators of actin organization in plants. Trends Plant Sci 2: 275-281

Steer MW and Steer JM (1989) Pollen tube tip growth. New Phytol 111: 323-358

Svitkina TM and Borisy GG (1999) Arp2/3 complex and actin depolymerizing factor/cofilin in dendritic organization and treadmilling of actin filament array in lamellipodia. J Cell Biol 145: 1009-1026

Tanaka I and Wakabayashi T (1992) Organization of the actin and microtubule cytoskeleton preceding pollen germination. An analysis using cultured pollen protoplasts of *Lilium longiflorum*. Planta 186: 473-482

Tang XJ, Hepler PK and Scordilis SP (1989) Immunochemical and immunocytochemical identification of a myosin heavy chain polypeptide in *Nicotiana* pollen tubes. J Cell Sci 92: 569-574

Taylor LP and Hepler PK (1997) Pollen germination and tube growth. Ann Rev Plant Physiol Plant Molec Biol 48: 461-491

Theriot JA and Mitchison TJ (1993) The three faces of profilin. Cell 75: 835-838

Thorn KS, Christensen HE, Shigeta R, Huddler D, Shalaby L, Lindberg U, Chua NH and Schutt CE (1997) The crystal structure of a major allergen from plants. Structure 5: 19-32

Tirlapur UK, Cai G, Faleri C, Moscatelli A, Scali M, Del Casino C, Tiezzi A and Cresti M (1995) Confocal imaging and immunogold electron microscopy of changes in distribution of myosin during pollen hydration, germination and pollen tube growth in *Nicotiana tabacum* L. Eur J Cell Biol 67: 209-217

Tiwari SC and Polito VS (1988a) Organization of the cytoskeleton in pollen tubes of *Pyrus communis:*a study employing conventional and freeze-substitution electron microscopy, immunofluorescence, and rhodamine-phalloidin. Protoplasma 147: 100-112

Tiwari SC and Polito VS (1988b) Spatial and temporal organization of actin during hydration, activation, and germination of pollen in *Pyrus communis* L.:a population study. Protoplasma 147: 5-15

Tominaga M, Morita K, Sonobe S, Yokota E and Shimmen T (1997) Microtubules regulate the organization of actin filaments at the cortical region in root hair cells of *Hydrocharis*. Protoplasma 199: 83-92

Tominaga M, Yokota E, Vidali L, Sonobe S, Hepler PK and Shimmen T (2000) Organization of actin cytoskeleton responsible for cytoplasmic streaming and for architecture of transvacuolar strand in root hair cells of *Hydrocharis*. Planta, In press

Valenta R, Duchêne M, Pettenburger K, Sillaber C, Valent P, Bettelheim P, Breitenbach M, Rumpold H, Kraft D and Scheiner O (1991) Identification of profilin as a novel pollen allergen; IgE autoreactivity in sensitized individuals. Science 253: 557-560

Vidali L and Hepler PK (1996) Purification and characterization of actin from *Lilium longiflorum* pollen. Mol Biol Cell 7: 2178 Suppl.

Vidali L and Hepler PK (1997) Characterization and localization of profilin in pollen grains and tubes of *Lilium longiflorum*. Cell Motil Cytoskeleton 36: 323-338

Vidali L, Yokota E, Cheung AY, Shimmen T and Hepler PK (1999) The 135 kDa actin-bundling protein from *Lilium longiflorum* pollen is the plant homologue of villin. Protoplasma 209: 283-291

Vitale ML, Rodriguez Del Castillo A, Tchakarov L and Trifaro JM (1991) Cortical filamentous actin disassembly and scinderin redistribution during chromaffin cell stimulation precede exocytosis, a phenomenon not exhibited by gelsolin. J Cell Biol 113: 1057-67

von Witsch M, Baluska F, Staiger CJ and Volkmann D (1998) Profilin is associated with the plasma membrane in microspores and pollen. Eur J Cell Biol 77: 303-312

Vos JW and Hepler PK (1998) Calmodulin is uniformly distributed during cell division in living stamen hair cells of *Tradescantia virginiana*. Protoplasma 201: 158-171

Vrtala S, Wiedemann P, Mittermann I, Eichler HG, Sperr WR, Valent P, Kraft D and Valenta R (1996) High-level expression in *Escherichia coli* and purification of recombinant plant profilins: comparison of IgE-binding capacity and allergenic activity. Biochem Biophys Res Comm 226: 42-50

Winter D, Podtelejnikov AV, Mann M and Li R (1997) The complex containing actin-related proteins Arp2 and Arp3 is required for the motility and integrity of yeast actin patches. Curr Biol 7: 519-529

Yang Z (1998) Signaling tip growth in plants. Curr Opin Plant Biol 1: 525-530

Yen L-F, Liu X and Cai S (1995) Polymerization of actin from maize pollen. Plant Physiol 107: 73-76

Yin HL, Janmey PA and Schleicher M (1990) Severin is a gelsolin prototype. FEBS Lett 264: 78-80

Yokota E and Shimmen T (1994) Isolation and characterization of plant myosin from pollen tubes of lily. Protoplasma 177: 153-162

Yokota E, McDonald AR, Liu B, Shimmen T and Palevitz BA (1995) Localization of a 170kDa myosin heavy chain in plant cells. Protoplasma 185: 178-187

Yokota E, Takahara K-i and Shimmen T (1998) Actin-bundling protein isolated from pollen tubes of lily. Biochemical and immunocytochemical characterization. Plant Physiol 116: 1421-1429

Yokota E, Muto S and Shimmen T (1999) Inhibitory regulation of higher-plant myosin by Ca^{2+} ions. Plant Physiol 119: 231-240

Yokota E and Shimmen T (1999) The 135-kDa actin-bundling protein from lily pollen tubes arranges F- actin into bundles with uniform polarity. Planta 209: 264-266

Yonezawa N, Nishida E and Sakai H (1985) pH control of actin polymerization by cofilin. J Biol Chem 260: 14410-14412

Chapter 19

ACTIN REARRANGEMENTS IN POLLEN TUBES ARE STIMULATED BY THE SELF-INCOMPATIBILITY (SI) RESPONSE IN *PAPAVER RHOEAS* L.

Benjamin N Snowman[1], Anja Geitmann[2], Anne Mie C Emons[2] and Vernonica E Franklin-Tong[1]

1 School of Biosciences, University of Birmingham, Edgbaston, B15 2TT, UK
2 Laboratory of Experimental Plant Morphology and Cell Biology, Department of Plant Sciences, Wageningen University, Arboretumlaan 4, 6703 BD Wageningen, The Netherlands

Key words: actin cytoskeleton, pollen, self-incompatibility, signal transduction

Abstract: Self-incompatibility (SI) is a genetically-controlled system used by many flowering plants to prevent self-fertilization. We have been investigating SI in *Papaver rhoeas* for a number of years, and have begun to build up a picture of some of the signalling events and mechanisms involved in this specific inhibition of pollen tube growth. Pollen tubes elongate by tip growth, which is dependent upon the actin cytoskeleton, the structure of which is controlled by actin-binding proteins. To date, no studies, to our knowledge, have investigated the effect that a self-incompatible response may have on actin organisation in pollen tubes. Our investigations provide clear evidence that the actin cytoskeleton of incompatible pollen tubes undergoing SI is dramatically altered. These data strongly suggest that the actin cytoskeleton is targetted by the signalling pathway stimulated in the SI response. These findings may have implications for our understanding of regulation of pollen tube growth.

1. INTRODUCTION

Self-incompatibility (SI) is one of the most important mechanisms used by flowering plants to prevent self-fertilization. By regulating the acceptance or rejection of pollen, fertilization by undesirable pollen is prevented. This acts to promote outbreeding. Genetical studies of a variety of SI systems suggest that they are generally under the control of genes residing at a single

C.J. Staiger et al. (eds.), Actin: A Dynamic Framework for Multiple Plant Cell Functions, 347–360.

multiallelic *S*-locus (reviewed by Franklin et al., 1995), although there are also SI systems under control of 2- and 4-loci. When a pollen grain lands on a pistil that also carries an *S*-allele that is genetically identical to that carried by itself, the hydration, germination and/or growth of the pollen are inhibited. Pollen carrying mutually exclusive *S*-alleles grow normally through the pistil and proceed to fertilize the ovules. SI is therefore an excellent example of precise cell-cell communication in response to explicit, defined signals.

SI in *Papaver rhoeas* L. (the field poppy) is under the control of a single multiallelic *S*-locus. Poppy SI is an example of a gametophytically controlled SI system, in which the *S*-phenotype of the pollen is controlled by its haploid genotype (Lawrence et al., 1978). We have cloned several alleles of the stigmatic *S* gene from this species (Foote et al., 1994; Walker et al., 1996; Kurup et al, 1998) and are able to reproduce the SI response *in vitro* using recombinant S proteins. Progress on the molecular basis of allelic specificity has been made, and we have recently established a requirement for certain residues that are crucial for the recognition of S_1 pollen (Kakeda et al, 1998; Jordan et al., 1999). Although we have identified an S protein-binding protein (SBP) in pollen (Hearn et al., 1996) and have evidence that SBP-binding of the stigmatic S protein is important for pollen inhibition (Jordan *et al*., 1999), we do not know if this is the S-receptor.

We have been investigating the signalling events and potential mechanisms involved in the SI response in pollen of *P. rhoeas*. Cytosolic free calcium (Ca^{2+}_i) acts as a second messenger and is established as playing an important role in the regulation of pollen tube growth (see Rudd & Franklin-Tong, 1999; Franklin-Tong, 1999a,b for recent reviews). There is good evidence from Ca^{2+}-imaging studies that the SI response in *P. rhoeas* activates a Ca^{2+}-mediated intracellular signalling pathway. Observed S-specific increases in $[Ca^{2+}]_i$ occur within seconds of SI challenge (Franklin-Tong et al., 1993, 1995, 1997), and we have recently obtained evidence that this involves the influx of Ca^{2+} in the shank of the pollen tube (Franklin-Tong et al., submitted). The rapidity of the Ca^{2+} response is consistent with the idea that the S-protein acts as a signal molecule that specifically triggers increases in $[Ca^{2+}]$. We propose that this is the first step in the signalling pathway.

These increases in $[Ca^{2+}]_i$ are thought to activate a calcium-dependent protein kinase (CDPK), resulting in S-specific increases in phosphorylation of a pollen protein, p26 (Rudd et al., 1996). Another phosphoprotein, p68, also exhibits increases in phosphorylation, but is not regulated by a calcium-dependent protein kinase (Rudd et al., 1997). We have recently cloned and characterized p26, and demonstrated that it has both homology to, and the activity of, a soluble inorganic pyrophosphatase (Rudd et al., submitted).

When phosphorylated, the pyrophosphatase activity of p26 is significantly reduced *in vitro*. The extremely energy demanding process of pollen tube growth, involving the synthesis of bio-polymers, such as those destined for new cell membrane and cell wall, will generate pyrophosphate as a by-product. Under normal growth conditions, an active pyrophosphatase will prevent the accumulation of PP_i through its hydrolysis to 2 P_i. This would drive biosynthesis and make catabolism less energetically feasible. This provides the basis for a model whereby phosphorylation of p26 contributes towards the inhibition of incompatible pollen tube growth. Inhibition of pyrophosphatase activity, resulting from the Ca^{2+}-dependent phosphorylation of p26, would result in very rapid and severe inhibition of pollen tube elongation as a result of the inhibition of anabolism. The arrest of growth is one of the classic markers of the observed SI phenotype in *Papaver*.

We have recently investigated later events occurring in *Papaver* pollen following an incompatible challenge, obtaining evidence for nuclear DNA fragmentation, which is a hallmark of programmed cell death (PCD), (Jordan et al., submitted). This feature is also observed when pollen is exposed to mastoparan at concentrations previously shown to increase $[Ca^{2+}]_i$ (Franklin-Tong et al., 1996). It therefore seems likely that the increases in $[Ca^{2+}]_i$ stimulated by both mastoparan and SI are responsible for eliciting a signaling cascade resulting in DNA fragmentation and possibly PCD.

The concept that there is likely to be more than one "failsafe" mechanism involved in the SI response clearly demonstrates the complexity of this system controlling the rejection of self-incompatible pollen in this species.

Pollen germination and tube growth are dependent upon the actin cytoskeleton (reviewed by Taylor & Hepler, 1997; Cai et al., 1997; Gibbon et al., 1999; Franklin-Tong, 1999a; Vidali & Hepler, this volume) which provides an intracellular framework, necessary for many essential processes. It provides physical support (Staiger et al. 1994, Valster et al. 1997) and acts as a track for myosin-based intracellular movement of organelles (Yokota et al., 1999a,b; Yokota & Shimmen, this volume). Tip growth of cells is dependent upon a typical configuration of the actin cytoskeleton (Geitmann & Emons 2000; Emons et al., this volume), which is controlled by actin-binding proteins. It is thought that the control of both pollen germination and tube growth involve a complex interplay between the cytoskeleton and signalling cascades, though there is little direct evidence for this at present.

So, what happens to the pollen tube cytoskeleton when the SI response is triggered? Since the SI response in *P. rhoeas* results in the rapid inhibition of pollen tube growth, it was thought likely that alterations in the actin cytoskeleton would take place as a consequence of an incompatible SI reaction. However, this key question has not, until recently, been investigated. In many cells, alterations in the cytoskeleton configuration are the first visible effects appearing in response to extracellular stimuli (Staiger, 2000), for example, in root hairs in response to Nod factors (Cárdenas et al., 1998; de Ruijter et al. 1999; Emons et al., this volume). Here we describe the identification of rapid structural changes to the actin cytoskeleton in pollen tubes of *P. rhoeas* undergoing the SI reaction (see also Snowman et al., 2000; Geitmann et al., 2000).

2. THE CYTOSKELETON IS RAPIDLY ALTERED IN THE SI RESPONSE

Incompatible and fully-compatible SI responses were induced in pollen tubes of *P. rhoeas* growing *in vitro* (see Kakeda *et al.*, 1998). The actin cytoskeleton was subsequently stabilised with *m*-maleimidobenzoyl N-hydroxysuccinimide ester (MBS) prior to chemical fixation, using the method modified from Doris & Steer (1996) and Miller et al. (1999). This method has been demonstrated to preserve actin cytoarchitecture well, in a manner similar to that using freeze-fixation prior to freeze substitution, which is well known to be superior to normal chemical fixation (Sonobe & Shibaoka, 1989; Miller et al., 1999). We have used fluorescently-labelled phalloidin to visualise actin filaments in these fixed pollen tubes before and after challenge with S proteins (Snowman et al., 2000; Geitmann et al., 2000; Snowman et al., in preparation).

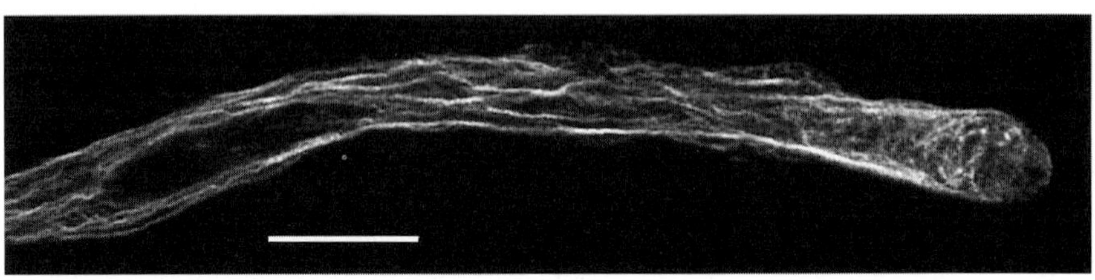

Figure 1. The cytoskeleton in normally growing pollen tubes. The F-actin cytoskeleton in a *P. rhoeas* pollen tube visualised with rhodamine-phalloidin and fluorescence confocal microscopy. Bar, 10 μm

Figure 1 shows a confocal section of a normally growing poppy pollen tube. In the apical region (generally the first 3–15 μm) of these pollen tubes, no heavily staining F-actin bundles were visible. This we believe demonstrates the absence of thick bundles of actin filaments. The lack of thick actin bundles in the apex has recently been demonstrated in both fixed and living pollen tubes (Miller et al., 1996; Kost et al., 1998; Gibbon et al., 1999).

Just behind this apical region, a region occupying ~10 μm of the sub-apex is observed. This intense staining by phalloidin indicates a concentrated network of F-actin (Fig. 1). This 'collar-shaped' arrangement has been demonstrated to be present in living pollen tubes, using a GFP-talin construct transiently expressed to detect actin filaments (Kost et al., 1998). Behind this region, and throughout the 'shank', bundles of actin filaments lying in a longitudinal or helical orientation are visible.

To investigate the effect of a SI response, identical samples of S protein were used to challenge pollen *in vitro*. This ensured that we demonstrated the S-specificity of the response, as the only difference between the two pollen samples was the *S*-genotype. Figure 2 shows representative examples of pollen tube actin cytoskeleton at different time points after challenge with S proteins. Figure 2a shows an incompatible pollen tube fixed immediately as S proteins were added. The organisation of the actin cytoskeleton seen here is indistinguishable from that of normally growing pollen tubes, and indicates that the pollen tubes at the time of challenge exhibited normal morphology.

Within 1–2 min. after S-protein perception, F-actin is seen at the extreme tip of the pollen tube (Fig. 2b). Following this apparent movement of F actin into the apex, an increasing number of F-actin bundles are detectable at the cortical region, close to the plasma membrane (Figs. 2b, 2c). Five min. after the induction of an incompatible SI challenge, the F-actin bundles appear to have further accumulated adjacent to the plasma membrane, as there is apparently continuous phalloidin staining there. Coincident with this is a decrease in the number and size of visible microfilament bundles in the lumen of the pollen tube (Fig. 2c).

Figures 2d and 2e show pollen tubes 10 and 20 min. after challenge with incompatible S proteins. The phalloidin staining has taken on a speckled appearance, indicating that the actin bundles have been drastically shortened. These fine fragments of actin are seen throughout the cytoplasm, but a distinct distribution is visualised at the cortex. By this stage, only a few actin filament bundles are visible in the cytoplasm. Further changes are detected later in the SI response. Figure 2f shows an incompatible pollen tube 1 h after SI induction, the fine punctate foci of F-actin appear to be further organised into larger aggregates. These appear qualitatively fewer in number, and more discernable as distinct structures. Again, these larger punctate foci of actin are distributed throughout the cytoplasm, but also display a prominent cortical localisation.

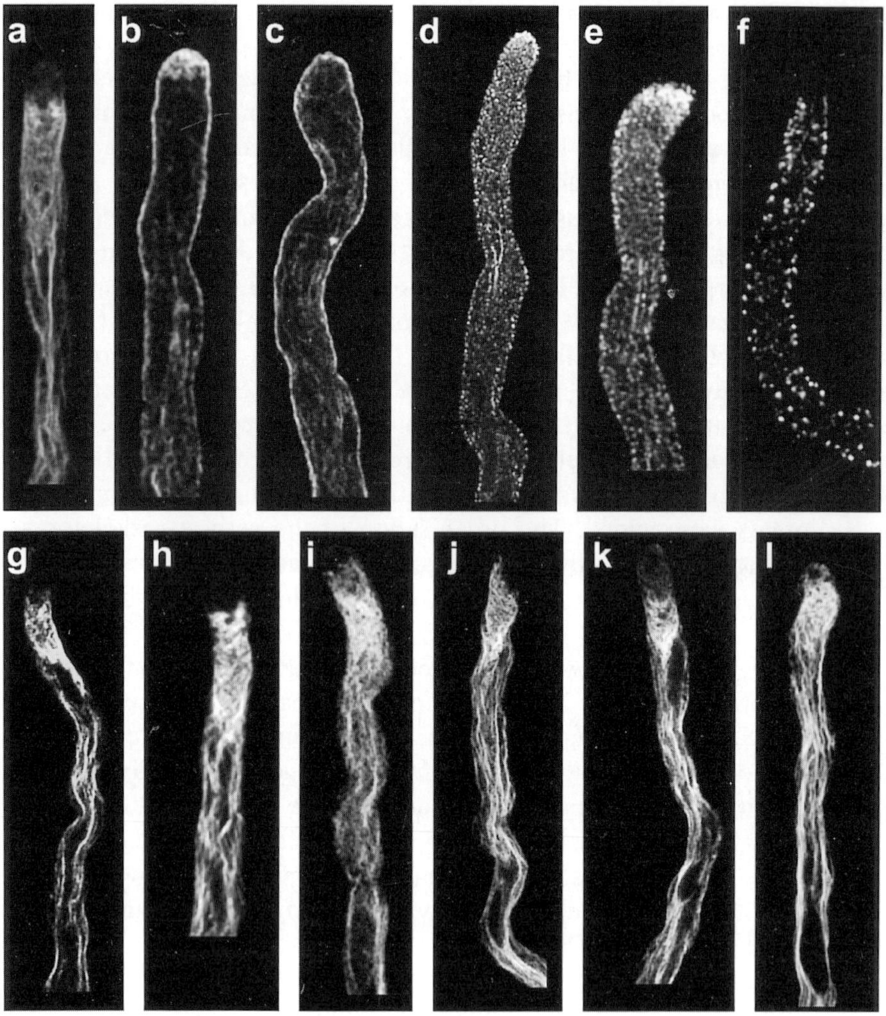

Figure 2. The cytoskeleton is rapidly altered in incompatible pollen in an S-specific manner. (**a-f**) Alterations in the F-actin cytoskeleton are induced by an incompatible SI response. (**a**) Normal F-actin cytoskeleton at t=0. Subsequent images show the actin cytoskeleton at the following time points after induction of an incompatible response: (**b**) 2 min, (**c**) 5 min, (**d**) 10 min, (**e**) 20 min, (**f**) 60 min. (**g–l**) No alterations in the F-actin cytoskeleton are detected in a compatible SI response. (**g**) 0 min, (**h**) 2 min, (**i**) 5 min, (**j**) 10 min, (**k**) 20 min, (**l**) 60 min. Pollen tubes are ~8 μm in diameter.

In contrast, and as expected, no alterations in F-actin organisation were observed when pollen tubes were challenged with fully-compatible S proteins (Figs. 2g–2l). These compatible pollen tubes displayed an F-actin cytoskeleton arrangement that was indistinguishable from the control pollen tubes that had not been challenged (Fig. 1). These data clearly demonstrate the absolute *S*-specificity of the response, as the only difference between the two sets of responses was the *S*-genotype of the pollen that was challenged. We have demonstrated that non-specific inhibition of growth does not result in similar actin rearrangements (Geitmann et al., 2000). This clearly suggests that the SI-induced alterations are triggered by the SI signalling cascade, rather than being merely due to the consequent inhibition of growth.

Analysis of the frequency of events occurring following exposure to incompatible S proteins defines the timing of events. This is shown in Figure 3. Pollen tubes were scored by the organization of their actin cytoskeleton following SI. Appearance of actin bundles at the tip, which was the first detectable change, was not scored for, as virtually all incompatible pollen tubes showed this within 1-2 min of SI challenge. The other phenotypes observed fell into three classes. In class I, phalloidin staining is as seen in normal growing pollen tubes. Class II have F actin localized adjacent to the plasma membrane. Class III is characterised by the appearance of membrane localised punctate foci of actin.

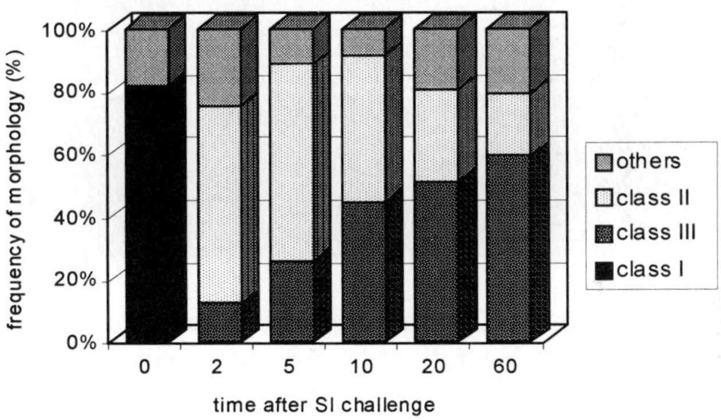

Figure 3. Quantification of events occurring to the actin cytoskeleton in pollen tubes undergoing an incompatible SI response. The histogram shows the percentages of pollen tubes displaying normal distribution of F-actin filaments (class I), membrane-localised F-actin filaments (class II) or the membrane-localised fragmented F-actin "punctate foci" (class III) following an incompatible challenge. "Others" indicate morphologies that do not fit these classes.

Following incompatible S-protein perception, the number of class I pollen tubes decreases very rapidly, and an increase in class II pollen tubes is detected. Over the subsequent time period, the number of class II pollen tubes decreases, and a proportional increase in pollen tubes of class III is seen. Following a fully compatible challenge 80% class I pollen tubes are seen (Geitmann et al., 2000). This shows that compatible pollen tubes challenged with the same S proteins grow normally and retain normal morphology.

These data clearly establish three things. First, they establish the S-specificity of this response. Second, they demonstrate that the cortical localization of actin, characterized by apparent continuous phalloidin staining, occurs prior to the formation of punctate foci of actin at the cell cortex. Third, they clearly indicate the speed of the response. Alterations to the actin organization, characterised by movement of actin bundles to the tip and marginalization of actin, are first detected within ~1 min. of stimulus perception. However, there are clearly other changes in the actin cytoskeleton that occur over a longer time period.

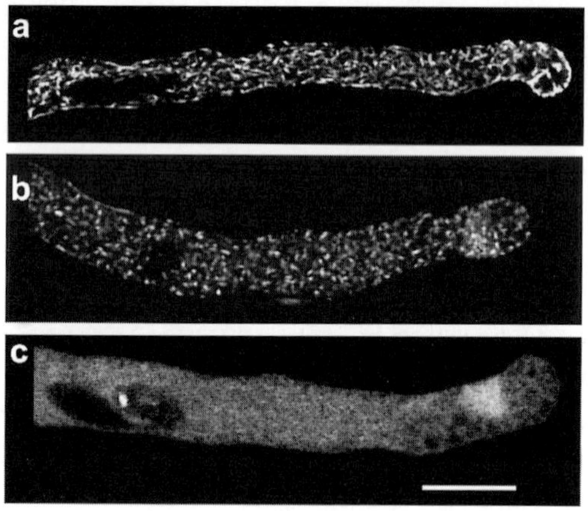

Figure 4. Effects of Latrunculin B (10 min treatment) on F-actin in pollen tubes of *P. rhoeas*. (a) 100 nM Lat B. (b) 1 μM Lat B. (c) 10 μM Lat B. Bar, 10 μm

Treatment of pollen tubes with Latrunculin B (Lat B), a potent actin depolymerizing agent, results in alterations in F-actin organisation (Fig. 4). Addition of 100 nM Lat B resulted in short microfilament bundles (Fig. 4a). Higher concentrations of Lat B (1 μM) resulted in visualisation of the presence of diffusely-stained actin (Fig. 4b). Still higher concentrations (10 μM) appeared to completely dissolve the bundles of actin filaments (Fig. 4c). These studies indicate that the F-actin patterns and distribution observed following treatment with Lat B is rather different to the changes in organization induced by SI, though the effect of Lat B has some similarity to the intermediate stages detected in incompatible pollen tubes. This suggests depolymerisation is likely to play a role in the SI response. However, other factors are likely also to be involved.

3. DISCUSSION

The data described here represent the first steps in reporting what happens structurally to the actin cytoskeleton of pollen tubes in the SI response (see also Snowman et al., 2000; Geitmann et al., submitted). We have described rapid and distinctive alterations to the actin cytoskeleton of incompatible pollen tubes in response to a SI reaction. Our data indicate that this response is S-specific, since fully compatible challenges showed actin cytoskeletal arrangements that were indistinguishable from those observed in normally growing pollen tubes. These data provide convincing evidence that the cytoskeleton is targeted by the signaling pathway(s) elicited in pollen tubes undergoing the SI response. Furthermore, as the onset of these events is so rapid, the inhibition of pollen tube growth may be (at least partly) a result of these cytoskeletal alterations. Since these results can be reproduced *in vivo* (Geitmann et al., 2000), we believe that the *in vitro* bioassay represents a reliable method to investigate the regulation of the actin cytoskeleton in response to S-proteins.

Our current imaging data for the SI-induced F-actin changes suggest that there is likely to be a decrease in the overall concentration of F-actin in incompatible pollen tubes since, qualitatively, the phalloidin signal appears to be reduced upon S-protein challenge. However, the comparison of the SI-induced alterations with the effect of latrunculin (LatB) suggest that, while depolymerisation may be partially responsible, the SI response has some distinctive features, namely, the apparent membrane localization and the formation of punctate aggregates of actin. Although there is little information on the regulation of actin polymer levels or dynamics in pollen, some insight into this has recently been provided by Gibbon et al., (1999). They studied the effects of Lat B on maize pollen and have demonstrated that pollen germination and tip growth are highly sensitive to perturbations in F-actin levels. In relating these observations with what is now known about SI in *Papaver*, one could surmise that SI-stimulated pollen tube inhibition, arises through rapid actin depolymerisation. Further studies are currently underway to investigate the nature of the alterations to the actin cytoskeleton more fully. A sensitive assay for quantifying levels of F-actin has recently been used by Gibbon et al., (1999) to measure changes in F-actin concentration in maize pollen. We have used this assay to determine if the levels of F-actin are altered in pollen tubes undergoing the SI response. Our data indicate that there is a rapid decrease in the level of F-actin in incompatible *Papaver* pollen tubes (Snowman et al., in preparation).

On the basis of what we have observed, there are several possibilities for how the SI-induced alterations might be achieved. Reductions in F-actin are consistent with the action of an actin-sequestering protein, like profilin, or stimulation of actin depolymerizing factor (ADF). F-actin breakdown is characteristic of ADF action (see reviews by Staiger et al., 1997; de Ruijter & Emons, 1999; Kovar & Staiger, this volume). ADF is known to cause the depolymerisation, and possibly fragmentation, of actin filaments (Moriyama & Yahara, 1999). Furthermore, its activity is conditional upon Ca^{2+}-dependent phosphorylation, PtdIns(4,5)P_2 binding and pH (Hawkins et al., 1993; Ressad et al., 1998; Smertenko et al., 1998). An incompatible SI response is known to induce increases in $[Ca^{2+}]_i$ and upregulate CDPK activity (Franklin-Tong et al., 1997; Rudd et al., 1997). These SI-induced increases in $[Ca^{2+}]_i$ would, therefore, be predicted to stimulate phosphorylation and inactivation of ADF. On face value, then, it would seem likely that ADF does not play a role in the SI-induced actin depolymerization that we observe. However, since ADF is localised to Ca^{2+}-high regions in root hairs and is thought to be active (Jiang et al., 1997), we cannot rule out the possibility of its involvement.

Profilin, another actin-binding protein present abundantly in pollen (Valenta et al., 1991; Staiger et al., 1993; Vidali & Hepler, 1997; Gibbon & Staiger, this volume), is also implicated as playing a possible role in these SI-induced changes in the pollen actin cytoskeleton. To date, analysis of plant profilin function *in vivo* points toward it acting as an actin monomer sequesterer. This was clearly demonstrated by Staiger et al. (1994) and Valster et al. (1997) who microinjected birch profilin into *Tradescantia* stamen hair cells, which resulted in F-actin depolymerisation. Perelroizen et al. (1996) have shown that high Ca^{2+} enhances actin sequestration by animal profilin. It has recently been demonstrated that Ca^{2+} increases the critical concentration for maize pollen actin assembly in the presence of profilin, indicating that profilin is a Ca^{2+}-activated sequestering factor (Kovar et al., 2000). This makes profilin a good candidate for playing a role in the SI-induced changes in actin observed in *Papaver* pollen.

The SI response in *Papaver* pollen is known to elevate $[Ca^{2+}]_i$ (Franklin-Tong et al., 1997), so we can begin to draw up a model for SI-induced changes in the actin cytoskeleton in *Papaver* pollen. Since the data on the effect of increased Ca^{2+} on ADF is equivocal (as discussed earlier), although ADF would appear to be an unlikely candidate as an effector, we cannot rule it out. Therefore, a model involving profilin appears more feasible. The presence of increased $[Ca^{2+}]_i$ could enhance G-actin sequestration by profilin, which would result in F-actin depolymerisation. Validitation of this model will depend upon measuring the cytosolic concentrations of G-actin, F-actin and profilin and the apparent K_d value for poppy pollen profilin binding to G-actin in the presence of varying concentrations of Ca^{2+}. However, it should be emphasized that ADF or other actin-depolymerising proteins could be activated in the SI response. Furthermore, we cannot rule out the possibility that F-actin severing and capping activities play a role, especially as there is circumstantial evidence for the existence of a gelsolin-like activity in plant cells (Kohno & Shimmen, 1987, 1988; Cárdenas et al., 1998; Yokota & Shimmen, this volume). In truth, it is likely that more than one actin-binding protein is involved in regulating the changes in actin organisation in these pollen tubes, since actin-binding proteins are known to act synergistically to effect cytoskeletal changes. Finally we have not attempted to explain the mechanism of the formation of punctate foci of actin induced in the SI response, since this would be pure speculation in the light of the lack of knowledge about this in plant cells at this point in time.

In conclusion, we have the concerted action of a complex set of signals and responses that are triggered as a consequence of the stigmatic S protein interaction with incompatible pollen during the SI response. The changes in actin organisation described here provide us with further information on the nature and mechanisms involved in the SI response in pollen of *P. rhoeas*. This contributes significantly to the overall picture of what we know about what appears to be an increasingly sophisticated mechanism for inhibiting pollen tube growth. Research is currently underway in our lab to identify the mechanisms involved.

ACKNOWLEDGMENTS

All images shown in this chapter were collected by BN Snowman as part of his PhD studies. We are indebted to John Martin and colleagues for their horticultural services. BNS gratefully acknowledges receipt of a BBSRC Studentship. A Geitmann has a TMR grant within the Marie Curie program of the European Union. VE Franklin-Tong's laboratory is funded by the Biotechnology and Biological Sciences Research Council (BBSRC). We gratefully acknowledge the input of CJ Staiger into recent aspects of this project.

REFERENCES

Cai G, Moscatelli A and Cresti M. (1997) Cytoskeletal organization and pollen tube growth. Trends Plant Sci 2: 86-91

Cárdenas L, Vidali L, Domínguez J, Pérez H, Sánchez F, Hepler PK and Quinto C (1998) Rearrangement of actin microfilaments in plant root hairs responding to *Rhizobium etli* nodulation signals. Plant Physiol 116: 871-877

de Ruijter NCA, Bisseling T and Emons AMC (1999) *Rhizobium* Nod factors induce an increase in subapical fine bundles of actin filaments in *Vicia sativa* root hairs within minutes. Mol Plant Microbe Inter 9: 829-832 de Ruijter NCA and Emons AMC (1999) Actin-binding proteins in plant cells. Plant Biol 1: 26-35

Didry D, Carlier M-F and Pantaloni D (1998) Synergy between actin depolymerizing factor cofilin and profilin in increasing actin filament turnover. J Biol Chem 273: 25602-25611

Doris FP and Steer MW (1996) Effects of fixatives and permeabilisation buffers on pollen tubes: implications for the localisation of actin microfilaments using phalloidin staining. Protoplasma 195: 25-36

Foote HG, Ride JP, Franklin-Tong VE, Walker EA, Lawrence MJ and Franklin FCH (1994) Cloning and expression of a novel self-incompatibility (S-)gene from *Papaver rhoeas* L. Proc Natl Acad Sci USA 91: 2265-2269

Franklin FCH, Lawrence MJ and Franklin-Tong VE (1995) Cell and molecular biology of self-incomaptibility in flowering plants. Int Rev Cytol 158: 1-64

Franklin-Tong VE (1999a) Signalling and the modulation of pollen tube growth. Plant Cell 11: 727-738

Franklin-Tong VE (1999b) Signalling in pollination. Curr Opin Plant Biol 2: 490-495

Franklin-Tong VE, Ride JP, Read ND, Trewavas AJ and Franklin FCH (1993) The self-incompatibility response in *Papaver rhoeas* is mediated by cytosolic free calcium. Plant J 4: 163-177

Franklin-Tong VE, Ride JP and Franklin FCH (1995) Recombinant stigmatic self-incompatibility (S-) protein elicits a Ca^{2+} transient in pollen of *Papaver rhoeas*. Plant J 8: 299-307

Franklin-Tong VE, Drøbak BK Allen AC, Watkins PAC and Trewavas AJ (1996) Growth of pollen tubes of *Papaver rhoeas* is regulated by a slow-moving calcium wave propagated by inositol 1,4,5-tris phosphate. Plant Cell 8: 1305-1321

Franklin-Tong VE, Hackett G and Hepler PK (1997) Ratio-imaging of $Ca^{2+}{}_i$ in the self-incompatibility response in pollen tubes of *Papaver rhoeas*. Plant J 12: 1375-1386

Franklin-Tong VE, Holdaway-Clarke T, Straatman KR, Kunkel JG and Hepler PK. Evidence for the involvement of calcium influx in the self-incompatibility response *Papaver rhoeas*. Submitted

Geitmann A and Emons AMC (2000) The cytoskeleton in plant and fungal cell tip growth. J Microsc, In press

Geitmann A, Snowman BN, Emons AMC and Franklin-Tong VE (2000) Alterations to the actin cytoskeleton of pollen tubes are induced by the self-incompatibility reaction in *Papaver rhoeas*. Plant Cell, in press

Gibbon BC, Kovar DR and Staiger CJ (1999) Latrunculin B has different effects on pollen germination and tube growth. Plant Cell 11: 2349-2363

Hawkins, M, Pope B, Maciver SK and Weeds AG (1993) Human actin depolymerising factor mediates a pH-sensitive destruction of actin filaments. Biochemistry 32: 9985-9993

Hearn MJ, Franklin FCH and Ride JP (1996) Identification of a membrane glycoprotein in pollen of *Papaver rhoeas* which binds stigmatic self-incompatibility (S-) proteins. Plant J 9: 467-475

Jiang CJ, Weeds AG and Hussey PJ (1997) The maize actin depolymerizing factor, ZmADF3, redistributes to the growing tip of elongating root hairs and can be induced to translocate into the nucleus with actin. Plant J 12: 1035-1043

Jordan ND, Kakeda K, Conner A, Ride JP, Franklin-Tong VE and Franklin FCH (1999) S-protein mutants indicate a functional role for SBP in the self-incompatibility reaction of *Papaver rhoeas*. Plant J 20: 119-125.

Jordan ND, Franklin FCH and Franklin-Tong VE. Evidence for programmed cell death triggered in the self-incompatibility response in pollen of *Papaver rhoeas*. Submitted

Kakeda K, Jordan ND, Conner A, Ride JP, Franklin-Tong VE and Franklin FCH (1998) Identification of residues in a hydrophilic loop of the *Papaver rhoeas* S protein that play a crucial role in recognition of incompatible pollen. Plant Cell 10: 1723-1731

Kost B, Speihofer P and Chua N-H (1998) A GFP-mouse talin fusion protein labels plant actin filaments *in vivo* and visualizes the actin cytoskeleton in growing pollen tubes. Plant J 16: 393-401

Kohno T and Shimmen T (1987) Ca^{2+}-induced fragmentation of actin filaments in pollen tubes. Protoplasma 141: 177-179

Kohno T and Shimmen T (1988) Mechanism of Ca^{2+}-inhibition of cytoplasmic streaming in lily pollen tubes. J Cell Sci 91: 501-509

Kovar DR, Drøbak BK and Staiger CJ (2000) Maize profilin isoforms are functionally distinct. Plant Cell, In press

Kurup S, Ride JP, Jordan ND, Fletcher G, Franklin-Tong VE and Franklin FCH (1998) Identification and cloning of related self-incompatibility S genes in *Papaver rhoeas* and *Papaver nudicaule*. Sex Plant Reprod 11: 192-198

Lawrence MJ, Afzal M and Kenrick J (1978) The genetical control of self-incompatibility in *Papaver rhoeas* L. Heredity 40: 239-285

Miller DD, Lancelle SA and Hepler PK (1996) Actin microfilaments do not form a dense meshwork in *Lilium longiflorum* pollen tube tips. Protoplasma 195: 123-132

Miller DD, de Ruijter NCA, Bisseling T and Emons AMC (1999) The role of actin in root hair morphosgenesis: studies with lipochito-oligosaccharide as a growth stimulator and cytochalasin as an actin-perturbing drug. Plant J 17: 141-154

Moriyama K and Yahara I (1999) Two activities of cofilin, severing and accelerating directional depolymerisation of actin filaments, are affected by mutations around the actin-binding helix. EMBO J 18: 6753-6761

Perelroizen I, Didry D, Christensen H, Chua N-H and Carlier M-F (1996) Role of nucleotide exchange and hydrolysis in the function of profilin in actin assembly J Biol Chem 271: 12302-12309

Ressad F, Didry D, Xia GX, Hong Y, Chua NH, Pantaloni D and Carlier M-F (1998) Kinetic analysis of the interaction of actin-depolymerizing factor (ADF)/cofilin with G- and F-actins. Comparison of plant and human ADFs and effect of phosphorylation. J Biol Chem 273: 20894-20902

Rudd JJ, Lord JM, Franklin FCH and Franklin-Tong VE (1996) Increased phosphorylation of a 26 kD pollen protein is induced by the self-incompatibility response in *Papaver rhoeas.* Plant Cell 8: 713-724

Rudd JJ, Franklin FCH and Franklin-Tong VE (1997) Ca^{2+}-independent phosphorylation of a 68 kD pollen protein is stimulated by the self-incompatibility response in *Papaver rhoeas.* Plant J 12: 507-514

Rudd JJ and Franklin-Tong VE (1999) Calcium signalling in plants. Cell Molec Life Sci 55: 214-232

Rudd JJ, Franklin-Tong VE and Franklin FCH. Molecular cloning and biochemical characterisation of p26: a pollen protein implicated in the self-incompatibility response in *Papaver rhoeas.* Submitted

Smertenko AP, Jiang C-J, Simmons NJ, Weeds AG, Davies DR and Hussey PJ (1998) Ser6 in the maize actin-depolymerizing factor, ZmADF3, is phosphorylated by a calcium-stimulated protein kinase and is essential for the control of functional activity. Plant J 14: 187-194

Snowman BN, Geitmann A, Clarke SR, Staiger CJ, Franklin FCH, Emons AMC and Franklin-Tong VE (2000) Signalling and the cytoskeleton of *Papaver rhoeas*. Ann Bot, In press

Sonobe S and Shibaoka H (1989) Cortical fine actin filaments in higher plant cells visualised by rhodamine phalloidin after treatment with m-maleimidobenzoyl-n-hydroxysuccinimide ester. Protoplasma 148: 80-86

Staiger CJ (2000) Signaling to the actin cytoskeleton in plants. Annu Rev Plant Physiol Plant Mol Biol 51: 257-288

Staiger CJ, Goodbody KC, Hussey PJ, Valenta R, Drøbak BK and Lloyd CW (1993) The profilin multigene family of maize: differential expression of three isoforms. Plant J 4: 631-641

Staiger CJ, Yuan M, Valenta R, Shaw PJ, Warn RM and Lloyd CW (1994) Microinjected profilin affects cytoplasmic streaming in plant cells by rapidly depolymerizing actin microfilaments. Curr Biol 4: 215-219

Staiger CJ, Gibbon BC, Kovar DR and Zonia LE (1997) Profilin and actin depolymerizing factor: modulators of actin organization in plants. Trends Plant Sci 2: 275-281

Taylor LP and Hepler PK (1997) Pollen germination and tube growth. Annu Rev Plant Physiol Plant Mol Biol 48: 461-491

Valenta R, Duchêne M, Pettenburger K, Sillaber C, Valent P, Bettelheim P, Breitenbach M, Rumpold H, Kraft D and Scheiner O (1991) Identification of profilin as a novel pollen allergen; IgE autoreactivity in sensitized individuals. Science 253: 557-560

Valster AH, Pierson ES, Valenta R., Hepler PK and Emons AMC (1997) Probing the plant actin cytoskeleton during cytokinesis and interphase by profilin microinjection. Plant Cell 9: 1815-1824

Vidali L and Hepler PK (1997) Characterization and localization of profilin in pollen grains and tubes of *Lilium longiflorum*. Cell Motil. Cytoskel 36: 323-338

Walker EA, Ride JP, Kurup S, Franklin-Tong VE, Lawrence MJ and Franklin FCH (1996) Molecular analysis of two functionally identical S_3 homologues of the self-incompatibility gene of *Papaver rhoeas* from two different populations. Plant Mol Biol 30: 983-994

Yokota E, Yukawa C, Muto S, Sonobe S and Shimmen T (1999a) Biochemical and immunocytochemical characterization of two types of myosins in cultured tobacco bright yellow-2 cells. Plant Physiol 121: 525-534

Yokota E, Muto S and Shimmen T (1999b) Inhibitory regulation of higher plant myosin by Ca^{2+} ion. Plant Physiol 119: 231-239

Chapter 20

THE ACTIN CYTOSKELETON DURING DIFFERENTIATION OF MICROSPORES TO MATURE POLLEN

The dynamic reorganization of actin arrays functions to integrate development

Laura Zonia

Institute of Experimental Botany, Academy of Sciences of the Czech Republic, Na Pernikárce 15, Prague 6, CZ–160 00, Czech Republic

Key words: actin inhibitors, asymmetrical mitosis, microspore development, nuclear migration, pollen development

Abstract: Development of the male gametophyte is a critical phase in the process of sexual reproduction in flowering plants. This developmental pathway proceeds through a complex sequence of events that include polarization, nuclear migration, asymmetrical mitosis, growth and differentiation. In the past few years, some of the signals and mechanisms that modulate pollen development have begun to be identified. A key role for actin cytoskeletal arrays has been documented. This review aims to summarize and supplement the current understanding of actin cytoskeletal stucture and function during the differentiation of microspores to mature pollen.

1. OVERVIEW OF MICROSPORE AND POLLEN DEVELOPMENT

The critical events during microspore and pollen development – cellular polarization, nuclear migration, asymmetrical mitosis and subsequent cellular differentiation – occur in all pollen and have been discussed in numerous reviews (Mascarenhas, 1989; McCormick, 1993; Bedinger et al., 1994; Twell, 1999). However, there are variations within this common framework that may reflect species-specific variations in the signaling pathways or variations in the mechanisms that coordinate development. This

C.J. Staiger et al. (eds.), Actin: A Dynamic Framework for Multiple Plant Cell Functions, 361–371.

section will highlight some of the main events that occur during formation of the male gametophyte for typical angiosperms.

1.1 Microspore development

The phase of microspore development follows the completion of meiosis of the microspore mother cell and proceeds until the start of the asymmetrical mitosis. The critical events are cellular growth and polarization. Following release from the tetrad, the microspore undergoes a rapid increase in cell volume and vacuole formation begins. Concurrently, microspores of some species undergo a change in cell shape from spheroidal to ellipsoidal (eg. *Nicotiana tabacum, Solanum tuberosum*), while in other species they remain spheroidal (eg. *Arabidopsis thaliana, Zea mays*). Microspore polarization is initiated with the onset of vacuole formation and cell shape change. During polarization, the nucleus migrates from a central position to an asymmetrical position in the cell. The pathway of nuclear migration as well as the final position of the nucleus prior to mitosis are species-specific. The microspores of some species have two nuclear migrations (eg. *N. tabacum, Tradescantia paludosa*), while others have only one (eg. *A. thaliana, Z. mays*). Irrespective of the mode of migration, the final asymmetrical positioning of the nucleus prior to the onset of the asymmetrical mitosis is considered to be an important determinant of the subsequent division symmetry and daughter cell fate (Zaki & Dickinson, 1990; Hause et al., 1993; Touraev et al., 1996; Gallagher & Smith, 1997). Blocking or disrupting the asymmetrical positioning of the nucleus with colchicine, a microtubule inhibitor, has been reported to switch the microspore mitosis to a symmetrical division and interfere with subsequent daughter cell fate determination (Zaki & Dickinson, 1991; Eady et al., 1995). It is possible to experimentally manipulate this developmental transition point and induce a switch from gametophytic development to a program of embryogenic development that can lead to the production of haploid plants (Zaki & Dickinson, 1990, 1991; Hause et al., 1993; Touraev et al., 1996, 1997).

1.2 Asymmetrical mitosis

With completion of cellular polarization in the late microspore, the necessary conditions and factors are in place to signal the start of the asymmetrical mitosis. Asymmetrical mitoses produce unequal daughter cells that will undergo different developmental fates and they occur at points of developmental differentiation (Gallagher & Smith, 1997; Scheres & Benfey, 1999). One of the interesting hallmarks of asymmetrical divisions is that the mitotic spindle is also asymmetrical, with one end more cone-shaped and one end more barrel-shaped (Heslop-Harrison, 1968; Brown & Lemmon, 1992; Terasaka & Niitsu, 1995; Zonia et al., 1999). Differentiation of the two unequal daughter nuclei becomes apparent immediately after telophase and cytokinesis. The generative nucleus is smaller and contains very condensed chromatin, while the vegetative nucleus is larger and its chromatin is more decondensed. During cytokinesis, a cell wall is formed between the generative and vegetative nuclei that eventually completely encloses the generative cell. The developmental fate of the generative cell is to undergo mitosis to produce the two sperm cells. The vegetative cell undergoes a different developmental fate, to germinate and produce the pollen tube that functions to deliver the non-motile gametes to the embryo sac.

1.3 Pollen development

Following the asymmetrical mitosis is the phase of pollen development. During the early stage of pollen development, generative cell wall formation is completed, the generative cell begins to move inside the vegetative cell, and will eventually be completely enclosed in a unique cell-within-a-cell arrangement. During the mid stage of pollen development, the generative cell and vegetative nucleus undergo a coordinated migration toward the center of the pollen grain. There is a high rate of metabolic activity in the vegetative cell during this stage, with the production of starch, proteins, and RNA transcripts. During the final stage leading to pollen maturity, the vegetative cell becomes metabolically quiescent. In some species that have tricellular pollen (eg. *A. thaliana, Z. mays*), the generative cell divides before final maturation to produce the two sperm cells. Immediately prior to release from the anther, the mature pollen grain undergoes a dehydration step that is necessary for the process of dehiscence and for maintenance of pollen viability within the environment once it is free from the plant.

2. ACTIN LOCALIZATION DURING MICROSPORE AND POLLEN DEVELOPMENT

The developmental events that occur during formation of the male gametophyte suggest a key role for the cytoskeleton. This section will discuss the localization and structural organization of actin arrays during microspore and pollen development in *S. tuberosum* (potato) and at three critical stages in *N. tabacum* (tobacco).

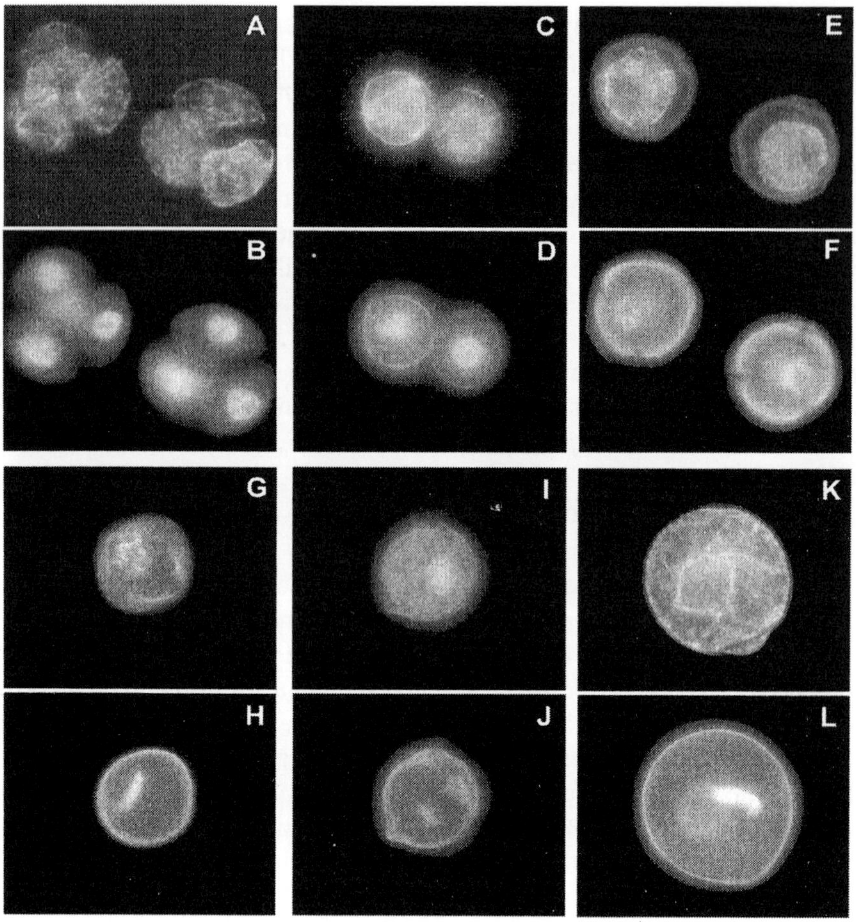

Figure 1. Actin localization during development of potato pollen. All cells were double-labeled with rhodamine phalloidin to visualize actin (**A, C, E, G, I, K**) and DAPI to visualize the nucleus (**B, D, F, H, J, L**). **A, B** tetrad, **C, D** early microspore, **E, F** late microspore, **G, H** metaphase, **I, J**, telophase, **K, L** mature pollen.

2.1 Tetrad

Actin has been observed in the tetrads of many species. In potato, they form an irregular cytoplasmic network that extends between the centrally-located nucleus and plasma membrane of each haploid cell (Figs. 1A, B). Similar actin arrays have been observed in the tetrads of *Lilium hybrida* (Sheldon & Hawes, 1988), *Brassica napus* (Gervais et al., 1994), tobacco (Zonia et al., 1999), *Gasteria verrucosa* (Van Lammeren et al., 1989), *Z. mays* (Staiger & Cande, 1991), *Magnolia soulangeana* (Dinis & Mesquita, 1993), and the orchid *Phalaenopsis* (Brown & Lemmon, 1991).

2.2 Free microspore

After release of the haploid cells from the tetrad, the free microspores undergo rapid reorganization of the actin arrays that were present during the tetrad stage. In tobacco, actin arrays are rapidly depolymerized and free microspores have few or no filaments (Zonia et al., 1999). Potato microspores have more extensive actin arrays at this stage (Figs. 1C, D). Some microspores have a spherical network centered around the nucleus and extending into the cytoplasm (Fig. 1C, right cell), and others have a triangular array (Fig. 1C, left cell). In *B. napus* microspores, actin forms a mesh in close association with the nucleus, and the cytoplasmic arrays are much finer (Hause et al., 1993; Gervais et al., 1994). Actin bundles were observed in the cytoplasm of microspores of *G. verrucosa* (Van Lammeren et al., 1989).

2.3 Late microspore and start of mitosis

At the late microspore stage, the nucleus has moved to an asymmetrical position within the cell and is ready to start the asymmetrical mitosis. In late tobacco microspores, the final position of the nucleus at the generative pole is extremely asymmetrical, as these cells change shape during polarization to become ellipsoidal (Fig. 2B). Once the nucleus is located at the pole, an actin network forms around it and tethers it to this asymmetrical position at the pole (Fig. 2A). Late microspores of potato and *B. napus* do not undergo such extreme cell shape changes, and also do not have such characteristically unique actin arrays. In potato, late microspores have a very dense mesh of actin filaments and bundles closely associated with the nucleus and the surrounding cytoplasmic domain. This array persists even after the chromatin condenses during prophase (Figs. 1E, F). In late microspores of *B. napus*, there is a dense actin array closely associated with the distal or vegetal pole of the nucleus (Gervais et al., 1994).

2.4 Asymmetrical mitosis

One of the interesting and unusual characteristics of the asymmetrical microspore mitosis is that it is mediated by an asymmetrical mitotic spindle as revealed by microtubule labeling. The asymmetrical mitosis represents the developmental transition point from microspore to pollen.

Actin filaments are aligned within the region of the mitotic spindle in potato (Fig. 1G, H), in tobacco (Zonia et al., 1999), and in the orchid *Phalaenopsis* (Brown & Lemon, 1992). Although actin filaments were present during metaphase in *B. napus*, they were not observed to be co-aligned within the mitotic spindle (Hause et al. 1993, Gervais et al., 1994).

Separation of the two unequal daughter cells begins with the assembly of the phragmoplast, the site of synthesis of the generative cell wall. Dense actin filaments are observed in the phragmoplast of potato (Figs. 1I, J), tobacco (Figs. 2C, D), and *B. napus* (Hause et al., 1992, 1993; Gervais et al., 1994). It is thought that the phragmoplast functions in the organization of vesicle trafficking and delivery of necessary building blocks to the expanding cell plate (Kakimoto & Shibaoka, 1988; see also Schmit, this volume). The phragmoplast expands outward along a vector perpendicular to the spindle long axis, and will eventually curve around and completely enclose the generative cell.

2.5 Early- to mid-pollen

The unequal daughter cells begin to differentiate immediately after mitosis. The generative cell is smaller, is located closer to the pole or cell perimeter, and its nucleus contains highly condensed chromatin. The nucleus of the vegetative cell is larger and contains decondensed chromatin. During the early pollen stage of tobacco, actin filaments remain associated with the generative cell wall and are organized in a dense mesh around the generative cell as it begins to move away from the pole (Zonia et al., 1999). During the mid-pollen stage, actin is organized as a dense interconnected mesh around both the vegetative nucleus and generative cell in tobacco (Figs. 2E, F) and in potato (not shown). This interlinked array functions to allow the generative cell and vegetative nucleus to move as one unit during their coordinated migration toward the center of the pollen grain. Differential localization patterns of actin filaments with the generative cell or vegetative nucleus during the early- to mid- pollen stages have also been observed in *B. napus* (Hause et al., 1992, 1993; Gervais et al., 1994).

2.6 Mature pollen

The final step in pollen maturation is an extensive dehydration process that prepares the grain for release from the plant and serves to sustain pollen viability. There have been different reports of actin organization in mature pollen depending on whether the pollen was harvested prior to or after this dehydration step. Actin in already dehydrated and dehisced pollen reveals an organization that has been called fibrillar, fusiform, or spiculate (Cresti et al., 1986; Heslop-Harrison et al., 1986; Tiwari & Polito, 1988; Heslop-Harrison & Heslop-Harrison, 1992). It is likely that during the dehydration step, actin filaments are organized into cross-linked bundles or 'storage forms'. In contrast, studies that analyze actin arrays in mature pollen prior to the dehydration step report dense cortical arrays as well as some cytoplasmic arrays in association with the vegetative nucleus (Hause et al., 1992; Gervais et al., 1994; Zonia et al., 1999). The actin arrays observed in potato pollen at this stage are shown in Fig. 1K. There are dense cortical and cytoplasmic arrays, and actin bundles are closely associated with the vegetative nucleus (Figs. 1K, L). Mature potato pollen also has an unusual organization of actin at the aperture, the site from which the pollen tube will eventually emerge during pollen germination. There is a patch of actin at the aperture tip and also slightly further back at the basal region of the aperture (Fig. 1K). These structures may prime the tube primordium for rapid elongation following hydration, and/or serve as organizational domains for the localization of enzymes and factors necessary during activation and tube emission.

3. FUNCTIONAL ANALYSIS OF ACTIN ARRAYS DURING MICROSPORE AND POLLEN DEVELOPMENT

The extensive changes in actin localization and structure that are observed at the critical developmental stages suggest a role for these arrays in integrating development. A recent report has provided evidence that the localization and organization of actin arrays is a fundamental determinant of pollen development in *N. tabacum* (Zonia et al., 1999). Exposing microspores or pollen to the actin inhibitors cytochalasin D (at 1 μM) or latrunculin B (at 5 nM) significantly blocked development at the following stages: late microspore and start of mitosis, progression through mitosis, and post-mitotic development. It was further demonstrated that development at these stages was blocked because of the targeted disruption of actin arrays by the inhibitors. The actin arrays at these stages during normal development in tobacco pollen are shown in Figures 2A–F, and the effect of 1 μM cytochalasin D on these arrays and on development are shown in Figures 2G–L.

At the late microspore stage, actin surrounds the nucleus and keeps it tethered to the generative pole prior to the asymmetrical mitosis (Figs. 2A, B). Cytochalasin D blocks formation of this array, and as a result the nucleus diffuses away from the pole (Figs. 2G, H). These cells are blocked from entering mitosis. During telophase, actin is localized in the phragmoplast (Figs. 2C, D). Cytochalasin disrupts the orientation of the phragmoplast and it is observed to curve away from the target site around the generative pole (Figs. 2I, J). Thus, during the asymmetrical micropsore mitosis, actin apparently has a role in guiding the orientation of synthesis of the generative cell wall. There have been previous reports of cytochalasin-induced mispositioning of the cell plate in other plant tissues (Palevitz & Hepler, 1974; Schmit & Lambert, 1988; Mineyuki & Gunning, 1990). During post-mitotic development, actin is associated with both the generative cell and vegetative nucleus and links them as a unit during the coordinated migration toward the center of the pollen grain (Figs. 2E, F). Treatment with cytochalasin D blocks formation of this linked array, and as a result the generative cell does not move from the pole, while the vegetative nucleus drifts to the opposite pole of the cell and becomes severely disorganized and degenerates (Figs. 2K, L).

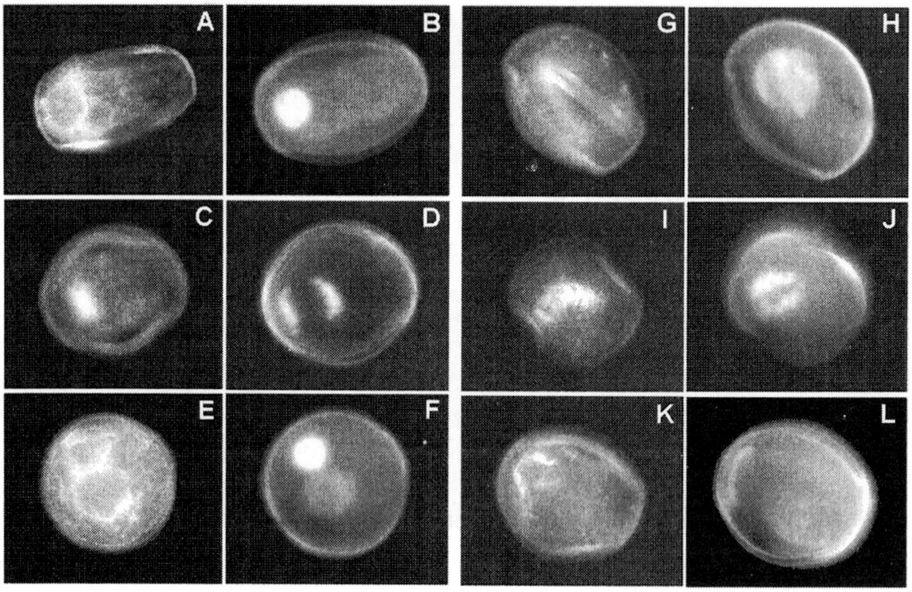

Figure 2. Actin localization during development of tobacco microspores and pollen (**A-F**) and the effect of 1 μM cytochalasin D (**G-L**). All cells were double-labeled with rhodamine-phalloidin to visualize actin (**A, C, E, G, I, K**) and DAPI to visualize the nucleus (**B, D, F, H, J, L**). **A, B, G, H** late microspore; **C, D, I, J** telophase; **E, F, K, L** mid-pollen.

Thus, at the critical stages of development – establishment of cellular polarity prior to the asymmetrical mitosis, progression through the asymmetrical mitosis, post-mitotic development – the observed actin arrays are demonstrated to have a functional role in mediating the cellular events that occur.

4. CONCLUSIONS

Formation of the male gametophyte is a useful system for addressing questions of cell biology, developmental regulation, and mechanisms of signal transduction in plants. The complex sequence of cellular events that must occur to produce the mature pollen grain suggest a role for the cytoskeleton. There is evidence that both actin and microtubule arrays have key roles in mediating the cellular events that occur during gametophyte development (Zonia et al., 1999). This review has discussed evidence for the role of actin arrays. Cytochemical localization studies have identified unique actin arrays at the critical stages of development, thus suggesting that the actin cytoskeleton is highly dynamic and rapidly reorganized. Functional analyses using inhibitors that target the actin cytoskeleton demonstrate that developmental progression and the precise orchestration of specific developmental events are dependent on the structure and localization of actin arrays. There is increasing evidence that actin reorganization during pollen development may be linked to phosphatidylinositol signal transduction cascades via associated actin-binding proteins and G-proteins (Drøbak et al., 1994; Lin et al., 1996; Lin & Yang, 1997; Staiger et al., 1997; von Witsch et al., 1998; Kost et al., 1999). Future work will more fully elucidate the signals that control actin reorganization and development in microspores and pollen.

ACKNOWLEDGEMENTS

The author is funded by the Grant Agency of the Czech Republic, Project Nr. 521/99/1354.

REFERENCES

Bedinger PA, Hardeman KJ and Loukides CA (1994) Travelling in style: the cell biology of pollen. Trends Cell Biol 4: 132-138

Brown RC and Lemmon BE (1991) Pollen development in orchids. 3. A novel generative pole microtubule system predicts unequal pollen mitosis. J Cell Sci 99: 273-281

Brown RC and Lemmon BE (1992) Pollen development in orchids. 4. Cytoskeleton and ultrastructure of the unequal pollen mitosis in *Phalaenopsis*. Protoplasma 167: 183-192

Cresti M, Hepler PK, Tiezzi A and Ciampolini F (1986) Fibrillar structures in *Nicotiana* pollen: changes in ultrastructure during pollen activation and tube emmision. In: Mulcahy DL, Bergamini-Mulcahy G, Ottaviano E (eds) Biotechnology and Ecology of Pollen. Berlin: Springer, pp 283-288

Dinis AM and Mesquita JF (1993) The F-actin distribution during microsporogenesis in *Magnolia soulangeana* Soul. (Magnoliaceae). Sex Plant Reprod 6: 57-63

Drøbak BK, Watkins PAC, Valenta R, Dove SK, Lloyd CW and Staiger CJ (1994) Inhibition of plant plasma membrane phosphoinositide phospholipase C by the actin-binding protein, profilin. Plant J 6: 389-400

Eady C, Lindsey K and Twell D (1995) The significance of microspore division and division symmetry for vegetative cell-specific transcription and generative cell differentiation. Plant Cell 7: 65-74

Gallagher K and Smith LG (1997) Asymmetric cell division and cell fate in plants. Curr Opin Cell Biol 9: 842-848

Gervais C, Simmonds DH and Newcomb W (1994) Actin microfilament organization during pollen development of *Brassica napus* cv. Topas. Protoplasma 183: 67-76

Hause B, Hause G, Pechan P and Van Lammeren AAM (1993) Cytoskeletal changes and induction of embryogenesis in microspore and pollen cultures of *Brassica napus* L. Cell Biol Intl 17: 153-168

Hause G, Hause B and Van Lammeren AAM (1992) Microtubular and actin filament configurations during microspore and pollen development in *Brassica napus* cv. Topas. Can J Bot 70: 1369-1376

Heslop-Harrison J (1968) Synchronous pollen mitosis and the formation of the generative cell in massulate orchids. J Cell Sci 3: 457-466

Heslop-Harrison J and Heslop-Harrison Y (1992) Cyclical transformations of the actin cytoskeleton of hyacinth pollen subjected to recurrent vapour-phase hydration and dehydration. Biol Cell 75: 245-252

Heslop-Harrison J, Heslop-Harrison Y, Cresti M, Tiezzi A and Ciampolini F (1986) Actin during pollen germination. J Cell Sci 86: 1-8

Kakimoto T and Shibaoka H (1988) Cytoskeletal ultrastructure of phragmoplast-nuclei complexes isolated from cultured tobacco cells. Protoplasma Suppl 2: 95-103

Kost B, Lemichez E, Spielhofer P, Hong Y, Tolias K, Carpenter C and Chua N-H (1999) Rac homologues and compartmentalized phosphatidylinositol 4,5-bisphosphate act in a common pathway to regulate polar pollen tube growth. J Cell Biol 145: 317-330

Lin Y and Yang Z (1997) Inhibition of pollen tube elongation by microinjected anti-Rop1Ps antibodies suggests a crucial role for Rho-type GTPases in the control of tip growth. Plant Cell 9: 1647-1659

Lin Y, Wang Y, Zhu J-K and Yang Z (1996) Localization of a Rho GTPase implies a role in tip growth and movement of the generative cell in pollen tubes. Plant Cell 8: 293-303

Mascarenhas JP (1989) The male gametophyte of flowering plants. Plant Cell 1 657-664

McCormick S (1993) Male gametophyte development. Plant Cell 5: 1265-1275

Mineyuki Y and Gunning BES (1990) A role for preprophase bands of microtubules in maturation of new cell walls, and a general proposal on the function of preprophase band sites in cell division in higher plants. J Cell Sci 97: 527-537

Palevitz B and Hepler PK (1974) The control of the plane of division during stomatal differentiation in *Allium*. II. Drug studies. Chromosoma 46: 327-341

Scheres B and Benfey PN (1999) Asymmetric cell division in plants. Annu Rev Plant Physiol Plant Mol Biol 50: 505-537

Schmit A-C and Lambert A-M (1988) Plant actin filament and microtubule interactions during anaphase-telophase transition: Effect of antagonist drugs. Biol Cell 64: 309-319

Sheldon JM and Hawes C (1988) The actin cytoskeleton during male meiosis in *Lilium*. Cell Biol Intl 12: 471-476

Staiger CJ and Cande WZ (1991) Microfilament distribution in maize meiotic mutants correlates with microtubule organization. Plant Cell 3: 637-644

Staiger CJ, Gibbon BC, Kovar DR and Zonia LE (1997) Profilin and actin-depolymerizing factor: modulators of actin organization in plants. Trends Plant Sci 2: 275-281

Terasaka O and Niitsu T (1995) The mitotic apparatus during unequal microspore division observed by a confocal laser scanning microscope. Protoplasma 189: 187-193

Tiwari SC and Polito VS (1988) Spatial and temporal organization of actin during hydration, activation, and germination of pollen in *Pyrus communis* L.: a population study. Protoplasma 147: 5-15

Touraev A, Pfosser M, Vicente O and Heberle-Bors E (1996) Stress as the major signal controlling the developmental fate of tobacco microspores: towards a unified model of induction of microspore/pollen embryogenesis. Planta 200: 144-152

Touraev A, Vicente O and Heberle-Bors E (1997) Initiation of microspore embryogenesis by stress. Trends Plant Sci 2: 297-302

Twell D (1999) Mechanisms of microspore polarity and differential cell fate determination in developing pollen. In: Cresti M, Cai G, Moscatelli A (eds) Fertilization in Higher Plants. Berlin: Springer, pp 201-215

Van Lammeren AAM, Bednara J and Willemse MTM (1989) Organization of the actin cytoskeleton during pollen development in *Gasteria verrucosa* (Mill.) H. Duval visualized with rhodamine-phalloidin. Planta 178: 531-539

von Witsch M, Baluska F, Staiger CJ and Volkmann D (1998) Profilin is associated with the plasma membrane in microspores and pollen. Eur J Cell Biol 77: 303-312

Zaki MAM and Dickinson HG (1990) Structural changes during the first divisions of embryos resulting from anther and free microspore culture in *Brassica napus*. Protoplasma 156: 149-162

Zaki MAM and Dickinson HG (1991) Microspore-derived embryos in *Brassica*: the significance of division symmetry in pollen mitosis I to embryogenic development. Sex Plant Reprod 4: 48-55

Zonia L, Tupy J and Staiger CJ (1999) Unique actin and microtubule arrays co-ordinate the differentiation of microspores to mature pollen in *Nicotiana tabacum*. J Exp Bot 50: 581-594

Chapter 21

ACTIN: A TARGET OF SIGNAL TRANSDUCTION IN ROOT HAIRS

Anne Mie C Emons and Norbert de Ruijter

Laboratory of Experimental Plant Morphology and Cell Biology, Department of Plant Sciences, Wageningen University, Arboretumlaan 4, 6703 BD Wageningen, The Netherlands

Key words: actin filament, legume, lipochito-oligosaccharide, Nod factor, plant-microbe interaction, *Rhizobium*, root hair, signal transduction, tip growth

Abstract: Signal transduction is a process that takes place primarily at the cell level. Therefore, root hairs are ideal cells to study cell biological aspects of signal transduction cascades, since they are easily accessible. A number of groups have reported changes in the actin cytoskeleton of root hairs induced by lipochito-oligosaccharides (LCOs). LCOs are signal molecules excreted by *Rhizobium* bacteria that induce root hair deformation in legume root hairs. In root hairs grown between glass slides, root hair deformation involves swelling of the hair tip and subsequent tip growth from one site in the swelling. Because this occurs in hairs that are terminating growth, root hair deformation implies a re-initiation of tip growth. Since modifications of the actin cytoskeleton are necessary for tip growth re-initiation, it was expected that the actin cytoskeleton is a downstream target of LCOs. We discuss the changes in the configuration of the actin cytoskeleton of root hairs of all developmental stages and after challenge with LCOs. Furthermore, we speculate about the possible mechanisms by which the signal may induce changes in the configuration of the actin cytoskeleton.

1. ROOT HAIR CYTOARCHITECTURE AND THE ACTIN CYTOSKELETON

Actin filament bundles are the backbones of cytoplasmic strands and, consequently, essentially organise the plant cytoarchitecture. For understanding actin dynamics in plant cells, knowledge of the cell architecture, the cytoarchitecture, is obligatory. Cytoplasmic strands traverse

C.J. Staiger et al. (eds.), Actin: A Dynamic Framework for Multiple Plant Cell Functions, 373–390.

the vacuole and, therefore, are bound by the vacuolar membrane. They are continuous with the cortical cytoplasm, which covers the whole plasma membrane. Together, transvacuolar cytoplasmic strands and cortical cytoplasm are one connected system. In fact, cytoplasmic strands without actin filaments in them do not exist. When the actin filaments are depolymerised by microinjection of the actin-binding protein (ABP) profilin, which binds to actin monomers, the cytoplasmic strands become thinner and snap back into the cortical cytoplasm, by which process a large central vacuole is formed. This has been shown in mature (Staiger et al., 1994) and growing (Valster et al., 1997) *Tradescantia* stamen hair cells. This can also be achieved by actin filament depolymerising drugs, such as the actin filament capping fungal toxin cytochalasin D (CD). At relatively high CD concentrations (1–5 μM), the increased depolymerisation of the actin filaments causes the formation of one large vacuole in growing root hairs of *Vicia sativa* (Esseling et al., 2000). However, bundles of actin filaments remain present in the cortical cytoplasm, both after profilin injection in *Tradescantia* stamen hair cells (Valster et al., 1997), and after CD treatment of root hairs (Miller et al., 1999, Esseling et al., 2000).

An obvious characteristic of cytoplasmic strands is the movement of the organelles inside them, known as cytoplasmic streaming. Actin filaments are the ***tracks*** along which this movement occurs, as shown by the above-mentioned experiments in which profilin was injected into *Tradescantia* stamen hair cells causing the cessation of cytoplasmic streaming (Staiger et al., 1994; Valster et al., 1997). In these experiments, a second injection with fluorescent phalloidin confirmed that depolymerisation of actin filaments causes the arrest of cytoplasmic streaming in cytoplasmic strands. Moreover, CD stops cytoplasmic streaming in *Vicia sativa* root hairs at concentrations above 5 μM (Miller et al., 1999).

Tip growing cells have "reverse-fountain" streaming, a term introduced by Iwanami (1956) for pollen tubes. Wide pollen tubes (*Lilium*), as well as wide root hairs (*Hydrocharis*, Tominaga et al., 1997, 1998) show an upward flow of cytoplasm near the plasma membrane and a downward flow in a transvacuolar cytoplasmic strand in the cell centre. Thinner root hairs, such as those of *Arabidopsis* (Ketelaar & Emons, unpublished), *Medicago truncatula* (Sieberer & Emons, personal communication) and *Vicia sativa* (Miller et al., 1997), as well as thinner pollen tubes (*Nicotiana tabacum*, de Win et al., 1999; *Papaver rhoeas*, Geitmann, unpublished), can have the downward streaming in the centre, but also near the plasma membrane. The most characteristic feature of reverse-fountain streaming is that the upward flow never reaches the plasma membrane at the very apex of a tip-growing cell. Its direction reverses just below the vesicle-rich region, an area at the root hair tip that almost exclusively contains vesicles. This characteristic organisation of the dynamic cytoplasm is necessary for tip growth to occur, which is local growth at one side of the cell, because it delivers the Golgi vesicles to the tip area. The site of insertion of new plasma membrane and exocytosis of cell wall material through Golgi vesicles (reviewed in Roberts, 1994) into a flexible cell wall determines the direction in which a plant cell grows. To achieve the delivery of Golgi vesicles to one side of the cell, growing root hairs, which develop from bulges on root epidermal cells (Fig. 1a), have a polar cytoarchitecture with a central vacuole in the basal part of the cell, a sub-apical cytoplasmically-dense region at the tip, and a thin smooth area bordering the tip plasma membrane (Fig. 1c). Electron microscopy has shown that the smooth 1–3 μm long area contains almost exclusively vesicles (Fig. 2, vetch) (chemical fixation: Bonnett & Newcomb, 1966, freeze-substitution: Emons, 1987; Ridge, 1988; Sherrier & VandenBosch, 1994; Galway et al., 1997). These vesicles are probably a mixture of exocytotic and endocytotic vesicles (Emons & Traas, 1986). During growth, the sub-apical cytoplasmic dense region, which is the area in which the direction of cytoplasmic streaming reverses, varies in length, while the vacuole increases in size in concord with cell elongation. The length of this sub-apical cytoplasmically-dense region in growing root hair tips deviates in *Medicago truncatula* between 3–30 μm in the cell cortex and between 3–15 μm in the cell centre.

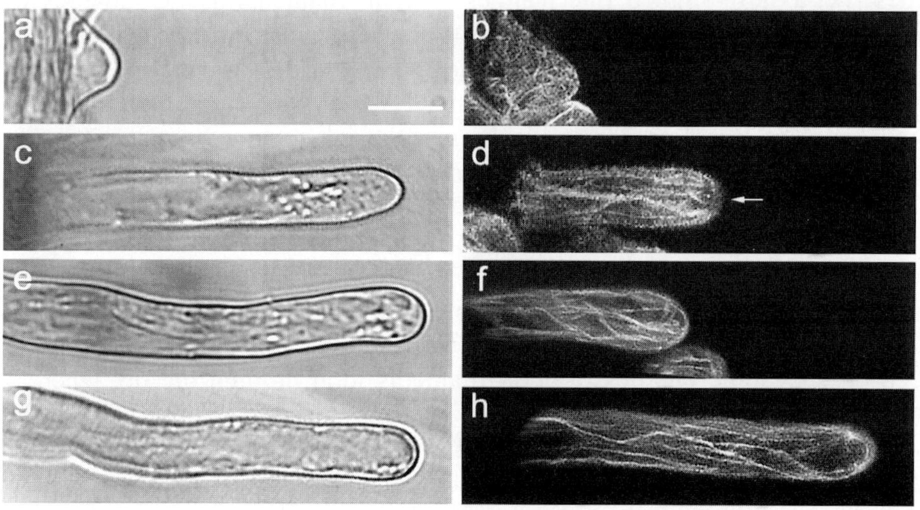

Figure 1. (**a–h**) Cytoarchitecture and actin filament configuration of subsequent developmental stages of *Vicia sativa* root hairs. (**a, c, e, g**) DIC images of living root hairs, representative for each developmental stage. (**b, d, f, h**) Representative fluorescence images showing the actin filament distribution in each developmental stage. Actin filaments were visualised by following the ester-aldehyde-choline (EAC) procedure (Miller et al., 1999).

Root hairs were prefixed for 3 min in 200 μM m-maleimido benzoyl N-hydroxysuccinimide ester in medium. Cytoplasmic streaming stopped within 1 minute. Roots were fixed for 20 min in a final concentration of 4% *p*-formaldehyde and 0.1% glutaraldehyde in Actin Stabilizing Buffer (ASB: 100 mM Pipes pH 6.8, 1 mM $MgCl_2$, 1 mM $CaCl_2$, 75 mM KCl) with 1 mM 4-2-aminoethylbenzene sulfonyl fluoride. Root hairs were permeabilised for 12 min with 100 μg/ml L-α-lyso-phosphatidylcholine in ASB. Actin filaments were stained with 0.33 μM fluorescein-phalloidin in ASB supplemented with 0.05% acetylated BSA. Roots were mounted in CITIFLUOR-glycerol. (**a, b**) Bulge containing a major central vacuole. Actin filament bundles are randomly oriented and pass through the peripheral cytoplasm at the tip. (**c, d**) Growing root hair with a smooth area at the very tip, and a subapical cytoplasmic dense region. The actin filament bundles are longitudinally oriented in the hair. In the subapical region, dense bundles of actin filaments are present (FB-actin), while the tip is devoid of actin filament bundles (*arrow*). (**e, f**) Root hair in which growth is terminating. Small vacuoles are present close to the tip with cytoplasmic strands between them. There is no tip area that is devoid of filamentous actin, and the subapical area with FB-actin is very short. (**g, h**) Full-grown root hair with one large vacuole and peripheral cytoplasm. Thick actin filament bundles are present in strands looping through the extreme tip. The magnification is the same in all images. Bar in (**a**), 20 μm.

In growing hairs, bundles of actin filaments are present in cytoplasmic strands around and traversing the central vacuole (Miller et al., 1999). These bundles lie longitudinally and helically in the root hair. In the sub-apical cytoplasmically-dense region, the filament bundles flare out into thinner and thinner bundles (Fig. 1c) and maybe even into single filaments. We call these net-axially-aligned fine bundles of actin filaments in the sub-apex of root hairs FB-actin (Miller et al., 1999). The very apex of the growing hair is devoid of actin filament bundles (arrow, Fig. 1d). In vetch, this actin filament-free region is 2–6 μm deep and 5–8 μm wide (Miller et al., 1999) and, more or less coincides with the vesicle-rich region. In root hairs that terminate growth the cytoplasmically-dense region, as well as the area with FB-actin, is very short (Fig. 1e). Furthermore, a region devoid of actin in the extreme tip can not be detected in the light microscope (Fig. 1f) (Miller et al., 1999; de Ruijter et al., 1999). In full-grown hairs, actin filament bundles in the sub-apical and tip area are situated in a thin peripheral layer of cytoplasm around a large central vacuole (Fig. 1g), just as elsewhere in the cell cortex. Some bundles of actin filaments loop through the tip, whereas others turn in the cortical cytoplasm before they reach the tip (Fig. 1h). FB-actin is not present in the sub-apex and there is no apical area devoid of filamentous actin. Comparison of the actin cytoskeleton of growing and non-growing root hairs and studies with CD have shown that the FB-actin cytoskeleton configuration and the tip area free of bundles of actin filaments are coupled phenomena and necessary for cell tip growth.

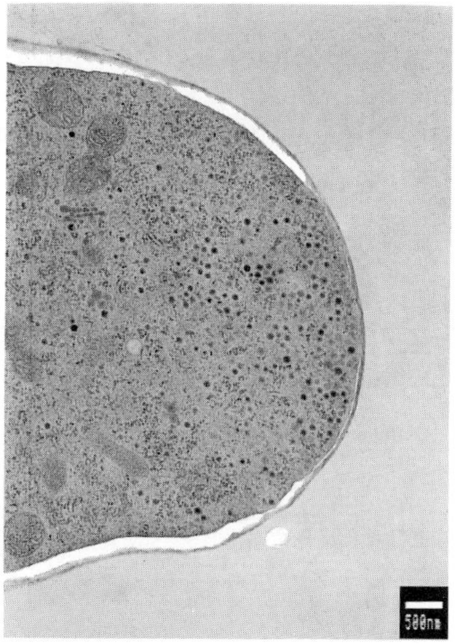

Figure 2. Electron microscopy photograph of the tip of a growing root hair of *Vicia sativa* with vesicles filling the tip cytoplasm.

2. NOD FACTOR-INDUCED CHANGES IN THE ACTIN CYTOSKELETON

When nitrogen-fixing *Rhizobium* bacteria encounter, and enter, their host legume plants, they can induce the formation of so-called root nodules, structures formed by the plant *de novo* in which the bacteria live and where nitrogen fixation takes place (Mylona et al., 1995). The bacteria enter the root via an infection thread, first formed inside the root hair (reviewed in Kijne, 1992). The bacteria excrete lipochito-oligosaccharides (LCOs), so-called Nod factors that, applied in concentrations as low as 10^{-12} M to liquid grown legume roots, induce several responses with a characteristic time course. In the first minutes after application, LCOs trigger physiological responses, including alkalinisation of the root hair (Ehrhardt et al., 1992; Felle et al., 1996), depolarisation of the root hair plasma membrane (Ehrhardt et al., 1992; Felle et al., 1996; Kurkdjian 1995) creation of a tip-focused calcium ion gradient (Cárdenas et al., 1999) and calcium spiking (Ehrhardt et al., 1996). Changes in the actin cytoskeleton precede the start of root hair deformation. These changes in the actin cytoskeleton occur early in the LCO-induced cascade, before gene expression can have taken place. Later, LCOs induce the expression of early nodulin genes, such as ENOD5, ENOD12, VsLb1 and Mtrip1 (Horvath et al., 1993; Journet et al., 1994;

Cook et al., 1995; Heidstra et al., 1997a). About 2–3 h after application of LCOs, root hair deformation is complete in 80–90% of the hairs that were terminating growth at the moment of application (Heidstra et al., 1994; de Ruijter et al., 1998). At that time point, cells in the root cortex start to divide (Mylona et al., 1995), the first step of nodule formation.

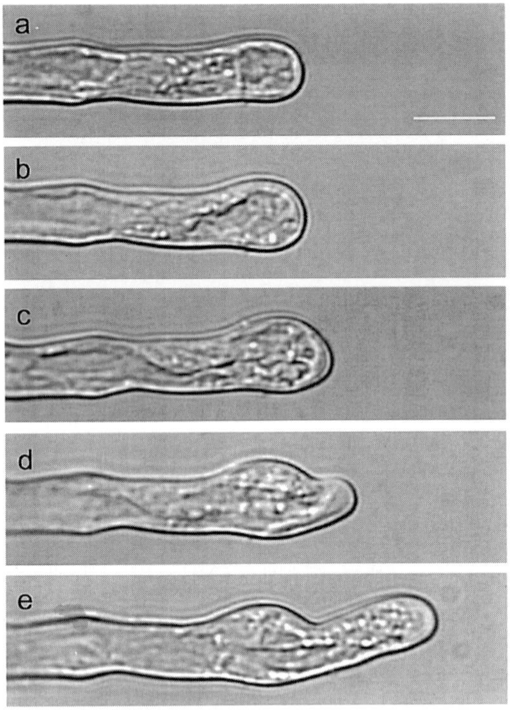

Figure 2. Root hair deformation after LCO application to roots of *Vicia sativa*. (**a–e**) A time-series of DIC images from one living root hair that was terminating growth, showing a swelling and reinitiation of a new outgrowth from that swelling after application of 10^{-10} M LCO. (**a**) Typical cytoarchitecture of a hair terminating growth. (**b**) An initial swelling of the tip at 1 h after LCO application. **c** At one side of the swelling cytoplasm accumulates at 1 h, 15 min after LCO application, a smooth area becomes visible, and tip-polarised growth is initiated. (**d**) Cytoplasmic strands orient towards the new tip at 1 h, 30 min after LCO application, and renewed tip growth is apparent. (**e**) The cytoarchitecture of the outgrowth (at t = 2 h, 30 min after LCO) has similar characteristics as a normal growing hair (see **Fig. 1c**). Magnification is the same for (**a–e**). Bar in (**a**), 20 μm.

To study the basic reactions of root hairs to LCOs, we have investigated the cellular changes that occur in vetch root hairs of roots that are growing between a coverslip and microscope slide. In this method, the hairs are easily accessible for the application of LCOs and visualisation by light microscopy. The hairs that react most visibly with root hair deformation are those that are terminating growth. Root hair deformation is characterised by a swelling of the root hair tip, followed by an outgrowth from that swelling (Fig. 3) (see also Heidstra et al., 1994). Outgrowth is reinitiation of tip growth, since it has all the characteristics of normal root hair tip growth (de Ruijter et al., 1998). The outgrowth has a tip-focused high concentration of cytosolic calcium, an accumulation of a spectrin-like epitope, and a vesicle-rich area at the tip, as well as a sub-apical cytoplasmically-dense area, and reverse-fountain streaming (de Ruijter et al., 1998). So, the cytoarchitecture of the outgrowth is exactly the same as the one described for growing root hairs.

Since the hairs that deform do not have the actin filament configuration necessary for tip growth, it was a logical step to study the actin cytoskeleton after Nod factor application. This could show that filamentous actin is not only a ***track*** or ***rail***, but also part of the plant cell ***signalling device***. The first reports about responses of the actin cytoskeleton to LCOs are from Allen et al. (1994) and Allen & Bennett (1996). They reported a fragmentation of the actin cytoskeleton in root hairs within the first 15 min after Nod factor application to *Medicago sativa* roots, seen as discontinuities in the longitudinal actin bundles. Later, Cárdenas et al. (1998) reported a fluorescent glow at the tip of *Phaseolus vulgaris* root hairs that had been injected with FITC-phalloidin, within 5 min after treatment with host-specific Nod factor. These authors interpreted this glow as actin filament fragmentation (Cárdenas et al., 1998). As discussed above, actin filament breakdown causes the disappearance of cytoplasmic strands. Thus, the results of Cárdenas et al. (1998) would suggest that actin filament fragmentation by Nod factor should cause the breakdown of cytoplasmic strands in root hairs. In *Vicia sativa* root hairs, however, actin filaments remain intact after Nod factor application (Miller et al., 1999), though they become more dynamic (Sieberer & Emons, personal communication). This is consistent with the idea that cytoplasmic strands can not exist without filamentous actin (Staiger et al., 1994, Valster et al., 1997). Cárdenas et al. (1998) reported also the formation of discrete foci of phalloidin staining in the outgrowth after Nod factor application to root hairs injected with FITC-phalloidin. We have not seen such foci in our actin visualisations of vetch root hairs, in which we used freeze-substitution as a technique alongside an optimised ester-aldehyde-choline (EAC) chemical-fixation procedure. The EAC procedure (see legend to Fig. 1) did not change the cytoarchitecture of the living cell and the actin cytoskeleton preserved in this way resembled the images of freeze-substituted cells (Miller et al., 1999).

These controversies made us examine in more detail the actin cytoskeleton in vetch root hairs challenged with LCOs. In a time series starting 3 min after LCO application, the roots with hairs were prepared with the EAC procedure and stained with fluorescein-phalloidin, and the number of actin filaments was counted at 0–50 μm from the tip. The number of fine bundles of actin filaments increased significantly in the sub-apical area of most root hairs within 3–15 min after Nod factor application (de Ruijter et al., 1999), and occurred in all stages of root hair growth (Figs. 4 & 5). Also those hairs that were in a developmental stage in which they do not respond to Nod factor with root hair deformation, did show the same response in the actin cytoskeleton.

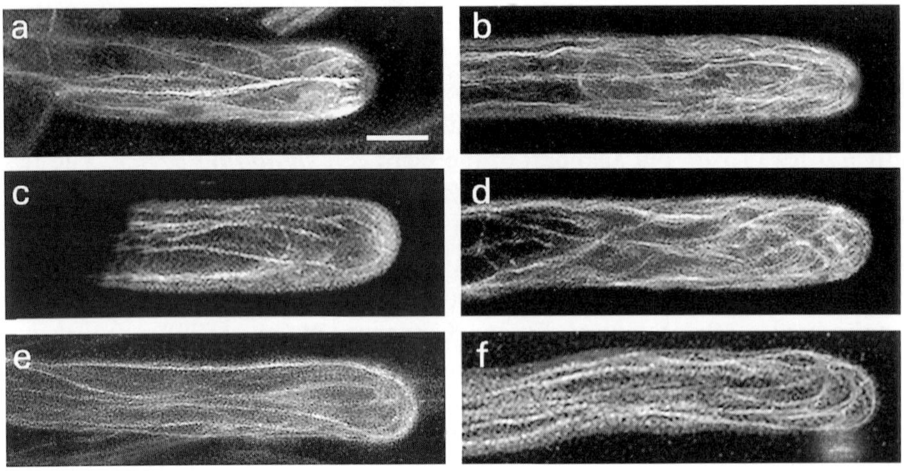

Figure 4. Increase in the density of bundles of actin filaments in *Vicia sativa* root hairs (**a, b**) Growing root hairs. Typical distribution and density of actin filament bundles (**a**) after replacing growth medium (control, t = 3 min), and (**b**) after application of LCOs (t = 15 min). Note the loss of thick bundles and increase of fine bundles of actin filaments. (**c, d**) Root hairs terminating growth. Typical distribution and density of actin filament bundles (**c**) after replacing growth medium (control, t = 6 min), and (**d**) after application of LCOs (t = 3 min). In hairs at this developmental stage, after challenge with LCOs, fine bundles of actin filaments often had a wider range of orientations in relation to the long axis of the hair, than growing hairs after LCO application. (**e, f**) Full-grown root hairs. Typical distribution and density of actin filament bundles (**e**) after replacing growth medium (control, t = 15 min), and (**f**) after application of LCOs (t = 15 min). To compare data, all settings during imaging and processing were similar. Confocal z-series were made at 1 μm steps. The magnification is the same in all images. Bar in (**a**), 10 μm.

Figure 4 shows examples of hairs in three developmental stages of growth, namely growing, growth terminating and full-grown, taken 3–15 min after Nod factor application or, as a control, after replacement of the plant growth medium only. In all three, the density of sub-apical net-axial fine bundles of actin filaments has significantly increased. After quantification (see legend to Fig. 5) of this increase in fine bundles of actin filaments, an interesting disparity appears. If we take the density of actin filament bundles in control growth terminating hairs as the threshold level necessary for growth, LCO application increased the density of fine bundles of actin filaments in growing hairs. Consequently, they kept on growing. The filamentous actin density in the growth terminating hairs increased to, or above, the threshold level for growth and reinitiated tip growth, after a lag period in which the hair tip swelled. We interpret this swelling of the root

Figure 5a: Density of actin filament bundles in control and LCO treated growing root hairs

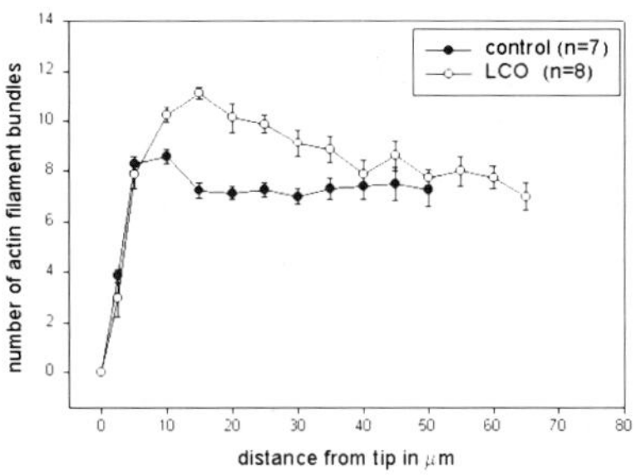

Figure 5b: Density of actin filament bundles in control and LCO treated root hairs that are terminating growth

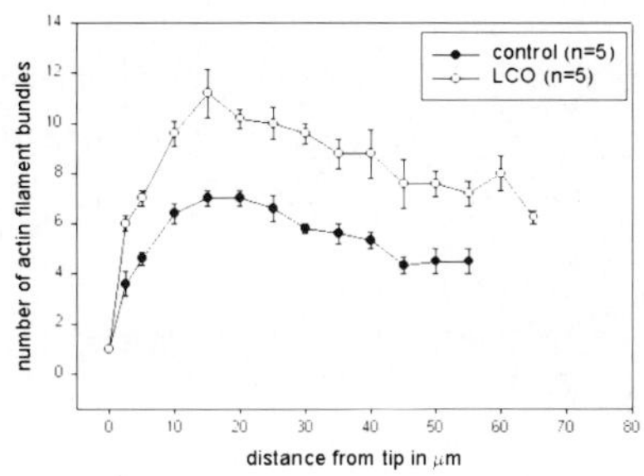

Figure 5c: Density of actin filament bundles in control and LCO treated full-grown root hairs

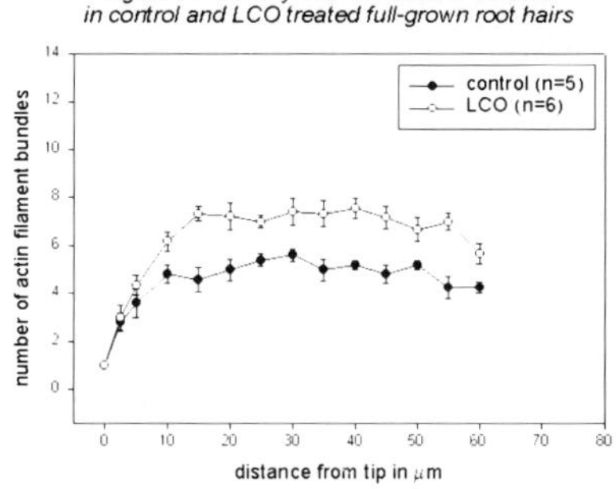

Figure 5. Mean densities of actin filament bundles in the subapical area of control and LCO-treated *Vicia sativa* root hairs. For **(a)** growing root hairs, **(b)** root hairs that are terminating growth, and **(c)** full-grown root hairs mean densities of actin filament bundles (y-axis) are displayed versus the distance from the tip (x-axis). In all root hairs, the density profile shows an increase in actin filament bundles after application of LCOs. Note that the increase in full-grown hairs (in **c**) does not exceed the critical FB-actin density that correlates with growth (control in **b**). A reproducible quantification procedure was done as follows. Random

sampling was done within each zone at 3–15 min after treatment. Confocal z-series were similarly processed. To quantify the number of actin filament bundles, a grid with equally spaced lines was superimposed onto the projected image of a root hair. The grid was perpendicular to the long axis of the hair and the zero-line was always positioned at the very tip. The number of all fine and thick actin filament bundles hit by these cross-lines was counted on a high-resolution screen and means of numbers of actin filaments were displayed. Error bars indicate SE.

hair tip as undirected growth (de Ruijter et al., 1998, Miller et al., 1999). The filamentous actin density in full-grown hairs, on the other hand, also increased after Nod factor treatment, but remained below the threshold level necessary for growth. Among other factors, like an already rigid secondary cell wall at tips of full-grown hairs, the limited increase of FB-actin in full-grown hairs may be the reason why those hairs do not respond with root hair deformation. Although the actin cytoskeleton responds to LCOs with reactivation of an actin filament configuration necessary for growth, there is a lag period of ~1 h before polarised tip growth starts. In this hour, the tip of the hair swells isodiametrically.

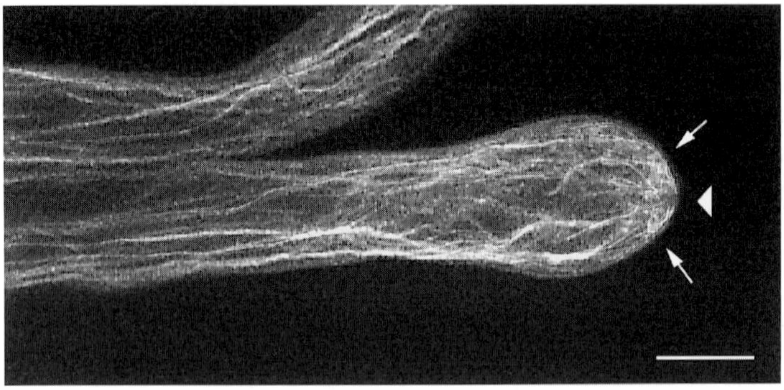

Figure 6. An immunofluorescence micrograph of the actin cytoskeleton, 75 min after LCO application, shows a short area with FB-actin pointing to a new outgrowth (arrows). Note that these bundles enclose an area at the plasma membrane devoid of actin filament bundles (arrowhead). The image is a z-projection of a series of confocal images over 10 μm depth at 0.5 μm intervals. Bar, 15 μm.

Growing hairs do not swell, or swell only slightly, because the tip growth apparatus is already present in abundance. During swelling, the FB-actin is less-focused than during polarised tip growth. Subsequently, the FB-actin focuses to a point in the swelling (Fig. 6). In the *in vivo* calcium imaging experiments with Indo-1 as vital and ratiometric calcium dye (de Ruijter et al., 1998) we have observed a spot of elevated $[Ca^{2+}]_c$ moving around in the peripheral cytoplasm of the swelling. Differential Interference Contrast (DIC) imaging of the same cell shows that this ‘hot spot’ of Ca^{2+} ions colocalises with an area of dense cytoplasm on top of dynamic cytoplasmic strands (de Ruijter et al., 1998). Such spots of dense cytoplasm in the peripheral cytoplasm of a swelling contain FB-actin (Fig. 6). Then, the hot spot of calcium and the FB-actin remain fixed at one site. From this site in the swelling, the outgrowth starts growing in a polarised fashion just like a normal root hair.

3. POSSIBLE MECHANISMS OF NOD FACTOR-INDUCED CHANGES IN THE ROOT HAIR ACTIN CYTOSKELETON

The LCO-induced increase in the number of fine bundles of actin filaments in the sub-apical area of root hairs suggests a signal transduction pathway involving the activation of ABPs. *In vitro*, the ABPs profilin and ADF/cofilin bind to PtdIns(4,5)P_2. This fact provides a link between the phosphatidylinositol signal transduction pathway and the actin cytoskeleton (Yonezawa et al., 1990; Aderem, 1992; Staiger et al., 1997). Therefore, it can be hypothesised that this signal transduction pathway is involved in Nod factor signalling. Recent experiments lend support to this hypothesis (Pingret et al., 1998). Inhibitors of phospholipase C (PLC) applied simultaneously with LCOs to *Medicago truncatula* roots inhibit the expression of a MtENOD12-GUS construct in the root epidermis. It is known that ENOD12 is induced by Nod factor in legume roots (Scheres et al., 1990, Pichon et al., 1992, Journet et al., 1994). Therefore, the inhibition of ENOD12 by PLC inhibitors indicates that PLC is part of the signal transduction pathway induced by Nod factors (Pingret et al., 1998). *In vitro*, PLC cleaves PtdIns(4,5)P_2 into inositol (1,4,5)-trisphosphate ($InsP_3$) and diacylglycerol (DAG) (Divecha & Irvine, 1995), suggesting a rapid increase in $InsP_3$ after Nod factor application. Whether PLC inhibitors block Nod-factor-induced responses is currently being studied in our laboratory.

Profilin and PtdIns(4,5)P_2 are present in tips of growing root hairs of *Zea mays*, *Lepidium sativum* and *Arabidopsis thaliana* (Braun et al., 1999). These authors suggest that profilin-actin interactions function in root hair tip growth. They report that "... *profilin, which is known to be a widely expressed, potent regulator of actin dynamics was specifically localised at tips of growing root hairs and co-distributed with a diffusely fluorescing cap of actin, but not with subapical actin microfilaments. Profilin and actin were present exclusively in the bulge of outgrowing root hairs; both disappeared when tip growth terminated, indicating a tip growth mechanism that involves profilin-actin interactions for the delivery and localized exocytosis of secretory vesicles*".

Unfortunately, in this work the actin filament bundles are not resolved well in the cell region close to the tip. Therefore, in the paper by Braun et al. (1999) it is difficult to draw conclusions on the absence of actin filament bundles in the extreme cell tip, or on the presence of sub-apical FB-actin. The thick bundles of actin filaments in the cytoplasmic strands in the vacuolated cell areas of root hairs are clear, but wrinkled, a sign of sub-optimal fixation. The brightly-fluorescing cap of diffusely-labelled actin in the growing root hairs, is comparable to what we see after inappropriate fixation and staining with fluorescent phalloidin (Esseling et al., 2000). Also in well-preserved freeze-substituted root hairs, details of the fine bundles of sub-apical actin filaments are difficult to visualise with anti-actin antibodies, since labelled G-actin may mask the FB-actin (Miller et al., 1999). We think that G-actin labelling is also the reason why in the freeze-substituted immunolabelled hairs of vetch the tip area free of actin filament bundles is smaller that in EAC fixed hairs (Miller et al., 1999). However, we can not exclude the possibility that the chemical fixation does not preserve the thinnest bundles, and that fluorescent phalloidin does not stain them. Braun et al., (1999), report an area in between the diffusely-labelled tip and the bundles in cytoplasmic strands in growing root hairs of *Lepidium* that has hardly any actin filaments which they observed to a lesser extent in root hairs of *Zea* and *Arabidopsis*. This is probably an artefact of the procedure, the more so since in all studied tip-growing cells this area contains dense cytoplasm which should need actin filaments as a backbone. Such a gap was also not reported by Miller et al. (1996) and by Kost et al. (1998) for living pollen tubes, the other well-known tip-growing cell of higher plants (see also Vidali & Hepler, and Kost et al., this volume). These authors used injection of fluorescently-labelled phalloidin (Miller et al., 1996) and transgenic pollen tubes expressing the GFP–talin actin binding domain (Kost et al., 1998) to visualise actin filaments in living pollen tubes. It is our experience that fixation of root hairs has to be performed under continuous observation in the DIC microscope and needs to be improved until the fixative and further chemicals used in the staining do not alter the cytoarchitecture of the living hair. For growing root hairs, this is a polarly-organised cytoarchitecture with a smooth, area in the extreme tip and a cytoplasmically-dense region in the sub-apex.

Regarding profilin in the work of Braun et al. (1999), one should, in general, be very cautious in interpretation of the results on the localisation of small, cytoplasmic molecules, as long as the accessible cell volume has not been taken into account. Vos and Hepler (1998) have shown this quite elegantly in a study on the localisation of calmodulin during cell division in

Tradescantia stamen hair cells. The profilin localisation reported by Braun et al. (1999) is exactly at the site of a large accessible volume in growing, full-grown, and CD-treated root hairs. Furthermore, such small proteins are difficult to fix and may relocate during the procedure. We are not suggesting that the interpretations are wrong, but point out that these localisations have not been proven yet.

The scheme shown in Figure 7 shows our current hypothesis for how ABPs such as profilin could induce actin rearrangements, but, we want to stress that we do not have evidence yet. Microinjection of signal transduction intermediates, fluorescent-tagged actin and ABPs, combined with fluorescence resonance energy transfer (FRET), and over/under-expression of the molecules involved, should help elucidate the molecular interactions in the signal transduction pathway, triggered by Nod factors.

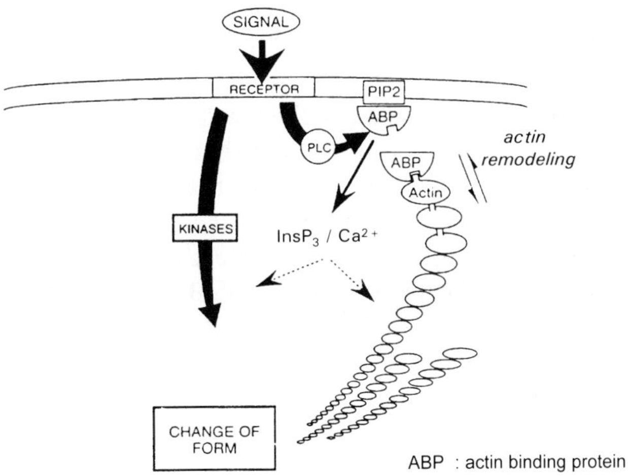

Figure 7. A scheme showing how actin remodeling and ABPs may be part of the signal transduction cascade after Nod factor application (signal) that leads to root hair deformation (*change of form*).

Nod factor causes an increase in the cytoplasmic pH in *Medicago sativa* root hairs (Ehrhardt et al., 1992; Felle et al., 1996). This may point to a regulation of the changes in the actin cytoskeleton by ABPs like actin depolymerizing factor (ADF)/cofilin (see also Kovar & Staiger, this volume). In LCO-treated *Medicago sativa* root hairs, the pH increases to levels (Ehrhardt et al., 1992; Felle et al., 1995, 1996) at which plant ADF is able to sever and depolymerise actin filaments (Staiger et al., 1997; Maciver et al., 1998). Moreover, in growing maize root hairs, the localisation of maize ADF3 (Jiang et al., 1997) coincides, essentially, with the region of

Vicia sativa root hairs in which an increase in fine bundles of actin filaments is seen shortly after Nod factor treatment (de Ruijter et al., 1999). However, for the ADF (immuno-)localisation (Jiang et al., 1997), the same concerns as outlined for profilin localisation can be raised, unless ADF is bound to filaments. We do not know the distribution of ADF in legume root hairs. However, it can be hypothesised that Nod factor-induced cytoplasmic pH changes influence the severing or depolymerising activity of ADF, which may result in a contribution to the increase of sub-apical fine bundles of actin filaments. Based on its activity, distribution and involvement in signal transduction pathways, ADF is also a good candidate for ABPs involved in the short-term actin rearrangements after Nod factor treatment. However, by analogy with ABP function in animal cells and yeast, we expect that additional ABPs are involved in the rearrangement of the actin cytoskeleton by Nod factor. For reviews on ABPs in plant cells, see de Ruijter and Emons (1999) and Chapters 2–8 of this volume.

Another factor should not be overlooked. That is the increase of $[Ca^{2+}]_c$ at the apex within 5 minutes after Nod factor application (Cárdenas et al., 1999). Calcium ions could act on actin via a number of ABPs (see also Yokota & Shimmen, and Gibbon & Staiger, this volume). Calcium may, among other things, dynamise existing filaments, stimulate filament polymerisation, and uncouple ABPs such as myosin from the filaments, and, therefore, uncouple the vesicles attached to these motor molecules from actin filaments, at the base of the vesicle-rich area. However, calcium ions could also directly depolymerise/dynamise the actin filaments and uncouple the exocytotic vesicles from them.

4. FOCUS FOR FUTURE RESEARCH OF SIGNAL - INDUCED CHANGES IN THE ACTIN CYTOSKELETON OF PLANT CELLS

To unravel the complex signal transduction pathways leading from a signal molecule such as the Nod factor to changes in cell morphogenesis, a bioassay is a useful tool. We have used the occurrence of root hair deformation by Nod factor challenge as a bioassay for studying the influence of external factors on this process (Heidstra et al., 1997b, Miller et al., 1999). Full root hair deformation involves swelling of the root hair tip followed by root hair outgrowth from that swelling. In vetch, this whole process takes 2–3 h. The imaging of calcium dynamics at the root hair tip after a challenge with LCOs (Cárdenas et al., 1999), and of calcium spiking (Ehrhardt et al., 1996) give us more refined tools for investigating the signal transduction cascades. In addition, the rapid, quantifiable increase of FB-actin in root hairs after Nod factor application (de Ruijter et al., 1999), provides a supplementary bioassay. Instead of waiting for 2 h to score root hair deformation, one can score the rapid effects of Nod factor on tip-focused calcium dynamics, the actin cytoskeleton, or calcium spiking. The use of drugs that affect the cytoskeleton and signal transduction cascades in combination with Nod factor, will make it possible to discriminate between the early factors in the signal transduction cascade that change the cytoskeleton, and later factors that need for instance calcium spiking and induction of new gene transcription. Ideally, whenever possible, this should be done in living cells. Therefore, GFP-technology and microinjection techniques will become important tools for studying the living cell, and will be used more extensively in combination with molecular genetic and biochemical approaches.

REFERENCES

Aderem A (1992) Signal transduction and the actin cytoskeleton: the roles of MARCKS and profilin. Trends Biochem Sci 17: 438-443

Allen NS, Bennett MN, Cox DN, Shipley A, Ehrhardt DW and Long SR (1994) Effects of Nod factors on alfalfa root hair Ca^{++} and H^{+} currents and on cytoskeletal behavior. In: Daniels MJ, Downie JA, Osbourn AE (eds) Advances in Molecular Genetics of Plant Microbe Interactions 3, Dordrecht, Netherlands: Kluwer Acad Pub, pp 107-113

Allen NS and Bennett MN (1996) Electro-optical imaging of F-actin and endoplasmic reticulum in living and fixed plant cells. Scanning Microsc Suppl: 10:177-187

Bonnett HT and Newcomb EH (1966) Coated vesicles and other cytoplasmic components of growing root hairs of radish. Protoplasma 62: 59-75

Braun M, Baluška F, von Witsch M and Menzel D (1999) Redistribution of actin, profilin and phosphatidylinositol 4,5-biphoshate in growing and maturing root hairs. Planta 209: 435-443

Cárdenas L, Vidali L, Domínguez J, Pérez H, Sánchez F, Hepler PK and Quinto C (1998) Rearrangement of actin microfilaments in plant root hairs responding to *Rhizobium etli* nodulation signals. Plant Physiol 116: 871-877

Cárdenas L, Feijó JA, Kunkel JG, Sánchez F, Holdaway-Clarke T, Hepler PK and Quinto C (1999) *Rhizobium* Nod factors induce increases in intracellular free calcium and extracellular calcium influxes in bean root hairs. Plant J 19: 347-352

Cook D, Dreyer D, Bonnet D, Howell M, Nony E and VandenBosch K (1995) Transient induction of a peroxidase gene in *Medicago truncatula* precedes infection by *Rhizobium meliloti*. Plant Cell 7: 43-55

de Ruijter NCA and Emons AMC (1999) Actin-binding proteins in plant cells. Plant Biol 1: 26-35

de Ruijter NCA, Rook MB, Bisseling T and Emons AMC (1998) Lipochito-oligosaccharides re-initiate root hair tip growth in *Vicia sativa* with high calcium and spectrin-like antigen at the tip. Plant J 13: 341-350 de Ruijter NCA, Bisseling T and Emons AMC (1999) *Rhizobium* Nod factors induce an increase in subapical fine bundles of actin filaments in *Vicia sativa* root hairs within minutes. Mol Plant Microbe Interact 12: 829-832

de Win AH, Pierson ES and Derksen J (1999) Rational analyses of organelle trajectories in tobacco pollen tubes reveal characteristics of the actomyosin cytoskeleton. Biophys J 76: 1648-1658

Divecha N and Irvine RF (1995) Phospholipid signalling. Cell 80: 269-278

Emons AMC (1987) The cytoskeleton and secretory vesicles in root hairs of *Equisetum* and *Limnobium* and cytoplasmic streaming in root hairs of *Equisetum*. Ann Bot 60: 625-632

Emons AMC and Traas JA (1986) Coated pits and coated vesicles on the plasma membrane of plant cells. Eur J Cell Biol 41: 57-64

Ehrhardt DW, Atkinson EM and Long SR (1992) Depolarization of alfalfa root hair membrane potential by *Rhizobium meliloti* Nod factors. Science 256: 998-1000

Ehrhardt DW, Wais R and Long SR (1996) Calcium spiking in plant root hairs responding to *Rhizobium* nodulation signals. Cell 85: 1-20

Esseling J, de Ruijter N and Emons AM (2000) The root hair actin cytoskelton as backbone, highway, morphogenetic instrument and target for signalling. In: Ridge RW, Emons AM (eds) Cell and Molecular Biology of Plant Root Hairs, Tokyo, Japan: Springer-Verlag. In press

Felle HH, Kondorosi É, Kondorosi Á and Schultze M (1995) Nod signal-induced plasma membrane potential changes in alfalfa root hairs are differently sensitive to structural modifications of the lipochito-oligosaccharide. Plant J 7: 939-947

Felle HH, Kondorosi É, Kondorosi Á and Schultze M (1996) Rapid alkalinization in alfalfa root hairs in response to rhizobial lipochito-oligosaccharide signals. Plant J 10: 295-301

Galway ME, Heckman Jr JW and Schiefelbein JW (1997) Growth and ultrastructure of *Arabidopsis* root hairs: the *rhd3* mutation alters vacuole enlargement and tip growth. Planta 201: 209-211

Heidstra R, Geurts R, Franssen H, Spaink HP, van Kammen A and Bisseling T (1994) Root hair deformation activity of nodulation factors and their fate on *Vicia sativa*. Plant Physiol 105: 787-797

Heidstra R, Nilsen G, Martinez-Abarca F, van Kammen A and Bisseling T (1997a) Nod factor-induced expression of leghemoglobin to study the mechanism of NH_4NO_3 inhibition on root hair deformation. Mol Plant Microbe Interact 10: 215-220

Heidstra R, Yang WC, Yalcin Y, Peck S, Emons AM, van Kammen A and Bisseling T (1997b) Ethylene provides positional information on cortical cell division but is not involved in Nod factor-induced root hair tip growth in *Rhizobium*-legume interaction. Development 124: 1781-1787

Horvath B, Heidstra R, Lados M, Moerman M, Spaink HP, Prome JC, van Kammen A and Bisseling T (1993) Lipo-oligosaccharides of *Rhizobium* induce infection-related early nodulin gene expression in pea root hairs. Plant J 4: 727-733

Iwanami Y (1956) Protoplasmic movement in pollen grains and pollen tubes. Phytomorphology 6: 288-295

Jiang C-J, Weeds AG and Hussey PJ (1997) The maize actin-depolymerizing factor, ZmADF3, redistributes to the growing tip of elongating root hairs and can be induced to translocate in the nucleus with actin. Plant J 12: 1035-1043

Journet EP, Pichon M, Dedieu A, de Billy F, Truchet G and Barker DG (1994) *Rhizobium meliloti* Nod factors elicit cell-specific transcription of the ENOD12 gene in transgenic alfalfa. Plant J 6: 241-249

Kijne JW (1992) The *Rhizobium* infection process. In: Stacey G, Burris RH, Evans HJ (eds) Biological Nitrogen Fixation, New York: Chapman and Hall, pp 349-398

Kost B, Spielhofer P and Chua N-H (1998) A GFP-mouse talin fusion protein labels plant actin filaments *in vivo* and visualizes the actin cytoskeleton in growing pollen tubes. Plant J 16: 393-401

Kurkdjian AC (1995) Role of differentiation of root epidermal cells in Nod factor (from *Rhizobium meliloti*)-induced root hair depolarization of *Medicago sativa*. Plant Physiol 107: 783-790

Maciver SK, Pope BJ, Whytock S and Weeds AG (1998) The effect of two actin depolymerizing factors (ADF/cofilins) on actin filament turnover: pH sensitivity of F-actin binding by human ADF, but not of *Acanthamoeba actophorin*. Eur J Biochem 256: 388-397

Miller DD, Lancelle SA and Hepler PK (1996) Actin microfilaments do not form a meshwork in *Lilium longiflorum* pollen tube tips. Protoplasma 195: 123-132

Miller DD, de Ruijter NCA and Emons AMC (1997) From signal to form: aspects of the cytoskeleton-plasma membrane-cell wall continuum in root hair tips. J Exp Bot 48: 1881-1896

Miller DD, de Ruijter NCA, Bisseling T and Emons AMC (1999) The role of actin in root hair morphogenesis: studies with lipochito-oligosaccharide as a growth stimulator and cytochalasin as an actin perturbing drug. Plant J 17: 141-154

Mylona P, Pawlowski K and Bisseling T (1995) Symbiotic nitrogen fixation. Plant Cell 7, 869-885

Pichon M, Journet EP, Dedieu A, De Billy F, Truchet G and Barker DG (1992) *Rhizobium meliloti* elicits transient expression of the early nodulin gene ENOD12 in the differentiating root epidermis of transgenic alfalfa. Plant Cell 4, 1199-1211

Pingret J-L, Journet E-P, Barker DG (1998) Rhizobium Nod factor signaling: Evidence for a G protein-mediated transduction mechanism. Plant Cell 10: 659-672

Ridge RW (1988) Freeze-substitution improves the ultrastructural preservation of legume root hairs. Bot Mag Tokyo 101: 427-441

Roberts K (1994) The plant extracellular matrix: in a new expansive mood. Curr Opin Cell Biol 6: 688-694

Scheres B, Van der Wiel C, Zalensky A, Horvath B, Spaink H, van Eck H, Zwartkruis F, Wolters A-M, Gloudemans T, van Kammen A and Bisseling T (1990) The ENOD12 gene product is involved in the infection process during pea-Rhizobium interaction. Cell 60: 281-294

Sherrier DJ and Van den Bosch KA (1994) Secretion of cell wall polysaccharides in *Vicia* root hairs. Plant J 5: 185-195

Staiger CJ, Yuan M, Valenta R, Shaw PJ, Warn RM and Lloyd CW (1994) Microinjected profilin affects cytoplasmic streaming in plant cells by rapidly depolymerizing actin microfilaments. Curr Biol 4: 215-219

Staiger CJ, Gibbon BC, Kovar DR and Zonia LE (1997) Profilin and actin-depolymerizing factor: modulators of actin organization in plants. Trends Plant Sci 2: 275-281

Tominaga M, Morita K, Sonobe S, Yokota E and Shimmen T (1997) Microtubules regulate the organization of actin filaments at the cortical region in root hair cells of *Hydrocharis*. Protoplasma 199: 83-92

Tominaga M, Sonobe S and Shimmen T (1998) Mechanism of inhibition of cytoplasmic streaming by auxin in root hair cells of *Hydrocharis*. Plant Cell Physiol 39: 1342-1349

Valster AH, Pierson ES, Valenta R, Hepler PK and Emons AMC (1997) Probing the plant actin cytoskeleton during cytokinesis and interphase by profilin microinjection. Plant Cell 9: 1815-1824

Vos JW and Hepler PK (1998) Calmodulin is uniformly distributed during cell division in living stamen hair cells of *Tradescantia virginiana*. Protoplasma 201: 158-171

Yonezawa N, Nishida E, Iida K, Yahara I and Sakai H (1990) Inhibition of the interactions of cofilin, destrin and deoxyribonuclease-I with actin by phosphoinositides. J Biol Chem 265: 8382-8386

Chapter 22

THE ROLE OF ACTIN DURING *ARABIDOPSIS* TRICHOME MORPHOGENESIS

Dan Szymanski

Department of Agronomy, Purdue University, West Lafayette, IN 47907 USA

Key words: actin, *Arabidopsis*, genetics, microtubule, morphogenesis, trichome

Abstract: *Arabidopsis* trichome (leaf hair) development is being used as a genetic system to study actin-dependent growth. *Arabidopsis* trichomes are unicellular structures that are derived from the epidermis. Trichome development has been established as a powerful model system to study the control of cell fate and pattern formation. Results from several recent papers suggest that a genetic analysis of trichome morphogenesis may yield insight into the role of the cytoskeleton during cellular morphogenesis. In particular, it is found that filamentous actin (F-actin) plays an important role during *Arabidopsis* trichome development. Specific rearrangements of the actin cytoskeleton occur during the development of the mature trichome cell. F-actin-disrupting drugs do not affect the establishment of polarity during trichome development; however, the maintenance and coordination of the normal pattern of cell growth are very sensitive to the same agents. In contrast, inhibitors that depolymerize microtubules severely inhibit cell polarization. Disruption of the actin cytoskeleton in growing trichomes phenocopies a class of mutations that cause a distorted trichome morphology. An analysis of cell shape and microfilament structure in wild-type, mutant, and drug-treated trichomes is consistent with a role for actin microfilaments in the maintenance and coordination of an established growth pattern during *Arabidopsis* trichome formation.

1. GENETIC ANALYSIS OF THE PLANT CYTOSKELETON

The actin cytoskeleton comprises an intricate filamentous network that is essential for the trafficking of organelles and vesicles in eukaryotic cells. This transport activity is thought to be necessary for the normal growth of

C.J. Staiger et al. (eds.), Actin: A Dynamic Framework for Multiple Plant Cell Functions, 391–410.

many plant cell types. The actin cytoskeleton is a dynamic filamentous structure. Understanding how the organization the actin cytoskeleton is regulated in response to both intrinsic and extrinsic cues is an important and difficult question in plant biology (reviewed in Staiger, 2000). A genetic analysis of this process may lead to important insights into the regulation of the actin cytoskeleton. This chapter describes the utility of *Arabidopsis* trichome genetics as a model system to identify genes that control actin organization.

To date, many advances have come from studies of the role of actin during pollen tube growth. In this system, a combination of approaches relate the localization and activity of actin and actin-binding proteins to pollen tube organization and expansion during tip growth (reviewed in Heslop-Harrison et al., 1986; Staiger et al., 1997; Taylor & Hepler, 1997; see also Vidali & Hepler, this volume). Because cytoskeletal function in most plant cell types is not amenable to biochemical approaches, it is important to develop genetic strategies to study cytoskeletal organization.

The power of a genetic approach to study the actin cytoskeleton has been established in organisms as diverse as yeasts, flies and worms (Ayscough, 1998). The feasibility of using genetic techniques to study the cytoskeleton and cell growth has also been demonstrated in plants. Reverse-genetic identification of mutations in *Arabidopsis* actin gene family members will provide important tools to study the function of actin (McKinney et al., 1995). Actin organization is important for normal coleoptile elongation in monocots (Thimann et al., 1992; Waller & Nick, 1997). The *Yin-Yang* mutant of rice displays an altered actin organization in coleoptile epidermal cells during auxin-stimulted cell elongation (Wang & Nick, 1998). Mutation of the *Arabidopsis* kinesin-like gene *ZWICHEL* (*ZWI*) leads to defects in trichome morphogenesis (Oppenheimer et al., 1997). The presence of more than 30 genes that affect different aspects of trichome morphogenesis suggests that a detailed analyses of this process will provide a unique experimental window into cytoskeletal organization (reviewed in Marks, 1997).

2. TRICHOME STRUCTURE AND FUNCTION

Trichomes, defined as hair-like appendages that extend from the epidermis of aerial tissues, are present on the surface of almost all terrestrial plants (Johnson, 1975). Trichomes comprise a very diverse set of structures, and many plants contain several types on a single leaf. These can be divided into two basic classes: 1) complex, multicellular structures comprised of several specialized cell types that have a secretory function. The aerial epidermis of 20–30% of vascular plants contain glandular trichomes (Duke, 1994). Secreted material often accumulates in the subcuticular space of the apical cell of the glandular trichome. And 2), simple multicellular or unicellular trichomes that extend from of the epidermal surface but lack a secretory anatomy.

It has been difficult to clearly demonstrate the function of plant trichomes, but several ideas have gained widespread acceptance. The presence of trichomes increases the boundary layer thickness between the epidermal tissue and the environment, and can protect against heat and water loss. In many species, trichomes are thought to protect the plant against insect or pathogen attack. These defensive qualities can be due to either chemical components that are secreted from trichomes or via the physical structure of the cell that limits insect access to or mobility on vegetative tissues. Figure 1 illustrates a glandular alfalfa trichome (Fig. 1A) and a *Mentzelia pumila* hooked trichome (Figs. 1B–E), both of which have the capacity to non-specifically immobilize insects at the leaf surface.

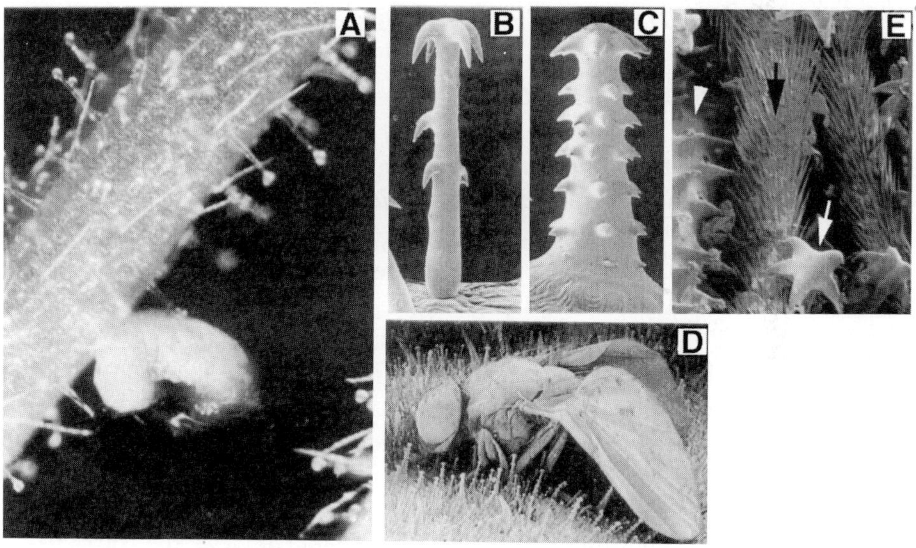

Figure 1. Scanning electron micrograph of trichomes. **(A)** An alfala weevil larvae attached to the sticky glandular exudate of an alfalfa trichome. Copyright (1975) RE Shade, reproduced with permission. **(B)** Type 1 *Mentzelia pumila* trichome. **(C)** Type 3 *M. pumila* trichome. **(D)** An agromyzid fly entangled in trichomes on the surface of a *M. pumila* plant. **(E)** High magnification of (D) showing the antennae of the fly wedged between type 1 and type 3 trichomes; white arrowhead, type 3 trichome; white arrow, type 1 trichome; black arrow, fly leg. (B–E) Copyright (1998) National Academy of Sciences USA, reproduced with permission.

Based on the potential economic importance of trichomes for the pharmaceutical, insecticidal, flavor and textile industries, considerable effort has been spent to understand the interplay between the cellular organization and function of plant trichomes. For example, the ultrastructure and vacuolar organization of the multicellular acid-secreting trichomes of cowpea have been examined in fixed specimens and in living cells (Lazzaro & Thomson, 1992, 1996). Cotton fibers are trichomes that develop from the outer integument cells of the ovule, and are an excellent model system to study cytoskeletal organization and cell wall synthesis. The microtubule and microfilament organization of developing cotton fibers has been examined in detail, (Seagull, 1986, 1990, 1992; Tiwari & Wilkins, 1995). The fine structure of trichomes from dozens of species of economic significance also has been documented (Rodriguiez et al., 1984). However, understanding how trichome structure and function is determined requires the use of molecular genetic techniques in conjunction with cell biological tools.

3. *ARABIDOPSIS* TRICHOME DEVELOPMENT

3.1 Trichome Initiation

Arabidopsis leaf trichomes are unicellular stellate trichomes that usually contain three branches. Historically, *Arabidopsis* trichome development has been used to address the question of how cell fate and pattern formation are regulated in the plane of the epidermis (reviewed in Larkin et al., 1997; Marks, 1997). Cell fate decision-making is complex and involves the concerted activity of at least seven genes that positively and negatively regulate trichome initiation. Many of the early acting genes encode putative transcriptional regulators (reviewed in Szymanski et al., 2000).

The ability of an epidermal cell to receive trichome differentiation signals is strictly regulated, and does not appear to be determined solely by the presence or absence of transcription factors. Cells that are mitotically cycling appear incapable of adopting the trichome cell fate (Pyke et al., 1991; Lloyd et al., 1994; Larkin et al., 1996). As leaf development progresses, leaf epidermal cells exit the mitotic cycle and undergo variable rounds of endoreduplication (DNA synthesis without cell division) (Szymanski & Marks, 1998). The ability of cells to receive trichome differentiation signals is limited to a narrow developmental window during the transition from the mitotic to the endomitotic cycle (Lloyd et al., 1994). Genes that alter the cell cycle also affect the spacing and tissue distribution of trichome formation, but the relationship between cell cycle status and the ability to receive trichome differentiation signals is not clear (Schnittger et al., 1998;

Szymanski & Marks, 1998). Trichome development represents a powerful experimental system to understand how transcription factors and cell cycle parameters interact with cytoskeletal components to alter the fate and morphogenesis of a cell.

3.2 Trichome Morphogenesis

Once an epidermal precursor enters the trichome pathway, it undergoes a complex morphogenetic program (Hülskamp et al., 1994; reviewed in Hülskamp et al., 1998). Based on morphological landmarks, trichome growth can be divided into discrete stages (Szymanski et al., 1998), as shown in Figure 2. The first sign of trichome formation is isodiametric expansion within the plane of the epidermis (stage 1). In cross section, the trichome precursor appears as an enlarged spherical cell. The cell nucleus is greatly enlarged, and it has been proposed that the trichome has undergone additional endoreduplication cycles relative to neighboring cells (Hülskamp et al., 1994). However, DNA content has not been measured in developing trichomes. During stage 1, neighboring socket cells form a tight association with the spherical trichome precursor. Stage 2 trichomes display obvious polarized elongation perpendicular to the leaf plane until the cell reaches a length of ~30 μm. The changes in cellular organization and shape that occur during stage 2 resemble tip-growing cells, but the nature of stage 2 growth has not been examined directly. During stage 3, branch initiation occurs sequentially on the developing stalk, often yielding a cell with three branches. The elongating branch buds initially have a blunt tip morphology (stage 4), but as the branches elongate, the tip morphology becomes more pointed (stage 5). The transition to stage 5 occurs very early in trichome development, and the vast majority of the cell volume is generated by diffuse growth during this phase. Once cell expansion has ceased, the cell wall acquires a papillate surface (stage 6). Although there is some variation between trichomes in the shape changes that occur during morphogenesis, each cell executes a similar developmental program.

Genetic and pharmacological experiments suggest that precise cytoskeletal function is required throughout trichome development. The *ZWI* gene encodes a minus-end directed kinesin-like motor protein that is required for normal branch formation and stalk elongation (Oppenheimer et al., 1997; Song et al., 1997). Further genetic analysis has identified many additional genes that are required for normal branch formation (Folkers et al., 1997; Krishnakumar & Oppenheimer, 1999; Luo & Oppenheimer, 1999; Perazza et al., 1999). An analysis of double mutant combinations for apparent loss-of-function branching mutants has uncovered a highly redundant branch initiation control pathway (Luo & Oppenheimer, 1999).

Pharmacological data suggest that microtubule-dependent function is required for stalk and branch initiation (see below). Drugs that affect actin organization have very distinct effects on trichome morphogenesis; polarized elongation and branch initiation are not obviously affected, while the coordinated expansion of branches and the stalk following stage 3 is severely inhibited (see below).

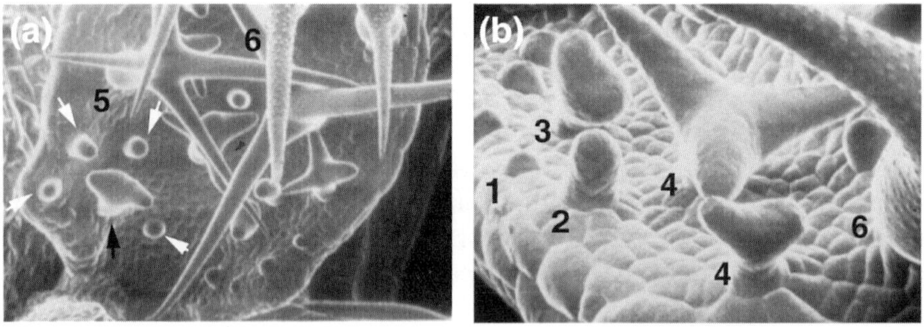

Figure 2. Stages of trichome development. Scanning electron micrographs of the adaxial surface of *Arabidopsis* leaves that illustrate aspects of trichome spacing and morphogenesis. **(a)** An example of secondary trichomes (white arrows) initiating around an existing central trichome (black arrows). Numbers to the left of each labeled trichome indicate the developmental stage. **(b)** An illustration of additional stages of trichome development: 1, isodiametric expansion in the plane of the epidermis; 2, stalk emergence and polar expansion; 3, branch initiation; 4, expansion of the stalk and branches with a blunt tip morphology; 5, trichome expansion with pointed branch tips; and 6, mature trichome with a papillate cell wall. Reproduced from Szymanski et al., 2000.

4. ACTIN FUNCTION AND CELL POLARITY IN OTHER SYSTEMS

Like all other polarized cell types, developing trichomes must define a specific cortical position for cell expansion and reinforce and maintain the asymmetry during morphogenesis (Drubin & Nelson, 1996). This hierarchical cascade of functional requirements has been studied in detail in many other experimental systems. In many cases, the actin cytoskeleton plays an essential role in the polarity establishment and maintenance.

The role of actin in cell polarity has been studied most thoroughly in the budding yeast *Saccharomyces cerevisiae*. Genetic analysis of bud formation has identified a clear hierarchy of regulation (Pringle et al., 1995). The GTPase cdc42 acts near the top of the regulatory cascade to locally regulate actin organization and the subsequent assembly of bud site-specific factors (Ayscough et al., 1997). F-actin is also essential for normal polarity establishment in the embryos of the brown algae, *Fucus* and *Pelvetia.* Pharmacological experiments with cytochalasins indicate that an unperturbed F-actin cytoskeleton is required for the normal onset of cell polarization in *Fucus* embryos (Quatrano, 1973; Brawley & Robinson, 1985). In *Pelvetia*, an F-actin patch in living embryos has been reported as an early marker of cell polarization and predicts the position of rhizoid emergence (Alessa & Kropf, 1999). Pharmacological evidence and F-actin localization also suggest that actin plays an important role during pollen germination (Mascarenhas & LaFountain, 1972; Tiwari & Polito, 1990). In the germinated pollen tube, a complex longitudinal array of actin bundles that terminate distal to the pollen tube tip likely plays an important role in vesicle delivery to the apical region of the cell (Picton and Steer, 1981; Lancelle et al., 1987, Miller et al., 1996). However, recent studies suggest that organization of specific membrane domains and actin-dependent function in the tip region of the tube are essential for normal pollen tube growth (Gibbon et al., 1999; Kost et al., 1999).

The cytoskeletal requirements for sensory bristle and hair formation in *Drosophila* are similar to those of *Arabidopsis* trichomes. In developing sensory bristles, a dense population of individual microtubules is surrounded by longitudinal hexagonal cortical bundles of microfilaments (Tilney et al., 1995). During early developmental stages of bristle development, the assembly of the core microtubules appears to precede microfilament bundle assembly (Tilney et al., 1996). Rearrangement of the microtubule and microfilament cytoskeletons also has been documented during hair formation on the *Drosophila* wing (Eaton et al., 1996; Turner & Adler, 1998). A pharmacological analysis suggests that microtubules are essential for wing hair initiation and morphogenesis, and the primary role of the actin cytoskeleton is to maintain the established growth pattern (Turner & Adler, 1998).

The specialized requirements for F-actin assembly during bristle development have been exploited to identify several genes that regulate microfilament organization. For example, removal of profilin or actin-capping protein genes is lethal in *Drosophila.* However, weak alleles of both genes yield bristle defects (Verheyen & Cooley, 1994; Hopmann et al., 1996). The *forked* and *singed* mutations cause bristle shape defects (Bender, 1960; Hoover et al., 1993). The predicted protein product of the *SINGED* gene shares amino acid similarity with the actin-bundling protein fascin (Cant et al., 1994). FORKED and SINGED appear to act sequentially to regulate actin bundle formation in developing bristles (Tilney et al., 1996; Wulfkuhle et al., 1998).

5. CYTOSKELETAL FUNCTION IN *ARABIDOPSIS* TRICHOME DEVELOPMENT

5.1 Pharmacology

Two recent publications provide preliminary data that *Arabidopsis* trichome formation is a useful genetic system to study actin-dependent growth (Mathur et al., 1999; Szymanski et al., 1999). Pharmacological data demonstrate that organized actin is not required for the initiation of polarized growth. Polarized growth during stage 2 and branch initiation during stage 3 are not noticeably affected by F-actin disrupting agents (Fig. 3A). However, subsequent cell expansion during stages 4 and 5 is disordered, and cell shape defects are apparent (Fig. 3B). F-actin-dependent function is required to coordinate cell expansion following the formation of a polarized stalk and branch structures.

The apparent unimportance of F-actin during the establishment of polarized growth contrasts with many other developmental processes in which F-actin is required. This is not due to an absence of F-actin during these stages of development. Using conventional fixation techniques coupled with the freeze-shattering permeabilization approach described by Wasteneys et al. (1997), anti-actin antibodies label intricate networks of F-actin in both stage 1 and late stage 2 cells (Szymanski et al., 1999). Perhaps the role of F-actin during stage 2 is to facilitate apical transport, but in its absence, diffusion or microtubule-dependent function are sufficient. It should be emphasized that while disruption of actin organization does not appear to affect the rate of cell elongation through stage 3, drug-treated cells at early stages of development are slightly swollen, and subtle kinetic effects on growth have not been examined (Szymanski et al., 1999).

Polar growth during stages 2 and 3 requires microtubule-dependent activities (Figs. 3C, 3D). Agents that either stabilize or depolymerize microtubules can inhibit the establishment of polarity (Mathur et al., 1999). This was clearly shown using a dexamethose-regulated form of the maize R gene to induce trichome formation in the presence or absence the microtubule-depolymerizing agent oryzalin (Szymanski et al., 1999). Application of high concentrations of oryzalin prior to initiation causes isotropic cell expansion without branch formation. However, at lower drug concentrations some polar expansion is observed (Mathur et al., 1999). Because staining for microtubule organization in *Arabidopsis* trichomes has not been conducted, the extent to which the microtubule cytoskeleton is disrupted in drug-treated cells is not known. Residual polar expansion of treated cells may be due to a drug-resistant population of microtubules. Cotton fiber elongation displays a similar sensitivity to microtubule-destabilizing agents (Tiwari & Wilkins, 1995), and a population of apparently functional drug-resistant microtubules have been described (Seagull, 1990).

Agents that disrupt F-actin organization severely disrupt trichome morphogenesis following branch initiation in stage 3 (Fig. 3B). In untreated cells, cell expansion during stages 4 and 5 is strictly regulated and includes diffuse growth of the stalk and branches, as well as potential tip-directed growth in the elongating branches. Drug-treated stage 4 and 5 trichomes display swollen stalks, twisted and swollen branches, and aborted branches. High concentrations of cytochalasin D (CD) do not stop cell growth. Leaves exposed to the drug for six days give rise to enlarged cells with a striking, distorted morphology (Fig. 3B). Interestingly, Mathur et al. (1999) show that the effects of F-actin stabilizing agents phalloidin and jasplakinolide are indistinguishable from those of CD and Latrunculin B, which tend to destabilize microfilaments. It was shown that each actin-binding drug severely disrupts cellular organization (Mathur et al., 1999). It is possible that the actin-dependent reinforcement of an organized growth pattern requires dynamic actin filaments or the activity of F-actin-binding proteins, both of which can be affected by different classes of actin-binding agents.

Even in mature trichomes that have completed the cell expansion program, the F-actin appears to be dynamic; CD rapidly fragments the cortical filaments. In stage 6 trichomes, the actin bundles are somewhat evenly spaced, and are oriented longitudinally in the stalk (Fig. 4A). After 30 min of CD treatment, the actin cytoskeleton is completely fragmented and disorganized (Fig. 4B). After 2 h of drug treatment, most stalks contain actin that is assembled into heavy rods (Fig. 4C), and after 24 h, bright punctate spots and heavy rods of F-actin are frequently detected (Fig. 4D). Surprisingly, cells in which the actin cytoskeleton has been completely depolymerized can recover from drug treatment and repolymerize oriented F-actin (Mathur et al., 1999). A direct analysis of actin dynamics in living cells would provide insight into the assembly of ordered F-actin arrays in trichomes.

The current data suggest that, at a very superficial level, there is a hierarchy of cytoskeletal control during trichome formation, with microtubule-dependent function acting before actin-dependent function. However, the temporal and functional requirments for microtubules are not entirely clear. Inhibitors of microtubule polymerization inhibit both polarized elongation during stage 2 and branch formation at stage 3 (Figs. 3C, 3D). Microtubule-dependent function is not limited to the onset of polarization. Branched wild-type trichomes treated with oryzalin display severe swelling and branch distortion during stages 4 and 5 (Fig. 3D).

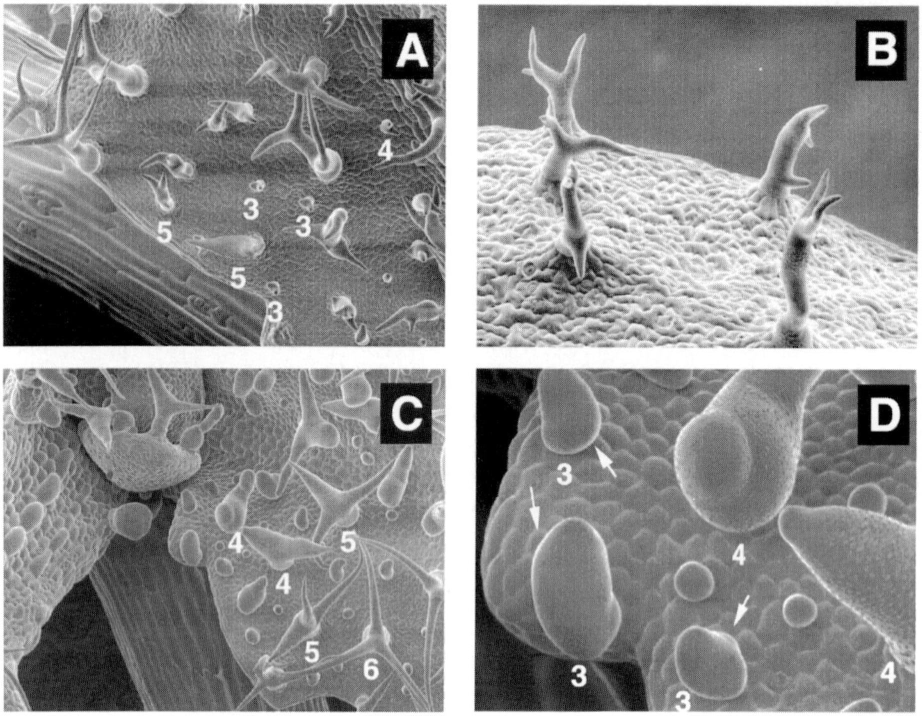

Figure 3. Effects of cytoskeletal inhibitors on trichome morphogenesis. Scanning electron micrographs of the adaxial leaf surface of drug-treated leaves. **(A)** Trichome morphology after 48 h of treatment with 50 μM Latrunculin B. **(B)** Trichome morphology after treatment with 50 μM cytochalasin D for 6 d. **(C)** Shape defects of cells treated with 100 μM oryzalin for 48 h. **(D)** High magnification of (C) demonstrating the swollen or distorted morphology of stage 3 and 4 oryzalin-treated cells. Numbers printed below each trichome indicate the developmental stage. Arrows indicate bulges that may represent defective trichome branches.

These data point to unique, but partially overlapping, roles for the microtubule and microfilament arrays during trichome development. The relationship between microtubule and F-actin organization is not clear. To begin to address this issue, Mathur et al. (1999) used a GFP–TALIN fusion protein as a probe for actin organization (Kost et al., 1998, 1999; see also Kost et al., this volume). Analysis was performed on living cells in which microtubules were disrupted with drugs and in *zwi* trichomes. In both cases F-actin was detected, suggesting that actin polymerization does not require native microtubule organization. However, based on the limited actin localization data in drug-treated and *zwi* trichomes, and the absence of microtubule localization data, it is difficult to conclude that microtubules do not influence the organization of microfilaments (see also Collings & Allen, this volume).

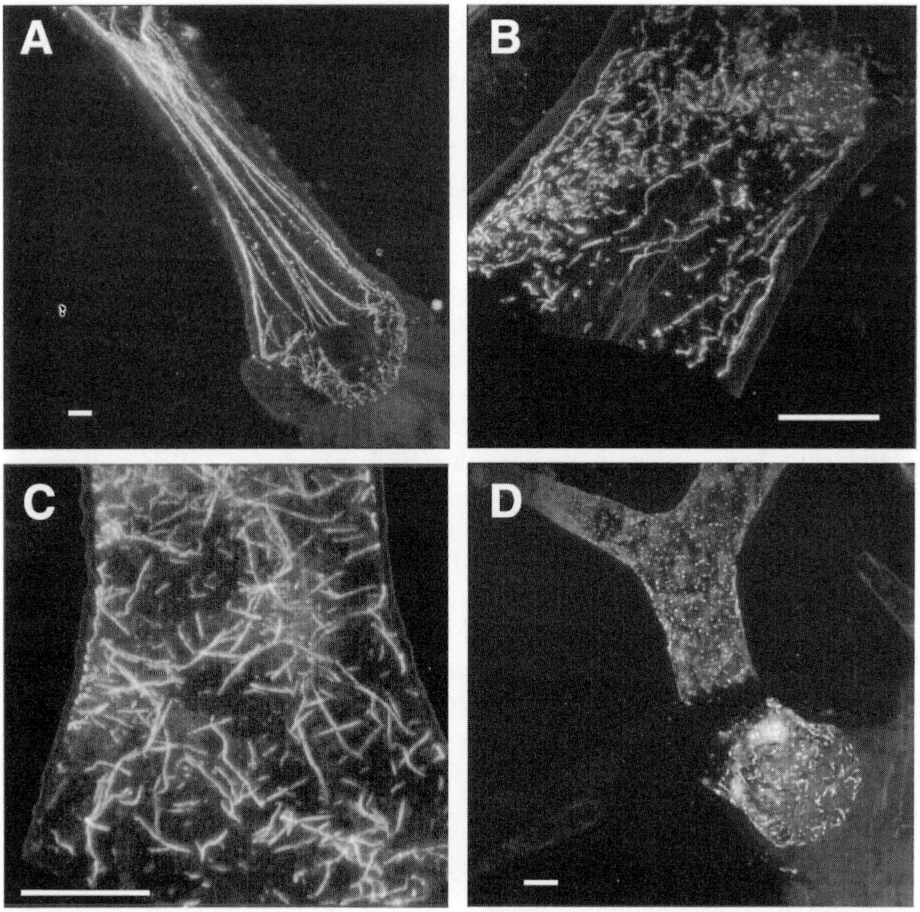

Figure 4. Time course of cytochalasin D (CD) effects on actin organization in mature trichomes. Each image is a maximum projection of confocal images of anti-actin immunofluorescence. **(A)** Typical wild-type trichome on a plant treated only with buffer. **(B)** Trichome treated for 30 min with 50 μM CD. **(C)** Trichome treated for 2 h with 50 μM CD. **(D)** Trichome treated for 24 h with 50 μM CD. Bar for (A–D), 10 μm.

5.2 Actin organization in developing trichomes

A survey of the actin organization in trichomes at each developmental stage has been conducted using both immunolocalization of actin in fixed samples (Szymanski et al., 1999) and the actin-binding domain of mouse TALIN fused to GFP (Mathur et al., 1999). In many cases the description of actin at specific developmental stages differed. For example, antibodies clearly detected F-actin structures in fixed stage 1 and 2 trichomes, however no similar structures were detected with GFP–TALIN. Clearly, both techniques have limitations. For example, the controversial dense actin patch at the tip of growing pollen tubes, that was detected using chemical-fixation methods, appears to be an artifact (Miller et al., 1996). Steady-state actin organization in pollen tubes reported with GFP-TALIN resembles the structures revealed with independent techniques (Kost et al., 1998, 1999). However, the effects of GFP–TALIN on plant cell growth are not known. GFP–actin fusions fail to complement actin mutants in yeast, and cause defects during cytokinesis in *Dictyostelium* (Doyle & Botstein, 1996; Westphal et al., 1997). GFP fusions to actin-binding proteins such as TALIN may display similar toxicity or may alter actin organization in living cells. Therefore, at this time, it is not clear which method provides the best description of actin organization in developing trichomes.

Nevertheless, both methods for actin localization reveal a similar general relationship between actin organization and morphological transitions in trichomes. In regions of the cell in which growth patterns are being established, such as branch buds and the apical regions of stage 2 stalks, the actin signal is diffuse (Szymanski et al., 1999; see Fig. 2D). The absence of extensive arrays of F-actin in these regions of the cell is similar to what is observed in the tips of growing pollen tubes (Miller et al., 1996; Kost et al., 1998, 1999). Subsequently, in other cellular domains in which the growth pattern is established, such as the central branch domains at stages 4 and 5, or the stage 6 stalk, F-actin dominates the anti-actin signal. A subset of the F-actin is aligned with the growth axis. This is consistent with a role for F-actin during the reinforcement and maintenance of cell growth pattern in trichome development.

The arrangement and location of F-actin and G-actin appear to be under strict spatial control within a given cell. This is most evident during the transition from stage 3 to stage 4. In stage 3 branch buds, actin signal is very strong and diffuse. Expanding stage 4 branches, which are located just 20 μm away, display a fine meshwork of polarized F-actin with a reduced diffuse actin signal at the tip. Similar spatial compartmentation of F-actin has been observed in tip-growing pollen tubes (Miller et al., 1996; Kost et al., 1998; see also Vidali & Hepler, this volume). In contrast to growing pollen tubes, stage 3 trichomes contain three separate elongation domains that appear to undergo similar rearrangements of F-actin. Each domain sequentially executes a highly regulated branch initiation and maintenance program. It is not known how the spatial control of actin polymerization and organization is achieved during branch growth, but almost certainly actin-binding proteins are involved.

Anti-actin antibodies detect actin filaments or fine bundles at each stage of trichome development. After stage 1, many of these bundles are aligned with the local axis of elongation. Similar aligned arrays of fine actin bundles have been observed in a variety of elongating cell types (Parthasarathy, 1985; Heslop-Harrison et al., 1986; Jackson & Heath, 1993). Even after cell expansion has ceased, stage 6 trichomes display fine, parallel arrays of evenly spaced F-actin bundles. Several groups have proposed that the presence of fine actin bundles, as opposed to heavy bundles of closely associated filaments, is associated with the ability to transport vesicles (Thimann et al., 1992; Foissner et al., 1996; Waller & Nick, 1997; Miller et al., 1999). The detection of actin-associated vesicle-like structures at all developmental stages and the observation of rapid longitudinal vesicle transport along the cortex of living stage 6 trichomes (Szymanski, unpublished) are consistent with this idea.

Regulated vesicle transport to the cell cortex is probably essential for the spatial control of cell expansion during stage 5. The length and diameter of stage 5 stalks and branches increase in a highly-ordered fashion. The actin cytoskeleton during stage 5 consists of an intricate network of actin filaments or bundles, some of which display clear longitudinal orientation (Fig. 5A). High magnification images of these structures reveal an intricate branched network of filaments or bundles (Fig. 5B).

This organization resembles the reticulating lattice of cortical F-actin in elongating cotton fibers (Seagull, 1990). Stage 5 trichomes are sensitive to F-actin disrupting agents, and display localized cell swelling and branch shape defects after 12 h of treatment. It is possible that disruption of actin organization in stage 5 cells leads to misregulated accumulation and fusion of vesicles at random positions within the cell. This could explain the variable shape defects observed in growing trichomes treated with CD.

5.3 The distorted trichome shape mutants

Precise actin-dependent function is required to coordinate and maintain cell expansion during later stages of trichome morphogenesis. The striking similarity of the distorted class of trichome shape mutants to CD-treated trichomes suggests that constrained cytoskeletal function during trichome morphogenesis may provide a useful genetic system to study actin function (Mathur et al., 1999; Szymanski et al., 1999). The *distorted1* (*dis1*) and *distorted2* trichome mutants have been used as visual markers for classical genetic mapping experiments for more than 20 years (Feenstra, 1978). Six additional mutants of this class were identified based on the aborted branch and swollen phenoytype, which reflect defects in maintaining a normal growth pattern. (Hulskamp et al. 1994).

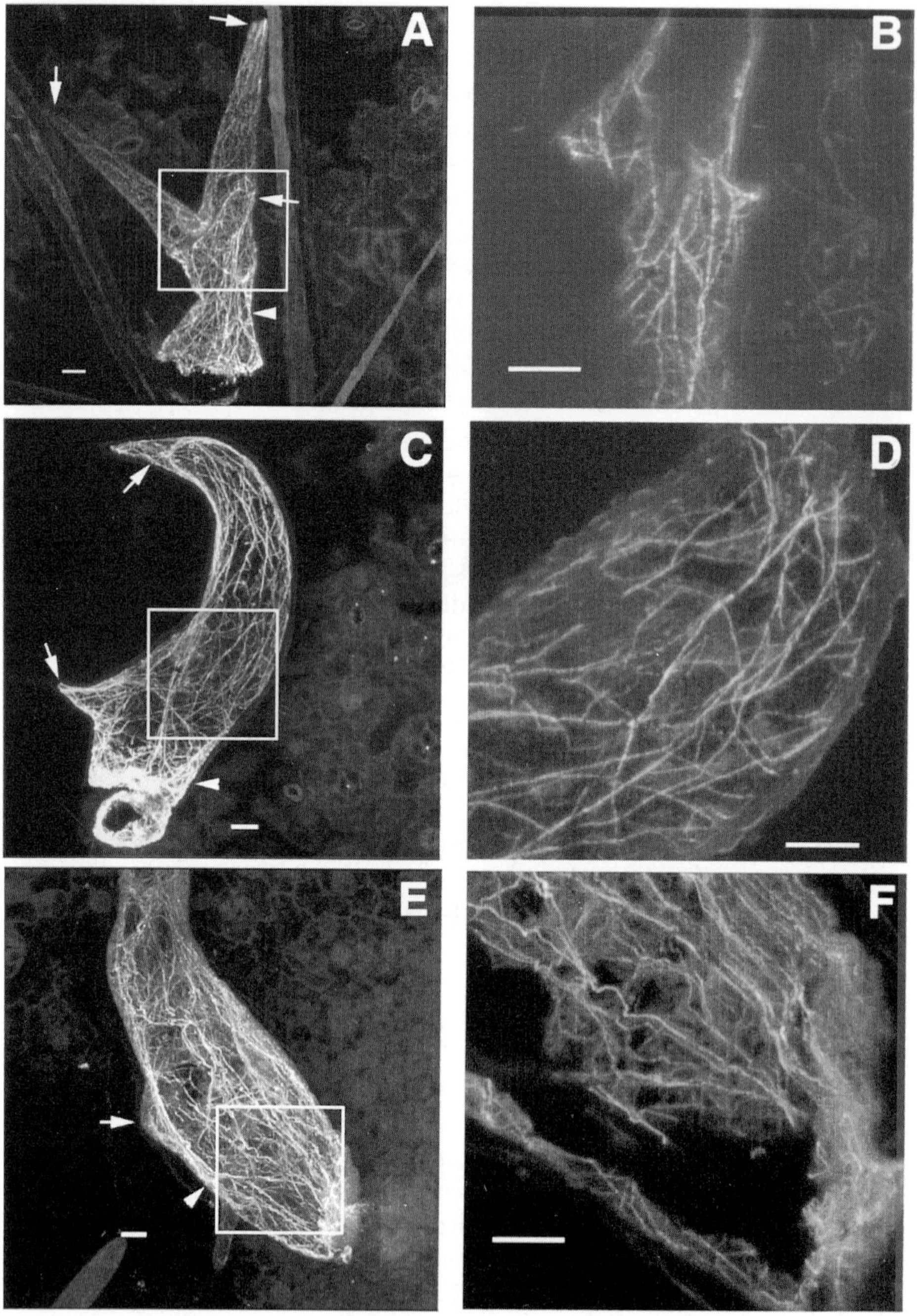

Figure 5. Actin organization in wild-type and distorted mutant trichomes. Each image is a maximum projection of confocal images of anti-actin immunofluorescence. **(A)** Three-branched stage 5 trichome. **(B)** High magnification of cortical actin signal derived from the

boxed region in (A). **(C)** Stage 6 *wrm* trichome. **(D)** High magnification of cortical actin signal derived from the boxed region in (C). **(E)** Stage 6 *crk* trichome. **(F)** High magnification of cortical actin signal derived from the boxed region in (E). Arrows mark branch tips. Arrowheads mark the stalk region. Bar for (A–F), 10 μm.

In both drug-treated and *gnarled* (*grl)* distorted trichomes, abnormal stalk swelling and expansion along the apical face of the cell often occur during the transition to stage 4 (Figs. 6A, 6B; see also Szymanski et al., 1999). Similar stage-specific defects are observed in *crooked* (*crk*) trichomes (Figs. 6C, 6D). The cellular reorganization during the transition to diffuse branch and stalk expansion is complex and likely requires several components. For example, reinforcement of an established cell growth pattern during stage 4 might require the regulated delivery or recycling of a specific class of vesicles. Any mutation that alters the timing, position, or specificity of vesicle transport could cause a breakdown in coordinated cell growth. The existence of at least eight trichome mutants with a similar distorted phenotype is consistent with the idea that multiple components are required to coordinate F-actin-dependent branch and stalk elongation after stage 4.

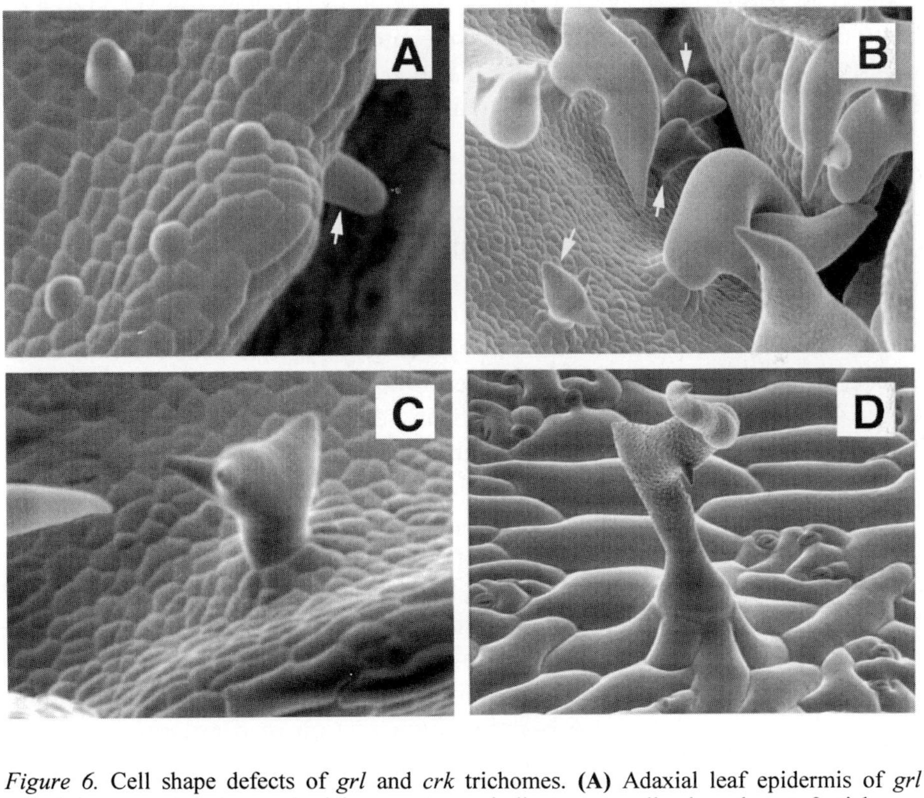

Figure 6. Cell shape defects of *grl* and *crk* trichomes. **(A)** Adaxial leaf epidermis of *grl* containing stage 1 and stage 2 trichomes. Arrow indicates normally-shaped stage 2 trichome.

(B) Epidermis of *grl* leaves containing stage 4 and stage 5 trichomes. Swollen stage 4 trichomes are labeled with arrows. **(C)** Stage 4 *crk* trichome. **(D)** Stage 6 *crk* trichome.

The actin organization of several distorted trichome mutants has been examined (Mathur et al., 1999; Szymanski et al., 1999). A detailed examination of actin organization in the *crk* and *grl* mutants detected altered F-actin organization in developing trichomes that was coincident with cell shape defects. These results do not prove that *CRK* and *GRL* directly affect actin organization, but demonstrate a close temporal link between misregulated F-actin organization and growth in the mutants. The actin organization in *alien*, *klunker*, *worm* (*wrm*), and *dis1* trichomes also were examined using GFP–TALIN (Mathur et al., 1999). In each case, the actin organization of stage 6 mutant trichomes differed from that of the wild type. An example of the actin immunolocalization in mature *wrm* trichome is shown in Figure 5. The steady state actin organization in stage 6 *wrm* (Figs. 5C, 5D) trichomes is similar to that of *grl* (Szymanski et al., 1999). Both display extensive branching of actin filaments in the stage 6 stalk, the extent of branching correlating with the severity of cell shape defects in almost all cases. The existence of heavily branched F-actin structures in the stalks of stage 6 trichomes is abnormal, however, similar structures are observed in wild-type stage 5 trichomes (Fig. 5B). The *grl* and *wrm* mutations may affect the spatial or temporal control of actin organization. Mathur et al. (1999) reported that the actin organization in *crk* trichomes appeared distinct from other distorted trichome mutants. Antibody signal from fixed specimens also consistently revealed an extensively cross-linked actin cytoskeleton in stage 6 *crk* trichomes (Fig. 5E). High magnification images of the cortical actin cytoskeleton in the base of an elongated *crk* stalk detected dense populations of extremely fine actin filaments or bundles, some of which display clear alignment with the long axis of the cell (Fig. 5F). It is possible that *CRK* affects F-actin bundling. The gene encoding the F-actin binding protein villin is transcribed at elevated levels in several elongating cell types in *Arabidopsis*, including trichomes (Klahre et al., 2000). Perhaps mutations in genes like villin will be identified in trichome morphology screens.

Although the shape of distorted mutants and CD-treated cells are indistinguishable, it is interesting that none of the distorted mutants have the actin organization of CD-treated cells. Perhaps F-actin-dependent function is highly constrained during the transition to the elongation and expansion phase of trichome growth (stages 4 and 5). Subtle defects in actin organization during this transition may have dramatic effects on trichome growth. Mutations that cause F-actin fragmentation or depolymerization in diverse cell types would most likely be lethal; however, additional screens for distorted trichome mutants may identify a trichome-specific gene with more dramatic effect on actin organization.

It is not known at present if mutations in distorted group genes affect actin organization directly or indirectly. It may be that F-actin, membrane recycling, and vacuole biogenesis are inter-dependent processes, and that distorted mutants may affect any one of them. Regardless, these cellular processes are fundamental to plant cell growth and are not well understood. A genetic analysis of distorted trichome mutants may identify both weak alleles of essential genes and mutations in genes that have a more specialized role in polarized cell elongation. Molecular and biochemical analysis of the distorted group gene products will lead to a more mechanistic understanding of plant cell growth control.

ACKNOWLEDGEMENTS

This work was made possible by support from David Marks and Sue Wick at the University of Minnesota, and the National Science Foundation Cytoskeleton Training Grant DBI 96002237. This project was also supported by NRI Competitive Grants Program/USDA Grant No. 99-35304-8525 to DBS.

REFERENCES

Alessa L and Kropf DL (1999) F-actin marks the rhizoid pole in living *Pelvetia compressa* zygotes. Development 126: 201-209

Ayscough KR (1998) In vivo functions of actin-binding proteins. Curr Opin Cell Biol 10: 102-111

Ayscough KR, Stryker J, Pokala N, Sanders M, Crews P and Drubin DG (1997) High rates of actin filament turnover in budding yeast and roles for actin in establishment and maintenance of cell polarity revealed using the actin inhibitor latrunculin-A. J Cell Biol 137: 399-416

Bender H (1960) Studies on the expression of various *singed* alleles in *Drosophila melanogaster*. Genetics 45: 867-883

Brawley SH and Robinson KR (1985) Cytochalasin treatment disrupts the endogenous currents associated wth cell polarization in fucoid zygotes: Studies of the role of F-actin in embryogenesis. J Cell Biol 100: 1173-1184

Cant K, Knowles BA, Mooseker MS and Cooley L (1994) *Drosophila* singed, a fascin homolog, is required for actin bundle formation during oogenesis and bristle extension. J Cell Biol 125: 369-380

Doyle T and Botstein D (1996) Movement of yeast cortical actin cytoskeleton visualized in vivo. Proc Natl Acad Sci USA 93: 3886-3891

Drubin, DG and Nelson, WJ (1996) Origins of cell polarity. Cell 84: 335-344

Duke SO (1994) Glandular trichomes: a focal point of chemical and structural interactions. Int J Plant Sci 155: 617-620

Eaton S, Wepf R and Simons K (1996) Roles for Rac1 and Cdc42 in planar polarization and hair outgrowth in the wing of *Drosophila*. J Cell Biol 135: 1277-1289

Eisner T, Eisner M and Hoebeke ER (1998) When defense backfires: detrimental effect of a plant's protective trichomes on an insect beneficial to the plant. Proc Natl Acad Sci USA 95: 4410-4414

Feenstra WJ (1978) Contiguity of linkage groups I and IV as revealed by linkage relationship of two newly isolated markers *dis-1* and *dis-2*. Arab Inf Serv 15: 35-38

Foissner I, Lichtscheidl IK and Wasteneys GO (1996) Actin-based vesicle dynamics and exocytosis during wound wall formation in characean internodal cells. Cell Motil Cytoskeleton 35: 35-48

Folkers U, Berger J and Hülskamp M (1997) Cell morphogenesis of trichomes in *Arabidopsis*: Differential control of primary and secondary branching by branch initiation regulators and cell growth. Development 124: 3779-3786

Gibbon, BC, Kovar, DR and Staiger, CJ (1999) Latrunculin B has different effects on pollen germination and tube growth. Plant Cell 11: 2349-2363.

Heslop-Harrison J, Heslop-Harrison Y, Cresti M, Tiezzi A and Ciampolini F (1986) Actin during pollen tube germination. J Cell Sci 86: 1-8

Hoover KK, Chien AJ and Corces VG (1993) Effects of transposable elements on the expression of the *forked* gene of *Drosophila melanogaster*. Genetics 135: 507-526

Hopmann R, Cooper JA and Miller KG (1996) Actin organization, bristle morphology, and viability are affected by actin capping protein mutations in *Drosophila*. J Cell Biol 133: 1293-1305

Hülskamp M, Folkers U and Grini PE (1998) Cell morphogenesis in *Arabidopsis*. Bioessays 20: 20-29

Hülskamp M, Misra S and Jürgens G (1994) Genetic dissection of trichome cell development in Arabidopsis. Cell 76: 555-566

Jackson SL and Heath IB (1993) The dynamic behavior of cytoplasmic F-actin in growing hyphae. Protoplasma 173: 23-34

Johnson HB (1975) Plant pubescence: an ecological perspective. Bot Rev 41: 233-258

Klahre U, Friederich E, Kost B, Louvard D and Chua N-H (2000) Villin-like actin-binding proteins are expressed ubiquitously in Arabidopsis. Plant Physiol 122: 35-48

Kost B, Spielhofer P and Chua N-H (1998) A GFP-mouse talin fusion protein labels plant actin filaments *in vivo* and visualizes the actin cytoskeleton in growing pollen tubes. Plant J 16: 393-401

Kost B, Lemichez E , Spielhofer P, Hong Y, Tolias K, Carpenter C and Chua N-H (1999) Rac homologues and compartmentalized phosphatidylinositol 4,5-bisphosphate act in a common pathway to regulate polar pollen tube growth. J Cell Biol 145: 317-330

Krishnakumar S and Oppenheimer DG (1999) Extragenic suppressors of the *Arabidopsis zwi-3* mutation identify new genes that function in trichome branch formation and pollen tube growth. Development 126: 3079-3088

Lancelle SA, Cresti M and Hepler PK (1987) Ultrastructure of the cytoskeleton in freeze-substituted pollen tubes of *Nicotiana alata*. Protoplasma 140: 141-150

Larkin JC, Marks MD, Nadeau J and Sack F (1997) Epidermal cell fate and patterning in leaves. Plant Cell 9: 1109-1120

Larkin JC, Young N, Prigge M and Marks MD (1996) The control of trichome spacing and number in *Arabidopsis*. Development 122: 997-1005

Lazzaro MD and Thomson WW (1992) Ultrastructural localization of calcium in the organic acid secreting trichomes of chickpea (*Cicer arietinum*). Can J Bot 70: 2319-2325

Lazzaro MD and Thomson WW (1996) The vacuolar-tubular continuum in living trichomes of chickpea (*Cicer arietinum*) provides a rapid means of solute delivery from base to tip. Protoplasma 193: 181-190

Lloyd AM, Schena M, Walbot V and Davis RW (1994) Epidermal cell fate determination in *Arabidopsis*: Patterns defined by a steroid-inducible regulator. Science 266: 436-439

Luo D and Oppenheimer DG (1999) Genetic control of trichome branch number in *Arabidopsis*: the roles of the *FURCA* loci. Development 126 5547-5557

Marks MD (1997) Molecular genetic analysis of trichome development in Arabidopsis. Annu Rev Plant Physiol Plant Mol Biol 48: 137-163

Mascarenhas JP and LaFountain J (1972) Protoplasmic streaming, cytochalasin B, and growth of the pollen tube. Tiss Cell 4: 11-14

Mathur J, Spielhofer P, Kost B and Chua N-H (1999) The actin cytoskeleton is required to elaborate and maintain spatial patterning during trichome cell morphogenesis in *Arabidopsis thaliana*. Development 126: 5559-5568

McKinney EC, Ali N, Traut A, Feldmann KA, Belostotsky DA, McDowell JM and Meagher RB (1995) Sequence-based identification of T-DNA insertion mutations in *Arabidopsis*: actin mutants *act2-1* and *act4-1*. Plant J 8: 613-622

Miller DD, de Ruijter NCA, Bisseling T and Emons AMC (1999) The role of actin in root hair morphogenesis: studies with lipochito-oligosaccharide as a growth stimulator and cytochalasin as an actin perturbing drug. Plant J 17: 141-154

Miller DD, Lancelle SA and Hepler PK (1996) Actin microfilaments do not form a dense meshwork in *Lilium longiflorum* pollen tubes. Protoplasma 195: 123-132

Oppenheimer DG, Pollock MA, Vacik J, Szymanski DB, Ericson B, Feldmann K and Marks MD (1997) Essential role of a kinesin-like protein in *Arabidopsis* trichome morphogenesis. Proc Natl Acad Sci USA 94: 6261-6266

Parthasarathy MV (1985) F-actin architecture in coleoptile epidermal cells. Eur J Cell Biol 39: 1-12

Perazza, D, Herzog M, Hulskamp M, Brown S, Dorne AM and Bonneville JM (1999) Trichome cell growth in *Arabidopsis thaliana* can be derepressed by mutations in at least five genes. Genetics 152: 461-476

Picton JM and Steer MW (1981) Determination of secretory vesicle production rates by dictysomes in pollen tubes of *Tradescantia* using cytochalasin D. J Cell Sci 49: 261-272

Pringle JR, Harkins HA, Zahner JE, De Virgilio C, Chant J, Corrado K and Fares H (1995) Establishment of cell polarity in yeast. Cold Spring Harb Symp Quant Biol 60: 729-744

Pyke KA, Marrison JL and Leech RM (1991) Temporal and spatial development of the cells of the expanding first leaf of *Arabidopsis thaliana* (L.) Heynh. J Exp Bot 42: 1407-1416

Quatrano RS (1973) Separation of processes associated with differentiation of two-celled *Fucus* embryos. Devel Biol 30: 209-213

Rodriguiez E, Healey PL and Mehta I (1984) Biology and Chemistry of Plant Trichomes, New York, Plenum Press

Schnittger A, Jürgens G and Hülskamp M (1998) Tissue layer and organ specificity of trichome formation are regulated by *GLABRA1* and *TRIPTYCHON* in *Arabidopsis*. Development 125: 2283-2289

Seagull RW (1986) Changes in microtubule organization an wall microfibril orientation during *in vitro* cotton fiber development: an immunofluorescent study. Can J Bot 64: 1373-1381

Seagull RW (1990) The effects of microtubule and microfilament disrupting agents on cytoskeletal arrays and wall deposition in developing cotton fibers. Protoplasma 159: 44-59

Seagull RW (1992) Quantitative electron microscopic study of changes in microtubule arrays and wall microfibril orientation during *in vitro* cotton fiber development. J Cell Sci 101: 561-577

Shade RE, Thompson TE and Campbell WR (1975) An alfalfa weevil larval resistance mechanism detected in *Medicago*. J Econ Entom 68: 399-404

Song H, Golovkin M, Reddy AS and Endow SA (1997) In vitro motility of AtKCBP, a calmodulin-binding kinesin protein of *Arabidopsis*. Proc Natl Acad Sci USA 94: 322-327

Staiger CJ (2000) Signaling to the actin cytoskeleton in plants. Annu Rev Plant Physiol Plant Mol Biol 51: 257-288

Staiger CJ, Gibbon BC, Kovar DR and Zonia LE (1997) Profilin and actin-depolymerizing factor: modulators of actin organzation in plants. Trends Plant Sci 2: 275-281

Szymanski DB, Lloyd AM and Marks DM (2000) Progress in the molecular genetic analysis of trichome initiation and morphogenesis in *Arabidopsis*. Trends Plant Sci, In press

Szymanski DB, Marks MD and Wick, SM (1999) Organized F-actin is essential for normal trichome morphogenesis in Arabidopsis. Plant Cell 11: 2331-2347

Szymanski DB, Jilk RA, Pollock SM and Marks MD (1998) Control of *GL2* expression in *Arabidopsis* leaves and trichomes. Development 125: 1161-1171

Szymanski, DB and Marks MD (1998) *GLABROUS1* overexpression and *TRIPTYCHON* alter the cell cycle and trichome cell fate in Arabidopsis. Plant Cell 10: 2047-2062

Taylor, LP and Hepler PK (1997) Pollen germination and tube growth. Annu Rev Plant Physiol Plant Mol Biol 48: 461-491

Thimann KV, Reese K and Nachmias VT (1992) Actin and the elongation of plant cells. Protoplasma 171: 153-166

Tilney LG, Connelly P, Smith S and Guild GM (1996) F-actin bundles in *Drosophila* bristles are assembled from modules composed of short filaments. J Cell Biol 135: 1291-1308

Tilney LG, Tilney MS and Guild GM (1995) F-actin bundles in *Drosophila* bristles. I. Two filament cross-links are involved in bundling. J Cell Biol 130: 629-638

Tiwari SC and Polito VS (1990) An analysis of the role of actin during pollen activation leading to germination in pear (*Pyrus communis* L.): treatment with cytochalasin D. Sex Plant Reprod 3: 121-129

Tiwari SC and Wilkins TA (1995) Cotton (*Gossypium hirsutum*) seed trichomes expand via diffuse growing mechanism. Can J Bot 73: 746-757

Turner CM and Adler PN (1998) Distinct roles for the actin and microtubule cytoskeletons in the morphogenesis of epidermal hairs during wing development in *Drosophila*. Mech Dev 70: 181-192

Verheyen EM and Cooley L (1994) Profilin mutations disrupt multiple actin-dependent processes during *Drosophila* development. Development 120: 717-728

Waller F and Nick P (1997) Response of actin microfilaments during phytochrome-controlled growth of maize seedlings. Protoplasma 200: 154-162

Wang Q-Y and Nick P (1998) The auxin response of actin is altered in the rice mutant *Yin-Yang*. Protoplasma 204: 22-33

Wasteneys GO, Willingale-Theune J and Menzel D (1997) Freeze shattering: A simple and effective method for permeabilizing higher plant cell walls. J Microsc 188: 51-61

Westphal M, Jungbluth A, Heidecker M, Muhlbauer B, Heizer C, Schwartz JM, Marriott G and Gerisch G (1997) Microfilament dynamics during cell movement and chemotaxis monitored using a GFP-actin fusion protein. Curr Biol 7: 176-83

Wulfkuhle JD, Petersen NS and Otto JJ (1998) Changes in the F-actin cytoskeleton during neurosensory bristle development in Drosophila: the role of singed and forked proteins. Cell Motil Cytoskeleton 40: 119-132

Chapter 23

ACTIN IN FORMATION OF STOMATAL COMPLEXES

Ann L Cleary

Plant Cell Biology Group, Research School of Biological Sciences, Australian National University, Canberra, ACT 2601, Australia

Key words: actin, *Allium*, guard mother cells, stomata, subsidiary mother cells, *Tradescantia*

Abstract: Stomata have long been recognised as a model system in which to study regulated division in plant cells because the precursor cells divide in predictable sequences and planes. The formation of functional stomatal complexes, whether a solitary guard cell pair or guard cells complexed with multiple subsidiary cells, requires precisely oriented cell divisions reliant on the correct functioning of the cytoskeleton. In particular, many stage-specific configurations of cortical actin form during the development of stomatal complexes. These may be critical to normal stomatal development and function. The distribution and function of actin during stomatogenesis is discussed.

1. STOMATAL COMPLEXES

Stomatal complexes are present on most plant organs, but are concentrated on the leaves where they have an essential function in gas exchange. The form of stomatal complexes is phylogenetically significant (e.g. Tomlinson, 1974). Irrespective of their simplicity, or complexity, production of stomatal complexes requires precisely positioned sequences of cell divisions. At the beginning of stomatal development asymmetric divisions in the leaf meristem form small guard mother cells (GMCs). In the simplest stomata there is one subsequent division; the GMC divides symmetrically to produce the guard cell pair. However, in many species, at some point after it is formed, the GMC begins to affect the neighbouring cells (Stebbins & Shah, 1960; Pickett-Heaps & Northcote, 1966a; Pickett-

C.J. Staiger et al. (eds.), Actin: A Dynamic Framework for Multiple Plant Cell Functions, 411–426.

Heaps, 1969a). In these more complex examples, epidermal cells surrounding the GMC, now designated subsidiary mother cells (SMCs), divide asymmetrically to form subsidiary cells. In all cases, the asymmetric divisions of stomatal precursor cells result in daughter cells of different shapes and sizes that have different developmental fates. Division of SMCs and GMCs occurs after the majority of meristematic divisions in the leaf epidermis have been completed and generally the sequence of divisions is predictable. Monocotyledons have an additional advantage for cell biologists in that cells are arranged in files and there is a developmental gradient of stomatal formation from the leaf base to the apex.

Two major categories of stomatal complex exist, based on the form of the guard cells: (i) the bone-, or dumb-bell, shaped guard cells of the grasses, and (ii) the kidney-, or elliptically, shaped guard cells of the non-grass monocotyledons and dicotyledons. Depending on the species, both forms of stomata may be found to lack, or develop, subsidiary cells, which can be present in a range of numbers. The first category is typified by grasses such as *Avena*, sugar cane, maize and winter rye. *Allium* is a classical example having elliptically shaped guard cells lacking subsidiary cells, while members of the Commelinaceae, such as *Tradescantia*, are representative of those containing subsidiary cells. Some of the earliest examinations of stomatogenesis were made in *Tradescantia* (Strasburger, 1866; Campbell, 1881) and *Allium* (Miehe, 1899, 1901–cited in Stebbins & Jain, 1960); two genera that remain amongst the most valuable today for unravelling the "mysteries" of stomatogenesis.

2. TECHNIQUES TO STUDY THE ROLE OF ACTIN

The formation of stomatal complexes is both spatially and temporally regulated and involves programmed changes in division plane – processes known to involve the action of microtubules (MTs). Until recently little was known of the distribution or dynamic behaviour of actin. The major impediments to the study of stomatal formation, and the role of actin in particular, are (i) the optical properties of the tissue to be studied and (ii) permeability of the experimental material to drugs and actin probes. Observations can be made in some intact tissues. However, the best results are obtained if the epidermal layer in which the stomata reside can be exposed in epidermal slices (e.g. *Allium* – Palevitz & Hepler, 1974a, 1974b) or as epidermal peels (e.g. *Tradescantia* – Cleary, 1995; Kennard & Cleary, 1997), and are conditional on the cells being maintained in a mitotically active state.

The role of actin in stomatal development was first inferred from studies with pharmacological agents, such as cytochalasin and phalloidin, that target actin structure and function (Palevitz & Hepler, 1974b; Palevitz, 1980). These, and more recently identified drugs (e.g. lantrunculin; Kandasamy & Meagher, 1999), remain essential tools for examining actin function in living cells. There are two types of drug studies: those where tissue is treated blindly (usually an intact tissue) and inferences made as to the process based on the outcome, and the more preferable method where the drug is applied to individual cells at known developmental stages and observations are made throughout the treatment, thereby revealing the subtleties of the disruptions to the cell division process. Advances in our understanding of actin function during stomatal formation will no doubt come from the injection or changed expression levels of plant actin or actin-associated proteins, such as profilin (Staiger et al., 1994) and actin depolymerizing factors (Jiang et al., 1997).

Despite general advances in techniques for preparing plant cells for light and electron microscopy, the preservation of actin remains problematical. There is a regrettable absence of ultrastructural reports of actin in developing stomatal complexes. Using variations on conventional fluorescence techniques for light microscopy, the actin cytoskeleton has been labelled in developing stomatal complexes of a few representative species: *Allium* (Mineyuki & Palevitz, 1990), winter rye (Cho & Wick, 1990, 1991), *Selaginella* (Cleary et al., 1992a), and *Tradescantia* (Cleary & Mathesius, 1996). These protocols utilise procedures involving fixation and subsequent labelling indirectly with anti-actin antibodies or directly with fluorescently tagged phalloidin. Importantly, not all features of actin organisation are revealed using any one of these techniques. The most comprehensive description of actin in a single species has been obtained in living stomatal cells following microinjection of phalloidin derivatives using pressure-injecting techniques (Cleary, 1995). An additional advantage of observing living cells is that changes in the actin cytoskeleton can be correlated directly to visible cytological changes occurring within the cells. The expression of GFP-talin bound to actin in living plant tissues (Kost et al. 1998) may facilitate the future non-invasive visualisation of actin and reveal more of its dynamic behaviour in stomata.

The first hurdle is to label the actin, but it is then equally important to be able to visualise the labelling. Most descriptions of actin organisation have been made using conventional epifluorescence microscopes, but significant insights have been made using confocal laser scanning microscopy to record stacks of adjacent single focal plane images that can be superimposed to make three-dimensional reconstructions (review Hepler & Gunning, 1998). In addition, the low levels of irradiation needed to detect fluorescence allow repeated observations of individual living, microinjected cells (Cleary et al.,

1992b; Cleary, 1995). The degree of actin filament bundling cannot be determined, even by the best available methods of fluorescence microscopy. Therefore, the fine structure of actin distribution awaits high quality ultrastructural preservation and examination by cryo-techniques. Using existing methods we are now aware that some configurations of actin are probably common to all meristematic cells, while others are specific to stomatal precursor cells.

3. DIVISION OF SUBSIDIARY MOTHER CELLS

Actin is essential to the successful production of subsidiary cells. The processes governing asymmetric, transverse divisions of GMC precursor cells are analogous to those found in terminal SMCs and they will also be considered here. The distribution and inferred role of actin can vary to some extent between species and also between lateral and terminal SMCs of the same complexes, but a general picture is emerging.

3.1 Nuclear migration

In preparation for the asymmetric division that gives rise to the GMC or subsidiary cell, the nucleus in each precursor cell migrates and establishes a polar axis providing an early visible sign that cell division will take place in a pre-determined off-centre location. In SMCs, nuclear migration is towards, and thought to be induced by, the adjacent GMC (Fig. 1) (Stebbins & Jain, 1960; Stebbins & Shah, 1960; Pickett-Heaps & Northcote, 1966a; Pickett-Heaps, 1969a). Premitotic nuclear migration occurs in the G1 phase of the SMC cell cycle (Sugimoto et al., 1993; Kennard & Cleary, 1997) and SMCs remain polarised for a significant period of time (throughout G1, S and G2 phases) prior to entering mitosis (Pickett-Heaps 1969a). The polarised state persists for at least 22 hours in *Tradescantia* SMCs (Kennard & Cleary, 1997).

Observation of living *Tradescantia* SMCs shows that the initial migration of spherically-shaped SMC nuclei is along cytoplasmic strands connected to the cortical region adjacent to the GMC (Kennard & Cleary, 1997). Actin filaments are abundant in these cytoplasmic strands and disruption of actin results in cessation of cytoplasmic streaming, breakage of the strands and total inhibition of nuclear migration (Kennard & Cleary, 1997). The involvement of actin filaments in pre-mitotic nuclear positioning has been shown also in GMC precursors of *Allium* (Mineyuki & Palevitz, 1990) and suggested for winter rye (Cho & Wick, 1991). In contrast, MTs mediate nuclear migration in the SMCs of cucumber hypocotyl epidermis (Kazama et al., 1995) and possibly also in SMCs of maize (Galatis et al., 1984).

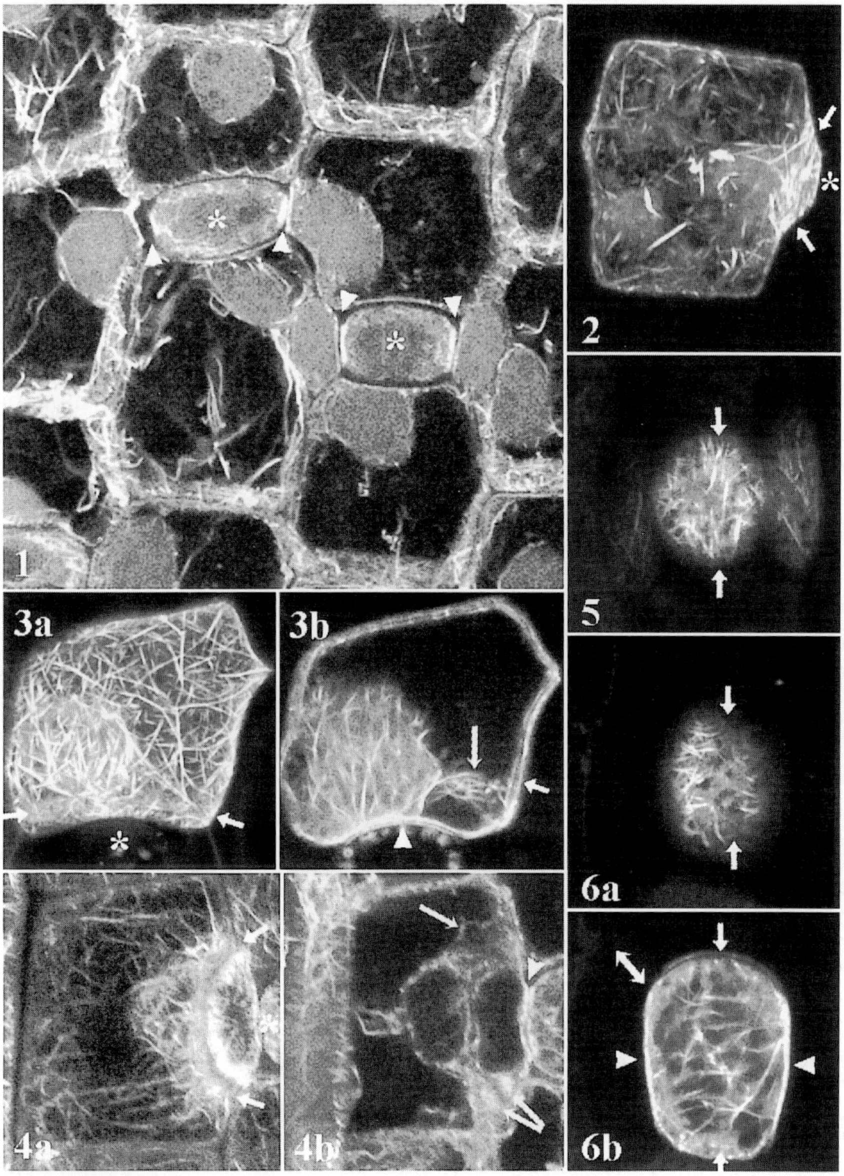

Actin in stomatal precursor cells.
Figure 1. Mid-plane of focus showing extensive cortical and cytoplasmic actin in GMCs (*) and adjacent SMCs. Cortical actin patches (arrowheads) are present in GMCs and polarised SMCs, but less obvious in subsidiary cells (right side complex).

Figure 2. Lateral SMC with a curved actin PPB (arrows) predicting the formation of a hemispherical subsidiary cell adjacent to the GMC (*).
Figure 3a. Cortical actin in a mitotic terminal SMC. The actin-depleted zone (ADZ; arrows) is very close to the terminal wall of the adjacent GMC (*).
Figure 3b. Mid-plane of focus showing extensive actin in the mitotic spindle, actin patch (arrowhead), and in the cytoplasmic strand (long arrow) connecting the spindle to the ADZ (short arrow).
Figure 4a. Cortical actin and ADZ (arrows) curving towards the GMC (*) in a lateral SMC.
Figure 4b. Mid-plane of focus with the phragmoplast containing actin (double arrow) approaching the cell cortex. The subsidiary cell nucleus is bounded by the phragmoplast and the actin patch (arrowhead). There are numerous other cytoplasmic strands (long arrow).
Figure 5. Prophase GMC with a longitudinally oriented actin PPB (arrows).
Figure 6a. Metaphase GMC with a longitudinally oriented ADZ (arrows).
Figure 6b. Mid-plane of focus showing cortical actin patches (arrowheads) and ADZ (arrows), and the axis of the oblique spindle marked by a double-headed arrow.

Nuclei remain motile after they reach the GMC, but movement diminishes concomitant with the nuclei becoming dome-shaped, and compressed against the GMC side of the SMC (Fig. 1). Nuclei with this morphology show some resistance to displacement by centrifugation (Kennard & Cleary, 1997). A mechanism exists to anchor SMC nuclei in position against centrifugation (Pickett-Heaps, 1969b, Kennard & Cleary, 1997), and we now know that the point of juxtaposition between the nucleus and the cell cortex is a site of pronounced actin accumulation (Cho & Wick, 1990; Cleary, 1995; Cleary & Mathesius, 1996).

3.2 Cortical actin patches

A cortical actin patch appears in the SMC against the wall adjacent to the GMC at some point after the SMC has become polarised (Fig. 1) (Cho & Wick, 1990; Cleary, 1995; Cleary & Mathesius, 1996; Gallagher & Smith, 1999). Actin patches are a general feature in lateral SMCs, but are less frequently observed in terminal SMCs (Cleary & Mathesius, 1996). Actin patches are composed of coaligned, thick actin bundles running perpendicular to the leaf surface (Cleary, 1995). They are resistant to treatment with cytochalasins, although fine cortical and cytoplasmic filaments extending from the patches are lost (Cho & Wick, 1990, 1991; Kennard & Cleary, 1997).

The function of the actin patch has not been directly demonstrated. However, the timing of the appearance of the patch and the resistance to displacement by centrifugation of nuclei with which the patch is associated suggests it has an anchoring role (Cho & Wick, 1990; 1991; Cleary, 1995; Cleary & Mathesius, 1996; Kennard & Cleary, 1997). A mechanism for securing nuclei may be important for maintaining SMC polarity during the long periods that SMC nuclei remain fixed in position prior to mitosis (Pickett-Heaps, 1969a; Kennard & Cleary, 1997). Cortical actin patches persist until the end of cytokinesis. They may have additional functions during mitosis when they are associated with one of the spindle poles (section 3.4).

3.3 Cortical actin and definition of the division site

A brief discussion of the major rearrangements of the cortical cytoskeleton known to occur prior to mitosis serves as a general introduction to stomatal divisions. One of the best known changes is the formation of the preprophase band (PPB), a circumferential band of parallel MTs and actin that marks the plane of cell division (Pickett-Heaps & Northcote, 1966b; Kakimoto & Shibaoka, 1987; Traas et al., 1987; review Mineyuki, 1999). Both cytoskeletal elements are essential for the establishment of a fully functional division site and to convey spatial information required during cytokinesis (Mineyuki & Gunning, 1990; Mineyuki & Palevitz, 1990). In prophase, the width of the PPB decreases as mitotic spindle MTs proliferate around the nucleus. When the MT PPB is at its narrowest, actin appears to be excluded from the centre of the PPB (Cleary and Smith, 1998). Both cytoskeletal components of the PPB disappear around the time of the breakdown of the nuclear envelope (Cleary et al., 1992b; Cleary & Smith, 1998; Hepler & Gunning, 1998; Mineyuki, 1999), leaving only cortical actin flanking the PPB site. The PPB site remains continuously marked throughout mitosis by this newly formed actin-depleted zone (ADZ) (Cleary et al., 1992b, Liu & Palevitz, 1992).

Microtubule PPBs have long been known to exist in all stomatal precursor cell types (Pickett-Heaps & Northcote, 1966a; Busby & Gunning, 1980). Formation of a transverse PPB in *Allium* GMC precursor cells precedes nuclear migration and is thus the first currently detectable indication of division polarity (Mineyuki & Palevitz, 1990). Actin has been identified in PPBs of *Allium* GMC precursors (Mineyuki and Palevitz, 1990). Conversely, the evidence from lateral SMCs is that nuclear migration precedes the formation of curved MT PPBs (Pickett-Heaps et al., 1969a; Kennard & Cleary, 1997). Nuclear anchoring by actin patches may be a pre-requisite for the subsequent establishment of the division site and its PPB (Cho & Wick, 1990; Cleary, 1995; Cleary & Mathesius, 1996).

There remains conflicting evidence for F-actin in SMC PPBs. Their presence in *Tradescantia* (Fig. 2) (Cleary, 1995), but their apparent absence in winter rye (Cho & Wick, 1990, 1991) SMCs, may result from natural variations between species or phylogeny, or result from differences in methodology (Cleary, 1995; Cleary & Mathesius, 1996). That curved actin

PPBs have been observed in *Tradescantia* lateral SMCs microinjected with phalloidin (Cleary, 1995), but not seen when the same cells are fixed (Cleary & Mathesius, 1996), indicates that the later option–differences in preparative technique–can result in the loss of this particularly sensitive component of the actin array (see discussion in Cleary & Mathesius, 1996). Direct evidence from one species, and the propensity for major cytoskeletal arrangements and cytokinetic mechanisms to be conserved within groups of plants (e.g. Pickett-Heaps et al., 1999), suggests that actin is a component of PPBs in all SMCs. Whereas the curved MT PPBs of lateral SMCs are thought to have unique nucleating sites in the cell cortex (Cho & Wick, 1989), nothing is known of the mode of formation of the curved actin PPBs.

As in other cell types, actin PPBs disappear from SMCs to be replaced by ADZs (Figs. 3 & 4). Direct observations of living SMCs show that actin PPBs and ADZs accurately mark the site of insertion of the new cell wall (Cleary, 1995). Persistent cortical actin, including actin patches, may play an essential role in guiding the expansion of the cell plate to the pre-defined division site at the ADZ.

3.4 Actin mechanisms guiding expansion of the cell plate

One of the persisting enigmas of plant cell biology is how the phragmoplast, responsible for forming the new cell wall, is guided to the area of the parental cell cortex defined earlier by the PPB and ADZ. It is generally accepted that a directional mechanism exists which guides the expanded phragmoplast once it is in close proximity to the parental wall (short distance guidance, see also Palevitz 1986; Wick, 1991). However, there is no direct evidence identifying the components of this orienting system or how the components interact to correct cell plate alignment. Cell plate realignments often represent the final stage in reaching the pre-determined division plane, but the cell plate may have to expand over large areas before it reaches and fuses with the parental wall at the division site (long distance guidance; see also Busby & Gunning, 1980). In extremely vacuolated cells, a phragmosome (a raft of cytoplasmic strands containing actin and MTs) forms in prophase and persists through telophase, connecting the edge of the cell plate to the cortical division site (Wick, 1991). However, phragmosomes are not seen in all vacuolated cells (e.g. SMCs) nor in any densely cytoplasmic cells (e.g. GMCs).

Stomatal morphogenesis requires precise control over phragmoplast expansion, a process that is subtly distinct for each of the stomatal precursor cell types. Terminal SMCs form slightly curved, transversely oriented cell plates, which cut off approximately rectangular terminal subsidiary cells, while lateral SMCs have highly curved cell plates partitioning off hemispherical lateral subsidiary cells. The effects of the actin disrupting agent, cytochalasin, are consistent with both the long and short distance guidance mechanisms of SMC cell plate expansion, as well as wall shaping, being facilitated by actin-dependent processes. Mitotic SMCs exhibit three distinct regions of cortical actin that together could provide a cortical "road map" to the division site (Figs. 3 & 4): (i) the actin patch on the subsidiary cell side of the division site, (ii) the ADZ marking the boundary of the two daughter cells, and (iii) residual cortical actin covering the remainder of the cell cortex (Cleary, 1995; Cleary & Mathesius, 1996). The interaction of actin in cytoplasmic strands, and/or actin associated with the mitotic spindle and phragmoplast, with components of the cortical actin array could function in the guidance of SMC cell plate expansion (Figs. 3b & 4b) (Cleary et al., 1992b; Cleary, 1995; see also Smith, 1999). Actin patches and a network of cortical actin filaments only are recorded for winter rye (Cho and Wick 1990). Difficulties with the complete preservation and visualisation of the cortical actin could explain the apparent absence of the ADZ in this species (see Cleary & Mathesius, 1996).

During mitosis, one SMC spindle pole remains anchored at the actin patch next to the GMC (Figs. 3b & 4b) (Cho & Wick, 1989; Cleary, 1995; Gallagher & Smith, 1999; Pickett-Heaps et al., 1999). The anchored pole acts as a pivot point for the spindle. The untethered spindle pole swings around placing one edge of the forming cell plate in close proximity to the division site (Cleary, 1996; Gallagher & Smith, 1999). Images consistent with this type of spindle rotation in SMCs are seen also in Stebbins and Jain (1960) and Cho and Wick (1989). Unlike algal zygotes (Kropf, 1992) and yeast cells (Palmer et al., 1992) where anchoring of spindles to a cortical patch of actin is essential to successful development, in the event of the SMC spindle becoming detached or skewed in position, cytokinesis is delayed while additional mechanisms reorient the spindle ensuring correct placement of the cell plate and normal subsidiary cell formation (Pickett-Heaps et al., 1999). Reorientation occurs when one edge of the cell plate is in close proximity to the cell cortex and hence potentially able to interact with the cortical actin (see also Palevitz, 1993). Thus, dislodging the spindle from the wall results in the initial misplacement of the cell plate, but it does not unduly disrupt final cell plate position or shape unless actin has been compromised (Pickett-Heaps et al., 1999). In the presence of cytochalasin, the spindle pole slowly detaches from the SMC wall and drifts away from the GMC indicating a role for the actin patch and/or its associated finer elements in maintaining spindle position and initial cell plate placement within the SMC (Cho & Wick, 1990; 1991; Pickett-Heaps et al., 1999). In addition, continued cytochalasin treatment of both lateral and terminal SMCs prevents any repositioning of mis-aligned cell plates, even when they are in close proximity to the cell cortex.

The formation of curved cell walls in SMCs is an example of directional control being exerted on growth of the cell plate before it comes close to the parental wall. It is more obvious in lateral than in terminal SMCs. The shape and position of the cell plate appear to be linked to the state of the reforming subsidiary cell nucleus in a manner similar to the curving of the cell plate around the generative nucleus during pollen grain mitosis (Brown & Lemmon, 1991). MTs and actin associated with the phragmoplast provide a physical link between the subsidiary cell nucleus and cell plate, so that as the subsidiary cell chromatin decondenses the cell plate is pulled in around it (Fig. 4b) (Cleary, 1995). In support of this view, cell plates fail to curve when they develop in the presence of cytochalasin (Cho & Wick, 1990; 1991; Pickett-Heaps et al., 1999). Additional support for the role of actin in cell plate guidance comes from the *discordia* mutants of maize. These are defective in asymmetrical divisions, including those of GMC precursor cells and SMCs (Gallagher & Smith, 1999). The authors propose that a defect in the cytoplasmic actin connection between the phragmoplast and the actin patch prevents the cell plate from being pulled around the GMC or subsidiary cell nucleus.

4. SYMMETRICAL DIVISION OF THE GMC

In most species, after the formation of any subsidiary cells, GMCs undergo DNA replication (S phase) (Kennard & Cleary, 1997), cortical actin realignments (Cleary & Mathesius, 1996) and division (see Stebbins & Jain, 1960). Divisions of GMCs are fairly straightforward in comparison to SMCs, but they remain highly programmed and dependent on actin in order to complete formation of functional stomatal complexes. A role for actin in stomatal formation, and in cytokinesis in general, was first elucidated for *Allium* GMCs (Palevitz & Hepler, 1974b; Palevitz, 1980).

4.1 Premitotic cortical actin

Before obvious cytological changes associated with their mitosis occur, GMCs accumulate actin on the lateral walls (Cho & Wick, 1990; Cleary, 1995; Cleary & Mathesius, 1996; Gallagher & Smith, 1999). The appearance of actin patches in GMCs occurs concurrently with the formation of patches in the lateral SMCs (Fig. 1) (Cleary & Mathesius, 1996). However, actin patches never form on the terminal walls even when the adjacent terminal SMCs have patches. As in SMCs, these actin patches are resistant to the depolymerising action of cytochalasins (Cho & Wick, 1991; Kennard & Cleary, 1997) and persist until the end of the GMC division, some time after the disappearance of patches from the subsidiary cells (Fig. 1). In comparison to SMCs, a role for the patches in maintaining premitotic nuclear position seems to be unnecessary in the small GMCs. Neither do they function in anchoring the spindle poles, which are remarkably mobile in these cells (section 4.3). However, they may be important for cell plate alignment during cytokinesis (Cho & Wick, 1990).

GMCs rearrange the cortical MTs on their periclinal walls prior to the establishment of the division site marked by the MT PPB. MTs in non-grass GMCs reorient from random or radial to longitudinal (Mineyuki et al., 1989; Cleary et al., 1992a). Grass GMCs go from a transverse interphase MT band, through radial or random, to a longitudinal array and MT PPB (Busby & Gunning, 1980; Cho & Wick, 1989; Cleary & Hardham, 1989; Galatis et al., 1982; Mullinax & Palevitz, 1989). From the few examples studied to date, actin goes through similar premitotic cortical rearrangements, although there are no data on the correlation of MTs and actin rearrangements at particular stages of GMC development. In *Selaginella*, actin reorients from random to a wide longitudinal band (Cleary et al., 1992a), while in *Tradescantia* cortical actin reorients from transverse, through radial to longitudinal (Cleary & Mathesius, 1996). The small size of GMCs limits the possibility of using microinjection of actin probes to examine such dynamic changes in actin distribution (Cleary, 1995). These reorientations of the cortical cytoskeleton may be necessary to place MTs and actin in the correct orientation to form PPBs and define the longitudinal division plane.

4.2 Cortical actin and the division site

PPBs of MTs were first observed in GMCs in ultrastructural studies (Pickett-Heaps & Northcote, 1966a). Illumination of the three dimensional structure of the MT PPBs, including connections to the spindle, followed the advent of immunofluorescence techniques (Mineyuki et al., 1989; Cho & Wick, 1989; Cleary et al., 1992a). The implementation of protocols known to improve its preservation revealed actin as a component of PPBs in *Selaginella* GMCs (Cleary et al., 1992a) and in forming PPBs (Fig. 5) and ADZs (Fig. 6) in *Tradescantia* GMCs (Cleary, 1995; Cleary & Mathesius, 1996).

4.3 Spindle rotation and cytokinesis

For cells in general, the mitotic spindle forms perpendicular to the division plane marked by the PPB (Mineyuki, 1999). GMCs are no exception (Mineyuki et al., 1988; Cho & Wick, 1989; Cleary & Hardham, 1989), but in many species they do provide an extreme example of a cell type in which the spindle, once formed, can undergo dynamic movements. GMC spindles can rotate freely up to 45° from the pre-programmed division plane (Fig. 6b) (Palevitz & Hepler, 1974b; Palevitz, 1986; Mineyuki et al., 1988; Cleary & Hardham, 1989). This morphological plasticity (Palevitz, 1993) is thought to overcome geometric constraints and maximise the space available for mitosis, particularly for the anaphase separation of chromosomes (Palevitz & Hepler, 1974a; Mineyuki et al. 1988). It is obvious that the role of the actin patches is not to anchor the prophase spindle poles, and that this flexibility in spindle position is not detrimental to the development of the guard cell pair.

The mis-orientation of the spindle is usually corrected in telophase by rotation of the nuclei-cell plate complex so that the cell plate becomes aligned in the pre-defined plane (Wick, 1991). As with SMCs, the final realignment occurs when the phragmoplast interacts with the plasma membrane (Palevitz, 1986). It is probable therefore that cortical actin patches and the ADZ define regions of the cell cortex that discourage or support, respectively, fusion of the cell plate (Fig. 6b). The ADZ anchors the cell plate, while the telophase nuclei come to rest adjacent to the actin patches. Realignment requires energy and is dependent on actin, as shown by the respective sensitivities to sodium azide (Palevitz & Hepler, 1974a; Palevitz, 1986) and cytochalasin (Palevitz and Hepler, 1974b; Cho & Wick, 1990).

5. CONCLUSION

Over the last 100 years we have learnt much about stomatal physiology and development; and in more recent years an understanding of the role of the cytoskeleton in these processes has been gained. As presented in this chapter, we are now aware that actin has fundamental roles. It is required during the earliest stages of cell polarisation through to the completion of cell division: (i) establishing division sites when cells prepare for a round of division, (ii) in providing persistent molecular markers for the division site during mitosis, and finally (iii) in guiding the growth and placement of the new cell wall. However, there remains a great deal to be learned and confirmed about the distribution and function of actin during stomatogenesis, particularly given the phylogenetic diversity in stomatal form.

Actin distribution is known only at the level of resolution of the light microscope. Ultrastructural investigations of stomatal actin are needed to identify the cyto-architecture of the actin arrays with a view to understanding how components of the actin arrays interact (e.g. phragmoplast and cortical actin during cytokinesis). The induction of SMC divisions by neighbouring GMCs was postulated in the 1960s (Pickett-Heaps & Northcote, 1966a; Stebbins & Jain, 1960; Stebbins & Shah, 1960). Since then no information has been gleaned as to the nature of the "inductive influence". Evidence that the inductive influence affects the actin cytoskeleton is apparent in the spatial and temporal distribution of the actin patches and the polarisation of cytoplasmic actin required for nuclear migration (see Cleary & Mathesius, 1996; Kennard & Cleary, 1997). The signal transduction pathways and the mechanisms controlling coordinated actin redistribution between cells of the complex are unknown, but could involve the phosphoinositol pathway acting via actin-binding proteins (e.g. Gibbon et al. 1997; Smertenko et al. 1998).

REFERENCES

Brown RC and Lemmon BE (1991) Pollen development in orchids. 5. A generative cell domain involved in spatial control of the hemispherical cell plate. J Cell Sci 100: 559-565

Busby CH and Gunning BES (1980) Observations on pre-prophase bands of microtubules in uniseriate hairs, stomatal complexes of sugar cane, and *Cyperus* root meristems. Eur J Cell Biol 21: 214-223

Campbell DH (1881) On the development of the stomata of *Tradescantia* and Indian corn. Amer Nat 15: 761-766

Cho S-O and Wick SM (1989) Microtubule orientation during stomatal differentiation in grasses. J Cell Sci 92: 581-594

Cho S-O and Wick SM (1990) Distribution and function of actin in the developing stomatal complex of winter rye (*Secale cereale* cv. Puma). Protoplasma 157: 154-164

Cho S-O and Wick SM (1991) Actin in the developing stomatal complex of winter rye: A comparison of actin antibodies and Rh-phalloidin labeling of control and CB-treated tissues. Cell Motil Cytoskel 19: 25-36

Cleary AL (1995) F-actin redistributions at the division site in living *Tradescantia* stomatal complexes as revealed by microinjection of rhodamine-phalloidin. Protoplasma 185: 152-165

Cleary A (1996) Regulation of cell division during formation of stomatal complexes: Importance of the cytoskeleton, cell plate realignments and the cell wall. In: The Symposium of the Construction of Cytoskeletal Arrays in Plant Cells. Osaka University, Osaka, pp 13-15

Cleary AL and Hardham AR (1989) Microtubule organization during development of stomatal complexes in *Lolium rigidum*. Protoplasma 149: 67-81

Cleary AL and Mathesius U (1996) Rearrangements of F-actin during stomatogenesis visualised by confocal microscopy in fixed and permeabilised *Tradescantia* leaf epidermis. Bot Acta 109: 15-24

Cleary AL and Smith LG (1998) The tangled1 gene is required for spatial control of cytoskeletal arrays associated with cell division during maize leaf development. Plant Cell 10: 1875-1888

Cleary AL, Brown RC and Lemmon BE (1992a) Establishment of division plane and mitosis in monoplastidic guard mother cells of *Selaginella*. Cell Motil Cytoskel 23: 89-101

Cleary AL, Gunning BES, Wasteneys GO and Hepler PK (1992b) Microtubule and F-actin dynamics at the division site in living *Tradescantia* stamen hair cells. J Cell Sci 103: 977-988

Galatis B, Apostolakos P, Katsaros C and Loukari H (1982) Pre-prophase microtubule band and local wall thickening in guard cell mother cells of some Leguminosae. Ann Bot 50: 779-791

Galatis B, Apostolakos P and Katsaros C (1984) Experimental studies on the function of the cortical cytoplasmic zone of the preprophase microtubule band. Protoplasma 122: 11-26

Gallagher K and Smith LG (1999) Discordia mutations specifically misorient asymmetric cell divisions during development of the maize leaf epidermis. Development 126: 4623-4633

Gibbon BC, Ren H and Staiger CJ (1997) Characterization of maize (*Zea mays*) pollen profilin function *in vitro* and in live cells. Biochem J 327: 909-915

Hepler PK and Gunning BES (1998) Confocal fluorescence microscopy of plant cells. Protoplasma 201: 121-157

Jiang C, Weeds AG and Hussey PJ (1997) The maize actin-depolymerizing factor, ZmADF3, redistributes to the growing tip of elongating root hairs and can be induced to translocate into the nucleus with actin. Plant J 12: 1035-1043

Kakimoto T and Shibaoka H (1987) Actin filaments and microtubules in the preprophase band and phragmoplast of tobacco cells. Protoplasma 140: 151-156

Kandasamy MK and Meagher RB (1999) Actin-organelle interaction: Association with chloroplast in *Arabidopsis* leaf mesophyll cells. Cell Motil Cytoskel 44: 110-118

Kazama H, Sugimoto K and Suzuki R (1995) Nuclear migration mechanism of the stomatal subsidiary mother cell of hypocotyl epidermis of *Cucumis sativus* L. J Cell Biochem 21A: 241

Kennard JL and Cleary AL (1997) Pre-mitotic nuclear migration in subsidiary mother cells of *Tradescantia* occurs in G1 of the cell cycle and requires F-actin. Cell Motil Cytoskel 36: 55-67

Kost B, Spielhofer P and Chua N-H (1998) A GFP-mouse talin fusion protein labels plant actin filaments *in vivo* and visualizes the actin cytoskeleton in growing pollen tubes. Plant J 16:393-401

Kropf DL (1992) Establishment and expression of cellular polarity in fucoid zygotes. Microbiol Rev 56: 316-339

Liu B and Palevitz BA (1992) Organization of cortical microfilaments in dividing root cells. Cell Motil Cytoskel 23: 252-264

Mineyuki Y (1999) The preprophase band of microtubules: Its function as a cytokinetic apparatus in higher plants. Int Rev Cytol 187: 1-49

Mineyuki Y and Gunning BES (1990) A role for preprophase bands of microtubules in maturation of new cell walls, and a general proposal on the function of preprophase band sites in cell division in higher plants. J Cell Sci 97: 527-537

Mineyuki Y and Palevitz BA (1990) Relationship between preprophase band organization, F-actin and the division site in *Allium*. Fluorescence and morphometric studies on cytochalasin-treated cells. J Cell Sci 97: 283-295

Mineyuki Y, Marc J and Palevitz BA (1988) Formation of the oblique spindle in dividing guard mother cells of *Allium*. Protoplasma 147: 200-203

Mineyuki Y, Marc J and Palevitz BA (1989) Development of the preprophase band from random cytoplasmic microtubules in guard mother cells of *Allium cepa* L. Planta 178: 291-296

Mullinax JB and Palevitz BA (1989) Microtubule reorganization accompanying preprophase band formation in guard mother cells of *Avena sativa* L. Protoplasma 149: 89-94

Palevitz BA (1980) Comparative effects of phalloidin and cytochalasin B on motility and morphogenesis in *Allium*. Can J Bot 58: 773-785

Palevitz BA (1986) Division plane determination in guard mother cells of *Allium*: Video time-lapse analysis of nuclear movements and phragmoplast rotation in the cortex. Dev Biol 117: 644-654

Palevitz BA (1993) Morphological plasticity of the mitotic apparatus in plants and its developmental consequences. Plant Cell 5: 1001-1009

Palevitz BA and Hepler PK (1974a) The control of the plane of division during stomatal differentiation in *Allium*. I. Spindle reorientation. Chromosoma 46: 297-326

Palevitz BA and Hepler PK (1974b) The control of the plane of division during stomatal differentiation in *Allium*. II. Drug studies. Chromosoma 46: 327-341

Palmer RE, Sullivan DS, Huffaker T and Koshland D (1992) Role of astral microtubules and actin in spindle orientation and migration in the budding yeast, *Saccharomyces cerevisiae*. J Cell Biol 119: 583-593

Pickett-Heaps JD (1969a) Preprophase microtubules and stomatal differentiation in *Commelina cyanea*. Aust. J. Biol. Sci. 22: 374-391

Pickett-Heaps JD (1969b) Preprophase microtubules and stomatal differentiation: some effects of centrifugation on symmetrical and asymmetrical cell division. J Ultrastruct Res 27: 24-44

Pickett-Heaps JD and Northcote DH (1966a) Cell division in the formation of the stomatal complex of the young wheat leaf. J Cell Sci 1: 121-128

Pickett-Heaps JD and Northcote DH (1966b) Organization of microtubules and endoplasmic reticulum during mitosis and cytokinesis in wheat meristems. J Cell Sci 1: 109-120

Pickett-Heaps JD, Gunning BES, Brown RC, Lemmon BE and Cleary AL (1999) The cytoplast concept in dividing plant cells: Cytoplasmic domains and the evolution of spatially organized cell division. Am J Bot 86: 153-172

Smertenko AP, Jiang C, Simmons NJ, Weeds AG, Davies DR and Hussey PJ (1998) Ser6 in the maize actin-depolymerizing factor, ZmADF3, is phosphorylated by a calcium-stimulated protein kinase and is essential for the control of functional activity. Plant Cell 14: 187-193

Smith LG (1999) Divide and conquer: Cytokinesis in plant cells. Curr Opin Plant Biol 2:447-453

Staiger CJ, Yuan M, Valenta R, Shaw PJ, Warn RM and Lloyd CW (1994) Microinjected profilin affects cytoplasmic streaming in plant cells by rapidly depolymerizing actin microfilaments. Curr Biol 4: 215-219

Stebbins GL and Jain SK (1960) Developmental studies of cell differentiation in the epidermis of monocotyledons. I. *Allium*, *Rhoeo,* and *Commelina*. Dev Biol 2: 409-426

Stebbins GL and Shah SS (1960) Developmental studies of cell differentiation in the epidermis of monocotyledons. II. Cytological features of stomatal development in the Gramineae. Dev Biol 2: 477-500

Strasburger E (1866) Ein beitrag zur Entwicklungsgeschichte der Spalltofrungen. Jahrb Wiss Bot 5: 297-342

Sugimoto K, Morita Y and Kazama H (1993) Red-light induced subsidiary cell differentiation in the hypocotyl epidermis of *Cucumis sativus* L. In: Proceedings of XV International Botanical Congress, Yokohama, p 435

Tomlinson PB (1974) Development of the stomatal complex as a taxonomic character in the monocotyledons. Taxon 23: 109-128

Traas JA, Doonan JH, Rawlins DJ, Shaw PJ, Watts J and Lloyd CW (1987) An actin network is present in the cytoplasm throughout the cell cycle of carrot cells and associates with the dividing nucleus. J Cell Biol 105: 387-395

Wick SM (1991) Spatial aspects of cytokinesis in plant cells. Curr Opin Cell Biol 3: 253-260

Chapter 24

STRUCTURE AND FUNCTION OF ACTIN FILAMENTS IN MATURE GUARD CELLS

Jae-Ung Hwang, Soon-Ok Eun and Youngsook Lee

Department of Life Science, Pohang University of Science and Technology, Pohang, 790–784, Republic of Korea

Key words: actin, guard cells, K^+ channels, signal transduction, stomata

Abstract: Recently, actin filaments in mature kidney-shaped guard cells of many plants have been shown using diverse methodologies. Interestingly, the arrangements of cortical actin filaments showed close similarities. Moreover, there has been evidence suggesting their roles in signal transduction. We review these recent data and draw a model for the function of cortical actin filaments in guard cells in daily stomatal movements.

1. INTRODUCTION

Plants have two contradicting needs for gaseous exchange with the environment. Photosynthesis requires an increase in gas exchange for the uptake of CO_2, but for conservation of water, the opposite is favoured. A pair of guard cells, which outlines the pore of a stoma, balances the two needs by precisely regulating stomatal aperture. To perform this critical function, guard cells have elaborate signaling systems to sense and respond to changes in the surrounding environment. During the signal transducing process of stomatal aperture change, many factors, such as calcium ions, cytosolic pH, protein kinases, protein phosphatases, G-proteins, phospholipases, etc., have been implicated (Assmann, 1993; MacRobbie, 1998). Recent studies have indicated that actin filaments act as a signal mediator in guard cells (Kim et al., 1995; Eun & Lee, 1997; Hwang et al., 1997; Liu & Luan, 1998). This role for actin filaments as a signal mediator is not new in non-plant systems, and evidence for similar schemes in several other plant systems is accumulating (Staiger, 2000).

C.J. Staiger et al. (eds.), Actin: A Dynamic Framework for Multiple Plant Cell Functions, 427–436.

In this article, we review the organization of actin filaments in mature kidney-shaped guard cells from various plant species and the changes in actin organization in response to physiological stimuli in *Commelina communis*. We also describe the experimental results that show the regulation of inward K^+ channels, a crucial factor for stomatal opening, through structural changes of actin. Finally, we discuss the possible control of actin reorganization and propose a model for its role during stomatal movements.

2. ACTIN ORGANIZATION IN MATURE GUARD CELLS

2.1 Organization of actin filaments in mature guard cells

C. communis is the most frequently used model plant for studying the physiology of guard cells. We first visualized actin in guard cells of stomata open under white light during the usual photoperiod. In live, mature guard cells of *C. communis*, microinjected fluorescein isothiocyanate (FITC)-phalloidin showed actin filaments radiating from the stomatal pore (Fig. 1A). However, the majority of injected guard cells lost their viability due to the difficulty of microinjection into mature guard cells having a thick cell wall and the fluorescence of the injected dye faded too soon to trace the dynamic changes of actin organization upon stimulation. As an alternative to this problem, cells were fixed and then labeled with actin antibody or rhodamine-phalloidin. The organization of actin filaments observed in the fixed cells was very similar to that seen in live guard cells (Figs. 1B, 1C), confirming the radial organization of cortical actin filaments in illuminated guard cells.

The radial arrangement of cortical actin filaments has been observed in kidney-shaped guard cells of other plant species as well. For example, in *Selaginella* and *Tradescantia*, actin filaments resembled the radial arrangement observed in *C. communis* (Cleary et al., 1996). In guard cells of dicotyledonous plants, *Vicia faba* and *Nicotiana plumbaginifolia*, radial actin filaments were detected, but were sparse in number and not so obvious as in *C. communis* amongst the additional cortical and subcortical actin filaments that were randomly distributed (Figs. 1D, 1E). The actin filaments in guard cells of *Arabidopsis thaliana*, transformed with a GFP-mouse talin reporter, also resembled those in guard cells of *Vicia* and *Nicotiana* (Kost et al., 1998). Thus, we can conclude that mature kidney-shaped guard cells have radially organized actin filaments at the cortex, although there are variations in the density and the orderliness depending on the plant species.

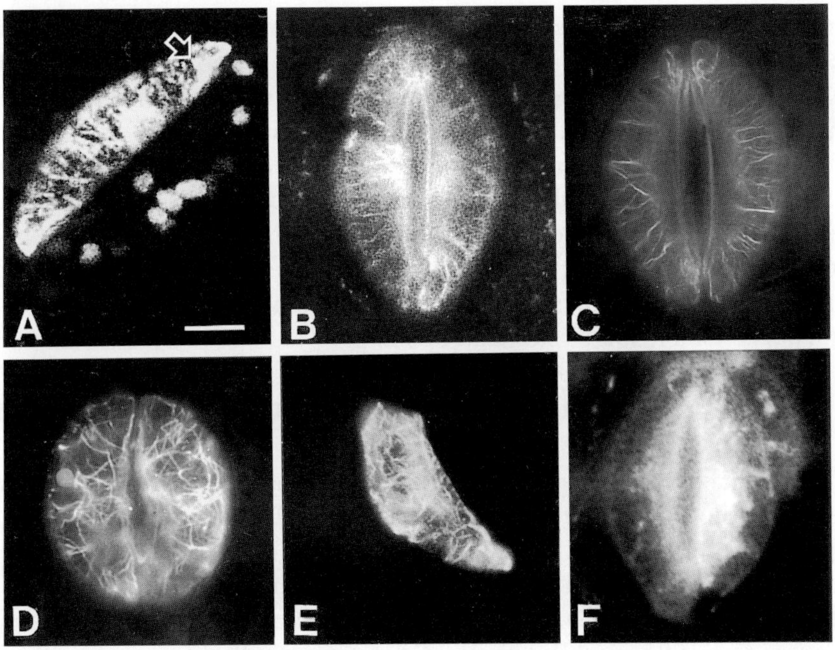

Figure 1. Actin in mature guard cells. (A, B, C & F) *C. communis.* (D) *N. plumbaginifolia.* (E) *V. faba.* (A) A live guard cell microinjected with FITC-phalloidin. The arrow points to the injection site. (B-F) Guard cells labeled with actin antibody (B, E & F) or rhodamine-phalloidin (C & D) after chemical fixation. Stomata opened under illumination (A-E) and closed under darkness (F). A is from Kim et al., 1995 and B & F are from Eun & Lee, 1997, with permission from the American Society of Plant Physiologists. Bar, 10 μm for A-F.

2.2 Signal-specific changes in the actin organization

In *C. communis*, the long radial actin filaments were observed in the guard cells of stomata open under daylight. However, under complete darkness, stomata closed and the radial organization of cortical actin filaments in guard cells completely disappeared or was replaced by randomly-oriented short fragments (Fig. 1F). Abscisic acid (ABA), which induces stomatal closure, depolymerized cortical actin filaments in guard cells as well. This response was detectable as early as 3 min, suggesting a possible involvement of actin in the early phase of signal transduction. However, this question remains unsolved since it is still not clear whether actin reorganization precedes stomatal closure (Eun & Lee, 1997).

In contrast to the common depolymerizing effects of two different closing signals, the opening stimuli showed different effects on actin organization. White light promoted the formation of long radial actin filaments near the plasma membrane. Circadian clock-controlled opening of stomata, 2–3 h before the beginning of photoperiod under high humidity, was also accompanied by the formation of cortical actin filaments (Eun & Lee, 1997). However, fusicoccin, a fungal toxin which promotes excessive stomatal opening, inhibited the formation of actin filaments and disrupted the already-formed filaments (Eun & Lee, 2000). Another opening stimulus, the hypotonic treatment which increases the guard cell volume and stomatal aperture, also induced depolymerization of actin filaments in *Vicia* guard cell protoplasts (Liu & Luan, 1998). The different effects of diverse opening signals on actin organization strongly indicate that formation of actin filaments is not a result of stomatal opening.

2.3 Effects of actin antagonists on stomatal movements

The signal-specific responses of actin filaments underline their role as a signaling molecule during stomatal movements. This possibility was further supported by the effects of actin antagonists, cytochalasin D (CD) and phalloidin, on the stomatal behaviour in two different plants (Kim et al., 1995; Hwang et al., 1997). CD completely depolymerized the actin filaments (Kim et al., 1995), while phalloidin promoted the formation of actin filaments and protected them from depolymerization even under conditions that normally caused depolymerization of actin filaments in guard cells (Kim et al., 1995; Eun & Lee, 2000). CD promoted the stomatal opening induced both by white light and by circadian clock at early photoperiod (Figs. 2A, 2B). In addition, it also enhanced stomatal closure in the presence of ABA (Fig. 2C) under conditions where the stomatal closure did not easily occur, such as low external CO_2 and high K^+ concentration in the medium. In the case of actin stabilization, stomatal movements were affected in the opposite direction (Figs. 2D, 2E). Phalloidin inhibited both the stomatal opening induced by white light or circadian clock and the stomatal closure induced by ABA. Phalloidin also showed a similar inhibitory effect on stomatal opening induced by fusicoccin, and reduced fusicoccin-induced depolymerization of actin filaments (Eun & Lee, 2000).

In short, CD made the guard cells more responsive to the stimuli while phalloidin had the opposite effect. In other words, when there are fewer actin filaments after CD treatment, guard cells change their volume to a larger extent in response to stimuli, whereas the increase in the number of actin filaments after phalloidin treatment reduces the extent of volume change. The actin antagonists had the same effects on stomatal movements in two different plants, *C. communis*, a monocotyledon, and *V. faba*, a dicotyledon. Therefore, we predict that actin filaments control the stomatal movements in a similar manner in all plants with kidney-shaped guard cells.

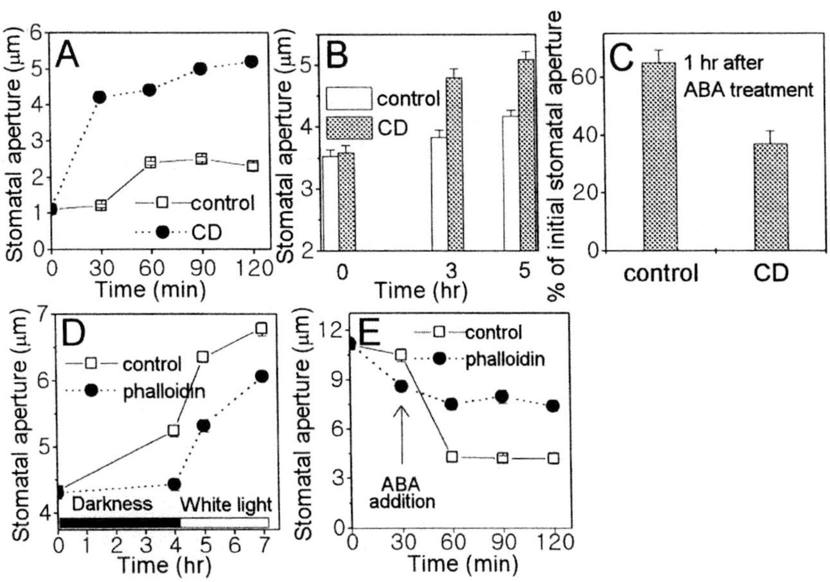

Figure 2. The effects of actin antagonists on stomatal movements. (A-C) CD (20 μM) effect on the circadian stomatal opening under darkness in *C. communis* (A), on the light-induced stomatal opening in *V. faba* (B), and on the ABA (10 μM)-induced stomatal closure in *C. communis* (C). In C, the incubation medium included 50 mM KCl and was bubbled with low CO_2 air. (D) Phalloidin (200 μM) effect on the circadian (h 0-4) and light-induced (h 4-7) stomatal opening in *V. faba.* (E) Phalloidin (100 μM) effect on the ABA-induced stomatal closure in *C. communis.* A & E are from Kim et al., 1995 and B & D are from Hwang et al., 1997, with permission from ASPP.

2.4 Independence of actin dynamics from microtubules

Actin filaments are often closely aligned with microtubules, and alteration in one results in reorganization of the other (Seagull, 1990). However, actin filaments and microtubules do not seem to have a close relationship either in structure or function in guard cells of *C. communis.* They did not colocalize, as clearly shown using the double-labeling technique, and furthermore, actin filaments were reorganized during stomatal movements, whereas microtubules retained an unaltered radial organization under all conditions tested (Eun & Lee, 1997).

3. EFFECTOR MOLECULES AND MODULATORS OF ACTIN REORGANIZATION

In animal cells, dynamic changes in actin filaments contribute to several signaling processes (Hall, 1998; Rijken et al., 1998). The rapid signal-induced reorganization of actin filaments observed in mature guard cells suggests that actin plays an important role in the signal transduction for stomatal movements. However, the mechanism by which actin filaments affect the volume changes of guard cells and the signal mediators that act up- or down-stream of actin in signaling cascades remains mostly unexplored.

3.1 The effector molecules of actin filaments

Ion movement proteins generate the driving force for guard cell volume changes and are the ultimate targets of signal cascades that lead to changes in stomatal aperture (MacRobbie, 1997). The cortical actin present in guard cells may regulate the ion movement proteins as in the volume adjustment of animal cells (Cantiello & Prat, 1996). This possibility was tested for the inward K^+ channels which are important for stomatal opening (Hwang et al., 1997; Liu & Luan, 1998).

CD potentiated the inward K^+ current in 50% of the cells examined in the whole cell configuration, whereas phalloidin decreased it in 76% of cells (Fig. 3). These effects of actin antagonists were detectable as early as 3 min after application of the drug and were confirmed in the outside-out single channel recording. Further analysis showed that the antagonists affected the opening probability of each channel and the number of active channels responsive to the stimulus but not the single channel conductance (Hwang *et al.*, 1997). We propose that as actin filaments depolymerize, the closed state of the channel becomes less stable and more channels become ready to respond to the hyperpolarized membrane potential, whereas long actin filaments stabilize the channels in the closed state and as a result, the channels become less responsive to hyperpolarization of membrane potential.

The effects of actin antagonists on inward K^+ channels correspond to their effects on stomatal opening; the enhancement of stomatal opening by CD may be mediated by its effect on inward K^+ channels and the opposite effect of phalloidin can be explained similarly. Since there are few cortical actin filaments in guard cells of a closed stoma, a probable scenario could be explained as follows. As a stoma opens, actin filaments may begin to form and inhibit the activation of inward K^+ channels and consequently slow down the opening process thus preventing the stoma from opening excessively. This path could be bypassed or enhanced with actin antagonists

(Fig. 2). Liu & Luan (1998) reported that the hypotonic condition depolymerized actin filaments as well as increased inward K^+ currents in guard cell protoplasts. This result is consistent with the hypothesis that actin depolymerization accelerates the increase of guard cell volume by activating the inward K^+ current.

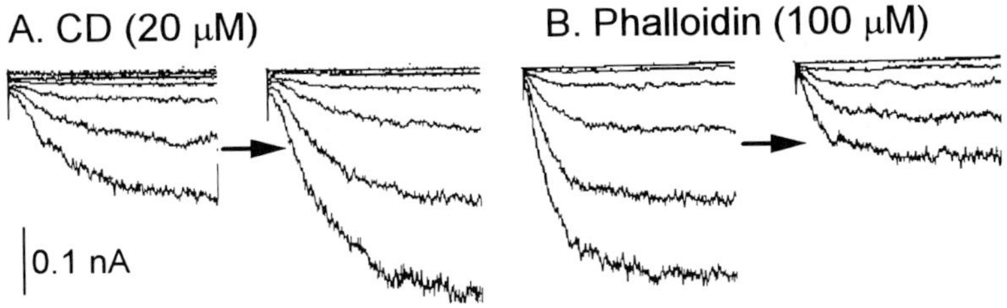

Figure 3. The effects of actin antagonists on inward K^+ currents of *V. faba* guard cells patch-clamped in the whole cell configuration (Hwang et al., 1997; with permission from ASPP). Membrane potential was stepped for 2 s from the holding potential of -60 mV to the test potential between -180 and -60 mV, in 20 mV-increments.

Then, how does actin interact with an ion channel? Some plant K^+ channels have a domain for binding to the cytoskeletal element (Fox & Guerinot, 1998) and many signaling molecules involved in the regulation of guard cell K^+ channels, such as protein kinases, phospholipases and G-proteins, were reported to be associated with actin filaments in other cell types (Putnam-Evans et al., 1989; Ibarrondo et al., 1995). Therefore, the structural changes of actin may affect the inward K^+ current in guard cells by actin-binding proteins or via regulating some signal mediators.

Actin filaments in guard cells may modulate other ion movement proteins and signaling factors required for the processes from signal perception to cellular responses. It has been demonstrated that near to, or at the plasma membrane, components for signal perception and transduction are located and form a complex (Okada, 1999; Trewavas, 1999). If this phenomenon holds true for guard cells, then cortical actin filaments, which also exist in the same area, may interact with other signaling molecules or ion movement proteins and so regulate their distribution and activity. In addition, during stomatal movements, the volume and surface area of a guard cell was known to change above the extent that the elasticity of plasma membrane can accommodate (Homman, 1998). Since the actin cytoskeleton is an essential player in the process of endocytosis and exocytosis (Battey et al, 1999), actin reorganization in guard cells may regulate the recruitment or retrieval process of plasma membrane during stomatal movements. Further studies are required to understand the detailed mechanisms by which actin modulates stomatal movements.

3.2 Actin modulating molecules

During the stomatal movements, what are the possible signal mediators that reorganize actin filaments? An obvious candidate includes the signal mediators in guard cells which, in other systems, have been demonstrated to regulate actin dynamics. Ca^{2+} is a potent second messenger which initiates the stomatal aperture changes (Assmann, 1993) and is known to regulate the activity of many actin-binding proteins (Forscher, 1989). Protein kinases and phosphatases are also important signal mediators of stomatal movements (MacRobbie, 1998) that have been reported to modulate actin organization in other plant and animal systems (Hartwig et al., 1992; Menzel et al., 1995). Cytosolic pH and phospholipases also belong to this group of possible actin modulators (Lin et al., 1997; Wakelam et al., 1997).

Many actin modulators found in diverse cell types have not been reported in guard cells yet. Actin-binding proteins, Rho family small G- proteins and inositol phospholipid kinases (Kandzari et al., 1996; Hall, 1998; Staiger, 2000) may exist in guard cells, and regulate the reorganization of actin filaments in response to various stimuli. Extensive studies on these molecules in relation to actin filaments in guard cells are likely to be highly rewarding.

4. SUMMARY AND PERSPECTIVE

Recent results reviewed in this article clearly show that actin structure in mature guard cells changes in response to stimuli, and that actin dynamics play a regulatory role in the process of stomatal movements such as the regulation of inward K^+ channel. It appears that actin dynamics are not the passive results of stomatal movements because actin reorganized rapidly and the direction of changes was independent from the final stomatal aperture. Moreover, the effects of actin antagonists on stomatal movements imply that actin reorganization is necessary for normal stomatal responses to physiological stimuli. Based on these results, we speculate that the function of actin reorganization in guard cells is as follows. Under darkness, the closed stomatal guard cells contain few cortical actin filaments, and therefore at the beginning of stomatal opening, a guard cell easily increases its volume in response to light. However, as stomata open, long actin filaments form near the plasma membrane and begin to control negatively the speed and extent of stomatal opening, where the modulation of inward

K^+ channels contributes. Without such a negative regulation, the stomata would open excessively. The presence of cortical actin filaments is predicted to be important in maintaining stomatal aperture at the optimum until stomata begin to close in response to signals from circadian clock, darkness, or ABA. When such signals are perceived, actin filaments are rapidly depolymerized and accelerate the process of stomatal closure. If actin filaments stayed polymerized, stomata would shut very slowly, leading to unnecessary water loss.

We conclude that actin reorganization in guard cells plays a regulatory role in daily stomatal movements for optimal gas exchange. There still remain many important questions to be addressed such as the precise pattern of diurnal change in actin organization and the specific changes of actin organization in response to other environmental and internal factors. In addition, studies on the possible modulation by actin of other ion movement proteins and the interaction between actin and other signaling molecules are necessary for obtaining a clear understanding of the role(s) of actin filaments in guard cell physiology.

ACKNOWLEDGEMENT

The authors thank Prof. IS Curtis (POSTECH, Korea) for reviewing this manuscript. Supported by the Science and Engineering Foundation of Korea (#98-0401-07-01-3) and by the Basic Science Research Fund of Pohang University of Science and Technology.

REFERENCES

Assmann SM (1993) Signal transduction in guard cells. Annu Rev Cell Biol 9: 345-375

Battey NH, James NC, Greenland AJ and Brownlee C (1999) Exocytosis and endocytosis. Plant Cell 11: 643-659

Cantiello HF and Prat AG (1996) Role of actin filament organization in ion channel activity and cell volume regulation. In: Nelson WJ (ed) Current Topics in Membrane, Vol 43: Membrane Protein-Cytoskeleton Interactions. San Diego: Academic Press, pp 373-396

Cleary AL and Mathesius U (1996) Rearrangements of F-actin during stomatogenesis visualised by confocal microscopy in fixed and permeabilised *Tradescantia* leaf epidermis Bot Acta 109: 15-24

Eun S-O and Lee Y (1997) Actin filaments of guard cells are reorganized in response to light and abscisic acid. Plant Physiol 115: 1491-1498

Eun S-O and Lee Y (2000) Stomatal opening by fusicoccin is accompanied by depolymerization of actin filaments in guard cells. Planta, in press

Forscher P (1989) Calcium and polyphosphoinositide control of cytoskeletal dynamics. Trends Neuro Sci 12: 468-474

Fox TC and Guerinot ML (1998) Molecular biology of cation transport in plants. Annu Rev Plant Physiol Plant Mol Biol 49: 669-96

Hall A (1998) Rho GTPases and the actin cytoskeleton. Science 279: 509-514

Hartwig JH, Thelen M, Rosen A, Janmey PA, Nairn AC and Aderem A (1992) MARCKS is an actin filament crosslinking protein regulated by protein kinase C and calcium-calmodulin. Nature 356: 618-622

Homann U (1998) Fusion and fission of plasma-membrane material accommodates for osmotically induced changes in the surface area of guard cell protoplasts. Planta 206: 329-333

Hwang J-U, Suh S, Yi H, Kim J and Lee Y (1997) Actin filaments modulate both stomatal opening and inward K^+-channel activities in guard cells of *Vicia faba* L. Plant Physiol 115: 335-342

Ibarrondo J, Joubert D, Dufour MN, Cohen-Solal A, Homburger V, Jard S and Guillon G (1995) Close association of the α subunits of G_q and G_{11} G proteins with actin filaments in WRK_1 cells: relation to G protein-mediated phospholipase C activation. Proc Natl Acad Sci USA 92: 8413-8417

Kandzari DE, Chen J and Goldshmidt-Clermont PJ (1996) Regulation of the actin cytoskeleton by inositol phospholipid pathways. Subcell Biochem 26: 97-114

Kim M, Hepler PK, Eun S-O, Ha KS and Lee Y (1995) Actin filaments in mature guard cells are radially distributed and involved in stomatal movement. Plant Physiol 109: 1077-1084

Kost B, Spielhofer P and Chua N-H (1998) GFP-mouse talin fusion protein labels plant actin filaments *in vivo* and visualizes the actin cytoskeleton in growing pollen tubes. Plant J 16: 393-401

Lin K-M, Wenegieme E, Lu P-J, Chen C-S and Yin HL (1997) Gelsolin binding to phosphatidylinositol 4,5-bisphosphate is modulated by calcium and pH. J Biol Chem 272: 20443-20450

Liu K and Luan S (1998) Voltage-dependent K^+ channels as targets of osmosensing in guard cells. Plant Cell 10: 1957-1970

MacRobbie EA (1998) Signal transduction and ion channels in guard cells. Philos Trans R Soc Lond B Biol Sci 353: 1475-1488

Menzel D, Vugrek O, Frank S and Elsner-Menzel C (1995) Protein phosphatase 2A, a potential regulator of actin dynamics and actin-based organelle motility in the green alga *Acetabularia*. Eur J Cell Biol 67: 179-187

Okada Y (1999) A scaffolding for regulation of volume-sensitive Cl^- channels. J Physiol (Lond) 520: 2

Putnam-Evans C, Harmon A, Palevitz BA, Fechheimer M and Cormier MJ (1989) Calcium-dependent protein kinase is localized with F-actin in plant cells. Cell Motil Cytoskel 12: 12-24

Rijken PJ, van Hall GJ, van der Heyden MA, Verkleij AJ and Boonstra J (1998) Actin polymerization is required for negative feedback regulation of epidermal growth factor-induced signal transduction. Exp Cell Res 243: 254-262.

Seagull RW (1990) The effects of microtubule and microfilament disrupting agents on cytoskeletal arrays and wall deposition in developing cotton fibers. Protoplasma 159: 44-59

Staiger CJ (2000) Signaling to the actin cytoskeleton in plants. Annu Rev Plant Physiol Plant Mol Biol 51: 257-288

Trewavas A (1999) How plants learn. Proc Natl Acad Sci USA 96: 4216-4218

Wakelam MJ, Hodgkin MN, Martin A and Saqib K (1997) Phospholipase D. Semin Cell Dev Biol 8: 305-310

Chapter 25

ACTIN DURING MITOSIS AND CYTOKINESIS

Structure and function of the F-actin cytoskeleton

Anne-Catherine Schmit

Institut de Biologie Moléculaire des Plantes, CNRS–ULP, 12 rue du Général Zimmer, F–67084 Strasbourg–Cedex, France

Key words: antibodies, cell division, drugs, phallotoxins, plant cells

Abstract: The improved methods of actin detection described in the first part of this review have clarified some controversies on the organization of the actin cytoskeleton in plant cells. Subtle differences in actin distribution are still found, suggesting that actin may have acquired different functions during evolution of plant cells. Such differences in the structure and function of the actin cytoskeleton during cell division are illustrated by various examples from the plant kingdom. Vacuolated and non-vacuolated higher plant cells as well as walled and wall-free cells are examined, and fern, algae and fungae cells including yeast are compared. The role of the actin cytoskeleton is discussed on the basis of drug studies on wild type cells and insights brought by studies of actin mutants.

1. INTRODUCTION

This review will focus on the organization and role of the actin cytoskeleton during cell division which represents one short part of the cell cycle. In the plant kingdom, although karyokinesis and cytokinesis are often chronologically linked, they are still independent processes. The mitotic spindle is built according to the division plane indicated by the preprophase band which in turn defines the cell plate position (review: Field et al., 1999).

During mitosis, the actin cytoskeleton is strongly linked to the microtubular cytoskeleton. Actin and tubulin form cytoplasmic complexes with the chaperone, TCP1, suggesting that a common folding pathway is involved in their regulation (Sternlicht et al., 1993; Ursic et al., 1994). Compelling evidence shows the interdependence between actin and tubulin polymers. They may either substitute for one another or act in synergy

C.J. Staiger et al. (eds.), Actin: A Dynamic Framework for Multiple Plant Cell Functions, 437–456.

(review: Gavin, 1997). Several motor proteins may fulfil equivalent functions and carry cargos on both actin filaments and microtubules (Karki & Holzbaum, 1999).

In conjunction with the microtubular cytoskeleton, the actin cytoskeleton is involved in many cellular events, including division and cytokinesis (Traas et al., 1987; Kengen & De Graaf, 1991; review: Gavin, 1997). The control of these events is of crucial importance, particularly in plant cells, where it influences both morphogenesis and development. Actin takes part in signal transduction and other regulatory pathways. Differences in actin organization will be described using examples chosen from the plant kingdom. The functional impact of actin during cell division will be discussed in the final part of this review.

2. METHODS FOR PLANT ACTIN IDENTIFICATION

2.1 Preservation of microfilament structure during labelling procedures

F-actin is notoriously sensitive to fixation and therefore difficult to preserve. Partial destruction, rearrangement and preferential stabilization can occur during labelling protocols. Consequently, the structural organization of F-actin in plant cells during mitosis and cytokinesis was, for a long time, subject to controversy. Differences were found depending on the techniques used for visualization and on the cell types analysed (review: Staiger & Schliwa, 1987). Due to technical progress made in the last few years, we now have a reliable idea of *in vivo* actin dynamics.

Various labelling protocols have been compared and adapted to each type of material. Fixation procedures have been improved and preserve the fine actin filaments present in the cortex. Actin filaments can be crosslinked by a short incubation in permeabilizing buffer containing 100 μM MBS (m-maleimidobenzoyl N-hydroxy-succinimide ester) prior to aldehyde fixation (Sonobe & Shibaoka, 1989) or by using 300 μM MBS in the extraction buffer, prior to direct labelling with fluorescent probe (Hush & Overall, 1992). Fluorescent phalloidin can be directly applied during the fixation procedure, added to non-fixed cells after permeabilization (Traas et al., 1987; Schmit & Lambert, 1990; Valster & Hepler, 1997) or following electroporation using pulses of direct current which create transient pores in the plasma membrane (Traas et al., 1987).

Such techniques have been used on permeabilized dividing cells including those of tobacco BY-2, (Kengen et al., 1995), fixed tissues such as maize root tips, (Baluška et al., 1997), and living cells such as blood lily endosperm (Schmit & Lambert, 1990; Endlé et al., 1998), Tradescantia stomatal complexes (Cleary, 1995) and Tradescantia stamen hairs (Valster & Hepler, 1997). Improved visualization of actin in *Acetabularia* was also obtained by microwave-accelerated fixation (Sawitzky et al., 1996).

2.2 Probes for actin visualization in plants

Heavy meromyosin decoration was the first probe used to demonstrate the presence of actin fibers in mitotic spindles (Forer & Jackson, 1975; 1976, 1979). New insights were obtained by using fluorescent phallotoxins as well as actin antibodies (Schmit & Lambert, 1985, 1987; Seagull et al., 1987; McCurdy et al., 1991). Phallotoxins are small peptides (of about 1 kDa) which diffuse instantaneously in the cytoplasm. Fluorescent phallotoxins label actin fibers in less than 5 min when injected into cells or when added in permeabilizing buffers. Although different types of plant F-actin differ in phalloidin affinity, similar images were obtained after immunolabelling using the various anti-actin antibodies listed in Table 1.

Table 1. Cross-reactivity of anti-actin antibodies on plant material

Source	Company	Materials	References
Smooth muscle α-actin	•Sigma A9172	Purified maize pollen actin	Liu & Yen 1992
Chicken gizzard	•Amersham N350	Pea stem extracts *Haemanthus*	Abe et al., 1992 Schmit & Lambert, 1988
Chicken gizzard (JLA 20)	•Developmental Studies Hybridoma Bank	*Nicotiana* pollen tubes	Lancelle & Hepler, 1991
Chicken gizzard (C4)	•ICN Immunobiologicals, Lisle, IL	*Nitella* *Chara* internodal cells *Allium* root tips	Lessard, 1988 Eleftheriou & Palevitz, 1992 Liu & Palevitz, 1992
Dictyostelium		*Chlamydomonas*	Detmers et al., 1985

Microinjection of fluorescently labelled actin could be developed for actin visualization *in vivo*. So far, it has not been possible to incorporate purified fluorescent actin from animal sources into the plant actin cytoskeleton. This could be due to divergence between plant and animal actins, a divergence which is greater than for tubulins. Recently, however, electroporation of fluorescent Dictyostelium actin in Dictyostelium amoebae led to successful incorporation of actin in the cytoskeleton (Yumura & Fukui, 1999).

Genetic engineering provides a new approach to actin visualization. The actin cytoskeleton is fluorescently labelled in plants by stably expressing a GFP-talin binding domain fusion protein which binds to actin *in vivo* (Kost et al., 1998). Similar expression of different GFP-fused proteins will be useful for *in vivo* labelling in plants. So far, GFP::ABP120 actin binding domain (Pang & Knecht, 1998) and GFP::actin (Westphal et al., 1997; Aizawa et al., 1997) have been used in Dictyostelium cells.

3. MICROFILAMENT DYNAMICS

Microinjection of fluorescent phalloidin revealed the dynamic redistribution of actin filaments *in vivo* during mitosis and cytokinesis. Although phallotoxins stabilize F-actin, low concentrations (5 - 30 nM) can be used (Schmit & Lambert, 1990; Endlé et al., 1998) or higher concentrations (> 250 nM) can be used with short observation times (Valster & Hepler, 1997). Comparative studies indicate that phalloidin-labelled F-actin reveals a cytoskeletal organization similar to that observed in untreated cells (Cleary et al., 1992). Significant differences observed in various plant species will be detailed below.

3.1 Higher plant cells

All eukaryotes possess an actin cytoskeleton, but the distribution in plant cells is strikingly different from that formed in animal cells. Although the plant actin cytoskeleton is present throughout the cell cycle, it is organized in different dynamic arrays (from thick cables to single actin filaments) which change during the cell cycle. Actin organization also varies depending on the tissue in the same species: endosperm, meristem or vacuolated cells.

3.1.1 Natural protoplasts

Endosperm cells from *Haemanthus katherinae* Bak. and *Clivia nobilis L.* have facilitated the study of the cytoskeleton dynamics during mitosis since they are large cells (about 100 μm in diameter) and they naturally lack a cell wall. Furthermore, this tissue forms a coenocyte in which cytokinesis is delayed with respect to karyokinesis. Cytokinesis occurs between non-sister as well as sister nuclei during the cellularization process, confirming the independence of both processes. Cell plate location is not indicated by a preprophase band since the tissue is not under polarized morphogenetic influence.

Actin filaments were first described at the structural level by Forer and Jackson (1975, 1976, 1979). Actin filaments have been found in close association with microtubules within the mitotic spindle.

Using fluorescent phallotoxins as probes, we visualize a plant-specific three-dimensional actin cytoskeleton in whole cells (Schmit & Lambert, 1985, 1987, 1988, 1990). Actin filaments form an elastic network composed of thin filaments in bundles. Like microtubules, actin polymers continuously assemble and disassemble. They form a dynamic cytoskeleton, present throughout interphase, mitosis and cytokinesis.

3.1.1.1 Interphase

Actin filaments intermingle with microtubules around the nucleus, throughout the cytoplasm, and at the cell cortex. Their distributions are similar but not identical. It is likely that both actin and microtubules participate in cell shape, in positioning of the nucleus at the center of the protoplast, and in cytoplasmic organelle movement (Schmit & Lambert, 1987).

3.1.1.2 Mitosis

A cortical basket of actin filaments forms a cage around the mitotic spindle during its formation (Fig. 1). This cage is stretched along the spindle axis at the equator and the mesh size decreases simultaneously at the polar regions during the metaphase-anaphase transition (Schmit & Lambert, 1987). Some fibers are also found within the microtubular spindle, close to the inter-polar and kinetochore microtubules (Forer, 1985; Schmit & Lambert, 1990). In late anaphase, a new population of actin filaments polymerize at the interzone. These filaments invade the cytoplasmic zone which is free of microtubules and where the phragmoplast will develop (Schmit & Lambert, 1985, 1988). This increase in the number of of actin filaments seems to be an active mechanism (Schmit & Lambert 1987; Endlé et al., 1998) rather than a passive invasion of polymers broken by rapid microtubule growth, as was suggested by Molé-Bajer and Bajer (1988).

3.1.1.3 Cytokinesis

During cell plate formation, there is a sudden increase in the number of very short actin filaments which run parallel to the neo-formed phragmoplast microtubules (Fig. 2). This suggests that an actin/tubulin assembly site is established and this site grows centrifugally. Thus, an actin wreath precedes and then intermingles with the microtubule ring until both reach the plasma membrane and cell plate maturation takes place (Endlé et al., 1998).

3.1.2 Walled cells

Spindle actin distribution is quite similar in walled cells and in endosperm cells. However, the organization of cortical fibers is specific to walled cells.

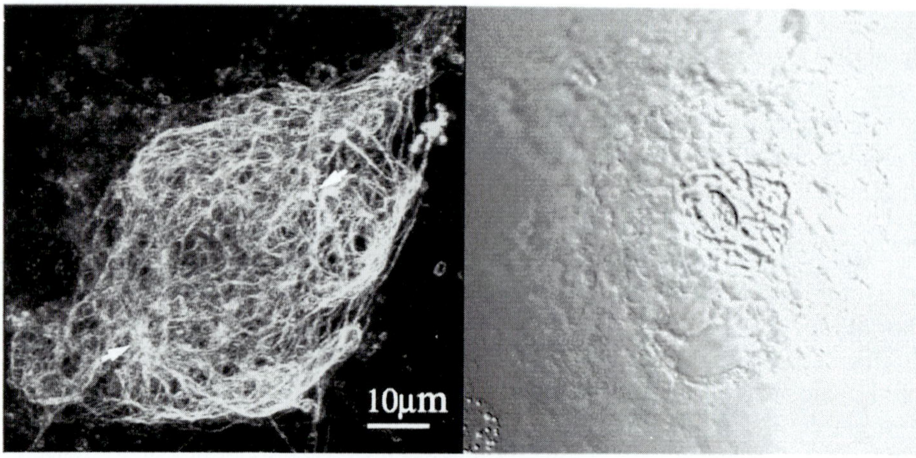

Figure 1. The cortical actin network in an endosperm prophase cell of *Haemanthus*. Three-dimensional reconstruction of 20 focal steps of 0.40 µm. Zeiss LSM 510 confocal laser scanning microscopy. (A) A cortical actin cage encircles the nucleus and some actin fibers forms at the forming poles (arrows). (B) DIC Nomarski image.

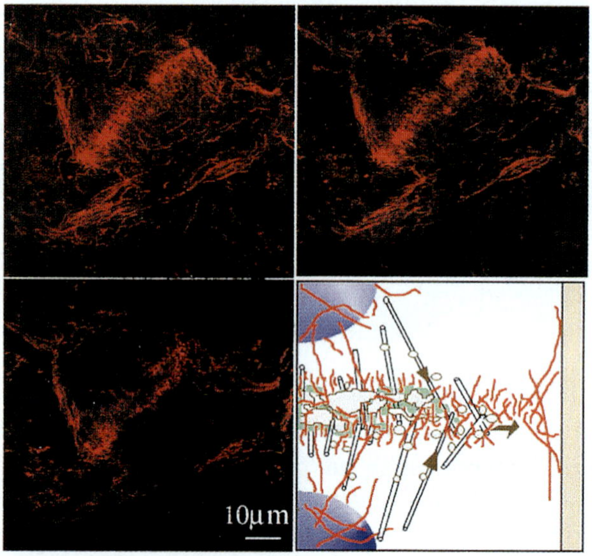

Figure 2. An endosperm cell of *Haemanthus* in telophase after perfusion of rhodamine-phalloidin in permeabilizing buffer. Analysis by confocal laser scanning microscopy (Zeiss LSM 510). (A-C) Three focal planes show the new polymerization of short actin filaments surrounding the forming cell plate. Long fibers persist in the cell cortex. (D) General model of cell plate growth in higher plants, adapted from Endlé et al. (1998). An expanding wreath of actin filaments (red) guide centrifugal phragmoplast expansion (large arrow). Golgi vesicles, transported along microtubules (thin arrows) fuse at the cell plate edges. Simultaneously,

vinculin-like antigens (green) are recruited at the cell plate. Nuclei are represented in blue, the mother cell wall appears in orange.

3.1.2.1 The phragmosome

Highly vacuolated cells re-enter the cell cycle after wounding. One of the first signs preceding mitosis is the centering of the nucleus. In G2, cytoplasmic strands radiate from the nuclear periphery to the cell cortex. They are oriented in two directions: along the future division plane corresponding to the phragmosome and along the future polar axis, suspending and polarizing the nucleus in preparation for division (Sinnott & Bloch, 1941; Traas et al., 1987; Staiger & Lloyd, 1991). These strands contain long actin cables and few microtubules which are both involved in the spatial control of the nucleus position, leading to cell polarization before mitosis. In asymmetrical divisions, like in subsidiary mother cells of *Tradescantia*, an accumulation of actin is observed at the anticlinal cell walls closest to the polarized nucleus, indicating the polar axis. Similarly, actin is important in symmetrical divisions (Cleary, 1995; Kennard & Cleary, 1997).

3.1.2.2 Actin cortical arrays and the preprophase band

Close to the plasma membrane, very thin cortical microfilament arrays are observed parallel to cortical microtubules (Sonobe & Shibaoka, 1989). During the G2/M transition, actin filaments are present in all cell types. A subset of cortical fibers, either recruited from the cortex or newly polymerized, coalign with the microtubular preprophase band to form an actin preprophase band. This latter band occupies a wider zone than microtubules. They mark the future site of the new cell wall insertion (Cleary, 1995). The formation of this barrel is critical for morphogenesis and it controls cytoskeleton signals involving both microtubules and actin filaments.

3.1.2.3 Spindle positioning and cytokinesis

Most of the actin preprophase band disappears from the cortical sites after nuclear envelope breakdown, leading to an actin-depleted zone. This negative template positions the cell plate correctly (Cleary, 1995). Simultaneously, other cytoplasmic actin arrays surround the mitotic spindle and cross the spindle (Seagull et al., 1987; Kengen et al., 1995). In late anaphase, two sets of actin filaments appear in the interzone. These actin filaments assemble *de novo,* close to the proximal surfaces of the reforming daughter nuclei (Zhang et al., 1993), with their (+) ends directed toward the cell plate (review: Staehelin & Hepler, 1996). In telophase, these actin filaments shorten and extend in wreath formation until they reach the mother cell walls. Cortical actin filaments link the edges of the phragmoplast, guiding lateral growth to the position previously occupied by the preprophase bands (Valster & Hepler, 1997).

3.2 Lower plant cells

3.2.1 Fern

Root tip cells of *Adiantum* show an actin distribution similar to that observed in higher plant cells, i.e., a pre-prophase actin band, an actin cage around the spindle, fibers near the kinetochore microtubules, phragmoplast actin fibers, and a persistent cortical network (Panteris et al., 1992).

3.2.2 Algae

In *Spirogyra crassa,* events of mitosis and cytokinesis overlap. Actin fibers, present as cytoplasmic tracks during interphase, vanish at mid-prophase. Concomitantly, actin increases around the nucleus and an actin band accumulates at the future division site. During cytokinesis, there is a centripetal in-growth of cross walls guided by actin filaments which are involved in positioning vesicles at the fusion sites. Cell wall closure involves both actin filaments and microtubules (Sawitzky & Grolig, 1995).

Chlamydomonas reinhardtii shows specific actin dynamics during the cell cycle (Harper et al., 1992). During interphase/prophase transition, diffuse perinuclear actin progressively relocalizes to form an angular structure, located close to the the basal apparatus. After duplication of the basal apparatus, a temporary cruciate structure curves at the anterior nuclear surface. During mitosis, a unique arc elongates between the spindle poles. This arc re-focuses to the site where the cleavage furrow will form and re-extends in anaphase to form a broad arc between the poles. During cytokinesis, actin localizes along the progressing cleavage furrow, starting from the anterior region of the cell and simultaneously encircling the daughter nuclei.

Unlike the mitotic spindles of *Mougeotia* (Grolig, 1990), *Spirogyra* or *Chlamydomonas*, those of *Oedogonium* show actin fibers extending from kinetochores parallel to the microtubule kinetochore fibers (Sampson et al., 1996)

All red alga families decribed so far have cortical actin arrays like those of green algae (McDonald et al., 1993).

3.2.3 Fungi

The cortical actin patches observed during interphase are involved in growth polarity and relocate prior to mitosis (Marks & Hyams, 1985;

Bretscher et al., 1994). Chromosome segregation during mitosis is intranuclear and occurs in the absence of detectable spindle actin. Actin filaments are exclusively cytoplasmic and form peripheral plaques involved in cell growth. These plaques are likely to be an actin storage system which is used until septum formation (Butt & Heath, 1988).

3.2.3.1 Schizosaccaromyces pombe

Fission yeast divides in a manner similar to animal cells. At the onset of mitosis, actin relocates to an equatorial ring indicating the future site of septum growth. Symetrical cleavage occurs via an actomyosin-based contractile ring which forms at the onset of mitosis. Septum formation is dependent on actin-coated vesicles carried by myosin motors (review: Robinow & Hyams, 1989). The actin ring contracts at the leading edge of the growing septum.

3.2.3.2 Saccharomyces cerevisiae

In the budding yeast, a ring of actin patches appears at the future bud emergence site in G1. Actin cables aligned toward the bud extend from cortical patches into the mother cell during S and G2. Actin patches redistribute to the surface of both daughter cells during mitosis. Before cytokinesis, cortical actin predominantly localizes at the neck of the bud. Actin cables play a role in the polarized transport of cell wall components (review: Bretscher et al., 1994).

3.2.3.3 Neozygites

In zygomycetes, one of the first signs of cytokinesis is the assembly of longitudinal actin filaments in a peripheral band at the equator. These filaments then shorten and form a transverse belt at the site of the future septum. The actin plaque concentrates on the invaginating plasma membrane during septum growth, suggesting that actin plays a role in the recruitment of wall precursors (Butt & Heath, 1988).

4. METHODS FOR INHIBITION OF ACTIN FUNCTION

To determinate the role of actin, various approaches have been used: (i) perfusion of pharmacological agents such as cytochalasins, latrunculins, phallotoxins or jasplakinolides, (ii) microinjection of actin-associated proteins, (iii) microinjection of antibodies, and (iv) mutagenesis.

To discriminate between actin-mediated and microtubule-mediated functions, microtubular depolymerizing drugs have been used.

While diverse cellular processes are affected by inhibiting actin activity, we will focus here on the effects on mitosis and cytokinesis.

4.1 Shift of the assembly/disassembly equilibrium

Cytochalasins specifically bind to actin (Ohmori & Toyama, 1992) by capping filament ends and thus preventing elongation (Cooper, 1987). Cytochalasin treatments (1 - 10 μM) lead to actin depolymerization. Fibers become shorter and specks or rods may be observed in plant material. Varying amounts of polar dots were also observed in meristematic cells (McCurdy et al., 1992).

Latrunculin A and B, marine toxins isolated from Red Sea sponge, are with 10 - 100-fold more potent in actin depolymerization than cytochalasins in leaf mesophyl cells of *Arabidopsis* (Kandasamy & Meagher, 1999) or in yeast (Theesfeld et al., 1999).

Profilin is a G-actin-associated protein which reduces or promotes actin assembly, depending on cellular conditions (review: Staiger et al., 1997). Profilin microinjection was used to prevent actin polymerization and to break down actin filaments (Staiger et al., 1994).

Phallotoxins are bicyclic peptides isolated from the mushroom *Amanita phalloides*. They shift the dynamic equilibrium to polymerization by lowering the critical concentration for polymerization up to 30-fold. Since these toxins are non-permeant, they are used on permeabilized cells, or injected. Phallotoxin-labelled actin filaments are stabilized at concentration above 300 nM, but below 50 nM the phallotoxin is still active and can be used to follow actin dynamics *in vivo*.

Jasplakinolides, novel macrocyclic peptides isolated from a sponge, induce microfilament polymerization. These peptides are permeant and can be perfused into living cells, but they compete with phalloidin for actin binding. Therefore, actin staining is done either by immunolocalization or by using transformed cells with a fluorescent actin network.

4.2 Immuno-blocking of actin activity

Microinjection of anti-actin antibodies has been used to block microfilament function *in vivo* in animal cells (Skalli et al., 1990). This approach could also be used in plant cells. Similarly, microinjection of antibodies directed against actin-associated proteins can also indirectly affect actin function, as described for myosin antibodies in *Acanthamoeba* (Sinard & Pollard, 1989), spectrin antibodies in *Amoeba proteus* (Choi & Jeon, 1992), or for talin antibodies in fibroblasts (Nuckolls et al., 1992).

4.3 Actin gene mutations

Cold-sensitive mutations in the actin locus have been identified in fission yeast. Four mutants analysed are due to single amino acid mutations and affect either actin patches and ring formation (*act1-48*) or actin stability (*act1-16, act1-32* and *act1-67*) (McCollum et al., 1999).

Budding yeast *act1* mutations disrupt cell shape, cell polarity, secretion, endocytosis, spindle orientation, nuclear migration, cytokinesis and mitochondrial distribution (review: Bretscher et al., 1994). The D157E mutant confers resistance to latrunculin A (Belmont et al., 1999).

A new tool for investigating actin is a yeast strain carrying an N-terminal, 6-histidine-tagged, plasmid-borne actin (Buzan et al., 1999). This strain allows selection of chromosomal actin mutants which might not grow otherwise.

A mutant strain of *Chlamydomonas* (*ida5*), deficient in the conventional actin gene, expresses a NAP protein which substitutes for actin in some cellular functions. Results obtained with this mutant give an insight into the highly regulated mechanisms in which actin participates (Kato-Minoura et al., 1998).

5. ROLES OF ACTIN DURING DIVISION-LINKED CELLULAR PROCESSES

A summary of actin functions during cell division is listed in Table 2.

Table 2. Involvement of actin during cell division

Genus and Species	Materials	Functions	References
Higher plants:			
Allium	Root tip	KK	McCurdy et al., 1991
Allium	Root tip	DPD	Mineyuki & Palevitz, 1990 Eleftheriou & Palevitz, 1992
Clivia	Endosperm	KK DPD	Schmit & Lambert, 1987-1990 Endlé et al., 1998
Daucus carota	Culture cells	DPD	Traas et al., 1987 Lloyd & Traas, 1988
Gasteria verrucosa	Pollen mother cells	KK	Van Lammeren et al., 1989
Haemanthus	Endosperm	KK DPD	Forer & Jackson, 1975-1979 Schmit & Lambert, 1987-1990 Endlé et al., 1998
Medicago sativa	Culture cells	KK	Seagull et al., 1987
Nicotiana tabaccum	Culture cells	KK	Kengen et al., 1995
Solanum melongea	Pollen mother cells	DPD	Traas et al., 1989
Tradescantia	Leaf epidermis stomatal complexes	DPD	Cleary, 1995

Tradescantia	Stomatal subsidiary mother cells	NM	Kennard & Cleary, 1997
Tradescantia	Root tip	DPD	Liu & Palevitz, 1992
Tradescantia	Stamen hair	DPD	Cleary et al., 1992 Valster & Hepler, 1997
Triticum	Root tip	KK	McCurdy et al., 1991
Vicia faba	Culture cells	KK	Seagull et al., 1987
Zea mays	Root cells	DPD	Baluška et al., 1997
Lower plants: *Adiantum*	Root tip	KK	Panteris et al., 1992
Chlamydomonas		CK Basal body movements	Harper et al., 1992
Oedogonium		KK	Sampson et al., 1996
Saccharomyces cerevisiae		Spindle orientation Nuclear migration DPD	Muhua et al., 1994 Bretscher et al., 1994
Schizosaccharomyces pombe		DPD	Marks & Hyams, 1996
Spirogyra		Nucleus centering DPD	Grolig, 1998 Sawitzky & Grolig, 1995

DPD: division plane determination and cytokinesis; KK: karyokinesis; NM: pre-mitotic nuclear migration

5.1 Cell shape

In endosperm cells which are natural protoplasts, actin has been proposed to be involved in cell shape determination (Schmit & Lambert, 1988). However, this is not the case in walled cells since actin disruption by drugs does not inhibit deposition of cellulose microfibrils (review: Staiger & Schliwa, 1987).

5.2 Organelle movement and position

Nuclear migration is the first visible manifestation of the establishment of division polarity. Subsidiary mother cells of stomatal complexes, which divide asymmetrically, constitute a model system to study the regulation of division. Pre-mitotic nuclear movement is actin-dependent and occurs during G1 in *Tradescantia* (Kennard & Cleary, 1997) and in *Spirogyra* (Grolig, 1998). In guard mother cells of *Allium*, actin is involved in nuclear movement following division (Palevitz, 1980). Drugs affecting the cytoskeleton (cytochalasin D 20 µM, phalloidin 5 µM, oryzalin 25 µM) distinctly alter nuclear movement, suggesting that actin plays a role in conveying tension along microtubules and integrating the translocation forces generated within the scaffold.

The dynamics of actin in *Chlamydomonas* suggests that actin plays a role in the separation of basal bodies and in the positioning of the spindle axis and cleavage furrow. However, results obtained using the mutant *bld2-1* which lacks basal bodies gives a different perspective. The positioning of the spindle and the furrow does not depend on actin but rather on centriole activity via the microtubules. In this strain, actin shows a variable and aberrant distribution which is not correlated with the spindle placement defect (Ehler et al., 1995).

5.3 Preprophase band organization, cell plate positioning, centrifugal expansion and stabilization

In Allium root tip cells treated with cytochalasins C or D, the preprophase band does not narrow. In addition, the preprophase bands which are already narrow at the time of treatment do not expand, suggesting that actin filaments control preprophase band morphology (Eleftheriou & Palevitz, 1992). Similar results were obtained by treating fern protonemata (Kadota & Wada, 1992). Triticum root cells are at least five times less sensitive to cytochalasin but show similar results (McCurdy et al, 1992). Such a variable sensitivity has not been explained so far.

Oryzalin depolymerizes microtubules in the preprophase band but the co-distributed actin arrays remain intact or disappear, depending on the material. Nevertheless, the actin preprophase band seems to be involved in recognizing the future cell-plate stabilization sites, as suggested by the analysis of asymmetric division. Recent results obtained using maize *discordia* mutant cells favor a role for actin in phragmoplast guidance. While phragmoplasts of asymmetrically dividing mutant cells are structurally normal, they are not guided to the position formerly occupied by the preprophase band. This effect mimicks that of cytochalasin on this process. (Gallagher & Smith, 1999).

Actin filaments are a major structural component of the cytokinetic apparatus. There is evidence that actin filaments play an active role in cell plate positioning since cytochalasin treatment induces misalignment (Palevitz, 1987b; Schmit & Lambert, 1988; Mineyuki & Gunning, 1990). In addition, there are structural actin links between the peripheral cell membrane and the growing cell plate (Schmit & Lambert, 1987; Valster & Hepler, 1997; Endlé et al., 1998). In dividing Tradescantia stamen hairs, profilin injection causes microfilament breakdown and leads to a block of cell plate formation (Staiger et al., 1994). This result also confirms a direct role for actin in maintaining cell architecture (Valster et al., 1997). Vinculin recruitment to the cell plate suggests that actin anchorage mechanisms act within the membrane as it forms (Endlé et al., 1998).

Drug studies on root cells of *Adiantum* suggest that microtubule and microfilament arrays are independant, since cytochalasin B (50-200 μg ml^{-1}) does not change microtubule organization and function during mitosis. On the other hand, most actin arrays disappear (except cortical ones) after oryzalin treatment (10 μM). This suggests that microtubules are involved in the organization and orientation of actin filaments and not *vice versa*.

5.4 Chromosome motion

Actin appears to be involved in the early interactions between the chromosomes and the spindle fibers and also in anaphase chromosome movement in meiotic crane fly spermatocytes (Forer & Pickett-Heaps, 1998). Actin is entangled between the chromosomes just after nuclear envelope breakdown. In both crane fly spermatocytes and Haemanthus endosperm cells, actin is found in chromosomal spindle fibers. These results suggest that actin is involved in chromosome contraction and chromosome motion (Czaban & Forer, 1992; Kengen et al., 1995). Since cytochalasin D (25 - 100 μg ml^{-1}) also blocks mitosis in *Oedogonium*, it is likely that actin plays a role in the attachment of chromosomes to the spindle (Sampson et al., 1996).

In previous studies on various plant material, chromosome movement was not disrupted by cytochalasins. This suggests that if actin participates in kinetochore anchoring to microtubules, other proteins may substitute for actin when cytochalasin is present (Palevitz, 1980; Schmit & Lambert, 1988).

5.5 Spindle integrity and orientation

The spindle-associated actin basket found in both endosperm cells and walled-cells could serve as an anchor and thus control lateral spindle movement (Schmit & Lambert, 1985, 1988; Hepler et al., 1993).

In budding yeast, actin depolymerization by latrunculin A showed that both spindle pole body migration and spindle orientation are sensitive to actin perturbations early in the cell cycle. However, maintenance of the correct spindle orientation in the G2/M phase does not depend on actin. These data suggest a role for actin cables either in guiding or anchoring astral microtubules into or within the bud (Palmer et al., 1992; Theesfeld et al., 1999). An actin-related protein of the Arp1 family is a major peptide of the dynactin complex. This complex may be involved in actin nucleation and in mediating the activity of dynein in spindle microtubule orientation in yeast (Muhua et al., 1994).

6. CONCLUSION

Development of new approaches to correlate *in vivo* actin organization with the changes taking place during mitosis and cytokinesis will be indispensable in furthering our understanding of actin function in higher plant cells. For example, methods used in *Dictyostelium* and could be adapted to plant cells : (i) electroporation (Yumura & Fukui, 1998) or microinjection of rhodamine-actin; and (ii) expression of transgenic proteins which label the actin cytoskeleton, such as GFP-actin (Westphal et al., 1997; Aizawa et al., 1997), or GFP-ABD, a 25 kDa actin-binding domain of the ABP-120 protein (Pang & Knecht, 1998). The actin binding domain of talin fused to GFP has already been used to label the actin cytoskeleton *in vivo* in tobacco cells (Kost et al., 1998).

A better knowledge of the relationship between actin and actin-associated proteins should provide new insights on microfilament function. A significant body of evidence on the cytokinetic pathway emerges from the study of yeast mutants. The division site is defined by the establishment of an actomyosin ring which results from a cascade of events involving numerous actin- and myosin-associated proteins. Actin recruitment and/or nucleation at this site is mediated by cdc12 formin-like proteins. IQGAP proteins and septins link signalling pathways to actin remodeling and form a scaffold to which other components can be recruited. Specific kinases and formin-like proteins are also involved in septation regulation (Field et al., 1999). In plant cells, actin is found in association with myosin, vinculin (Endlé et al., 1998) and a troponin T epitope (Lim et al., 1986). Changes in actin dynamics and regulation are also under the control of capping proteins like profilin (Staiger et al., 1997), twisting factors like ADF1/cofilin (McGough et al., 1997), and binding proteins like ABP1. Dissection of the structure-function relationship between actin and associated proteins is underway. Development of new tools adapted specifically to plant cells will undoubtedly help in understanding actin activity *in vivo* and in exploring the relationship between the actin and microtubular cytoskeleton during both mitosis and cytokinesis.

ACKNOWLEDGMENT

I would like to thank Jean Canaday who helped me with her critical comments and her valuable linguistic advice.

REFERENCES

Abe S, Ito Yand Davies E (1992) Co-sedimentation of actin, tubulin and membranes in the cytoskeleton fractions from peas and mouse 3T3 cells. J Exp Bot 43: 941-949

Aizawa H, Sameshima M and Yahara I (1997) A green fluorescent protein-actin fusion protein dominantly inhibits cytokinesis, cell spreading and locomotion in *Dictyostelium.* Cell Struct Funct 22: 335-345

Baluška F, Vitha S, Barlow PW and Volkmann D (1997) Rearrangements of F-actin arrays in growing cells of intact maize root apex tissues: A major developmental switch occurs in the postmitotic transition region. Eur J Cell Biol 72: 113-121

Belmont LD, Patterson GM and Drubin DG (1999) New actin mutants allow further characterization of the nucleotide binding cleft and drug binding sites. J Cell Sci 112: 1325-1336

Bretscher A, Drees B, Harsay E, Schott D and Wang T (1994) What are the basic functions of microfilaments? Insights from studies in budding yeast. J Cell Biol 126: 821-825

Butt TM and Heath IB (1988) The changing distribution of actin and nuclear behavior during the cell cycle of the mite-pathogenic fungus *Neozygites* sp. Eur J Cell Biol 46: 499-505

Buzan J, Du J, Karpova T and Frieden C (1999) Histidine-tagged wild-type yeast actin: Approach for obtaining yeast actin mutants. Proc Natl Acad Sci USA 96: 2823-2827

Choi EY and Jeon KW (1992) Role of spectrin in *Amoeba proteus*, as studied by microinjection of anti-spectrin monoclonal antibodies. Exp Cell Res 199: 174-178

Cleary AL, Gunning BES, Wasteneys GO and Hepler PK (1992) Microtubule and F-actin dynamics at the division site in living *Tradescantia* stamen hair cells. J Cell Sci 103: 977-988

Cleary AL (1995) F-actin redistributions at the division site in living Tradescantia stomatal complexes as revealed by microinjection of rhodamin-phalloidin. Protoplasma 185: 152-165

Cooper JA (1987) Effects of cytochalasin and phalloidin on actin. J Cell Biol 105: 1473-1478

Czaban BB and Forer A (1992) Rhodamine-labelled phalloidin stains components in the chromosomal spindle fibers of crane-fly spermatocytes and Haemanthus endosperm cells. Biochem Cell Biol 70: 664-676

Detmers PA, Goodenough UW and Condeelis J (1985) Localization of actin in *Chlamydomonas* using anti-actin and NBD-phallicidin. Cell Motil Cytoskel 5: 415-430

Ehler LL, Holmes JA and Dutcher SK (1995) Loss of spatial control of the mitotic spindle apparatus in a *Chlamydomonas reinhardtii* mutant strain lacking basal bodies. Genetics 141: 945-960

Eleftheriou EP and Palevitz BA (1992) The effect of cytochalasin D on preprophase band organization in root tip cells of *Allium*. J Cell Sci 103: 989-998

Endlé MC, Stoppin V, Lambert AM and Schmit AC (1998) The growing cell plate of higher plants is a site of both actin assembly and vinculin-like antigen recruitment. Eur J Cell Biol 77: 10-18

Field C, Li R and Oegema K (1999) Cytokinesis in eucaryotes: A mechanistic comparison Curr Opin Cell Biol 11: 68-80

Forer A (1985) Does actin produce the force that moves a chromosome to the pole during anaphase? Can J Cell Biol 63: 585-598

Forer A and Jackson WT (1975) Actin in the higher plant *Haemanthus katherinae* Baker. Cytobiologie 10: 217-226

Forer A and Jackson WT (1976) Actin filaments in the endosperm mitotic spindles in a higher plant, *Haemanthus katherinae* Baker. Cytobiologie 12: 199-214

Forer A and Jackson WT (1979) Actin in spindles of Haemanthus katherinae endosperm. I. General results using various glycerination methods. J Cell Sci 37: 323-347

Forer A and Pickett-Heaps JD (1998) Cytochalasin D and latrunculin affect chromosome behaviour during meiosis in crane-fly spermatocytes. Chromosome Res 6: 533-549

Gallagher K and Smith LG (1999) *discordia* mutations specifically misorient asymmetric cell divisions during development of the maize leaf epidermis. Development 126: 4623-4633

Gavin RH (1997) Microtubule – microfilament synergy in the cytoskeleton Int Rev Cytol 173: 207-242

Grolig F, Weigang-Köhler K and Wagner G (1990) Different extent of F-actin bundling in walled cells and protoplasts of *Mougeotia scalaris*. Protoplasma 157: 225-230

Grolig F (1998) Nuclear centering in *Spirogyra*: Force integration by microfilaments along microtubules. Planta 204: 54-63

Harper JDI, McCurdy DW, Sanders MA Salisbury JL and John PCL (1992) Actin dynamics during the cell cycle in *Chlamydomonas reinhardtii*. Cell Motil Cytoskel 22: 117-126

Hepler PK, Cleary AL, Gunning BES, Wadsworth P, Wasteneys GO and Zhang DH (1993) Cytoskeletal dynamics in living plant cells. Cell Biol Int 17: 127-142

Hush JM and Overall RL (1992) Re-orientation of cortical F-actin is not necessary for wound-induced microtubule re-orientation and cell polarity establishment. Protoplasma 169: 97-106

Kakoda A and Wada M (1992) The circular arrangement of cortical microtubules around the subapex of tip-growing fern protonemata is sensitive to cytochalasin B. Plant Cell Physiol 33: 99-102

Kandasamy MK and Meagher RB (1999) Actin-organelle interaction: Association with chloroplast in Arabidopsis leaf mesophyll cells. Cell Motil Cytoskel 44: 110-118

Karki S and Holzbaur ELF (1999) Cytoplasmic dynein and dynactin in cell division and intracellular transport. Curr Opin Cell Biol 11: 45-53

Kato-Minoura T, Uryu S, Hirono M and Kamiya R (1998) Highly divergent actin expressed in a Chlamydomonas mutant lacking the conventional actin gene. Biochem Biophys Res Comm 251: 71-76

Kengen HMP and De Graaf BHJ (1991) Microtubules and actin filaments co-localize extensively in non-fixed cells of tobacco. Protoplasma 163: 62-65

Kengen HMP, Eygensteyn J and van Amstel TNM (1995) F-actin in mitotic spindles of synchronized suspension culture cells of tobacco visualized by confocal laser scanning microscopy. Cell Biol Int 19: 585-592

Kennard JL and Cleary AL (1997) Pre-mitotic nuclear migration in subsidiary mother cells of *Tradescantia* occurs in G1 of the cell cycle and requires F-actin. Cell Motil Cytoskel 36: 55-67

Kost B, Spielhofer P and Chua N (1998) A GFP-mouse talin fusion protein labels plant actin filaments in vivo and visualizes the actin cytoskeleton in growing pollen tubes. Plant J 16: 393-401

Laine RO, Zeile W, Kang F, Purich DL and Southwick FS (1997) Vinculin proteolysis unmasks an ActA homolog for actin-based Shigella motility. J Cell Biol 138: 1255-1264

Lancelle SA and Hepler PK (1988) Cytochalasin induced ultrastructural alteration in Nicotiana pollen tubes. Protoplasma Suppl 2: 65-75

Lessard JL (1988) Two monoclonal antibodies to actin: One muscle selective and one generally reactive. Cell Motil Cytoskel 10: 349-362

Lim SS, Hering GE and Borisy GG (1986) Widespread occurrence of anti-troponin T crossreactive in non-muscle cells. J Cell Sci 85: 1-19

Liu XO and Yen LF (1992) Purification and characterization of actin from maize pollen. Plant Physiol 99: 1151-1155

Liu B and Palevitz BA (1992) Organization of cortical microfilaments in dividing root cells. Cell Motil Cytoskel 23: 252-264

Marks and Hyams JS (1985) Localization of F-actin through the cell division cycle of *Schizosaccharomyces pombe*. Eur J Cell Biol 39: 27-32

McCollum D, Balasubramanian M and Gould K (1999) Identification of cold-sensitive mutations in the Schizosaccharomyces pombe actin locus. FEBS Lett 451: 321-326

McCurdy DW, Palevitz BA and Gunning BES (1991) Effect of cytochalasins on actin in dividing root tip cells of *Allium* and *Triticum*: A comparative immunocytochemical study. Cell Motil Cytoskel 18: 107-112

McDonald AR, Garbary DJ and Duckett JG (1993) Rhodamine-phalloidin staining of F-actin in rhodophyta. Biotech Histochem 68: 91-98

McGough A, Pope B, Chiu W and Weeds A (1997) Cofilin changes the twist of F-actin: Implications for actin filament dynamics and cellular function. J Cell Biol 138: 771-781

Mineyuki Y and Gunning BES (1990) A role for preprophase bands of microtubules in maturation of new cell walls, and a general proposal on the function of preprophase band sites in cell division in higher plants. J Cell Sci 97: 527-537

Molé-Bajer J, Bajer AS and Inoué S (1988) Three-dimensional localization and redistribution of F-actin in higher plant mitosis and cell plate formation. Cell Motil Cytoskel 10: 217-228

Muhua LI, Karpova TS and Cooper JA (1994) A yeast actin-related protein homologous to that in vertebrate dynactin complex is important for spindle orientation and nuclear migration. Cell 78: 669-679

Nuckolls GH, Romer LH and Burridge K (1992) Microinjection of antibodies against talin inhibits the spreading and migration of fibroblasts. J Cell Sci 102:753-762

Ohmori H and Toyama S (1992) Direct proof that the primary site of action of cytochalasin on cell motility processes is actin. J Cell Biol 116: 933-941

Palevitz BA (1980) Comparative effects of phalloidin and cytochalasin B on motility and morphogenesis in *Allium*. Can J Bot 58: 773-785

Palevitz BA (1987a) Actin in the preprophase band of *Allium cepa*. J Cell Biol 104: 1515-1519

Palevitz BA (1987b) Accumulation of F-actin during cytokinesis in *Allium*. Correlation with microtubule distribution and the effect of drugs. Protoplasma 141: 24-32

Palmer RE, Sullivan DS, Huffaker T and Koshland D (1992) Role of astral microtubules and actin in spindle orientation and migration in the budding yeast, *Saccharomyces cerevisiae*. J Cell Biol 119: 583-593

Pang KM and Knecht DA (1998) Use of a fusion protein between GFP and an actin-binding domain to visualize transient filamentous-actin structures. Curr Biol 8: 405-408

Panteris E, Apostolakis P and Galatis B (1992) The organization of F-actin in root tip cells of *Adiantum capillus veneris* throughout the cell cycle. Protoplasma 170: 128-137

Robinow CF and Hyams JS (1989) General cytology of fission yeasts. In: Nasim A, Young PG and Johnson BF (eds) The Molecular Biology of Fission Yeast. New York: Academic Press, pp 273-331

Sampson K, Pickett-Heaps JD and Forer A (1996) Cytochalasin D blocks chromosomal attachment to the spindle in the green alga *Oedogonium*. Protoplasma 192: 130-144

Sawitzky H and Grolig F (1995) Phragmoplast of the green alga *Spirogyra* is functionally distinct from the higher plant phragmoplast. J Cell Biol 130: 1359-1371

Sawitzky H, Willingale-Theune J and Menzel D (1996) Improved visualization of F-actin in the green alga *Acetabularia* by microwave-accelerated fixation and simultaneous FITC-phalloidin staining. Histochem J 28: 353-360

Schmit AC and Lambert AM (1985) F-actin dynamics is associated to microtubule function during cytokinesis in higher plants. A revised concept. In: de Brabander M, de Mey J (eds) Microtubule and Microtubule Inhibitors. Amsterdam: Elsevier Science Publishers B.V., pp 243-252

Schmit AC and Lambert AM (1987) Characterization and dynamics of cytoplasmic F-actin in higher plant endosperm cells during interphase, mitosis and cytokinesis. J Cell Biol 105: 2157-2166

Schmit AC and Lambert AM (1988) Plant actin filament and microtubule interactions during anaphase-telophase transition: effects of antagonist drugs. Biol Cell 64: 309-319

Schmit AC and Lambert AM (1990) Microinjected fluorescent phalloidin in vivo reveals the F-actin dynamics and assembly in higher plant mitotic cells. Plant Cell 2: 129-138

Seagull RW, Falconer MM and Weerdenburg CA (1987) Microfilaments: Dynamic arrays in higher plant cells. J Cell Biol 104: 995-1004

Sinard JH and Pollard TD (1989) Microinjection into *Acanthamoeba castellanii* of monoclonal antibodies to myosin-II slows but does not stop cell locomotion. Cell Motil Cytoskel 12: 42-52

Sinnott EW and Bloch R (1941) Division in vacuolated plant cells. Amer J Bot 28: 225-232

Skalli O, Gabbiani F and Gabbiani G (1990) Action of general and alpha-smooth muscle-specific actin antibody microinjection on stress fibers of cultured smooth muscle cells.Exp Cell Res 187: 119-125

Sonobe S and Shibaoka H (1989) Cortical fine actin filaments in higher plant cells visualized by rhodamine-phalloidin after pretreatment with m-maleimidobenzoyl N-hydroxy-succinimide ester. Protoplasma 148: 80-86

Staehelin LA and Hepler PK (1996) Cytokinesis in higher plants. Cell 84: 821-824

Staiger CJ and Lloyd CW (1991) The plant cytoskeleton. Curr Opin Cell Biol 3: 33-42

Staiger CJ and Schliwa M (1987) Actin localization and function in higher plants. Protoplasma 141: 1-12

Staiger CJ, Yuan M, Valenta R, Shaw PJ, Warn RM and Lloyd CW (1994) Microinjected profilin affects cytoplasmic streaming in plant cells by rapidly depolymerizing actin microfilaments. Curr Biol 4: 215-219

Staiger CJ, Gibbon BC, Kovar DR and Zonia LE (1997) Profilin and actin-depolymerizing factor: Modulators of actin organization in plants. Trends Plant Sci 2: 275-281

Sternlicht H, Farr GW, Sternlicht ML, Driscoll JK, Willison K and Yaffe MB (1993) The t-complex polypeptide 1 complex is a chaperonin for tubulin and actin *in vivo*. Proc Natl Acad Sci USA 90: 9422-9426

Traas JA, Doonan JH, Rawlins D, Shaw PJ, Watts J and Lloyd CW (1987) An actin network is present in the cytoplasm throughout the division cycle of carrot cells: actin co-distributes with the four microtubule arrays. J Cell Biol 105: 387-395

Theesfeld CL, Irazoqui JE, Bloom K and Lew DJ (1999) The role of actin in spindle orientation changes during the Saccharomyces cerevisiae cell cycle. J Cell Biol 146: 1019-1032

Ursic D, Sedbrook JC, Himmel KL and Culbertson MR (1994) The essential yeast Tcp1 protein affects actin and microtubules. Mol Biol Cell 5: 1065-1080

Valster AH and Hepler PK (1997) Caffeine inhibition of cytokinesis: Effect on the phragmoplast cytoskeleton in living Tradescantia stamen hair cells. Protoplasma 196: 155-166

Valster AH, Pierson ES, Valenta R, Hepler PK and Emons AMC (1997) Probing the plant actin cytoskeleton during cytokinesis and interphase by profilin microinjection. Plant Cell 9: 1815-1824

Van Lammeren AAM, Bednara J and Willemse MTM (1989) Organization of the actin cytoskeleton during pollen development in *Gasteria verrucosa* (Mill.) visualized with rhodamine-phalloidin. Planta 178: 531-539

Yumura S and Fukui Y (1998) Spatiotemporal dynamics of actin concentration during cytokinesis and locomotion in *Dictyostelium*. J Cell Sci 111: 2097-2108

Westphal M, Jungbluth A, Heidecker M, Mühlbauer B, Heizer C, Schwartz JM, Marriott G and Gerisch G (1997) Microfilament dynamics during cell movement and chemotaxis monitored using a GFP-actin fusion protein. Curr Biol 7: 176-183

Zhang D, Wadsworth P and Hepler PK (1993) Dynamics of microfilaments are similar, but distinct from microtubules during cytokinesis in living, dividing plant cells. Cell Motil Cytoskel 24: 151-155

Chapter 26

ACTIN AND MYOSIN VIII IN DEVELOPING ROOT APEX CELLS

Development- and tissue-specific distributions with possible relevance for diverse root cell functions

František Baluška[1], Peter W Barlow[2] and Dieter Volkmann[1]

1 Botanisches Institut der Universität Bonn, Kirschallee 1, D–53115 Bonn, Germany
2 IACR–Long Ashton Research Station, Department of Agricultural Sciences, University of Bristol, Long Ashton, Bristol BS41 9AF, UK

Key words: actin, cell elongation, root apex, root cap, root hair, transition zone

Abstract: Root apices represent an ideal model object for studies on plant cell growth and development. We have exploited this opportunity for detailed analysis of the actin-based cytoskeleton in cells of various root tissues throughout their cellular development. During mitosis, cells re-distribute their actin filaments (AFs) and myosin VIII molecules from the cytoplasm to the cell periphery where they accumulate at putative AF-organizing centres (AFOCs) facing the spindle poles. Postmitotic root cap columella cells differentiate first into gravity-sensing statocytes which are unique among postmitotic root cells due to the lack of any distinct cables of AFs. Later, statocytes, as well as peripheral cap cells, transform into secretory cells equipped with dense AF networks distributed throughout their cytoplasm. They retain abundant AFs after being shed from the root. Intriguingly, however, all root cap cells lack myosin VIII at their periphery. By contrast, all postmitotic cells of the root body, as they traverse the transition zone, show myosin VIII localized at their periphery. Myosin VIII localizes especially at the plasmodesmata in the non-growing cross walls. In cells of the transition zone, unique AF bundles develop which are proposed to participate in the onset of rapid cell elongation. These AF bundles are initiated at the nuclear peripheries and are organized via myosin VIII-enriched cross-walls, these two sites obviously act as the major AFOCs of postmitotic root-body cells. Treatment of roots with latrunculin B reveals that dynamic AFs are essential for both vacuome-driven cell elongation and root hair formation. In the transition zone and elongation region, cells of the inner cortex localize plant myosin VIII molecules abundantly at their pit-fields. These distinctive subcellular sites, like cross-walls and root hair apices, represent powerful AFOCs capable of organizing abundant AFs.

C.J. Staiger et al. (eds.), Actin: A Dynamic Framework for Multiple Plant Cell Functions, 457–476.

1. INTRODUCTION

An actin cytoskeleton plays a central role in diverse intracellular processes, the list of which is far from complete (for recent reviews, see Volkmann & Baluška, 1999; Staiger, 2000; other chapters in this volume). One typical role of the actin-based cytoskeleton is that its actin filaments (AFs) serve as tracks along which myosins (for plant cell myosin, see Yokota & Shimmen; and Reichelt & Kendrick-Jones, this volume) transport Golgi-based vesicles (Mollenhauer & Morré, 1976) and other diverse organelles, including small vacuoles (Verbelen & Tao, 1998), whole Golgi complexes (Boevink et al., 1998; Nebenführ et al., 1999), mitochondria (Olyslaegers & Verbelen, 1998) and endoplasmic reticulum (ER) elements (e.g., Kachar & Reese, 1988; Lichtscheidl et al., 1990; Knebel et al., 1990; Liebe & Menzel, 1995; Lichtscheidl & Baluška, this volume). In short, myosins are membrane-associated and actin-activated molecular machines which hydrolyse ATP and use its energy for driving intracellular movements along bundles of AFs (Titus, 1993; for plant cells, see Asada & Collings, 1997). Nevertheless, it is postulated that this picture reverses in situations where plant myosins are firmly anchored at the plasma membrane and their motor activities are directed toward dragging AFs along and beneath the plasma membrane. Actually, such a situation can occur experimentally *in vitro* when F-actin elements are induced to move along non-biological surfaces coated with myosins and myosin-containing extracts (Higashi-Fujime, 1991; Kohno et al., 1991; Shimmen & Yokota, 1994). This alternative scenario seems to be valid for the plant unconventional myosin VIII (Knight & Kendrick-Jones, 1993; Reichelt et al., 1997) which localises to cellular peripheries in the form of distinct dots (Reichelt et al., 1999; Volkmann & Baluška, 1999.

Movements of AFs beneath and along the plasma membrane, which is equipped with plant myosin VIII molecules (for the analogous situation with myosin I, see Zot et al., 1992), can be expected to exert local tension on the plasma membrane (e.g., Wolfe & Steponkus, 1993). Similarly, cytoplasmic strands are known to exert pulling forces on the plasma membrane causing its local distortion (Hahne & Hoffman, 1984). Such stretching of a stress/strain-loaded plasma membrane might indirectly stimulate the activity of stretch-activated ion channels (Lee et al., 1999; for plant cells, see Ding & Pickard, 1993; Garrill et al., 1996; Ramahaleo et al., 1996). Moreover, physical linkages between ion channels and elements of plasma-membrane-linked AF meshworks might also more directly control the activities of these channels (Watson et al., 1992; Cantiello & Prat, 1996; Prat et al., 1996; Hwang et al., 1997; Glogauer et al., 1998; Liu & Berg, 1999). As vacuome-driven plant cell elongation requires activities of both ion and water channels, one could propose a testable hypothesis: that actomyosin-based forces are essential for the execution of vacuome-driven cell elongation (see Waller et al., this volume) through direct and indirect impacts of actomyosin forces on ion channel activities. Root apices represent a unique assemblage of different plant cell types encompassing diverse developmental pathways and contrasting cellular fates. The anatomical simplicity of root apices makes them one of the most useful experimental plant systems (e.g., Barlow, 1989, 1994; Barlow & Baluška, 2000; Silk, 1992). Moreover, the suitability of root apices for the Steedman's wax embedding technique (e.g., Baluška et al., 1992; Vitha et al., 2000; this volume) allows tissue- and development-specific visualization of actin and myosin VIII, as well as other antigens, using specific antibodies applied on sections taken from the whole root apex (Baluška et al., 1997a; Vitha et al., 1997, 2000; Reichelt et al., 1999).

2. MITOTIC CELLS REDISTRIBUTE ACTIN AND MYOSIN VIII

Throughout interphase, the whole cytoplasmic interior is homogeneously pervaded by AFs. Dense AF networks are arranged as perinuclear baskets from which individual AFs extend towards the cell periphery and associate with the plasma membrane (Figs. 1, 2). In contrast, myosin VIII is present within the cytoplasm in low amounts only, whereas it is abundant at the cell periphery (Fig. 3), presumably in association with the plasma membrane (Reichelt et al., 1999; Volkmann & Baluška, 1999). This feature is less prominent in quiescent centre cells and their immediate descendents, as well as in all epidermis cells. The myosin VIII-positive dots at the cell periphery have been identified as plasmodesmata and pit-fields (Reichelt et al., 1999).

The above situation changes dramatically during mitosis when fine AFs extend from those peripheral sites, which may be putative AF-organizing centres (AFOCs), and face towards the cytoplasmic domains that are adjacent to spindle poles (Figs. 1C-E). Myosin VIII is similarly restricted to the actin-enriched cell periphery domains where it forms prominent continuous labelling (Figs. 3C, D). This latter feature suggests that, in addition to plasmodesmata, other plasma membrane portions, including perhaps the plasmodesmata-associated cortical ER elements, recruit myosin VIII molecules released from the cytoplasm during mitosis. Mitotic re-organisation of AFs and myosin VIII, and especially the enrichment of actin and myosin VIII at cell-periphery domains facing spindle poles, appear to be involved in orienting, or anchoring, the mitotic spindles (Baluška et al., 2000). This enrichment of AFs at peripheral sites facing the spindle pole, and also the depletion of AFs at pre-prophase band-marked domains, persist during cytokinesis, whereas myosin VIII re-distributes further and accumulates at the callosic cell plates during the final stages of cytokinesis (Reichelt et al., 1999; Volkmann & Baluška, 1999.

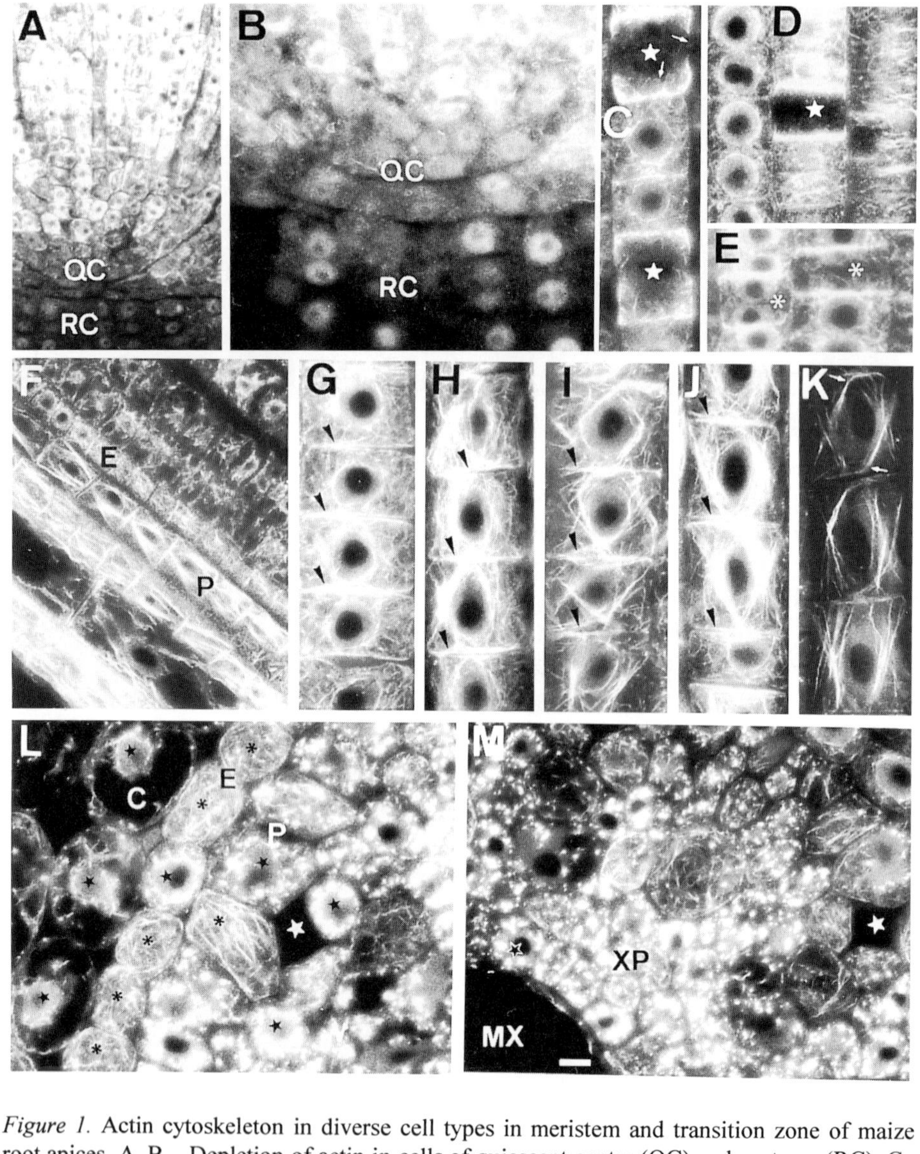

Figure 1. Actin cytoskeleton in diverse cell types in meristem and transition zone of maize root apices. A, B – Depletion of actin in cells of quiescent centre (QC) and root cap (RC). C–E – In the meristem, interphase cells show AFs organized as dense networks radiating from nuclear surfaces towards cell peripheries. This pattern changes dramatically in mitotic cells; actin becomes depleted from spindle regions (stars in C, D) and accumulates at cell-periphery domains facing spindle poles (lower arrow in C). In contrast, those cell periphery domains which are pre-disposed for cell plate fusion are actin-depleted (upper arrow in C). This unique distribution of the actin cytoskeleton persists also during cytokinesis (asterisks in E). F – In

the transition zone, actin is organized in the form of prominent bundles organized via putative AFOCs assembled at nuclear surfaces and cross-walls. This is especially prominent in cells of pericycle (P). Note the prominent actin in cells of the outer stele (lower part of F), whereas cells of inner cortex show less prominent actin cytoskeleton (upper part of F). G–K – Developmental sequence of actin bundle formation in cells of pericycle around centrally positioned nuclei and near cross-walls (black arrowheads in G–J). Note that, after leaving the nuclear region, bundles align along cross-walls (white arrows in K – this image is under-exposed to allow visualization of AF bundles alignment along the plasma membrane). L, M – Cross-sections through the transition zone of the root apex reveal abundant longitudinal (with respect of root axis) bundles around nuclei (black stars). Some cells are sectioned at their peripheries and so reveal cortical AFs near their plasma membranes which are arranged transversely to the longitudinal AF bundles (black snowflakes). Note that vacuolated cells of the cortex have fewer AFs, while phloem (white stars in L, M) and metaxylem (MX in M) are almost devoid of AFs. RC – root cap, QC – quiescent centre, E – endodermis, P – pericycle, C – cortex, MX – metaxylem, XP – xylem parenchyma. Bar = 17 μm (A) and 7 μm (B–M).

3. ROOT CAP CELLS: GRAVIPERCEPTION VERSUS CELL SEPARATION

Among all root tissues, the root cap is the most intriguing one due to the progression of most of its cells through three distinct developmental stages which culminate in a dramatic disintegration event. After only a limited number of mitotic cycles, the postmitotic root-cap columella cells develop into approximately isodiametric gravity-perceiving cells, also called statocytes. These peripheral cells contain only few small vacuoles. Towards the exterior of the cap, the cells transform into elongated and more vacuolate secretory cells. Cells at the root cap periphery produce large amounts of extracellular mucilage which covers the whole root apex and performs multiple functions critical for root growth (Vermeer & McCully, 1982). Subsequently, development of root-cap cells terminates when they are shed from the cap periphery (e.g., Hawes et al., 1998). These specific cellular fates clearly distinguish root cap cells from all other root body cells and are closely associated with unique re-organizations of the actin cytoskeleton.

The actin cytoskeleton is dramatically depleted from the cytoplasm of meristematic and early postmitotic root cap cells, as well as of quiescent centre cells (Figs. 1A, B). Interestingly, it is precisely these cells which show the highest endogenous levels of auxin (Sabatini et al., 1999). Statocytes are devoid of any AF bundles which make them quite distinct from all other postmitotic root cells (Baluška et al., 1997b). Together with the absence of dense ER networks deeper in the cytoplasm, this feature of AF depletion is critical in allowing sedimentation of amyloplasts, or statoliths (Baluška & Hasenstein, 1997), which is the basis of statolith-mediated gravisensing (Sack, 1997). Sedimented amyloplasts are, however, enmeshed in a fine network of short AFs interlinking individual statoliths to each other as well as to the cortical ER elements and the plasma membrane. This highly specialised population of AFs is supposed to play a critical role in the perception and transduction of gravity (Sievers et al., 1991; Volkmann et al., 1999; Volkmann & Baluška, 1999; Volkmann & Baluška, this volume).

In contrast, secretory cells of the cap are equipped with prominent networks of AFs and numerous ER elements, both of which support the cells' specialization for production of root cap mucilage. Both statocytes and secretory root cap cells lack myosin VIII at their peripheries (not shown but see Volkmann & Baluška, this volume), and this feature seems to be relevant for the progression of the cells towards tissue disintegration. Intriguingly, cells shed from the root cap periphery retain abundant AFs, indicating their viability and active role in modulating the root-surface micro-environment (e.g., Vermeer & McCully, 1982; Hawes et al., 1998).

4. TRANSITION ZONE: ACTOMYOSIN-BASED GROWTH REGION WHERE CELLS PREPARE FOR F-ACTIN DEPENDENT AND VACUOME-DRIVEN CELL ELONGATION

After leaving the apical meristem, root cells enter the transition growth zone which serves as a preparatory region for their further growth via the vacuome-driven rapid cell elongation (Baluška et al., 1996a). Cells traversing the transition zone are, in many features, more similar to meristematic cells (although they do not divide any more) than to elongating cells (Baluška et al., 1994). Among these features, the most evident is that nuclei are still organizing active 'cell bodies' (Baluška et al., 1998). This is obvious from the approximately central positions of the nuclei which are supported by numerous cytoskeleton-based cytoplasmic strands, radiating from the nuclei towards the cellular peripheries. In contrast, the nuclei of cells performing vacuome-driven cell elongation are settled against the longitudinal (extending) cell walls. These nuclei are depleted of perinuclear AFs and microtubules, do not organize cytoplasmic strands, and are indicative of inactive plant 'cell bodies' (Baluška et al., 1998). They can, however, be re-activated either by external signals (e.g., wounding – Mews et al., 1996) or by internal signals (e.g., initiation of root hairs – Baluška et al., 1998, 2000, and of lateral root primordia – Baluška, in preparation).

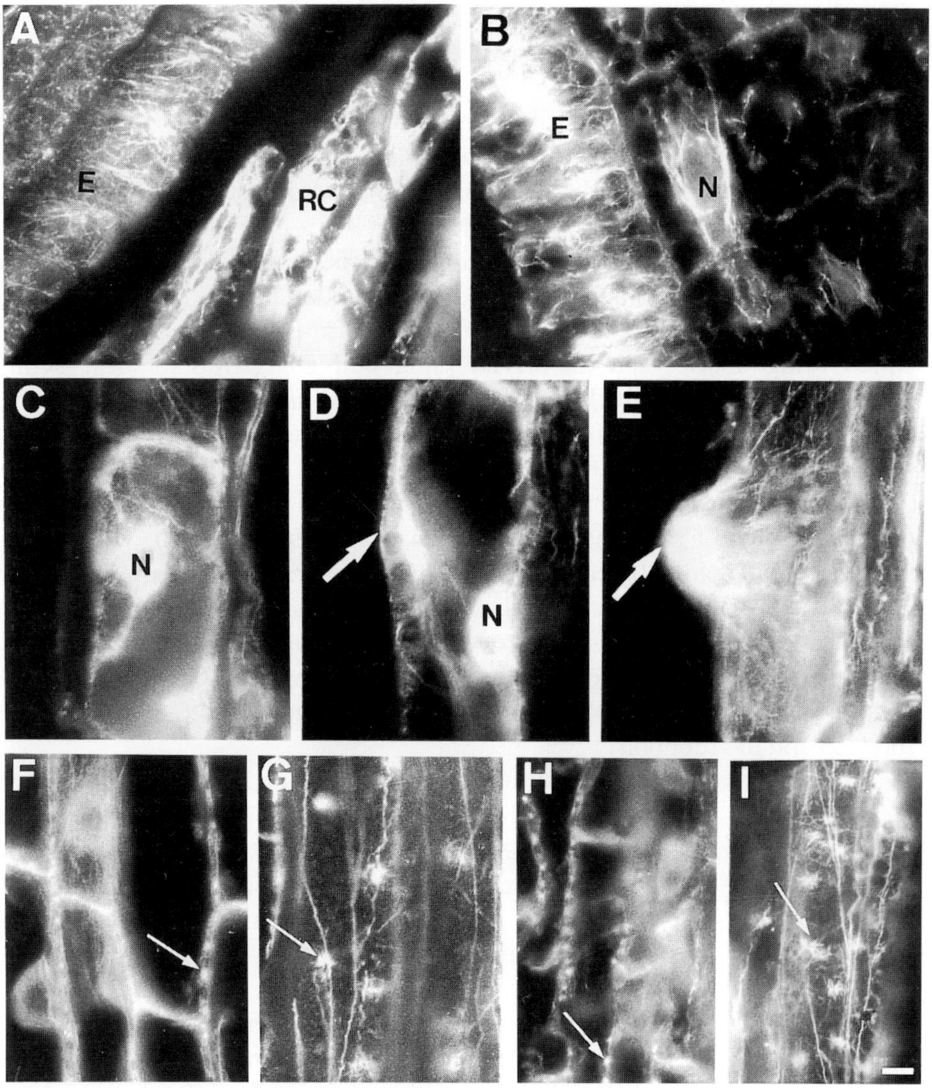

Figure 2. Distributions of AFs in elongating cells of maize root apices. A – Root cap cells (RC), when they are shed from the root periphery, are equipped with prominent AF bundles. In the root body, epidermis cells (E) show numerous cortical AFs arranged transversely with respect to the root axis. B – Transverse arrangement of AF bundles is typical of epidermis cells in the distal part of the transition zone. Adjacent outer cortex cells are already at the stage of early elongation and AF bundles 'embrace' the nuclei and anchor at cross-walls. C-E – Elongating trichoblasts initiating root hairs re-arrange their AFs when the bulging domains (arrows in D and E) become the major AFOCs and attract most AFs. F-I – Pit-fields of the inner cortex (arrows) represent another powerful AFOCs of elongating cells, as revealed by jasplakinolide (10^{-5}, 4 h; F, G) and mild salt-stress (NaCl 200 mM for 1 h; H, I) treatments. Bar = 8 µm

Root cells located within the transition zone are unique with respect to their calcium and auxin responsiveness (Ishikawa & Evans, 1992, 1993; Baluška et al., 1994). Moreover, transition-zone cells initiate diverse root tropisms (for root gravitropism see, e.g., Baluška et al., 1996b) and are the target for thigmostimulation and aluminium toxicity, both of which are phenomena impinge on calcium- and auxin-related processes (Ishikawa & Evans, 1992, 1993; Sivaguru & Horst, 1998; Sivaguru et al., 1999; Kollmeier et al., 2000). The transition-zone cells are proposed to act as sensors of both environmental and developmental factors (Baluška et al., 1994, 1996a). They transmit information towards endogenous processes which bring about vacuome-driven rapid cell elongation in an F-actin-dependent fashion (Baluška et al., 1997a; Volkmann & Baluška, 1999; Šamaj et al., 2000). In many situations, these responses can show different timings at the opposite sides of root apices. This induces differential cell growth in the apical part of the cell elongation region and results ultimately in a range of root tropisms (e.g. Baluška et al., 1994, 1996b). The above-mentioned cap columella cells also pass through a transition phase of cellular development immediately after they leave the meristem, as evidenced by their cell shapes and the absence of a central vacuole. Thus, one could generalize that the transition phase of plant cell development is well suited for the sensing of diverse exogenous and endogenous factors, in both the root cap and the root body.

Postmitotic root body cells traversing the transition growth zone perform a prominent re-organization of the actin cytoskeleton with the perinuclear arrays of AFs transforming into axially arranged, thick bundles (Baluška et al., 1997a). The final outcome of this re-organization, which is characteristic in a more-or-less prominent form in all elongating plant cells (see below), is a population of longitudinally arranged AF bundles which are anchored at non-growing cross walls (for leaf cells, see Masuda et al., 1991; Ryu et al., 1995, 1997). The first sign of this reconstruction of the actin cytoskeleton, which is prominent especially in the densely cytoplasmic cells of the outer stele, pericycle and endodermis, is a sudden proliferation of nuclear-envelope-associated AFs which then bundle together (Baluška et al., 1997a). These nucleus-based AF bundles apparently have an inherent polarity (see also Yokota & Shimmen, 1999, this volume). Moreover, after growing out from perinuclear sites and approaching the non-growing cross walls, they exhibit contrasting architectures (for schematic overview see Fig. 4). One AF bundle end succeeds in interacting with the plasma membrane and then grows further along the plasma membrane. This anchors the whole AF bundle at a specific position. The opposite end of the AF bundle fails to interact with the plasma membrane and becomes clearly splayed in appearance. However, it might physically interact with endomembranes.

This suggests that AF-bundling factors (Bartles, 2000; for plant cells see Vidali et al., 1999; Klahre et al., 2000) are either absent or not working properly at this splayed bundle end. Similarly splayed are F-actin tails which propel bacteria of *Listeria monocytogenes* when the actin-binding protein, α-actinin, was depleted in an in vitro movement assay (Loisel et al., 1999; Machesky & Cooper, 1999). Importantly, lateral surfaces of centrally positioned nuclei act as some kind of reference point for these AF bundles, which keep their nuclear contact areas intact, even after their ends approach the non-growing cross walls. Typically, four AF bundles are visible in 10-μm-thick median sections (for schematic overview of two such bundles see Fig. 4A). Altogether, our data identify nuclear surfaces and non-growing myosin VIII-enriched cross-walls as the major AFOCs of postmitotic cells of the transition zone. Other unconventional myosins have also been proposed to anchor AFs at the plasma membrane (for myosin I, see Heintzelman & Mooseker, 1992; for myosin VI, see Self et al., 1999).

One of the most characteristic features of these unique AF bundles is their straight appearance, suggesting that they are under tension. This feature is lost immediately after the onset of rapid cell elongation. The longitudinal AF bundles then acquire a crinkled appearance (Baluška et al., 1997a). This indicates that these AF bundles, and their interactions with endomembranes and the plasma membrane, are directly related to the developmental transition from the relatively slow cytoplasmic growth of meristematic and transition zone cells to the vacuome-driven growth characteristic of the zone of rapid cell elongation. One possibility is that both ends of these nucleus-associated AF bundles are growing, and, after their approach to the cross walls, they exert a pushing forces on the plasma membrane (reviewed by Borisy & Svitkina, 2000). In addition, one end of these bundles laterally associates, perhaps via membrane-associated myosin VIII molecules, with the plasma membrane. The combined actin polymerization and actomyosin-based forces might thus contribute to a mechanical stretching of the plasma membrane. As root cells traverse the transition zone, the tensional stress imposed on the plasma membrane is expected to increase gradually. Our working hypothesis is that after reaching a critical threshold value, putative stretch-activated channels are switched on, allowing an influx of osmotically-active ions and water molecules which initiate and drive rapid cell elongation. Due to this feature, the vacuome-driven rapid cell elongation starts suddenly (Kubica et al., 1991; Ivanov & Maximov, 1999). The pushing forces of the actin nano-machinery (see Borisy & Svitkina, 2000) may also help overcome the yield threshold of cell growth (the variable, Y, in the Lockhart growth equation – see Barlow, 1989).

Latrunculins represent a new generation of powerful actin drugs which bind to G-actin monomers, hence precluding their polymerization into F-actin (see Gibbon et al., 1999 for plant cells). Due to the natural dynamicity of AFs, latrunculins rapidly 'precipitate' the whole cellular complement of G-actin, leading to total disintegration of AFs. This property makes these drugs for ideal tools in revealing which processes are F-actin dependent and, conversely, which processes proceed unhindered in the absence of F-actin arrays. Our experiments have shown that exposing growing maize root apices to latrunculin B for 2 h (which disintegrates all AFs, not shown) increases the number of cells in the transition zone, indicating an F-actin dependence for the onset of vacuome-driven cell elongation. After a longer (12 h) absence of F-actin, root cortex cells fail to execute rapid cell elongation, although cell divisions continue. As a consequence, the whole of the former elongation region becomes filled with short postmitotic cells. Similar, but milder, effects were induced with the general inhibitor of myosin motor activity, 3,4 butanedione monoxime (Baluška et al., 1997a; Volkmann & Baluška, 1999; Šamaj et al., 2000), indicating that, besides actin polymerization, also myosin activities are involved in the initiation of vacuome-driven cell elongation.

5. ASSOCIATION OF MYOSIN VIII AND ACTIN WITH PIT-FIELDS IN ELONGATING INNER CORTEX CELLS

In contrast to cells of the meristem and transition zone, which have myosin VIII associated preferentially with their peripheries, elongating cells have most of their myosin VIII molecules localized in the cytoplasm. The only exceptions to this are the pit-fields of the inner cortex cells which show abundant myosin VIII at their side walls in the transition zone and distal portion of the elongation region. As phloem unloading is accomplished in these root zones, one interpretation of this might be that plasmodesmata-associated myosin VIII plays some role in symplasmic transport (see Overall et al, this volume) and in this respect redresses the deficiency of sucrose input which was estimated on the basis of passive acropetal symplasmic transport (Bret-Harte & Silk, 1994). For instance, those plasmodesmata which are expected to be more active in symplasmic transport of sucrose into the meristem also show more myosin VIII associated with them. The most dramatic examples of this phenomenon are the myosin VIII-enriched sieve elements in the root zone of phloem unloading (Fig. 3O). Moreover, all root cortex cells accumulate abundant myosin VIII at the plasmodesmata/pit-fields areas in root apices exposed to compacted soil (Baluška, Bengough, Volkmann, in preparation). These root apices exhibit an increased lateral force, generated via excessive vacuome formation based on an additional unloading of osmotically active sucrose from phloem elements, which helps to form cracks within compact soil in advance of the root apex.

Although myosin VIII-rich pit-fields of the inner cortex cells do not show abundant AFs using the ICN antibody (clone C4), AFs can be visualized with this antibody at these pit-fields after exposing the roots to mild osmotic stress (200 mM NaCl for 1 h), or to mastoparan (data not shown) and jasplakinolide treatments. However, a polyclonal antibody raised against maize pollen actin (gift of Chris Staiger) is able to recognize AFs at pit-fields in the inner cortex of control root apices (Baluška, preliminary data). All these findings identify the plasmodesmata and pit-fields as potential AFOCs.

6. TRICHOBLASTS: ROOT HAIR FORMATION IS F-ACTIN DEPENDENT

Of all the elongating cells of the root, those of the epidermis have the most abundant AFs (Baluška et al., 1997a, 2000b). This feature corresponds well with the dramatic increase of cytoplasmic streaming in prospective trichoblasts and is one of the first signs of root hair initiation (Baluška & Lichtscheidl, unpublished). During the early bulge outgrowth, longitudinal AF bundles re-organize and an actin-rich domain is assembled in the vicinity of the bulge (Figs. 2C–E; Baluška et al., 2000). At this early stage of root hair formation, AF cables in the trichoblast become oriented transversely with respect to the root axis (Miller et al., 1999; Braun et al., 1999; Baluška et al., 2000) and are typically associated with the nuclei moving into the bulge (Baluška et al., 1998, 2000). This subcellular bulge domain ultimately transforms into the tip-growing root-hair apex. It represents a powerful AFOC as it not only attracts to itself most of the newly assembled AF cables (Fig. 2E; Baluška et al., 2000) but it also obviously polarizes the whole exocytosis apparatus. Interesting in this respect is the accumulation of myosin VIII at the extreme apices of emerging root hairs (Fig. 2L).

From the 'cell body' perspective, tip-growing root hairs can be considered as active plant 'cell bodies' (Baluška et al., 1998), polarized by the actin-rich cell-cortex domains of the root hair tips (Braun et al., 1999; Baluška et al., 2000) which acts as a potent AFOCs. In addition to the vacuome-driven rapid cell elongation, root hair formation has been identified as another F-actin-dependent growth process (Baluška et al., 2000). In the absence of AFs, root hair formation is stopped after the bulge formation, indicating that this latter event results perhaps from localized cell wall loosening (discussed in Baluška et al., 2000). Analysis of living, latrunculin-

B treated trichoblasts during their initiation of root hairs has revealed an instantaneous vacuolation of the bulge-zone which then fails to initiate tip growth (Ovečka et al., 2000).

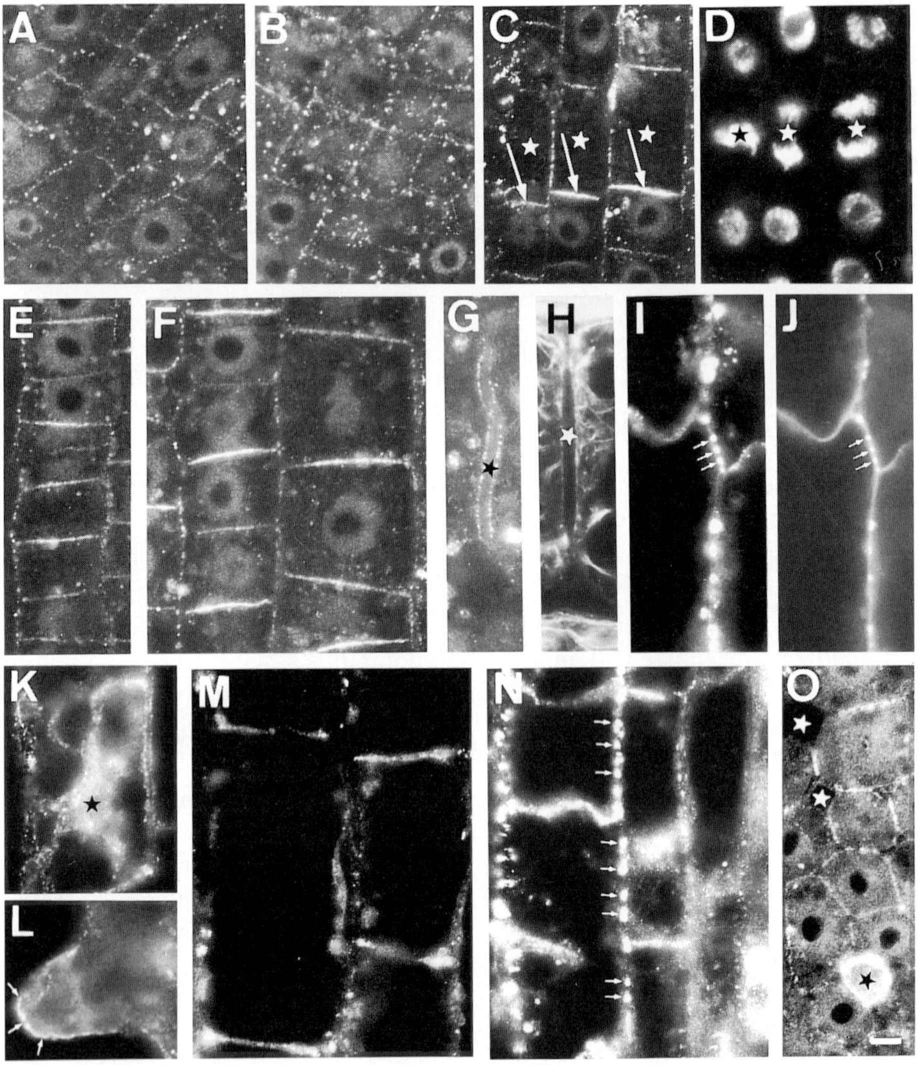

Figure 4. Localization of myosin VIII in cells of maize root apices. A–D – Meristematic cells just behind the quiescent centre (A) show numerous fluorescent dots at cellular peripheries and less fluorescence in the cytoplasm. Further back from the root tip (B), the size of these peripheral dots increases while mitotic cells (stars in C and D) accumulate abundant myosin VIII at cell walls facing the spindle poles (arrows in C and D). In the transition zone, cross-walls of inner cortex cells (E and F) show continuous labelling while side-walls are decorated with myosin VIII-positive dots. G, H – Myosin VIII-positive dots (G) and actin (H) at the cross-wall (star) of a metaxylem cell. I, J – In elongating inner cortex cells, myosin VIII-

positive pit-fields (arrows in I) are enriched with callose (arrows in J), as identified by aniline blue staining. K – Abundant distributions of myosin VIII in postmitotic epidermal cells (star indicates position of nucleus). L – Myosin VIII accumulates at the plasma membrane of emerging root hair apices (arrows). M – In the elongation region, cells in outer and middle cortex show only faint labelling with myosin VIII antibody. N – In the inner cortex, elongating cells accumulate abundant myosin VIII at pit-fields (arrows). O – Cross-section showing that myosin VIII localizes at cell-cell contacts whereas it is depleted from cell peripheries facing extracellular spaces (white stars). Abundant myosin VIII is associated with phloem elements (black star). Bar = 8 μm

A root hair mutant of *Arabidopsis thaliana*, which is unable to transform the outgrowing bulges into tip-growing root hairs, has been cloned and identified as defective for a potassium ion channel (Liam Dolan, personal communication). This finding indicates a close relationships between the actin cytoskeleton and vacuome formation, supporting the notion that both rapid cell elongation and tip growth depend on intact F-actin because this is essential for accomplishment of extensive vacuome formation. Intriguingly, on the other hand, most plant morphogenetic processes seem to proceed normally during the long-term absence of F-actin, just as they can when microtubules (the other major element of the plant cytoskeleton) are disordered (Traas et al., 1995). The absence of actin results in dwarfism, and this phenotype resembles those of various genetic dwarfs of *Arabidopsis*. This dwarfism occurs when seeds are imbibed and germinated in latrunculin (Baluška, Jásik, Edelmann, Salajová, Volkmann, submited).

7. ACTIN FILAMENT ORGANIZING CENTRES

Our data strongly imply the existence of putative AFOCs in plant cells. Distinct domains at the nuclear envelope and plasma membrane obviously support AF assembly. Among these domains, we can mention the plasma membrane domains which face the spindle poles in mitotic cells, the cross-walls in the postmitotic cells of the transition zone (Fig. 4C) and of the elongation region (Fig. 4B, C), as well as the bulging domains and apices of root hairs. In addition, myosin VIII-enriched pit-fields of longitudinal walls of the inner cortex cells also behave as potentional AFOCs (Fig. 4B). Because cross-walls are densely populated with primary plasmodesmata, one could speculate that plasmodesmata, irrespective of whether they are primary or secondary, are relevant for the organization of AFs. In accordance with this notion, both plasmodesmata and pit-fields are known to be depleted in cortical microtubules (reviewed by Baluška et al., 2000) and are prominently enriched with myosin VIII (Reichelt et al., 1999; Reichelt & Kendrick-Jones, this volume).

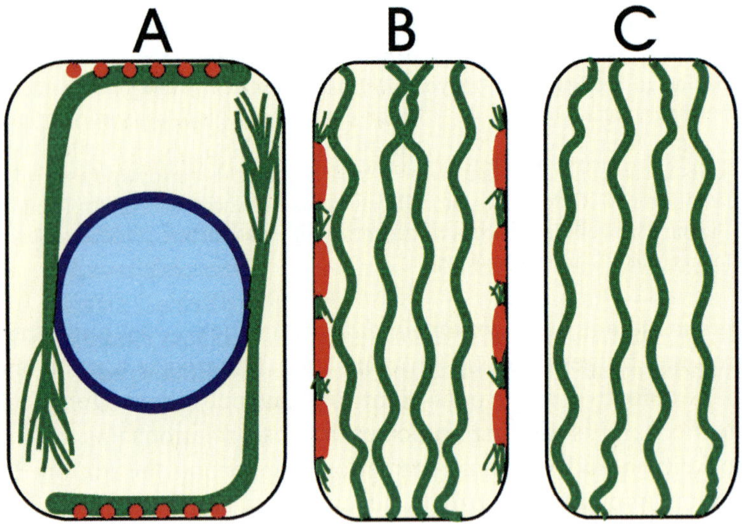

Figure 5. Schematic overview of distributions of AFs (green) and myosin VIII (red) in postmitotic maize root cells. A – Cells of the transition zone develop straight AF bundles which attach laterally to nuclear surfaces and are symmetrically organized from both cross-walls thus matching the growth axes of the cell and of the root. These AF bundles show an inherent polarity: while one end of a bundle is anchored at a myosin VIII-enriched cross-wall, the other end, with a loosely-splayed configuration, resides in the cytoplasm. For simplicity, only two such AF bundles are depicted (compare with Fig. 1K), but up to eight bundles may develop in any one cell. B, C – More proximal, elongating cells are equipped with undulating, axially arranged AF cables. These are typical also of elongated plant cells. In the case of inner cortex cells (B), myosin VIII-enriched pit-fields (red structures) act as additional AF-organizing sites. Other elongating root cells (C) lack such myosin-enriched sites.

Studies on yeast and animal cells identified several putative components of AFOCs (e.g., Shariff & Luna, 1992; Miki et al., 1996; Frazier & Field, 1997; Moreau & Way, 1998; Machesky & Gould, 1999; Yuan & Chia, 1999; Suetsugu et al., 1999; Defacque et al., 2000; Raucher et al., 2000). Our present data on root cells indicate that the unconventional myosin VIII might be involved in conferring AFOC-like properties to discrete portions of the plasma membrane in root cells. This is in accordance with data from animal cells where unconventional myosins have been identified as molecules regulating AF assembly and organization (reviewed by Soldati et al., 1999; Wu et al., 2000). Moreover, myosin I of yeast supports local actin assembly (Goodson et al., 1996) via its interactions with components of the Arp2/3 complex (Machesky, 2000; Evangelista et al., 2000; Lechler et al., 2000), which is crucial for the nucleation of new AFs (Machesky & Gould, 1999). Our preliminary data on maize root cells (Baluška, von Witsch, Šamaj, and Volkmann, in preparation) imply putative plant homologues of Arp2/3 proteins (see Klahre & Chua, 1999) as further AFOC-relevant components of myosin VIII-enriched plasmodesmata and pit-fields.

8. CONCLUSIONS

The actin cytoskeleton has been analyzed in cells of all root apex tissues throughout their diverse pathways of cellular development. For this, we applied indirect immunofluorescence, using ICN actin antibody (clone C4), to Steedman's wax sections taken from whole root apices. In combination with conventional epifluorescence microscopy, this powerful technique (see also Vitha et al., 1997, 2000, this volume) allows viewing of all root apex cells arranged within the context of three-dimensionally organized tissues. This is critical for analysis of tissue- and development-specific organizations of AFs and for assessing their relationships to subcellular distributions of myosin VIII.

Our analysis allows us to reach seven tentative conclusions, listed below, which pave solid avenues to arrive at an understanding of the involvement of AFs and actomyosin-based forces in plant cell growth and development.

1. Unique organizations of the actin cytoskeleton are typical of cells of diverse tissues and for various phases of cellular development.
2. During mitosis, AFs become depleted from spindle regions and from domains marked by preprophase bands, the AFs are then recruited to cell cortex domains facing the spindle poles. The same redistribution pattern is true also for the unconventional myosin VIII.
3. Dramatic re-organization of AFs is accomplished in postmitotic root-body cells of the transition zone in preparation for their rapid cell elongation. Here, perinuclear AF networks transform into distinctive AF bundles.
4. In meristematic and transition zone cels, myosin VIII molecules are localized preferentially at plasmodesmata and pit-fields. In the elongation region, myosin VIII abundantly localizes to pit-fields only in the inner cortex cells. In contrast, all root cap cells lack myosin VIII at their peripheries.
5. Latrunculin B efficiently depolymerizes all AFs in root apices. Root cells devoid of F-actin continue to divide, but then fail to accomplish rapid cell elongation.
6. Roots devoid of F-actin do not form root hairs, although the first stage of hair formation, the bulge outgrowth, proceeds normally.
7. Distinct subcellular domains of the plasma membrane act as AFOCs in root cells. Plant molecules responsible for this domain-specific

organization of AFs await identification but our data suggest that the unconventional myosin VIII is a strong candidate for this role.

REFERENCES

Asada T and Collings D (1997) Molecular motors in higher plants. Trends Plant Sci 2: 29-37

Baluška F and Hasenstein KH (1997) Root cytoskeleton: Its role in perception of and response to gravity. Planta 203: S69-S78

Baluška F, Parker JS and Barlow PW (1992) Specific patterns of cortical and endoplasmic microtubules associated with cell growth and tissue differentiation in roots of maize (*Zea mays* L.). J Cell Sci 103: 191-200

Baluška F, Barlow PW and Kubica Š (1994) Importance of the post-mitotic 'isodiametric' growth (PIG) region for growth and development of roots. Plant and Soil 167: 31-42

Baluška F, Volkmann D and Barlow PW (1996a) Specialized zones of development in roots: View from the cellular level. Plant Physiol 112: 3-4

Baluška F, Hauskrecht M, Barlow PW and Sievers A (1996b) Gravitropism of the primary root of maize: A complex pattern of differential cellular growth in the cortex independent of the microtubular cytoskeleton. Planta 198: 310-318

Baluška F, Kreibaum A, Vitha S, Parker JS, Barlow PW and Sievers A (1997a) Central root cap cells are depleted of endoplasmic microtubules and actin microfilament bundles: Implications for their role as gravity-sensing statocytes. Protoplasma 196: 212-223

Baluška F, Vitha S, Barlow PW and Volkmann D (1997b) Rearrangements of F-actin arrays in growing cells of intact maize root apex tissues: A major developmental switch occurs in the postmitotic transition region. Eur J Cell Biol 72: 113-121

Baluška F, Barlow PW, Lichtscheidl IK and Volkmann D (1998) The plant cell body: A cytoskeletal tool for cellular development and morphogenesis. Protoplasma 202: 1-10

Baluška F, Volkmann D and Barlow PW (2000) Actin-based domains of 'cell periphery complex' and their associations with polarized 'cell bodies' in higher plants. Plant Biol, in press

Barlow PW (1989) Anatomical controls of root growth. Aspects Appl Biol 22: 57-66

Barlow PW (1994) Cell divisions in meristems and their contribution to organogenesis and plant form. In: Shape and Form in Plants and Fungi. Ingram DS and Hudson A (eds), London: Academic Press, pp 169-193

Barlow PW and Baluška F (2000) Cytoskeletal perspectives on root growth and morphogenesis. Annu Rev Plant Physiol Plant Mol Biol 51: 289-322

Bartles JR (2000) Parallel actin bundles and their multiple actin-bundling proteins. Curr Opin Cell Biol 12: 72-78

Boevink P, Oparka K, Santa Cruz S, Martin B, Betteridge A and Hawes C (1998) Stacks on tracks: The plant Golgi apparatus traffics on an actin/ER network. Plant J 15: 441-447

Borisy GG and Svitkina TM (2000) Actin machinery: Pushing the envelope. Curr Opin Cell Biol 12: 104-112

Braun M, Baluška F, von Witsch M and Menzel D (1999) Redistribution of actin, profilin and phosphatidylinositol-4,5-bisphosphate (PIP_2) in growing and maturing root hairs. Planta 209: 435-443

Bret-Harte MS and Silk WK (1994) Nonvascular, symplasmic diffusion of sucrose cannot satisfy the carbon demands of growth in the primary root tip of *Zea mays* L. Plant Physiol 105: 19-33

Cantiello HF and Prat AG (1996) Role of actin filament organization in ion channel activity and cell volume regulation. In: Membrane Protein-Cytoskeleton Interactions. Nelson WJ (ed), San Diego: Academic Press, pp 373-396

Ding JP and Pickard BG (1993) Mechanosensory calcium selective channels in onion epidermis. Plant J 3: 83-110

Defacque H, Egeberg M, Habermann A, Diakonova M, Roy C, Mangeat P, Voelter W, Marriott G, Pfannstiel J, Faulstich H and Griffiths G (2000) Involvement of ezrin/moezin in de novo actin assembly on phagosomal membranes. EMBO J 19: 199-212

Evangelista M, Klebl BM, Tong AHY, Webb BA, Leeuw T, Leberer E, Whiteway M, Thomas DY and Boone C (2000) A role for myosin-I in actin assembly through interactions with Vrp1, Bee1p, and the Arp2/3 complex. J Cell Biol 148: 353-362

Frazier JA and Field CM (1997) Actin cytoskeleton: Are FH proteins local organizers? Curr Biol 7: R414-R417

Garrill A, Findlay GP and Tyerman SD (1996) Mechanosensitive ion channels. In: Membranes: Specialized Functions in Plants. Smallwood M, Knox JP and Bowles DJ (eds), Oxford: BIOS Scientific Publishers, pp 247-260

Gibbon BC, Kovar DR and Staiger CJ (1999) Latrunculin B has different effects on pollen germination and tube growth. Plant Cell 11: 2349-2364

Glogauer M, Arora P, Chou D, Janmey PA, Downey GP and McCulloch CAG (1998) The role of actin-binding protein 280 in integrin-dependent mechanoprotection. J Biol Chem 273: 1689-1698

Goodson H, Anderson B, Warrick H, Pon L and Spudich J (1996) Synthetic lethality screen identifies a novel yeast myosin I gene (*MYO5*): Myosin I proteins are required for polarization of the actin cytoskeleton. J Cell Biol 133: 1277-1291

Hahne G and Hoffman F (1984) The effect of laser microsurgery on cytoplasmic strands and cytoplasmic streaming in isolated plant protoplasts. Eur J Cell Biol 33: 175-179

Hawes MC, Brigham LA, Wen F, Woo HH and Zhu Y (1998) Function of root border cells in plant health: Pioneers in the rhizosphere. Annu Rev Phytopathol 36: 311-327

Heintzelman MB and Mooseker MS (1992) Assembly of the intestinal brush border cytoskeleton. Curr Top Dev Biol 26: 93-122

Higashi-Fujime S (1991) Reconstitution of active movement *in vitro* based on the actin-myosin interaction. Int Rev Cytol 125: 95-138

Hwang J-U, Suh S, Yi H, Kim J and Lee Y (1997) Actin filaments modulate both stomatal opening and inward K^+-channel activities in guard cells of *Vicia faba* L. Plant Physiol 115: 335-342

Ishikawa H and Evans ML (1992) Induction of curvature in maize roots by calcium or by thigmostimulation. Role of the postmitotic isodiametric growth zone. Plant Physiol 100: 762-768

Ishikawa H and Evans ML (1993) The role of the distal elongation zone in the response of maize roots to auxin and gravity. Plant Physiol 102: 1203-1210

Ivanov VB and Maximov VN (1999) The change in the relative rate of cell elongation along the root meristem and the apical region of the elongation zone. Russ J Plant Physiol 46: 73-82

Kachar B and Reese TS (1988) The mechanism of cytoplasmic streaming in Characean algal cells: Sliding of endoplasmic reticulum along actin filaments. J Cell Biol 106: 1545-1552

Klahre U and Chua N-H (1999) The *Arabidopsis* ACTIN-RELATED PROTEIN2 (AtARP2) promoter directs expression in xylem precursor cells and pollen. Plant Mol Biol 41: 65-73

Klahre U, Friederich E, Kost B, Louvard D and Chua N-H (2000) Villin-like actin-binding proteins are expressed ubiquitously in *Arabidopsis*. Plant Physiol 122: 35-47

Knebel W, Quader H and Schnepf E (1990) Mobile and immobile endoplasmic reticulum in onion bulb epidermis cells: Short- and long-term observations with a confocal laser scanning microscope. Eur J Cell Biol 52: 328-340

Knight A and Kendrick-Jones J (1993) A myosin-like protein from a higher plant. J Mol Biol 231: 148-154

Kohno T, Okagaki T, Kohama K and Shimmen T (1991) Pollen tube extract supports the movement of actin filaments *in vitro*. Protoplasma 161: 75-77

Kollmeier M, Felle HH and Horst WJ (2000) Genotypical differences in aluminum resistance of maize are expressed in the distal part of the transition zone. Is reduced basipetal auxin flow involved in inhibition of root elongation by aluminum? Plant Physiol 122: 945-956

Kubica Š, Baluška F and Hauskrecht M (1991) Elemental growth rate and rRNA transcript maturation exhibit the same pattern in individual tissues of the maize root apex. Ann Bot 68: 387-391

Lechler T, Shevchenko A, Shevchenko A and Li R (2000) Direct involvement of yeast type I myosins in Cdc42-dependent actin polymerization. J Cell Biol 148: 363-373

Lee J, Ishihara A, Oxford G, Johnson B and Jacobson K (1999) Regulation of cell movements is mediated by stretch-activated calcium channels. Nature 400: 382-386

Lichtscheidl IK, Lancelle SA and Hepler PK (1990) Actin-endoplasmic reticulum complexes in *Drosera*. Their structural relationship with the plasmalemma, nucleus, and organelles in cells prepared by high pressure freezing. Protoplasma 155: 116-126

Liebe S and Menzel D (1995) Actomyosin-based motility of endoplasmic reticulum and chloroplasts in Vallisneria mesophyll cells. Biol Cell 85: 207-222

Liu Q-S and Berg DK (1999) Actin filaments and the opposing actions of CaM kinase II and calcineurin in regulation of α7-containing nicotinic receptors on chick ciliary ganglion neurons. J Neurosci 19: 10280-10288

Loisel TP, Boujemaa R, Pantaloni D and Carlier ME (1999) Reconstitution of actin-based motility of *Listeria* and *Shigella* using pure proteins. Nature 401: 613-616

Machesky LM and Gould KL (1999) The Arp2/3 complex: A multifunctional actin organizer. Curr Opin Cell Biol 11: 117-121

Machesky LM (2000) The tails of two myosins. J Cell Biol 148: 219-221

Machesky LM and Cooper JA (1999) Bare bones of the cytoskeleton. Nature 401: 542-543

Masuda Y, Takagi S and Nagai R (1991) Protease-sensitive anchoring of microfilament bundles provides tracks for cytoplasmic streaming in *Vallisneria*. Protoplasma 162: 151-159

Mews M, Baluška F and Volkmann D (1996) Tissue- and development-specific distribution of PSTAIR-proteins in cells of control and wounded maize root apices. J Exp Bot 47: 819-829

Miki H, Miura K and Takenawa T (1996) N-WASP, a novel actin-depolymerizing protein regulates the cortical cytoskeletal rearrangement in a PIP_2-dependent manner downstream of tyrosine kinases. EMBO J 15: 5326-5335

Miller DD, de Ruijter NCA, Bisseling T and Emons AMC (1999) The role of actin in root hair morphogenesis: Studies with lipochito-oligosaccharide as a growth stimulator and cytochalasin as an actin perturbing drug. Plant J 17: 141-154

Mollenhauer HH and Morré DJ (1976) Cytochalasin B, but not colchicine, inhibits migration of secretory vesicles in root tips of maize. Protoplasma 87: 39-48

Moreau V and Way M (1998) Cdc42 is required for membrane dependent actin polymerization *in vitro*. FEBS Lett 427: 353-356

Nebenführ A, Gallagher LA, Dunahay TG, Frohlick JA, Mazurkiewicz AM, Meehl JB and Staehelin LA (1999) Stop-and-go movements of plant Golgi stacks are mediated by the acto-myosin system. Plant Physiol 121: 1127-1141

Olyslaegers G and Verbelen J-P (1998) Improved staining of F-actin and co-localization of mitochondria in plant cells. J Microsc 192: 73-77

Ovečka M, Baluška F, Nadubinská M and Volkmann D (2000) Actomyosin and exocytosis inhibitors alter root hair morphology in *Poa annua* L. Biológia, in press

Prat AG, Holtzman EJ, Brown D, Cunningham CC, Reisin IL, Kleyman TR, McLaughlin M, Jackson GR, Lydon J and Cantiello HF (1996) Renal epithelial protein (Apx) is an actin cytoskeleton-regulated Na^+ channel. J Biol Chem 271: 18045-18053

Ramahaleo T, Alexandre J and Lassalles J-P (1996) Stretch activated channels in plant cells. A new model for osmoelastic coupling. Plant Physiol Biochem 34: 327-334

Raucher D, Stauffer T, Chen W, Shen K, Guo S, York JD, Sheetz MP and Meyer T (2000) Phosphatidylinositol 4,5-bisphosphate functions as a second messenger that regulates cytoskeleton – plasma membrane adhesion. Cell 100: 221-228

Reichelt S, Knight AE, Hodge TP, Baluška F, Volkmann D and Kendrick-Jones J (1997) Characterization of a plant myosin. Mol Biol Cell 7 (Suppl.) 216a

Reichelt S, Knight AE, Hodge TP, Baluška F, Šamaj J, Volkmann D and Kendrick-Jones J. (1999) Characterization of the unconventional myosin VIII in plant cells and its localization at the post-cytokinetic cell wall. Plant J 19: 555-569

Ryu J-H, Takagi S and Nagai R (1995) Stationary organization of the actin cytoskeleton in *Vallisneria*: The role of stable microfilaments at the end walls. J Cell Sci 108: 1531-1539

Ryu J-H, Mizuno K, Takagi S and Nagai R (1997) Extracellular components implicated in the stationary organization of the actin cytoskeleton in mesophyll cells of *Vallisneria*. Plant Cell Physiol 38: 420-432

Sabatini S, Beis D, Wolkenfelt H, Murfett J, Guilfoyle T, Malamy J, Benfey P, Leyser O, Bechtold N, Weisbeek P and Scheres B (1999) An auxin-dependent distal organizer of pattern and polarity in the Arabidopsis root. Cell 99: 463-472

Sack F (1997) Plastids and gravitropic sensing. Planta 203: S63-S68

Šamaj J, Peters M, Volkmann D and Baluška F (2000) Effects of myosin ATPase inhibitor 2,3-butanedione 2-monoxime on distributions of myosins, F-actin, microtubules, and cortical endoplasmic reticulum in maize root apices. Plant Cell Physiol, in press

Self T, Sobe T, Copeland NG, Jenkins NA, Avraham KB and Steel KP (1999) Role of myosin VI in the differentiation of cochlear hair cells. Dev Biol 214: 331-341

Shariff A and Luna EJ (1992) Diacylglycerol-stimulated formation of actin nucleation sites at plasma membrane. Science 256: 245-247

Sievers A, Buchen B, Volkmann D and Hejnowicz Z (1991) Role of the cytoskeleton in gravity perception. In: The Cytoskeletal Basis of Plant Growth and Form. Lloyd CW (ed), London: Academic Press, pp 169-182

Silk WK (1992) Steady form from changing cells. Int J Plant Sci 153: S49-S58

Shimmen T and Yokota E (1994) Physiological and biochemical aspects of cytoplasmic streaming. Int Rev Cytol 155: 97-139

Sivaguru M and Horst WJ (1998) The distal part of the transition zone is the most aluminum-sensitive apical root zone of maize. Plant Physiol 116: 155-163

Sivaguru M, Baluška F, Volkmann D, Felle H and Horst WJ (1999) Impacts of aluminum on cytoskeleton of maize root apex: Short-term effects on distal part of transition zone. Plant Physiol 119: 1073-1082

Soldati T, Schwarz EC and Geissler H (1999) Unconventional myosins at the crossroad of signal transduction and cytoskeleton remodeling. Protoplasma 209: 28-37

Staiger CJ (2000) Signaling to the actin cytoskeleton in plants. Annu Rev Plant Physiol Plant Mol Biol 51: 257-288

Suetsugu S, Miki H and Takenawa T (1999) Distinct roles of profilin in cell morphological changes: Microspikes, membrane ruffles, stress fibers, and cytokinesis. FEBS Lett 457: 470-474

Titus MA (1993) Myosins. Curr Opin Cell Biol 5: 77-81

Traas J, Bellini C, Nacry P, Kronenberger J, Bouchex D and Caboche M (1995) Normal differentiation patterns in plants lacking microtubular preprophase bands. Nature 357: 676-677

Verbelen J-P and Tao W (1998) Mobile arrays of vacuole ripples are common in plant cells. Plant Cell Rep 17: 917-920

Vermeer J and McCully ME (1982) The rhizosphere in *Zea*: New insight into its structure and development. Planta 156: 45-61

Vidali L, Yokota E, Cheung AY, Shimmen T and Hepler PK (1999) The 135 kDa actin-bundling protein from *Lilium longiflorum* pollen is the plant homologue of villin. Protoplasma 209: 283-291

Vitha S, Baluška F, Mews M and Volkmann D (1997) Immunofluorescence detection of F-actin on low melting point wax sections from plant tissues. J Histochem Cytochem 45: 89-95

Vitha S, Baluška F, Braun M, Šamaj J, Volkmann D and Barlow PW (2000) Comparison of cryofixation and aldehyde fixation for plant actin immunocytochemistry: Aldehydes do not destroy F-actin. Histochem J, in press

Volkmann D and Baluška F (1999) The actin cytoskeleton in plants: From transport networks to signaling networks. Microsc Res Tech 47: 135-154

Volkmann D, Baluška F, Lichtscheidl IK, Driss-Ecole D and Perbal G (1999) Statoliths motions in gravity-perceiving plant cells: Does actomyosin counteract gravity? FASEB J 13 (Suppl.): S143-S147

Watson AJM, Levine S, Donowitz M and Montrose MH (1992) Serum regulates Na^+/H^+ exchange in Caco-2 cells by a mechanism which is dependent on F-actin. J Biol Chem 267: 956-962

Wolfe J and Steponkus PL (1993) Mechanical properties of the plasma membrane of isolated plant protoplasts. Plant Physiol 71: 276-285

Wu X, Jung G and Hammer JA III (2000) Functions of unconventional myosins. Curr Opin Cell Biol 12: 42-51

Yokota E and Shimmen T (1999) The 135-kDa actin-bundling protein from lily pollen tubes arranges F-actin into bundles with uniform polarity. Planta 209: 264-266

Yuan A and Chia CP (1999) Co-loss of profilin I, II and cofilin with actin from maturing phagosomes in *Dictyostelium discoideum*. Protoplasma 209: 214-225

Zot HG, Doberstein SK and Pollard TD (1992) Myosin-I moves actin filaments on a phospholipid substrate: Implications for membrane targeting. J Cell Biol 116: 367-376

Chapter 27

ACTIN AND SIGNAL-CONTROLLED CELL ELONGATION IN COLEOPTILES

Frank Waller, Qi-Yan Wang and Peter Nick

Institut für Biologie II, Schänzlestr. 1, D–79104 Freiburg, Germany

Key words: auxin, coleoptiles, cytochalasin, phytochrome

Abstract: The role of actin for cell growth has been investigated in great detail for tip-growing cells such as root hairs or pollen tubes. The role of actin for the growth of interphase cells within a tissue context is much less understood and has therefore been analyzed for signal-controlled cell elongation in Graminean coleoptiles using phytochrome and auxin as triggers. Growth inhibition is found to be correlated to actin bundling, whereas a stimulation of cell elongation is accompanied by the formation of a more delicate actin meshwork especially in the region of the cell poles. The response of actin to light is found to be rapid and to precede changes in growth rate. The analysis of the *Yin-Yang* mutant in rice, where actin disassembles in response to auxin indicates the presence of a highly dynamic actin ppulation that is important for cell growth. The experimental findings are discussed in the framework of three alternative models for a role of actin in the elongation of cells within a tissue context: 1. A mechanical limitation of cell expansion by the rigour of the actin meshwork, 2. A control of cell-wall extensibility via an actin-controlled orientation of cortical microtubules, and 3. A control of actin-guided vesicle transport towards sites of wall synthesis. The chapter ends with an outlook on approaches that will distinguish between these alternatives.

1. SCOPE OF THE STUDY

The role of actin filaments has been analyzed in great detail for tip-growing cells such as root hairs, pollen tubes, trichomes, protonemata, fungal hyphae or rhizoids (other chapters of this volume) uncovering a polar transport of vesicles along actin filaments towards the growing tip of those cells, where the vesicles unload their cargo leading to intussusception of new cell-wall material into the growing areas that are confined to the very tip of these cells.

C.J. Staiger et al. (eds.), Actin: A Dynamic Framework for Multiple Plant Cell Functions, 477–496.

In contrast, the function of actin during the expansion of cells that are integrated into a tissue context, is much less understood. It seems that in those cells growth is not confined to a single site, but extends over larger areas. This more diffuse localisation of growth activity is accompanied by a different mode of expansion – rather than increasing wall area via intussusception of new cell-wall material, those cells expand by extending the distance between cellulose microfibrils and by apposing new microfibrils to the inner layers of the cell wall. In this growth mode, the regulation of a growth axis seems to be intimately linked to the orientation of cellulose microfibrils. The synthesis of new microfibrils is guided by underlying microtubules: Transverse orientation of microtubules and, consequently microfibrils, supports cell elongation, whereas a loss of this transverse orientation causes a shift of cell expansion towards lateral thickening (for reviews see Giddings & Staehelin, 1991; Williamson, 1991; Nick, 1998). Interestingly, the orientation of cortical microtubules can change in response to a range of signals such as light, gravity, or various plant hormones. These responses are rapid (Himmelspach et al., 1999), and they are accompanied by corresponding changes in cell expansion (for review see Nick, 1998). Whereas cellular morphogenesis of tip-growing cells has been discussed in terms of signal-dependent changes of polar vesicle transport, it thus seems that, in tissues, it is signal-triggered microtubule reorientation that drives morphogenetic responses to the environment.

A closer look uncovers, however, several inconsistencies between microtubule responses and growth (Nick et al., 1991; Nick & Schäfer, 1994). Moreover, although in many cases the microtubular response itself is rapid enough to precede the growth response (Nick et al., 1990; Himmelspach et al., 1999), one has to take into account that at least a certain part of the cellulose fibers has to be deposited in the new direction, before a change of growth rate is to be expected. These considerations indicate that there must be rapid mechanisms that are independent of microtubules and that can inhibit growth. These might include events that cause a general decrease of wall-extensibility such as cross-links between cellulose microfibrils or lignification (e.g. Müsel et al., 1997). However, the stimuli that induce growth inhibition, very often also change the proportionality of the cell (Baskin & Bivens, 1995), which means that this non-microtubular mechanism must be based on some kind of directionality.

The prime candidate for such a directional mechanism, capable of rapid responses and connected to signaling is very well known from tip-growing cells: it is the actin cytoskeleton. In elongating tissues, actin is usually organized in interconnected, long strands that are aligned with the axis of cell growth. These strand have been found to drive cytoplasmic streaming in cells, where this effect is prominent (Nagai, 1979; Shimmen et al., 1995). However, they are also observed in tissues that are not characterized by a dramatic extent of cytoplasmic streaming. The function of these actin strands that are present in a broad range of tissues remains obscure. Thus, the question seems reasonable, whether it is these actin strands that are responsible for microtubule-independent rapid growth responses of cells within a tissue context.

2. ADVANTAGES OF THE COLEOPTILE SYSTEM

A system, where these questions can be addressed, should ideally fulfill the following conditions:

1. Growth should be based exclusively on cell expansion, not on cell division to allow correlations between growth rate, cellular and biochemical events.
2. Growth should by accessible to rapid and consistent control by external signals.
3. The cells should be large enough to allow for cell biological studies.
4. The system should develop synchronously to provide a homogenous population for biochemical studies.
5. The physiology of growth should be well established.

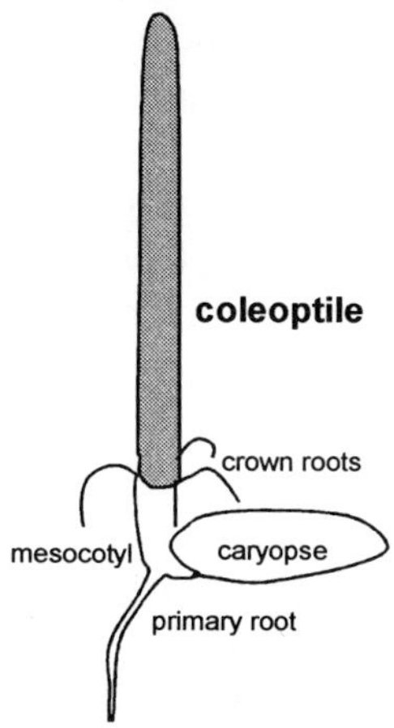

Figure 1. Typical morphology of a Graminean seedling. The coleoptile is separated from the mesocotyl by the crown roots that form the main rooting system in the adult plant. In the Graminea, the primary root does not persist, but dies a few weeks after germination. The apical meristem forming the primary leaves is hidden in the coleoptile and is located slightly above the the node. The primary leaves pierce the coleoptiles a few days after germination depending on the lighting conditions.

These requirements are met by the Graminean coleoptile (Fig. 1). This organ develops as outgrowth of the scutellum (the Graminean homologue of the cotyledon) during late embryogenesis and ensheaths the apical meristem and the embryonic leaves. Following germination it grows rapidly. Growth is carried exclusively by cell elongation. Germination and development can be easily synchronized to high homogeneity by soaking the caryopses in water. The biological role of the coleoptile is to guide the primary leaves through the soil to the surface. As soon as it has reached the surface, it ceases to grow further and is pierced by the primary leaves. It continues to live, however, for several weeks and seems to play a role in the organization of crown-roots, the major rooting system in adult grasses (Nick, 1995).

As adaptation to its biological function, coleoptiles possess an astounding ability to adjust growth with respect to external and internal stimuli:

1. Growth of older coleoptiles is inhibited by red- and far-red light, perceived by the plant photoreceptor phytochrome as well as by blue light, perceived by the plant photoreceptor cryptochrome. This is biologically meaningful, because light indicates that the coleoptile has reached the surface and fulfilled its task. In contrast, the growth of very young coleoptiles is stimulated by phytochrome, but inhibited by blue light. This might be an adaptation to the changed spectral composition of light that penetrates into the soil - whereas blue light is scattered strongly and cannot penetrate, red and especially far-red light can penetrate more deeply. A high red to blue ratio thus indicates that the coleoptile is approaching the surface, but has not yet reached it - an acceleration of growth is therefore a meaningful response (Du Buy & Nuernbergk, 1930).
2. In addition to light, the direction of coleoptile growth is guided by gravity. If maize coleoptiles are displaced from a vertical orientation, they respond by gravitropic curvature with a delay time of less than 30 min (Nick & Schäfer, 1988; Himmelspach et al., 1999), the response of rice coleoptiles is even faster (Godbolé et al., 2000).
3. The coleoptile responds to asymmetries in the light distribution by phototropic curvature and the formation of a stable transverse polarity that guides the development of the crown roots such that they are preferentially emerging at the shaded side (Nick, 1995).
4. These growth responses are carried by changes in the direction and the extent of auxin fluxes from the coleoptile tip (where auxin is produced) towards the base of the coleoptile. The inhibition of growth by red light is triggered by a phytochrome-induced inhibition of auxin transport from the tip into the elongating zone of the coleoptile (Furuya et al., 1969). Gravitropic and phototropic bending are caused by a redistribution of auxin across the coleoptile towards the lower or the shaded flank (Cholodny, 1927; Dolk, 1936; Went, 1928).

The target of these signals seems to be the epidermis that poses a mechanical constraint to the extension of the inner tissues. When the epidermis is removed, the inner tissues extend spontaneously, whereas the epidermis shrinks (Kutschera et al., 1987). By application of auxin, this shrinkage is reduced, whereas the extension of the inner tissue is maintained. This means that, in the intact coleoptile, growth is limited by a reduced extensibility of the epidermis and that auxin can stimulate growth by increasing this extensibility. Electron microscopical studies (Bergfeld et al., 1988) show that it is the outer surface of epidermal cells, where this constraint of growth is located. Thus it is possible to explain the signal response of an entire organ (the coleoptile) in terms of signal-triggered changes in the extensibility of the outer epidermal cell wall. The coleoptile epidermis consists of well-polarized, relatively large cells that are accessible to a range of cell-biological and histological approaches and is thus an ideal system to study the role of actin for the growth of cells that are integrated into a tissue context.

3. LIGHT-INDUCED BUNDLING OF ACTIN IN MAIZE COLEOPTILES

To obtain insight into the possible role of actin in cell growth, the structure of actin filaments was analyzed in epidermal cells of maize coleoptiles under different light conditions. In this system, cell growth is tightly controlled by light perceived by the phytochrome system: When maize seedlings are grown in darkness, the mesocotyl is elongating strongly whereas the coleoptile shows only reduced development. Grown in continuous far-red light, a light quality perceived only by the phytochrome system, mesocotyl growth is suppressed but growth of the coleoptile and primary leaf is promoted. The growth kinetic differs in these tissues: Coleoptile elongation is accelerated by far-red light, leading to a length of about 40 mm within 4 days after germination. In darkness, the final length is identical, but is accomplished only after 6 days. In contrast, the mesocotyl in dark-grown seedlings is more than five times longer compared with that of seedlings grown in far-red light (Waller & Nick, 1997).

3.1 The light response of the coleoptile is based on changed cell elongation

The analysis of cell length and cell number in epidermal cells of the coleoptile during the growth period from 2 to 6 days after germination reveals a close correlation between coleoptile length and cell length. The growth curve of epidermal cells and of the whole organ closely resemble each other and cell numbers remain constant. This suggests that growth of the coleoptile is accomplished exclusively by cell elongation without any contribution of cell division (Waller & Nick, 1997). Epidermal cells in dark-grown mesocotyls also show an elongation during the first days after germination, but from day 5 to day 6 after germination, average cell length declines drastically and pairs of adjacent short cells can be observed (Waller & Nick, 1997). Thus, in contrast to the coleoptile, cell elongation cannot account completely for the growth response of the mesocotyl.

3.2 The cells that control growth possess specific actin filaments

Upon staining with fluorescence-labelled phalloidin and analysis by fluorescence microscopy and laser scanning confocal microscopy a fine meshwork of actin filaments can be visualized in many cell types. This pattern is observed in epidermal cells of primary leafs, in primary leaf companion cells and in subepidermal cells of coleoptiles. In contrast to the cells of the inner tissues, epidermal cells of the coleoptile (and of the mesocotyl) are characterized by bundles of actin filaments spanning the longitudinal axis of the cell, forming an array of finer filaments at the poles of the cell only. This difference is especially striking, if epidermal cells are compared to the immediately neighboured subepidermal cells: Thick longitudinal bundles of actin filaments are only observed in epidermal cells whereas neighbouring subepidermal cells do not exhibit this pattern (Waller & Nick, 1997). This difference is significant if one takes into account that it is the epidermis that limits and regulates coleoptile growth (Kutschera et al., 1987), whereas the inner tissues play a more passive role in the regulation of growth.

3.3 Condensed actin filament arrays correlate with reduced cell elongation

When the behaviour of actin filaments in the coleoptile epidermis is followed over time in dark- and light-grown seedlings, a significant change in the state of longitudinal bundles is observed: In young seedlings (2.5 to 3 days after germination) that had been raised under far-red light, the longitudinal arrays of actin filaments consist of a number of fine strands spanning the longitudinal axis of the cells. In addition, a fan of fine filaments can be observed near the cell poles. When growth slows down after day 4, these filaments become increasingly bundled and the fine polar strands disappear. In this stage most of the longitudinal actin filaments are concentrated in two to four thick longitudinal actin bundles.

In dark-grown seedlings, longitudinal filaments are comparatively bundled up to day 3 after germination along with a slow growth of the coleoptile. They loosen into arrays with more and finer longitudinal filaments and fine strands near the poles between days 4 and 5 when dark-grown coleoptiles undergo rapid elongation. Eventually, the actin filaments recondense into thick bundles from 6 days after germination, accompanied by a rapid decline in growth rate.

These observations can be summarized as follows (Waller & Nick, 1997): During periods of fast elongation (2 to 3 days after germination for far-red treated seedlings; 4 to 5 days after germination for dark-grown seedlings) of the coleoptile longitudinal actin filaments in the epidermal cell layer are organized into a number of fine strands, whereas prior or subsequent to the phase of strong elongation these filaments are observed to form few thick bundles.

3.4 The light response of actin filaments is fast

In epidermal cells of the mesocotyl, a similar correlation between cell elongation and the organization of longitudinal actin filaments has been observed. These actin filaments respond rapidly to changes in growth rate: When dark grown maize seedlings are transferred to far-red light, mesocotyl growth slows down by about 70% within one hour after onset of irradiation. This growth inhibition is accompanied by a rapid and dramatic response of actin filaments: The originally fine strands are replaced by dense bundles that have, in addition, approached the lateral cell walls.

The main findings of this study on the relation between actin filaments and phytochrome-controlled cell elongation can be summarized as follows:

1. The actin filament arrangement of epidermal cells, where the control of growth is located, differs significantly from the underlying tissue.
2. Epidermal actin filaments occur in two configurations: massive longitudinal bundles or a loose array consisting of fine longitudinal strands.
3. The bundled configuration is observed in cells, where elongation is blocked, the loose array is observed in cells, where elongation is elevated.

The transition from the loose array to the bundled array in response to light is rapid and parallel to an inhibition of cell elongation.

4. AUXIN-INDUCED DYNAMICS OF ACTIN IN RICE COLEOPTILES

The response of Graminean coleoptiles to phytochrome seems to be controlled by changes in auxin transport (Furuya et al., 1969). This plant hormone that is produced by a very limited region in the very tip of the coleoptile, is transported in basipetal direction and this transport can be regulated by a number of signals such as red light (Furuya et al., 1969), blue light (Went, 1928; Nick et al., 1992), and gravity (Godbolé et al., 2000). By limiting the flow of auxin into a given region of the coleoptile, cell growth in this region can be controlled allowing, in case of auxin asymmetries, for curving responses such as photo- and gravitropism.

4.1 Screen for cytoskeletal rice mutants

On this background, mutants were selected in rice, where the response of the cytoskeleton to auxin as a key signal was disturbed (Nick et al., 1994). Such mutations should become manifest as alterations of coleoptile growth, when the cytoskeleton is differentially eliminated depending on its dynamics. This is possible for drugs that inhibit the addition of monomers to cytoskeletal polymers. Cytoskeletal elements with a high level of dynamic equilibrium between assembly and disassembly should be more susceptible to such compounds as compared to more inert structures with low rates of monomer addition and polymer disassembly. In other words: it is possible to screen for mutants with reduced cytoskeletal dynamics by searching for seedlings, where coleoptile elongation has acquired resistance to cytoskeletal assembly inhibitors such as the aryl carbamates, the dinitroaniline herbicides (in case of microtubules) or the cytochalasins (in case of actin filaments). Since these compounds cause a loss of growth axiality, the growth of susceptible plants is dramatically blocked, accompanied by lateral swelling, whereas resistance becomes manifest as unimpaired coleoptile elongation.

4.2 *Yin-Yang*, a mutant with elevated resistance to aryl carbamates

Among mutants that were found to be resistant to aryl carbamates such as phenyl urethane or propham, *Yin-Yang* was isolated as a recessive mutant due to a moderately increased resistance of seedling growth. It was found to be cross-resistant to drugs with a different site of action such as colchicine, indicating that the mutation was not based on alterations of the drug-binding site of tubulins (Wang & Nick, 1998). A physiological analysis revealed that cell number was increased in the epidermis of mutant coleoptiles, which was compensated by a reduced cell length. Microtubules were found to be oriented strictly transverse at a time, when they had already assumed a longitudinal orientation in the wild type, accompanied by a prolonged period of coleoptile elongation.

During experiments on the gravitropic response of rice coleoptiles, a striking 'misbehaviour' was observed for the *Yin-Yang* mutant (Fig. 2): The gravitropic lag phase (around 30 min for the wild type) was drastically reduced, the gravitropic response became detectable virtually immediately after the onset of stimulation.

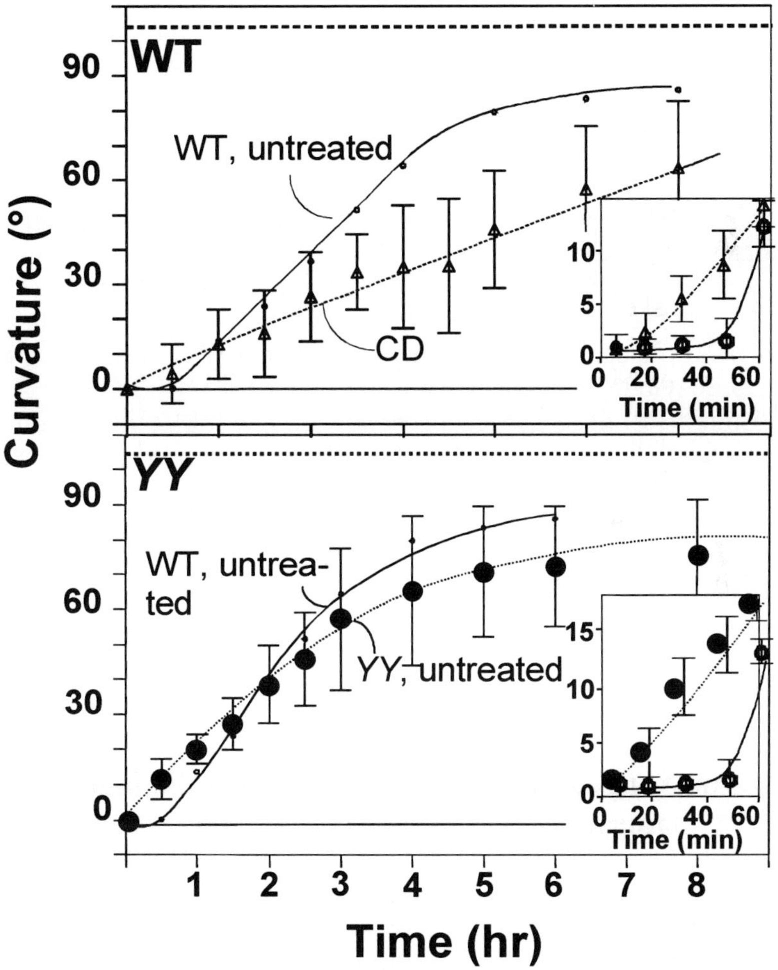

Figure 2. Gravitropic response of the *Yin-Yang* mutant. Rice seedlings were placed horizontally and the gravitropic curvature of the coleoptiles was followed over time in untreated coleoptiles versus plants that had been pretreated for 30 min with 2 μM of

cytochalasin D. The response of the wild type is shown in the upper, the response of the *Yin-Yang* mutant in the lower panel. The inset shows a blow-up of the early time course to show the elimination of the gravitropic lag-phase in the mutant (lower panel) or by treatment with cytochalasin D in the wild type (upper panel). The dashed line in the upper panel shows the final curvature reached by wild-type coleoptiles after treatment with cytochalasin D, and the dotted line in the lower panel the final curvature reached by untreated *Yin-Yang* coleoptiles. In both cases. a significant overbending is observed.

On the other hand, the rate of curving was reduced as compared to the wild type resulting in a response that initiated earlier, but required a longer time until the coleoptile was completely vertically again. Moreover, the curving response did not stop when the coleoptile had recovered its originally vertical orientation, but continued leading to a significant overresponse in the mutant. These alterations of gravitropism were accompanied by a changed microtubular response in the bundle-sheath cells of the coleoptile, where the gravitropic stimulus is perceived: In the wild type, microtubules are found to be strictly transverse in these cells, but reorient rapidly in response to gravitropic stimulation by a reorientation into longitudinal arrays (Godbolé et al., 2000). In the mutant, microtubules are found to be longitudinal before the onset of gravitropic orientation (Nick et al., 1997), which has been correlated with a redistribution of auxin fluxes from basipetal transport to lateral transport towards the lower flank of the rice coleoptile (Godbolé et al., 2000).

Interestingly, all these aspects of the altered gravitropic response in *Yin-Yang* mutants could be mimicked in the wild type by pretreatment with the actin-polymerization blocker cytochalasin D: Microtubules in the gravisensing cells assumed a longitudinal orientation in the absence of gravity (Nick et al., 1997), the gravitropic lag phase was drastically reduced, and the rate of bending was reduced accompanied by significant overbending (Wang & Nick, 1998). This raised the possibility that the mutation was only indirectly related to changes in microtubular dynamics, and that the phenotype might be caused by alterations of the actin filament system.

To test this assumption more directly, the *Yin-Yang* coleoptiles were analyzed with respect to the response of cell elongation to auxin (Wang & Nick, 1998). Coleoptile segments were incubated in the presence of 5 μM auxin, a concentration that represents the optimum of elongation growth, and increasing concentrations of the actin-assembly blocker cytochalasin D were added into the assay to obtain a dose-response relation. The resulting curve revealed that auxin-induced elongation became inhibited in wild-type coleoptiles with concentrations exceeding 0.2 μM of cytochalasin D. In the *Yin-Yang* mutant, the threshold of inhibition was lowered by a factor of 10. Interestingly, if the experiment was repeated at lower auxin concentrations that were suboptimal for cell elongation, the elevated cytochalasin-D sensitivity in the mutant disappeared. In order to observe this increased sensitivity, a minimal concentration of 200 nM of auxin had to be added into the assay. This type of study was repeated for different temperatures to obtain different growth rates and produced three major results:

1. The *Yin-Yang* mutant is more sensitive to cytochalasin D.
2. This elevated sensitivity is induced by auxin.
3. This elevated sensitivity does not depend on growth rate.

Stimulated by these physiological studies, the *Yin-Yang* mutant was further analyzed with respect to the structure of actin filaments in the coleoptile epidermis. In coleoptile segments, where the endogenous auxin had been depleted by incubation in water, the actin filaments of the wild type were organized in dense bundles as was expected from these cells that showed a low rate of elongation (see above). Upon addition of auxin, these bundles became loosened and dispersed into finer strands that were parallel to each other. Again, this was to be expected from the observations in maize, where such a loose array was characteristic for cells with stimulated elongation. The situation in the *Yin-Yang* mutant differed conspicuously: upon auxin depletion, the actin filaments still maintained a loose configuration similar to that observed in the wild type after addition of auxin. After addition of auxin, these fine actin strands disappeared rapidly and were replaced by basket-like networks of short filaments that formed around the nucleus. The formation of such nuclear baskets could be mimicked in the wild type by treatment with cytochalasin D. As observed already for growth, the *Yin-Yang* mutant exhibited an elevated sensitivity of actin filament integrity with respect to cytochalasin D. In the rice mesocotyl that is under control of gibberellic acid rather than auxin (Toyomasu et al., 1994), the formation of nuclear baskets could be induced by gibberellic acid in the *Yin-Yang* mutant, but not in the wild type.

Since cytochalasin D inhibits the addition of actin monomers to growing actin filaments, the sensitivity of a given actin filament to this drug depends on its rate of turnover (Cooper, 1987): Dynamic actin filaments with a high dynamic equilibrium between assembly and disassembly are eliminated by low concentrations of cytochalasin D, whereas inert actin filaments with slow turnover are expected to be relatively resistant. The elevated sensitivity of the *Yin-Yang* mutant to cytochalasin D could thus be a consequence of an elevated dynamics of actin filaments.

Auxin (and gibberellin in case of the mesocotyl) seems to stimulate the dynamics of actin filaments, because the sensitivity of cell elongation (and of actin filament integrity) to cytochalasin D is elevated in the presence of auxin. This effect of auxin is very dramatic in the *Yin-Yang* mutant, but can be observed in a less pronounced fashion in the wild type as well (Wang & Nick, 1998). This stimulation of actin dynamics by auxin must occur in the wild type in a tight balance between elevated disassembly and elevated assembly, because the actin filaments would disappear otherwise. In the mutant, this balance seems to be disturbed in the mutant with disassembly being stimulated by auxin, whereas the assembly seems to lag behind, leading to an increased pool of monomeric G-actin that is then repartitioned into the nuclear meshwork characteristic for auxin-treated *Yin-Yang* cells. This situation can be mimicked in the wild type by treatment with cytochalasin D. Again, the resulting elevated level of G-actin is organized into the nuclear basket. The observation that, in *Yin-Yang*, the actin filaments are found in the loose configuration even upon depletion of endogenous auxin suggests that the mutation might partially decouple actin dynamics from auxin signalling. Auxin perception and early signalling seems to be intact, because auxin sensitivity (as assessed from the dose-response relationship) is not affected in the mutant. The target of the mutation must be located further downstream, either in late steps of auxin signalling or in factors that regulate the dynamics of the actin system.

The major findings of the mutant analysis can be summarized into the following statements:

1. Several aspects of the *Yin-Yang* mutant; as different as gravitropic response, cell elongation, microtubule orientation in the gravity-sensing cells of the bundle sheath, or formation of nuclear actin baskets; can be mimicked in the wild type by treatment with cytochalasin D.
2. The sensitivity to cytochalasin D is elevated in the *Yin-Yang* mutant with respect to cell elongation.
3. This sensitivity is induced by auxin (or by gibberellin in the mesocotyl).
4. Auxin induces a higher dynamic equilibrium of actin filament assembly and disassembly.
5. The pool of G-actin is tightly controlled in the wild type.
6. This tight control is maintained in the mutant for low levels of actin filament dynamics. It is increasingly lost, when the dynamics is increased by auxin or gibberellin.
7. The early events of auxin signalling are not affected in the mutant.

Upon depletion of auxin the actin filaments in the *Yin-Yang* mutant behave as if auxin is still present.

5. IDEAS ABOUT THE ROLE OF ACTIN IN CELL GROWTH

The work described in this chapter suggests a role of actin filaments for the control of cell elongation by signals such as light and auxin. The light stimulus activates the phytochrome system which causes changes of auxin transport and, in consequence, changes of intracellular auxin content. Actin filaments respond to these signals (light or auxin, respectively) by passing between two configurations: a loose configuration composed of fine longitudinal strands that seem to consist of highly dynamic actin filaments, and a bundled configuration of lower dynamics. The loose configuration is closely correlated with elevated cell elongation, whereas the bundled configuration is found in cells that elongate slowly. The actin filament response is confined to the epidermis, the tissue that limits coleoptile elongation.

Results obtained with the *Yin-Yang* mutant demonstrate that specific alterations of actin filament organization (i.e. induction of nuclear baskets by auxin) are accompanied by corresponding changes in the behaviour of cell elongation (reduced cell elongation). Moreover, when these alterations in the actin filament system are mimicked in the wild type by treatment with cytochalasin D, this results in the same effect, namely, an inhibition of cell elongation (Wang & Nick, 1998). The signal response of the epidermal actin array is fast – at least as fast as the response of cell elongation (Waller & Nick, 1997).

Thus, actin filaments are necessary to maintain cell elongation in epidermal cells of Graminean coleoptiles. They can fulfill this function, when they are organized in loose arrays made up of fine strands, they cannot fulfill this function, when they are bundled in response to auxin depletion.

However, this conclusion does not explain, how actin filaments control cell elongation. One can imagine three ideas to answer this question:

Idea 1: The rigour of the actin filament system could mechanically constrain cell elongation (Fig. 3A). This idea is supported by the following evidence:

A. Actin filaments form fan-like arrays near the poles of epidermal cells (Waller & Nick, 1997).
B. The actin filament system of adjacent cells is coupled through the plasmodesmata (White et al., 1994).
C. Mild treatment with cytochalasin D that still preserves actin filaments as such, detaches the actin cytoskeleton from the epidermal poles in such a way that it collapses, very much like a spring that had been allowed to contract (Wang & Nick, 1998).
D. In soy-bean cells, the rigour of the actin system had been estimated by means of a laser forceps (Grabski & Schindler, 1996) and found to be tightly correlated with cell growth. Auxin makes the actin cytoskeleton relax and stimulates elongation, whereas cytokinin or aluminum that increase the rigour of actin filaments, inhibit cell elongation (Grabski & Schindler; 1995, 1996).

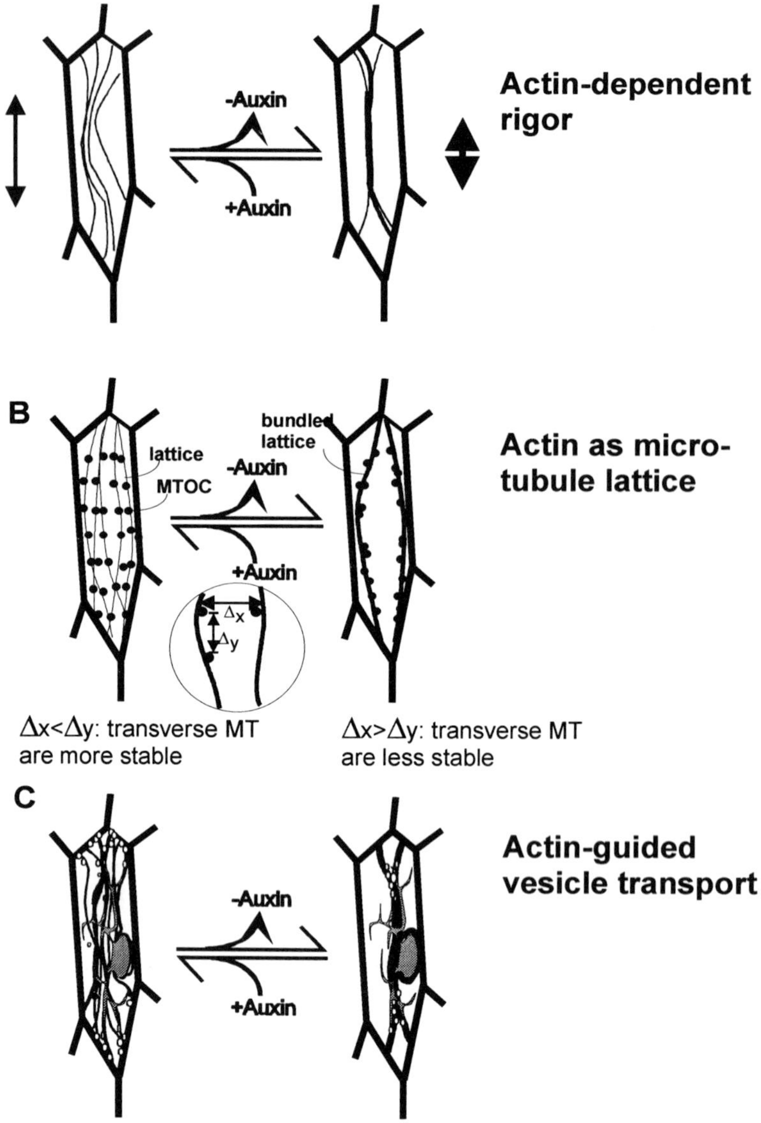

Figure 3. Models for the potential role of actin filaments in the control of auxin-dependent elongation growth. **A** Bundling of actin filaments in response to auxin depletion increases the rigour of the actin cytoskeleton and reduces cellular extensibility in longitudinal direction (double-headed arrow). **B** Bundling of the actin lattice causes a repatterning of microtubule-organizing centers (MTOC) increasing the minimal distance between neighboured MTOC in transverse direction over that in longitudinal direction. In consequence, longitudinal microtubules become more stable as compared to transverse microtubules and new microtubules reassemble more frequently in longitudinal direction causing a net reorientation of microtubules into a longitudinal array that causes a longitudinal deposition of cellulose and thus a loss of cell elongation. **C** Bundling of the actin lattice inhibits transport of vesicles to

the polar regions of the cell impairing the supply of membrane components that are necessary to maintain cell elongation.

E. Bundling of actin filaments is confined to the epidermis, and coleoptile growth is limited by the epidermis (Waller & Nick, 1997).
F. Bundled actin filaments are correlated with inhibited cell elongation, both in maize and in rice coleoptiles (Waller & Nick, 1997; Wang & Nick, 1998).

If actin filaments impede growth due to their rigour, the elimination of actin filaments by cytochalasin D should remove this growth constraint, and elongation should be stimulated. However, the opposite has been observed (Thimann et al., 1992; Wang & Nick, 1997).

Idea 2: Actin could control cell elongation by determining microtubule orientation (Fig. 3B).

Actin filaments and microtubules are intimately linked. The transverse microtubule arrays characteristic of elongating cells are spontaneously lost in response to cytochalasin D (Nick et al., 1997; Wang, unpublished results). The reorientation of microtubules in response to a signal has been shown to be based on disassembly of discordant microtubules and reassembly in a new direction (Himmelspach et al., 1999). Disassembly and reassembly are biochemical processes that do not convey directional information *per se*, but require some kind of either lattice or field that provides directional cues that are amplified into a direction-dependent stability. These directional cues would then organize microtubule nucleation and microtubule elongation in space. This lattice or field has to embody some kind of axis and it must respond to signals. Actin filaments would meet both conditions. If microtubule-organizing centers (MTOC) would be attached to actin filaments they could by dynamically reshuffeled between a disperse distribution (maintained by the loose actin filament configuration), and an aggregated distribution (maintained by the bundled actin filament configuration). By modulating the degree of actin filament bundling, the distances between the points of the MTOC-lattice could be tuned in such a way that the minimal distance between two MTOC would be different in a direction-dependent manner. As long as the actin filaments remain in the loose configuration, the minimal distance would be smaller in transverse direction, larger in longitudinal direction. This would result in a higher stability of transverse microtubules as compared to longitudinal microtubules. Bundling of actin filaments would increase the minimal distance between neighbouring MTOC in transverse direction, such that the stability of longitudinal microtubules would exceed that of transverse microtubules. All the experimental data reported above are consistent with this actin-MTOC-lattice model, moreover this model would explain, why the microtubules in the gravity-sensitive inner cells of *Yin-Yang* spontaneously assume a longitudinal array (Nick et al., 1997), and why the *Yin-Yang* mutant is cross-resistant to antimicrotubular drugs with different binding sites (Wang & Nick, 1998).

Idea 3: Actin could guide vesicle transport to those areas, where cell wall is produced in a way similar to that found in tip-growing cells (Fig. 3C).

The loose actin filament configuration would support this function, the bundled configuration would impede the movement of vesicles along actin resulting in growth inhibition. The advantage of the vesicle model over the actin-rigour model (Fig. 3A) consists in the explanation, why cytochalasin D is expected to block cell elongation. If the tracks for vesicle transport are affected, this should result in impaired transport and impaired cell elongation. Further support for this model comes from pharmacological studies of root expansion growth, where a loss of growth axis was observed in response to drugs that affect the actin-myosin system or vesicle secretion (Baskin & Bivens, 1995). The cargo of these vesicles might be components of the cell wall or of the plasma membrane.

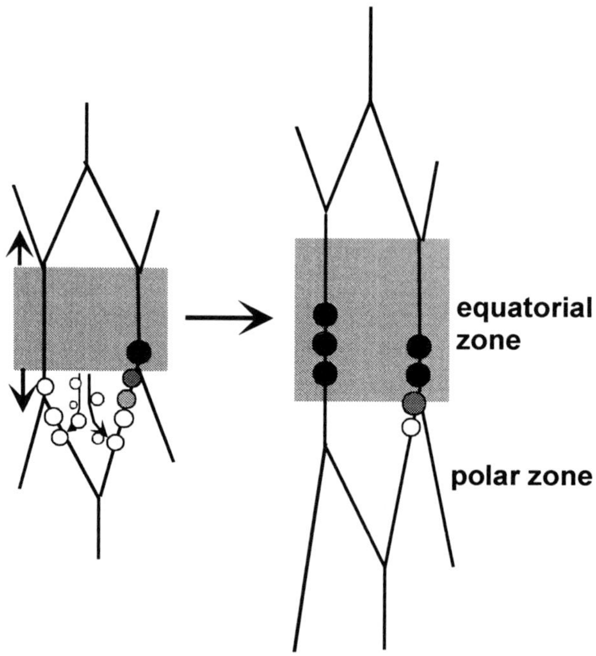

Figure 4. Elongation of an epidermal cell. Elongation is mainly brought about by extension of the central equatorial zone (gray) towards the two poles. Cell-wall components that are necessary to maintain the growth of the equatorial zone (circles) could be transported to the polar zone and would shift to the equatorial zone, when it migrates over the respective

position of the plasma membrane. Thus membrane components of the equatorial zone are expected to be have experienced a longer history (symbolized by the gray and black colour of the circles) as compared to the newly inserted membrane components in the polar regions (symbolized by the white colour of the circles).

At this point, it is necessary to emphasize a principal difference with the mechanism that drives growth in tip-growing cells: The growth of cells in a tissue does not occur by intussusception of new cell-wall material into the cell wall, but by apposition of new layers to the inner face of the preexisting cell wall. Cell expansion occurs by changing the distance between preexisting cellulose microfibrils, and by apposition of new microfibrils that allow for further expansion. As pointed out in the beginning of this chapter, this growth mode is intimately linked to the orientation of cortical microtubules. A guided transport of vesicles loaded with cell-wall material might be needed to maintain the synthesis of cellulose, an inhibition of this transport would gradually cause a depletion of monomer supply and thus cellulose synthesis. From this, an inhibition of growth might be expected. However, auxin-induced coleoptile growth can be maintained under conditions, where cellulose synthesis has been severely inhibited (Edelmann & Fry, 1992). This indicates that microfibril apposition is not an important target for actin-guided vesicle trafficking.

If growth of cells in a tissue were exclusively driven by extension of lateral walls supported by a mechanism dependent on microtubules and microfibrils, one would expect more or less cylindrical cells that are long in case of rapid elongation and short in case of growth inhibition. What is observed under the microscope, is something else: epidermal cells are more appropriately seen as extended rhomboids that are separated by a more or less cylindrical zone in the equator (Fig. 4). In case of elongated cells the equatorial zone is large in relation to the rhomboid polar regions, such that the cell as a whole resembles a cylinder. When growth is suppressed and the cells remain short, their true shape can be seen more clearly, because the equatorial region is short in relation to the rhomboid poles. This is especially striking in the cells of the coleoptile tip prior to expansion. This shape suggests that the polar region of the cells grow by a different mechanism than the equatorial zone. The prominence of rhomboid cells in the coleoptile apex suggests that growth in the polar regions precedes that in the equator. When these cells elongate by a factor of 20 or more, this means that there is a difference in terms of maturation between cell wall and plasma membrane in the very equator as compared to the polar regions: The nearer they are located to the equator, the more mature they are expected to be.

This leads to a model, where in the polar region components are inserted into the young cell wall and into the young plasma membrane that are necessary to maintain intercalar growth once the expanding equatorial zone has reached the respective region of the cell wall (Fig. 4). Such components might be either elements of the cellulose-synthetizing complexes residing in the plasma membrane (Giddings & Staehelin, 1991), or membrane proteins that link the cortical microtubules to the cell wall or just basic components of primary wall and plasma membrane.

In such a model, elongating cells within a tissue would grow based on a two-phase mechanism: In the early phase of cell-wall formation in the polar regions of the cell, actin filaments would act to guide vesicles to the cell poles, whereas microtubules would control the subsequent phase, when the matured cell wall extends and grows by apposition. According to the actin-vesicle model of growth control, the loose actin filament configuration would be necessary to maintain growth because it provides the tracks (and, via myosins, the motile force) to support the immense flow of membrane and wall material necessary for the rapid expansion of epidermal cells. The extent of plasma-membrane flow in these cells is astonishingly high – the complete plasma membrane has been reported to turn over within three hours (Steer, 1988). This means that a block of vesicle-trafficking caused by the bundling of actin filaments should impair growth within minutes, consistent with the fast response of both, epidermal filaments and cell elongation that has been observed after induction of the phytochrome system (Waller & Nick, 1997).

6. OUTLOOK

To distinguish between the three models discussed above, it is necessary to find experimental answers to the following questions:

1. Where do epidermal cells expand? Is growth completely diffuse over the entire cell, or are there distinct regions, where membrane formation takes place preferentially?
2. What is the role of actin-driven motors (i.e. myosins) in auxin-induced cell elongation?
3. How is auxin-signalling coupled to actin-bundling?
4. How is the polarity of vesicle transport (from the cell center towards the cell periphery) controlled?

REFERENCES

Baskin TI and Bivens NJ (1995) Stimulation of radial expansion in *Arabidopsis* roots by inhibitors of actomyosin and vesicle secretion but not by various inhibitors of metabolism. Planta 197: 514-521

Bergfeld R, Speth V and Schopfer P (1988) Reorientation of microfibrils and microtubules at the outer epidermal wall of maize coleoptiles during auxin-mediated growth. Bot Acta 101: 57-67

Cholodny N (1927) Wuchshormone und Tropismen bei Pflanzen. Biol Zentralbl 47: 604-626

Cooper JA (1987) Effects of cytochalasin and phalloidin on actin. J Cell Biol 105: 1473-1478

Dolk HE (1936) Geotropism and the growth substance. Rec Trav Bot Néerl 33: 509-585

du Buy HG and Nuernbergk E (1929) Über das Wachstum der Koleoptile und des Mesokotyls von *Avena sativa* unter verschiedenen Außenbedingungen. Proc. K. Ned Akad Wetenschappen 32: 614-624

Edelmann HG and Fry SC (1992) Effect of cellulose synthesis inhibition on growth and the integration of xyloglucan into pea internode cell walls. Plant Physiol 100: 993-997

Giddings TH and Staehelin A (1991) Microtubule-mediated control of microfibril deposition: A re-examination of the hypothesis. In: The Cytoskeletal Basis of Plant Growth and Form. Lloyd CW (ed), London: Academic Press, pp 85-99

Godbolé R, Michalke W, Nick P and Hertel R (2000) Cytoskeletal drugs and gravity-induced lateral auxin transport in rice coleoptiles. Plant Biol 2: 1-6

Furuya M, Pjon Ch-J, Fujii T and Ito M (1969) Phytochrome action in *Oryza sativa* L. III. The separation of photoperceptive site and growing zone in coleoptiles, and auxin transport as effector system. Dev Growth Differ 11: 62-76

Grabski S and Schindler M (1995) Aluminum induces rigor within the actin network of soybean cells. Plant Physiol 108: 897-901

Grabski S and Schindler M (1996) Auxins and cytokinins as antipodal modulators of elasticity within the actin network of plant cells. Plant Physiol 110: 965-970

Himmelspach R, Wymer CL, Lloyd CW and Nick P (1999) Gravity-induced reorientation of cortical microtubules observed *in vivo*. Plant J 18: 449-453

Kutschera U, Bergfeld R and Schopfer P (1987) Cooperation of epidermal and inner tissues in auxin-mediated growth of maize coleoptiles. Planta 170: 168-180

Müsel G, Schindler T, Bergfeld R, Ruel K, Jacquet G, Lapierre C, Speth V and Schopfer P (1997) Structure and distribution of lignin in primary and secondary cell walls of maize coleoptiles analyzed by chemical and immunological probes. Planta 201: 146-156

Nagai R (1979) Cytoplasmic streaming in plant cells. What's New in Plant Physiology 10: 45-48

Nick P and Schäfer E (1988) Spatial memory during the tropism of maize (*Zea mays* L.) coleoptiles. Planta 175: 380-388

Nick P, Furuya M and Schäfer E (1991) Do microtubules control growth in tropism? Experiments with maize coleoptiles. Plant Cell Physiol 32: 873-880

Nick P, Schäfer E and Furuya M (1992) Auxin redistribution during first positive phototropism in corn coleoptiles. Microtubule reorientation and the Cholodny-Went theory. Plant Physiol 91: 1302-1308

Nick P and Schäfer E (1994) Polarity induction versus phototropism in maize: Auxin cannot replace blue light. Planta 195: 63-69

Nick P (1995) Phototropic stimulation can shift the gradient of crown root emergence in maize. Bot Acta 110: 291-291

Nick P, Godbolé R and Wang, Q.Y. (1997) Probing rice gravitropism with cytoskeletal drugs and cytoskeletal mutants. Biol Bull 192: 141-143

Nick P (1998) Signalling to the microtubular cytoskeleton in plants. Int Rev Cytol 184: 33-80

Shimmen T, Hamatani M, Saito S, Yokota E, Mimura T, Fusetani N, and Karaki H (1995) Roles of actin filaments in cytoplasmic streaming and organization of transvacuolar strands in root hair cells of *Hydrocharis*. Protoplasma 185: 188-193

Steer MW (1988) Plasma membrane turnover in plant cells. J Exp Bot 39: 987-996
Thimann KV, Reese K and Nachmikas VT (1992) Actin and the elongation of plant cells. Protoplasma 171: 151-16692
Toyomasu T, Yamane H, Murofushi N and Nick P (1994) Phytochrome inhibits the effectiveness of gibberellins to induce cell elongation in rice. Planta 194: 256-263
Waller F und Nick P (1997) Response of actin microfilaments during phytochrome-controlled growth of maize seedlings. Protoplasma 200: 154-162
Wang QY and Nick P (1998) The auxin response of actin is altered in the rice mutant *Yin-Yang*. Protoplasma 204: 22-33
Went RW (1928) Wuchsstoff und Wachstum. Rec Trav Bot Neerl 25: 1-116
Williamson RE (1991) Orientation of cortical microtubules in interphase plant cells. Int Rev Cytol 129: 135-206
White RG, Badelt K, Overall RL and Vesk M (1994) Actin associated with plasmodesmata. Protoplasma 180: 169-184

Chapter 28

ACTIN AND MYOSIN IN PLASMODESMATA

Robyn L Overall, Rosemary G White[1,2], Leila M Blackman and Janine E Radford[2]

School of Biological Sciences, University of Sydney, NSW, 2006, Australia
1 Current address: CSIRO Plant Industry, Canberra, ACT, 2601 Australia
2 Department of Biological Sciences, Monash University, Melbourne, Victoria, 3800, Australia

Key words: actin, cell-to-cell communication, centrin, intercellular transport, myosin, plasmodesmata, plants, regulation, ultrastructure

Abstract: The motile proteins actin and myosin appear to be integral components of plasmodesmata, the channels through plant cell walls that provide a direct link between the cytoplasm of adjacent cells. Antibodies to both actin and myosin label the entire length of plasmodesmata in lower and higher plants. In combination with ultrastructural evidence, we speculate that actin could be arranged as a spiral around the central strand of endoplasmic reticulum (ER) within each plasmodesma, and myosin may form the observed links between the ER and the surrounding tube of plasma membrane that is also continuous from cell to cell. Inhibitor studies suggest that both proteins are important in maintaining the structural integrity of plasmodesmata. Actin and myosin disruptors also affect intercellular transport in some species, implying that actomyosin contraction may regulate transport either at the plasmodesma neck or by altering bulk flow from cell to cell. These findings suggest that, together with other contractile proteins, actin and myosin play a key role in regulating plasmodesma structure and intercellular transport in plants.

1. CELL-CELL COMMUNICATION VIA PLASMODESMATA

Plasmodesmata are plasma–membrane–lined channels that cross the cell wall, providing a pathway for transport between the cytoplasm of neighbouring plant cells. They contain a central element of endoplasmic reticulum (ER) known as the desmotubule (Figs. 1a,b) which connects the ER of neighbouring cells (Overall et al., 1982). The structure and regulation

C.J. Staiger et al. (eds.), Actin: A Dynamic Framework for Multiple Plant Cell Functions, 497–515.

of these channels have been the subject of several recent reviews (Lucas et al., 1993; Overall & Blackman 1996; McLean et al., 1997; Ding, 1998, Ding et al., 1999; Lucas, 1999; Overall, 1999; van Bel et al., 1999). Until recently, plasmodesmata were considered to be simple conduits for the passive transport of small molecules up to 1 kDa (Robards & Lucas, 1990). However, recent advances have shown that plasmodesmata are highly dynamic structures which under certain conditions, allow the intercellular transport of much larger molecules (Lucas et al., 1993; Ding, 1999; Lucas, 1999). Despite these recent findings, we know little about the molecular architecture of plasmodesmata, nor the exact mechanism of transport through them, or how this transport is regulated.

Transport through plasmodesmata and its dynamic regulation might be expected to involve motility or contraction generated by the cytoskeleton. In this chapter, we discuss the evidence showing that cytoskeletal proteins form an integral part of plasmodesma structure and that these proteins may be involved in intercellular transport or its regulation.

2. ACTIN AND MYOSIN ARE FOUND IN PLASMODESMATA

Actin bundles labelled with the specific F-actin stain, rhodamine-phalloidin, occasionally focus on pit fields in epidermal tissue, and faint strands can be seen crossing cell walls that contain only individual plasmodesmata not concentrated in pit fields (White et al., 1994). Pit fields themselves are sometimes brightly stained by rhodamine-phalloidin (Fig. 2a) (Hush & Overall, 1992; White et al., 1994). Antibodies to animal or plant actin label a single band at approximately 43 kDa on Western blots of whole plant extracts (Metcalf et al., 1980; White et al., 1994; Blackman & Overall, 1998). These anti-actin antibodies labeled plasmodesmata in young and mature cell walls of *Hordeum* (Figs. 2b,c), *Nicotiana* (White et al., 1994), *Zea* (Reichelt et al., 1999) and *Chara* (Figure 2d) (Blackman & Overall, 1998) at the transmission electron microscopy (TEM) level.

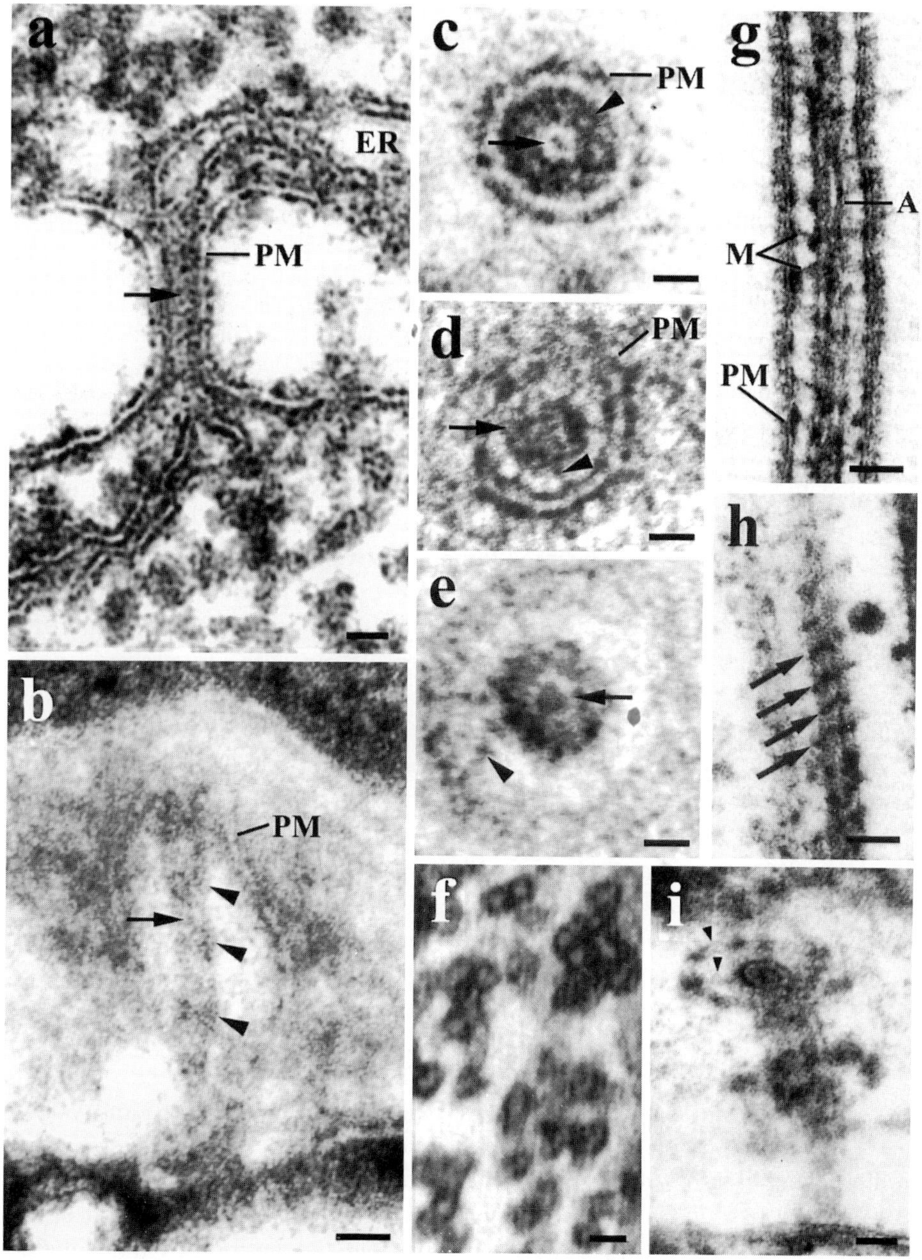

Figure 1. Electron micrographs comparing plasmodesmata with actin and brush border microvillae. (a-b) Longitudinal images of plasmodesmata showing the plasma membrane (PM), endoplasmic reticulum (ER) and desmotubule (arrow). The desmotubule is surrounded by a spiral of electron dense material (arrowheads). (a) Azolla root conventionally fixed in the presence of tannic acid (from Overall et al., 1982). Scale bar = 25 nm. (b) Freeze substituted

barley root. (R. G. White, unpublished) Scale bar = 50 nm. (c-d) Transverse sections of plasmodesmata delimited by the plasma membrane and containing a desmotubule (arrow). The desmotubule is surrounded by a 'mottled layer' of electron-lucent particles which are particularly clear in (c) (arrowhead). In (d-e), an electron-lucent cytoplasmic lumen is seen between the 'mottled layer' and the plasma membrane (PM). This region is traversed by electron-dense spokes (arrowheads). (c) Azolla root (from Overall et al., 1982). Scale bar = 10 nm. (d) Barley root (S. J. Brett and R. G. White, unpublished). Scale bar = 10 nm. (e) *Egeria densa* leaves (supplied by J. E. Radford). Scale bar = 10 nm. (f) Cross section of a bundle of negatively stained actin filaments (from Maciver et al., 1991). Scale bar = 20 nm. (g) Longitudinal section through a brush border microvillus. The core bundle of actin filaments (A) is connected laterally to the plasma membrane (PM) by myosin cross filaments (M). (h) Longitudinal section through a de-membranated brush border microvillus in which the cross filaments retain their striped pattern. (g-h) From Matsudaira and Burgess (1982). Scale bars = 50 nm. (i) Oblique section of plasmodesma with extracellular filaments (arrowheads) and particles, fixed in the presence of tannic acid (from Badelt et al., 1994). Scale bar = 25 nm.

Since plasmodesmata were labeled along their entire length and at the neck where they open into the cytoplasm, we conclude that actin most probably lines the cytoplasmic lumen between the desmotubule and surrounding plasma membrane, linking the cytoskeleton of adjacent cells. Actin integrity and antigenicity is compromised during preparation for TEM, and these results were obtained only after all measures to preserve actin (omission of osmium, minimal chemical fixation or freeze-fixation, low temperatures during polymerization) were taken.

Antibodies to animal and plant myosin label bands in whole plant extracts consistent with either complete myosin or its breakdown products (Parke et al., 1986; Qiao et al., 1994; Miller et al., 1995; Blackman & Overall, 1998; Radford & White, 1998; Reichelt et al., 1999). Although myosin was extremely difficult to preserve intact for Western blot analysis (Radford & White, 1998), the antigenic sites targeted by the antibodies used appear relatively robust, and plasmodesmata can be stained by both immunofluorescence and immuno-EM (Blackman & Overall, 1998; Radford & White, 1998; Reichelt et al., 1999). Immunofluorescence images show intense labelling of pit fields (Figs. 3a-d) (Radford & White, 1998; Reichelt et al., 1999) and faint labeling of single plasmodesmata (Radford and White, 1998) by antibodies to animal or plant myosins. TEM images revealed that antibodies against myosin labeled the entire length of plasmodesmata (Figs. 3e-h) (Blackman & Overall, 1998; Radford & White, 1998; Reichelt et al., 1999). Radford and White (1998) showed that the antibodies that labeled plasmodesmata also labeled the surface of the ER, strands of cytoplasm and other organelle surfaces in cells, in locations where myosin would be expected (Grolig et al., 1988; Qiao et al., 1994; Miller et al., 1995). Myosin antibodies also labeled the modified central cavities of mature plasmodesmata in Chara cell walls (Figure 3h), indicating that, as for actin, the myosin cytoskeleton appears to be continuous from cell to cell in lower and higher plants.

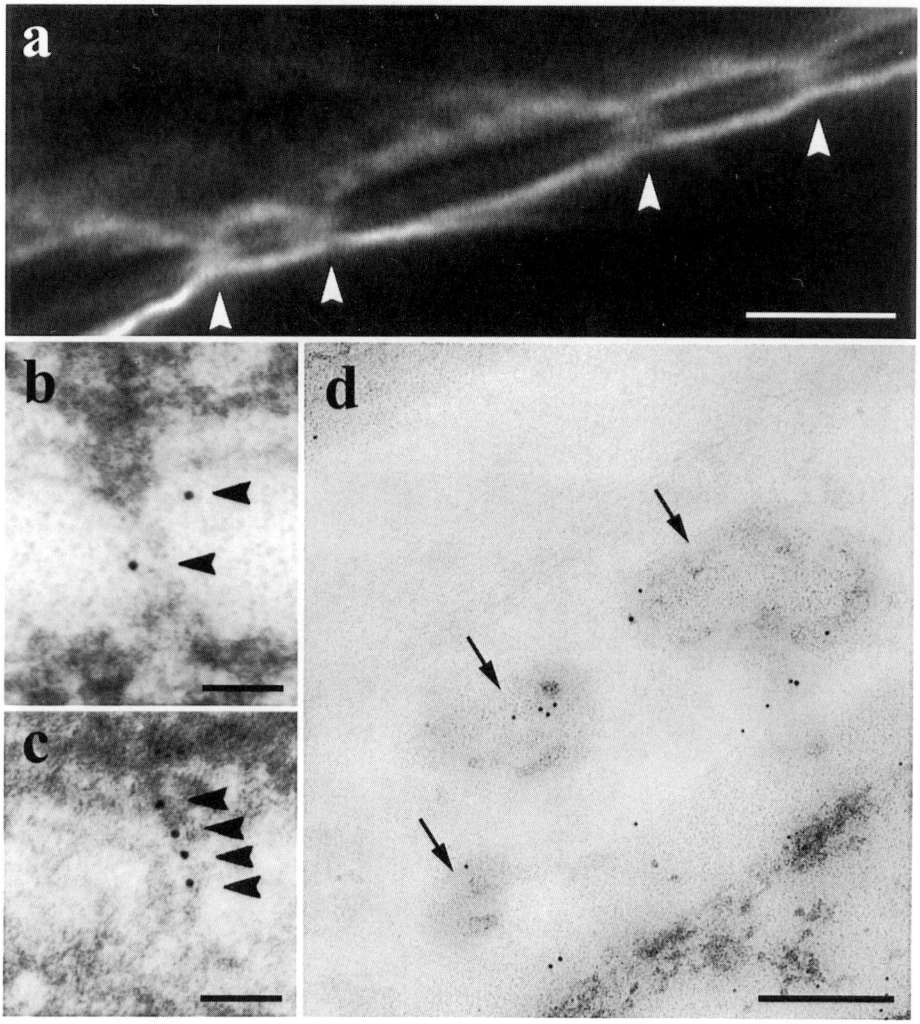

Figure 2. Localisation of actin to plasmodesmata. (a) Confocal laser scanning microscope images of inner epidermal peels of *Hordeum vulgare* stained with rhodamine-phalloidin, showing actin filaments and fluorescent pit fields (arrowheads). Scale bar = 5 μm (from White et al., 1994). (b-d) Electron micrographs showing labelling of plasmodesmata by anti-actin antibodies. (b-c) Gold label (arrowheads) on plasmodesmata from *Hordeum vulgare*. Scale bars = 50 nm. (from White et al., 1994). (d) Plasmodesmata branches (arrows) from *Chara corallina*. (L. M. Blackman, unpublished) Scale bar = 100 nm.

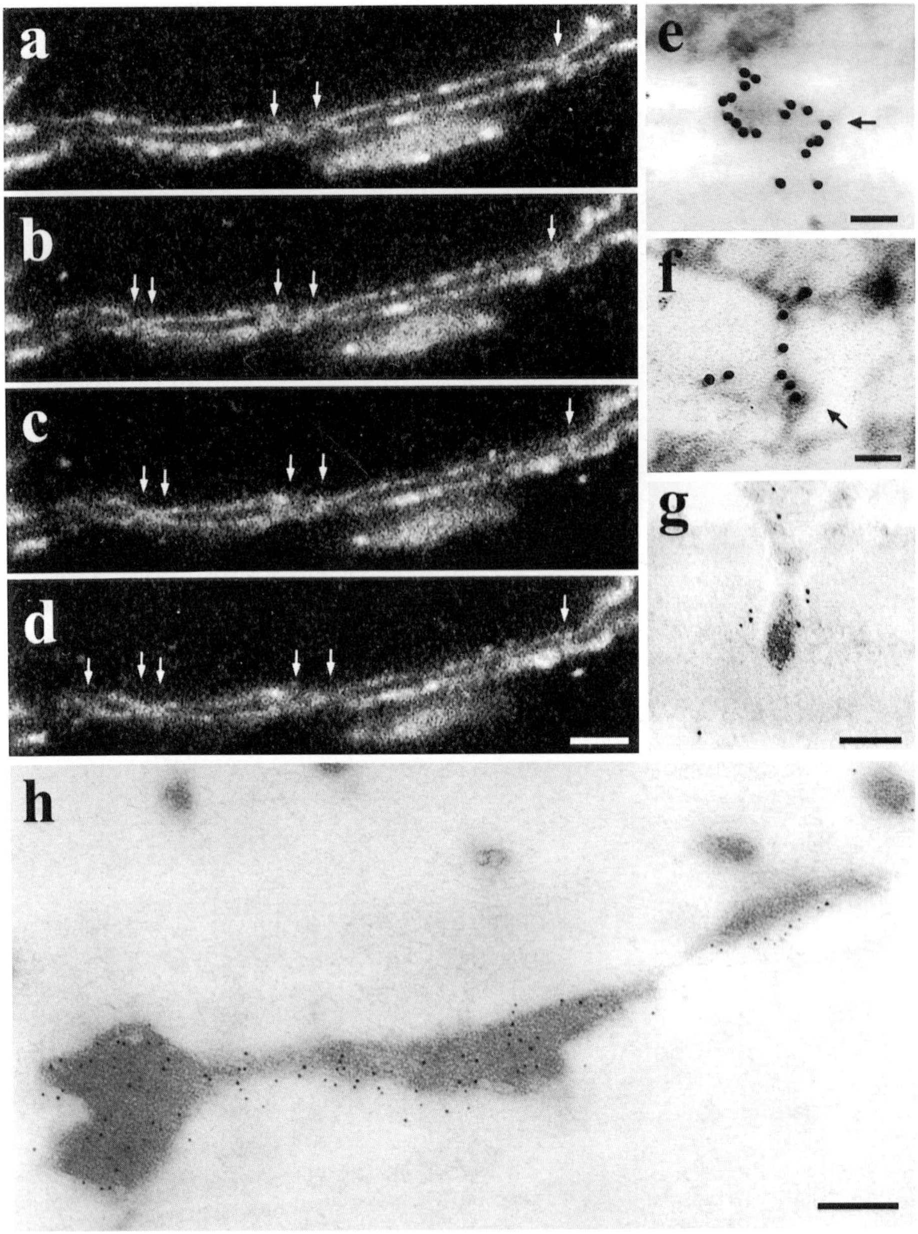

Figure 3. Immunolocalisation of myosin to plasmodesmata. (a-d) Adjacent optical sections of a cell wall of *Zea mays* immunofluorescently labelled with anti-myosin antibodies, showing fluorescence traversing the wall at pit-fields (arrow). Scale bar = 3 μm. (e-h) Immuno-EM of anti-myosin antibodies on plasmodesmata. (e-f) *Z. mays* plasmodesmata (arrows). Scale bar = 50 nm (a-f from Radford & White, 1998). (g) Immature plasmodesmata from *Chara corallina*. Scale bar = 100 nm. (h) Mature plasmodesmata from *Chara corallina* showing labelled middle cavity (Blackman, unpublished). Scale bar = 100 nm.

Fifteen families of myosins have been identified so far (Titus & Gilbert, 1999) but only some of these have been identified in plants (Knight and Kendrick-Jones, 1993; Kinkema & Schiefelbein, 1994; Kinkema et al., 1994; Yamamoto et al., 1999). Interestingly, the antibodies raised against the tail region of a specific plant myosin VIII labeled only the plasma membrane and plasmodesmata but no other structures in the cells (Reichelt et al., 1999). In addition, the antibody used to localize myosin in Chara plasmodesmata (Blackman & Overall, 1998) probably recognizes myosin V (Plazinski et al., 1997). Since different types of myosin are involved in different functions (Baker & Titus, 1998), the identification of the types of myosin in plasmodesmata will be critical in determining the function of actin and myosin in cell-to-cell communication.

3. ARRANGEMENT OF ACTIN AND MYOSIN IN PLASMODESMATA

The large size of primary and secondary antibodies, each of which is about 8 nm (Roth, 1982), means that immuno-EM does not have sufficient resolution to localize actin or myosin to specific structures seen in ultrastructural images of plasmodesmata. Therefore, the exact location and arrangement of actin and myosin can only be speculated upon.

One approach to this speculation is to compare the staining patterns and dimensions of components of the plasmodesmata with that for actin and myosin. Surrounding the desmotubule is a 'mottled layer' which may abut directly onto the plasma membrane (Figs. 1a,c) or there may be an electron lucent region between this 'mottled layer' and the cell membrane (Figs. 1 b,d,e). This electron-lucent region may contain electron-dense spokes (Burgess, 1971; Tilney et al., 1991; Ding et al., 1992; Schulz, 1995; Cook et al., 1997) connecting the 'mottled layer' to the cell membrane (Figs. 1d,e). These images have been variously interpreted (for review see van Bel, 1999), but one interpretation outlined in Overall (1999) is that the 'mottled layer' consists of negatively stained electron lucent particles arranged in a spiral (Fig. 1b). This spiral is connected to the cell membrane via electron-dense spokes traversing the electron-lucent region or cytoplasmic lumen.

Actin filaments that have been fixed in the presence of tannic acid appear in cross section as electron-lucent particles (Fig. 1f) (Maciver et al., 1991). These particles are of similar dimensions and staining patterns to the electron-lucent particles in the 'mottled layer' of plasmodesmata in tissue fixed in the presence of tannic acid (compare Figs. 1c and 1f). The negatively stained actin in Figure 1f is a little larger than a pure actin filament as it has been co-polymerised with alpha actinin and actaphorin.

This suggests that the spiral of electron-lucent particles around the desmotubule is actin. In transverse sections of plasmodesmata, there is usually only one site per plasmodesma labeled with the anti-actin antibody (White et al, 1994). Presumably, only one antigenic site was at the surface of the section, as would be expected if it were in a spiral arrangement.

Similarly, the electron dense spokes in the plasmodesmata are also comparable in appearance and dimensions to brush border myosin I (BBM) in epithelial microvilli (compare Figs. 1d,e and 1g). Here, actin is bundled to form a stable structural element in the microvillus, and around this bundle is a 'finger' of cell membrane which is anchored to the actin via regularly spaced BBM molecules (Fig. 1g) (Matsudaira & Burgess, 1982). When the membrane has been removed, the myosin retains a transverse striping pattern around the core of actin (Fig. 1h), reminiscent of the patterning around the desmotubule (Fig. 1b). The electron-dense spokes in plasmodesmata may well be myosin.

There are other filamentous structures associated with plasmodesmata that could also comprise elements of the cytoskeleton. For example, there is an electron dense connection between the plasma membrane and the endoplasmic reticulum around the neck region of plasmodesmata (Overall et al., 1982) and a variety of electron-dense filaments and particles surrounding the plasmodesmata in material treated with tannic acid (Fig. 1i) (Badelt et al., 1994). The position of these structures suggests that they could be involved in contraction to regulate intercellular transport pathways so that the involvement of actin or myosin in these structures is also a possibility.

4. ROLE OF ACTIN AND MYOSIN IN MAINTENANCE OF STRUCTURE OF PLASMODESMA

One possible role for actin and myosin in plasmodesmata could be similar to that of actin and its associated BBM in the intestinal epithelium of animal cells (Matsudaira & Burgess, 1982), that is, maintenance of structure. Nevertheless, this myosin retains some dynamic behaviour in that calcium, and other regulatory molecules, can alter the attachment to actin and the tightness of binding to the plasma membrane (for review see Hammer, 1994). There are several other animal myosins whose main role in cells is to provide structural integrity, such as ninaC (Hicks et al., 1996), and similarly, both actin and myosin may serve to maintain plasmodesma structure. Plant myosins are generally small compared to the large, motile myosin II of animal cells, and some of these may well have a primarily structural role, as is postulated for the unconventional small myosin VIII (Reichelt et al.,1999).

A structural role is further supported by the effects of the actin-disrupter, cytochalasin D, which caused plasmodesmata in the fern, *Nephrolepis exultata*, to open wide at their necks and to lose their extracellular structures, although little effect was seen in plasmodesmata from higher plants (White et al., 1994). One functional approach to assessing structural integrity is to determine the largest size of molecules that can freely move through plasmodesmata, termed the size exclusion limit (SEL). Ding et al. (1996) showed that in tobacco mesophyll cells, application or injection of the actin antagonists, cytochalasin D or profilin, increased the SEL of plasmodesmata, pointing to a role for actin in maintaining structural integrity. Further evidence for the role of actin and myosin in maintaining structure and control over the SEL comes from studies in which ATP levels have been experimentally altered, leading to a change in the SEL (Reid & Overall, 1992; Tucker, 1993; Cleland et al., 1994). A myosin antagonist, 2,3-butanedione monoxime, caused plasmodesma closure rather than opening, in *Zea mays* and *Allium cepa* (Radford & White, 1998). Tilney et al. (1991) and Turner et al. (1994) observed drastic loss of structure in plasmodesmata treated with proteases, emphasizing the role of proteins in maintenance of plasmodesma structure.

5. ROLE OF ACTIN AND MYOSIN IN INTERCELLULAR TRANSPORT

Intercellular transport is thought to occur mainly through the cytoplasmic lumen between the desmotubule and the plasma membrane, but the ER lumen and membrane are also possible routes (Overall, 1999). Actin and myosin may be involved in orchestrating or regulating this transport through plasmodesmata. Certainly, components necessary for a functional actomyosin system, such as adenosine triphosphate (ATP) and calcium, appear to be involved. ATPase activity has been localized at plasmodesmata (van Steveninck, 1976; Belitser et al., 1982; Franceschi & Lucas, 1982; Nougarède et al., 1985; Chauhan et al., 1991). In barley roots, this ATPase activity has been shown to be calcium activated (Belitser et al., 1982). Higher than normal intracellular levels of calcium will halt intercellular movement (Erwee & Goodwin, 1983, Tucker, 1990; Lew, 1994; Holdaway-Clarke et al., 2000).

5.1 Non-selective transport in the cytoplasmic lumen

Transport through plasmodesmata is defined as non-selective when substances below the normal SEL pass without requiring any special transport mechanism or modulation of the plasmodesma structure (for review see Schulz, 1999). The non-selective transport can vary between species, tissue types and developmental stages. For example, while the usual SEL for plasmodesmata is in the order of 1 kDa (Robards & Lucas, 1990), free green fluorescent protein (GFP), which is a 27 kDa protein (Imlau et al., 1999: Oparka et al., 1999), and GFP fusion proteins as large as 50 kDa (Oparka et al., 1999), can move from cell to cell in sink tissue.

Transport may take place by diffusion through the cytoplasmic lumen in the plasmodesmata. In *Chara*, for example, the cell-to-cell movement of radioactive chloride (Bostrom & Walker, 1975) and rubidium (Ding & Tazawa, 1989) occurs by diffusion. Disruption of actin filaments by cytochalasin E in *Chara* does not inhibit the cell-to-cell diffusion of rubidium (Ding & Tazawa, 1989).

However, it is also possible that there is an actomyosin-powered bulk flow through the plasmodesmata, essentially allowing for cytoplasmic streaming to be continuous between adjacent cells. In support of this idea is that movement of carbon isotopes in *Nitella flexilis* (Zawadzki & Fensom, 1986a,b) and *N. translucens* occurs by an active mechanism (Dale et al., 1983). Carbon transport rates in *Chara* and *N. translucens* show a polarity (Dale et al., 1983; Ding et al., 1991) which is not explained by differences in photosynthetic activity (Ding et al., 1991). Treatments which decrease the ATP concentration inhibit both cytoplasmic streaming and cell-to-cell transport in *N. flexilis* (Zawadzki & Fensom, 1986a) and *Chara* (Reid & Overall, 1992). The disruption of streaming in *N. translucens* by N-ethylmaleimide, which inhibits the action of myosin, also stops cell-to-cell movement of radioactive carbon tracers (Dale et al., 1983). Cell-to-cell transport in *S. purpurea* is inhibited by the secondary messengers IP_2 and IP_3 (Tucker, 1988). Interestingly, the injection of CaBAPTA, a calcium-loaded calcium chelator, stops cytoplasmic streaming in adjacent cells and cell-to-cell transport (Tucker, 1990). These findings predict that actin and myosin may be involved in the intercellular transport in some species.

A final possibility is that since the actomyosin system is contractile, perhaps it could play a role in pumping materials through plasmodesmata, maybe even against a concentration gradient. If pumping were co-ordinated at each end, a peristaltic type of contraction could force cytoplasmic contents from one cell to the next. There is some evidence for asymmetry in the electrical resistance of plasmodesmata (Overall & Gunning, 1982), but dye injection studies have, so far, provided no evidence for asymmetric transport of material from one tissue or cell to the next.

5.2 Selective transport in the cytoplasmic lumen

Transport of large molecules through plasmodesmata, made possible only by modulation of the structure of plasmodesmata or a specific transport mechanism, is defined as selective (for review see Schulz, 1999).

One such example may occur in the plasmodesmata between companion cells and sieve elements in the phloem. These plasmodesmata appear to allow the movement of specific proteins into the enucleate sieve element that have no protein synthesis machinery (Fisher et al, 1992). Up to 200 different proteins, with molecular sizes between 10 and 200 kDa, including actin and profilin (Schobert et al., 1998), have been found in phloem exudates (Fisher et al., 1992; Sakuth et al., 1993; Nakamura et al., 1993; Balachandran et al., 1997). The sucrose-transporter protein, SUT1, and *SUT1* mRNA are also found in the sieve element, indicating that both the protein and mRNA undergo cell-to-cell transport into the sieve element (Kuhn et al., 1997).

Complex molecular interactions occur between viruses and plasmodesmata during cell-to-cell (for review see Reichel et al., 1999; also see McLean & Zambryski, this volume) and long-distance movement of plant viruses (for review see Carrington et al., 1996; Gilbertson & Lucas, 1996). Viruses such as tobacco mosaic virus (TMV) temporarily modify plasmodesmata (Oparka et al., 1997) and move from cell to cell in an uncapsulated form. Essential for cell to cell movement of TMV is a virus-encoded 30 kDa movement protein (MP) (Deom et al., 1987). The MP increases the SEL (Citovsky & Zambryski, 1991), binds to the TMV mRNA (Citovsky et al., 1990) and chaperones unfolded viral RNA through the enlarged plasmodesma (Citovsky et al., 1992). Recent findings show that these functions are not restricted to viral MPs. For example, a plant protein which can modify plasmodesmata and transport sense and anti-sense mRNA has been identified in *Cucurbita maxima* (Xoconostile-Cázares et al., 1999). Interestingly, the TMV MP also co-localises with actin filaments (McLean et al., 1995) and microtubules (Heinlein et al., 1995; McLean et al., 1995), indicating a role for the cytoskeleton in cell-to-cell transport (also see McLean & Zambryski, this volume).

During development, regulation of the permeability of plasmodesmata occurs, creating symplastic domains (Goodwin & Lyndon, 1983; Duckett et al., 1994; Rinnie et al., 1998; Gisel et al., 1999) and presumably triggering organ development in a similar manner to that seen in animal systems (Kalimi & Lo, 1988). Also critical to normal development is non-cell autonomous action of some transcription factors (for review see Jackson & Hake, 1997). The shoot meristem consists of three distinct layers, the epidermal layer, the subepidermal layer and the inner core (Szymkowiak & Sussex, 1992; Carpenter & Coen, 1995). Through the use of chimeras, it has been shown that signals from one layer can affect the development of the other layers (Szymkowiak & Sussex, 1992). For example, the 45 kDa transcription factor, KNOTTED1, is found in all layers of the meristem, however the mRNA is only ever found in the inner core of cells (Jackson et al., 1994). Microinjected fluorescently labelled KNOTTED1 protein has recently been shown to increase the SEL of plasmodesmata and catalyse its own movement, and that of the *KNOTTED1* mRNA, through the plasmodesmata (Lucas et al., 1995; Mezitt & Lucas, 1996). Other examples of non-cell automonous acting transcription factors include the Antirrhinum floral homeotic genes *FLORICAULA* (Carpenter & Coen, 1995; Hantke et al., 1995), *DEFICIENS* and *GLOBOSA*, the Arabidopsis trichome patterning gene, *TRIPTYCHON* (Schnittger et al., 1999) and *LIGULELESS-1* from maize (Becraft et al., 1990). Central to our understanding of plant development will be the exploration of plasmodesmata function and regulation.

We propose that actin and myosin may be directly responsible for the specific movement of viral and plant proteins and mRNA. In yeast (Long et al., 1997; Takizawa et al., 1997) and *Caenorhabditis elegans* (Guo & Kemphues, 1996), actin is directly responsible for the correct distribution of mRNA and hence for subsequent developmental events.

5.3 Transport via the endoplasmic reticulum

In general, the ER component of plasmodesmata, as viewed in TEM images, is tightly constricted and is only rarely seen dilated (for example, see Fig. 1e, Overall & Blackman, 1996). Most studies have assumed that intercellular transport would not occur through this compartment (Gunning & Overall, 1983). However, a recent micro-injection study has shown that the ER lumen can provide another intercellular transport pathway. Fluorescent probes up to 3 kDa injected into the ER of tobacco and 10 kDa into Torenia stem epidermal cells can move into the ER and nucleus of the adjacent cell (Cantrill et al., 1999). In contrast, probes of 3 kDa and above will not move into the adjacent cell if the injection is cytoplasmic. In addition, Gamalei et al., (1994) have shown that a contraction of the ER/plasmodesmata system of the intermediary cells is associated with decreased phloem loading at low temperatures.

Studies using fluorescent lipid and phospholipid analogs have shown that the ER membrane can also form a dynamic pathway through plasmodesmata (Grabski et al., 1993). In addition, fusion proteins of GFP and MPs from TMV (Más & Beachy, 1998; Reichel & Beachy, 1998) and alfalfa mosaic virus (AMV) (Huang & Zhang, 1999) localize with the ER. The biochemical studies of the MPs of AMV and TMV indicate that they behave as integral membrane proteins (Moore et al., 1992; Schaad et al., 1997; Huang &

Zhang, 1999). Perhaps the ER membrane provides a pathway for intercellular movement of these proteins.

These intercellular movements via the lumen and membrane of the ER, may be simply diffusive. However, it is possible that there is some selective filtering within the ER lumen. Indeed, an antibody against calreticulin, a calcium-sequestering protein found specifically in the ER lumen, is localised to plasmodesmata, whereas antibodies against two other ER proteins, the HDEL retention signal and BiP (immunoglobulin binding protein), did not localize to plasmodesmata (Baluška et al., 1999).

It is a tantalizing possibility that the ER may actually move through plasmodesmata. In the cytoplasm, actin (Goosen-de Roo et al., 1983; Quader et al., 1987; Lichtscheidl & Url, 1990) and myosin (Grolig et al., 1988; Qiao et al., 1994), which are intimately associated with the ER, act together to provide constant rearrangement and movement of the ER (Grolig et al., 1988; Kachar & Reese, 1988; Knebel et al., 1990). If ER does move through plasmodesmata, which way would it travel? Is all of the actin in a plasmodesma oriented parallel, and is the orientation within a cell wall aligned? Perhaps the recently characterized myosin VI, which moves contrary to other myosins (Wells et al., 1999) would allow for bidirectional transport. Before we can answer these questions, we need to determine whether actin or myosin antagonists affect this ER trafficking. In addition, intercellular movement of the ER would need to be reconciled with the suggestion that ER can be divided into a number of domains, including that within the plasmodesmata (Staehelin, 1997).

6. OTHER COMPONENTS OF THE CYTOSKELETON AT PLASMODESMATA

Centrin has been localised to the cytoplasmic opening of individual plasmodesmata in a number of species and is particularly prominent in the forming cell plate when primary plasmodesmata are forming in the new cell wall (Blackman et al., 1999). Centrin may be responsible for regulation at the neck region, since it contracts in response to calcium and relaxes in response to ATP. This work correlates with several studies showing effects of ATP on cell-to-cell transport. At present, it is unclear how this mechanism would be coordinated with a structural or functional role for actin and myosin.

Interestingly, tubulin was found in protein extracts of walls containing plasmodesmata and not in extracts from walls without plasmodesmata from *Chara* (Blackman & Overall, 1998). It may be that microtubules are involved in bringing macromolecules such as viruses (Heinlein et al., 1995; McLean et al., 1995) to plasmodesmata where actin and myosin are involved in the movement of molecules through plasmodesmata. Such systems, where microtubules provide long-distance transport and actin provides short-range movement, have been identified in the transport of vesicles in animal cells (Langford, 1995; Allan & Schroer, 1999).

7. CONCLUDING REMARKS

We conclude that, although the evidence is only beginning to emerge, actin, myosin and their associated regulatory and structural proteins, will prove to be essential components of many types of plasmodesmata. It now remains to clarify what role these proteins play in the operation of plasmodesmata.

REFERENCES

Allan VJ and Schroer TA (1999) Membrane motors. Curr Opin Cell Biol 11: 476-482

Badelt K, White RG, Overall RL and Vesk M (1994) Ultrastructural specializations of the cell wall sleeve around plasmodesmata. Amer J Bot 81: 1422-1427

Baker JP and Titus MA (1998) Myosins: Matching functions with motors. Curr Opin Cell Biol 10: 80-86

Balachandran S, Xiang Y, Schobert C, Thompson GA and Lucas WJ (1997) Phloem sap proteins from *Cucurbita maxima* and *Ricinus communis* have the capacity to traffic cell to cell through plasmodesmata. Proc Natl Acad Sci USA 94: 14150-14155

Baluška F, Šamaj J, Napier R and Volkmann D (1999) Maize calreticulin localizes preferentially to plasmodesmata in root apex. Plant J 19: 481-488

Becraft PW, Bongard-Pierce DK, Sylvester AW, Poethig RS, Freeling M, Becraft PW and Freeling M (1990) The *liguleless-1* gene acts tissue specifically in maize leaf development. Dev Biol 141: 220-232

Belitser NV, Zaalishvili GV and Sytnianskaja NP (1982) Ca^{2+}-binding sites and Ca^{2+}-ATPase activity in barley root tip cells. Protoplasma 111: 63-78

Blackman LM and Overall RL (1998) Immunolocalization of the cytoskeleton to plasmodesmata of *Chara corallina*. Plant J 14: 733-741

Blackman LM, Harper JDI and Overall RL (1999) Localization of a centrin-like protein to higher plant plasmodesmata. Eur J Cell Biol 78: 297-304

Bostrom TE and Walker NA (1975) Intercellular transport in plants. I. The rate of transport of chloride and the electric resistance. J Exp Bot 95: 767-782

Burgess J (1971) Observations on structure and differentiation in plasmodesmata. Protoplasma 73: 83-95

Cantrill LC, Overall RL and Goodwin PB (1999) Cell-to-cell communication via plant endomembranes. Cell Biol Int 23: in press

Carpenter R and Coen ES (1995) Transposon induced chimeras show that *floricaula*, a meristem identity gene, acts non-autonomously between cell layers. Development 121: 19-26

Carrington JC, Kasschau KD, Mahajan SK and Schaad MC (1996) Cell-to-cell and long-distance transport of viruses in plants. Plant Cell 8: 1669-1681

Chauhan E, Cowan DS and Hall JL (1991) Cytochemical localization of plasma membrane ATPase activity in plant cells. Protoplasma 165: 27-36

Citovsky V, Knorr D, Schuster G and Zambryski P (1990) The P30 movement protein of tobacco mosaic virus is a single-strand nucleic acid binding protein. Cell 60: 637-647

Citovsky V, Wong ML, Shaw AL, Prasad BVV and Zambryski P (1992) Visualization and characterisation of tobacco mosaic virus movement protein binding to single-stranded nucleic acids. Plant Cell 4: 397-411

Citovsky V and Zambryski P (1991) How do plant virus nucleic acids move through intercellular connections? BioEssays 13: 373-379

Cleland RE, Fujiwara T and Lucas WJ (1994) Plasmodesmal mediated cell to cell transport in wheat roots is modulated by anaerobic stress. Protoplasma 178: 81-85

Cook ME, Graham LE, Botha CEJ and Lavin CA (1997) Comparative ultrastructure of plasmodesmata of *Chara* and selected bryophytes: Toward an elucidation of the evolutionary origin of plant plasmodesmata. Amer J Bot 84: 1169-1178

Dale N, Lunn G and Fensom DS (1983) Rates of axial transport of ^{11}C and ^{14}C in Characean cells: Faster than visible streaming? J Exp Bot 34: 130-143

Deom CM, Oliver MJ and Beachy RN (1987) The 30-kilodalton gene product of tobacco mosaic virus potentiates virus movement. Science 237: 389-393

Ding B (1998) Intercellular protein trafficking through plasmodesmata. Plant Mol Biol 38: 279-310

Ding B, Itaya A and Woo Y.M (1999) Plasmodesmata and cell-to-cell communication in plants. Int Rev Cytol 190: 251-252

Ding B, Kwon M.-O and Warnberg L (1996) Evidence that actin filaments are involved in controlling the permeability of plasmodesmata in tobacco mesophyll. Plant J 10: 157-164

Ding B, Turgeon R and Parthasarathy M.V (1992) Substructure of freeze-substituted plasmodesmata. Protoplasma 169: 28-41

Ding D-Q, Amino S, Mimura T, Nagata T and Tazawa M (1991) Intercellular transport and subcellular distribution of photoassimilates in *Chara corallina*. J Exp Bot 42: 1392-1398

Ding D-Q and Tazawa M (1989) Influence of cytoplasmic streaming and turgor pressure gradient on transnodal transport of rubidium and electrical conductance in *Chara corallina*. Plant Cell Physiol 30: 739-748

Duckett CM, Oparka KJ, Prior DAM, Dolan L and Roberts K (1994) Dye-coupling in the root epidermis of *Arabidopsis* is progressively reduced during development. Development 120: 3247-3255

Erwee MG, and Goodwin PB (1983) Characterisation of *Egeria densa* Planch. leaf symplast. Inhibition of the intercellular movement of fluorescent probes by group II ions. Planta 158: 320-328

Fisher DB, Wu Y and Ku MSB (1992) Turnover of soluble proteins in the wheat sieve tube. Plant Physiol 100: 1433-1441

Franceschi VR and Lucas WJ (1982) The relationship between the charasome to chloride uptake in *Chara corallina*: Physiological and histochemical investigations. Planta 154: 525-537

Gamalei YV, van Bel AJE, Pakhomova M, V and Sjutkina AV (1994) Effects of temperature on the conformation of the endoplasmic reticulum and on starch accumulation in leaves with the symplasmic minor-vein configuration. Planta 194: 443-453

Gilbertson RL and Lucas WJ (1996) How do viruses traffic on the 'vascular highway'? Trends Plant Sci 1: 260-268

Gisel A, Barella S, Hempel FD and Zambryski PC (1999) Temporal and spatial regulation of symplastic trafficking during development in Arabidopsis thaliana apices. Development 126: 1879-1889

Goodwin PB and Lyndon RF (1983) Synchronisation of cell division during transition to flowering in Silene apices not due to increased symplast permeability. Protoplasma 116: 219-222

Goosen-de Roo L, Burggraaf PD and Libbenga KR (1983) Microtubule bundles associated with tubular endoplasmic reticulum in fusiform cells in the active cambial zone of *Fraxinus excelsior* L. Protoplasma 116: 204-208

Grabski S, de Feijter AW and Schindler M (1993) Endoplasmic reticulum forms a dynamic continuum for lipid diffusion between contiguous soybean root cells. Plant Cell 5: 25-38

Grolig F, Williamson RE, Parke J, Miller C and Anderton BH (1988) Myosin and Ca^{2+}-sensitive streaming in the alga *Chara*: Detection of two polypeptides reacting with a monoclonal anti-myosin and their localization in the streaming endoplasm. Eur J Cell Biol 47: 22-31

Gunning BES and Overall RL (1983) Plasmodesmata and cell-to-cell transport in plants. BioScience 33: 260-265

Guo S and Kemphues KJ (1996) A non-muscle myosin required for embryonic polarity in *Caenorhabditis elegans*. Nature 382: 455-458

Hammer JA (1994) The structure and function of unconventional myosins: A review. J Muscle Res Cell Motil 15: 1-10

Hantke SS, Carpenter R and Coen ES (1995) Expression of *floricaula* in single cell layers of periclinal chimeras activates downstream homeotic genes in all layers of floral meristems. Development 121: 27-35

Heinlein M, Epel BL, Padgett HS and Beachy RN (1995) Interaction of tobamovirus movement proteins with the plant cytoskeleton. Science 270: 1983-1985

Hicks JL Liu XR and Williams DS (1996) Role of the ninaC proteins in photoreceptor cell structure–ultrastructure of ninaC deletion mutants and binding to actin filaments. Cell Motil Cytoskel 35: 367-379

Holdaway-Clarke TL, Walker NA, Hepler PK and Overall RL (2000) Physiological elevations in cytoplasmic free calcium by cold or ion injection result in transient closure of higher plant plasmodesmata. Planta 210: 329-335

Huang M and Zhang L (1999) Association of the movement protein of alfalfa mosaic virus with the endoplasmic reticulum and its trafficking in epidermal cells of onion bulb scales. Mol Plant Microbe Interact 12: 680-690

Hush JM and Overall RL (1992) Re-orientation of cortical F-actin is not necessary for wound-induced microtubule re-orientation and cell polarity establishment. Protoplasma 169: 97-106

Imlau A, Truernit E and Sauer N (1999) Cell-to-cell and long-distance trafficking of the green fluorescent protein in the phloem and symplastic unloading of the protein into sink tissues. Plant Cell 11: 309-322

Jackson D and Hake S (1997) Morphogenesis on the move: Cell-to-cell trafficking of plant regulatory proteins. Curr Opin Gen Dev 7: 495-500

Jackson D, Veit B and Hake S (1994) Expression of maize *KNOTTED1* related homeobox genes in shoot apical meristem predicts patterns of morphogenesis in the vegetative shoot. Development 120: 405-413

Kachar B and Reese TS (1988) The mechanism of cytoplasmic streaming in Characean algal cells: Sliding of endoplasmic reticulum along actin filaments. J Cell Biol 106: 1545-1552

Kalimi GH and Lo CW (1988) Communication compartments in the gastrulating mouse embryo. J Cell Biol 107: 241-255

Kinkema M and Schiefelbein J (1994) A myosin from a higher plant has structural similarities to class V myosins. J Mol Biol 239: 591-597

Kinkema M, Wang H and Schiefelbein J (1994) Molecular analysis of the myosin gene family in *Arabidopsis thaliana*. Plant Mol Biol 26: 1139-1153

Knebel W, Quader H and Schnepf E (1990) Mobile and immobile endoplasmic reticulum in onion bulb epidermis cells: Short- and long-term observations with a confocal laser scanning microscope. Eur J Cell Biol 52: 328-340

Knight AE and Kendrick-Jones J (1993) A myosin-like protein from a higher plant. J Mol Biol 231: 148-154

Kuhn C, Franceschi VR, Schulz A, Lemoine R and Frommer WB (1997) Macromolecular trafficking indicated by localization and turnover of sucrose transporters in enucleate sieve elements. Science 275: 1298-1300

Langford GM (1995) Actin- and microtubule-dependent organelle motors: interrelationships between the two motility systems. Curr Opin Cell Biol 7: 82-88

Lew RR (1994) Regulation of electrical coupling between Arabidopsis root hairs. Planta 193: 67-73

Lichtscheidl IK and Url WG (1990) Organisation and dynamics of cortical endoplasmic reticulum in inner epidermal cells of onion bulb scales. Protoplasma 157: 203-215

Long RM, Singer RH, Meng XH, Gonzalez I, Nasmyth K and Jansen RP (1997) Mating type switching in yeast controlled by asymmetric localization of *ASH1* mRNA. Science 277: 383-387

Lucas WJ (1999) Plasmodesmata and the cell-to-cell transport of proteins and nucleoprotein complexes. J Exp Bot 50: 979-987

Lucas WJ, Bouchepillon S, Jackson DP, Nguyen L, Baker L, Ding B and Hake S (1995) Selective trafficking of KNOTTED1 homeodomain protein and its mRNA through plasmodesmata. Science 270: 1980-1983

Lucas WJ, Ding B and van der Schoot C (1993) Plasmodesmata and the supracellular nature of plants. New Phytol 125: 435-476

McLean BG, Hempel FE and Zambryski PC (1997) Plant intercellular communication via plasmodesmata. Plant Cell 9: 1043-1054

McLean BG, Zupan J and Zambryski PC (1995) Tobacco mosaic virus movement protein associates with the cytoskeleton in tobacco cells. Plant Cell 7: 2101-2114

Maciver SK, Wachsstock DH, Schwarz WH and Pollard TD (1991) The actin filament severing protein actophorin promotes the formation of rigid bundles of actin filaments crosslinked with α-actinin. J Cell Biol 115: 1621-1628

Más P and Beachy RN (1998) Distribution of TMV movement protein in single living protoplasts immobilized in agarose. Plant J 15: 835-842

Matsudaira PT and Burgess DR (1982) Organization of the cross-filaments in intestinal microvilli. J Cell Biol 92: 657-664

Metfcalf TN, Szabo LJ, Schubert KR and Wang JL (1980) Immunochemical identification of an actin-like protein from soybean seedlings. Nature 285: 171-172

Mezitt LA and Lucas WJ (1996) Plasmodesmal cell-to-cell transport of proteins and nucleic acids. Plant Mol Biol 32: 251-273

Miller DD, Scordilis SP and Hepler PK (1995) Identification and localization of three classes of myosins in pollen tubes of *Lilium longiflorum* and *Nicotiana alata*. J Cell Sci 108: 2549-2563

Moore PJ, Fenczik CA, Deom CM and Beachy RN (1992) Developmental changes in plasmodesmata in transgenic tobacco expressing the movement protein of tobacco mosaic virus. Protoplasma 170: 115-127

Nakamura S, Hayashi H, Mori S and Chino M (1993) Protein phosphorylation in the sieve tubes of rice plants. Plant Cell Physiol 34: 927-933

Nougarède A, Landré P, Rembur J and Niebla Hernandez M (1985) Des variations d'activités de la 5'-nucléotidase et de l'adénylate-cyclase sont-elles des composantes de la levée d'inhibition du bourgeon cotylédonaire du pois? Can J Bot 63: 309-323

Oparka KJ, Roberts AG, Boevink P, Santa Cruz S, Roberts L, Pradel KS, Imlau A, Kotlizky G, Sauer N and Epel B (1999) Simple but not branched plasmodesmata allow the nonspecific trafficking of proteins in developing tobacco leaves. Cell 97: 743-754

Oparka KJ, Prior DAM, Santa Cruz S, Padgett HS and Beachy RN (1997) Gating of epidermal plasmodesmata is restricted to the leading edge of expanding infection sites of tobacco mosaic virus (TMV). Plant J 12: 781-789

Overall RL (1999) Substructure of plasmodesmata. In: van Bel AJE and van Kesteren WJP (eds) Plasmodesmata: Structure, Function, Role in Cell Communication. Berlin: Springer-Verlag, pp 129-148

Overall RL and Blackman LM (1996) A model of the macromolecular structure of plasmodesmata. Trends Plant Sci 1: 307-311

Overall RL and Gunning BES (1982) Intercellular communication in Azolla roots: II. Electrical coupling. Protoplasma 111: 151-160

Overall RL, Wolfe J and Gunning BES (1982) Intercellular communication in Azolla roots. I. Ultrastructure of plasmodesmata. Protoplasma 111: 134-150

Parke J, Miller C and Anderton BH (1986) Higher plant myosin heavy-chain using a monoclonal antibody. Eur J Cell Biol 41: 9-13

Plazinski J, Elliott J, Hurley UA, Burch J, Arioli T and Williamson RE (1997) Myosins from angiosperms ferns and algae. Amplification of gene fragments with versatile PCR primers and detection of protein products with a monoclonal antibody to a conserved head epitope. Protoplasma 196: 78-86

Qiao L, Jablonsky PP, Elliott J and Williamson RE (1994) A 170 kDa polypeptide from mung bean shares multiple epitopes with rabbit skeletal myosin and binds ADP-agarose. Cell Biol Int 18: 1035-1047

Quader H, Hofmann A and Schnepf E (1987) Shape and movement of the endoplasmic reticulum in onion bulb epidermis cells: possible involvement of actin. Eur J Cell Biol 44: 17-26

Radford JE and White R.G (1998) Localization of a myosin-like protein to plasmodesmata. Plant J 14: 743-750

Reichel C and Beachy RN (1998) Tobacco mosaic virus infection induces severe morphological changes of the endoplasmic reticulum. Proc Natl Acad Sci USA 95: 11169-11174

Reichel C, Más P and Beachy RN (1999) The role of the ER and cytoskeleton in plant viral trafficking. Trends Plant Sci 4: 458-462

Reichelt S, Knight AE, Hodge TP, Baluška F, Šamaj J, Volkmann D and Kendrick-Jones J (1999) Characterization of the unconventional myosin VIII in plant cells and its localization oat the post-cytokinetic cell wall. Plant J 19: 555-567

Reid RJ Overall RL (1992) Intercellular communication in *Chara*: Factors affecting transnodal electrical resistance and solute fluxes. Plant Cell Environ 15: 507-517

Rinne PLH and van der Schoot C (1998) Symplasmic fields in the tunica of the shoot apical meristem coordinate morphogenetic events. Development 125: 1477-1485

Robards AW and Lucas WJ (1990) Plasmodesmata. Annu Rev Plant Physiol Plant Mol Biol 41: 369-419

Roth J (1982) The protein A-gold (pAg) technique: A quantitative approach for antigen localization on thin sections. In: Bullock GR and Perutz P (eds) Techniques in Immunocytochemistry. New York: Academic Press, pp 15-57

Sakuth T, Schobert C, Pecsvaradi A, Eichholz A, Komor E and Orlich G (1993) Specific proteins in the sieve-tube exudate of Ricinus communis L. seedlings: Separation, characterization and in-vivo labelling. Planta 191: 207-13.

Schaad MC, Jensen PE and Carrington JC (1997) Formation of plant RNA virus replication complexes on membranes: Role of an endoplasmic reticulum-targeted viral protein. EMBO J 16: 4049-4059

Schnittger A, Folkers U, Schwab B, Jürgens G and Hulskamp M (1999) Generation of a spacing pattern: The role of *TRIPTYCHON* in trichome patterning in *Arabidopsis*. Plant Cell 11: 1105-1116

Schobert C, Baker L, Szederkényi J, Grossmann P, Komor E, Hayashi H, Chino M and Lucas WJ (1998) Identification of immunologically related proteins in sieve-tube exudate collected from monocotyledonous and dicotyledonous plants. Planta 206: 245-252

Schulz A (1995) Plasmodesmal widening accompanies the short-term increase in symplasmic phloem unloading in pea root tips under osmotic stress. Protoplasma 188: 22-37

Schulz A (1999) Physiological control of plasmodesmal gating. In: van Bel AJE and van Kesteren WJP (eds) Plasmodesmata: Structure, Function, Role in Cell Communication. Berlin: Springer-Verlag, pp 173-204

Staehelin LA (1997) The plant ER: A dynamic organelle composed of a large number of discrete functional domains. Plant J 11: 1151-1165

Szymkowiak EJ and Sussex IM (1992) The internal meristem layer (L3) determines floral meristem size and carpel number in tomato periclinal chimeras. Plant Cell 4: 1089-1100

Takizawa PA, Sil A, Swedlow JR, Herskowitz I and Vale RD (1997) Actin-dependent localization of an RNA encoding a cell-fate determinant in yeast. Nature 389: 90-93

Tilney LG, Cooke TJ, Connelly PS and Tilney MS (1991) The structure of plasmodesmata as revealed by plasmolysis detergent extraction and protease digestion. J Cell Biol 112: 739-747

Titus MA and Gilbert SP (1999) The diversity of molecular motors: An overview. Cell Mol Life Sci 56: 181-183

Tucker EB (1988) Inositol bisphosphate and inositol trisphosphate inhibit cell-to-cell passage of carboxyfluorescein in staminal hairs of *Setcreasea purpurea*. Planta 174: 358-363

Tucker EB (1990) Calcium-loaded 1,2-bis(2-aminophenoxy)ethane -N,N,N',N'-tetraacetic acid blocks cell-to-cell diffusion of carboxyfluorescein in staminal hairs of *Setcreasea purpurea*. Planta 182: 34-38

Tucker EB (1993) Azide treatment enhances cell-to-cell diffusion in staminal hairs of *Setcreasea purpurea*. Protoplasma 174: 45-49

Turner A, Wells B and Roberts K (1994) Plasmodesmata of maize root tips: Structure and composition. J Cell Sci 107: 3351-3361

van Bel AJE, Günther S and van Kesteren WJP (1999) Plasmodesmata a maze of questions. In: van Bel AJE and van Kesteren WJP (eds) Plasmodesmata: Structure, Function, Role in Cell Communication. Berlin: Springer-Verlag, pp 1-26

van Steveninck FM (1976) Cytochemical evidence for ion transport through plasmodesmata. In: Gunning BES and Robards AW (eds) Intercellular Communication in Plants: Studies on Plasmodesmata. Berlin: Springer-Verlag, pp 131-147

Wells AL, Lin AW, Chen L-Q, Safer D, Cain SM, Hasson T, Carragher BI, Milligan RA and Sweeney HL (1999) Myosin VI is an actin-based motor that moves backwards. Nature 401: 505-508

White RG, Badelt K, Overall RL and Vesk M (1994) Actin associated with plasmodesmata. Protoplasma 180: 169-184

Xoconostle-Cázares B, Yu X, Ruiz-Medrano R, Wang HL, Monzer J, Yoo BC, McFarland KC, Franceschi VR and Lucas WJ (1999) Plant paralog to viral movement protein that potentiates transport of mRNA into the phloem. Science 283: 94-98

Yamamoto K, Hamada S and Kashiyama T (1999) Myosins from plants. Cell Mol Life Sci 56: 227-232

Zawadzki T and Fensom DS (1986a) Transnodal transport of ^{14}C in *Nitella flexilis*. I. Tandem cells without applied pressure gradients. J Exp Bot 37: 1341-1352

Zawadzki T and Fensom DS (1986b) Transnodal transport of ^{14}C in *Nitella flexilis*. II. Tandem cells with applied pressure gradients. J Exp Bot 37: 1353-1363

Chapter 29

INTERACTIONS BETWEEN VIRAL MOVEMENT PROTEINS AND THE CYTOSKELETON

B Gail McLean[1,2] and Patricia C Zambryski[3]

1 Department of Biochemistry and Molecular Biology, University of Maryland School of Medicine, Baltimore, MD 21201 USA
2 USDA/NRICGP, Stop 2241, 1400 Independence Ave SW, Washington, DC 20250 USA
3 Department of Plant and Microbial Biology, University of California, Berkeley, CA 94720 USA

Key words: actin, myosin, movement protein, plant virus, plasmodesmata, tobacco mosaic virus (TMV)

Abstract: The cytoskeleton is involved in many diverse processes, from cell structure to signal transduction and transport. In recent years, the study of plant virus movement has suggested a role for the plant cytoskeleton in virus transport. Much of this research focuses on the association of the tobacco mosaic virus movement protein with plant cytoskeletal elements. Future studies will determine the role of this association in intracellular and intercellular or plasmodesmal transport of plant viruses. Furthermore, as viruses tend to exploit host cellular processes, interactions between viral components and the cytoskeleton may play vital roles during all stages of virus infection in plant cells.

Typically, when a virus infects a plant cell, it replicates in and moves from the initially infected cell. This cell-to-cell or local movement occurs through plasmodesmata, cytoplasmic bridges that traverse the cell walls and provide cytoplasmic continuity among plant cells. The virus eventually reaches the vascular system where it then moves long distance in the plant. During the past decade, the study of plant virus movement has played a crucial role in characterizing plasmodesmal transport in plant cells (Citovsky, 1993, 1999; Lazarowitz & Beachy, 1999). Now, recent research suggests that the study of plant viruses may also be a tool for investigation of cytoskeletal function in plant cells (reviewed in Lazarowitz, 1999).

C.J. Staiger et al. (eds.), Actin: A Dynamic Framework for Multiple Plant Cell Functions, 517–540.

Interaction of the virus with cytoskeletal elements may occur at several steps in infection. An obvious function for cytoskeletal elements is in virus transport, both intracellular and intercellular. Indeed, the recent observation that a viral protein required in cell-to-cell movement colocalizes with cytoskeletal elements suggests a role for the cytoskeleton in virus transport. As work with plant viruses and the plant cytoskeleton is in its earliest stages, this chapter will first describe known plant virus-cytoskeleton interactions and then speculate on other possible virus-cytoskeleton interactions and functions.

1. ASSOCIATION OF VIRAL PROTEINS WITH THE CYTOSKELETON—IS THERE A ROLE IN VIRAL MOVEMENT?

One of the most obvious questions in the study of viral infection is how does the virus move? Movement of plant viruses commonly occurs in two modes: local movement from cell to cell, and long distance movement through vascular tissue (Carrington et al., 1996; Maule, 1991). While the mechanisms of long distance movement are not yet well characterized, the process of local transport has been studied extensively. Most plant virus families encode a protein dedicated solely to cell-to-cell movement, called movement protein (MP) (Carrington et al., 1996; Citovsky, 1999; Ding, 1998; Ghoshroy et al., 1997; Lartey and Citovsky, 1997; Lucas and Gilbertson, 1994; McLean et al., 1997). The best studied MP is that of tobacco mosaic virus (TMV). TMV itself is undoubtedly the best characterized plant virus and its study significantly advanced molecular biology and biotechnology (Creager et al., 1999; Fraenkel-Conrat, 1986; Turpen, 1999). TMV encodes at least 4 proteins, two that function in replication and two, the coat protein and the MP, that function in movement. Coat protein is required for both encapsidating the viral genome and moving TMV long distance through the vascular tissue. The MP of TMV potentiates its cell-to-cell movement.

TMV MP as well as several other plant viral MPs binds single strand nucleic acids to form a protein/nucleic acid complex (Citovsky et al., 1990, 1992; Fujiwara et al., 1993; Osman et al., 1992; Schoumacher et al., 1992). MPs also increase, or gate, the size exclusion limit of plasmodesmata to allow intercellular transport of large molecules, including MPs themselves (Angell et al., 1996; Derrick et al., 1992; Ding et al., 1995; Fujiwara et al., 1993; Noueiry et al., 1994; Poirson et al., 1993; Vaquero et al., 1994; Waigmann et al., 1994; Waigmann & Zambryski, 1995; Wolf et al., 1989).

During cell-to-cell transport, the MP binds the viral RNA, forming a ribonucleoprotein (RNP) complex; the cooperative binding of TMV MP to TMV RNA unfolds the RNA to form a thin, elongated complex (Citovsky et al., 1992). This complex then moves through the cell to interact with the plasmodesmata and mediate its transport into the neighboring cell. Since neither virion assembly nor capsid formation is required, cell-to-cell movement is essentially genome transport (Carrington et al., 1996).

TMV replicates in the cytoplasm where the MP-viral RNA complex consequently forms. The protein content and organized nature of the cytoplasm creates a very viscous environment and thus restricts diffusion of large molecular complexes (Luby-Phelps, 1993, 1994; Pickett-Heaps et al., 1999). Movement of large RNP MP-viral RNA complexes through the cytoplasm probably is not diffusion driven but instead is dependent on active transport. How then does the virus move through the cytoplasm to reach the plasmodesmata at the cell periphery? Cytoskeletal elements permeate the highly structured cytoplasm, helping define the architecture of the cell. Potentially, the cytoskeletal filaments, and their associated motor proteins, provide tracks upon which large molecules such as RNPs are actively propelled through the cytoplasm. Numerous studies in both plant and animal cells suggest that the cytoskeleton acts as a trafficking system for intracellular transport, translocating vesicles, and organelles, protein and even mRNA (Bassell & Singer, 1997; Cole & Lippincott-Schwartz, 1995; Hesketh, 1994; Langford, 1995; Oleynikov & Singer, 1998; Vale, 1987; Wilhelm & Vale, 1993; Williamson, 1986). Thus, as pathogens such as viruses tend to exploit existing cellular pathways and processes, viruses may commandeer an endogenous, cytoskeleton-mediated transport mechanism to move in and between cells.

Indeed, studies of bacterial and viral infections in animal cells demonstrate the importance of the cytoskeleton for spread of infection (Cudmore et al., 1997; Dramsi & Cossart, 1998). Many viral and bacterial pathogens disrupt or rearrange the actin cytoskeleton upon entry and internalization within the host animal cell. A recent, exciting finding has been the discovery that the vaccinia virus and bacteria from different genera actively move in the host cell cytoplasm using polarized actin (Cudmore et al., 1995; Dramsi & Cossart, 1998; Ireton & Cossart, 1998; Strauss, 1996; Theriot, 1995). Actin is not the only cytoskeletal network exploited by animal pathogens. The microtubule network also plays an important role in viral protein trafficking and distribution. Viruses such as adenovirus, reovirus, and HSV-1 bind and interact with microtubules (Luftig, 1982; Penfold et al., 1994; Sodeik et al., 1997; Topp et al., 1994). For instance, cytosolic transport of HSV capsids is mediated by microtubules. Incoming capsids are proposed to bind microtubules and then use dynein to move from the cell periphery to the nucleus, thus pirating machinery responsible for retrograde organelle transport (Sodeik et al., 1997; Topp et al., 1994). The viral matrix protein of vesicular stomatitus virus also interacts with tubulin *in vitro* and *in vivo*, and data from *in vitro* studies show that this viral protein binds to both polymerized and unpolymerized tubulin (Melki et al., 1994).

A possible interaction between plant viruses and the cytoskeleton was originally described in EM studies of comoviruses, suggesting viral particles aligned with microtubules (Kim & Fulton, 1975). In a different study, the 65-kD HSP70-related protein of beet yellows closterovirus was found to bind microtubules *in vitro*. This protein apparently functions in cell-to-cell movement, suggesting microtubule association may be important for movement (Agranovsky et al., 1998; Karasev et al., 1992). Another plant viral protein, the aphid transmission factor (ATF) from cauliflower mosaic virus (CaMV), also colocalizes with microtubules when expressed in either plant cells or insect cells (Blanc et al., 1996). While the function of this interaction is unknown, the authors speculate that ATF may bind to a tubulin-like receptor on the aphid feeding apparatus or that ATF may interact with the plant cytoskeleton in early stages of CaMV infection.

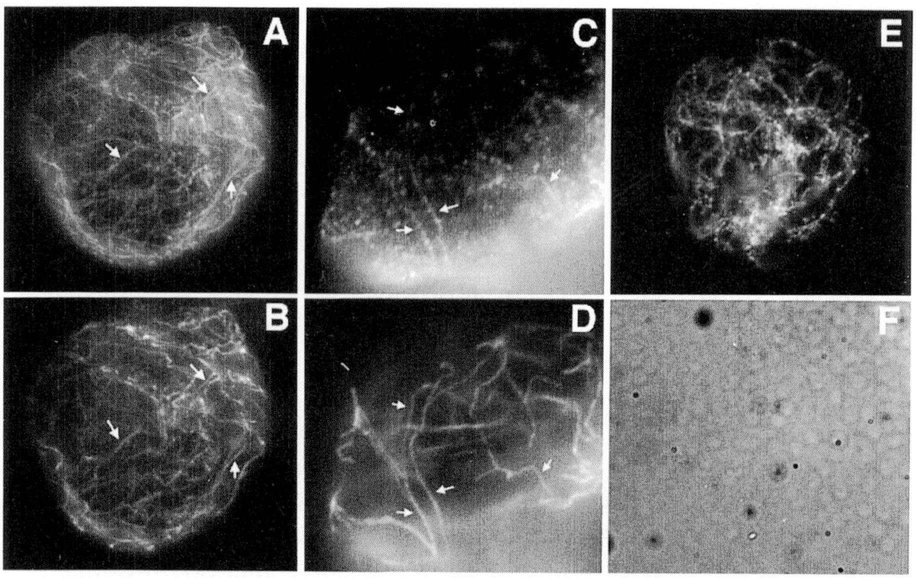

Figure 1. Colocalization of TMV MP with microtubules. Panels **A** and **B** show that TMV MP alone (in the absence of other viral proteins) colocalizes with microtubules. Panels **C** and **D** confirm that, during infection with wild-type TMV, MP colocalizes with microtubules. Panels **E** and **F** show the purification of TMV MP-GFP with microtubule cytoskeletons, suggesting TMV MP binds tightly to microtubules. **A**. TMV MP expression in tobacco protoplasts transfected with a plasmid encoding TMV MP, detected by MP antibody and a

fluorescent secondary antibody. **B**. Microtubule cytoskeleton in transfected protoplasts, detected with tubulin antibody and a fluorescent secondary antibody. **C**. TMV MP expression in TMV infected tobacco protoplasts, detected by MP antibody and a fluorescent secondary antibody. **D**. Microtubule cytoskeleton in TMV infected protoplasts, detected with tubulin antibody and a fluorescent secondary antibody. **E**. TMV MP-GFP fluorescence associated with isolated microtubule cytoskeletons from tobacco protoplasts transfected with plasmid encoding TMV MP-GFP. **F**. Light micrograph of cytoskeleton in panel E demonstrating cell wall and membranes are no longer present.

A number of recent studies focus on the interaction of TMV MP and the cytoskeleton. The MP of TMV was shown to colocalize in tobacco protoplasts primarily with microtubules (Fig. 1; Heinlein et al., 1995; McLean et al., 1995) and to a lesser extent with actin filaments (McLean et al., 1995). The colocalization of MP with microtubules was observed both when MP was immunologically detected in protoplasts infected with wildtype TMV and in protoplasts transiently expressing MP alone in the absence of virus (McLean et al., 1995). That MP alone in the absence of other viral components colocalized with microtubules indicates that this cytoskeletal association is a property of the MP itself and is not due to interaction with other viral proteins. Experiments utilizing a MP-GFP fusion demonstrated that MP colocalizes with microtubules in living protoplasts (Heinlein et al., 1995; McLean et al., 1995). Interestingly, over time, the MP-GFP localization changed, with the fluorescence moving from microtubules to the edge of the cell. The interaction of MP with microtubules was confirmed by several different methods. Treatment of protoplasts with microtubule disrupting agents also disrupted MP filaments, showing an interaction between microtubules and MP (Heinlein et al., 1995; McLean et al., 1995). MP also copurifies with isolated microtubule cytoskeletons or ‘ghosts’, suggesting a strong interaction between microtubules and MP (our unpublished data). Finally, *in vitro* experiments demonstrate that MP can bind directly to both polymerized and unpolymerized tubulin; thus, MP may be able to bind microtubules directly *in vivo* (McLean et al., 1995).

Transient expression of MP alone in protoplasts also indicated the interaction of MP with actin filaments (McLean et al., 1995). The interaction of MP with actin filaments *in vivo* was seen at a much lower frequency than the interaction with microtubules and was disrupted by treatment of the cell with cytochalasin to depolymerize the actin filaments. *In vitro*, MP bound both polymerized and unpolymerized actin, suggesting MP may bind actin directly *in vivo*. The infrequent detection of an interaction between MP and actin filaments may result from the plant actin cytoskeleton being very diverse in regard to gene family and isoforms (Meagher, 1990; Volkmann & Baluska, 1999; Meagher et al., this volume). Possibly, MP interacts with a specific isoform of actin or with an actin that is transiently expressed at a particular time or structure in the cell. Association of MP with other viral proteins may limit when MP can interact with actin, perhaps allowing interaction at a very specific cellular structure. Thus, expression of MP alone may favor detection of the actin-MP interaction. Alternatively, a less likely possibility is that the fusion of GFP to MP may hinder the interaction with actin. TMV MP is not the only plant MP that interacts with actin. A recent *in vitro* study of cucumber mosaic virus (CMV) MP suggests that it also interacts with actin (Wang et al., 1999). CMV MP appears to inhibit actin polymerization and in some cases seems to decrease the amount of filamentous actin. As yet, the specific effects of TMV MP on actin have not been determined. It would be interesting to see if TMV MP also can induce depolymerization of actin filaments and if this is a common property of MPs that move cell to cell through plasmodesmata.

The MP-GFP protein fusion is a valuable tool for observing MP localization over extended time periods (Oparka et al., 1996, 1997). Detailed analyses of MP-GFP intracellular localizations in both protoplasts and leaf tissues indicate that the dynamic distribution and morphologies of TMV MP-GFP expression are similar in both systems (Heinlein et al., 1998; Mas & Beachy, 1998; Oparka et al., 1997; Reichel & Beachy, 1998). In cells infected with TMV carrying a MP-GFP fusion, MP-GFP associates with microtubules during mid–late stages of TMV infection. Early in infection, MP-GFP is first detected in leaves at the cell wall, near or in plasmodesmata, and in protoplasts, at punctate sites at or near the surface, possibly at sites where the endoplasmic reticulum (ER) attaches to the plasma membrane. When the MP-GFP appears in the cytoplasm, it associates first with cortical ER. Additionally, in this early–mid stage of TMV infection, the ER is severely disrupted and forms large aggregates in which both MP-GFP and TMV replicase are detected (Reichel & Beachy, 1998). Only in mid–late stages of TMV infection does MP-GFP appear as filaments that colocalize with cortical and noncortical microtubules. Finally, late in infection, the MP-GFP is again observed near the plasma membrane, at or near plasmodesmata in leaves and at distinct punctate sites at the plasma membrane in protoplasts. At this time the ER returns to a normal organized cortical reticulum and the large aggregates disappear (Reichel & Beachy, 1998). Biochemical analysis of isolated microsomes indicates that MP-GFP behaves like an integral ER membrane protein and may be exposed on the cytoplasmic face of ER (Reichel & Beachy, 1998). Thus, there is clear evidence that MP associates with both microtubules and ER in plant cells. Future studies of this dual association will determine if both the microtubules and ER play a role in intracellular trafficking of TMV.

These time course studies of MP-GFP illustrate that MP associates with microtubules at a specific stage during TMV infection. In addition to the filamentous colocalization of MP with microtubules, the punctate MP-GFP sites (punctae) in some cells also appear to coincide with microtubules (Heinlein et al., 1998; Reichel & Beachy, 1998). Microtubules are postulated to attach at punctate cytoskeleton-plasma membrane-wall adhesion sites in plant cells and have been observed at sites where the cortical ER and plasma membrane are anchored to the cell wall (Pickard & Ding, 1993). When microtubules in these cells are disrupted by oryzalin, the MP-GFP fluorescent punctae enlarge (Heinlein et al., 1998); this result is similar to that observed for sites where the cortical ER and plasma membrane are anchored to the cell wall (Knebel et al., 1990; Lichtscheidl and Url, 1990). When the cells were treated with cytochalasin D to disrupt actin filaments, the distribution and structure of fluorescent punctae was not changed but the intensity of the fluorescence was reduced (Heinlein et al., 1998). Interestingly, actin and microtubules appear to interact at or near the plasma membrane in tobacco cells, and the localization and organization of microtubules may establish the arrangement of cortical actin arrays (Collings et al., 1998; see also Collings & Allen, this volume). Thus, an interaction between actin filaments and microtubules may help form or anchor the MP-GFP punctae at the cell periphery.

2. MP-RNP MOVEMENT: ANALOGIES WITH mRNA TRAFFICKING

TMV MP, both alone and with other viral proteins, interacts with the cytoskeleton. How this interaction occurs is unknown. Since TMV MP is a tenacious single strand nucleic acid binding protein, it may bind cytoskeletal elements as an RNP complex. Possibly, transport of the viral RNP may mimic the mRNA and RNP transport described in animal cells. Indeed, as viruses tend to exploit host mechanisms, TMV-RNP cell-to-cell movement may pirate a host RNP trafficking pathway.

Transport and consequent localization of mRNA in animal cells illustrate that the cytoskeleton plays an active role in transport and/or anchoring several mRNA transcripts, such as *Drosophila BICOID*, an anteriorly-localized RNA, or *OSKAR*, a posteriorly-localized RNA, to specific intracellular locations (Bassell & Singer, 1997; Jansen, 1999; Oleynikov & Singer, 1998; Singer, 1992; St Johnston, 1995; Wilhelm & Vale, 1993). Typically, the transported mRNAs are complexed with other proteins to form RNP particles, and these RNP particles are detected in transport and bound to cytoskeletal filaments. Thus, an RNA binding protein such as the

TMV MP may link RNA to the cytoskeleton for active transport to, or anchoring at, specific cellular locations.

That TMV MP interacts with both microtubules and actin may signal the use of these filaments for viral RNA transport and/or anchoring in the plant cell. Studies in animal cells show that mRNA/RNP transport and localization can occur by either microtubules or actin filaments. Some protein components involved in mRNA trafficking actually participate in both microtubule- and actin-dependent localization pathways (Jansen, 1999; Oleynikov & Singer, 1998). Apparently, these RNA binding proteins can switch from one type of cytoskeletal element to another. For example, when bound to one type of RNA, *Drosophila* STAUFEN utilizes microtubule-dependent pathways (Ferrandon et al., 1994); when bound to a different RNA, STAUFEN utilizes actin-dependent pathways (Broadus et al., 1998). Two almost identical proteins, Vg1 and ZBP-1, interact with different cytoskeletal elements in different cells, suggesting that each of these RNA binding proteins can associate with several types of cytoskeletal filaments (Deshler et al., 1997, 1998; Elisha et al., 1995; Havin et al., 1998). Interestingly, the Vg1 protein behaves similarly to TMV MP in that it binds RNA, microtubules, and ER membranes. The association of Vg1 with both microtubules and ER led to the hypothesis that Vg1 attaches mRNA to a specific subcompartment of ER that binds microtubules (Deshler et al., 1997).

Similarly, recent studies in plants suggest that some mRNAs which localize to the ER also bind cytoskeletal components (Muench et al., 1998; Wu et al., 1998). In rice endosperm cells, the mRNAs encoding prolamine and glutelin storage proteins localize to morphologically-distinct membranes of the ER. The prolamine polysomes attach to cytoskeletal elements and to ER membranes which bind prolamine protein bodies (Wu et al., 1998). The ribosome-free prolamine mRNA binds the cytoskeleton, indicating the prolamine mRNA associates with the cytoskeleton before translation (Muench et al., 1998). Thus, the RNA molecule itself contains a signal for cytoskeletal interaction. Studies in animal cells also show that the ribosome is not essential for RNA-cytoskeleton interaction (Bag & Pramanik, 1987; Bassell et al., 1994; Taneja et al., 1992).

A combination of immunostaining and *in situ* hybridization suggests that TMV RNA colocalizes with microtubules in TMV-infected protoplasts (unpublished data described in Reichel et al., 1999). When the cytoskeleton was disrupted with pharmacological agents, ER-associated complexes containing TMV MP and viral RNA remained in the perinuclear region of the cell and did not spread through the cytoplasm. Reichel et al. (1999) suggest that TMV RNA is transported intracellularly on microtubules; whether the interaction of RNA with microtubules occurs through the RNA or MP is not stated. As purified MP binds to tubulin *in vitro*, the binding of MP to microtubules *in vivo* does not appear to be via RNA. RNA binding, however, may alter the MP interaction with microtubules and initiate transport of the RNP along the cytoskeleton. For instance, both viral RNA and MP may be targeted to microtubules. Upon binding RNA, the MP changes conformation to interact with a microtubule motor for active intracellular movement. Specific cellular localizations of MP-GFP were disrupted or lost when TMV MP deletion mutants were used, suggesting MP conformation is responsible for different intracellular localizations and possibly transport (Kahn et al., 1998).

Another function for MP-cytoskeleton interaction may be to prevent movement of RNA until a 'mature' MP-RNA complex is formed. Citovsky et al. (1990) speculate that the MP-RNA complex represents a specific pool of RNA molecules that are targeted for movement and excluded from replication. Indeed, TMV MP blocks translation of viral RNA *in vitro* (Karpova et al., 1997). MP may bind RNA and then anchor the translationally-repressed MP-RNA complexes on microtubules in the cytoplasm, the complete or 'mature' complexes being released only in response to a signal for cell-to-cell, plasmodesmal transport. In this case, interaction with the microtubules is a regulatory rather than a transport function. Similar mechanisms have been observed in animal cells. The phosphoprotein TB-RBP attaches specific translationally-repressed RNAs to microtubules (Han et al., 1995), and the transcription factor NF-kappa B is anchored in the cytoplasm until a change in the cytoskeleton (depolymerization of microtubules) releases the protein for import into the nucleus (Rosette & Karin, 1995).

3. MP INTERACTIONS WITH CYTOSKELETAL ELEMENTS AND PLASMODESMATA: POTENTIAL TRANSPORT AND GATING FUNCTIONS

Unlike animal viruses which can move from cell to cell by budding, plant viruses must cross a cell wall. Consequently many plant viruses, such as TMV, interact with plasmodesmata, cytoplasmic bridges which traverse the cell wall. Plasmodesmata provide continuity between plant cells by allowing movement of small molecules while regulating movement of large molecules (Ding, 1998; Ghoshroy et al., 1997; Lucas, 1995; McLean et al., 1997; Oparka, 1993; Oparka et al., 1999; Zambryski, 1995). This transport through intercellular channels, rather than by release into the extracellular space followed by attachment to cell surface receptors, is potentially the most significant evolutionary adaptation distinguishing plant viruses from animal viruses (Carrington et al., 1996). TMV MP and MP-GFP localize to plasmodesmata (Ding et al., 1992; Moore et al., 1992; Oparka et al., 1997; Padgett et al., 1996; Tomenius et al., 1987), and several viral MPs gate plasmodesmata to allow movement of large molecules (Angell et al., 1996; Derrick et al., 1992; Ding et al., 1995; Fujiwara et al., 1993; Noueiry et al., 1994; Poirson et al., 1993; Vaquero et al., 1994; Waigmann et al., 1994; Waigmann & Zambryski, 1995; Wolf et al., 1989). As a consequence of this ability to functionally modify plasmodesmata, viral MPs have been used extensively to study plasmodesmal transport.

Structurally, plasmodesmata consist of an outer sheath which is contiguous with the plasma membrane and a central core of modified ER termed the desmotubule (Ding et al., 1992; Overall & Blackman, 1996; Robards & Lucas, 1990; Tilney et al., 1991; Turner et al., 1994). While the specific components of plasmodesmata are not yet identified, two cytoskeletal elements, actin and myosin, have been identified as possible plasmodesmata components (Blackman & Overall, 1998; Radford & White, 1998; Reichelt et al., 1999; White et al., 1994; see also Overall et al., this volume). Both actin and myosin localize to the plasmodesmal opening and along plasmodesmal channels. Actin-myosin interactions may determine the size of the opening (the size exclusion limit) or the dimensions of the presumed transport channels extending along the length of the plasmodesmata. Recently, calreticulin, a highly-conserved calcium-sequestering protein residing in the ER lumen, was found to localize to to the plant cell periphery predominantly at or near plasmodesmata (Baluska et al., 1999). Calreticulin may regulate plasmodesmal dynamics by its calcium-buffering capacity, and intriguingly, myosin VIII, a plasmodesmal-associated myosin, is also regulated by calcium (Reichelt et al., 1999). Thus, plants may control plasmodesmal transport by calcium-mediated actin-myosin contraction or relaxation. Another possibility is that myosin may help move either the ER, or membrane proteins, via underlying actin filaments to and through plasmodesmata (Pollard et al., 1991). As the TMV MP is closely associated with the ER, then, the MP RNP complex may be transported to or through plasmodesmata by the ER and an actin-myosin interaction. Similar to TMV MP, the alfalfa mosaic virus MP also behaves as an integral membrane protein and has been observed to be closely associated with the ER membrane component of plasmodesmata (Huang and Zhang, 1999). Thus, during plasmodesmal transport, myosin-generated movement along actin filaments may traffic proteins either as a direct cytoskeletal interaction or as an ER membrane-associated protein. Recent experiments, however, suggest that proteins do not use a direct actin filament interaction to move through plasmodesmata channels (Crawford & Zambryski, unpublished).

A role for actin filaments in regulation of plasmodesmal gating has been clearly demonstrated. Treatment of cells with cytochalasin, an agent that depolymerizes actin filaments, causes the plasmodesmal opening to appear greatly enlarged (Ding et al., 1996). Moreover, microinjection of substances that perturb actin filaments, such as cytochalasin or the actin-binding protein, profilin, dramatically increases the leaf mesophyll plasmodesmal size exclusion limit (Ding et al., 1996). A recent study with CMV MP further suggests a role for actin in plasmodesmal transport by virus MPs (Wang et al., 1999). CMV MP, like many other MPs, is thought to play a role in gating plasmodesmata for cell-to-cell movement of large molecules and has been localized to plasmodesmata (Ding et al., 1995; Itaya et al., 1997). When cells were pretreated with phalloidin to stabilize actin filaments, the cell-to-cell transport of large molecules by CMV MP was dramatically reduced (Wang et al., 1999). CMV MP itself inhibits actin polymerization and occasionally seems to decrease the amount of actin filaments, suggesting that MPs may gate plasmodesmata by destabilizing actin filaments (Wang et al., 1999).

Thus, as actin and myosin are associated with plasmodesmata, the interaction of MP with actin may gate and transport MP-RNP through plasmodesmata via an actin-myosin interaction. As data suggest MP does not interact with actin filaments to move through the plasmodesmal channel, MP may interact with actin to remove actin filaments and gate plasmodesmata (Crawford & Zambryski, unpublished data). Further, the interaction of MP with a plasmodesmal component may alter the conformation of the MP, perhaps causing MP-RNP complex to move from microtubules onto actin filaments associated with plasmodesmata. An intriguing finding is that plasmodesmal transport of the TMV-MP-RNP complex converts the nontranslatable MP-RNP into a form that is translated and replicated in the next cell (Karpova et al., 1997). Thus, interaction of the TMV MP with actin at the plasmodesmata may convert MP RNP into a translatable complex.

4. HOW MIGHT THE CYTOSKELETON FUNCTION IN TMV MP TRANSPORT?

Based on characterization of RNP transport in animal cells, the colocalization of TMV MP with the plant cytoskeleton suggests that the MP or MP-RNP complex may be actively transported in the cytoplasm (Heinlein et al., 1995; McLean et al., 1995). There are several possible mechanisms for cytoskeletal mediated movement of TMV MP (Fig. 2).

First, the MP-RNP complex could use a linked system of microtubules and actin filaments for active transport (Model 1; McLean et al., 1995). This system is similar to models proposed for some animal systems, such as for squid giant axon organelles, in which microtubules provide tracks for extended intracellular movement and actin filaments direct short, targeted movement to local sites (Atkinson et al., 1992; Kuznetsov et al., 1992; Langford, 1995). In one scenario, after MP-RNP formation at an ER-associated replication site, the MP-RNP initially moves on microtubules for extended, possibly bi-directional, movement in the cytoplasm (Heinlein et al., 1998; McLean et al., 1995). This microtubule-dependent transport may propagate new replication complexes on the ER within the cell in addition to actively directing the MP RNP to plasmodesmata at the cell periphery (Heinlein et al., 1998; Lazarowitz & Beachy, 1999; Reichel & Beachy, 1998). At the cell periphery, the MP-RNP moves from the microtubules onto actin filaments associated with the plasmodesmata (McLean et al., 1995). Association with plasmodesmal actin filaments then would provide targeted, unidirectional movement to the plasmodesmata. Interaction of MP with actin at the plasmodesmal opening may also gate the plasmodesmata to allow movement of the large RNP (McLean et al., 1995).

Another possibility is that the MP is transported as an ER associated protein to and through plasmodesmata (Model 2; Oparka et al., 1999; Padgett et al., 1996; Reichel & Beachy, 1998; Crawford & Zambryski, unpublished). Unlike animal cells where organization and movement of the ER depends on microtubules, in plant cells it appears to be governed by an actomyosin system (Staehelin, 1997). In this scenario, the interaction of the MP with microtubules would anchor the MP close to replication complexes so that specific viral RNA-MP complexes are formed. The interaction of MP with actin then may enable MP to gate plasmodesmata.

Numerous other models can also be proposed. For example, the MP may primarily use actin filaments instead of the ER for transport; the association with microtubules may anchor the MP close to replication complexes as described above. Another possibility is that the MP is anchored on the ER until a specific conformational change occurs, perhaps one involving the binding of the viral genome. Then the MP interacts with microtubules for active transport to the cell periphery. Finally, transport may occur primarily via the ER, and the localization of MP to microtubules may target the MP for degradation (Padgett et al., 1996; Reichel & Beachy, 1998). As microtubules are required for movement of RNA in TMV infected cells, a role for microtubules in targeting MP degradation seems unlikely.

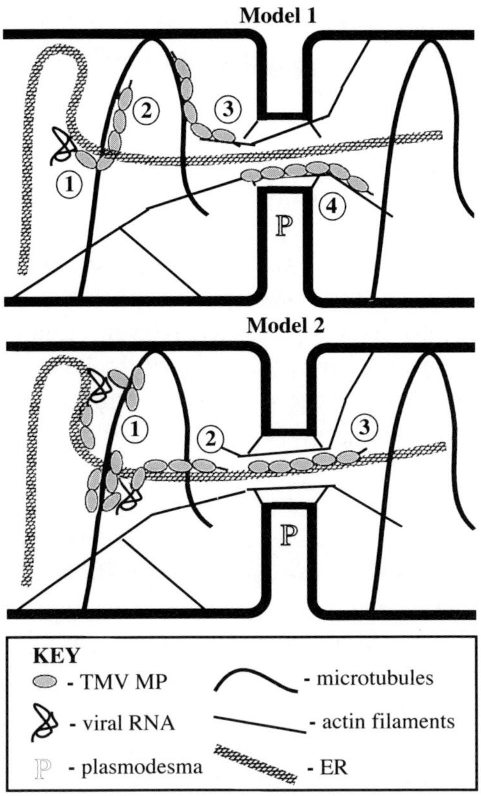

Figure 2. Models for TMV MP transport. As noted in the text, there are many possible mechanisms for transport of TMV MP. Two models are depicted here. In one, TMV MP is transported by cytoskeletal elements, and in the second, TMV MP is transported by the ER.
Model 1. In this scenario, TMV MP uses microtubules for extended transport and actin filaments for targeting and/or plasmodesma modification. 1. TMV MP associates with viral RNA produced in replication complexes on the ER. 2. The MP-RNA complex (RNP) travels through the cytoplasm via active transport on microtubules. 3. The RNP moves from microtubules onto plasmodesmal-associated actin filaments for targeting to the plasmodesma. Interaction of MP with actin at the plasmodesmal opening may also gate the plasmodesma. 4. The RNP is transported through plasmodesma possibly by an actinomyosin mechanism.
Model 2. In this scenario, the ER transports TMV MP, and the interaction of the MP with microtubules provides a regulatory function. 1. The TMV MP is anchored on microtubules until MP binds the viral RNA at the replication complex on the outer membrane of the ER. Once the MP-RNA complex (RNP) forms, MP is released from microtubules and associates with the ER. 2. The RNP moves through the cytoplasm via the ER. 3. The RNP is transported through plasmodesma by associating with the ER of the desmotubule. Interaction of the MP with actin at the plasmodesmal opening would gate the plasmodesmata for RNP transport.anchored on microtubules until MP binds the viral RNA at the replication complex on the outer membrane of the ER. Once the MP-RNA complex (RNP) forms, MP is released from microtubules and associates with the ER. 2. The RNP moves through the cytoplasm via

the ER. 3. The RNP is transported through plasmodesma by associating with the ER of the desmotubule. Interaction of the MP with actin at the plasmodesmal opening would gate the plasmodesmata for RNP transport.

5. CYTOSKELETAL INTERACTIONS—PERHAPS NOT JUST FOR VIRAL TRANSPORT

The recent studies of TMV MP focus on the possible role of the cytoskeleton and ER in the trafficking of the MP-RNA complex. In animal cells, however, the cytoskeleton is also involved in virus replication (Avalos et al., 1997; Ben-Ze'ev et al., 1983; Li et al., 1998; Topp et al., 1994), and in both plant and animal cells, components of protein synthesis machinery are associated with the cytoskeleton (Condeelis, 1995; Davies et al., 1991; Durso & Cyr, 1994; Hesketh, 1994; Hovland et al., 1996; Volkmann & Baluska, 1999). Moreover, RNP complexes sometimes also contain components of the protein synthesis apparatus (Barbarese et al., 1995; reviewed in Jansen, 1999). Viruses, and specifically viral RNPs, may interact with the cytoskeletal components for translation of proteins necessary for replication and movement. Another potential research area for plant virus-cytoskeleton interaction, then, is the role of the cytoskeletal elements in virus replication and protein synthesis.

The role of the cytoskeleton during replication and protein synthesis may be quite varied. Actin is implicated in several stages of the replicative cycle of baculovirus AcMNPV (Lanier & Volkman, 1998; Ohkawa and Volkman, 1999). For some animal viruses, nucleocapsids associate with actin only after termination of RNA synthesis. This specific interaction may guarantee that only the mature, 'transcriptionally-silent' nucleocapsids are transported by actin filaments to the plasma membrane for budding (Moyer et al., 1990). During HIV-1 infection, the cytoskeleton is thought to promote both establishment of the reverse transcriptase complex in the host cell and subsequent reverse transcription in the complex (Bukrinskaya et al., 1998). In fact, actin and tubulin stimulate transcription of different viruses (De et al., 1991; Hamaguchi et al., 1985; Huang et al., 1993; Moyer et al., 1986, 1990). Thus, cytoskeletal elements are important for propagation as well as movement of viruses.

Similarly, the importance of endomembranes, such as the ER, in RNA virus replication is well established in both animal and plant cells (Bienz et al., 1994; Egger et al., 1996; Heinlein et al., 1998; Restrepo-Hartwig & Ahlquist, 1996; Schaad et al., 1997; reviewed in Buck, 1996). Viral RNA replication and protein synthesis occur in the same complex on the ER (Beachy & Zaitlin, 1975; TMV replication reviewed in Buck, 1999). Both

TMV replicase and TMV MP, which has no role in replication, colocalize to the ER, suggesting that viral RNA synthesis, protein synthesis, and assembly of MP complexes occur in close proximity (Heinlein et al., 1998; Reichel & Beachy, 1998; Reichel et al., 1999). As TMV MP binds single-strand nucleic acid in a sequence-nonspecific manner, the interaction of the MP and replicase complex may confer specificity on RNA binding and consequently RNA movement (Buck, 1999). The cytoskeletal components actin and tubulin are closely associated with these TMV replication bodies (Ding et al., 1998; Heinlein et al., 1998). The protein synthesis elongation factor 1α (EF1α) also colocalizes with the TMV replicase complex in tobacco (Ding et al., 1998), and in various cell types, EF1α has been shown to associate with either actin filaments or microtubules (Clore et al., 1996; Collings et al., 1994; Condeelis, 1995; Durso and Cyr, 1994; Durso et al., 1996; Wu et al., 1998; Yang et al., 1990). In both animal and plant cells, association of factors such as EF1α with replication complexes is thought to reflect a role in protein synthesis (Buck, 1999; Condeelis, 1995). Interestingly, both purified TMV and brome mosaic virus RNA polymerase contain host proteins related to different subunits of the translation factor eIF3 (Osman & Buck, 1997; Quadt et al., 1993).

Components of the protein translation machinery (polysomes, protein factors, mRNAs) associate with cytoskeleton in both animal and plant cells (Davies et al., 1991, 1993; Hesketh, 1994; see also Stankovic et al., this volume). Many polysomes, instead of being free in the cytosol, evidently are associated with the cytoskeleton, and some polysomes, termed cytomatrix-bound, appear to associate with both cytoskeletal elements and membranes (Davies et al., 1991, 1993; Ito et al., 1994; Pachter, 1992; Stankovic et al., 1993). Exactly how polysomes interact with the cytoskeleton remains unclear. In some cases, the ribosome or nascent polypeptide is suggested to anchor the polysome to the cytoskeleton (Davies et al., 1993; Stankovic et al., 1993), but in other cases, the polysomes appear to be anchored via mRNA and not the ribosome (Bag & Pramanik, 1987). Thus, the close association of host proteins such as ribosomes, translation initiation factors, and elongation factors with cytoskeletal elements suggests that the cytoskeleton may compartmentalize protein synthesis machinery in the cytoplasm (Condeelis, 1995; Hesketh, 1994; Hovland et al., 1996). An interaction between viral proteins and the cytoskeleton would bring the virus to areas of the cytoplasm equipped with machinery for viral protein synthesis. The association of the MP with cytoskeletal filaments, then, may be required not only for transport but also for protein synthesis.

6. CONCLUSIONS

Studies of both plant and animal cytoskeletons show that cytoskeletal elements are involved in many diverse processes, from cell structure and cytokinesis to signal transduction and transport. Furthermore, the coordinated interplay between microtubules and actin filaments appears to be vital for many cell processes. Both microtubules and actin filaments may play important roles during plant virus infection. The many properties of the TMV MP—RNA binding, intracellular transport, plasmodesmata gating, intercellular transport—may all involve or require a cytoskeletal interaction. Whether cytoskeletal association plays a primary role in driving intracellular and intercellular transport of the MP/MP RNP complex or in anchoring the MP in the cytoplasm for nucleoprotein complex formation still remains to be determined. Further, cytoskeletal interactions have primarily focused on the MPs of viruses that replicate in the cytoplasm. It will be interesting to determine if cytoskeletal elements also play a role in transport of nuclear replicating viruses. That the plant nuclear localization signal receptor importin alpha associates with cytoskeletal elements suggests a role for the cytoskeleton in nuclear transport (Smith & Raikhel, 1998). As yet, however, no interaction between the cytoskeleton and proteins from nuclear replicating viruses has been reported.

Thus, the many aspects of viral infection, from RNP transport and plasmodesmal gating to protein synthesis and replication, provide an opportunity for many cytoskeletal interactions. An intriguing observation from the studies of the TMV MP is the association of the MP with both the cytoskeleton and the ER. Future studies will reveal if, and how, interplay between cytoskeletal elements and ER membranes directs viral transport within and through cells. Interactions between viral proteins and cytoskeletal elements provide new challenges to plant virologists and plant cell biologists in the dual quest to characterize virus infection and cytoskeletal functions.

ACKNOWLEDGEMENTS

Studies of viral movement and plasmodesmal transport in the Zambryski laboratory are supported by National Institutes of Health (# GM 45244).

REFERENCES

Agranovsky AA, Folimonov AS, Folimonova S, Morozov S, Schiemann J, Lesemann D and Atabekov JG (1998) Beet yellows closterovirus HSP70-like protein mediates the cell-to-cell movement of a potexvirus transport-deficient mutant and a hordeivirus-based chimeric virus. J Gen Virol 79: 889-895

Angell SM, Davies C and Baulcombe DC (1996) Cell-to-cell movement of potato virus X is associated with a change in the size-exclusion limit of plasmodesmata in trichome cells of *Nicotiana clevelandii*. Virology 216: 197-201

Atkinson SJ, Doberstein SK and Pollard TD (1992) Moving off the beaten tracks. Curr Biol 2: 326-328

Avalos RT, Yu Z and Nayak DP (1997) Association of influenza virus NP and M1 proteins with cellular cytoskeletal elements in influenza virus-infected cells. J Virol 71: 2947-2958

Bag J and Pramanik S (1987) Attachment of mRNA to the cytoskeletal framework and translational control of gene expression in rat L6 muscle cells. Biochem Cell Biol 65: 565-575

Baluska F, Samaj J, Napier R and Volkmann D (1999) Maize calreticulin localizes preferentially to plasmodesmata in root apex. Plant J 19: 481-488

Barbarese E, Koppel DE, Deutscher MP, Smith CL, Ainger K, Morgan F and Carson JH (1995) Protein translation components are colocalized in granules in oligodendrocytes. J Cell Sci 108: 2781-2790

Bassell G and Singer RH (1997) mRNA and cytoskeletal filaments. Curr Opin Cell Biol 9: 109-115

Bassell GJ, Singer RH and Kosik KS (1994) Association of poly(A) mRNA with microtubules in cultured neurons. Neuron 12: 571-582

Beachy RN and Zaitlin M (1975) Replication of tobacco mosaic virus. VI. Replicative intermediate and other viral related RNAs associated with polyribosomes. Virology 63: 84-97

Ben-Ze'ev A, Abulafia R and Bratosin S (1983) Herpes simplex virus assembly and protein transport are associated with the cytoskeletal framework and the nuclear matrix in infected BSC-1 cells. Virology 129: 501-507

Bienz K, Egger D and Pfister T (1994) Characteristics of the poliovirus replication complex. Arch Virol Suppl 9: 147-157

Blackman LM and Overall RL (1998) Immunolocalization of the cytoskeleton to the plasmodesmata of *Chara corallina*. Plant J 14: 733-741

Blanc S, Schmidt I, Vantard M, Scholthof HB, Kuhl G, Esperandieu P, Cerutti M and Louis C (1996) The aphid transmission factor of cauliflower mosaic virus forms a stable complex with microtubules in both insect and plant cells. Proc Natl Acad Sci USA 93: 15158-15163

Broadus J, Fuerstenberg S and Doe CQ (1998) Staufen-dependent localization of prospero mRNA contributes to neuroblast daughter-cell fate. Nature 391: 792-795

Buck KW (1996) Comparison of the replication of positive-stranded RNA viruses of plants and animals. Adv Virus Res 47: 159-251

Buck KW (1999) Replication of tobacco mosaic virus RNA. Phil Trans R Soc Lond B 354: 613-627

Burinskaya A, Brichacek B, Mann A and Stevenson M (1998) Establishment of a functional human immunodeficiency virus type 1 (HIV-1) reverse transcription complex involves the cytoskeleton. J Exp Med 188: 2113-2125

Carrington JC, Kasschau KD, Mahajan SK and Schaad MC (1996) Cell-to-cell and long-distance transport of viruses in plants. Plant Cell 8: 1669-1681

Citovsky V (1993) Probing plasmodesmal transport with plant viruses. Plant Physiol 102: 1071-1076

Citovsky V (1999) Tobacco mosaic virus: a pioneer of cell-to-cell movement. Phil Trans R Soc Lond B 354: 637-643

Citovsky V, Knorr D, Schuster G and Zambryski P (1990) The P30 movement protein of tobacco mosaic virus is a single strand nucleic acid binding protein. Cell 60: 637-647

Citovsky V, Wong ML, Shaw A, Prasad BVV and Zambryski P (1992) Visualization and characterization of tobacco mosaic virus movement protein binding to single-stranded nucleic acids. Plant Cell 4: 397-411

Clore A, Dannenhoffer J and Larkins B (1996) EF1α is associated with a cytoskeleton network surrounding protein bodies in maize endosperm cells. Plant Cell 8: 2003-2014

Cole NB and Lippincott-Schwartz J (1995) Organization of organelles and membrane traffic by microtubules. Curr Opin Cell Biol 7: 55-64

Collings DA, Asada T, Allen NS and Shibaoka H (1998) Plasma membrane-associated actin in bright yellow 2 tobacco cells. Evidence for interaction with microtubules. Plant Physiol 118: 917-928

Collings DA, Wasteneys GO, Miyazaki M and Williamson RE (1994) Elongation factor 1 alpha is a component of the subcortical actin bundles of characean algae. Cell Biol Int 18: 1019-1024

Condeelis J (1995) Elongation factor 1α, translation and the cytoskeleton. Trends Biochem Sci 20: 169-170

Creager AN, Scholthof KB, Citovsky V and Scholthof HB (1999) Tobacco mosaic virus. Pioneering research for a century. Plant Cell 11: 301-308

Cudmore S, Cossart P, Griffiths G and Way M (1995) Actin-based motility of vaccinia virus. Nature 378: 636-638

Cudmore S, Reckmann I and Way M (1997) Viral manipulations of the actin cytoskeleton. Trends Microbiol 5: 142-148

Davies E, Comer EC, Lionberger JM, Stankovic B and Abe S (1993) Cytoskeleton-bound polysomes in plants. III. Polysome-cytoskeleton-membrane interactions in corn endosperm. Cell Biol Int 17: 331-340

Davies E, Fillingham BD, Oto Y and Abe S (1991) Evidence for the existence of cytoskeleton-bound polysomes in plants. Cell Biol Int Rep 15: 973-981

De BP, Lesoon A and Banerjee AK (1991) Human parainfluenza virus type 3 transcription in vitro: role of cellular actin in mRNA synthesis. J Virol 65: 3268-3275

Derrick PM, Barker H and Oparka KJ (1992) Increase in plasmodesmatal permeability during cell-to-cell spread of tobacco rattle tobravirus from individually inoculated cells. Plant Cell 4: 1405-1412

Deshler JO, Highett MI, Abramson T and Schnapp BJ (1998) A highly conserved RNA-binding protein for cytoplasmic mRNA localization in vertebrates. Curr Biol 8: 489-496

Deshler JO, Highett MI and Schnapp BJ (1997) Localization of Xenopus Vg1 mRNA by Vera protein and the endoplasmic reticulum. Science 276: 1128-1131

Ding B (1998) Intercellular protein trafficking through plasmodesmata. Plant Molec Biol 38: 279-319

Ding B, Haudsenshield JS, Hull RJ, Wolf S, Beachy RN and Lucas WJ (1992) Secondary plasmodesmata are specific sites of localization of the tobacco mosaic virus movement protein in transgenic tobacco plants. Plant Cell 4: 915-928

Ding B, Kwon M-O and Warnberg L (1996) Evidence that actin filaments are involved in controlling the permeability of plasmodesmata in tobacco mesophyll. Plant J 10: 157-164

Ding B, Li Q, Nguyen L, Palukaitis P and Lucas WJ (1995) Cucumber mosaic virus 3a protein potentiates cell-to-cell trafficking of CMV RNA in tobacco plants. Virology 207: 345-353

Ding B, Turgeon R and Parthasarathy MV (1992) Substructure of freeze-substituted plasmodesmata. Protoplasma 169: 28-41

Ding XS, Bao Y, Carter SA and Nelson RS (1998) Host factors involved in virus replication and cell-to-cell movement In: 1997 Annual Report of the Samuel Roberts Noble Foundation. Ardmore, OK: Noble Foundation Communications, pp 94-96

Dramsi S and Cossart P (1998) Intracellular pathogens and the actin cytoskeleton. Annu Rev Cell Dev Biol 14: 137-166

Durso NA and Cyr RJ (1994) Beyond translation: elongation factor-1α and the cytoskeleton. Protoplasma 180: 99-105

Durso NA, Leslie JD and Cyr RJ (1996) In situ immunocytochemical evidence that a homolog of protein translation elongation factor EF-1α is associated with microtubules in carrot cells. Protoplasma 190: 141-150

Egger D, Pasamontes L, Bolten R, Boyko V and Bienz K (1996) Reversible dissociation of the poliovirus replication complex: functions and interactions of its components in viral RNA synthesis. J Virol 70: 8675-8683

Elisha Z, Havin L, Ringel I and Yisraeli JK (1995) Vg1 RNA binding protein mediates the association of Vg1 RNA with microtubules in Xenopus oocytes. EMBO J 14: 5109-5114

Ferrandon D, Elphick L, Nusslein-Volhard C and St Johnston D (1994) Staufen protein associates with the 3'UTR of bicoid mRNA to form particles that move in a microtubule-dependent manner. Cell 79: 1221-1232

Fraenkel-Conrat H (1986) Tobacco mosaic virus: The history of tobacco mosaic virus and the evolution of molecular biology. In: van Regenmortel MVH, Fraenkel-Conrat H (eds) The Plant Viruses. New York, NY: Kluwer Academic/Plenum Press, pp 5-18

Fujiwara T, Giesman-Cookmeyer D, Ding B, Lommel SA and Lucas WJ (1993) Cell-to-cell trafficking of macromolecules through plasmodesmata potentiated by the red clover necrotic virus movement protein. Plant Cell 5: 1783-1794

Ghoshroy S, Lartey R, Sheng J and Citovsky V (1997) Transport of proteins and nucleic acids through plasmodesmata. Annu Rev Plant Physiol Plant Mol Biol 48: 27-49

Hamaguchi M, Nishikawa K, Toyoda T, Yoshida T, Hanaichi T and Nagai Y (1985) Transcriptive complex of Newcastle disease virus. II. Structural and functional assembly associated with the cytoskeletal framework. Virology 147: 295-308

Han JR, Yiu GK and Hecht NB (1995) Testis/brain RNA-binding protein attaches translationally repressed and transported mRNAs to microtubules. Proc Natl Acad Sci USA 92: 9550-9554

Havin L, Git A, Elisha Z, Oberman F, Yaniv K, Schwartz SP, Standart N and Yisraeli JK (1998) RNA-binding protein conserved in both microtubule- and microfilament-based RNA localization. Genes Dev 12: 1593-1598

Heinlein M, Epel BL and Beachy RN (1995) Interaction of tobamovirus movement proteins with the plant cytoskeleton. Science 270: 1983-1985

Heinlein M, Padgett HS, Gens JS, Pickard BG, Casper SJ, Epel BL and Beachy RN (1998) Changing patterns of localization of the tobacco mosaic virus movement protein and replicase to the endoplasmic reticulum and microtubules during infection. Plant Cell 10: 1107-1120

Hesketh J (1994) Translation and the cytoskeleton: a mechanism for targeted protein synthesis. Mol Biol Rep 19: 233-243

Hovland R, Hesketh JE and Pryme IF (1996) The compartmentalization of protein synthesis: Importance of cytoskeleton and role in mRNA targeting. Int J Biochem Cell Biol 28: 1089-1105

Huang M and Zhang L (1999) Association of the movement protein of alfalfa mosaic virus with the endoplasmic reticulum and its trafficking in epidermal cells of onion bulb scales. Mol Plant Micr Int 12: 680-690

Huang YT, Romito RR, De BP and Banerjee AK (1993) Characterization of the in vitro system for the synthesis of mRNA from human respiratory syncytial virus. Virology 193: 862-867

Ireton K and Cossart P (1998) Interaction of invasive bacteria with host signaling pathways. Curr Opin Cell Biol 10: 276-283

Itaya A, Hickman H, Bao Y, Nelson R and Ding B (1997) Cell-to-cell trafficking of cucumber mosaic virus movement protein:green fluorescent protein fusion produced by biolistic bombardment in tobacco. Plant J 12: 1221-1230

Ito Y, Abe S and Davies E (1994) Co-localization of cytoskeleton proteins and polysomes with a membrane fraction from peas. J Exp Bot 45: 253-259

Jansen RP (1999) RNA-cytoskeletal associations. FASEB J 13: 455-466

Kahn RW, Lapidot N, Heinlein M, Reichel C, Cooper B, Gafny R and Beachy RN (1998) Domains of the TMV movement protein involved in subcellular localization. Plant J 15: 15-25

Karasev AV, Kashina AS, Gelfand VI and Dolja VV (1992) HSP70-related 65 kDa protein of beet yellows closterovirus is a microtubule-binding protein. FEBS Lett 304: 12-14

Karpova OV, Ivanov KI, Rodionova NP, Dorokhov YL and Atabekov JG (1997) Nontranslatability and dissimilar behavior in plants and protoplasts of viral RNA and movement protein complexes formed *in vitro*. Virology 230: 11-21

Kim KS and Fulton JP (1975) An association of plant cell microtubules and virus particles. Virology 64: 560-565

Knebel W, Quader H and Schnepf E (1990) Mobile and immobile endoplasmic reticulum in onion bulk epidermis cells: Short- and long-term observations with a confocal laser scanning microscope. Eur J Cell Biol 52: 328-340

Kuznetsov SA, Langford GM and Weiss DG (1992) Actin-dependent organelle movement in squid axoplasm. Nature 356: 722-725

Langford GM (1995) Actin- and microtubule-dependent organelle motors: interrelationships between the two motility systems. Curr Opin Cell Biol 7: 82-88

Lanier LM and Volkman LE (1998) Actin binding and nucleation by *Autographa california* M nucleopolyhedrovirus. Virology 243: 167-177

Lartey R and Citovsky V (1997) Nucleic acid transport in plant-pathogen interactions. Genet Eng 19: 201-214

Lazarowitz SG (1999) Probing plant cell structure and function with viral movement proteins. Curr Opin Plant Biol 2: 332-338

Lazarowitz SG and Beachy RN (1999) Viral movement proteins as probes for intracellular and intercellular trafficking in plants. Plant Cell 11: 535-548

Li E, Stupack D, Bokoch GM and Nemerow GR (1998) Adenovirus endocytosis requires actin cytoskeleton reorganization mediated by Rho family GTPases. J Virol 72: 8806-8812

Lichtscheidl IK and Url WG (1990) Organization and dynamics of cortical endoplasmic reticulum in inner epidermal cells of onion bulb scales. Protoplasma 157: 203-215

Luby-Phelps K (1993) Effect of cytoarchitecture on the transport and localization of protein synthetic machinery. J Cell Biochem 52: 140-147

Luby-Phelps K (1994) Physical properties of cytoplasm. Curr Opin Cell Biol 6: 3-9

Lucas WJ (1995) Plasmodesmata: intercellular channels for macromolecular transport in plants. Curr Opin Cell Biol 7: 673-680

Lucas WJ and Gilbertson RL (1994) Plasmodesmata in relation to viral movement within leaf tissues. Annu Rev Phytopathol 32: 387-411

Luftig RB (1982) Does the cytoskeleton play a significant role in animal virus replication? J Theor Biol 99: 173-191

Mas P and Beachy RN (1998) Distribution of TMV movement protein in single living protoplasts immobilized in agarose. Plant J 15: 835-842

Maule AJ (1991) Virus movement in infected plants. Crit Rev Plant Sci 9: 457-473

McLean BG, Hempel FD and Zambryski PC (1997) Plant intercellular communication via plasmodesmata. Plant Cell 9: 1043-1054

McLean BG, Zupan J and Zambryski PC (1995) Tobacco mosaic virus movement protein associates with the cytoskeleton in tobacco cells. Plant Cell 7: 2101-2114

Meagher RB (1990) Divergence and differential expression of actin gene families in higher plants. Int Rev Cytol 125: 139-163

Melki R, Gaudin Y and Blondel D (1994) Interaction between tubulin and the viral matrix protein of vesicular stomatitis virus: possible implications in the viral cytopathic effect. Virology 202: 339-347

Moore PJ, Fenczik CA, Deom CM and Beachy RN (1992) Developmental changes in plasmodesmata in transgenic tobacco expressing the movement protein of tobacco mosaic virus. Protoplasma 170: 115-127

Moyer SA, Baker SC and Horikami SM (1990) Host cell proteins required for measles virus reproduction. J Gen Virol 71: 775-783

Moyer SA, Baker SC and Lessard JL (1986) Tubulin: a factor necessary for the synthesis of both Sendai virus and vesicular stomatitis virus RNAs. Proc Natl Acad Sci USA 83: 5405-5409

Muench DG, Wu Y, Coughlan SJ and Okita TW (1998) Evidence for a cytoskeleton-associated binding site involved in prolamine mRNA localization to the protein bodies in rice endosperm tissue. Plant Physiol 116: 559-569

Noueiry AO, Lucas WJ and Gilbertson RL (1994) Two proteins of a plant DNA virus coordinate nuclear and plasmodesmal transport. Cell 76: 925-932

Ohkawa T and Volkman LE (1999) Nuclear F-actin is required for AcMNPV nucleocapsid morphogenesis. Virology 264: 1-4

Oleynikov Y and Singer RH (1998) RNA localization: different zipcodes, same postman? Trends Cell Biol 8: 381-383

Oparka KJ (1993) Signalling via plasmodesmata–the neglected pathway. Semin Cell Biol 4: 131-138

Oparka KJ, Boevink P and Santa Cruz S (1996) Studying the movement of plant viruses using green fluorescent protein. Trends Plant Sci 1: 412-418

Oparka KJ, Prior DA, Santa Cruz S, Padgett HS and Beachy RN (1997) Gating of epidermal plasmodesmata is restricted to the leading edge of expanding infection sites of tobacco mosaic virus (TMV). Plant J 12: 781-789

Oparka KJ, Roberts AG, Boevink P, Santa Cruz S, Roberts I, Pradel KS, Imlau A, Kotlizky G, Sauer N and Epel B (1999) Simple, but not branched, plasmodesmata allow the nonspecific trafficking of proteins in developing tobacco leaves. Cell 97: 743-754

Oparka KJ, Roberts AG, Santa Cruz S, Boevink P, Prior DAM and Smallcombe A (1997) Using GFP to study virus invasion and spread in plant tissues. Nature 388: 401-402

Osman TA and Buck KW (1997) The tobacco mosaic virus RNA polymerase complex contains a plant protein related to the RNA-binding subunit of yeast eIF-3. J Virol 71: 6075-6082

Osman TAM, Hayes RJ and Buck KW (1992) Cooperative binding of the red clover necrotic mosaic virus movement protein to single-stranded nucleic acids. J Gen Virol 73: 223-227

Overall RL and Blackman LM (1996) A model of the macromolecular structure of plasmodesmata. Trends Plant Sci 1: 307-311

Pachter JS (1992) Association of mRNA with the cytoskeletal framework: its role in the regulation of gene expression. Crit Rev Euk Gene Exp 2: 1-18

Padgett HS, Epel BL, Kahn TW, Heinlein M, Watanabe Y and Beachy RN (1996) Distribution of tobamovirus movement protein in infected cells and implications for cell-to-cell spread of infection. Plant J 10: 1079-1088

Penfold ME, Armati P and Cunningham AL (1994) Axonal transport of herpes simplex virions to epidermal cells: evidence for a specialized mode of virus transport and assembly. Proc Natl Acad Sci USA 91: 6529-6533

Pickard BG and Ding JP (1993) The mechanosensory calcium-selective ion channel: Key component of a plasmalemmal control center? Aus J Plant Physiol 20: 439-459

Pickett-Heaps JD, Gunning BES, Brown RC, Lemmon BE and Cleary AL (1999) The cytoplast concept in dividing plant cells: cytoplasmic domains and the evolution of spatially organized cell division. Am J Bot 86: 153-172

Poirson A, Turner AP, Giovane C, Berna A, Roberts K and Godefroy-Colburn T (1993) Effects of the alfalfa mosaic virus movement protein expressed in transgenic plants on the permeability of plasmodesmata. J Gen Virol 74: 2459-2461

Pollard TD, Doberstein SK and Zot HG (1991) Myosin-I. Annu Rev Physiol 53: 653-681

Quadt R, Kao CC, Browning KS, Hershberger RP and Ahlquist P (1993) Characterization of a host protein associated with brome mosaic virus RNA-dependent RNA polymerase. Proc Natl Acad Sci USA 90: 1498-1502

Radford JE and White RG (1998) Localization of a myosin-like protein to plasmodesmata. Plant J 14: 743-750

Reichel C and Beachy RN (1998) Tobacco mosaic virus infection induces severe morphological changes of the endoplasmic reticulum. Proc Natl Acad Sci USA 95: 11169-11174

Reichel C, Mas P and Beachy RN (1999) The role of the ER and cytoskeleton in plant viral trafficking. Trends Plant Sci 4: 458-462

Reichelt S, Knight AE, Hodge TP, Baluska F, Samaj J, Volkmann D and Kendrick-Jones J (1999) Characterization of the unconventional myosin VIII in plant cells and its localization at the post-cytokinetic cell wall. Plant J 19: 555-567

Restrepo-Hartwig MA and Ahlquist P (1996) Brome mosaic virus helicase- and polymerase-like proteins colocalize on the endoplasmic reticulum at sites of viral RNA synthesis. J Virol 70: 8908-8916

Robards AW and Lucas WJ (1990) Plasmodesmata. Annu Rev Plant Physiol Plant Mol Biol 41: 369-419

Rosette C and Karin M (1995) Cytoskeletal control of gene expression: depolymerization of microtubules activates NF-kappa B. J Cell Biol 128: 1111-1119

Schaad MC, Jensen PE and Carrington JC (1997) Formation of plant RNA virus replication complexes on membranes: role of an endoplasmic reticulum-targeted viral protein. EMBO J 16: 4049-4059

Schoumacher F, Erny C, Berna A, Godefroy-Colburn T and Stussi-Garaud C (1992) Nucleic acid binding properties of the alfalfa mosaic virus movement protein produced in yeast. Virology 188: 896-899

Singer R (1992) The cytoskeleton and mRNA localization. Curr Opin Cell Biol 4: 15-19

Smith HMS and Raikhel NV (1998) Nuclear localization signal receptor importin α associates with the cytoskeleton. Plant Cell 10: 1791-1799

Sodeik B, Ebersold MW and Helenius A (1997) Microtubule-mediated transport of incoming herpes simplex virus 1 capsids to the nucleus. J Cell Biol 136: 1007-1021

St Johnston D (1995) The intracellular localization of messenger RNAs. Cell 81: 161-170

Staehelin LA (1997) The plant ER: a dynamic organelle composed of a large number of discrete functional domains. Plant J 11: 1151-1165

Stankovic B, Abe S and Davies E (1993) Co-localization of polysomes, cytoskeleton, and membranes with protein bodies form corn endosperm. Protoplasma 177: 66-72

Strauss EJ (1996) Intracellular pathogens: A virus joins the movement. Curr Biol 6: 504-507

Taneja KL, Lifshitz LM, Fay FS and Singer RH (1992) Poly(A) RNA codistribution with microfilaments: Evaluation by in situ hybridizaiton and quantitative digital imaging microscopy. J Cell Biol 119: 1245-1260

Theriot JA (1995) The cell biology of infection by intracellular bacterial pathogens. Annu Rev Cell Dev Biol 11: 213-239

Tilney LG, Cooke TJ, Connelly PS and Tilney MS (1991) The structure of plasmodesmata as revealed by plasmolysis, detergent extraction, and protease digestion. J Cell Biol 112: 739-747

Tomenius K, Clapham D and Meshi T (1987) Localization by immunogold cytochemistry of the virus coded 30K protein in plasmodesmata of leaves infected with tobacco mosaic virus. Virology 160: 363-371

Topp KS, Meade LB and LaVail JH (1994) Microtubule polarity in the peripheral processes of trigeminal ganglion cells: Relevance for the retrograde transport of herpes simplex virus. J Neurosci 14: 318-325

Turner A, Wells B and Roberts K (1994) Plasmodesmata of maize root tip–structure and composition. J Cell Sci 107: 3351-3361

Turpen TH (1999) Tobacco mosaic virus and the virescence of biotechnology. Phil Trans R Soc Lond B 354: 665-673

Vale RD (1987) Intracellular transport using microtubule-based motors. Annu Rev Cell Biol 3: 347-378

Vaquero C, Turner AP, Demangeat G, Sanz A, Serra MT, Roberts K and Garcia-Luque I (1994) The 3a protein from cucmber mosaic virus increases the gating capacity of plasmodesmata in transgenic tobacco plants. J Gen Virol 75: 3193-3197

Volkmann D and Baluska F (1999) Actin cytoskeleton in plants: from transport networks to signaling networks. Micros Res Tech 47: 135-154

Waigmann E, Lucas WJ, Citovsky V and Zambryski P (1994) Direct functional assay for tobacco mosaic virus cell-to-cell movement protein and identification of a domain involved in increasing plasmodesmal permeability. Proc Natl Acad Sci USA 91: 1433-1437

Waigmann E and Zambryski P (1995) TMV movement protein mediated protein transport between trichome cells. Plant Cell 7: 2069-2079

Wang X-C, Long M and Ming Y (1999) Actin is involved in the cell-to-cell trafficking of cucumber mosaic virus movement protein. In: Proceedings of the 1999 ASPP Plant Biology Meeting. Baltimore, MD. p 64

White RG, Badelt K, Overall RL and Vesk M (1994) Actin associated with plasmodesmata. Protoplasma 180: 169-184

Wilhelm JE and Vale RD (1993) RNA on the move: the mRNA localization pathway. J Cell Biol 123: 269-274

Williamson R (1986) Organelle movements along actin filaments and microtubules. Plant Physiol 82: 631-634

Wolf S, Deom CM, Beachy RN, and Lucas WJ (1989) Movement protein of tobacco mosaic virus modifies plasmodesmatal size exclusion limit. Science 246: 377-379

Wu Y, Muench DG, Kim Y-T, Hwang Y-S and Okita TW (1998) Identification of polypeptides associated with an enriched cytoskeleton-protein body fraction from developing rice endosperm. Plant Cell Physiol 39: 1251-1257

Yang J, Demma M, Warren V, Dharmawardhane S and Condeelis J (1990) Identification of an actin-binding protein from *Dictyostelium* as elongation factor 1α. Nature 347: 494-496

Zambryski P (1995) Plasmodesmata: Plant channels for molecules on the move. Science 270: 1943-1944

Chapter 30

INTERACTIONS BETWEEN THE ACTIN CYTOSKELETON AND AN AUXIN TRANSPORT PROTEIN

Gloria K Muday

Department of Biology, Wake Forest University, Winston–Salem NC, 27109 USA

Key words: actin, auxin transport, F-actin affinity chromatography, NPA-binding protein, plasma membrane, polarity

Abstract: In shoots, polar auxin transport is basipetal (i.e., from the shoot apex toward the base), and is driven by the basal localization of the auxin efflux carrier complex. One mechanism by which this efflux carrier complex could be localized to the basal membrane is through attachment to the actin cytoskeleton. The efflux carrier protein complex is believed to consist of several polypeptides, including a regulatory subunit that binds auxin transport inhibitors such as naphthylphthalamic acid (NPA). Several lines of experimentation have been used to determine whether the NPA-binding protein interacts with actin filaments. The NPA-binding protein has been shown to partition with the actin cytoskeleton during detergent extraction. Agents that specifically alter the polymerization state of the actin cytoskeleton also change the amount of NPA-binding protein and actin recovered in these cytoskeletal pellets. Actin affinity columns were prepared with polymers of actin purified from zucchini hypocotyl tissue. NPA-binding activity was eluted in a single peak from the actin filament column. Cytochalasin D, which fragments the actin cytoskeleton, was shown to reduce polar auxin transport in zucchini hypocotyls. The interaction of the NPA-binding protein with the actin cytoskeleton may localize it in one plane of the plasma membrane, and thereby control the polarity of auxin transport.

The mechanisms by which cells and tissues develop and maintain polarity is a growing area of study. In mammalian systems, a number of proteins have been examined to understand how asymmetric cellular localization is established and maintained. In asymmetric cells, such as epithelial or nerve cells, there are mechanisms both for initially sorting proteins into different membrane domains and for maintaining the

C.J. Staiger et al. (eds.), Actin: A Dynamic Framework for Multiple Plant Cell Functions, 541–556.

localization after initial sorting is complete (Drubin & Nelson, 1996). For several proteins, it has been shown clearly that an attachment to the actin cytoskeleton, either directly or through large protein complexes, mediates the localization (Froehner, 1993). Disruption of the actin cytoskeleton can randomize the distribution of these proteins and destroy the cellular asymmetry.

In plant cells, the best understood example of a protein with asymmetric cellular localization is the auxin efflux carrier, which controls polar auxin transport. Auxins, of which indole-3-acetic acid (IAA) is the predominant naturally occurring hormone, move through plants by a unique polar transport mechanism (reviewed in Goldsmith, 1977; Lomax et al., 1995). This polar movement of auxin is from the shoot meristem towards the base of stems, and is a cell-to-cell movement. Polar auxin transport results in an auxin gradient down the length of the plant, with the highest auxin concentrations found in the regions of greatest elongation (Ortuno et al., 1990). There are two protein complexes that control auxin movement into and out of cells. These protein complexes, the auxin uptake carrier and the auxin efflux carrier, respectively, are shown in Figure 1. IAA can move into cells both passively, as it is hydrophobic when protonated, and through an uptake carrier (Lomax et al., 1995). The amount and direction of polar auxin transport are thought to be controlled by the auxin efflux carrier. Basal localization of an auxin efflux carrier in cells of the plant stem has also been proposed to determine the polarity of IAA transport (Rubery & Sheldrake, 1974; Jacobs & Gilbert, 1983; Müller et al., 1998).

This review focuses on examination of the data which indicate that a subunit of the auxin efflux carrier interacts with the actin cytoskeleton. This actin association may fix this protein in one plane of the plasma membrane. It should also be noted that the initial establishment of the polarity of auxin efflux carrier may require the localized targeting of vesicles. Morris and Robinson (1998) have shown that monesin and Brefeldin A, inhibitors of Golgi vesicle secretion, also inhibit auxin efflux carrier activity. Additionally, treatments with Brefeldin A perturb the development of embryo polarity and localization of the PIN1 protein (a subunit of the efflux carrier; see below) in developing embryos (Steinmann et al., 1999). Together, these results demonstrate that polypeptides of the efflux carrier move through the Golgi vesicle transport pathway and that this pathway could differentially target proteins to the basal membrane. The interaction of the efflux carrier with the actin cytoskeleton could then function to maintain this asymmetric localization.

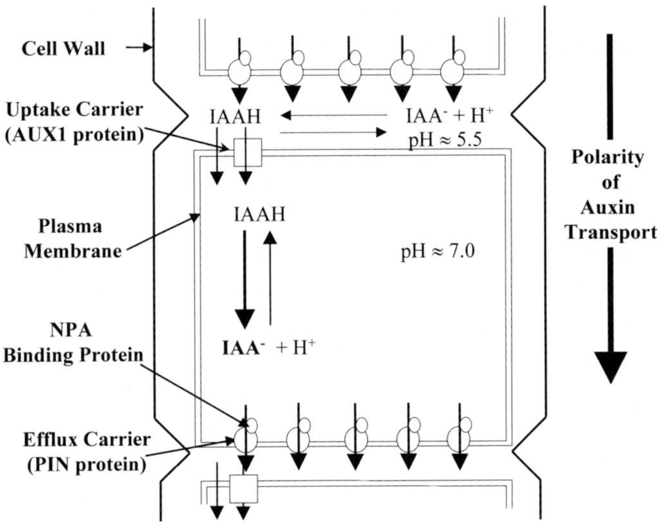

Figure 1. Schematic model of the chemiosmotic hypotheis for polar auxin transport. The protein complexes that mediate auxin movement are shown and the names of individual polypeptides are in parentheses, when known. (Modified from Lomax et al., 1995).

1. BIOCHEMICAL CHARACTERIZATION OF THE AUXIN EFFLUX CARRIER

A biochemical dissection of the auxin efflux carrier will increase our understanding of the regulation of this protein complex, and how its localization to the basal plasma membrane controls the polarity of auxin movement. The efflux carrier complex appears to be composed of more than one polypeptide: an integral membrane transporter encoded by the one of the members of the PIN gene family, an NPA-binding protein that may act as a regulatory polypeptide (Dixon et al., 1996), and perhaps a third, rapidly turned over protein that connects these two subunits (Morris et al., 1991). Several members of the PIN gene family in *Arabidopsis* have been identified (Gälweiler et al., 1998; Müller et al., 1998), indicating that there are multiple auxin efflux carriers with distinct expression patterns. Plants with mutations in either of two genes of this family have phenotypes consistent with tissue-specific alterations in auxin transport (Okada et al., 1991; Gälweiler et al., 1998; Müller et al., 1998) and have alterations in auxin transport in the affected tissues (Okada et al., 1991; Chen et al., 1998; Rashotte et al., 2000).

PIN genes encode proteins with 10 membrane-spanning domains and similarity to other membrane transport proteins (Chen et al., 1999). The protein products of these genes show an asymmetric localization in the plasma membrane that is consistent with controlling the polarity of auxin movement (Gälweiler et al., 1998; Müller et al., 1998). It has been suggested, therefore, that the PIN genes encode one polypeptide of the auxin efflux carrier.

Until the recent identification of the PIN proteins, most studies of the auxin efflux carrier focused on the NPA-binding protein. The activity of this protein can be followed using a binding assay with [^{3}H]-NPA. For these assays, radiolabeled NPA is incubated with membrane vesicles or solublilized protein. The protein-ligand complexes are recovered by filtration or centrifugation. The NPA ligand binds with high affinity to a single class of NPA-binding proteins associated with the zucchini plasma membrane (Muday et al., 1993). Therefore, this assay has allowed extensive biochemical characterization of the NPA-binding protein.

Several lines of evidence suggest that the protein that binds inhibitors of auxin efflux, including NPA, is distinct from the PIN gene products. Treatments with inhibitors of protein translation and protein processing in the Golgi apparatus reduce the regulation of auxin transport by NPA, without altering the amount of NPA-binding activity (Morris et al., 1991; Wilkinson & Morris, 1994; Morris & Robinson, 1998). These results suggest that the NPA-binding and auxin-efflux activities are on separate proteins, and support the idea that a third protein may connect them (Morris et al., 1991).

1.1 The NPA-binding protein is a peripheral membrane protein

Additionally, it appears that the NPA-binding protein is peripherally associated with the plasma membrane. Treatment of plasma membrane vesicles with potassium iodide (KI) or sodium bicarbonate released the NPA-binding protein into the supernatant after ultracentrifugation, suggesting that NPA binds to a peripheral protein (Cox & Muday, 1994). Furthermore, the NPA-binding protein is still active in detergent-insoluble pellets. These pellets should be almost free of lipids, yet the majority of NPA-binding activity was recovered in these pellets suggesting that the NPA-binding protein does not require a lipophilic environment for activity (Cox & Muday, 1994; Butler et al., 1998). Therefore, our current model is that the NPA-binding protein is a peripheral membrane, regulatory protein. This model is consistent with the results of Morris et al. (1991) and

Wilkinson and Morris (1994) that indicate that NPA-binding activity and auxin-efflux activity are on two distinct polypeptides.

1.2 The NPA-binding protein is localized to the cytoplasmic face of the membrane

Biochemical evidence suggests that NPA-binding activity is localized to the cytoplasmic face of the plasma membrane. Several investigators have examined the protease sensitivity of NPA-binding activity in plasma membranes isolated from zucchini hypocotyls. Treatment of intact right-side-out vesicles with protease does not lead to loss of NPA-binding activity (Dixon et al., 1996, Bernasconi et al., 1996), which is consistent with a cytoplasmic localization. In contrast, disruption of membranes by detergent, followed by protease treatment, results in a total loss of NPA-binding activity (Bernasconi et al., 1996). Furthermore, plasma membrane vesicles have been subjected to several different treatments that should convert them to an inside-out orientation (Hertel et al., 1983; Dixon et al., 1996; Bernasconi et al., 1996), although the effectiveness of these treatments was only verified by analysis of marker enzymes in one case (Dixon et al., 1996). In two of these reports, including the example in which the ability of treatments to convert vesicles to an inside-out orientation was verified, inside-out vesicles had increases in both NPA-binding activity and protease sensitivity of the binding activity (Hertel et al., 1983; Dixon et al., 1996). Therefore, the NPA-binding site appears to be localized to the cytoplasmic face of the membrane and poised for interaction with the cytoskeleton.

1.3 NPA-binding protein partitions with actin during detergent extraction

The first step of purification of integral membrane proteins is the treatment with detergent to solubilize the protein and release it from the membrane. As it was initially assumed that the NPA-binding activity and the auxin efflux carrier activity were localized to the same polypeptide, most investigators initiated experiments using detergent solubilization with the goal of releasing NPA-binding activity. Although several reports in the literature indicate that NPA-binding activity can be released from the membrane by detergent treatment, all of these procedures resulted in very low yields of soluble NPA-binding activity (Sussman & Gardner, 1980; Jacobs & Gilbert, 1983; Cox & Muday, 1994; Bernasconi et al., 1996). In two of these reports, the amount of NPA-binding activity in the detergent-insoluble pellet was quantified and, in both cases, the majority of the activity was in the detergent-insoluble pellet (Sussman & Gardner, 1980; Cox & Muday, 1994). The insolubility of the NPA-binding protein during detergent extraction may be due to interaction with the cytoskeleton, as cytoskeletal-associated proteins show this behavior (Carraway, 1992).

The first study that addressed whether the detergent insolubility of the NPA binding protein was due to association with the cytoskeleton was performed by Cox and Muday (1994). After treating purified zucchini plasma membranes with Triton X-100, the presence of both NPA-binding activity and actin and tubulin polypeptides were examined in the pellet and supernatant fractions following ultracentrifugation (Cox & Muday, 1994). Actin, tubulin and NPA-binding activity all partitioned preferentially into the detergent-insoluble pellet. Treatment of the detergent-insoluble or cytoskeletal pellet with cytochalasin B, a drug that fragments the filamentous form of actin (F-actin), released [^{3}H]-NPA-binding activity into the supernatant after ultracentrifugation. Use of this drug *in vitro* caused the release of both actin and tubulin cytoskeletal fragments, so these experiments could not differentiate between association with actin or tubulin (Cox & Muday, 1994).

Although this initial study supported the argument that the NPA-binding protein interacts with the cytoskeleton, the use of purified plasma membranes with treatments designed to alter actin polymerization was not optimal. Rather, the use of fresh and relatively crude extracts of zucchini hypocotyl proteins proved to be a better method for determining whether the NPA-binding protein was interacting with the cytoskeleton, and for determining which cytoskeletal polymer was the site of interaction (Butler et al., 1998). Butler et al. (1998) also found that NPA-binding activity and actin partitioned into the cytoskeleton pellet after detergent extraction of fresh extracts. There was very little tubulin polypeptide found in these extracts, suggesting an interaction between the NPA-binding protein and actin.

As a more direct test for the interaction between the NPA-binding protein and actin, intact zucchini hypocotyls were treated *in vivo* with three drugs that alter cytoskeletal organization (Butler et al., 1998). After drug treatment, extracts were prepared, treated with detergent, and the amount of NPA-binding activity in the detergent-insoluble cytoskeletal pellets was measured. Phalloidin and cytochalasin D act on the actin cytoskeleton to stabilize or to fragment polymers, respectively. Treatment with phalloidin increased both the amount of pelletable actin and NPA-binding activity, while cytochalasin D treatment decreased both pelletable actin and NPA-binding activity. In contrast, taxol treatment stabilized microtubules, resulting in an increase in pelletable tubulin after detergent solubilization, but not an increase in pelletable NPA-binding activity (Butler et al., 1998).

Butler et al. (1998) used one additional treatment to depolymerize actin. The buffer Tris, at high concentrations, has been reported previously to lead to actin depolymerization (Pinder et al., 1995). Treatment of detergent-insoluble pellets from either fresh extracts or plasma membranes isolated from zucchini hypocotyls with Tris lead to a dose-dependent decrease in both pelletable actin and NPA-binding activity (Butler et al., 1998). Initial experiments using Tris did not result in a concomitant increase in NPA-binding activity in the detergent supernatant, so it was not clear whether Tris was releasing NPA-binding activity or denaturing the NPA-binding activity. In order to stabilize the NPA-binding protein during its release, the NPA ligand was included during the detergent extraction and Tris treatment. The Tris was removed and the pH lowered to a level more optimal for NPA binding, resulting in the recovery of NPA-binding activity in the supernatant (Butler et al., 1998). Therefore, it appeared that Tris released NPA-binding activity, and did not denature the protein (Butler et al., 1998). Together, these results were consistent with an association of the NPA-binding protein with the actin cytoskeleton, although the interactions were only demonstrated indirectly. The next step in demonstration of this interaction was to show that the NPA-binding protein could bind *in vitro* to homogenous and purified actin filaments.

1.4 Interaction of the NPA-binding protein with actin filaments *in vitro*

The first approach to demonstrate the interaction of the NPA-binding protein with actin filaments was to subject detergent-insoluble cytoskeletal pellets to rounds of polymerization and depolymerization. Throughout these cycles of actin polymerization, both the location of actin polypeptide and NPA-binding activity were followed. Initially, actin and NPA-binding activity were recovered in the detergent-insoluble pellet. Upon treatment with KI, both actin and NPA-binding activity moved into the supernatant, as filaments were depolymerized (Cox & Muday, 1994). The KI was removed to allow actin filaments to reform and the resulting sample subjected to centrifugation. After this treatment, both actin filaments and NPA-binding activity were found predominantly in the pellet. Although NPA-binding activity was preferentially partitioned into the samples that were enriched in actin, there was very low recovery. Furthermore, both actin and tubulin were enriched in these samples, suggesting that this procedure was not a very specific way to recover actin polymers (Cox & Muday, 1994). Therefore, an alternative approach was developed to more directly examine actin interactions.

To test the interaction of the NPA-binding protein with purified actin filaments directly, F-actin affinity columns were prepared and their ability to retain NPA-binding activity was examined. In all plants examined to date, actin is encoded by a multigene family (see Meagher et al., this volume). In *Arabidopsis*, there are eight expressed genes, for which there are distinct tissue-specific expression patterns (Kandasamy et al., 1999). Therefore, it was critical to obtain actin isoforms from the tissues that are known to transport auxin and to possess NPA-binding activity, in case variations in actin–actin-binding protein interactions exist.

There are no procedures in the literature for purification of actin that is competent for polymerization from hypocotyls although procedures to purify maize pollen actin have been published (Liu & Yen, 1992; Ren et al., 1997). For purification of animal and lower eukaryote actin, several reports document the use of affinity columns prepared with the enzyme DNase I coupled to a solid support (Sheterline et al., 1998; Zechel, 1980). DNase I binds to G-actin with a 1:1 ratio and a high affinity, is commercially available, and has been used previously to partially purify actin from pea roots (Andersland et al., 1992). Therefore, this approach was chosen to purify zucchini hypocotyl actin.

Actin was purified to electrophoretic homogeneity from zucchini hypocotyls using DNase I chromatography followed by ultracentrifugation (Hu et al., 2000). Since actin was eluted from the DNase I resin with formamide, which can denature proteins, it was particularly critical to demonstrate that this actin was native. First, the profilin-binding activity of purified zucchini hypocotyl actin was compared to purified and native maize pollen actin isolated according to Ren et al. (1997). The profilin-binding ability was compared using two isoforms of maize profilin, one expressed in pollen (ZmPRO1) and one expressed predominantly in vegetative tissues (ZmPRO5). The resulting K_d values for these two actin pools were not statistically different under these conditions (Hu et al., 2000). The native structure of the actin was also confirmed by the ability of the purified actin to bind and inhibit DNase I activity (Hu & Muday, unpublished results).

The ability of purified zucchini hypocotyl actin to form filaments was demonstrated by sedimentation of F-actin during ultracentrifugation, decreased mobility of F-actin on native gels, and by examination of actin filaments with electron microscopy (Hu et al., 2000). Ultrastructural examination of *in vitro* polymerized actin showed helical filaments with a width of 6.8 nm (Hu et al., 2000), consistent with the conformation of maize actin filaments assembled *in vitro* (Ren et al., 1997). Together, these results demonstrated that our protocol can be used to purify native zucchini hypocotyl actin that is competent for polymerization.

Purified, native zucchini hypocotyl actin was then used to prepare both G- and F-actin columns. BSA was used to create a third affinity matrix to test for non-specific protein interactions (Hu et al., 2000). The selectivity of the F-actin column was demonstrated by examination of binding of vertebrate α-actinin to the F-actin column. Purified α -actinin was shown to bind tightly to the F-actin, but weakly to the G-actin column (Hu et al., 2000).

Since the NPA-binding protein is associated with the plasma membrane, isolated plasma membranes were used as the starting sample for chromatography on the actin columns. Plasma membranes were treated with Triton X-100 and Tris, resulting in the recovery of NPA-binding activity in the supernatant after ultracentrifugation. This soluble sample was applied to the actin or BSA columns. Eluted protein samples were analysed for NPA-binding activity and were examined by silver stain after SDS-PAGE. NPA-binding activity was retained by the F-actin column and reproducibly eluted with high salt concentrations. In five separate experiments, NPA-binding activity was localized to one or two fractions eluted from the F-actin column and was significantly greater than the activity eluted from a BSA column or an F-actin column to which no solubilized proteins were applied (Hu et al., 2000).

The elution of NPA-binding activity from the F-actin column is the strongest evidence to date indicating the association of the NPA-binding protein with the actin cytoskeleton. A second question was whether these columns would retain sufficient quantities of the NPA-binding protein to allow isolation of this protein for amino acid sequence analysis. In order to examine the pattern of proteins retained by the column, the samples were subjected to SDS-PAGE followed by silver staining. There was no consistent band found in samples that contained NPA-binding activity (Hu & Muday, unpublished results). In contrast, there were routinely two proteins that eluted from the F-actin column at salt concentrations that were higher than required to elute NPA-binding activity (Hu et al., 2000). These two proteins of 30–35 kDa were recognized by annexin antisera (Hu et al., 2000). As annexins are plasma membrane proteins that have been shown to interact with the actin cytoskeleton (Calvert et al., 1996), this result further supports the ability of these columns to retain F-actin binding proteins in a specific fashion.

2. DRUGS THAT FRAGMENT THE ACTIN CYTOSKELETON REDUCE POLAR AUXIN TRANSPORT

The interaction of the NPA-binding protein with the actin cytoskeleton may be necessary for either movement of auxin across the membrane or for the polar localization of the efflux carrier complex. If either of these hypotheses are correct, then disruption of the actin cytoskeleton would be predicted to reduce polar auxin transport. Treatment of either corn coleoptiles (Cande et al., 1973) or zucchini hypocotyls (Butler et al., 1998) with cytochalasins have been shown to reduce auxin transport.

The effect of cytochalasin D on auxin transport in zucchini hypocotyls was measured using a modification of previously published assays (Butler et al., 1998). Zucchini hypocotyl segments treated with and without cytochalasin D are simultaneously loaded with [^{3}H]-IAA and [^{14}C]-benzoic acid, and the amount of radioactivity transported out of each end of the segment is recovered in agar blocks. When zucchini hypocotyls are treated with cytochalasin D, there is a statistically significant reduction (as judged by student's *t* test) in basipetal auxin transport, as shown in Table 1 (data from Butler et al., 1998). The average and standard error of 12 separate experiments are reported. This reduction in transport is not at the level of diffusion, as there are no changes in the amount of either basipetal benzoic acid movement or acropetal auxin transport. As this assay measures passive diffusion from the segment as well as polar transport, each measurement contains some background diffusion. The level of background diffusion can be assessed by examining the percentage of either acropetal auxin movement or benzoic acid diffusion. Basipetal IAA transport can be normalized by subtracting the amount of diffusion (calculated by averaging the background values), and subtracted from the value for basipetal auxin transport. The magnitude of the effect of cytochalasin D treatment increases to two-fold when the normalized values are compared (Table 1).

Table 1. Cytochalasin D Reduces Polar Auxin Transport.

	% Transport		
	– Cytochalasin D	+ Cytochalasin D	*P* value
Basipetal IAA	19.0 ± 1.2	13.3 ± 1.1	< 0.005
Acropetal IAA	8.0 ± 0.4	8.8 ± 0.8	> 0.02
Basipetal BA	7.0 ± 0.9	7.0 ± 0.8	> 0.02
Normalized Basipetal IAA	11.3	5.6	

If an intact actin cytoskeleton is required for localization of the auxin efflux carrier complex, should fragmentation of actin filaments with cytochalasin lead to a total loss of auxin transport? This question can be considered by examination of the model in Figure 2. If the efflux carriers are totally randomized, then transport should be at the level of diffusion. Basipetal IAA transport is not reduced to the level of diffusion by cytochalasin. However, the cytochalasin D treatment was for 1 hour, and this may not have been of sufficient time to allow all the efflux carriers to randomize. Additionally, there may have been partial recovery of polar auxin transport capacity during the 1.5 hour transport period. These results are consistent with the model shown in Figure 2, although more complex possibilities cannot yet be eliminated.

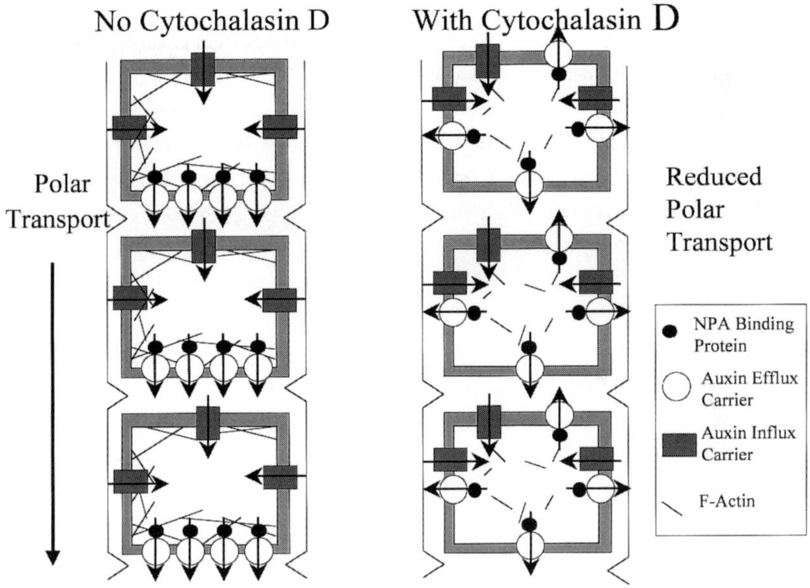

Figure 2. Model for the effect of cytochalasin D on polar auxin transport. A file of untreated cells (left) is compared to cells treated with cytochalasin D (right). The fragmentation of actin filaments by cytochalasin D is shown, as well as the predicted randomization of the auxin efflux carrier complex as a result of loss of actin structure.

3. AUXIN TRANSPORT MAY INFLUENCE THE DEVELOPMENT OF ORGAN POLARITY

The experimental results discussed above suggest a mechanism for the maintenance of the auxin efflux carrier asymmetric cellular localization through attachment to the actin cytoskeleton. It is also possible that this auxin efflux carrier asymmetry may play an important role in the development of asymmetry on the organ level, as well. Numerous studies have found that treatment of developing embryos with auxin transport inhibitors leads to abnormal pattern formation (Schiavone & Cooke, 1987; Liu et al., 1993; Fischer & Neuhaus, 1996). Very similar developmental alterations can be observed in *Arabidopsis* mutants that have alterations in auxin transport, as are seen when *Brassica* embryos are treated with auxin transport inhibitors (Hadfi et al., 1998). Additionally, treatment of the single-celled symmetric zygotes of the brown algae, *Fucus*, with auxin transport inhibitors perturbs the normal developmental program and the ability of zygotes to form polarity in response to environmental gradients (Basu & Muday, unpublished observations). Changes in the organization of the actin cytoskeleton have been implicated in the early stages of *Fucus* polarity development. Therefore, these results suggest that the interactions between the actin cytoskeleton and auxin transport proteins could be important in the initial stages of polarity development as embryos go from one symmetric cell to two unequal and polar cells.

4. CONCLUSIONS

The results from these studies indicate that a regulatory subunit of the auxin efflux carrier, the NPA-binding protein, binds directly to actin filaments. The actin association of the NPA-binding protein may localize the auxin efflux carrier complex in one plane of the plasma membrane and thereby control the polarity of auxin transport. Gälweiler et al. (1998) and Müller et al. (1998) have used antibodies that recognize two different isoforms of an integral membrane protein of the auxin efflux carrier to show that these proteins are localized to one plane of the plasma membrane. Additionally, Butler et al. (1998) and Cande et al. (1973) have shown that treatment of zucchini hypocotyls and corn coleoptiles, respectively, with cytochalasins reduce polar auxin transport, consistent with the actin cytoskeleton being required to maintain the polar distribution of auxin transport proteins. Our studies have shown that a key regulatory subunit of the auxin efflux carrier, the NPA-binding protein, binds actin filaments (Hu et al., 2000), and suggest that this interaction may be one mechanism that controls the polarity of auxin movement.

Other studies in both plants and animals have demonstrated the importance of the actin cytoskeleton in establishment and maintenance of cell polarity. In yeast and the brown alga *Fucus*, initial establishment of cell polarity requires an intact actin cytoskeleton (Goodner & Quantrano, 1993; Li et al., 1995) and is preceded by changes in the organization of the actin cytoskeleton (Ayscough & Drubin, 1996; Kropf et al., 1989; Alessa & Kropf, 1999; Chant, 1999). In order to maintain cellular polarity, a number of plasma membrane proteins that have asymmetric localization maintain their distribution by attachment to the actin cytoskeleton. In the developing zygotes of *Fucus*, the dihydropyridine receptor has been shown to develop asymmetric localization that also requires an intact actin cytoskeleton (Shaw & Quatrano, 1996). Both the acetylcholine receptor of neurons, and the Na^+, K^+-ATPase of epithelial cells have polar distributions that are required for their function (Froehner, 1993; Apel & Merlie, 1995). Evidence for both of these protein complexes indicates that attachment to the actin cytoskeleton controls their localization (Nelson & Hammerton, 1989; Froehner, 1993). The acetylcholine receptor is a particularly interesting example, since one of the proteins in the complex is a 43-kDa peripheral membrane protein (rapsyn) which is associated with the actin cytoskeleton. In mice deficient for this protein, the acetylcholine receptor fails to localize properly (Gautam et al., 1995). The NPA-binding protein may function in a similar way to localize the efflux carrier.

Differential actin association of the NPA-binding protein may also provide a level of regulation of auxin transport. Morris and Johnson (1990) suggest that in tissues with reduced polar auxin transport, the reduction in transport may be due to a randomization of the location of the efflux carrier, rather than to changes in abundance of the efflux carrier. This could be mediated by a decrease in actin association of the NPA-binding protein. Additionally, auxin transport is reduced in response to ethylene treatment (Suttle, 1988) and this reduction may be controlled by changes in cytoskeletal attachment of the NPA-binding protein (Ebenezer, 1997). The ethylene induced reduction in auxin transport is accompanied by a statistically significant ($P < 0.01$) decrease in the amount of NPA-binding activity in detergent-insoluble cytoskeletal pellets, although the abundance of the protein in crude extracts is constant (Ebenezer, 1997). Consequently, it may be that alteration in cytoskeletal association of the NPA-binding protein acts to control the amount of polar auxin transport.

Together, these results suggest that the NPA-binding protein is associated with the actin cytoskeleton *in vitro*, and that this association is required for maximal auxin transport and its regulation by NPA *in vivo*. Although the significance of this association is not yet clear, it may be that attachment of the NPA-binding protein to the actin cytoskeleton serves to localize this protein in one plane of the plasma membrane and to thereby control the polarity of auxin transport.

ACKNOWLEDGEMENTS

We appreciate the helpful comments during the preparation of this manuscript from members of the Muday lab and Jennifer Waters Shuler. This work was supported by grants to GKM from the NSF (IBN-9318250) and NASA (NAG2-1203), as well as by the NASA Specialized Center for Research and Training at North Carolina State University.

REFERENCES

Alessa L and Kropf DL (1999) F-actin marks the rhizoid pole in living *Pelvetia compressa* zygotes. Development 126: 201-209

Andersland JM, Jagendorf AT and Parthasarathy MV (1992) The isolation of actin from pea roots by DNase I affinity chromatography. Plant Physiol 100: 1716-1723

Apel ED and Merlie JP (1995) Assembly of the postsynaptic apparatus. Curr Opin Neurobiol 5: 62-67

Ayscough KR and Drubin DG (1996) Actin: general principles from studies in yeast. Annu Rev Cell Dev Biol 12: 129-160

Bernasconi P, Bhavesh CP, Reagan JD and Subramanian MV (1996) The N-1-naphthylphthalamic acid-binding protein is an integral membrane protein. Plant Physiol 111: 427-432

Butler JH, Hu S, Brady SR, Dixon MW and Muday GK (1998) In vitro and in vivo evidence for actin association of the naphthylphthalamic acid-binding protein from zucchini hypocotyls. Plant J 13: 291-301

Calvert CM, Gant SJ and Bowles DJ (1996) Tomato annexins p34 and p35 bind to F-actin and display nucleotide phosphodiesterase activity inhibited by phospholipid binding. Plant Cell 8: 333-342

Cande WZ, Goldsmith MHM and Ray PM (1973) Polar auxin transport and auxin-induced elongation in the absence of cytoplasmic streaming. Planta 111: 279-296

Carraway C (1992) Association of cytoskeletal proteins with membranes. In: Carraway K, Carraway C (eds) The Cytoskeleton: A Practical Approach. Oxford: IRL Press, pp 123-150

Chant J (1999) Cell polarity in yeast. Annu Rev Cell Dev Biol 15: 365-391

Chen R, Hilson P, Sedbrook J, Rosen E, Caspar T and Masson PH (1998) The *Arabidopsis thaliana AGRAVITROPIC1* gene encodes a component of the polar-auxin-transport efflux carrier. Proc Natl Acad Sci USA 95: 15112-15117

Chen R, Rosen E and Masson PH (1999) Gravitropism in higher plants. Plant Physiol 120: 343-350

Cox DN and Muday GK (1994) NPA binding activity is peripheral to the plasma membrane and is associated with the cytoskeleton. Plant Cell 6: 1941-1953

Dixon MW, Jacobson JA, Cady CT and Muday GK (1996) Cytoplasmic orientation of the naphthylphthalamic acid-binding protein in zucchini plasma membrane vesicles. Plant Physiol 112: 421-432

Drubin DG and Nelson WJ (1996) Origins of cell polarity. Cell 84: 335-344

Ebenezer C (1997) Regulation of auxin transport through changes in the cytoskeletal association of the NPA binding protein. MS Thesis. Wake Forest University, Winston-Salem, NC

Fischer C and Neuhaus G (1996) Influence of auxin on the establishment of bilateral symmetry in monocots. Plant J 9: 659-669

Froehner SC (1993) Regulation of ion channel distribution at synapses. Annu Rev Neurosci 16: 347-368

Gälweiler L, Guan C, Müller A, Wisman E, Mendgen K, Yephremov A and Palme K (1998) Regulation of polar auxin transport by AtPIN1 in *Arabidopsis* vascular tissue. Science 282: 2226-2230

Gautam M, Noakes PG, Mudd J, Nichol M, Chu GC, Sanes JR and Merlie JP (1995) Failure of postsynaptic specialization to develop at neuromuscular junctions of rapsyn-deficient mice. Nature 377: 232-236

Goldsmith MHM (1977) The polar transport of auxin. Annu Rev Plant Physiol 28: 439-478.

Goodner B and Quantrano RS (1993) *Fucus* embryogenesis: a model to study the establishment of polarity. Plant Cell 5: 1471-1481

Hadfi K, Speth V and Neuhaus G (1998) Auxin-induced developmental patterns in *Brassica juncea* embryos. Development 125: 879-887

Hertel R, Lomax TL and Briggs WR (1983) Auxin transport in membrane vesicles from *Cucurbita pepo* L. Planta 157: 193-201

Hu S, Brady SR, Kovar D, Staiger CJ, Clarke GB, Roux SJ and Muday G (2000) Identification of plant F-actin-binding proteins by F-actin chromatography. Plant J, In review

Jacobs M and Gilbert SF (1983) Basal localization of the presumptive auxin transport carrier in pea stem cells. Science 220: 1297-1300

Kandasamy M, McKinney E and Meagher R (1999) The late pollen-specific actins in angiosperms. Plant J 18: 681-691

Kropf DL, Berge SK and Quatrano RS (1989) Actin localization during *Fucus* embryogenesis. Plant Cell 1: 191-200

Li R, Zheng Y and Drubin DG (1995) Regulation of cortical actin cytoskeleton assembly during polarized cell growth in budding yeast. J Cell Biol 128: 599-615

Liu C-M, Xu Z-H and Chua N-H (1993) Auxin polar transport is essential for the establishment of bilateral symmetry during early plant embryogenesis. Plant Cell 5: 621-630

Liu X and Yen L-F (1992) Purification and characterization of actin from maize pollen. Plant Physiol 99: 1151-1155

Lomax TL, Muday GK and Rubery P (1995) Auxin transport. In: Davies PJ (ed) Plant Hormones: Physiology, Biochemistry, and Molecular Biology. Norwell: Kluwer Academic Press, pp 509-530

Morris DA and Johnson CF (1990) The role of auxin efflux carriers in the reversible loss of polar auxin transport in the pea (*Pisum sativum* L.) stem. Planta 181: 117-124

Morris DA and Robinson JS (1998) Targeting of auxin carriers to the plasma membrane: differential effects of brefeldin A on the traffic of auxin uptake and efflux carriers. Planta 205: 606-612

Morris DA, Rubery PH, Jarman J and Sabater M (1991) Effects of inhibitors of protein synthesis on transmembrane auxin transport in *Cucurbita pepo* L. hypocotyl segments. J Exp Bot 42: 773-783

Muday GK, Brunn SA, Haworth P and Subramanian M (1993) Evidence for a single naphthylphthalamic acid binding site on the zucchini plasma membrane. Plant Physiol 103: 449-456

Müller A, Guan C, Gälweiler L, Tanzler P, Huijser P, Marchant A, Parry G, Bennett M, Wisman E and Palme K (1998) *AtPIN2* defines a locus of *Arabidopsis* for root gravitropism control. EMBO J 17: 6903-6911

Nelson WJ and Hammerton RW (1989) A membrane-cytoskeletal complex containing Na^+,K^+-ATPase, ankyrin, and fodrin in Madin-Darby canine kidney (MDCK) cells: implications for the biogenesis of epithelial cell polarity. J Cell Biol 108: 893-902

Okada K, Ueda J, Komaki MK, Bell CJ and Shimura Y (1991) Requirement of the auxin polar transport system in early stages of Arabidopsis floral bud formation. Plant Cell 3: 677-684

Ortuno A, Sanchez-Bravo J, Moral JR, Acosta M and Sabater F (1990) Changes in the concentration of indole-3-acetic acid during the growth of etiolated lupin hypocotyls. Physiol Plant 78: 211-217

Pinder JC, Sleep JA, Bennett PM and Gratzer WB (1995) Concentrated Tris solutions for the preparation, depolymerization, and assay of actin; application to erythroid actin. Anal Biochem 225: 291-295

Rashotte A, Brady S, Reed R, Ante S and Muday G (2000) Basipetal auxin transport is required for gravitropism in roots of Arabidopsis. Plant Physiol 122: 481-490

Ren H, Gibbon BC, Ashworth SL, Sherman DM, Yuan M and Staiger CJ (1997) Actin purified from maize pollen functions in living plant cells. Plant Cell 9: 1445-1457

Rubery PH and Sheldrake AR (1974) Carrier-mediated auxin transport. Planta 118: 101-121

Schiavone FM and Cooke TJ (1987) Unusual patterns of somatic embryogenesis in the domesticated carrot: developmental effects of exogenous auxins and auxin transport inhibitors. Cell Differ 21: 53-62

Shaw SL and Quatrano RS (1996) Polar localization of a dihydropyridine receptor on living *Fucus* zygotes. J Cell Sci 109: 335-342

Sheterline P, Clayton J and Sparrow JC (1998) Actin. Protein Profile 4: 1-272

Steinmann T, Geldner N, Grebe M, Mangold S, Jackson C, Paris S, Gälweiler L, Palme K and Jurgens G (1999) Coordinated polar localization of auxin efflux carrier PIN1 by GNOM ARF GEF. Science 268: 316-318

Sussman MR and Gardner G (1980) Solubilization of the receptor for N-1-napthylphthalamic acid. Plant Physiol 66: 1074-1078

Suttle JC (1988) Effect of ethylene treatment on polar IAA transport, net IAA uptake and specific binding of N-1-naphthylphthalamic acid in tissues and microsomes isolated from etiolated pea epicotyls. Plant Physiol 88: 795-799

Wilkinson S and Morris DA (1994) Targeting of auxin carriers to the plasma membrane: effects of monensin on transmembrane auxin transport in *Cucurbita pepo* L. tissue. Planta 193: 194-202

Zechel K (1980) Isolation of polymerization-competent cytoplasmic actin by affinity chromatography on immobilized DNase I using formamide as eluant. Eur J Biochem 110: 343-348

Chapter 31

ACTIN CYTOSKELETON RELATED TO GRAVISENSING IN HIGHER PLANTS

Dieter Volkmann and František Baluška

Botanisches Institut der Universität Bonn, Kirschallee 1, D–53115 Bonn, Germany

Key words: actin, amyloplasts, gravisensing, root cap, statocytes, statoliths

Abstract: Gravisensing cells (statocytes) from plant root caps are characterized by a polar arrangement of organelles and sedimented amyloplast-based statoliths. Immunofluorescence microscopy fails to visualize prominent actin filaments in statocytes but indicates a highly dynamic cytoskeletal network, composed at least of actin, myosin-like proteins and profilin, surrounding sedimented statoliths. Experiments under microgravity demonstrated that the positioning of statoliths depends on the external gravitational force and on endocellular cytoskeleton-based forces exerted on their surfaces. Accepting the amyloplast-based statolith hypothesis, these results strongly suggest that gravisensing occurs in a close vicinity of statolith surfaces. Experiments with grass nodes revealed transient changes of the signalling molecule IP_3 within few seconds after gravistimulation. The importance of mutants for dissecting the gravity-related signal transduction chains is highlighted.

1. INTRODUCTION

Gravisensing belongs to the general category of mechanosensing processes which encompass also the perception of sound and pressure in animals, and touch and osmosis in plants. The most outstanding feature of mechanosensory perception is the extreme velocity of the process, having latencies of less than one millisecond (Kernan & Zuker, 1995). By this characteristic, the mechanosensing process is faster than any other biological process transforming environmental stimuli into diverse cellular responses, including photosensing (Hargrave & McDowell, 1993). This special situation is explained by the idea that mechanical forces might be directly transduced at mechanosensitive ion channels of the plasma membrane (e.g.

C.J. Staiger et al. (eds.), Actin: A Dynamic Framework for Multiple Plant Cell Functions, 557–571.

Sachs, 1997; Spencer et al., 1999) via cytoskeletal elements (e.g. Glogauer et al., 1998), thereby bypassing a need for special receptor molecules (Kernan & Zuker, 1995). Growing root apices proved to be sensitive towards extremely weak mechanical stimuli (Monshausen & Sievers, 1998).

Gravity-oriented growth of higher plant organs, like roots, is mainly under the control of two signal transduction chains (Volkmann & Tewinkel, 1998). The first one is initiated in specialized gravity-sensing cells (known as statocytes) where gravity-induced stimuli transform into a physiological response (for a review, see Sack, 1997). The second one ends up in cells of the transition zone (Baluška et al., 1996) which initiates the differential cell elongation (Zieschang & Sievers, 1991). Thus, the transition zone acts as the target tissue which receives signals transmitted from the site of gravisensing. The transition zone cells are uniquely sensitive towards auxin and calcium (Ishikawa & Evans, 1992, 1993) and are induced into their differential elongation at the opposite root flanks, resulting in the gravity-induced re-orientation of the whole root apex (Evans & Ishikawa, 1997). We would like to highlight putative roles of the actomyosin-based cytoskeleton at the gravisensing sites represented by amyloplast surfaces (see below). Currently, a large body of evidence is emerging for the involvement of diverse molecules related to the phosphoinositide metabolism in both signal perception and transduction (for plant cells, see Munnik et al., 1998). Finally, we stress the importance of mutants relevant for the dissection and exploration of signal transduction chains during root gravisensing.

2. SITES OF GRAVISENSING

Since the pioneering surgical experiments performed by Juniper et al. (1966) with maize roots, the root cap is well established as the site of gravisensing for the underground parts of plants (for genetic evidence, see Tsugeki & Fedoroff, 1999). For above-ground plant organs mutants suggest a role of the endodermis as site of perception (e.g. Fukaki et al., 1998; Tasaka et al., 1999). In the root cap centre, highly specialized columella cells are characterized by a polar arrangement of cell organelles whereby the nucleus is mainly localized at the proximal cell pole, whereas amyloplasts are sedimented mostly on cortical ER membranes at the distal pole (for review, see Volkmann & Sievers, 1979). In spite of recent controversal discussions of 'starchless mutants' in relation to the first step of stimulus transformation, it is accepted that sedimentable particles (statoliths) are involved in an intracellular perception mechanism (for a review, see Sack, 1997; for controversal discussion compare Pickard & Ding, 1992; Staves et al., 1997a,b). Recently, Blancaflor et al. (1998) specified the central root cap cells of the Arabidopsis roots as the most important gravisensing site.

Experiments using centrifuges (Wendt et al., 1987) and microgravity (Perbal et al., 1987; for review, see Perbal et al., 1997) ruled out the plasma membrane and ER membranes as cell structures directly involved in the process of early stimulus transformation. On the other hand, dynamic cytoskeletal elements, especially those based on the actomyosin complex, have recently come into discussion as sensors and mediators of various environmental signals (for reviews, see Volkmann & Baluška, 1999; Staiger, 2000). In general, actomyosin-based networks represent hot candidates for transducing the mechanical force (e.g. Bähler, 1996; Mermall et al., 1998; Volkmann & Baluška, 1999; Soldati et al., 1999; Staiger, 2000; Wu et al., 2000) transformed at the statolith surfaces and passed towards membranes at the cell periphery (Sievers et al., 1989; Sievers et al., 1991). Thus, they might be implicated not only in signal perception but also in propagation of perceived stimuli (for review, see Baluška & Hasenstein, 1997).

3. CYTOSKELETAL ELEMENTS AS A BASIS FOR STATOLITH SHORT-RANGE MOTIONS

Sedimentation of statoliths is obviously facilitated by the specific cytoskeletal status of root cap statocytes which differ from all other root cells (Baluška et al., 1997). Until now, endoplasmic microtubules (MTs) and bundles of actin filaments (AFs) have not been observed in this unique type of root cap cells. Statolith sedimentation, however, is not just a simple one-way process executed progressively during development. Rather, as can be observed using high resolution video microscopy, it represents a complex form of endocellular motility accomplished via continuous short-range motions (Sack et al., 1986; Volkmann et al., 1999). Dynamic statoliths occur often in clusters (Smith et al., 1997), typically showing saltatory movements (Sack et al., 1986). This situation has also been reported for barium sulfate-filled statoliths in unicellular graviresponding rhizoids of the green alga *Chara* (Hejnowicz & Sievers 1981). Statoliths motions are not randomized, as is typical of non-sedimentable amyloplasts, but are preferentially correlated according to the direction of the gravity vector. Thus, in developing statocytes, short-range motions of amyloplasts are progressively channelled into directed motions which ultimately culminate in their sedimentation. In other words, oriented gravity-dependent statolith motions overcome the noise of their constitutive random motions (Volkmann et al., 1999).

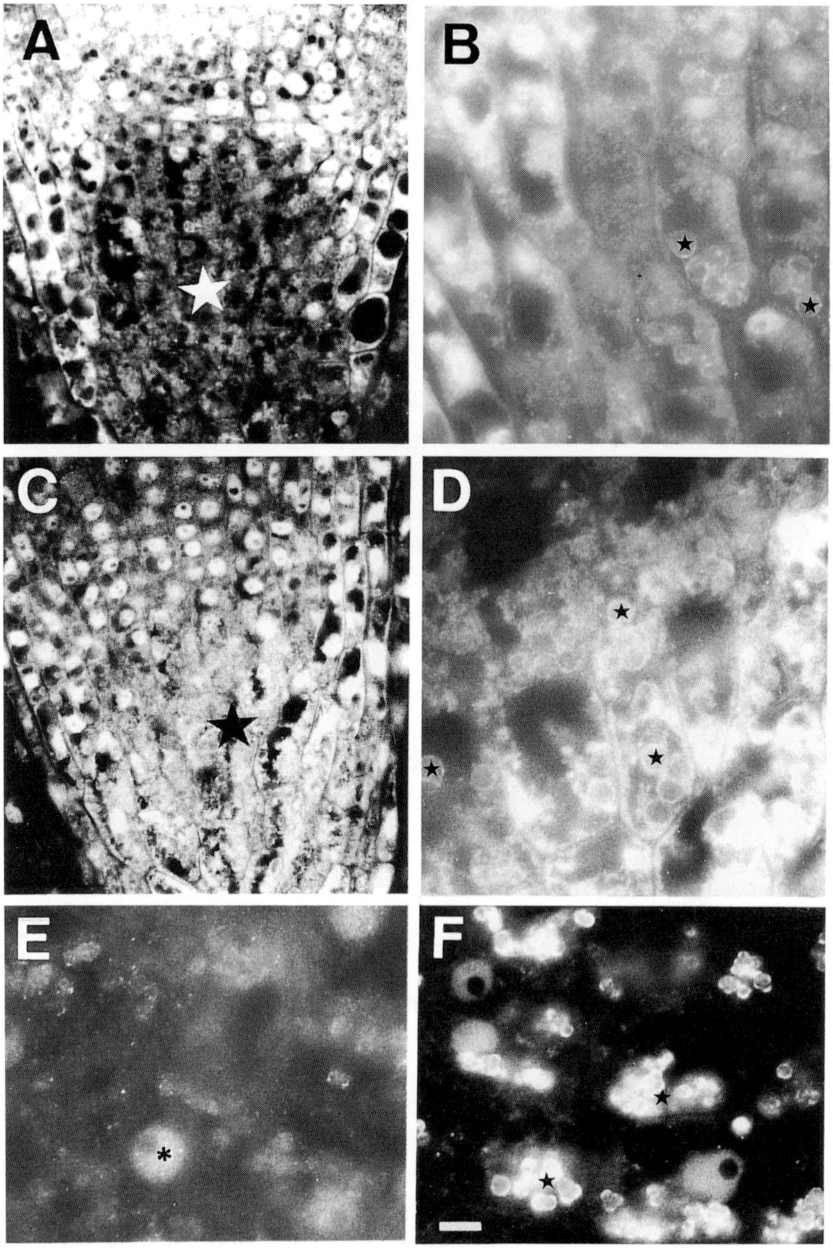

Figure 1. Distributions of actin (A,B), profilin(s) (C,D) reactive to the ZmPRO5 antibody (gift from Chris Staiger), myosin VIII (E), and myosin-like protein (F) reactive to the myosin II antibody (Sigma, M7648) in statocytes (big stars in A, C) of alfalfa (A-D) and maize (E, F) root apices. Note accumulation of actin, profilins, and myosin-like proteins around surfaces of sedimented amyloplasts (small stars in B, D, F). Myosin VIII does not accumulate around statoliths (E). Asterisk in E indicates nucleus. Bar = 18 μm for A, C; and 10 μm for B, D-F.

Actomyosin-driven transport of cell structures and organelles, especially of chloroplasts, is well documented for algae such as *Nitella* or *Acetabularia* (Williamson, 1993; Menzel, 1994; Grolig & Pierson, this volume; Takagi, this volume) as well as for ferns and higher plants (Sato et al., 1999; Kandasamy & Meagher, 1999). Using monospecific antibodies, it can be demonstrated that root cap cells, especially statocytes, possess a highly specific status in comparison to cells from other root tissues concerning the expression of actin and its related proteins, myosin and profilin (Fig. 1). In spite of intensive studies using different immunocytological techniques, visualization of prominent AFs in root statocytes is still lacking (for recent literature compare Baluška & Hasenstein, 1997). In contrast, other cell types of the root proper have been shown to be rich in filamentous actin, distinctly organized in different cell types (Baluška et al., 1997; Vitha et al. 1997; Vitha et al., 2000; Baluška et al., this volume; Vitha et al., this volume). Recently, Blancaflor and Hasenstein (this volume) were able to visualize F-actin networks within isolated statocytes of Arabidopsis roots using a special technique combining phalloidin fluorochromes with glycerol incubation. On the other hand, prominent F-actin cables were observed in statocytes of coleoptiles (White & Sack, 1990) and hypocotyls (Volkmann et al., 1993) after phalloidin labelling. However, these statocytes of above-ground plant organs are distinguished by large central vacuoles and by cytoplasmic streaming which is driven by the actomyosin system (Grolig & Pierson, this volume). The latter situation is completely different to cytoplasmic-rich root cap statocytes lacking any cytoplasmic streaming. A diffuse actin signal in root statocytes surrounds sedimented statoliths. This suggests dense, but short, dynamic F-actin elements surrounded by abundant G-actin in these cells (compare Fig. 1). Interestingly, when statocytes of maize and barley root caps have been labelled for F-actin after their enzymatic release from cap tissues, then distinct AF networks were found (White & Sack, 1990). This latter finding indicates that the actin cytoskeleton in root cap statocytes is capable of rapidly assembling a more robust system if these cells are treated with cell wall digesting enzymes which thereby change the cellular tensegrity system (Ingber, 1997; Chicurel et al., 1998). A powerful tool for visualization of the plant actin cytoskeleton has been developed using a Green Fluorescence Protein (GFP)-mouse talin construct which visualizes even actin oligomers (Kost et al., 1998; Kost et al., this volume). Using this in vivo approach, dense meshworks of presumably oligomeric AFs are implicated around sedimented statoliths (Mathur Jaideep, personal communication).

Concerning the motor molecule, myosin, essential information is relatively rare for higher plants. Using heterologous antibodies against myosin II from chicken muscle, prominent labelling was localized in the close vicinity of statoliths of cress root statocytes (Wunsch & Volkmann, 1993; compare also Fig. 1). Similar observations have been reported for statoliths of maize root caps (Baluška & Hasenstein, 1997) and Chara rhizoids (Braun, 1996). Profilin, the actin-binding protein which is essential for the organization of F-actin (Staiger et al., 1997; Gibbon & Staiger, this volume), is detectable around statoliths (Fig. 1). All this a suggests that dynamic actin-based cytoskeleton surrounds the statolith envelopes and probably the structural basis for continuous saltatory motions of statoliths, even in their sedimented state (Sack et al., 1986; for review, compare Volkmann et al., 1999).

4. STATOLITHS BEHAVIOUR UNDER MICROGRAVITY CONDITIONS: ACTOMYOSIN COUNTERACTS GRAVITY

Experiments under microgravity showed that the actual position of statoliths in statocytes depends on two forces, the external gravitational force and the endocellular cytoskeletal forces, both of which impinge on the statoliths (Lorenzi & Perbal, 1990; Volkmann et al., 1991). In these experiments, the possibility to switch from higher **g** levels to microgravity has proven to be very important. Experiments investigating cress and lentil roots (Perbal et al., 1986; Volkmann et al., 1986; Perbal & Driss-Ecole, 1989; Lorenzi & Perbal, 1990; Laurinavi□ius et al., 1996) indicated that statoliths under microgravity did not show random distributions as might be expected, but showed remarkable shifts in direction towards the proximal statocyte pole, thus implicating active motion of these organelles. Direct evidence for actively driven movements of statoliths came from experiments on rockets when, after launch accelerations of some **g**, gravitational conditions changed immediately to microgravity. Under these experimental conditions, statoliths moved in the opposite direction to the originally acting gravity vector within a few minutes (Volkmann et al., 1991). These results were confirmed by observations made during long term experiments (Perbal & Driss-Ecole, 1994; Laurinavi□ius et al., 1996; Smith et al., 1997; for review, see Volkmann et al., 1999). Corresponding behaviour of statoliths from the Chara rhizoid was directly observed in orbit by tele-communication (Volkmann et al., 1991). In the presence of cytochalasin D, this motion did not occur and statoliths remained in their launch position (Buchen et al., 1993; for review, see Sievers et al., 1996). Interestingly, one actin isoform strongly decreased under short-term microgravity conditions (Janßen et al., 1996).

5. THRESHOLD VALUES AS INDICATIONS FOR STIMULUS TRANSFORMATION IN CLOSE VICINITY OF STATOLITH ENVELOPES

Microgravity experiments offered possibilities to study important parameters, such as threshold values, under controlled conditions of sensor physiology. For plant material cultivated entirely under microgravity conditions, the minimum dose under continuous stimulation (presentation time, tp; for details compare Perbal et al., 1997) has been estimated to be 20–30 **g**×sec for cress roots (Volkmann & Tewinkel, 1996), and by extrapolation from microgravity data to be 27 **g**×sec for lentil roots (Perbal & Driss-Ecole, 1994). Correlation of this threshold value with the position of statoliths show that these moved very slightly in the direction of the stimulating gravity vector, generally less than 1 µm. Thus, one must conclude that stimulus transformation occurs in the close vicinity of the statolith envelopes. On the basis of these results, it can be hypothesized that gravity perceiving cells are transforming the gravity stimulus into endocellular signalling pathways by a cytoskeleton-mediated measurement of the actual positions of continuously repositioning statoliths.

6. SECOND MESSENGER MOLECULES INVOLVED IN GRAVITROPIC SIGNAL TRANSDUCTION CHAINS

Important molecules related to signalling are second messengers like phosphoinositides and calcium (e.g. Yang, 1996). Components of the phosphoinositol signalling pathway have recently been investigated in maize nodes after their gravistimulation (Perera et al., 1999). Within 10 s after gravistimulation, inositol 1,4,5-trisphosphate (IP_3) increases transiently in the faster-growing lower half. Additionally, the activity of phosphatidyl-inositol 4,5 bisphosphate (PIP_2) kinase increased transiently within 10 min. These results indicate up-regulation of PIP_2 biosynthesis during the gravisensing–graviresponding process. At the moment, however, it is unclear to which signal transduction chain these events are pertinent. Intriguingly, plastids are unique with respect to some lipidic molecular species (e.g. Miège & Maréchal, 1999) which might be relevant for signalling associated with amyloplast-based gravisensing.

Regulation of calcium channels via cytoskeletal elements and second messengers is becoming increasingly evident in plant signal transduction chains (Thuleau et al., 1998). Changes of cytosolic free calcium have been documented for Arabidopsis roots after stimulation by touch (Legué et al., 1997) and gravity (Davies et al., 1999). The same is true for calmodulin (see Braam et al., 1997). Cyclopiazonic acid, a specific inhibitor of ER-Ca^{2+}-ATPases, inhibited the graviresponse of cress roots but not their growth (Sievers & Busch, 1992). For tip-growing pollen tubes, there is general agreement that the two second-messenger molecules, calcium and IP_3, are essential for signalling-based tip growth (Franklin-Tong et al., 1996; Staiger, 2000).

Convincing data are accumulating that mitogen activated protein (MAP) kinases, another typical component of signal transduction chains and well known from yeast to animals, play crucial roles in the sensing of diverse environmental factors in plants (e.g. Yang, 1996; Bögre et al., 1996). For instance, specific MAP kinase (MMK4) becomes activated within one minute already after two seconds of mechanical stimulation of alfalfa leaves (Bögre et al., 1996). MAP kinases have been identified as universal, multipurpose signalling tools performing important functions, among others, as mediators of mechano-sensitive responses (e.g. Heberle-Bors & Hirt, 1994; Bögre et al., 1996; Hirt, 1997; Jonak et al., 1999). However, the crucial question again remains open; to which of the signal-transduction chains these molecules are related, and how they interact with the cytoskeleton.

7. ROLE OF MUTANTS IN DISSECTING AND EXPLORING GRAVITY-RELATED SIGNAL TRANSDUCTION CHAINS

Recent genetical evidence for participation of cytoskeletal elements in gravisensing came from an Arabidopsis mutant showing altered response to gravity (ARG, Sedbrook et al., 1999; Rosen et al., 1999). The authors found that the *ARG1* locus encodes for a 45 kDa protein (DnaJ-like protein) containing a coiled-coil region homologous to coiled-coils found in cytoskeleton-interacting proteins. Interestingly, this gene is expressed in all plant organs investigated so far, and the encoded protein is related to a conserved molecule involved in signal transduction in *Caenorhabditis elegans*. Therefore, *ARG1* gene might code for a phylogenetically primitive component of the signal transduction chains associated with the process of gravity sensing in plants (for evolutionary considerations, compare Barlow, 1995). The authors (Rosen et al., 1999) do, however, not exclude a more general role of the DnaJ-like protein. For instance, it might be a component of macromolecular chaperone complexes having direct impacts on protein folding, trafficking and degradation, but only indirect effects on gravisensing and signal-transduction processes. A number of agravitropic mutants are related to auxin physiology suggesting a global role of these genes in graviresponses (Chen et al., 1998; Luschnig et al., 1998; Godbolé et al., 1999; Marchant et al., 1999), but their involvement in gravisensing is unclear. In this respect, mutants of the PIN family dissecting the auxin transport mechanisms and pathways (e.g. Müller et al., 1998; Palme & Gälweiler, 1999) are becoming highly important.

The recent view of plasmodesmata as gateable channels (Ding, 1998; Overall et al., this volume), including the plasmodesmal localization of molecules like actin (White et al., 1994; Baluška et al., this volume), myosin VIII (Reichelt et al., 1999; Reichelt & Kendrick-Jones, this volume; Baluška et al., this volume), epitopes reactive to heterologous myosin II antibody (Radford & White, 1998), and ER-based calreticulin (Baluška et al., 1999), sheds some light on intercellular signal transmission. Interestingly, root cap statocytes lack myosin VIII and calreticulin at their plasmodesmata, but these molecules can by recruited there during plasmolysis (Baluška & Volkmann, in preparation). The possible relevance of this phenomenon for root cap gravisensing remains to be explored.

8. CONCLUSIONS AND OUTLOOK

If we accept the starch statolith hypothesis, then it must be concluded that the transformation of a gravity stimulus, i.e. gravisensing, has to occur in close vicinity to the statolith envelopes which are surrounded by a highly dynamic cytoskeletal network based putatively on oligomeric F-actin. First evidences emerge that metabolites of the phosphatidylinositol pathway are involved in the gravity-related signal transduction chains. In addition, mechanosensation of gravity might be tightly linked to those MAP kinases which are activated via mechanical stimuli. All this calls for more intensive investigations of interactions among components of the actin cytoskeleton, diverse phosphoinositide signalling molecules, calcium, and MAP kinases. These complex investigations might include the following approaches:

— identification of functional actin isovariants of the highly diverse plant actin gene family (Meagher et al., 1999; this volume);

— identification of additional relevant cytoskeletal elements of plants, like actin-related proteins (Arps), which are organizers of the actin dynamics (Machesky & Gould, 1999). The Arp2 gene has already been identified in the Arabidopsis genome (Klahre & Chua, 1999);

— testing the amyloplast-based statoliths as putative excitable organelles capable of generating and conveying calcium signals, and their envelopes as suitable assembly sites of protein complexes performing gravity perception: putative gravisensing signalosome (for signalosomes and transducisomes related to MAPK cascades, see Burack and Shaw, 2000; for conserved COP9 signalosome, see Freilich et al., 1999 and Mundt et al., 1999; for signalplex in relation to the phototransduction cascade, see Montell 1999);

— exploration of possible participation of dynamic stroma-filled plastid tubules (stromules) which are abundant around chlorophyll-free plastids of roots (Köhler et al., 1997; Köhler & Hanson, 2000);

— selection of mutants which are directly dissecting gravity-related signal transduction chains.

ACKNOWLEDGEMENT

Work reported here has been supported via AGRAVIS (Bonn) by Deutsche Agentur für Raumfahrtangelegenheiten (DARA, Bonn) and Ministerium für Wissenschaft und Forschung (MWF, Düsseldorf). Authors would like to thank Andreas Sievers for helpful comments on the manuscript.

REFERENCES

Bähler M (1996) Myosins on the move to signal transduction. Curr Opin Cell Biol 8: 18-22

Baluška F and Hasenstein KH (1997) Root cytoskeleton: Its role in perception of and response to gravity. Planta 203: S69-S78

Baluška F, Volkmann D and Barlow PW (1996) Specialized zones of development in roots: View from the cellular level. Plant Physiol 112: 3-4

Baluška F, Šamaj J, Napier, R and Volkmann D (1999) Maize calreticulin localizes to plasmodesmata in root apex. Plant J 19: 481-487

Baluška F, Vitha S, Barlow PW and Volkmann D (1997) Rearrangements of F-actin arrays in growing cells of intact maize root apex tissues: A major developmental switch occurs in the postmitotic transition region. Eur J Cell Biol 72: 113-121

Baluška F, Kreibaum A, Vitha S, Parker JS, Barlow PW and Sievers A (1997) Central root cap cells are depleted of endoplasmic microtubules and actin filament bundles: Implications for their role as gravity-sensing statocytes. Protoplasma 196: 212-223

Barlow PW (1995) Gravity perception in plants: A multiplicity of systems derived by evolution. Plant Cell Environ 18: 951-952

Blancaflor EB, Fasano JM and Gilroy S (1998) Mapping the functional roles of cap cells in the response of *Arabidopsis* primary roots to gravity. Plant Physiol 116: 213-222

Bögre L, Ligterink W, Heberle-Bors E and Hirt H (1996) Mechanosensors in plants. Nature 383: 489-490

Braam J, Sistrunk ML, Polisensky DH, Wei X, Purugganan M, Antosiewicz DM, Campbell P and Johnson KA (1997) Plant responses to environmental stress: Regulation and functions of the *Arabidopsis TCH* genes. Planta 203: S35-S41

Braun M (1996) Immunocytolocalization of myosin in rhizoids of *Chara globularis* Thuill. Protoplasma 191: 1-8

Buchen B, Braun M and Sievers A (1993) Statoliths pull on microfilaments. Experiments under microgravity. Protoplasma 172: 38-42

Burack WR and Shaw AS (2000) Signal transduction: Hanging on a scaffold. Curr Opin Cell Biol 12: 211-216

Chen R, Hilson P, Sedbrook J, Rosen E, Caspar T and Masson PH (1998) The *Arabidopsis thaliana AGRAVITROPIC1* gene encodes a component of the polar-auxin-transport efflux carrier. Proc Natl Acad Sci USA 95: 15112-15117

Chicurel ME, Chen CS and Ingber DE (1998) Cellular control lies in the balance of forces. Curr Opin Cell Biol 10: 232-239

Davies E, Shimps B, Brown K and Stankovi□ B (1999) Gravity, stress, calcium and gene expression. J Grav Physiol 6: P21-P22

Ding B (1998) Intercellular protein trafficking through plasmodesmata. Plant Mol Biol 38: 279-310

Evans ML and Ishikawa H (1997) Cellular specificity of the gravitropic motor response in roots. Planta 203: S115-S122

Franklin-Tong VE, Drøbak B, Allan AC, Watkins PAC and Trewavas AJ (1996) Growth of pollen tubes of *Papaver rhoeas* is regulated by a slow-moving calcium wave propagated by inositol 1,4,5-trisphosphate. Plant Cell 8: 1305-1321

Freilich S, Oron E, Kapp Y, Nevo-Caspi Y, Orgad S, Segal D and Chamovitz DA (1999) The COP9 signalosome is essential for development of *Drosophila melanogaster*. Curr Biol 9: 1187-1190

Fukaki H, Wysocka-Diller J, Kato T, Fujisawa H, Benfey PN and Tasaka M (1998) Genetic evidence that the endodermis is essential for shoot gravitropism in *Arabidopsis thaliana*. Plant J 14: 425-430

Glogauer M, Arora P, Chou D, Janmey PA, Downey GP and McCulloch CAG (1998) The role of actin-binding protein 280 in integrin-dependent mechanoprotection. J Biol Chem 273: 1689-1698

Godbolé R, Takahashi H and Hertel R (1999) The *Lazy* mutation in rice affects a step between statoliths and gravity-induced lateral auxin transport. Plant Biol 1: 379-381

Hargrave PA and McDowell JH (1993) Rhodopsin and phototransduction. Int Rev Cytol 137: 49-97

Heberle-Bors CJC and Hirt H (1994) MAP kinases: Universal multipurpose signaling tools. Plant Mol Biol 24: 407-416

Hejnowicz Z and Sievers A (1981) Regulation of the position of statoliths in *Chara* rhizoids. Protoplasma 108: 117-137

Hirt H (1997) Multiple roles of MAPkinases in plant signal transduction. Trends Plant Sci 2: 11-15

Ingber DE (1997) Tensegrity: The architectural basis of cellular mechanotransduction. Annu Rev Physiol 59: 575-559

Ishikawa H and Evans ML (1992) Induction of curvature in maize roots by calcium or by thigmostimulation. Role of the postmitotic isodiametric growth zone. Plant Physiol 100: 762-768

Ishikawa H and Evans ML (1993) The role of the distal elongation zone in the response of maize roots to auxin and gravity. Plant Physiol 102: 1203-1210

Janßen M, Hunte C, Schulz M and Schnabl H (1996) Tissue specification and intracellular distribution of actin isofoms in *Vicia faba* L. Protoplasma 191: 158-163

Jonak C, Ligterink W and Hirt H (1999) MAP kinases in plant signal transduction. Cell Mol Life Sci 55: 204-213

Juniper BE, Groves S, Landau-Schachar B and Audus LJ (1966) Root cap and the perception of gravity. Nature 209: 93-94

Kandasamy MK and Meagher RB (1999) Actin-organelle interaction: Association with chloroplast in Arabidopsis leaf mesophyll cells. Cell Motil Cytoskel 44: 110-118

Kernan M and Zuker Ch (1995) Genetic approaches to mechanosensory transduction. Curr Opin Neurobiol 5: 443-448

Klahre U and Chua N-H (1999) The *Arabidopsis* ACTIN-RELATED PROTEIN2 (AtARP2) promoter directs expression in xylem precursor cells and pollen. Plant Mol Biol 41: 65-73

Köhler RH and Hanson MR (2000) Plastid tubules of higher plants are tissue-specific and developmentally regulated. J Cell Sci 113: 81-89

Köhler RH, Cao J, Zipfle WR, Webb WW and Hanson MR (1997) Exchange of protein molecules through connections between higher plant plastids. Science 276: 2039-2042

Kost B, Spielhofer P and Chua N-H (1998) A GFP-mouse talin fusion protein labels plant actin filaments *in vivo* and visualizes the actin cytoskeleton in growing pollen tubes. Plant J 16: 393-401

Laurinavi□ius R, Sto□kus A, Buchen B and Sievers A (1996) Structure of cress root statocytes in microgravity (BION-10 mission). Adv Space Res 17: 91-94

Legué V, Blancaflor E, Wymer C, Perbal G, Fantin D and Gilroy S (1997) Cytoplasmic free Ca^{2+} in Arabidopsis roots changes in response to touch but not gravity. Plant Physiol 114: 789-800

Lorenzi G and Perbal G (1990) Root growth and statocyte polarity in lentil seedling roots grown in microgravity or on a slowly rotating clinostat. Physiol Plant 78: 532-537

Luschnig C, Gaxiola RA, Grisafi P and Fink GR (1998) EIR1, a root-specific protein involved in auxin transport, is required for gravitropism in *Arabidopsis thaliana*. Genes Dev 12: 2175-2187

Machesky LM and Gould KL (1999) The ARP2/3 complex: A multifunctional actin organizer. Curr Opin Cell Biol 11: 117-121

Marchant A, Kargul J, May ST, Müller P, Delbarre A, Perrot-Rechenmann C and Bennett MJ (1999) AUX1 regulates root gravitropism in *Arabidopsis* by facilitating auxin uptake within root apical tissues. EMBO J 18: 2066-2073

Meagher RB, McKinney EC and Kandasamy MK (1999) Isovariant dynamics expand and buffer the responses of complex systems: The diverse plant actin gene family. Plant Cell 11: 995-1005

Menzel D (1994) Tansley Review No. 77. Cell differentiation and the cytoskeleton in *Acetabularia*. New Phytol 128: 369-393

Mermall V, Post PL and Mooseker MS (1998) Unconventional myosins in cell movement, membrane traffic, and signal transduction. Science 279: 527533

Miège C and Maréchal É (1999) 1,2-*sn*-Diacylglycerol in plant cells: Product, substrate and regulator. Plant Physiol Biochem 37: 795-808

Montell C (1999) Visual transduction in *Drosophila*. Annu Rev Cell Dev Biol 15: 231-268

Monshausen G and Sievers A (1998) Weak mechanical stimulation causes hyperpolarisation in root cells of *Lepidium*. Bot Acta 111: 303-306

Müller A, Guan C, Gälweiler L, Tänzler P, Huijser P, Marchant A, Parry G, Bennett M, Wisman E and Palme K (1998) *AtPIN2* defines a locus of *Arabidopsis* for root gravitropism control. EMBO J 23: 6903-6911

Mundt KE, Porte J, Murray JM, Brikos C, Christensen PU, Caspari T, Hagan IM, Millar JBA, Simanis V, Hofmann K and Carr AM (1999) The COP9/signalosome complex is conserved in fission yeast and has a role in S phase. Curr Biol 9: 1427-1430

Munnik T, Irvine RF and Musgrave A (1998) Phospholipid signalling in plants. Biochim Biophys Acta 1389: 222-272

Palme K and Gälweiler L (1999) PIN-pointing the molecular basis of auxin transport. Curr Opin Plant Biol 2: 375-381

Perbal G and Driss-Ecole D (1989) Polarity of statocytes in lentil seedling roots grown in space (Spacelab D1 Mission). Physiol Plant 75: 518-524

Perbal G and Driss-Ecole D (1994) Sensitivity to gravistimulus of lentil seedling roots grown in space during the IML1 Mission of Spacelab. Physiol Plant 90: 313-318

Perbal G, Driss-Ecole D, Rutin J and Sallé G (1987) Graviperception of lentil seedling roots grown in space (Spacelab D1 Mission). Physiol Plant 70: 119-126

Perbal G, Driss-Ecole D, Sallé G and Raffin F (1986) Perception of gravity in the lentil root. Naturwissenschaften 73: 444-446

Perbal G, Driss-Ecole D, Tewinkel M and Volkmann D (1997) Statocyte polarity and gravisensitivity in seedling roots grown in microgravity. Planta 203: S57-S62

Perera IY, Heilmann I and Boss WF (1999) Transient and sustained increases in inositol 1,4,5-trisphosphate precede the differential growth response in gravistimulated maize pulvini. Proc Natl Acad Sci USA 96: 5838-5843

Pickard BG and Ding JP (1992) Gravity sensing by higher plants. In: Comparative Aspects of Mechanoreceptor Systems, Pharis RP and Reid DM (eds), Heidelberg: Springer-Verlag, pp 193-281

Radford JE and White RG (1998) Localization of a myosin-like protein to plasmodesmata. Plant J 14: 743-750

Reichelt S, Knight AE, Hodge TP, Baluška F, Šamaj J, Volkmann D and Kendrick-Jones J (1999) Characterization of the unconventional myosin VIII in plant cells and its localization at the post-cytokinetic cell wall. Plant J 19: 555-567

Rosen E, Chen R and Masson PH (1999) Root gravitropism: A complex response to a simple stimulus? Trends Plant Sci 4: 407-412

Sachs F (1997) Mechanical transduction by ion channels: How forces reach the channel. In: Cytoskeletal Regulation of Membrane Function, Froehner SC (ed), New York: Rockefeller University Press, pp 209-218

Sack F (1997) Plastids and gravitropic sensing. Planta 203: S63-S68

Sack FD, Suyemoto M and Leopold AC (1986) Amyloplast sedimentation and organelle saltation in living columella cells. Am J Bot 73: 1692-1698

Sato Y, Kadota A and Wada M (1999) Mechanically induced avoidance response of chloroplasts in fern protonemal cells. Plant Physiol 121: 37-44

Sedbrook JC, Chen R and Masson PH (1999) ARG1 (altered response to gravity) encodes a DnaJ-like protein that potentially interacts with the cytoskeleton. Proc Natl Acad Sci USA 96: 1140-1145

Sievers A and Busch MB (1992) An inhibitor of the Ca^{2+}-ATPase in the sarcoplasmic and endoplasmic reticula inhibits transduction of the gravity stimulus in cress roots. Planta 188: 619-622

Sievers A, Buchen B and Hodick D (1996) Gravity sensing in tip-growing cells. Trends Plant Sci 1: 273-279

Sievers A, Buchen B, Volkmann D and Hejnowicz Z (1991) Role of the cytoskeleton in gravity perception. In: Lloyd CW (ed) The Cytoskeletal Basis of Plant Growth and Form. London: Academic Press, pp 169-182

Sievers A, Kruse S, Kuo-Huang LL and Wendt M (1989) Statoliths and microfilaments in plant cells. Planta 179: 275-278

Smith JD, Todd P and Staehelin LA (1997) Modulation of statolith mass and grouping in white clover (*Trifolium repens*) grown in 1-g, microgravity and on the clinostat. Plant J 12: 1361-1373

Soldati T, Schwarz EC and Geissler H (1999) Unconventional myosins at the crossroad of signal transduction and cytoskeleton remodeling. Protoplasma 209: 28-37

Spencer RH, Chang G and Rees DC (1999) 'Feeling the pressure': Structural insights into a gated mechanosensitive channel. Curr Opin Struct Biol 9: 448-454

Staiger CJ (2000) Signaling to the actin cytoskeleton in plants. Annu Rev Plant Physiol Plant Mol Biol 51: 257-288

Staiger CJ, Gibbon BC, Kovar DR and Zonia LE (1997) Profilin and actin-depolymerizing factor: Modulators of actin organization in plants. Trends Plant Sci 2: 275-281

Staves MP, Wayne R and Leopold AC (1997a) Cytochalasin D does not inhibit gravitropism in roots. Am J Bot 84: 1530-1535

Staves MP, Wayne R and Leopold AC (1997b) The effect of the external medium on the gravitropic curvature of rice (*Oryza sativa*, Poaceae) roots. Am J Bot 84: 1522-1529

Tasaka M, Kato T and Fukaki H (1999) The endodermis and shoot gravitropism. Trends Plant Sci 4: 103-107

Thuleau P, Schroeder JI and Ranjeva R (1998) Recent advances in the regulation of plant calcium channels: Evidence for regulation by G-proteins, the cytoskeleton and second messengers. Curr Opin Plant Biol 1: 424-427

Tsugeki R and Fedoroff NV (1999) Genetic ablation of root cap cells in *Arabidopsis*. Proc Natl Acad Sci USA 96: 12941-12946

Vitha S, Baluška F, Mews M and Volkmann D (1997) Immunofluorescence detection of F-actin on low melting wax sections from plant tissue. J Histochem Cytochem 45: 89-95

Vitha S, Baluška F, Braun M, Šamaj J, Volkmann D and Barlow PW (2000) Comparison of cryofixation and aldehyde fixation for plant actin immunocytochemistry: Aldehydes do not destroy F-actin. Histochem J, in press

Volkmann D and Baluška F (1999) The actin cytoskeleton in plants: From transport networks to signaling networks. Microsc Res Tech 47: 135-154

Volkmann D and Sievers A (1979) Graviperception in multicellular organs. In: Encyclopedia of Plant Physiology, NS, vol. 7, Physiology of Movements. Haupt W and Feinleib ME (eds), Berlin: Springer-Verlag, pp 573-600

Volkmann D and Tewinkel M (1996) Gravisensing of cress roots: Investigations of threshold values under specific conditions of sensor physiology in microgravity. Plant Cell Environm 19: 1195-1202

Volkmann D and Tewinkel M (1998) Gravisensitivity of cress roots. Adv Space Res 21: 1209-1217

Volkmann D, Behrens HM and Sievers A (1986) Development and gravity sensing of cress roots under microgravity. Naturwissenschaften 73: 438-441

Volkmann D, Winn-Börner U and Waberzeck K (1993) Graviresponsiveness of cress seedlings and structural status of presumptive statocytes from the hypocotyl. J Plant Physiol 142: 710-716

Volkmann D, Baluška F, Lichtscheidl IK, Driss-Ecole D and Perbal G (1999) Statoliths motions in gravity-perceiving plant cells: Does actomyosin counteract gravity? FASEB J 13: S143-S147

Volkmann D, Buchen B, Hejnowicz Z, Tewinkel M and Sievers A (1991) Oriented movement of statoliths studied in a reduced gravitational field during parabolic flights of rockets. Planta 185: 153-161

Wendt M, Kuo-Huang LL and Sievers A (1987) Gravitropic bending of cress roots without contact between amyloplasts and complexes of endoplasmic reticulum. Planta 172: 321-329

White RG and Sack FD (1990) Actin microfilaments in presumptive statocytes of root caps and coleoptiles. Am J Bot 77: 17-26

White RG, Badelt K, Overall RL and Vesk M (1994) Actin associated with plasmodesmata. Protoplasma 180: 169-184

Williamson RE (1993) Organelle movements. Annu Rev Plant Physiol Plant Mol Biol 44: 181-202

Wu X, Jung G and Hammer JA III (2000) Functions of unconventional myosins. Curr Opin Cell Biol 12: 42-51

Wunsch C and Volkmann D (1993) Immunocytological detection of myosin in the root tip cells of *Lepidium sativum*. Eur J Cell Biol Suppl 61: p 46

Yang Z (1996) Signal transduction proteins in plants: An overview. In: Signal Transduction in Plant Growth and Development. Verma DPS (ed), Wien/New York: Springer-Verlag, pp 1-37

Zieschang HE and Sievers A (1991) Graviresponse and the localization of its initiating cells in *Phleum pratense* L. Planta 184: 468-477

Chapter 32

PLANT ACTIN CYTOSKELETAL RESPONSES TO ATTACK AND INVASION BY PATHOGENIC FUNGI

Issei Kobayashi[1] and Yuhko Kobayashi[2]

1 Laboratory of Plant Pathology, Faculty of Bioresources, Mie University, Japan
2 Center for Molecular Biology and Genetics, Mie University, Japan

Key words: actin isoforms, cytoskeleton, epidermal cell, fungal pathogen, GTPase

Abstract: A cytoskeletal network is involved in the temporal and spatial regulation of information in all eukaryotic cells. Recently, we reported that the plant actin cytoskeleton plays an important role in the expression of nonhost resistance during early stages of fungal infection. When plant cells are attacked by nonpathogenic fungi, actin microfilaments gather beneath the sites of penetration. This remodeling of microfilament arrays is closely related to polarization of defense-related responses and interruption of fungal penetration. Treatments with actin inhibitors suppress polarization of defense-related responses and facilitate fungal penetration. To determine the molecular mechanisms that underlie remodeling of the cytoskeleton during fungal penetration, we investigated gene expression patterns for several actin isoforms and Rac small GTPases. When barley and tobacco leaves were inoculated with the nonpathogen, *Erysiphe pisi*, expression of *HvAct3* and *NtRac2* mRNAs was increased specifically. These results suggest that specific actin isoforms and Rac GTPases may be closely linked to remodeling of the actin cytoskeleton during defense responses. Finally, the possibility of artificial regulation of general resistance expression is discussed.

1. INTRODUCTION

The actin cytoskeleton is a highly conserved subcellular component whose organization changes dramatically during the cell cycle, development and adaptation to external stimuli. In plant cells, it is well known that the cytoskeleton participates in diverse cellular processes such as cytoplasmic

C.J. Staiger et al. (eds.), Actin: A Dynamic Framework for Multiple Plant Cell Functions, 573–585.

streaming (Kamiya, 1981) and motility of organelles (Williamson, 1986). Furthermore, when plant cells perceive a variety of external stimuli such as low temperature (Quader et al., 1989) and wounding (La Claire, 1989), the actin cytoskeleton is rapidly reorganized. Because plants are sedentary, the ability to quickly respond to stimuli through the temporal and spatial control of cellular organization, which is mediated by the cytoskeleton, must be essential to survive under changeable environmental conditions. Although microbial invasion also provides serious stress to attacked plants, little is known about the contribution of the cytoskeleton to plant defense responses. Recently, we demonstrated a dynamic reorganization of microfilaments during fungal invasion and, by using several actin inhibitors, provided evidence for the involvement of the actin cytoskeleton in defense against fungal invasion. These results reveal the importance of temporal and spatial control of the resistance machinery. Furthermore, these findings not only provide new prospects for understanding plant defense responses, but also suggest novel approaches for genetic engineering to improve crop plant properties. In this chapter, we summarize our recent work and advances on this topic. Also, we briefly describe prospects for the enhancement of plant disease resistance through the artificial regulation of the actin cytoskeleton.

2. INVOLVEMENT OF THE ACTIN CYTOSKELETON IN BLOCKING FUNGAL PENETRATION

When plants are attacked by pathogenic microbes, various defense mechanisms are activated. For example, fungal invasion induces a number of visible changes, such as cytoplasmic aggregation, wall apposition and hypersensitive cell death. Many defense-related, morphologically detectable responses are accompanied by reorganization of the cytoplasm. Therefore, it is reasonable to assume that the cytoskeleton plays a role in the defense-related responses, because cytoskeletal elements, such as microfilaments and microtubules, provide the major force for organelle and cytoplasmic movements (Kamiya 1981; Seagull 1989; Williamson 1986). In this section, we discuss the involvement of actin microfilaments in plant defense responses.

2.1 Visible changes in plant cells during fungal penetration

Visible cytoplasmic responses are induced by an attempted penetration, including movement of the nucleus towards an encounter site (Pappelis et al., 1974; Gross et al., 1993; Škalamera and Heath 1998), cytoplasmic aggregation (Bushnell & Bergquist, 1975; Kunoh et al., 1985a,b; Takemoto et al., 1999; Furuse et al., 1999), formation of a cell wall apposition (papilla) beneath the penetration site (Aist 1976), rearrangement of cytoplasmic strands (Kitazawa et al., 1973; Kobayashi et al., 1993) and changes in the velocity of cytoplasmic streaming (Kobayashi et al., 1990; Tomiyama, 1956). In several cases, the arrangement of microfilaments and microtubules at the contact sites between plant cells and fungal infection structures undergoes marked changes (Kobayashi et al., 1992, 1994; Gross et al., 1993; Škalamera & Heath 1998; Xu et al., 1998; McLusky et al., 1999). Kobayashi et al. (1992) found that the microfilaments and microtubules become radially arranged beneath the contact site in barley coleoptile cells during attempts to penetrate by the nonpathogenic powdery mildew, *Erysiphe pisi*. Although a similar reorganization is observed in coleoptile cells inoculated with a pathogenic powdery mildew fungus, *E. graminis* f. sp. *hordei*, the incidence of reorganization is much lower than that during *E. pisi* infection. These results suggest that cytoskeletal rearrangements in barley cells might be closely associated with a failure of attempted penetration by the nonpathogen.

The rearrangement of microfilaments is observed not only in barley coleoptile cells inoculated with *E. pisi* but also in many plant–nonpathogenic fungus interactions (Fig. 1). Gross et al. (1993) report that when cultured parsley cells are attacked by a nonpathogen, *Phytophthora infestans*, microtubules are depolymerized around penetration sites and a radial array of microfilaments is formed at the penetration sites. Xu et al. (1998) showed that the rice blast fungus, *Magnaporthe grisea*, is able to trigger the accumulation of autofluorescent compounds and rearrangements of actin in onion epidermal cells, independent of penetration by the fungus. Kobayashi et al. (1994) report that a similar reorganization of the cytoskeleton occurs in a genetically specified race–cultivar relationship. In incompatible interactions of flax and the flax rust fungus, microfilaments and microtubules are reorganized to form radial arrays focused on the contact sites between flax mesophyll cells and flax rust infection structures. Such dynamic reorganization is rarely observed in compatible race–cultivar interactions (Kobayashi et al., 1994). These results strongly suggest that the cytoskeleton is involved not only in general resistance but also in race–specific resistance that follows the gene-for-gene hypothesis.

2.2 Effects of actin inhibitors on defense responses

To investigate the involvement of microfilaments in general resistance of plants against pathogenic fungi, partially dissected coleoptiles that were inoculated with the nonpathogen, *E. pisi*, were treated with several actin inhibitors (Kobayashi et al., 1997a,b). If coleoptiles are not treated with inhibitors, the nonpathogen always fails to penetrate.

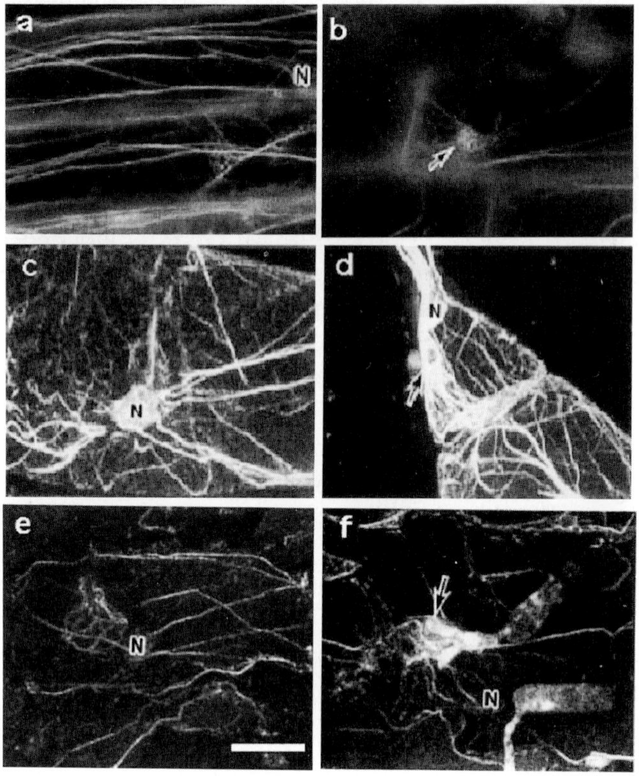

Figure 1. Rearrangements of microfilaments in wheat coleoptile cells (**a,d**), tobacco leaf trichome cells (**b,e**) and tomato epidermal cells (**c,f**) which were inoculated with *Erysiphe pisi*. Microfilaments were stained with rhodamine-phalloidin following fixation with 1 mM MBS and 1% formaldehyde. (**a**) Microfilaments in an uninoculated coleoptile cell were arranged longitudinally or obliquely with respect to the longitudinal axis of the cell. (**b,c**) In uninoculated trichome (**b**) and epidermal cells (**c**), microfilaments were distributed throughout the cell. (**d,e,f**) A radial array of microfilaments was observed around an incipient papilla (arrows) beneath the appressorium of the nonpathogen 15 h after inoculation. Bar, 50 μm. N, nucleus

However, when coleoptiles are treated with actin inhibitors, *E. pisi* is able to penetrate successfully and to form haustoria in nonhost barley cells (Kobayashi et al., 1997a). Actin polymerization inhibitors, especially cytochalasin A and E, are effective in causing an increase in penetration efficiency of *E. pisi*, from 0 to over 60%. Furthermore, cytochalasin treatment allows other nonpathogens, *C. lagenarium* and *A. alternata*, to penetrate successfully into the nonhost barley cells. These effects of cytochalasins on general resistance are observed in other plants such as wheat, rye, cucumber, tomato and tobacco (Kobayashi et al., 1997b). When leaf tissues of these nonhost plants are treated with cytochalasins, several nonpathogenic fungi, which normally fail or have very low efficiency of penetration, also are able to penetrate the cells of these plants.

Previous microscopic and histochemical investigations reveal that when *E. pisi* begins to penetrate into barley coleoptile cells, defense-related reactions such as a massive cytoplasmic aggregation, deposition of papillae, accumulation of callose, protein, carbohydrate and autofluoresent compounds (Kunoh et al., 1982, 1985a, 1985c; Kobayashi et al., 1997a) are observed at sites of fungal penetration. Furthermore cytoplasmic streaming of the attacked cell is activated (Kobayashi et al., 1991) and cytoplasmic strands gather beneath appressoria of *E. pisi* (Kobayashi et al., 1993). Studies with fluorescent probes, such as $DiOC_7$ and BODIPY-ceramide, show that mitochondria and Golgi gather beneath fungal penetration sites (Takemoto & Kobayashi, unpublished data). Actin inhibitors totally suppress the polarization of these defense-related responses and organelles (Kobayashi et al., 1997a; Takemoto & Kobayashi, unpublished data). Inhibitory effects of actin inhibitors on defense responses are also observed in other plant–fungus interactions (Gross et al., 1993; Hazen & Bushnell, 1983; Takemoto et al., 1999).

2.3 Possible roles of actin cytoskeleton in defense mechanisms against fungal invasion

The most likely role of the cytoskeleton in defense response expression is polarization (mobilization) of defense-related reactions to appropriate subcellular locations. Many authors have reported the translocation of organelles (Gross et al., 1993; Hazen & Bushnell, 1983; Jahnen & Hahlbrock, 1988; Snyder & Nicholson, 1990), defense-related compounds such as phytoalexins (Snyder & Nicholson, 1990), enzymes involved in phytoalexin synthesis (Jahnen & Hahlbrock, 1988), and mRNAs encoding PR proteins (Gross et al., 1993) during attempted penetration by the fungus. These responses might be caused by *de novo* synthesis following gene activation or by modification of general metabolism due to recognition of specific or nonspecific signals associated with the fungal penetration attempt (Boyd et al., 1994; Gregersen et al., 1990; Lamb, 1994; Lamb et al., 1989; Renelt et al., 1993). The fact that the polarization of defense-related reactions occurs coincident with the formation of a radial arrangement of the cytoskeleton at the same site led us to suggest that the cytoskeleton may be involved in polarization of defense-related materials (Kobayashi et al. 1992, 1997a). The results described above strongly support this possibility. The recent work of McLusky et al. (1999) shows that autofluorescence material at the penetration site of *Botrytis allii* is caused by the accumulation of feruloyl-3'methoxytyramine and feruloyltyramine in onion epidermal cells and is associated with a striking polarization of actin microfilaments. Thus, polarization of organelles and/or defense-related compounds is one possible function of the cytoskeleton in expression of resistance at the time of attempted penetration by the fungus.

Alternatively, the cytoskeleton rearrangements may directly relate to defense gene expression. Takemoto et al. (1999) reported that induction of phenylalanine ammonia-lyase mRNA, production of a potato phytoalexin, rishitin, and accumulation of PR proteins is suppressed or delayed by treatment with cytochalasin D in elicitor-treated potato tuber slices (Takemoto et al., 1999; Furuse et al., 1999). In interactions between pea and the pea pathogen, *Mycosphaerella pinodes*, cytochalasin A specifically suppresses a ribosome-bound form of phenylalanine ammonia-lyase mRNA in pea tissues treated with a fungal elicitor, whereas no change occurs in the house-keeping gene histone H1 (Sugimoto et al., 2000).

Another possible function of the actin cytoskeleton in plant defense mechanisms, is that it transmits physical and/or chemical signals from the fungal penetration site to the nucleus. Fungal pathogens break through plant cell walls and invade cells rather quickly. Therefore, rapid responses to fungal attack are required by plant cells. In eukaryotic cells, external signals are received by receptors at the surface of cells, and the signals are transmitted to an intracellular target through a cascade of second messengers. The important modulator protein for phosphatidylinositol (PI)-mediated signal transduction, PI 4-kinase, is closely associated with the cytoskeleton in carrot cells (Xu et al., 1992). Yang et al. (1993) reported that the amino acid sequence of a PI 4-kinase activator protein from carrot shows high homology with an actin-binding protein from *Dictyostelium.* The plant cytoskeleton may, therefore, provide an information superhighway for the emergency call of defense responses against pathogenic microbes.

3. PUTATIVE MOLECULAR MECHANISMS OF ACTIN REORGANIZATION DURING FUNGAL PENETRATION

As described above, the dynamic rearrangement of microfilaments may play an important role in the expression of resistance mechanisms during fungal attack by temporally and spatially regulating plant defense responses. In this context, control of microfilament organization seems to be a critical determinant of the success or failure of fungal invasion. Determination of the regulatory system(s) for microfilament organization in the expression of defense responses may, therefore, greatly enhance our understanding of plant disease resistance and may provide new molecular targets for enhancing disease resistance. In this section, we describe our recent observations that specific isoforms of actin and Rac GTPase are activated or suppressed in barley and tobacco plants during the early fungal infection process.

3.1 Differential expression of actin isoforms

In general, a large number of actin genes are found in each plant genome and form a gene family. Even in the small genome of *Arabidopsis*, eight expressed actin genes exist (McDowell et al., 1996b). The occurence of multiple genes coding for divergent actin isoforms raises the possibility that different isoforms can have distinct roles *in vivo* (Meagher, 1991; see also Meagher et al., this volume). The *Arabidopsis* actin gene, *ACT7*, is the only isotype to respond to several external stimuli including hormone treatments, light regime and wounding (McDowell et al., 1996a). We found that expression of a specific actin gene was activated in barley leaf after challenge with the nonpathogenic powdery mildew fungus, *E. pisi* (Fig. 2).

Normally three actin genes, *HvAct1–HVAct3,* were expressed predominantly in barley cells. Only *HvAct3* gene transcripts accumulated 6 to 12 hours after inoculation, whereas two other actin genes were not activated by fungal invasion. Dynamic rearrangement of microfilaments is observed during this same period (Kobayashi et al., 1991), corresponding to fungal appressorium maturation and penetration. Interestingly, the amount of total actin gene transcripts did not show any marked change. These results suggest that specific actin isoforms may be involved in defense responses and reorganization of actin cytoskeleton. In animal cells, actin related proteins (Arp) are thought to be required for actin filament nucleation (Machesky & Gould, 1999). Although specific actin isoforms may participate in actin assembly as Arps, the function of specific plant actin isovariants remains to be investigated.

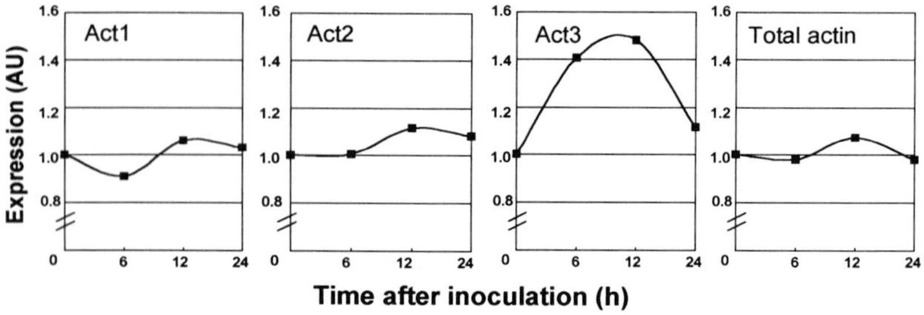

Figure 2. Expression pattern of actin mRNA in barley leaves inoculated with the nonpathogen, *Erysiphe pisi*. Representative results are shown. Values represent relative values to 0 time as 1.

3.2 Rac GTPases

Recently, it has been shown that members of the Rho family of small GTPases are key regulators of the actin cytoskeleton and, through their interaction with multiple target proteins, they ensure coordinated control of other cellular activities, such as membrane dynamics, gene expression and cell cycle progression (Hall, 1998). Three distinct members, Rho, Rac and Cdc42, comprise the Rho GTPase family. Only Rac genes, however, have been cloned from higher plants (Borg et al., 1997; Winge et al., 1997; Yang and Watson, 1993). Little is known about plant Rac function, but recent work suggests that some Rac isoforms may play a role in controlling cell division and polarization of tip growing pollen tubes (Li et al., 1999; Kost et al., 1999; Lin et al., 1996; see also Li & Yang, this volume). These results lead us to the hypothesis that plant Rac GTPases may be involved in reorganization of microfilaments and polarization of defense responses against fungal invasion. To investigate this hypothesis, we cloned tobacco Rac genes and analyzed their expression in plants inoculated with pathogenic (*Erysiphe cichoracearum*) and nonpathogenic (*E. pisi*) powdery mildews (Fig. 3).

Transcripts for two Rac isoforms, *NtRac1* and *NtRac3,* were detectable in tobacco leaf with Northern analysis. When the leaves were inoculated with a nonpathogen, both *NtRac1* and *NtRac3* transcript levels were increased slightly. On the other hand, the attack of pathogenic *E. cichoracearum* resulted in decreased *NtRac1* transcript levels, although *NtRac3* didn't show any change. Transcription of certain defense genes is activated by nonpathogen or incompatible pathogen attack, whereas compatible pathogen suppresses expression of the same defense genes (Bell et al., 1986; Lo et al., 1999).

```
                                I                                                      II
NtRac1   --MSASRFIKCVTVGDGAVGKTCLLISYTSNTFPTDYVPTVFDNFSANVVVDGSTVNLGLWDTAGQEDYNRLRPLSYRGADVFILAFSLISKASYENVSKKWIPELR
NtRac2   --MSASRFIKCVTVGDGAVGKTCLLISYTSNTFPTDYVPTVFDNFSANVVVNGSTVNLGLWDTAGQEDYNRLRPLSYRGADVFILAFSLISKASYENVSKKWIPELK
NtRac3   MASSASRFIKCVTVGDGAVGKTCMLICYTSNKFPTDYVPTVFDNFSANVVVEGTTVNLGLWDTAGQEDYNRLRPLSYRGADVFVLAFSLVSRASYENILKKWIPELQ
NtRac4   --MSGSRFIKCVTVGDGAVGKTCLLISYTSNTFPSDYVPTVFDNFSANVVVDGSTINLGLWDTAGQEDYNRLRPLSYRGADVFILAFSLISKASHENVFKKWISELR
MsRac1   --MQA---IKCVVVGDGAVGKTCLLISYTTNAFPGEYIPTVFDNYSANVMVDGKPVNLGLWDTAGQEDYDRLRPLSYPQTDVFLICFSLVSPASFENVRAKWYPEVR
           ..    ****.**********:**.**:* ** :*:******:****:*:*..:**************:*******  :***::.***:* **.**:  ** .*::
                        III                                      IV                                V
NtRac1   HYAPGVPIILVGTKLDLREDKQFF--LDHPGAVPLTTAQGEELRKSIGASAYIECSAKTQQNVKAVFDAAIKVVLQPPKQKKKKKRKGQKACSIL----------
NtRac2   HYAPGCPIVLVGTKLDLRDDKQFF--IDHPGAVPITTAQGEELRKTIGAPSNNECSSKTQENVKAVFDAAIKVVLQPPK-AKKKKGKSQRACSIL----------
NtRac3   HYAPGIPVVLAGTKLDLREDKHFL--ADHPGLVPVTTAQGEELRKQIGAAYYIECSSKTQQNVKAVFDAAIKVVIKPPQ-KQKEKKKQRRGCLMNVMCGRKLVCLK
NtRac4   HYAPGVPIILVGTKLDLREDKQFF--LDHPGAVPITTTQGEDLRKSVGAAAYIECSAKTQQNVKAVFDAAIKVVLQPPKQKKKKKRKGQKACSIL----------
MsRac1   HHCPNTPIILVGTKLDLRDDKDTIEKLKEKKLTPITYPQGLAMAKEIGAVKYLECSALTQRGLKTVFDEAIRAVLCPPP-VKKRKRK----CLLL----------
         *:.*. *::*.*******:**. :    ..   .*:* .**  : * :**    ***: **..:*:*** **:.*: **   :*.* *    * :
```

Figure 3. Multiple alignment of amino acid sequences for 4 tobacco *Rac* genes. The sequence of mouse *Rac1* gene is included for comparison. Note that 5 essential motifs of Rac proteins are well conserved. I, ATP/GTP-binding site motif A (P-loop); II, DXXG motif; III, NKXD motif; IV, EXSAX motif; V, geranylgeranyl group binding site (CXXL box). Asterisks indicate positions which have a single, fully conserved residue. Colons indicate that one of the following 'strong' groups is fully conserved: STA, NEQK, NHQK, NDEQ, QHRK, MILV, MILF, HY and FYW. Dots indicate that one of the following 'weaker' groups is fully conserved: CSA, ATV, SAG, STNK, STPA, SGND, SNDEQK, NDEQHK, NEQHRK, FVLIM, HFY.

The expression pattern of *NtRac1* is similar to such defense genes. Borg et al. (1997) reported *Rac* mRNA accumulation in *Lotus japonicus* during nodule development. These results suggest that Rac GTPases may be involved generally in stress responses during the microbial infection process. However, the roles of plant Rac proteins in responses against biotic stresses are unclear, and whether the GTPases participate in the combat against pathogenic fungus through microfilament reorganization remains to be elucidated.

4. CONCLUSIONS

It is becoming increasingly evident that the actin cytoskeleton is involved in plant defense responses against fungal pathogens, through the regulation of temporal and spatial arrangement of defense related molecules and organelles during the early stages of infection. In addition to the inhibitor studies described above, our preliminary experiments showed that transient overexpression of the actin-binding protein, profilin, in barley coleoptile cells caused actin depolymerization, similar to a report by Staiger et al. (1994). Furthermore, overexpression of profilin resulted in the suppression of resistance to fungal penetration (Kobayashi & Furuhashi, unpublished result). These results strongly suggest that the actin cytoskeleton is involved in blocking fungal penetration.

Although little is known about the regulation of microfilament reorganization, further investigations of actin isoforms and Rac GTPases may lead to breakthroughs in our understanding of the molecular mechanisms underlying cytoskeletal regulation. Furthermore, determination of the molecular basis of this regulatory mechanism will provide a possible target for improvement of crop plants with molecular means. Recently, overexpression of a constitutively active form (i.e. a GTP-bound form) of Rac protein was shown to enhance reactive oxygen species production and hypersensitivity in rice plants inoculated with rice blast fungus (Kawasaki et al., 1999). This result suggests that activation of Rac GTPases may cause constitutively expressed disease resistance in plants, although the main target of the mutant Rac protein in this case may be NADPH oxidase assembly (Bokoch, 1994; Lamb & Dixon, 1997). Alteration of the actin cytoskeleton and other defense mechanisms should be investigated in this transgenic rice line. To determine the molecular mechanism of actin regulation and its involvement in defense responses, transgenic plants that possess modified actin-related genes will also be powerful tools.

ACKNOWLEDGEMENT

The work reported here has been supported in part by Grant-in-Aid for Scientific Research (No.10306004 and 11660045) from the Ministry of Education, Science and Culture of Japan.

REFERENCES

Aist JR (1976) Papillae and related wound plugs of plant cells. Annu Rev Phytopathol 14:145-163

Bell JN, Ryder TB, Wingate VP, Bailey JA and Lamb CJ (1986) Differential accumulation of plant defense gene transcripts in a compatible and an incompatible plant-pathogen interaction. Mol Cell Biol 6: 1615-1623

Bokoch, GM (1994) Regulation of the human neutrophil NADPH oxydase by the Rac GTP-binding proteins. Curr Opin Cell Biol 6: 212-218

Borg S, Brandstrup B, Jensen TJ and Poulsen C (1997) Identification of new protein species among 33 different small GTP-binding proteins encoded by cDNAs from *Lotus japonicus*, and expression of corresponding mRNAs in developing root nodules. Plant J 11: 237-250

Boyd LA, Smith PH, Green RM and Brow KM (1994) The relationship between the expression of defense-related genes and mildew development in barley. Mol Plant Microbe Interact 7: 401-410

Bushnell WR and Bergquist S (1975) Aggregation of host cytoplasm and the formation of papillae and haustoria in powdery mildew of barley. Phytopathology 65: 310-318

Furuse K, Takemoto D, Doke N and Kawakita K (1999) Involvement of actin filament association in hypersensitive reactions in potato cells. Physiol Mol Plant Pathol 54: 51-61

Gregersen PL, Collinge DB and Smedegaard-Petersen V (1990) Early induction of new mRNAs accompanies the resistance reaction of barley to the wheat pathogen, *Erysiphe graminis* f sp *tritici.* Physiol Mol Plant Pathol 36: 471-481

Gross P, Julius C, Schmelzer E and Hahlbrock K (1993) Translocation of cytoplasm and nucleus to fungal penetration sites is associated with depolymerization of microtubules and defense gene activation in infected, cultured parsley cells. EMBO J 12: 1735-1744

Hall A (1998) Rho GTPases and the actin cytoskeleton. Science 279: 509-514

Hazen BE and Bushnell WR (1983) Inhibition of the hypersensitive reaction in barley to powdery mildew by heat shock and cytochalasin B. Physiol Plant Pathol 23: 421-438

Jahnen W and Hahlbrock K (1988) Cellular localization of nonhost resistance reactions of parsley (*Petroselinum crispum*) to fungal infection. Planta 173: 197-204

Kamiya N (1981) Physical and chemical basis of cytoplasmic streaming. Annu Rev Plant Physiol 32: 205-236

Kawasaki T, Henmi K, Ono E, Hatakeyama S, Iwano M, Satoh H and Shimamoto K (1999) The small GTP-binding protein Rac is a regulator of cell death in plants. Proc Natl Acad Sci USA 96: 10992-10926

Kitazawa K, Inagaki H and Tomiyama K (1973) Cinephotomicrographic observation on the dynamic responses of protoplasm of a potato plant cell to infection by *Phytophtora infestans.* Phytopath Z 76: 80-86

Kobayashi I, Komura T, Sakamoto Y, Yamaoka N and Kunoh H (1990) Recognition of a pathogen and a nonpathogen by barley coleoptile cells. (I) Cytoplasmic responses to the nonpathogen, *Erysiphe pisi*, prior to its penetration. Physiol Mol Plant Pathol 37: 479-490

Kobayashi I, Kobayashi Y, Yamaoka N and Kunoh H (1991) An immunofluorescent cytochemical technique applying micromanipulation to detect microtubules in plant tissues inoculated with fungal spores. Can J Bot 69: 2634-2636

Kobayashi I, Kobayashi Y, Yamaoka N and Kunoh H (1992) Recognition of a pathogen and a nonpathogen by barley coleoptile cells. III. Responses of microtubules and actin filaments in barley coleoptile cells to penetration attempts. Can J Bot 70: 1815-1823

Kobayashi I, Kobayashi Y and Hardham AR (1994) Dynamic reorganization of microtubules and microfilaments in flax cells during the resistance response to flax rust infection. Planta 195: 237-247

Kobayashi Y, Kobayashi I and Kunoh H (1993) Recognition of a pathogen and a nonpathogen by barley coleoptile cells. (II) Alteration of cytoplasmic strands in coleoptile cells caused by the pathogen, *Erysiphe graminis*, and *E pisi*, prior to their penetration. Physiol Mol Plant Pathol 43: 243-254

Kobayashi Y, Kobayashi I, Funaki Y, Fujimoto S, Takemoto T and Kunoh H (1997a) Dynamic reorganization of microfilaments and microtubules is necessary for the expression of non-host resistance in barley coleoptile cells. Plant J 11: 525-537

Kobayashi Y, Yamada M, Kobayashi I and Kunoh H (1997b) Actin microfilaments are required for the expression of nonhost resistance in higher plants. Plant Cell Physiol 38: 725-733

Kost B, Lemichez E, Spielhofer P, Hong Y, Tolias K, Carpenter C and Chua N-H (1999) Rac homologues and compartmentalized phosphatidylinositol 4,5-bisphosphate act in a common pathway to regulate polar pollen tube growth. J Cell Biol 145: 317-330

Kunoh H, Aist JR and Hayashimoto A (1985a) The occurrence of cytoplasmic aggregates induced by *Erysiphe pisi* in barley coleoptile cells before the host cell walls are penetrated. Physiol Plant Pathol 26: 199-207

Kunoh H, Hayashimoto A, Harui M and Ishizaki H (1985b) Induced susceptibility and enhanced resistance at the cellular level in barley coleoptiles. I. The significance of timing of fungal invasion. Physiol Plant Pathol 27: 43-54

Kunoh H, Kuno K and Ishizaki H (1985c) Cytological studies of the early stages of powdery mildew in barley and wheat. XI. Autofluorescence and haloes at penetration sites of appressoria of *Erysiphe graminis hordei* and *Erysiphe pisi* on barley coleoptiles. Can J Bot 63: 1535-1539

Kunoh H, Yamamori K and Ishizaki H (1982) Cytological studies of early stages of powdery mildew in barley and wheat. VIII. Autofluorescence at penetration sites of *Erysiphe graminis hordei* on living barley coleoptiles. Physiol Plant Pathol 21: 373-379

La Claire JW III (1989) Actin cytoskeleton in intact and wounded coencytic green algae. Planta 177: 47-57

Lamb CJ (1994) Plant disease resistance genes in signal perception and transduction. Cell 76: 419-422

Lamb CJ and Dixon RA (1997) The oxidative burst in plant disease resistance. Annu Rev Plant Physiol Plant Mol Biol 48: 251-275

Lamb CJ, Lawton MA, Dron M and Dixon RA (1989) Signal and transduction mechanisms for activation of plant defenses against microbial attack. Cell 56: 215-224

Li H, Lin Y, Heath RM, Zhu MX and Yang Z (1999) Control of pollen tube tip growth by a Rop GTPase dependent pathway that leads to tip-localized calcium influx. Plant Cell 11: 1731-1742

Lin Y, Wang Y, Zhu J-K and Yang Z (1996) Localization of a Rho GTPase implies a role in tip growth and movement of the generative cell in pollen tubes. Plant Cell 8: 293-303

Lo SC, Hipskind JD and Nicholson RL (1999) cDNA cloning of a sorghum pathogenesis-related protein (PR-10) and differential expression of defense-related genes following inoculation with *Cochliobolus heterostrophus* or *Colletotrichum sublineolum.* Mol Plant Microbe Interact 12: 479-489

Machesky LM and Gould KL (1999) The Arp2/3 complex: a multifunctional actin organizer. Curr Opin Cell Biol 11: 117-121

McDowell J M, An Y-Q, Huang S, McKinney EC and Meagher RB (1996a) The Arabidopsis ACT7 actin gene is expressed in rapidly developing tissues and responds to several external stimuli. Plant Physiol 111: 699-711

McDowell J M, Huang S, McKinney E C, An Y-Q and Meagher R B (1996b) Structure and evolution of actin gene family in *Arabidopsis thaliana.* Genetics 142: 587-602

McLusky SR, Bennett MH, Beale MH, Lewis MJ, Gaskin P and Mansfield JW (1999) Cell wall alterations and localized accumulation of feruloyl-3'-methoxytyramine in onion epidermis at sites of attempted penetration by *Botrytis allii* are associated with actin polymerization, peroxidase activity and suppression of flavonoid biosynthesis. Plant J 17: 523-534

Meagher RB (1991) Divergence and differential expression of actin gene families in higher plants. Int Rev Cytol 125: 139-163

Pappelis AJ, Pappelis GA and Kulfinski FB (1974) Nuclear orientation in onion epidermal cells in relation to wounding and infection. Phytopathology 64: 1010-1012

Quader H, Hofmann A and Schnepf E (1989) Reorganization of the endoplasmic reticulum in epidermal cells of onion bulb scales after cold stress: Involvement of cytoskeletal elements. Planta 177: 273-280

Renelt A, Collong C, Hahlbrock K, Nürnberger T, Parker JE, Sacks WR and Scheel D (1993) Studies on elicitor recognition and signal transduction in plant defense. J Exp Bot 44: 257-268

Seagull RW (1989) The plant cytoskeleton. CRC Critic Rev Plant Sci 8: 131-167

Škalamera D and Heath MC (1998) Changes in the cytoskeleton accompanying infection-induced nuclear movements and the hypersensitive response in plant cells invaded by rust fungi. Plant J 16: 191-200

Snyder BA and Nicholson RL (1990) Synthesis of phytoalexins in sorghum as a site-specific response to fungal ingress. Science 248: 1637-1638

Staiger CJ Yuan M Valenta R Shaw PJ Warn RM and Lloyd CW (1994) Microinjected profilin affects cytoplasmic streaming in plant cells by rapidly depolymerizing actin microfilaments. Curr Biol 4: 215-219

Sugimoto M, Toyoda K, Ichinose Y, Yamada T and Shiraishi T (2000) Cytochalasin A inhibits the binding of phenylalanine ammonia-lyase mRNA to ribosomes during induction of phytoalexin in pea seedlings. Plant Cell Physiol 41: 234-238

Takemoto D, Maeda H, Yoshioka H, Doke N and Kawakita K (1999) Effects of cytochalasin D on defense responses of potato tuber discs treated with hyphal wall components of *Phytophthora infestans.* Plant Sci 141: 219-226

Tomiyama K (1956) Cell physiological studies on the resistance of potato plant to *Phytophthora infestans.* Annu Phytopath Soc Japan 21: 54-62

Williamson RE (1986) Organelle movements along actin filaments and microtubules. Plant Physiol 82: 631-634

Winge P, Brembu T and Bones AM (1997) Cloning and characterization of rac-like cDNAs from *Arabidopsis thaliana.* Plant Mol Biol 35: 483-495

Xu P, Lloyd CW, Staiger CJ and Drøbak BK (1992) Association of phosphatidylinositol 4-kinase with the plant cytoskeleton. Plant Cell 4: 941-951

Xu J-R, Staiger CJ and Hamer JE (1998) Inactivation of the mitogen-activated protein kinase Mps1 from the rice blast fungus prevents penetration of host cells but allows activation of plant defense responses. Proc Natl Acad Sci USA 95: 12713-12718

Yang W, Burkhart W, Cavallius J, Merrick WC and Boss WF (1993) Purification and characterization of a phosphatidylinositol 4-kinase activator in carrot cells. J Biol Chem 268: 392-398

Yang Z and Watson JC (1993) Molecular cloning and characterization of rho, a ras-related small GTP binding protein from garden pea. Proc Natl Acad Sci USA 90: 8732-8736

Chapter 33

ACTIN IN THE SECONDARY VASCULAR SYSTEM OF WOODY PLANTS

Nigel Chaffey and Peter W Barlow

IACR–Long Ashton Research Station, Department of Agricultural Science, University of Bristol, Long Ashton, Bristol BS41 9AF, UK

Key words: cambium, cytoskeleton, trees, wood-formation

Abstract: A brief historical survey of ultrastructural work on actin filaments (AFs) in secondary vascular differentiation is presented. The role(s) of AFs during secondary vascular differentiation in angiosperm trees are considered in the light of recent immunofluorescent studies. AFs are present in all cells of the vascular cambium at all stages of its seasonal cycle, and in all cambial derivatives throughout their differentiation. The predominantly net axial arrangement of AFs (parallel to the cell's long axis) in all cambial derivatives suggests a role in cytoplasmic streaming and delivery of materials to the cell tips (sites of wall synthesis and modification). The presence of other arrangements of AFs – such as rings and ellipses – particularly in differentiating xylem cells implies an involvement in the wall modifications of these cells. A recently described cytoskeletal feature of trees is the axially oriented AFs and microtubules within the long-lived xylem ray cells. The role of this cytoskeletal association in xylem—phloem exchange is discussed.

1. INTRODUCTION

No two woods are anatomically identical, the differences between them being a reflection of the relative proportions of the various cell types inherent to wood, the structural characteristics of the individual cells, and the genetic structure of the species. That wood is the product of a secondary meristem marks it out from tissues studied by the majority of plant biologists, who tend to concentrate upon the primary plant body (see other contributions in this volume). Moreover, in the context of applied science, an understanding of the process of wood-cell differentiation is essential to the

C.J. Staiger et al. (eds.), Actin: A Dynamic Framework for Multiple Plant Cell Functions, 587–600.

proper targeting of efforts concerned with molecular-biological manipulation of tree physiology and productivity.

Wood (secondary xylem) and bark (secondary phloem) are produced by two basic cell types in the vascular cambium, the fusiform initials and the ray initials. The various differentiating derivatives of these cells then become distributed in characteristic patterns within the two tissues. Some of the most fundamental changes that occur during formation of secondary xylem and secondary phloem relate to the modification of cell shape and wall structure. Of primary importance here, is the uniform orientation of the relatively inflexible cellulose microfibrils within the cell walls. The direction of cell expansion is consequently constrained and only anisotropic growth is permitted (e.g. Green, 1969). Ever since the first ultrastructural report of the parallel co-alignment of microtubules (MTs) in the cortical cytoplasm and microfibrils in the nearby cell wall (Ledbetter & Porter, 1963), this relationship between MTs and cell wall structure has been confirmed by others (reviewed by Giddings & Staehelin, 1991). Although this hypothesis of MT-microfibril co-operation has yet to receive support from all workers (see discussion in Barnett et al., 1998; Emons & Mulder, 2000), it has nevertheless focused attention on the possible role(s) of the cytoskeleton in plant-cell differentiation.

In view of the large number of actin genes in plant species and the importance attached to actin in the evolution of plants (Meagher et al., 1999), it would be surprising if actin had not in some way participated in the development of the arboreal form. This form is characterised by extensive growth in height and girth of the shoot coupled with regular branching of the growing axes. It is the result of the sustained production of anisotropic cells from the primary meristems. To uphold this shoot form, a woody support system, which also incorporates a solute conduction system, has evolved. These two systems, along with the vascular cambium from which they are derived, are collectively termed the 'secondary vascular system' (SVS). The development of the SVS requires not only the division and anisotropic growth of fusiform cambial cells, but also the elongation of fibre cells, as well as the elaboration of complex wall structures (vessel and sieve tube end-wall perforations, pits and tertiary wall thickenings). As we shall show, actin, especially in the form of actin filaments (AFs), participates in all these formative processes.

2. SURVEY OF EARLIER WORK ON ACTIN FILAMENTS AND SECONDARY TISSUE DEVELOPMENT IN TREES

From the mid-1960s onwards, ultrastructural observations slowly fostered the concept of the plant cytoskeleton. Although much work was done during this period on the ultrastructure of the SVS, rather little of it focussed on this universal feature of the cytoplasm. Where it did so (see discussion in Chaffey et al., 1996, 1997b), the emphasis was mainly on recording the MT component of the cytoskeleton to the apparent neglect of the AFs. The principal conclusion from these studies (Table 1) is that the AFs in interphase fusiform cambial cells exist in bundles and these are oriented longitudinally, parallel to the cell's long axis.

Table 1. Ultrastructural studies of actin filaments within the cambium of tree species

Aspect studied	Taxon	Authors
Active cambium (Hardwood)	*Acer pseudoplatanus*	Catesson, 1974
	Aesculus hippocastanum	Rao, 1985; Chaffey et al., 1997a, c
	Fraxinus americana	Srivastava, 1996
	F. excelsior	Goosen-de Roo et al., 1983, 1984
	Salix alba and *S. eleagnos*	Parthasarathy & Mühlethaler, 1972
Active cambium (Softwoods)	*Picea pongeus* and *Zamia* sp.	Parthasarathy & Pesacreta, 1980
	Pinus strobus	Srivastava & O'Brien, 1966
	Several gymnosperm spp.	Tsuda, 1975a
Seasonal cambium (Hardwoods)	*Aesculus hippocastanum*	Chaffey et al., 1998
	Robinia pseudoacacia	Farrar & Evert, 1997
	Tectona grandis	Rao & Dave, 1983

Unfortunately, a similar consensus view is not possible regarding the AFs of ray cambial cells. There appear to be no studies devoted exclusively to these cells even though, numerically, they are of similar frequency to fusiform cells. When they are mentioned, it is usually to contrast them with the 'more interesting' fusiform cambial cells. If the ray cells are not carefully distinguished from the fusiform initials, there can be confusion. For example, when the orientation of AFs is described as 'longitudinal', it is not certain what this means in the context of a cuboid ray cell – is the orientation with respect to the longest side of the ray cell, to the axis of the ray tissue, or to the organ axis? Parthasarathy and Mühlethaler (1972) considered AF bundles to be rare in cambial ray initials and in those secondary vascular cells which do not undergo elongation. From our own TEM work on shoots and roots of horse-chestnut (Chaffey et al., manuscript in preparation) and shoots of hybrid aspen (Chaffey, Barlow & Sundberg, manuscript in preparation), we can also state that AF bundles are less frequently seen in ray cells compared to fusiform cells, but where they are seen they have no preferred orientation (i.e. they are apparently arranged at random).

Since the idea was first mooted by O'Brien and Thimann (1966) for oat coleoptile cells, the AFs in active cambial cells have generally been considered to be involved in cytoplasmic streaming (Parthasarathy & Mühlethaler, 1972; Catesson, 1974, 1990; Tsuda, 1975a, Parthasarathy & Pesacreta, 1980). But, as stressed by Catesson (1990), further clarification of this putative role is required. However, because AFs are present in dormant cambial cells, and sometimes are found only during this dormant stage of the seasonal cycle (e.g. Srivastava, 1966; Srivastava & O'Brien, 1966; Catesson, 1974; Rao, 1985), AFs have been suggested as playing a role in protein storage (Catesson, 1974, 1990). This latter interpretation appears particularly attractive in the case of the cambial cells in *Acer pseudoplatanus* where AFs were found from October (beginning of cambial dormancy) but were absent from the end of March (beginning of spring reactivation) (Catesson, 1974).

A TEM study of Goosen-de Roo et al. (1983) examined the distribution of AFs and tubular endoplasmic reticulum (ER) in the SVS of *Fraxinus excelsior*. Although they found ER and AFs in both active fusiform and ray cambial cells, axially-aligned AF bundles and tubular ER were found only in the fusiform cells. This association was thought to relate to the pronounced longitudinal cytoplasmic streaming occuring in these cells and to their high elongation growth rate. A relationship between streaming and growth is strengthened by their presence during the first stages of fibre differentiation (these cells undergo the most elongation in the SVS), and by their absence from ray cambial cells where cytoplasmic streaming was observed to be slower and more varied in direction.

Goosen-de Roo et al. (1984) also described co-oriented MTs and AFs in association with division wall formation in the fusiform cambial cells of *F. excelsior*. Here, the MTs and AFs were found within the phragmosome and were oriented parallel to its long axis. The phragmosomal MTs were considered to represent a new category of MT, involved in guiding the cell plate towards the site of the former preprophase band of MTs at the parental cell periphery. However, the authors were silent on the role of AFs in cell division.

The few TEM studies which concern AFs in differentiating fusiform cambial derivatives include those of Parthasarathy and Mühlethaler (1972) on immature phloem derivatives of two *Salix* spp., and of Tsuda (1975b) on the tracheids of *Pinus densiflora* and *Cryptomeria japonica.* The latter author conceded that AFs might be ubiquitous in developing tracheids of gymnosperms. Tsuda also recorded AFs in the phloem ray cells of *Tsuga sieboldii.* TEM observations on the taproot and shoot of *Aesculus hippocastanum* showed that, during the earliest stages of differentiation of axial secondary xylem and secondary phloem cells, AFs retained their axial orientation, despite the orientation of MTs changing markedly in these cells (Chaffey et al., 1997a).

In all the TEM studies considered above, the term 'AF' is used largely as a morphological descriptor (and for conformity with the other contributions in this volume). The presence of actin in these bundles did not become evident until the use of fluorescent-phalloidin revealed that F-actin comprised the AFs of root vascular tissues in the gymnosperms, *Chamaecyparis obtusa* and *Pseudotsuga menziesii* (Pesacreta et al., 1982). Subsequently, an actin nature of plant AFs has been widely demonstrated (Parthasarathy et al., 1985). However, although fluorescent-phalloidin labelling is a popular method of visualising F-actin in plant cells (Tewinkel et al., 1989), we have been unable to label AFs in fixed hardwood tissues by this method. This approach has also been unsuccessful so far in softwoods (Dr R. Funada, personal communication). The study of the SVS cytoskeleton of trees has, nevertheless, been advanced by the technique of indirect immunofluorescence (IIF), using anti-actin and anti-tubulin antibodies. Commencing with the study of Uehara and Hogetsu (1993), most IIF work has concentrated on the MTs, but now it has also revealed the actin component of the cytoskeleton (Chaffey et al., 1997d), as well as myosin and other cytoskeleton-related proteins. It is gratifying that, in our own work on the SVS of trees, the TEM images of AF bundles and the pattern of immunostaining with anti-actin antibody are so similar that we can confidently infer that the AF bundles contain actin (Chaffey, 2000a; Chaffey et al., 2000b). The crucial confirmation of this conclusion by immuno-staining at the EM level is to be expected shortly.

3. CYTOSKELETAL ACTIN AND SECONDARY VASCULAR TISSUE DEVELOPMENT

3.1 Visualization of actin filaments by indirect immunofluorescence microscopy

Our current method of revealing, by IIF, the actin cytoskeleton of woody material is to use the animal-derived actin C4 monoclonal antibody [ICN]. The material is pre-fixed in 3-maleimidobenzoyl N-hydroxysuccinimide ester, fixed in 3.7% formaldehyde (± 0.2% glutaraldehyde, ± 10% dimethyl sulphoxide), ethanol-dehydrated, and embedded in butyl methylmethacrylate resin which is cured with UV at –20 °C. After removal of the resin, the 6–10 μm sections are immuno-stained and examined in either an epifluorescence or confocal laser scanning microscope. The protocol is more fully described in Chaffey et al. (1997d) and Chaffey (1998, 2000c).

3.2 Special problems associated with the fixation of cambium and its derivatives

Cambial cells are thin-walled, highly vacuolate, and sandwiched between secondary xylem and phloem cells with much thicker walls. Because of this, the cambium, when it is in its active, cell-production phase, tears easily during sampling (e.g. Chaffey, 1997). Moreover, the SVS contains many different cell types which are at different physiological stages of differentiation, and an optimum fixation regime for one cell type need not be optimum for another (Kidwai & Robards, 1969).

4. ACTIN FILAMENTS IN THE SECONDARY TISSUES OF ANGIOSPERM TREES

AFs prepared for both TEM and IIF in the way described above were examined in developing secondary tissues of roots and shoots of *Aesculus hippocastanum* L. (Hippocastanaceae) (horse-chestnut) during the first 5 years of growth. Except where indicated, results for roots and shoots are the same. Where appropriate, they are compared with a second angiosperm tree, the hybrid aspen *Populus tremula* L. x *P. tremuloides* Michx (Salicaceae).

4.1 Anatomical background

From outside to inside of the organ (shoot or root), the secondary vascular tissues consist of secondary phloem, cambium, and secondary xylem (Fig. 1A). True cambial initials (see Catesson, 1974) are difficult to identify and hence the term 'cambium' is used to include the putative initial cells and their dividing derivatives; the latter may also be embarking on the earliest stages of cell differentiation. Fusiform cambial cells are thin-walled, highly vacuolate, and elongated parallel to the organ axis. In *Aesculus*, their xylem derivatives widen and differentiate as fibres, vessel elements, and axial parenchyma cells (Chaffey et al., 1999). Ultimately, the fibres are longer, axial parenchyma cells are shorter, and vessel elements are about the same length as the fusiform cambial cells. At maturity, the walls of all these differentiated cells are variously pitted, modified at their ends, and show secondary thickening and lignification.

Phloem consists of companion cells, sieve elements, axial parenchyma and fibre-sclereids fibres (Chaffey et al., 2000a), all cells being wider and having thicker walls than the fusiform cambial cells. Relative to the fusiform initials, fibre-sclereids are considerably longer and appear similar to xylem fibres. Sieve elements are about the same length as the initials, whereas companion cells and parenchyma cells are considerably shorter. The end walls of sieve elements are perforated as sieve plates.

Ray cambial cells are cuboid, thin-walled and vacuolate. Concomitant with organ widening, their derivatives elongate perpendicularly to the organ axis. Xylem ray cells develop thicker walls which lignify and become variously pitted, but are alive at maturity. They are further classified as either 'contact' cells which have large-diameter pits in the walls that border vessel elements, or as 'isolation' cells which do not make contact with vessels and bear only simple pits. Phloem ray cells, by contrast, seem to be a single cell type. Although they also have thickened walls, these are neither lignified nor elaborated, as are those of the xylem ray cells.

4.2 Actin filaments in the secondary vascular system

4.2.1 The cambium

In interphase, fusiform cambial cells display bundled AFs. These bundles are frequently branched and are longitudinally oriented (Fig. 1B), although they may also appear as a series of parallel helices of low pitch. Apparently, they extend from one end of the cell to the other. Similar arrangements are also seen in fusiform cambial cells of other angiosperm (willow – Chaffey et al., 1997d; hybrid aspen – Chaffey, Barlow & Sundberg, manuscript in preparation) and gymnosperm trees (*Pinus pinea* – unpublished observation; several other species – Funada et al., 2000). In ray cambial cells, by contrast, AF bundles are less frequent and are randomly arranged. These AF arrangements are maintained throughout the seasonal cycle in both cambial cell types.

4.2.2 Axial elements of the secondary phloem

Extensive searching reveals that the predominant AF array in these cells is the axial-helical bundle (Fig. 1C), which is common in secondary xylem fibres. Ellipses of AFs are seen in association with the development of simple pits in phloem fibres, but no specific arrays of AFs have been seen in connection with the developing sieve areas of the sieve elements.

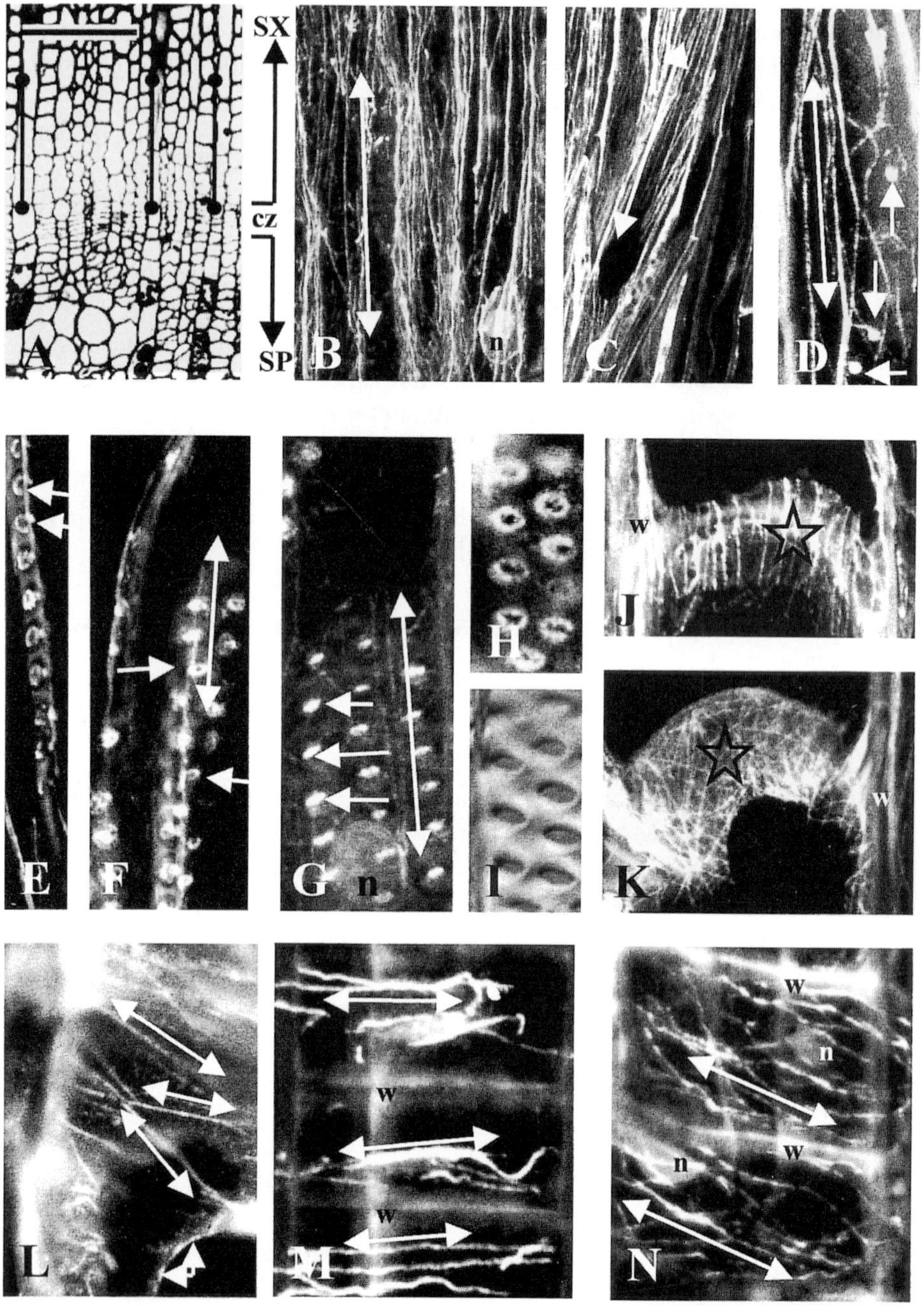

Figure 1. Aspects of secondary vascular differentiation in *Aesculus hippocastanum* (A, D-J, L and N) and hybrid aspen (B, C, K, M). Confocal (B, C, E-G, J, M, N) and epifluorescence (D, H, L) images of longitudinal sections stained with anti-actin or anti α-tubulin (N) primary antibody and FITC-labelled secondary antibody. In all images, orientation of actin filaments is indicated by bi-directional arrows. A, transverse section showing tissues of secondary

vascular system; rays indicated by headed lines; B, fusiform cambial cells; C, phloem fibres; D, xylem fibres, arrows indicate sites of pit formation; E-G, stages in bordered pit formation (arrows) in vessel elements: rings of actin filaments (E), then narrower rings/ellipses (F), and finally, ellipses; H, circular array of actin fluorescence at periphery of developing contact pits; I, Nomarski image of H; J, K, meshwork (☆) of actin filaments at the developing perforation plate of vessel elements; L, actin filaments associated with tertiary-thickening, and development of perforation (arrows) in a vessel element; M, axial actin filament cables (cf. thinner bundles in F and G) in xylem ray cells; N, axially aligned microtubule cables in xylem ray cells. Key: cz, cambial zone; n, nucleus; SP, secondary phloem; SX, secondary xylem; w, wall. Scale bar = 90 μm (A), 30 μm (B, C), 20 μm (D-L), 10 μm (M, N).

4.2.3 Axial elements of the secondary xylem

At early and late stages of cell differentiation, the AFs of all axial secondary xylem cells retain their net axial orientation (Fig. 1D). However, in late differentiation additional AF arrangements occur in all cell types. Ellipses of AFs are associated with sites of pit formation in fibres and axial parenchyma cells.

Vessel elements possess several distinct AF arrays: (i) helical AF bundles extending the length of the cell (Fig. 1F, G); (ii) AF rings associated with early stages of bordered pit formation and the developing pit aperture (Fig. 1E, F, G); (iii) circular AF arrays associated with the development of contact pits (Fig. 1H, I); (iv) an AF-meshwork upon the perforation plate itself (Fig. 1J, K); (v) a circle of AFs at the periphery of the developing end-wall perforations (Fig. 1L), and, (vi), linear AF arrays associated with tertiary thickening and which lie parallel to the developing wall-ridge (Fig. 1L) and also bridge the region between adjacent wall ridges.

4.2.4 Radial elements of the secondary vascular system

As soon as the ray cell derivatives commence elongation, AF bundles become more readily observable by IIF, appearing either axially aligned (parallel to the axis of cell elongation), in the xylem, or helically arranged in the phloem. Additionally, in the xylem rays, AF rings are seen in association with the contact pits of both developing and mature contact cells. AF ellipses are associated with the simple pits of isolation ray cells. In mature, fully differentiated xylem, axially aligned AFs are seen (Fig. 1M); however, the AFs here are appreciably thicker (cable-like) than in differentiating ray cells.

5. ACTIN FILAMENT–MICROTUBULE INTERACTIONS

From the images obtained by IIF, it seems likely that there are situations where AFs and MTs co-localise, suggesting that they may co-operate in some processes. In other cases, the two structures are independent. In the SVS of *Ae. hippocastanum*, we find both actin and tubulin at the sites of developing bordered pits (vessel elements), contact pits (vessel elements and contact rays), and at the periphery of the perforation plate in vessels prior to the plate's dissolution. Co-orientation of AFs and MTs is also evident at sites of tertiary wall thickening, but not during secondary wall formation. In all other situations, MTs and AFs appear to be spatially independent, although the net axial orientation of MTs in mature xylem ray cells (Fig. 1N) is similar to that of the AFs in such cells (Fig. 1M).

6. ACTIN FILAMENTS IN CONIFERS

To our knowledge, the only other IIF study of AFs in trees is that of Funada et al. (2000) which concerns the active fusiform cambial cells and their xylem derivatives in conifers. Axial AF bundles were found in the cambial cells and in tracheids during primary and secondary wall formation. Transverse or oblique AFs were found in some tracheids at the final stages of their differentiation. This alteration in the orientation of AFs seems to be a feature common to many maturing, elongated cells (Seagull et al., 1987; Hasezawa et al., 1989). Unlike the situation described above for hardwoods, no AFs were reported at sites of developing bordered pits, even though circular MTs have been seen there (e.g. Uehara & Hogetsu, 1993; Funada et al., 2000). Although these authors did not speculate on the role of AFs in these cells, they concluded that the orientation of AF bundles was independent of the rearrangements that affected the MTs in these differentiating cells (cf. Chaffey et al., 1997a).

7. MULTIFUNCTIONAL ACTIN FILAMENTS

What is the role of the AF cytoskeleton in secondary vascular differentiation in trees? AF bundles have often been inferred to be involved in cytoplasmic streaming (see above). Our observations with IIF might support such a view, particularly for fusiform cambial cells, their derivatives, and for ray cells. A directed cytoplasmic flow is likely to be particularly important for the tip growth of fibres and fibre-sclereids, and for cells which undergo elaboration of their end-walls – principally, sieve and vessel elements. In this context, the AF–ER associations recorded by Goosen-de Roo et al. (1983) can probably be seen as an early indication of the phenomenon, recently demonstrated by Boevink et al. (1998), where Golgi apparatus are seen to move along actin-ER strands.

Wherever AFs and MTs are co-oriented in the same cytoplasmic domain, e.g., sites of pit and perforation formation, and tertiary thickening in xylem cells, a co-operative role is suggested (cf. Seagull, 1993). There is little indication of the nature of this co-operation. However, in view of the fact that no secondary wall deposition takes place within the region bounded by the MT–AF association, we suggest that it here contributes to the establishment and maintenance of MT-free areas at the plasmalemma. In the case of bordered pits, there is evidence for a definite functional role for the actin component of the AF–MT association. When wall material is deposited to form the pit border, the diameters of both the MT and AF rings which mark the pit aperture diminish (Chaffey et al., 1999, 2000b). Having recently localised myosin to this aperture region (Chaffey & Barlow, manuscript in preparation), we infer that an actomyosin complex may mediate the contraction of the MT ring in a manner akin to that of muscle contraction (e.g. Jackson, 1982). In this way, the shape of the pit border is maintained and, at the same time, the border can continue to thicken.

It is more problematic to suggest roles for AFs in mature, fully differentiated cells. In the tree SVS this is even more acute in the case of the long-lived xylem ray cells, which may survive for many months, or even years. For a long time it has been suggested that these cells provide a symplasmic pathway for the two-way exchange of materials between the xylem vessels and the phloem sieve tubes (e.g. van Bel, 1990). Further, the xylem ray cells are a major site of deposition of storage materials, such as starch, protein and fat (e.g. Sauter & Kloth, 1986). These materials are accumulated at the end of the growth season and remobilised during the following spring. To be useful for renewed cambial activity, these reserves need to be transported from the ray cells to the cambium. The tangential walls of xylem ray cells contain relatively high numbers of plasmodesmata (Sauter & Kloth, 1986), and are hence inferred to represent a relatively low-resistance route for symplasmic transport. However, their cytoplasm contains a large number of organelles (such as protein bodies, mitochondria, plastids, oleosomes) which are likely to increase considerably the free-diffusion pathway from one end of the cell to the other.

We therefore propose that the axially oriented AF cables (and also the similarly-oriented MT cables) in the xylem ray cells facilitate intracellular transport of materials which assist in promoting and sustaining the resumption of cambial activity after dormancy has broken. Then, during the growing season, secondary metabolites are transported centripetally through the sapwood and deposited in the oldest woody cells, giving them the characteristic property of heartwood. Such directed, long-distance transport of secondary metabolites may also require AF cables for its efficient accomplishment. Of course, it is not unappreciated that, in addition to modulating the intracellular flux towards plasmodesmata, actin filaments may also control the permeability of these structures (Ding et al., 1996). Hence, actin could regulate the complete intercellular flow-path of ray cells.

ACKNOWLEDGEMENTS

IACR - Long Ashton Research Station receives grant-aided support from the BBSRC.

REFERENCES

Barnett JR, Chaffey NJ and Barlow PW (1998) Cortical microtubules and microfibril angle. In: Butterfield BG (ed) Microfibril Angle in Wood. Christchurch, New Zealand: IAWA/IUFRO, pp. 253-271

Boevink P, Oparka K, Cruz SS, Martin B, Betteridge A and Hawes C (1998) Stacks on tracks: the plant Golgi apparatus traffics on an actin/ER network. Plant J 15: 441-447

Catesson A-M (1974) Cambial cells. In: Robards AW (ed) Dynamic Aspects of Plant Ultrastructure. London: McGraw-Hill Book Co., pp. 358-390

Catesson A-M (1990) Cambial cytology and chemistry. In: Iqbal M (ed) The Vascular Cambium. Taunton, UK: Research Studies, pp. 63-112

Chaffey NJ (1997) Big *is* beautiful: science (and art) in wood microscopy. Microsc Today November 1997: 12-14

Chaffey NJ (1998) A protocol for immunostaining of microtubules and microfilaments in the secondary vascular tissues of hardwood trees for fluorescence microscopy. Microsc Today January 1998: 20

Chaffey NJ (2000a) Cytoskeleton, cell walls and cambium: new insights into secondary xylem differentiation. In: Savidge R, Barnett J, Napier R (eds) Cell and Molecular Biology of Wood Formation. Oxford: Bios, in press

Chaffey NJ (2000b) An introduction to the problems of working with the secondary vascular system of trees. In: Chaffey NJ (ed) Wood Formation in Trees. Developmental Cell Biology Techniques. The Netherlands: Harwood Academic Publishers, in press

Chaffey NJ (2000c) Immunolocalisation of the cytoskeleton in the secondary vascular system of angiosperm trees and its visualization using epifluorescence microscopy. In: Chaffey NJ (ed) Wood Formation in Trees. Developmental Cell Biology Techniques. The Netherlands: Harwood Academic Publishers, in press

Chaffey NJ, Barlow PW and Barnett JR (1996) Microtubular cytoskeleton of vascular cambium and its derivatives in roots of *Aesculus hippocastanum* L. (Hippocastanaceae). In: Donaldson LA, Butterfield BG, Singh PA, Whitehouse LJ (eds) Recent Advances in

Wood Anatomy. Rotorua, New Zealand: New Zealand Forest Research Institute, pp. 171-183

Chaffey NJ, Barlow PW and Barnett JR (1997a) Cortical microtubules rearrange during differentiation of vascular cambial derivatives, microfilaments do not. Trees 11: 333-341

Chaffey NJ, Barlow PW and Barnett JR (1998) A seasonal cycle of cell wall structure is accompanied by a cyclical rearrangement of cortical microtubules in fusiform cambial cells within taproots of *Aesculus hippocastanum* (Hippocastanaceae). New Phytol 139: 623-635

Chaffey NJ, Barlow PW and Barnett JR (2000a) Structure-function relationships during secondary phloem development in *Aesculus hippocastanum*: Microtubules and cell walls. Tree Physiol (in press)

Chaffey NJ, Barlow PW and Barnett JR (2000b) A cytoskeletal basis for wood formation: The involvement of microfilaments. Planta (in press)

Chaffey NJ, Barnett JR and Barlow PW (1997b) Cortical microtubule involvement in bordered pit formation in secondary xylem vessel elements of *Aesculus hippocastanum* L. (Hippocastanaceae): A correlative study using electron microscopy and indirect immunofluorescence microscopy. Protoplasma 197: 64-75

Chaffey NJ, Barnett JR and Barlow PW (1997c) Endomembranes, cytoskeleton and cell walls: aspects of the ultrastructure of the vascular cambium of taproots of *Aesculus hippocastanum* L. (Hippocastanaceae). Int J Plant Sci 158: 97-109

Chaffey NJ, Barnett JR and Barlow PW (1997d) Visualization of the cytoskeleton within the secondary vascular system of hardwood species. J Microsc 187: 77-84

Chaffey NJ, Barnett JR and Barlow PW (1999) A cytoskeletal basis for wood formation in angiosperm trees: the involvement of cortical microtubules. Planta 208: 19-30

Ding B, Kwon M-O and Warnberg L (1996) Evidence that actin filaments are involved in controlling the permeability of plasmodesmata in tobacco mesophyll. Plant J 10: 157-164

Emons AMC and Mulder BM (2000) How the deposition of cellulose microfibrils builds cell wall architecture. Trends Plant Sci 5: 35-40

Farrar JJ and Evert.RF (1997) Seasonal changes in the ultrastructure of the vascular cambuim of *Robinia pseudoacacia*. Trees 11: 191-202

Funada R, Furusawa O, Shibagaki M, Miura H, Miura T, Abe H and Ohtani J (2000) The role of cytoskeleton in secondary xylem differentiation in conifers. In: Savidge R, Barnett J, Napier R (eds) Cell and Molecular Biology of Wood Formation. Oxford: Bios (in press)

Giddings TH, Jr and Staehelin LA (1991) Microtubule-mediated control of microfibril deposition: a re-examination of the hypothesis. In: Lloyd CW (ed) The Cytoskeletal Basis of Plant Growth and Form. London: Academic Press, pp. 85-99

Goosen-de Roo L, Bakhuizen R, van Spronsen PC and Libbenga KR (1984) The presence of extended phragmosomes containing cytoskeletal elements in fusiform cambial cell so *Fraxinus excelsior* L. Protoplasma 122: 145-152

Goosen-de Roo L, Burggraaf PD and Libbenga KR (1983) Microfilament bundles associated with tubular endoplasmic reticulum in fusiform cells in the active cambial zone of *Fraxinus excelsior* L. Protoplasma 116: 204-208

Green PB (1969) Cell morphogenesis. Annu Rev Plant Physiol 20: 365-394

Hasezawa S, Hogetsu T and Syono K, (1989) Changes in actin filaments and cellulose fibrils in elongating cells derived from tobacco protoplasts. J Plant Physiol 134: 115-119

Jackson WT (1982) Actomyosin. In: Lloyd CW (ed) The Cytoskeleton in Plant Growth and Development. London: Academic Press, pp. 3-29

Kidwai P and Robards AW (1969) The appearance of differentiating cells after fixation in different solutions. J Exp Bot 20: 664-670

Ledbetter MC and Porter KR (1963) A "microtubule" in plant cell fine structure. J Cell Biol 19: 239-250

Meagher RB, McKinney EC and Vitale AV (1999) The evolution of new structures. Clues from plant cytoskeletal genes. TIG 15: 278-284

O'Brien TP and Thimann KV (1966) Intracellular fibers in oat coleoptile cells and their possible significance in cytoplasmic streaming. Proc Natl Acad Sci USA 56: 888-894

Parthasarathy MV and Mühlethaler K (1972) Cytoplasmic microfilaments in plant cells. J Ultrastruct Res 38: 46-62

Parthasarathy MV, Perdue TD, Witztum A and Alvernaz J (1985) Actin network as a normal component of the cytoskeleton in many vascular plant cells. Amer J Bot 72: 1318-1323

Parthasarathy MV and Pesacreta TC (1980) Microfilaments in plant vascular cells. Can J Bot 58: 807-815

Pesacreta TC, Carley WW, Webb WW and Parthasarathy MV (1982) F-actin in conifer roots. Proc Natl Acad Sci USA 79: 2898-2901

Rao KS (1985) Seasonal ultrastructural changes in the cambium of *Aesculus hippocastanum* L. Ann Sci Nat Bot Paris 13e sér 7: 213-228

Rao KS and Dave YS (1983) Ultrastructure of active and dormant cambial cells in teak (*Tectona grandis* L. f). New Phytol 93: 447-456

Sauter JJ and Kloth S (1986) Plasmodesmatal frequency and radial translocation rates in ray cells of poplar (*Populus* x *canadensis* Moench 'robusta'). Planta 168: 377-380

Seagull RW (1989) The plant cytoskeleton. Crit Rev Plant Sci 8: 131-167

Seagull RW, Falconer MM and Weerdenburg CA (1987) Microfilaments. Dynamic arrays in higher plants. J Cell Biol 104: 995-1004

Srivastava LM (1966) On the fine structure of the cambium of *Fraxinus americana* L. J Cell Biol 31: 79-93

Srivastava LM and O'Brien TP (1966) On the ultrastructure of cambium and its vascular derivatives. I. Cambium of *Pinus strobus* L. Protoplasma 61: 257-276

Tewinkel M, Kruse S, Quader H, Volkmann D and Sievers A (1989) Visualization of actin filament patterns in plant cells without pre-fixation: A comparison of differently modified phallotoxins. Protoplasma 149: 178-182

Tsuda M (1975a) The ultrastructure of the vascular cambium and its derivatives in coniferous species. I. Cambial cells. Bull Tokyo Univ For 67: 158-226

Tsuda M (1975b) The ultrastructure of the vascular cambium and its derivatives in coniferous species. II. Xylem cells and phloem parenchyma cells. Bull Tokyo Univ For 68: 25-89

Uehara K and Hogetsu T (1993) Arrangement of cortical microtubules during formation of bordered pits in the tracheids of *Taxus*. Protoplasma 172: 145-153

van Bel AJE (1990) Xylem-phloem exchange via the rays: The undervalued route of transport. J Exp Bot 41: 631-644

Chapter 34

METHODS FOR DETECTION AND IDENTIFICATION OF F-ACTIN IN FIXED AND PERMEABILIZED PLANT TISSUES

Elison B Blancaflor[1] and Karl H Hasenstein[2]

1 Plant Biology Division, The Samuel Roberts Noble Foundation, 2510 Sam Noble Parkway, Ardmore, OK 73401, USA
2 Department of Biology, P.O. Box 42451, The University of Louisiana, Lafayette, LA 70504, USA

Key words: actin, antibody labeling, immunofluorescence, phalloidin staining

Abstract: The fixation, staining, and visualization of F-actin in plant roots are explained largely based on the comparisons between maize and flax roots. Suitable fixation includes formaldehyde and glutaraldehyde containing Pipes buffers that typically result in reliable staining with antibodies and sometimes with phalloidin-linked fluorochromes. Since cytochalasins appear to alter phalloidin binding rather than destroying F-actin, more actin-specific antibodies may be required to confirm F-actin depolymerization. F-actin staining is confounded by the fact that treatments that improve actin antibody staining, such as preincubation in methanol, are detrimental for phalloidin-based protocols. In contrast, incubation of phalloidin-labeled fluorochromes with <2% glycerol in living cells can produce accurate and reliable staining of F-actin that cannot be evaluated otherwise.

1. INTRODUCTION

Our ability to visualize the organization of the cytoskeleton in plant cells has improved dramatically during the last few years due to the advances in green fluorescent protein (GFP) technology, the refinement of microinjection techniques, the development of new antibodies, and better immunostaining protocols. Both microtubules and actin filaments in living plant cells have recently been observed using GFP fused to microtubule- and actin-binding proteins (Marc et al., 1998; Kost et al., 1998; Vitha et al., this volume). Successful microinjection of fluorescently labeled tubulin (Yuan et al., 1994;

C.J. Staiger et al. (eds.), Actin: A Dynamic Framework for Multiple Plant Cell Functions, 601–618.

Wasteneys et al., 1993; Himmelspach et al., 1999) and phalloidin (Schmit and Lambert, 1990; Ren et al., 1997) into plant cells has also allowed investigators to monitor the dynamics of both these cytoskeletal elements *in vivo*.

Despite the advances in GFP technology and of microinjection technique, these methods are limited to a few cells. Especially the study of the cytoskeleton on a tissue scale requires reliable visualization of the cytoskeletal organization in plants in fixed plant tissues (Hepler & Gunning 1998). This is in large part due to limitations in our ability to image the cytoskeleton in cells located in the interior of the plant. So far, the information on the cytoskeleton in living plant cells derived from GFP and microinjection techniques has been limited to isolated cells or cells located on the plant surface. Therefore, standard fixation and sectioning techniques will continue to be an important tool for studies on the plant cytoskeleton.

In contrast to microtubules, methods for preserving actin in plant cells and tissues for subsequent labeling with fluorescent probes have been problematic. Several protocols ranging from varying the concentrations and components of the fixative, or cross-linking to modifying permeabilization techniques have been tried in order to obtain a better representation of F-actin organization in plants (Traas et al., 1987; Sonobe & Shibaoka, 1989; Doris & Steer, 1996; Olyslaegers & Verbelen, 1998; Wasteneys et al., 1997). Many of these techniques have led to substantial improvements in our ability to image actin filaments in fixed plant cells and tissues at the light microscope level. However, results appear to be specific for the cell type used, or whether antibodies or phallotoxin derivatives were employed to label actin. General reports on immunostaining methods in fixed plant cells for microscopy provide details on routine preparation for immunolabeling (Wick, 1993; Brown & Lemmon, 1995). This article will focus primarily on procedures and associated problems of labeling actin filaments in sectioned plant material for confocal microscopy. The methods presented are mostly based on our experience with vibratome-sectioned root tissues which we have used successfully in identifying cytoskeletal organization under various conditions and treatments (Blancaflor & Hasenstein, 1993; Blancaflor & Hasenstein 1995a,b, Blancaflor & Hasenstein, 1997; Blancaflor et al., 1998; Hasenstein et al., 1999). However, the techniques, observations and recommendations presented in this chapter apply to other tissues as well. Alternative fixation, permeabilization and staining techniques developed by other researchers are included to provide several options in designing the best strategy for F-actin visualization.

2. BASIC PROTOCOLS

2.1 Indirect immunofluorescence microscopy

Although immunofluorescence techniques have been used routinely for the localization of microtubules in plant tissues (Brown & Lemmon, 1995; Wick, 1993), analogous attempts to localize F-actin have been difficult because different protocols result in a different arrangement or staining pattern of F-actin. Several investigators attribute this problem to aldehyde fixation, which reportedly destroys actin filaments in plant cells (Parthasarathy et al., 1985; Traas et al., 1987; Wick, 1993). However, reports claiming poor F-actin preservation after aldehyde fixation have come mainly from studies employing fluorescent derivatives of phalloidin, a fungal metabolite that binds preferentially to F-actin (Wulf et al., 1979, see discussion below). An increasing number of reports of successful visualization of F-actin in plant cells rely on commercially available anti-actin antibodies such as the C4 mouse monoclonal anti-actin antibody from ICN (Liu & Palevitz, 1992; Cho & Wick, 1991; McCurdy & Gunning, 1990; Baluška et al., 1997a,b; Vitha et al., 1997; Sivaguru et al., 1999; Szymanski et al., 1999) and the N350 mouse IgM antibody from Amersham (Wasteneys et al.,1997; Szymanski et al., 1999). Improved labeling of actin using indirect immunofluorescence has also been demonstrated with monoclonal and polyclonal antibodies raised against plant actins (Andersland et al., 1994; Collings et al., 1998a,b; Blancaflor et al., 1998). We have obtained good results in the staining of F-actin in vibratome-sectioned root tissues by indirect immunofluorescence using an antibody against pea actin (designated mAb3H11, Andersland et al. 1994). The details of the methods are presented below.

2.1.1 Fixation

We routinely use PHEM buffer for stabilizing microtubules in root tissues and have also been successful with this buffer in preserving actin filaments (Blancaflor & Hasenstein, 1997; Blancaflor et al., 1998). PHEM buffer consists of 60 mM PIPES, 25 mM HEPES, 2mM $MgCl_2$, and 10 mM EGTA, pH 6.9 (Brown & Lemmon, 1995). Other investigators use PME buffer (50 mM PIPES, 2 mM $MgSO_2$, 5 mM EGTA) for stabilizing actin filaments in other plant tissues, such as shoots and onion epidermal cells (Collings et al., 1998a,b; Wasteneys et al., 1997), and this buffer works for root tissue as well. Despite the potential drawbacks of aldehyde fixation, staining of F-actin with antibodies benefits from fixation in a combination of fresh 1% glutaraldehyde and 3 - 4% formaldehyde (prepared from 37%

formalin solution), or 3% paraformaldehyde and 5% dimethyl sulfoxide (DMSO, v/v) in either PHEM or PME buffer. After excision of the desired tissue with a sharp razor blade, the isolated tissue is simply immersed into a glass vial containing the prepared fixative. The open vials with fixative and tissue are placed in a dessicator and a light vacuum is applied. The vacuum is released after 5 to 10 min when all tissues have settled to the bottom of the vials and the tissue is left in fixative for an additional 1 - 2 hrs. Thicker plant organs may need longer fixation, but it is advantageous to divide larger organs into smaller pieces thereby optimizing the penetration of the fixative. After fixation, the samples are washed by rinsing in either PHEM or PME buffer 3 × 5 min.

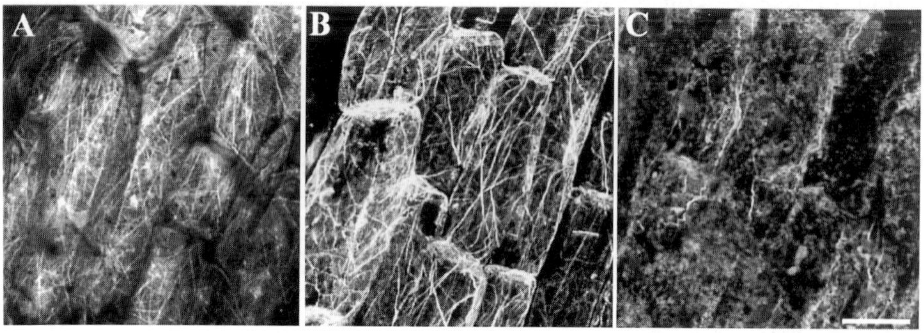

Figure 1. F-actin in the elongation zone of longitudinal sections of flax (*Linum usitatissimum*) roots. Roots were fixed in 3.7% formaldehyde and 1% glutaraldehyde and processed for indirect immunofluorescence using a monoclonal anti-actin antibody against pea actin (mAB3H11, A & B) or stained with Texas red-phalloidin (C). Images were captured using a Bio-Rad MRC 1024 confocal laser scanning microscope. 14 optical sections at 1 μm intervals were merged into one image. Roots not pretreated with $NaBH_4$ show high cytoplasmic autofluorescence, which interferes with the actin signal (A). In order to reduce free aldehydes, glutaraldehyde fixed specimen were pretreated with $NaBH_4$ (1 mg/ml) prior to processing, which significantly reduces this background and reveals fine actin filaments (B). Although processed similarly as the samples in panel A and B, F-actin was poorly stained and fragmented after staining with Texas red-phalloidin (C). All images show equal magnification, Bar = 30 μm.

Fixatives containing glutaraldehyde are notorious for their tendency to induce cytoplasmic autofluorescence (Wick, 1993). To minimize this effect, it may be useful to treat glutaraldehyde-fixed tissues with 1 mg/ml sodium borohydride ($NaBH_4$) for 10 min and reduce free aldehydes and hence the background fluorescence (Wick, 1993; Braun & Wasteneys, 1998). However, aldehyde-induced autofluorescence may be a plant-specific problem. In particular, flax (*Linum usitatissimum*) root tissue fixed with 1% glutaraldehyde exhibited a strong diffuse fluorescence in the cytoplasm without $NaBH_4$ pretreatment. The enhanced background fluorescence resulted in a reduction in both the signal-to-noise ratio and the information that could be obtained from the F-actin signal (Fig. 1A). Pretreatment with $NaBH_4$ eliminated the interfering cytoplasmic autofluorescence and allowed the documentation of delicate actin filaments (Fig. 1B). In contrast to the detailed staining with the mAB3H11 antibody, the fluorescent signal obtained after Texas red-phalloidin was fragmented and of poor quality (Fig. 1C).

In maize (*Zea mays* L.) roots, the cytoplasmic autofluorescence was not as intense or even absent after fixation in 1% glutaraldehyde and the pretreatment with $NaBH_4$ could be omitted (data not shown). The tissue and plant specificity for F-actin visualization is further emphasized by the possibility of obtaining reliable and detailed staining in the absence of glutaraldehyde (Blancaflor & Hasenstein, 1997). The varying requirement for glutaraldehyde in the fixative probably indicates differences in cytoplasmic composition, redox state or the presence of specific actin isoforms. Since a generic recipe for the chemical fixation of F-actin is not available, cryofixation and high-pressure freeze-fixation are valuable alternatives for the preservation of antigenicity and ultrastructure (Raudakoski et al., 1991; Hess et al., 1995; He & Wetzstein, 1995; Baskin et al., 1996; Thijssen et al., 1998).

2.1.2 Sectioning

To preserve the positional relationship between cells and provide access for antibodies, sectioning is necessary. Various embedding media for plant material suitable for sectioning have been tested for both F-actin and microtubule labeling, among them Steedman's wax (Brown & Lemmon, 1995; Vitha et al., 1997; Sivaguru et al., 1999; Vitha et al., this volume), polyethylene glycol (Brown et al., 1989) and butyl methyl methacrylate (Baskin et al., 1992; Baskin et al., 1996). However, embedding prior to immunolabeling requires dehydration and infiltration, and potentially reduces the quality of staining due to incomplete removal of the embedding media or masking of epitopes. Hand-sections of plant tissues have also produced excellent images of F-actin (Hush & Overall, 1992; Uetake & Peterson, 1997), but obtaining good longitudinal sections, especially of smaller plant organs, is almost impossible with this technique.

We have routinely sectioned plant material with a Vibratome-1000 (Technical Products International) for immunolabeling cytoskeletal components in root tissues (Blancaflor et al. 1998; Hasenstein et al., 1999 and references therein). A vibratome uses a vibrating blade to slice tissues without the need for embedding or freezing and provides better control over section thickness compared to hand sectioning. However, the vibratome is useful only for relatively thick plant organs. With practice, the vibratome can be used for sectioning tissue that is at least 300 μm thick. However, due to the need for some type of support, thin plant material, such as leaves or Arabidopsis roots, is difficult to section with such an instrument and therefore require embedding (Kobayashi et al., 1997; Baskin at al., 1996), or extended periods of exposure to wall digesting enzymes to allow access of antibodies (Harper et al., 1996; Wymer et al., 1999).

For vibratome sectioning, the tissue is first attached to an (aluminum) block, which has been coated with a thin film of cyanoacrylate (super glue). Attaching the tissue to the block with this technique requires patience and some practice to avoid using excessive amounts of glue. Too much glue will adversely affect the tissue properties for sectioning and subsequent labeling. After attaching the tissue to the block, the block is secured with a clamp in the vibratome bath that has been filled with either distilled water or buffer. Vibratome sectioning with half-width, double-edged razor blades (0.4 μm thickness) can routinely provide 50 μm sections. Sections less than ca. 30 - 40 μm thick break easily and are not workable. Sections are retrieved from the vibratome bath using the wide end of a Pasteur pipette and transferred to a Petri dish half-filled with PHEM or PME buffer and then carefully transferred to glass slides coated with 1 mg/ml poly-L-lysine. Alternatively, sections can be secured onto glass cover slips using a thin film of agarose (for details see Brown & Lemmon, 1995). We found the latter method to be more successful in preventing the loss of sections during processing. Furthermore, agarose prevents drying of sections during processing and minimizes collapse and distortion of the sections (Brown & Lemmon, 1995). Then the sections are allowed to dry partially. Excessive drying increases background fluorescence and causes tissue shrinkage.

2.1.3 Antibody staining and mounting of sections

After the sections are secured on either glass slides or cover slips, the tissue is slightly digested in a cocktail of wall degrading enzymes in PHEM buffer (1% cellulase; 0.1% pectolyase, typically in the presence of 0.1% BSA to reduce proteolytic activity) followed by 3 washes in PHEM buffer. After enzyme treatment, sections are incubated in 1% Triton X-100 in PHEM buffer for 15 - 20 min followed by another 3 washes in PHEM buffer. Treatment with this non-ionic detergent solubilizes lipids and membrane-related proteins thereby helping to reduce the autofluorescence (Wick, 1993). Detergent also facilitates penetration of aqueous antibody preparations. The sections are incubated in the primary antibody for 1 - 2 h, washed in buffer (3 x 5 min) followed by 1 - 2 h incubation in the secondary antibody conjugated to a suitable fluorochrome. After 3 additional washes in buffer to remove the secondary antibody, sections are mounted in 20%

Mowiol 4-88 (Calbiochem) with the anti-fading reagent para-phenylene-diamine (0.1%) in PBS, pH 8.5.

Some primary antibodies (e.g., C4) require pretreatment of the sections in organic solvents, e.g., cold methanol (- 20 °C for 10 min) or acetone for successful labeling (Wick, 1993; Williamson et al., 1987; Tang et al., 1989). Staining with the monoclonal anti pea actin antibody also benefits from the methanol incubation (Andersland et al., 1994). Presumably, pretreatment with organic solvents denatures proteins and facilitates binding to the proper epitope by the antibody (Tang et al., 1989; Wick, 1993). Vitha et al. (1997) reported that methanol pretreatment was not necessary to obtain F-actin labeling in wax embedded maize roots. They claim that the ethanol and acetone used during the dewaxing step could substitute for methanol. We tried research grade ethanol as a substitute for methanol in processing vibratome-sectioned maize roots. Despite some F-actin fluorescence, the labeling was not as extensive as in methanol-pretreated roots. In addition, fluorescent aggregates were visible in different regions of the ethanol-treated tissue (see Figs. 2A and B). Furthermore, actin strands in all cells types were fragmented and no actin labeling was observed in the elongation zone. It therefore appears that methanol treatment is optimal for the anti pea actin antibody. However, the basis for the enhanced staining by methanol is still poorly understood (Tang et al., 1989; Wick, 1993).

3. LABELING F-ACTIN WITH FLUORESCENT-PHALLOIDIN

Most studies on actin in plant cells relied on labeling actin with fluorescent derivatives of phalloidin, low molecular weight fungal metabolites that cross the plasma membrane readily and bind preferentially to F-actin (Wulf et al., 1979). Methods for sectioning and permeabilizing plant tissues for F-actin staining using fluorescently-labeled phalloidin is similar to the procedures described for indirect immunofluorescence. However, instead of adding antibodies after detergent extraction, the sections are incubated for 2 - 3 hrs in fluorescently conjugated phalloidin. Our experience with maize roots shows that it is beneficial to reduce the formaldehyde concentration to about 2% to stain F-actin with phalloidin, particularly in epidermal and cortical cells of the elongation and maturation zone (Blancaflor & Hasenstein, 1997; Blancaflor et al.,1998). Formaldehyde concentrations above 3% resulted in diffuse staining and the appearance of fragmented actin. A similar observation was also reported for maize shoots (Waller & Nick, 1997).

3.1 Problems associated with phalloidin labeling

As mentioned earlier, reports claiming the sensitivity of F-actin to aldehyde fixation have come from labeling studies employing phalloidin (Parthasarathy et al., 1985; Traas et al., 1987; Sonobe & Shibaoka, 1989; Blancaflor & Hasenstein, 1997). Based on the diffuse and sometimes fragmented appearance, it is possible that the actin becomes disrupted after aldehyde fixation (Wick, 1993; Traas et al., 1987). However, it is equally possible that aldehyde fixation could diminish phalloidin binding to F-actin resulting in poor labeling rather then causing fragmentation of F-actin (Tang et al., 1989). The latter possibility is supported by our work comparing indirect immunofluorescence labeling of actin in maize and flax roots. Although roots fixed in 3.7% formaldehyde and 1% glutaraldehyde and labeled with a pea anti-actin antibody resulted in excellent preservation of actin filaments in both species (Figs. 1B, 2A), roots that were fixed similarly but stained with Texas red-phalloidin showed poor F-actin labeling (Figs. 1C, 2C.). Furthermore, a recent study reported that paraformaldehyde concentrations as high as 8% still resulted in successful labeling of F-actin when processed by indirect immunofluorescence using polyclonal anti-actin antibodies (Collings et al., 1998a).

Concerns about the efficiency of phalloidin as an indicator of F-actin have also come from studies comparing the effect of cytochalasin B (CB) on F-actin organization. Tang et al. (1989) treated pollen tubes with CB and examined F-actin organization after labeling with either rhodamine-phalloidin or with actin antibodies. Their results showed that F-actin was still recognizably intact in CB-treated pollen tubes stained with actin antibodies, while rhodamine-phalloidin stained pollen tubes showed only a diffuse staining pattern and fragmented actin strands. Their results support the argument that CB alters the binding of phalloidin to F-actin rather then causing actin depolymerization. Therefore, previous claims that CB destroys F-actin need to be re-examined with antibodies.

The difficulties of staining of F-actin with phalloidin are also well demonstrated by comparing actin staining of vibratome sections from maize and flax roots. Although after mild fixation (2% formaldehyde) F-actin in maize roots is easily visualized by rhodamine-phalloidin (see Blancaflor & Hasenstein, 1997; Fig. 2D), the same protocol did not stain F-actin in flax roots (Fig. 2E) with the exception of some signal in the vascular cells (Fig. 2F). Despite several variations in buffer and fixative components, satisfactory labeling of F-actin in flax roots with phalloidin was not possible. However, the extensive network of F-actin in flax roots could be visualized with the monoclonal antibody against pea actin described above (Andersland et al., 1994; see Figs. 1A, B). The C4 antibody from ICN, a commonly used antibody for plant actin localization, did not reveal the intricate structure of actin filaments that we observed using the mAb3H11 antibody. Based on the comparisons between antibody and phalloidin staining, it seems that the latter is more generic than antibody staining. The variability in the F-actin staining, either in the genuine form or after fixation, suggests a range of configurations or structures of F-actin that are sensitive to the fixation and incubation medium.

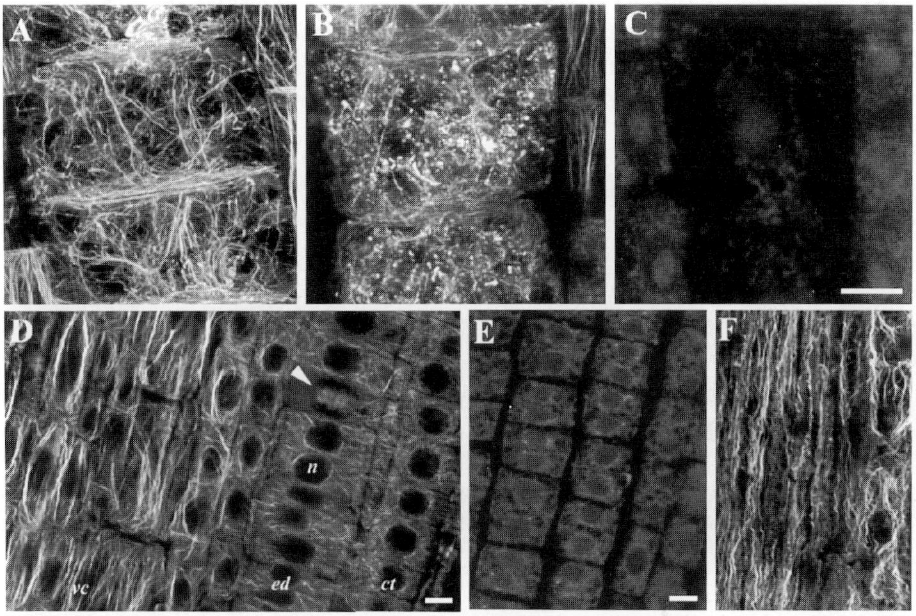

Figure 2. Metaxylem cells in a maize root stained with mAB3H11 actin antibody after pretreatment with methanol. An extensive array of longitudinal and random actin filaments could be visualized (A). Metaxylem cells in a maize root stained with the same actin antibody after pretreatment with ethanol. The ethanol pretreatment produces large fluorescent aggregates and fewer actin filaments than methanol pretreatment (B). Metaxylem cells in a methanol-treated root similar to panel A but stained with Texas red-phalloidin shows complete absence of actin staining (C). The lack of an F-actin signal indicates that the steps that improve staining with antibody diminish staining with phalloidin. Cells of maize roots fixed in 2% formaldehyde close to the vascular region stained with Texas red-phalloidin show uniform F-actin labeling (D) but flax roots fixed and stained similarly as roots in panel D show no F-actin labeling (E). Only some cells in the vascular region of flax roots showed F-actin staining stained with Texas red-phalloidin (F). All images were collected using a Bio-Rad MRC 1024 confocal microscope. Panels A and B are projections of 15 optical sections. Panels C to F are single optical sections. Phragmoplast (arrowhead). Pro-endodermal cells (*ed*), Cortex (*ct*), vascular cells (*vc*). Bar in C = 20 μm for panels A-C. Bar in D and E = 10 μm for panels D-F.

3.2 Comparison between actin antibody and phalloidin staining of fixed root sections

Using the optimal protocol for staining F-actin in maize roots with either phalloidin or antibodies, we compared the organization of F-actin in various regions of the root. Both phalloidin and antibody staining revealed transverse F-actin in epidermal cells (Figs. 3A, B) and pro-endodermal cells (Figs. 2D, 3C). Parenchyma cells close to and within the vascular region and cortical cells in the elongation zone exhibited longitudinal actin cables (Figs. 3D-F) and both methods also showed actin associated with the phragmoplasts (Figs. 2D, 3G, 3H).

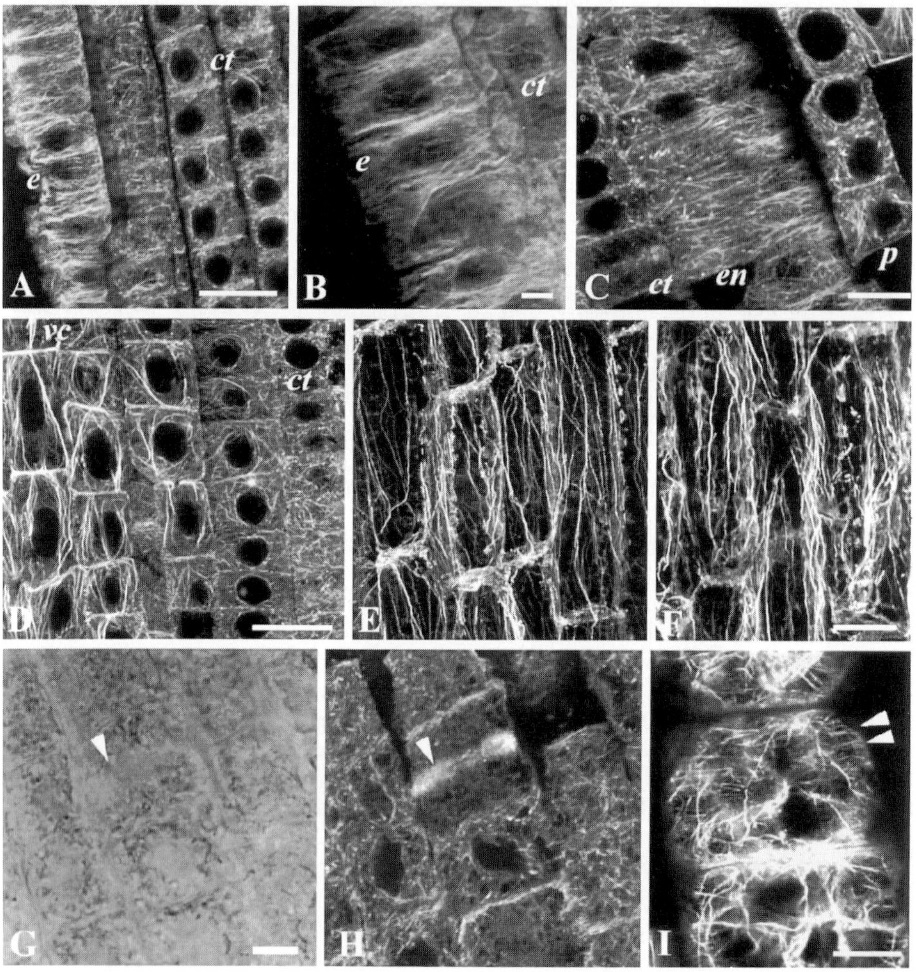

Figure 3. Comparison between indirect immunofluorescence and Texas red-phalloidin labeling of F-actin in maize (A-F, I) and flax (G-H) roots. F-actin was stained using a

monoclonal anti pea actin antibody (A, C, D, E, H, I) or Texas red-phalloidin (B, F). Images were captured using a Bio-Rad MRC 1024 confocal microscope. Panels A, B, C, D and H are single optical sections. Panels E, F and I are projections of 8-12 optical sections. Epidermal cells (*e*) and cortical cells (*ct*) in the apical region of maize roots after staining with actin antibody (A) and Texas red-phalloidin (B) revealed transverse cortical actin filaments in epidermal cells. Transverse actin filaments were also observed in proendodermal (*en*) cells after antibody staining which was similar to the pattern observed after phalloidin staining; Pericycle (*p*), cortex (*ct*). (C, compare with Fig. 2D and Blancaflor and Hasenstein, 1997). Longitudinal bundles in vascular cells (*vc*) were observed after both antibody and phalloidin staining (see also Fig. 2D) and therefore likely represent the true organization of F-actin; Randomly oriented and weaker stained actin filaments were seen in the inner cortical cells (*ct*) (D). Longitudinal actin cables in the outer cortex of the elongation zone after antibody (E) and phalloidin labeling (F). While the brightfield image shows no detail of the developing cell plate (arrowhead, G), the corresponding fluorescence image of F-actin (H) in the meristematic region of flax roots reveals random cytoplasmic and cortical actin strands and actin association with the phragmoplast (arrowhead). F-actin in developing metaxylem of maize roots stained with actin antibody (I). In addition to irregular actin cables, fine transverse arrays of actin filaments could be seen (arrowheads). Bars in A, D, F, I = 20 μm, bars in B, C, G = 10 μm.

Although the organization of F-actin was similar in many regions of the root after phalloidin or antibody staining, there were some distinct advantages in using the mAb3H11 antibody. The first advantage was the ability to image F-actin in a wider variety of tissue such as flax roots (see Figs. 1, 3H, as discussed above). Second, the background fluorescence in roots stained with the actin antibody, particularly in the meristematic region, was significantly less, therefore finer actin strands were visible. Fine transverse cortical actin strands were observed in developing metaxylem elements, which were not visible in phalloidin-stained roots (Fig. 3I).

3.3 Optimizing F-actin staining using fluorescent phalloidin derivatives

3.3.1 Crosslinking with succidimyl derivatives

To avoid the reported pitfalls of aldehyde fixation, Sonobe and Shibaoka (1989) developed a protocol using the protein cross-linking agent m-maleimodobenzoyl N-hydroxysuccinimide ester (MBS) in order to reduce the problems associated with aldehyde fixation. Their protocol involves pretreating the cells with 100 - 300 μM MBS for 30 min to 1 h followed by standard formaldehyde fixation. For tobacco BY-2 cells, this method yielded images of fine cortical actin filaments in cells fixed with formaldehyde but not with glutaraldehyde (Sonobe & Shibaoka, 1989).

Attempts to apply this technique to plant tissues have led to conflicting reports. Leaf epidermal cells of *Transdescantia* pretreated with 200 μM

MBS for 1h followed by aldehyde fixation revealed a dense network of fine actin filaments as well as thicker bundles not observed without MBS pretreatment (Cleary & Mathesius, 1996). Fresh longitudinal hand sections of unfixed pea roots showed good F-actin preservation with rhodamine-phalloidin after pretreatment with 300 μM MBS (Hush & Overall, 1992). Brighter and more abundant actin filaments after MBS pretreatment of pollen tubes prior to fixation with aldehydes indicate that MBS could protect actin filaments against aldehyde damage (Doris & Steer, 1996). However, this method was reported to be ineffective in onion root cell squashes (Liu & Palevitz, 1992) and maize shoots (Waller & Nick, 1997), but no report indicated that MBS might have adverse effects on F-actin staining. Therefore, the decision to use MBS as part of the staining protocol should be evaluated on a case-by-case basis.

3.3.2 Glycerol permeabilization of non-fixed tissues

Olyslaegers and Verbelen (1998) recently developed a glycerol permeabilization procedure to stain F-actin filaments with rhodamine-phalloidin in non-fixed onion epidermal cells and cultured tobacco cells. This technique allowed improved visualization of even the finer actin strands in both cell types. Adapting this method to vibratome sections of maize roots resulted in excellent images of F-actin in various regions of the root. Many of the finer actin filaments, particularly in the elongation zone, could now be observed (Fig. 4) that were not previously visible in fixed maize root sections (Blancaflor & Hasenstein, 1997; Blancaflor et al., 1998; Vitha et al., 1997; Baluška et al., 1997a). A similar observation was also made in epidermal sheets of rye. Direct labeling of unfixed leaf tissues with rhodamine-phalloidin showed actin filaments in greater detail, including fine filaments not visible after fixation (Cho & Wick, 1991).

For glycerol permeabilization of maize root sections, non-fixed roots were excised and attached to the vibratome as described above. Excision and securing the root to the vibratome bath should be done as fast as possible to prevent drying of the root. Sections 50 - 70 μm thick are collected and immediately immersed in a small volume (20 μl) of PME buffer containing 1 - 2% glycerol (w/v), 300 mM mannitol, and 0.2 μM Texas red-phalloidin. A coverslip is gently placed on top of the sections making sure that the sections do not fold over. This preparation can now be observed under the epifluorescence or confocal microscope. Very faint staining could be detected within 5 min of phalloidin application. The staining intensified after 10 min, and distinct actin bundles as well as finer actin strands became visible after about 20 min.

Although this method of labeling yielded striking images of F-actin in maize roots, there are several disadvantages and potential criticisms to this technique:

1) The fluorescence of F-actin labeled in this manner is less intense than in fixed preparations. Therefore, imaging F-actin in such sections requires increasing the laser power of the confocal microscope and aperture to allow more light to excite the sample. This resulted in rapid bleaching of the sample and loss of resolution.

2) The applied phalloidin could induce polymerization of actin. Therefore, sections should be observed immediately after phalloidin application.

3) The staining pattern was not as uniform as staining in fixed root sections, particularly in the elongation zone and root cap. Although it was possible to locate areas in the elongation zone that had several cells labeled (Fig. 4A), often only single cells were labeled.

4) Not all plant tissues are amenable to this type of staining. Vibratome-sectioned flax roots showed weak actin labeling with this technique and emphasizes the problems of phalloidin binding to (flax) actin isoforms that was discussed previously.

Despite the potential drawbacks to this technique, the glycerol-facilitated visualization of actin is promising. Optical sectioning by confocal microscopy and 3D reconstruction revealed an extensive network of actin filaments in cells that was not previously observed (Figs. 4A-D). In addition to the thick actin bundles, fine transverse actin filaments could be seen in elongating cells (Figs. 4A, B) and randomly oriented fine actin filaments were observed in the transition zone (Fig. 4C) and inner cortex of the elongation zone (Fig. 4D). Fine cortical and cytoplasmic actin strands were also observed in meristematic cells, pro-endodermal cells and developing metaxylem elements (data not shown).

The pro-endodermal cells showed transverse actin filaments with an organization similar for phalloidin and antibody staining in fixed roots (see Figs. 2D, 3C). Therefore, the glycerol permeabilization technique reliably produces the same staining patterns observed in fixed roots, but yields additional information about actin organization. Because of the quality of images obtained, refinement of this technique should be pursued further. With adequate controls (Traas et al., 1987) this method should be a valuable tool in visualizing F-actin in plant tissues and facilitate our understanding of the role of the actin cytoskeleton in plant roots.

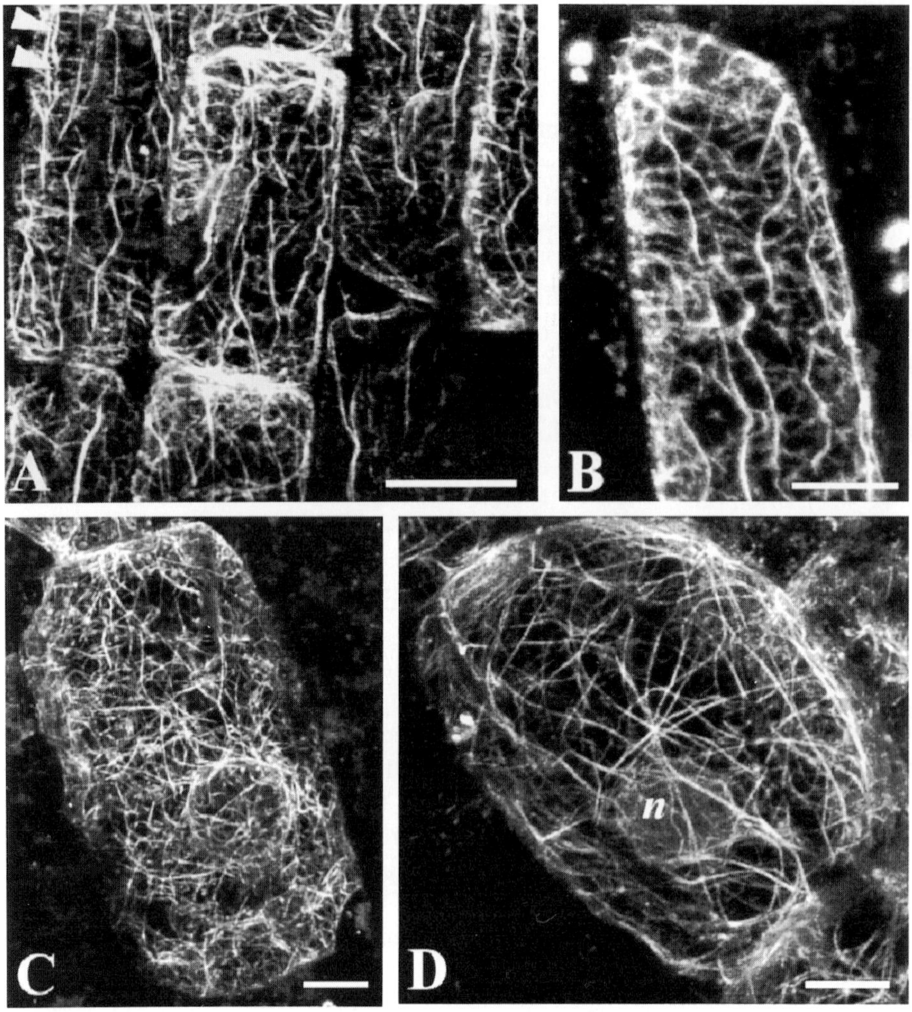

Figure 4. F-actin in non-fixed longitudinal vibratome sections of maize roots permeabilized with glycerol and incubated in Texas red-phalloidin exhibit better staining than fixed sections. All images were acquired with a Bio-Rad 1024 confocal microscope. All panels are projections of 15-25 optical sections. The fluorescence image of F-actin in the elongation zone reveals longitudinal, transverse (arrowheads) and irregular F-actin bundles (A), which are easier seen at higher magnification (B). Outer cortical cells (C) in the transition zone and inner cortical cells (D) in the elongation zone reveal an extensive network of actin filaments not previously documented in fixed root tissue. The structure and organization of the filaments show that the nuclear domain (*n*) is not a center of F-actin distribution. Bar in A = 25 μm; All other bars = 10 μm.

4. CONCLUSIONS: IMPLICATIONS FOR IDENTIFYING F-ACTIN ORGANIZATION IN ROOTS

In addition to reports that described the F-actin organization in fixed roots after various treatments and under various environmental conditions (Hush & Overall, 1992; Uetake and Peterson, 1997; Blancaflor & Hasenstein, 1997; Staves et al., 1997; Blancaflor et al., 1998; Sivaguru et al., 1999), there are reports that speculated on the correlation between the specific type of F-actin organization and development in roots (Vitha et al., 1997; Baluška et al., 1997a; Baluška et al., 1998; Baluška & Hasenstein, 1997; Vitha et al., this volume). The information presented in this chapter demonstrates the importance of carefully evaluating the labeling protocols used to visualize the organization of F-actin and probably other antigens in plant tissues. Because of the variability in results obtained from different techniques, poorly fixed or stained material can easily lead to misinterpretations.

Despite the differences between phalloidin and antibody staining, good staining with at least one label is necessary to illustrate differential actin labeling (e.g., the reduced actin staining in columella cells of fixed corn roots stained with phalloidin, Blancaflor & Hasenstein, 1997). Although the mAB3H11 antibody does not stain the central columella region, the glycerol technique does stain this region with phalloidin (Blancaflor & Collings, unpublished). This peculiar staining pattern in columella cells either reflects a diminished amount of F-actin (Baluška et al., 1997b; Baluška & Hasenstein, 1997), a change in the prevalent form of actin in these cells, or the presence of other proteins that cover the immunogenic actin epitope(s). Since F-actin has been implicated in the gravisensing system of plants (Sievers et al., 1991; Volkmann & Baluška, this volume), the reduced level of staining in fixed columella cells could indicate that the gravisensing system might rely on a subset of F-actin. Regardless of the implications of this particular problem, the characterization of F-actin should ideally be confirmed with two independent methods of staining.

Because of the difficulties associated with actin staining described above, the real organization, turnover and stability of the actin cytoskeleton in plant cells is still uncertain. Currently there are more than 10 isoforms of actin identified in plants with different expression patterns (Meagher et al., 1999; Meagher et al., this volume). The development of antibodies to specific actin isoforms will help in understanding the function of actin in plant development and clarify the differences in actin organization between tissue types. Furthermore, a combination of labeling techniques in both live and fixed plant tissues is likely to improve our understanding of the organization and function of actin in plants.

ACKNOWLEDGMENT

This work was supported by the Samuel Roberts Noble Foundation Inc. (E.B.B.) and the National Aeronautics and Space Administration (grant NAG10-0190 to K.H.H.). We thank Dr. Xin Shun Ding for comments on the manuscript. We are also grateful to Deb Fisher and Dr. Richard J. Cyr (Pennsylvania State University) for providing the anti pea actin antibody.

REFERENCES

Andersland JM, Fisher DD, Wymer CL, Cyr RJ and Parthasarathy MV (1994) Characterization of a monoclonal antibody prepared against plant actin. Cell Motil Cytoskel 29: 339-344

Baluška F, Vitha S, Barlow PW and Volkmann D (1997a) Rearrangements of F-actin arrays in growing cells of intact maize root apex tissues: A major developmental switch occurs in the postmitotic transition region. Eur J Cell Biol 72: 113-121

Baluška F, Kreibaum AK, Vitha S, Parker JS, Barlow PW and Sievers A (1997b) Central root cap cells are depleted of endoplasmic microtubules and actin microfilament bundles: Implications for their role as gravity-sensing statocytes. Protoplasma 196: 212-223

Baluška F and Hasenstein KH (1997) Root cytoskeleton: its role in perception of and response to gravity. Planta 203: S69-S78

Baluška F, Barlow P, Lichtscheidl IK and Volkmann D (1998) The plant cell body: A cytoskeletal tool for cellular development and morphogenesis. Protoplasma 202: 1-10

Baskin TI, Busby CH, Fowke LC, Sammut M and Gubler F (1992) Improvements in immunostaining samples embedded in methacrylate: Localization of microtubules and other antigens throughout developing organs in plants of diverse taxa. Planta 187: 405-413

Baskin TI, Miller DD, Vos JW, Wilson JE and Hepler PK (1996) Cryofixing single cells and multicellular specimens enhances structure and immunocytochemistry for light microscopy. J Microsc 182: 149-161

Blancaflor EB and Hasenstein KH (1993) Organization of cortical microtubules in graviresponding maize roots. Planta 191: 231-237

Blancaflor EB and Hasenstein KH (1995a) Time course and auxin sensitivity of cortical microtubule reorientation in maize roots. Protoplasma 185: 72-82

Blancaflor EB and Hasenstein KH (1995b) Growth and microtubule orientation of maize roots subjected to osmotic stress. Int J Plant Sci 156: 794-802

Blancaflor EB and Hasenstein KH (1997) The organization of the actin cytoskeleton in vertical and graviresponding primary roots of maize. Plant Physiol 113: 1147-1455

Blancaflor EB, Jones DL and Gilroy S (1998) Alterations in the cytoskeleton accompany aluminum-induced growth inhibition and morphological changes in primary roots of maize. Plant Physiol 116: 157-172

Braun M and Wasteneys GO (1998) Distribution and dynamics of the cytoskeleton in graviresponding protonemata and rhizoids of *Characean* algae: Exclusion of microtubules and a convergence of actin filaments in the apex suggest an actin-mediated gravitropism. Planta 205: 39-50

Brown RC and Lemmon BE (1995) Methods in plant immunolight microscopy. Meth Cell Biol 49: 85-107

Brown RC, Lemmon BE and Mullinax J (1989) Immunofluorescent staining of microtubules in plant tissues: improved embedding and sectioning techniques using polyethylene glycol (PEG) and Steedman's wax. Bot Acta 102: 54-61

Cho SO and Wick SM (1991) Actin in the developing stomatal complex of winter rye - a comparison of actin antibodies and rh-phalloidin labeling of control and CB-treated tissues. Cell Motil Cytoskel 19: 25-36

Cleary AL and Mathesius U (1996) Rearrangement of F-actin during stomatogenesis visualized by confocal microscopy in fixed and permeabilized *Tradescantia* leaf epidermis. Bot Acta 109: 15-24

Collings DA, Winter H, Wyatt SE and Allen NS (1998a) Growth dynamics and cytoskeleton during stem maturation and gravity-induced stem bending in *Zea mays* L. Planta 207: 246-258

Collings DA, Asada T, Allen NS and Shibaoka H (1998b) Plasma-membrane-associated actin in bright yellow 2 tobacco cells. Plant Physiol 118: 917-928

Doris FP and Steer MW (1996) Effects of fixatives and permeabilization buffers on pollen tubes: implications for localisation of actin microfilaments using phalloidin staining. Protoplasma 195: 25-36

Harper JDI, Holdaway NJ, Brecknock SL, Busby CH and Overall RL (1996) A simple and rapid technique for the immunofluorescence confocal microscopy of intact *Arabidopsis* root tips. Cytobios 87: 71-78

Hasenstein KH, Blancaflor EB and Lee JS (1999) The microtubule cytoskeleton does not integrate auxin transport and gravitropism in maize roots. Physiol Plant 105: 729-738

He Y and Wetzstein H (1995) Improved structural preservation and immunolocalization of the cytoskeleton in animal and plant cells by freeze-substitution fixations. Mol Biol Cell 6: 2150-2150

Hepler PK and Gunning BES (1998) Confocal fluorescence microscopy of plant cells. Protoplasma 201: 121-157

Hess M, Mittermann I, Luschnig C and Valenta R (1995) Immunocytochemical localization of actin and profilin in the generative cell of angiosperm pollen - TEM studies on high-pressure frozen and freeze-substituted *Ledebouria socialis* Roth (Hyacinthaceae). Histochem Cell Biol 104: 443-451

Himmelspach R, Wymer CL, Lloyd CW and Nick P (1999) Gravity-induced reorientation of cortical microtubules observed *in vivo*. Plant J 18: 449-453

Hush JM and Overall RL (1992) Re-orientation of F-actin is not necessary for wound-induced microtubule reorientation and cell polarity establishment. Protoplasma 169: 97-106

Kobayashi Y, Kobayashi I, Funaki Y, Fujimoto S, Takemoto T and Kunoh H (1997) Dynamic reorganization of microfilaments and microtubules is necessary for the expression of non-host resistance in barley coleoptile cells. Plant J 11: 525-537

Kost B, Spielhofer P and Chua N-H (1998) A GFP-mouse talin fusion protein labels plant actin filaments *in vivo* and visualizes the actin cytoskeleton in growing pollen tubes. Plant J 16: 393-401

Liu B and Palevitz BA (1992) Organization of cortical microfilaments in dividing root cells. Cell Motil Cytoskel 23: 252-264

Marc J, Granger CL, Brincat J, Fisher DD, KaoT-H, McCubbin AG and Cyr RJ (1998) A GFP-MAP4 reporter gene for visualizing cortical microtubule rearrangements in living epidermal cells. Plant Cell 10: 1927-1939

Meagher RB, McKinney EC and Kandasamy MK (1999) Isovariant dynamics expands and buffers the responses of complex systems: The diverse plant actin gene family. Plant Cell 11: 1-12

McCurdy DW and Gunning BES (1990) Reorganization of cortical actin microfilaments and microtubules at preprophase and mitosis in wheat root-tip cells: A double label immunofluorescence study. Cell Motil Cytoskel 15: 76-87

Olyslaegers G and Verbelen JP (1998) Improved staining of F-actin and co-localization of mitochondria in plant cells. J Microsc 192: 73-77

Parthasarathy MV, Perdue TD, Witzum A and Alvernaz J (1985) Actin network as a normal component of the cytoskeleton in many vascular plant cells. Amer J Bot 72: 1318-1323

Raudaskoski M, Rupes I and Timonen S (1991) Immunofluorescence microscopy of the cytoskeleton in filamentous fungi after quick-freezing and low-temperature fixation. Exp Mycol 15: 167-173

Ren H, Gibbon BC, Ashworth SL, Sherman DM, Yuan M and Staiger CS (1997) Actin purified from maize pollen functions in living plant cells. Plant Cell 9: 1445-1457

Schmit AC and Lambert AM (1990) Microinjected fluorescent phalloidin *in vivo* reveals F-actin dynamics and assembly in higher plant mitotic cells. Plant Cell 2: 129-138

Sievers A, Buchen B, Volkmann D, Hejnowicz Z (1991) Role of the cytoskeleton in gravity perception. In: Lloyd CW (ed) The Cytoskeletal Basis of Plant Growth and Form. London: Academic Press, pp 169-182

Sivaguru M , Baluška F, Volkmann D, Felle HH and Horst WJ (1999) Impacts of aluminum on the cytoskeleton of the maize root apex. Short-term effects on the distal part of the transition zone. Plant Physiol 119: 1073-1082

Sonobe S and Shibaoka H (1989) Cortical fine actin filaments in higher plant cells visualized by rhodamine-phalloidin after pretreatment with m-maleimidobenzoyl N-hydroxysuccinimide ester. Protoplasma 148: 80-86

Staves MP, Wayne R and Leopold AC (1997) Cytochalasin D does not inhibit gravitropism in roots. Amer J Bot 84: 1530-1535

Szymanski DB, Marks MD and Wick SM (1999) Organized F-actin is essential for normal trichome morphogenesis in *Arabidopsis*. Plant Cell 11: 2331-2347

Tang X, Lancelle SA and Hepler PK (1989) Fluorescence microscopic localization of actin in pollen tubes: Comparison of actin antibody and phalloidin staining. Cell Motil Cytoskel 12: 216-224

Thijssen M, Mittempergher F, Van Aelst A and Van Went J (1997) Improved ultrastructural preservation of Petunia and Brassica ovules and embryo sacs by high pressure freezing and freeze substitution. Protoplasma 197: 199-209

Traas JA, Doonan JH, Rawlins DJ, Shaw PJ, Watts J and Lloyd CW (1987) An actin network is present in the cytoplasm throughout the cell cycle of carrot cells and associates with the dividing nucleus. J Cell Biol 105: 387-395

Uetake Y and Peterson RL (1997) Changes in actin filament arrays in protocorm cells of the orchid species, *Spiranthes sinensis*, induced by the symbiotic fungus *Ceratobasidium cornigerum*. Can J Bot 75: 1661-1669

Vitha S, Baluška F, Mews M and Volkmann D (1997) Immunofluorescence detection of F-actin on low melting wax sections from plant tissues. J Histochem Cytochem 45: 89-95

Waller F and Nick P (1997) Response of actin microfilaments during phytochrome-controlled growth of maize seedlings. Protoplasma 200: 154-162

Wasteneys GO, Gunning BES and Hepler PK (1993) Microinjection of fluorescent brain tubulin reveals dynamic properties of cortical microtubules in living plant cells. Cell Motil Cytoskel 24: 205-213

Wasteneys GO, Willingale-Theune J and Menzel D (1997) Freeze shattering: A simple and effective method for permeabilizing higher plant cells walls. J Microsc 188: 51-61

Wick SM (1993) Immunolabeling of antigens in plant cells. Meth Cell Biol 37: 171-200

Williamson RE, McCurdy DW, Hurley UA and Perkin JL (1987) Actin of *Chara* giant internodal cells. Plant Physiol 85: 268-272

Wulf E, Deboben A, Baatz FA, Faulstich H and Wieland T (1979) Fluorescent phallotoxin, a tool for the visualization of cellular actin. Proc Natl Acad Sci USA 76: 4498-4502

Wymer CL, Beven AF, Boudonck K and Lloyd CW (1999) Confocal microscopy of plant cells. Meth Mol Biol 122: 103-130

Yuan M, Shaw PJ, Warn R and Lloyd CW (1994) Dynamic reorientation of cortical microtubules from transverse to longitudinal in living plant cells. Proc Natl Acad Sci USA 91: 6050-6053

Chapter 35

STEEDMAN'S WAX FOR F-ACTIN VISUALIZATION

Stanislav Vitha[1,2], František Baluška[3], Jan Jasik[4], Dieter Volkmann[3] and Peter W Barlow[5]

1 Institute of Plant Molecular Biology, Czech Academy of Sciences, České Budějovice, CZ–37005 Czech Republic
2 Department of Biochemistry, University of Nevada, Reno, NV 89557, USA
3 Botanisches Institut der Universität Bonn, Kirschallee 1, D–53115 Bonn, Germany
4 Department of Plant Physiology, Comenius University, SK–84215 Bratislava, Slovakia
5 IACR–Long Ashton Research Station, Department of Agricultural Sciences, University of Bristol, Long Ashton, Bristol BS41 9AF UK

Key words: cryofixation, embedding, formaldehyde fixation, immunofluorescence, polyester wax

Abstract: Actin filaments are visualised by means of indirect immunofluorescence in plant tissues that were fixed in formaldehyde, embedded in low-melting polyester wax and sectioned on a microtome. The technique described here avoids usage of detergents and organic solvents and is also compatible with immunolocalization of many other antigens.

1. INTRODUCTION

An actin cytoskeleton in plant cells can be visualised by several methods. Depending on the questions to be answered, the investigator can choose to study actin *in vivo*, using transgenic plants expressing GFP fusion proteins (see Kost et al., this volume) or via microinjection of fluorescently labelled proteins (e.g. Hepler & Hush, 1996). On the other hand, where the dynamism of F-actin network is not the main object of study, or if the cells of interest are deep inside a large organ or piece of tissue and are not easily accessible for the above mentioned techniques, it may be preferable to work with tissues that were fixed and further processed. Again, fluorescence is the visualisation method of choice because of its high sensitivity and spatial

C.J. Staiger et al. (eds.), Actin: A Dynamic Framework for Multiple Plant Cell Functions, 619–636.

resolution. The two most common approaches for actin filament staining are through the use of: 1) fluorescently labelled phalloidin, a toxin that has high affinity towards F-actin (affinity cytochemistry), and 2) anti-actin antibodies followed by a fluorescently labelled secondary antibody (indirect immunofluorescence).

Despite its simplicity, the phalloidin technique is not without problems (Blancaflor & Hasenstein, this volume). In some cell types, the phalloidin binding sites may be masked with other actin-binding proteins (e.g. Ao & Lehrer, 1995; Jiang et al., 1997; Nishida et al., 1987). Furthermore, there are often problems with the high background fluorescence and rapid fading of the signal (e.g. La Claire, 1989; McCurdy & Gunning, 1990). The use of antibodies is therefore preferable.

This chapter deals with immunofluorescence detection of F-actin in cells of plant tissue that has been fixed, embedded in wax and sectioned on a microtome. The procedure has four stages:

1. Fixation
2. Tissue processing
3. Immunostaining
4. Microscopy

1.1 Fixation

The aim of fixation is to kill the cell rapidly and to immobilise proteins in the same state and location as they were *in vivo*. Chemical fixatives commonly employ aldehydes as crosslinking agents. Formaldehyde is preferred for immunofluorescence at the light microscope level. It penetrates the tissue quickly but has only limited crosslinking capabilities. Glutaraldehyde is used less often. It is a stronger crosslinker, but penetrates slowly and causes strong tissue autofluorescence which hinders its use for immunofluorescence unless special precautions are taken. Moreover, glutaraldehyde induces immediately synthesis of callose in plant cells (Hughes & Gunning, 1980). There is a concern that the slowness of chemical fixation leaves time for abnormal re-arrangements of the cytoplasm and cytoskeleton (Doris & Steer, 1996; He & Wetzstein, 1995; Mersey & McCully, 1978). This is especially critical in the case of higher plant cells encased within robust cellulosic cell walls. Moreover, F-actin is traditionally considered to be extremely sensitive to aldehydes (Doris & Steer, 1996; Parthasarathy et al., 1985; Seagull et al., 1987; Traas et al., 1987). To stabilise F-actin, Sonobe and Shibaoka (1989) introduced treatment with MBS (n-maleimidobenzoic acid N-hydroxysuccinimide ester) prior to formaldehyde fixation. However, our and other authors' data (e.g. Blancaflor & Hasenstein, this volume) indicate that fine F-actin network can be preserved with formaldehyde fixation. In our material (*Zea mays* and

Arabidopsis thaliana) MBS did not improve F-actin preservation. Just the opposite, incubation with MBS (>15 min) actually prevented immunolabelling of phragmoplasts (Vitha et al., 1997).

One of the alternatives to chemical fixation is rapid freezing followed by freeze-substitution (further referred to as cryofixation; e.g. Baskin et al., 1995; Roy et al., 1997). Cryofixation is fast-acting, thus preventing pre-fixation artefacts, which is an advantage for immunolocalization of epitopes that do not survive chemical fixation. Maneta-Peyret et al. (1999) also found cryofixation to be superior for EM immunocytochemistry of lipids. Cryofixation, however, may not work well for larger objects because the relatively slow cooling rates inside the tissue can cause ice-crystal damage. Also, since the proteins in cryofixed cells are not crosslinked, redistribution of some small proteins may occur. As emphasised by Melan and Sluder (1992), unless the fixative completely immobilises soluble proteins, further sample preparations and differential extraction can lead to artefactual redistribution and/or extraction of the antigens. All this is not an issue for electron microscopy where the embedding resin is not removed prior to immunolabelling. However, procedures for light microscopy usually include de-embedding and rehydration in order to make tissue epitopes accessible for antibodies. False localization is obviously not a problem with F-actin networks and the same localization patterns were seen in roots that were either cryofixed or formaldehyde fixed and then embedded in Steedman's wax (Vitha et al., 2000). However, researchers should be aware of this potential for localization artefacts if other proteins are being simultaneously immunolocalised in the same tissue. Various fixatives and their mode of action are discussed in depth by Pearse (1991). With respect of actin immunofluorescence, it is well known that fine meshworks are extremely difficult to visualize. Unless they are effectively cross-linked and/or bundled, their fine nature makes them sensitive towards fixation and permeabilization (Small et al., 1999). In contrast, robust bundles of actin filaments are more resistant even towards effective fixation and heavy permeabilization. For instance, dense meshworks at growing apices of root hair tips can be visualized using mild permeabilization (Brown et al., 1999) but not after heavy permeabilization (Miller et al., 1999).

1.2 Tissue processing

The purpose of this step is to enable penetration of antibodies and also to allow observation of tissue structure and the fluorescence signal with minimum background. The tissue that has been fixed can be either stained in toto, or macerated, hand-sectioned or vibratome-sectioned, or as described in this chapter, embedded in wax and sectioned on a microtome. Embedding and sectioning allows study of individual cells in the context of the surrounding tissue, and in the case of root tips, for example, to observe the succession of developmental stages. The choice of embedding medium is critical for successful immunostaining. Paraffin wax, the classical embedding medium of histology and anatomy, is unsuitable here because its relatively high melting-point destroys F-actin and renders many other proteins non-immunoreactive. Methacrylate-based resins can be used at low temperatures and polymerised by UV light. Although they provide good structural preservation of the tissue, the preservation of antigenicity may not be ideal and application of organic solvents to remove resin from tissue sections compromises antigenicity. Unless de-embedding steps are included, only the surface of the sections is available for immunostaining, and the signal is weak. Moreover, it is not possible to cut sections more than several micrometers thick and special equipment is needed, such as ultramicrotome or retracting microtome, in order to section resin blocks with glass knives. Unlike paraffin or Steedman's wax, resin sections do not form ribbons; serial sectioning is thus less convenient and more time consuming.

Steedman (1957) introduced a new, ribbon-forming, sectioning medium – low melting-point polyester wax (today also known as Steedman's wax). This wax has a melting point of 35-37°C, is soluble in ethanol, and has sectioning properties very similar to paraffin wax. Some of the technical conveniences of Steedman's wax compared to paraffin were noted by Sidman et al. (1961): „Sections of polyester wax fragment less. Hard tissues... commonly split away from a paraffin ribbon, but not from one of polyester wax. Polyester ribbons do not generate static electricity." Although Steedman's wax never replaced paraffin for routine histology, its suitability for immunocytochemistry had been noticed early on (Baird, 1967; Sidman et al., 1961) and it has been used for detection of various antigens, including cytoskeletal, secretory and membrane-associated proteins (Oke & Suarez-Quian, 1993), BrDU-labeled DNA (Hanazono et al., 1990), as well as for *in situ* hybridization (Schneitz et al., 1998; Shuttlesworth & Mills, 1995). In plant tissues, Steedman's wax has proved suitable for the immunofluorescence detection of microtubules (Brown et al., 1989; Baluška et al., 1992; Timmers et al., 1999), actin filaments (Baluška et al., 1997; Vitha et al., 1997, 2000, this volume), plasma membrane H^+-ATPases (Jahn et al., 1998), PSTAIR proteins (Mews et al., 1996), cyclins (Mews et al., 1997), arabinogalactan-protein (AGP) epitopes (Šamaj et al., 1998, 1999), HDEL proteins and calreticulin (Baluška et al., 1999a), profilins (von Witsch et al., 1998), as well as cadherin- and catenin-like proteins (Baluška et al., 1999b). Importantly, we routinely immunolocalize these and other antigens with the same procedure as for F-actin.

1.3 Immunostaining

After the microtome sections, which are attached to glass slides, have been dewaxed and rehydrated, antibodies have direct access to their antigens. Indirect methods, using a primary antibody that is then detected with labelled secondary antibodies, are preferred because of their high sensitivity.

1.3.1 Antibodies

We use the monoclonal C-4 anti-actin antibody (Lessard, 1988) from ICN Pharmaceuticals (Costa Mesa, CA, USA) for staining of all classes of F-actin in various plant species, including *Zea mays*, *Arabidopsis thaliana, Lepidium sativum, Oryza sativa* and *Medicago sativa.* Secondary antibodies conjugated with various fluorochromes are available from many suppliers, FITC conjugates being the most common.

1.3.2 Controls

Proper controls are essential in order to show that the immunofluorescence signal reflects localization of the target protein. The controls include:

a) Both antibodies omitted: This reveals the autofluorescence of the tissue.
b) Primary antibody omitted: In combination with the previous control, this control indicates non-specific binding of the secondary antibody.
c) Use of pre-immune serum (ideally from the same animal used to raise the primary antibody) or normal immunoglobulin, at the same dilution or protein concentration as the primary antibody: This control is similar to b), but also checks for non-specific binding of the primary antibody. Immunoglobulins often bind to plastids and starch granules.
d) Competition: Primary antibody is pre-incubated for 30 min with the target protein before it is applied to the sections. This control should not give any significant fluorescent staining. It is used to check for non-specific binding of the primary antibody.
e) Use of non-related primary antibody: This is expected to produce different localization pattern, e.g. anti-tubulin antibody, and is therefore a check for the distribution of the antigen in question.

1.3.3 Autofluorescence

Many tissues exhibit autofluorescence and this is associated especially with their cell walls and chloroplasts. Glutaraldehyde fixatives are known to induce autofluorescence and induce callose (Hughes & Gunning, 1980). Therefore, they should be avoided, or special steps should be taken to quench the induced fluorescence, e.g. by treatment with $NaBH_4$ or with Schiff's reagent followed by reduction with $NaBH_4$ (Tagliaferro et al., 1997).

Several steps can be taken to minimise natural autofluorescence. Choice of fluorescence filters which closely match the excitation and emission spectra of the fluorochrome. Since autofluorescence is usually of a broad spectrum, this improves the signal-to-noise ratio. We have good experience with FITC-conjugated antibodies viewed with filter set containing a bandpass emission filter at 535 ± 20 nm (e.g. XF100 filter set, Omega Optical, Inc., Brattleboro, VT , USA). The added bonus is that such filter can be used for viewing GFP in fresh tissue since it eliminates red autofluorescence of chloroplasts. If FITC-conjugated secondary antibodies are used, labelled sections may be stained with Toluidine Blue (see Table 3) to diminish the autofluorescence. The sections should not be overstained, because the specific immunofluorescence signal could be quenched as well. Note that Toluidine Blue staining may not be used with some multi-color filter sets as it emits strong reddish fluorescence.

1.3.4 Non-specific staining

Similar to autofluorescence can be the effects of non-specific staining resulting from too high a concentration of either the primary or secondary antibodies and/or insufficient washing. The antibody dilutions given in this protocol (see Methods section 3.3) serve as a starting point. Optimal dilutions should be determined for a given material by pilot experiments with several dilutions of the primary and secondary antibodies. A comparison of control slides where the first or both antibodies were omitted will indicate whether the secondary antibody binds in a non-specific manner. If that is the case, the antibody can be pre-absorbed with tissue powder from corresponding plant material (see Section 2.2). Non-specific binding of antibodies is also reduced by pre-incubating slides with a blocking solution (SB – see section 2.3, containing 0.5–3% BSA or non-immune serum from the same species as used to raise the secondary antibody) and by diluting antibodies in the blocking solution. To remove any aggregates that may have formed during storage, microcentrifuge the diluted secondary antibody before use for 5 min at 4°C at maximum speed. A mild detergent such as 0.1% v/v Tween 20 increases the efficiency of washing, following the incubation with antibodies.

1.4 Fluorescence Microscopy

Because the tissue has been sectioned into relatively thin slices, standard epifluorescence microscopy can be used for rapid screening and photography. When working with thicker sections and when three-dimensional arrays of F-actin are to be documented, confocal microscopy or computational deconvolution microscopy may be used. Discussion of these techniques is, however, outside the scope of this chapter. The choice of excitation and emission wavelengths is dictated by the fluorochrome used. Longpass emission (barrier) filters give brighter fluorescence signal, but for multiple labelling and/or in case of strong autofluorescence, bandpass barrier filters should be used (see Section 1.3.3). High quality fluorescence filters can be obtained, e.g., from Omega Optical (http://www.omegafilters.com). A database of excitation and emission spectra of many fluorochromes has been made available on-line by BioRad (http://fluorescence.bio-rad.com).

2. MATERIALS

2.1 Steedman's wax

Steedman's wax can be purchased ready to use under the name of 'Polyester wax' from Ted Pella (Redding, CA, USA) or Electron Microscopy Sciences (Fort Washington, PA, USA), or it can be easily and less expensively mixed in the laboratory. Among the different suppliers of the individual components, we were best satisfied with those obtained from Aldrich (Milwaukee, WI, USA).

To prepare the Steedman's wax from the individual components, melt 900 g polyethylene glycol 400 distearate (Aldrich, cat. # 30,541-3) and 100 g 1-hexadecanol (Aldrich cat. # 25,874-1) in a large beaker in an incubator at 65 °C. When completely melted, stir the wax very thoroughly for several minutes, using a stirring bar. Pour the wax into a tray lined with aluminium foil and leave at room temperature to harden. Alternatively, pour the wax in 50 ml plastic conical tubes and close. This prevents contamination by dust and other particles which can cause problems during sectioning. The wax can be stored at room temperature indefinitely. Do not use a microwave oven to melt the wax for embedding as it seems to affect its sectioning properties. For embedding, melt only the necessary amount of wax in an Erlenmeyer flask at 37°C. If using a water bath to melt the wax, ensure that the container is closed to keep out moisture.

2.2 Tissue powder

Use the same tissue type as the one under investigation (e.g. leaves, flowers). Fix the tissue and dehydrate through an ethanol series (see Sections 3.1 and 3.2), air-dry and then grind with a pestle and mortar in liquid nitrogen. The resulting powder can be stored at –20°C. To pre-absorb the secondary antibody, add 1.5 ml diluted antibody to 0.2 ml tissue powder in a microcentrifuge tube, vortex and incubate at 4°C for 24 h. Vortex again, spin for 5 min at 4°C at maximum speed and use the supernatant. This powder (after Schneitz et al., 1998) works better than the acetone powders often recommended by immunocytochemical handbooks (Tanya Wagner, personal communication).

2.3 Buffers and solutions

a) **Stabilising buffer (SB)**: This is the same buffer as used in protocols for tubulin immunofluorescence (e.g. Brown et al., 1989). The buffer consists of 50 mM PIPES, 5 mM $MgSO_4$ and 5 mM EGTA. PIPES-Na salt is soluble in water, whereas if the 'free acid' PIPES is used, KOH pellets must be added in order that the PIPES will be dissolved. Finally, adjust the pH to 6.9 with KOH or H_2SO_4. Store at 4°C for up to several weeks.
b) **Formaldehyde fixative**: Suspend paraformaldehyde powder (e.g. Sigma, St. Louis, MO, USA; Catalogue No. P-6148) in SB to 3.7% w/v. Heat in a water bath at 65°C and mix occasionally until it dissolves completely. Allow the solution to cool. Prepare fresh before use.
c) **PBS**: 0.14 M NaCl, 2.7 mM KCl, 6.5 mM Na_2HPO_4, 1.5 mM KH_2PO_4, pH 7.3. It is convenient to prepare 10× concentrated solution that can be autoclaved or preserved by 3.0 mM NaN_3. Caution: NaN_3 is toxic.
d) **1% BSA/SB stock**: 1% w/v Bovine serum albumin (BSA, preferably fatty acid free) in SB. Store 1 ml aliquots at –20°C.
e) **Primary antibody**: Monoclonal anti-actin C-4 antibody (ICN) diluted 1:200 in 1% BSA/SB stock. Mix fresh before use.
f) **Secondary antibody**: Anti-mouse IgG-FITC conjugate (e.g. Sigma, Catalogue No. F-5262) diluted 1:100 in 1% BSA/SB stock. Mix fresh before use, spin 5 min at 15 000 rpm at 4°C to remove any aggregates. If pre-absorption with tissue powder is necessary, see Section 2.2.
g) **DAPI stock**: 4,6-diamidino-2-phenylidone (Sigma, Catalogue No. D-9542) 0.1 mg/ml in water, store at 4°C in the dark.
h) **Toluidine Blue solution**: 0.01% w/v Toluidine Blue O in PBS.
i) **Anti-fade mounting medium with DAPI**: Use either commercially available anti-fade mountants, such as ProLong, SlowFade (Molecular

Probes, Eugene, OR, USA) or Vectashield (Vector Labs, Burlingame, CA, USA), and add DAPI stock to 1 μg/ml. Alternatively, prepare the medium as follows. Dissolve 50 mg p-phenylendiamine in 5 ml phosphate-saline (0.01 M phosphate buffer pH 7.4, 0.15 M NaCl) and adjust the pH to 8.0 with carbonate/bicarbonate buffer (consists of 4 ml 0.2 M Na_2CO_3 + 46 ml 0.2 M $NaHCO_3$, pH 9.2). To this solution, add 45 ml glycerol and 0.5 ml of DAPI stock. Mix thoroughly. Aliquot into 1.5 ml microcentri-fuge tubes and store in darknes at –20°C. Good for a year or until it turns dark.

3. METHODS

3.1 Fixation, dehydration and embedding

Cut pieces of tissue into glass vials containing the fixative. Vacuum infiltrate in a desiccator for 10 min or until no air bubbles come out of the tissue. Leaves are often more difficult to infiltrate (see Technical Note 3). Slowly raise the pressure back to normal and incubate for an additional time to complete the 1 h fixation. Wash with SB and PBS (see Table 1). Change all solutions using a glass Pasteur pipette with a rubber suction bulb. For small objects, such as suspension culture cells, it is advantageous to enclose the pipette tip in a fine nylon mesh (pore size of 50 μm or smaller) to prevent samples being sucked into the pipette and lost.

Table 1. Fixation, dehydration and embedding in Steedman's wax

Treatment	Duration
Fixation with an initial vacuum infiltration	1 h
SB wash	2 x 10 min
PBS wash	2 x 10 min
30, 50, 70, 90% ethanol/PBS, 97% ethanol	30 min each at 4°C
0.1% Toluidine Blue in ethanol	10 min
97% ethanol wash	10 min
Move samples to 37°C incubator	Allow samples to warm up
Wax + ethanol 1+1	Overnight
100% Wax	2 x 1 h

Either analytical or technical grade (97%) ethanol can be used for tissue dehydration; it is not necessary to use dried (100%) ethanol. Dehydration should be performed at 4°C to minimize possible protein extraction. To be able to see the samples better, the dehydrated tissue is stained with Toluidine Blue (about 0.1% in 97% ethanol). Afterwards, samples are moved to 37°C and all further steps are performed at this temperature. The duration of the steps depends on specimen size. The times given here are used for embedding maize root tips (about 7 mm long). Increase the times for larger specimens. For small seedlings and suspension culture cells, the times may be shortened by 50%. See also Technical Note 4.

To make the wax+ethanol mixture, simply add molten (37°C) wax to the vials containing samples in ethanol. At the next step, just exchange the mixture for a pure wax. For the final stage of embedding, use a warm plate (37°C); it is very important not to overheat the samples. If no warm plate with a fine temperature control is avaliable, work quickly at room temperature on a bench. Pre-warm a small Petri dish and the silicon embedding molds to 37°C. Fill the mold cavities with molten wax. Pour the samples from the glass vial into the pre-warmed Petri dish and transfer samples into the mold cavities. For embedding large pieces of tissue, Petri dishes or Peel-A-Way® disposable histology molds may be used.

Samples can be picked directly from the vial, but sometimes the specimens stick together or are tangled (as with Arabidopsis seedlings) and are hard to separate from each other.

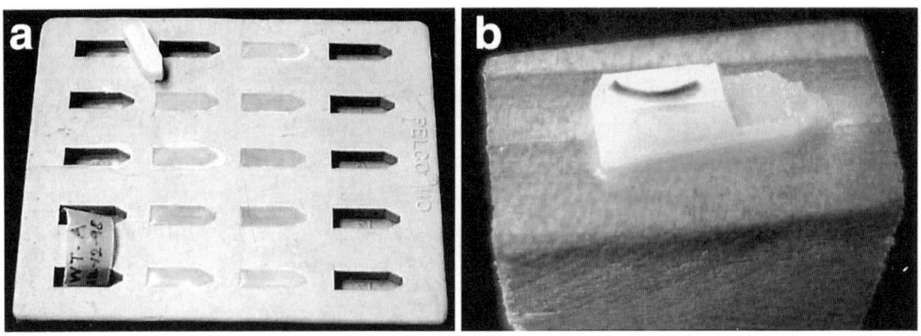

Figure 1. Wax blocks with embedded specimens in the embedding mold (a), and after mounting and trimming for sectioning (b).

Add more wax to the embedding molds so that the surface is convex. The wax will shrink somewhat upon solidification. Move the embedding mold away from the warm plate, leave at room temperature to solidify overnight (Fig. 1a). Blocks can then be used for sectioning immediately after hardening (the wax fully hardens 12 — 24 h after embedding) or can be stored in a dry place (avoid refrigerators for their humidity) at room temperature for several months without loss of antigenicity (even for years).

3.2 Sectioning

Mount the wax blocks on wooden stubs. The top of a new stub should be first soaked with molten wax. Warm a razor blade briefly in a flame and lay it on top of the wooden block. Place the wax block on the blade and, as its base starts to melt, slide it off the blade on the wood. Do not use too much heat otherwise the whole wax block will melt. Avoid touching the top of the wax block. Allow the wax to harden at room temperature for about 30 min, then trim the sides of the block around the specimen with a razor blade, creating a pyramidal shape with straight, parallel edges (Fig. 1b).

Clamp the block into the specimen holder of a rotary microtome. Make sure the top face of the wax block is aligned with the knife. High cutting speeds (about 4 cuts per second) give better sections in our hands, but other authors prefer cutting at very low speed. As soon the wax block starts cutting and a ribbon of sections begins to form, pick up the end of the ribbon with a fine-tipped paint brush and lift the ribbon up so that it does not slide and crumple upon the knife surface (Fig. 2a).

Ribbons as long as 30 cm can easily be obtained. Use a second paint brush to detach the ribbon from the knife edge and transfer it on a clean woodboard, cardboard or sheet of paper (see Technical Note 5). Cut the ribbon into shorter pieces with a razor blade and place them on a glass slide.

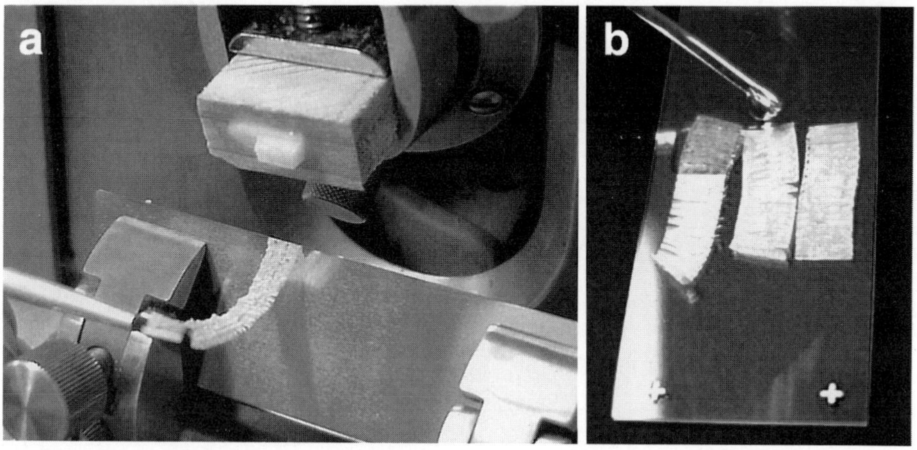

Figure 2. Sectioning and stretching of sections upon a slide. a) The end of the ribbon is being held by a paint brush during sectioning. b) A drop of water is added to the top of the ribbons on a glass slide, excess water is then removed from the lower end.

Note that sections will have been compressed during sectioning but will expand by 30–50% at the next step. Holding the slide with one end raised about 45°, add a small drop of water (at room temperature) to the upper end of the ribbons (Fig. 2b). The ribbons start to expand immediately. As soon as the water reaches the bottom of the ribbon, blot excess fluid with a piece of filter paper. Keep slides at an angle and let dry at room temperature for several hours or overnight. Slides can then be used for immunostaining or stored at 4°C for later use.

Sections sometimes do not stick well to slides and are lost during the immunostaining procedure. Several improvements of the initial protocol were accordingly proposed (see Technical Notes 1 and 2). For our routine work with sections of root tips, leaves and shoot apices, the standard protocol was satisfactory, when using slides coated with Mayer's glycerol-albumen or poly-L-lysine, or with Superfrost Plus® charged slides.

3.3 Immunostaining

Most steps are done in Coplin staining jars. Slides are transferred from jar to jar with forceps. Technical grade ethanol (97%) is sufficient for dewaxing, but it was noticed that if it contains acetone above 1%, the tissue-antigenicity vanishes and autofluorescence of the tissue is somewhat increased.

Take the rehydrated slide and wipe excess fluid from around the sections. Apply 100 μl of the diluted primary antibody and rock the slide forth and back to distribute the antibody evenly over all tissue sections. Cover the sections gently with a piece of Parafilm® (American National Can, Chicago, IL, USA).

Table 2. Dewaxing and rehydration of sections. The steps indicated with * are optional, see Technical Note 6.

Treatment	Duration
97 % ethanol	3 × 10 min (Thick sections may need longer)
90% ethanol/PBS	10 min
50 % ethanol/PBS	10 min
PBS	10 min
SB*	30 min
100% methanol chilled to – 20°C *	10 min at –20°C
SB	90 min

This will prevent the antibody from evaporating and also maintain a uniform layer of solution on the sections. Place the slides in a humid chamber (e.g., in a large Petri dish lined with wet filter paper), and incubate at room temperature. Both signal intensity and F-actin preservation in *Zea mays* are improved if incubation is done at room temperature rather than at 37°C (Vitha et al., 1997), but in *Arabidopsis*, 0.5 h at 37°C gives somewhat better results. The immunostaining procedure is outlined in Table 3.

After the last wash, the labeled slides are ready for mounting. Put a drop (20–30 μl) of the anti-fade mounting medium on the sections, rock the slide forth and back in order to distribute the medium evenly, and cover with a coverslip. Blot excess fluid with filter paper and seal with nail polish. The slides can be stored at +4°C for few days and at –20°C for 3–4 weeks, after which time the fluorescence will have faded too much.

Table 3. F-Actin immunostaining procedure. For steps indicated with *, see Technical Note 7.

Treatment	Duration
Primary antibody*	1 h at room temperature in a humid chamber
0.1% w/v Tween 20 in SB*	2 × 10 min
SB*	10 min
Secondary antibody*	1 h at room temperature, in a humid chamber, in the dark
0.1% w/v Tween 20 in PBS	10 min
PBS	10 min
Toluidine Blue 0.01% in PBS	10 min
PBS	10 min

Examples of F-actin immunofluorescence in shoot and root tissues are shown in Figures 3 and 4.

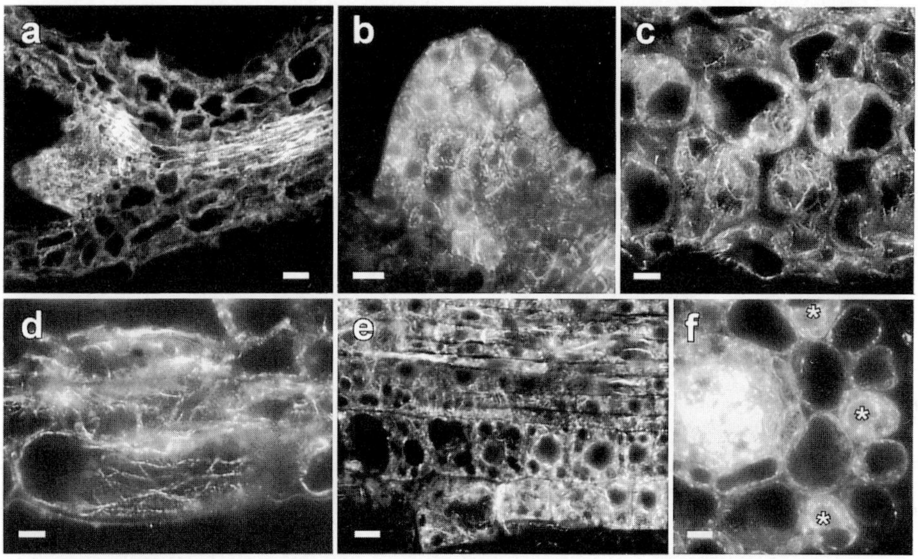

Figure 3. Actin immunofluorescence in six days-old Arabidopsis seedlings in longitudinal (a-e) and cross sections (f). Bars = 50 μm (a) or 10 μm (b-f). a) General view of a hypocotyl, apical meristem and basal parts of cotyledons. Note the intense F-actin fluorescence in the meristem and stele. b) Higher magnification view of the meristem from a. c) Cotyledon, paradermal section. d) Epidermis of the hypocotyl. e) Longitudinal section showing F-actin in epidermis, cortex and stele of the root elongation zone. f) Cross section of the root elongation zone. Note the difference in actin immunofluorescence between atrichoblasts and trichoblasts (asterisks) in the epidermis.

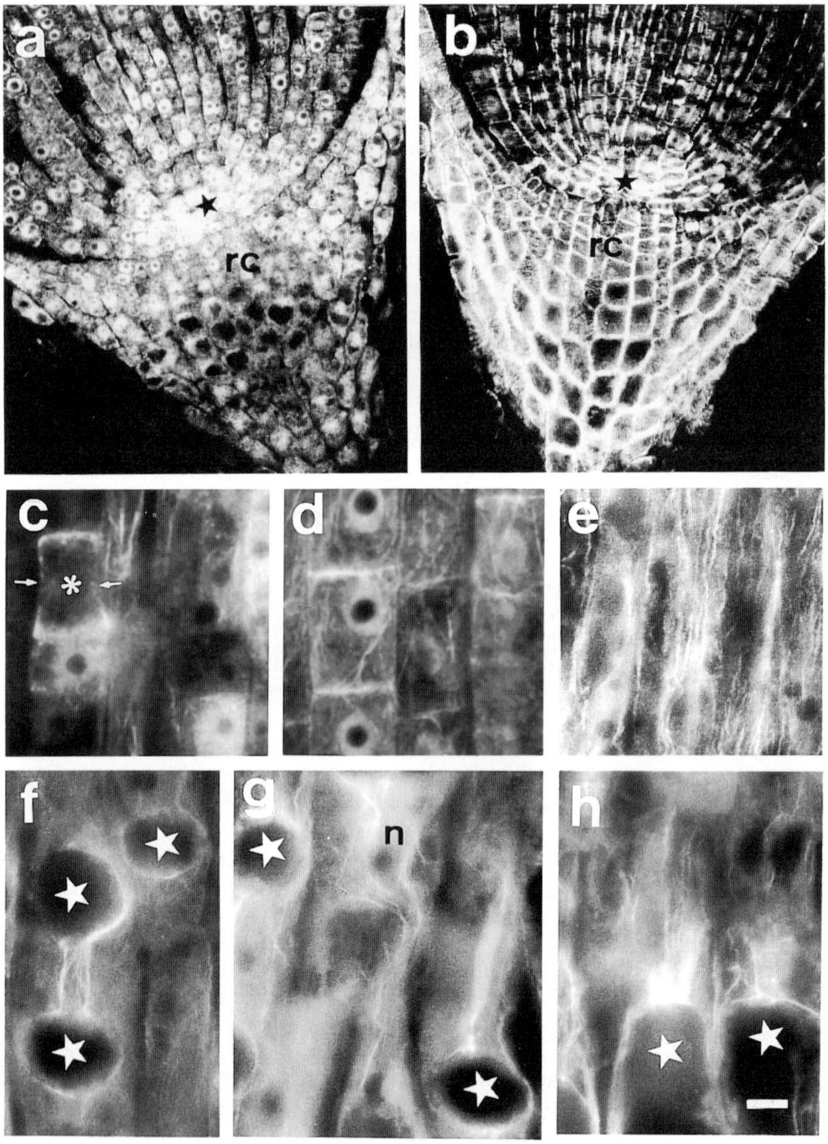

Figure 4. Actin in root cells of *Oryza sativa* (a,c,d,e), and *Vicia faba* (f,g,h). a) Root apex of rice shows accumulation of actin in quiescent centre cells (star) and depletion of actin in central cells of the root cap (rc). b) The same object but labeled with the anti-tubulin antibody. Note that the central cells lack endoplasmic microtubules bur are equipped with abundant cortical microtubules. c) Dividing cell (asterisk) in the rice root apex showing actin depletion at cell periphery domains marked by pre-prophase band (arrows). d) Cells of the transition zone accumulate actin at cross-walls and assemble perinulclear AF bundles. e) Elongating root cells with longitudinal AF cables. f,g,h). Root cells of *Vicia faba* show accumulation of actin at peripheries of vacuoles (stars) which organize longitudinal AF cables. n = nucleus. Bar = 30 μm (a,b) or 10μm (c-h).

TECHNICAL NOTES

1. To improve section adherence for *in situ* hybridisation, Shuttlesworth and Mills (1995) developed the following protocol: 0.5 g bacteriological grade bovine gelatin (EM Science, Gibbstown, NJ, USA) is dissolved in 80 ml of deionized water to which is added 0.5 ml of Hipure liquid gelatin (Norland Products, North Brunswick, NJ, USA). After mixing the two gelatins and cooling the solutions, 0.1 g of chrome alum ($CrK(SO_4)_2 \cdot 12H_2O$) dissolved in 20 ml of deionized water is added. Thoroughly pre-cleaned microscope slides are then subbed with this solution. A small drop (about 1–2 μl) of the subbing mixture is put onto a slide and, using gloved finger, the solution is rubbed thoroughly over the slide. Allow to dry. Once the wax sections are attached to the slide and dried overnight, they can be stored at 4°C until used. Dewax the tissue in two changes of 100% ethanol for 10 min each, air-dry for 15 min. Again treat with fresh 100% ethanol, air dry 15 min, rehydrate twice in 90% ethanol and twice in 70% ethanol.
2. Section adherence to slides for immunohistochemistry and *in situ* hybridization of plant material can be also improved simply by 15 min of air-drying after dewaxing in ethanol (Richardson & Dym, 1994) without any special coating of the slides (using SuperFrost Plus® charged slides; Tanya Wagner, personal communication). After air drying, continue with the standard protocol: rehydration in 90%, 50% ethanol, etc.
3. Leaves are difficult to infiltrate with the fixative, since they float on the surface and the cuticle is water-repellent. For such specimens, vacuum infiltration in a syringe can be performed: take a 10-ml syringe, remove the piston, stop the outlet of the syringe with a gloved finger and pour in several ml of the fixative with pieces of tissue. Insert the piston and invert the syringe so that the outlet is upwards. Purge all air and excess fixative from the syringe, leaving only about 2-3 ml liquid. Stop the outlet with the thumb and pull the piston back. This will create a vacuum. Many small bubbles appear on the surface of the leaves, as the air trapped inside the issue expands and escapes. Shake the syringe vigorously, still holding the piston back and keeping the thumb on the outlet. The goal is to shake the bubbles off the leaf surface so that the air does not return when the syringe is pressurised again. Samples will often get stuck on the syringe walls. Check this occasionally and keep shaking. When all specimens are in the fixative, release the piston. The pressure inside reverts to normal and fixative is pushed inside the tissue, replacing the air that escaped. Leaves which are infiltrated with the fixative turn darker green and are semi-translucent. Repeat the infiltration 2–3 times. Finally, remove the piston and transfer the samples into fresh fixative and fix for the desired time. It is still a good idea to put them under vacuum in a desiccator for 10 min or so, to complete the infiltration.
4. For small objects, such as suspension cultures and organs of *Arabidopsis* seedlings, a faster dehydration/embedding protocol may be used with dehydration steps of 15 min each. Warm the samples in ethanol to 37°C and add wax without mixing. Wax settles to the bottom, while the samples float upwards. Once the specimens have sunk (after about 1 h), remove the upper ethanol layer and part of the wax, and add more wax. When the samples sink again, change the old wax for fresh wax and after 30 min transfer to embedding molds and let harden at room temperature.
5. For cutting the ribbon of sections, a paper, cardboard or wooden surface is preferred, as the wax does not stick as much as it does to a plastic or metal surface.
6. Treatment of sections with cold methanol and the preceding wash with SB are optional. F-actin immunofluorescence of maize root sections is stronger without these steps, but background staining is higher (Vitha et al., 1997). In *Arabidopsis* seedlings and *Medicago* suspension culture cells, the immunostaining was satisfactory without the methanol step.
7. PBS can be used instead of SB for diluting antibodies and washing, but SB gives slightly better preservation of F-actin and stronger fluorescence labelling. The use of detergent

(Tween 20) for washing helps reduce background staining, but in tissues with older, highly vacuolated cells, the cytoplasm is sometimes washed away and lost.

ACKNOWLEDGMENTS

We thank Dr. Tanya Wagner (Duke University, Durham, NC) for helpful discussion.

REFERENCES

Ao XL and Lehrer SS (1995) Phalloidin unzips nebulin from thin filaments in skeletal myofibrils. J Cell Sci 108: 3397-4303

Baird IL (1967) Polyester wax as an embedding medium for serial sectioning of decalcified specimens. Am J Med Techn 33: 393-396

Baluška F, Parker JS and Barlow PW (1992) Specific patterns of cortical and endoplasmic microtubules associated with cell growth and tissue differentiation in roots of maize (*Zea mays* L.). J Cell Sci 103: 191-200

Baluška F, Barlow PW and Volkmann D (1996) Complete disintegration of the microtubular cytoskeleton precedes its auxin-mediated reconstruction in postmitotic maize root cells. Plant Cell Physiol 37: 1013-1021

Baluška F, Šamaj J and Volkmann D (1999a) Proteins reacting with cadherin and catenin antibodies are present in maize showing tissue-, domain-, and development-specific associations with endoplasmic-reticulum membranes and actin microfilaments in root cells. Protoplasma 206: 174-187

Baluška F, Šamaj J, Napier R and Volkmann D (1999b) Maize calreticulin localizes preferentially to plasmodesmata in root apex. Plant J 19: 481-488

Baluška F, Vitha S, Barlow PW and Volkmann D (1997) Rearrangements of F-actin arrays in cells of intact maize root apex tissues: A major developmental switch occurs in the postmitotic transition region. Eur J Cell Biol 72: 113-121

Baskin TI, Miller DD, Vos JW, Wilson JE and Hepler PK (1995) Cryofixing single cells and multicellular specimens enhances structure and immunocytochemistry for light microscopy. J Microsc 182: 149-161

Braun M, Baluška F, von Witsch M and Menzel D (1999) Redistribution of actin, profilin and phosphatidylinositol-4,5-bisphosphate (PIP_2) in growing and maturing root hairs. Planta 209: 435-443

Brown RC, Lemmon BE and Mullinax JB (1989) Immunofluorescent staining of microtubules in plant tissues: Improved embedding and sectioning techniques using polyethylene glycol (PEG) and Steedman's wax. Bot Acta 102: 54-61

Doris FP and Steer MW (1996) Effects of fixatives and permeabilisation buffers on pollen tubes: implications for localization of actin microfilaments using phalloidin staining. Protoplasma 195: 25-36

Hanazono M, Yoshiki A, Ôta K, Kitoh J and Kusakabe M (1990) Immunohistochemical detection of DNA replication in mouse uterine cells by bromdeoxyuridine labeling of wax- and resin-embedded tissue sections. Stain Techn 65: 139-149

He Y and Wetzstein HY (1995) Fixation induces differential tip morphology and immunolocalization of the cytoskeleton in pollen tubes. Physiol Plant 93: 757-763

Hepler P and Hush J (1996) Behavior of microtubules in living plant cells. Plant Physiol 112: 455-461

Hughes JE and Gunning BES (1980) Glutaraldehyde-induced deposition of callose. Can J Bot 58: 250-258

Jahn T, Baluška F, Michalke W, Harper JF and Volkmann D (1998) Plasma membrane H^+-ATPase in the root apex: Evidence for strong expression in xylem parenchyma and asymmetric localization within cortical and epidermal cells. Physiol Plant 104: 311-316

Jiang C-J, Weeds AG and Hussey PJ (1997) The maize actin-depolymerizing factor, ZmADF3 redistributes to the growing tip of elongating root hairs and can be induced to translocate into the nucleus with actin. Plant J 12: 1035-1043

La Claire JW (1989) Actin cytoskeleton in intact and wounded coenocytic green algae. Planta 177: 47-57

Lessard JL (1988) Two monoclonal antibodies to actin: one muscle selective and one generally reactive. Cell Mot Cytoskel 10: 349-362

Maneta-Peyret L, Compère P, Moreau P, Goffinet G and Cassagne C (1999) Immunocytochemistry of lipids: Chemical fixatives have dramatic effects on the preservation of tissue lipids. Histochem J 31: 541-547

McCurdy DW and Gunning BES (1990) Reorganization of cortical actin microfilaments and microtubules at preprophase and mitosis in wheat root-tip cells: A double label immunofluorescence study. Cell Mot Cytoskel 15: 76-87

Melan MA and Sluder G (1992) Redistribution and differential extraction of soluble proteins in permeabilized cultured cells. Implications for immunofluorescence microscopy. J Cell Sci 101: 731-743

Mersey B and McCully ME (1978) Monitoring the course of fixation of plant cells. J Microsc 114: 49-76

Mews M, Baluška F and Volkmann D (1996) Tissue- and development-specific distribution of PSTAIR-proteins in cells of control and wounded maize root apices. J Exp Bot 47: 819-829

Mews M, Sek FJ, Moore R, Volkmann D, Gunning BES and John PCL (1997) Mitotic cyclin distribution during maize cell division: Implications for the sequence diversity and function of cyclins in plants. Protoplasma 200: 128-145

Miller DD, de Ruijter NCA, Bisseling T and Emons AMC (1999) The role of actin in root hair morphogenesis: Studies with lipochito-oligosaccharide as a growth stimulator and cytochalasin as an actin perturbing drug. Plant J 17: 141-154

Nishida E, Iida K, Yonezawa N, Koyasu S, Yahara I and Sakai H (1987) Cofilin is a component of intranuclear and cytoplasmic actin rods induced in cultured cells. Proc Natl Acad Sci USA 84: 5262-5266

Oke BO and Suarez-Quian CA (1993) Localization of secretory, membrane-associated cytoskeletal proteins in rat testis using an improved immunocytochemical protocol that employs polyester wax. Biol Reprod 48: 621-631

Parthasarathy MV, Perdue TD, Wiztum A and Alvernaz J (1985) Actin network as a normal component of the cytoskeleton in many vascular plant cells. Am J Bot 72: 1318-1323

Pearse AGE (1991) Histochemistry: Theoretical and Applied, 4th Ed, Vol 1. Edinburgh: Churchill Livingstone

Richardson LL and Dym M (1994) Improved adhesiveness of polyester wax sections for immunocytochemistry. Biotechniques 17: 846-848

Roy S, Eckard KJ, Lancelle S, Hepler PK and Lord EM (1997) High-pressure freezing improves the ultrastructural preservation of in vivo grown lily pollen tubes. Protoplasma 200: 87-98

Šamaj J, Baluška F and Volkmann D (1998) Cell-specific expression of two arabinogalactan protein epitopes recognized by monoclonal antibodies JIM8 and JIM13 in maize roots. Protoplasma 204: 1-12

Šamaj J, Braun M, Baluška F, Ensikat H, Tsumuraya Y and Volkmann D (1999) Specific localization of arabinogalactan-protein epitopes at the surface of maize root hairs. Plant Cell Physiol 40: 874-883

Schneitz K, Baker SC, Gasser CS and Redweik A (1998) Pattern formation and growth during floral organogenesis: HUELLENLOS and INTEGUMENTA are required for the formation of the proximal region of the ovule primordium in *Arabidopsis thaliana*. Development 125: 2555-2563

Seagull RW, Falconer MM and Weerdenburg CA (1987) Microfilaments: Dynamic arrays in higher plant cells. J Cell Biol 104: 995-1004

Shuttlesworth GA and Mills N (1995) Section adherence to glass slides from polyester wax-embedded tissue for *in situ* hybridization. Biotechniques 18: 948-950

Sidman RL, Mottla PA and Feder N (1961) Improved polyester wax embedding for histology. Stain Techn 36: 279-284

Small J-V, Rottner K, Hahne P and Anderson KI (1999) Visualising the actin cytoskeleton. Micr Res Techn 47: 3-17

Sonobe S and Shibaoka H (1989) Cortical fine actin filaments in higher plant cells visualized by rhodamine phalloidin after pretreatment with n-maleimidobenzoyl N-hydroxysuccimide ester. Protoplasma 148: 80-86

Steedman HF (1957) A new ribboning embedding medium for histology. Nature 179: 1345

Tagliaferro P, Tandler CJ, Ramos AJ, Saavedra JP and Brusco A (1997) Immunofluorescence and glutaraldehyde fixation. A new procedure based on the Schiff-quenching method. J Neurosci Meth 77: 191-197

Timmers ACJ, Auriac M-C and Truchet G (1999) Refined analysis of early symbiotic steps of the *Rhizobium-Medicago* interaction in relationship with microtubular cytoskeleton rearrangements. Development 126: 3617-3628

Traas JA, Doonan JH, Rawlins DJ, Watts J and Lloyd CW (1987) An actin network is present in the cytoplasm throughout the cell cycle of carrot cells and associates with the dividing nucleus. J Cell Biol 105: 387-395

Vitha S, Baluška F, Mews M and Volkmann D (1997) Immunofluorescence detection of F-actin on low melting point wax sections from plant tissues. J Histochem Cytochem 45: 89-95

Vitha S, Baluška F, Braun M, Šamaj J, Volkmann D and Barlow PW (2000) Comparison of cryofixation and aldehyde fixation for plant actin immunocytochemistry: Aldehydes do not destroy F-actin. Histochem J, in press

von Witsch M, Baluška F, Staiger CJ and Volkmann D (1998) Profilin is associated with the plasma membrane in microspores and pollen. Eur J Cell Biol 77: 303-312

Chapter 36

NON-INVASIVE F-ACTIN VISUALIZATION IN LIVING PLANT CELLS USING A GFP-MOUSE TALIN FUSION PROTEIN

Benedikt Kost[1], Pius Spielhofer[2,3], Jaideep Mathur[1], Chun-Hai Dong[1] and Nam-Hai Chua[2]

1 Laboratory of Plant Cell Biology, Institute of Molecular Agrobiology, National University of Singapore, 1 Research Link, 117604 Singapore
2 Laboratory of Plant Molecular Biology, Rockefeller University, 1230 York Avenue, New York, NY 10021–6399, USA
3 present address: Institute of Plant Physiology, Universität Bern, Altenbergrain 21, 3013 Bern, Switzerland

Key words: actin visualization, GFP, mouse talin I/LWEQ domain

Abstract: The actin cytoskeleton has essential functions in the survival, the division, the growth and the morphogenesis of plant cells, and in the development of plant organs. A key approach to investigating these functions is the observation of the organization and the dynamic behavior of filamentous actin structures in plants cells. Visualizing plant actin filaments has been notoriously difficult, because they are highly sensitive to standard fixation and cell permeabilization procedures. This chapter demonstrates that the expression in plant cells of an optimized mutant version of the green fluorescent protein (GFP) from *Aequorea victoria* fused to the well characterized F-actin binding domain of mouse talin labels the actin cytoskeleton and allows non-invasive observation of its dynamic organization by fluorescence microscopy.

1. ACTIN VISUALIZATION IN PLANT CELLS

Filamentous actin structures in plant cells are more sensitive to fixation and cell permeabilization procedures than microtubules. As a consequence, the widespread presence of actin filaments in different types of plant cells throughout the cell cycle was established only about a decade ago (for reviews see Lloyd, 1988; Staiger & Schliwa, 1987), whereas the essential

C.J. Staiger et al. (eds.), Actin: A Dynamic Framework for Multiple Plant Cell Functions, 637–659.

features of the structure of the plant microtubular cytoskeleton during interphase and mitosis have been known since the 1960s (Ledbetter & Porter, 1963; Pickett-Heaps & Northcote, 1966). Although actin filaments in plant cells have been detected earlier in a few studies based on electron microscopy and decoration with heavy meromyosin (e.g. Palevitz et al., 1974), systematic analysis of F-actin organization in plant cells began only in the 1980s after methods have been developed that allow the visualization of plant actin filaments using fluorescent phallotoxins or antibodies (Kakimoto & Shibaoka, 1987a; McCurdy et al., 1988; Palevitz, 1987; Parthasarathy et al., 1985; Seagull et al., 1987; Traas et al., 1987). Since then, fluorescence microscopy has contributed the bulk of the current knowledge on the structure of the plant actin cytoskeleton. Electron microscopy was employed occasionally to confirm results obtained with fluorescence microscopy (e.g. Lancelle et al., 1987; Tiwari & Polito, 1988) or to analyze very small structures at maximal resolution (White et al., 1994).

Because of the simplicity of this technique, labeling with fluorescent derivatives of phallotoxins has been most commonly applied to observe actin structures in plant cells. Phallotoxins are heptapeptides that are produced by the poisonous mushroom, *Amanita phalloides*. They bind to F-actin much more tightly than to G-actin and stabilize actin filaments by reducing the threshold G-actin concentration required for actin polymerization (Andersland & Parthasarathy, 1992; Cooper, 1987). Incubation with fluorescent FITC or rhodamine conjugates of phalloidin, or other phallotoxins at low concentrations, can be employed to visualize filamentous actin structures in animal (Wulf et al., 1979) and in plant cells (Tewinkel et al., 1989). This technique has allowed numerous important observations, although it is not free of caveats. Phallotoxin conjugates may have variable affinities to the different actin isoforms known to be present in plant cells (Meagher et al., 1999) and may therefore not label the entire actin cytoskeleton in certain cell types. Also, phallotoxin conjugates can potentially induce artificial stabilization and elongation of actin filaments. To avoid these problems, indirect immunofluorescence techniques using antibodies against animal or plant actin isoforms have been employed to visualize plant actin structures in a few studies (Baluška et al., 1997; McCurdy et al., 1988; Miller et al., 1999). Direct comparison of immunolabeling with phallotoxin staining has indicated that the former technique can produce more reliable and qualitatively better results (Cho & Wick, 1991; Tang et al., 1989).

F-actin staining using both phallotoxin conjugates and anti-actin antibodies generally works well with isolated cells but is more difficult to apply to intact plant tissues. Plant cell walls hinder the passage of any molecular probe and are impenetrable for bulky immunoglobulins. Two principal approaches have been developed to permeabilize plant cell walls in preparation for actin labeling: enzymatic degradation and physical removal. Treatment with cell wall degrading enzymes often disrupts tissue organization and is likely to affect the organization especially of the cortical cytoskeleton. Better results can be obtained when the cell interior is exposed to actin probes by physical sectioning after embedding (Blancaflor & Hasenstein, 1997, this volume). However, to avoid problems with insufficient preservation of actin structures and antigenicity, the embedding and sectioning procedures have to be very carefully optimized (Vitha et al., 1997, this volume). Alternatively, "freeze-shattering" can be employed to make the interior of plant cells accessible for actin probes. In this process, cells that have been rapidly frozen in liquid nitrogen are fractured under pressure to remove cell walls. Good preservation of cellular structure and actin antigenicity has been achieved using this relatively simple technique (Wasteneys et al., 1997).

Staining of actin filaments in plant cells with either phallotoxin conjugates or antibodies generally requires fixation and/or membrane permeabilization. The plant actin cytoskeleton has been reported to be very sensitive to classical osmium tetroxide (Maupin-Szamier & Pollard, 1978; Staiger & Schliwa, 1987) and aldehyde (Lehrer, 1981; Parthasarathy et al., 1985) fixation procedures. Pre-treatments with lysine (Kakimoto & Shibaoka, 1987a), tropomyosin (Kakimoto & Shibaoka, 1987b) or MBS (m-maleimidobenzoyl N-hydroxysuccinimide ester; Sonobe & Shibaoka, 1989) have been shown to help the preservation of actin filaments during fixation in certain types of plant cells, but may affect actin organization. Completely avoiding fixation and directly introducing phallotoxin conjugates into cells after membrane permeabilization by electroporation or by extraction with DMSO or a detergent, has been reported to result in improved actin staining in different types of plant cells (Heslop-Harrison & Heslop-Harrison, 1991; Pierson, 1988; Traas et al., 1987). However, the risk of artificial actin polymerization induced by phallotoxin conjugates appears to be particularly high in unfixed cells. Physical fixation by rapid freeze-substitution has been described to result in excellent preservation of the entire structure of plant cells, including the actin cytoskeleton, for both electron and fluorescence microscopy (Baskin et al., 1996; Lancelle et al., 1986). Unfortunately, this technique is labor-intensive and has been employed only occasionally to study plant actin structures (e.g. Emons & Derksen, 1986; Lancelle & Hepler, 1992).

With the introduction of techniques that allow microinjection of phallotoxin conjugates, it became possible to visualize actin organization in living plant cells (Schmit & Lambert, 1990; Zhang et al., 1993). Actin structures observed in living cells that show normal morphology and behavior are likely to be unaffected by the visualization procedure. Also, in vivo actin staining methods greatly facilitate the analysis of dynamic changes in actin organization during cellular processes. Microinjection of phallotoxin conjugates has provided new insights into actin organization and dynamics in plant cells during cell division (Cleary, 1995; Endlé et al., 1998; Valster & Hepler, 1997), stomatal movement (Kim et al., 1995), tip growth (Miller et al., 1996) and root nodulation (Cardénas et al., 1998). However, to prevent effects on actin organization in living cells, it is crucial to inject only minute amounts of phallotoxin conjugates which may not be sufficient for the visualization of certain actin structures (Valster & Hepler, 1997). Furthermore, microinjection is technically demanding, labor-intensive and its application is restricted to certain types of plant cells.

We have found recently that transient or stable expression of the green fluorescent protein (GFP) fused to the F-actin binding domain of mouse talin can be used to non-invasively visualize filamentous actin structures in a range of different types of living plant cells and tissues (Figs. 1-8; Kost et al., 1998). This technique avoids many of the difficulties that are associated with the use of traditional methods for actin visualization. It is expected to allow easy and reliable *in vivo* observation of actin organization in all plant cells and tissues that are amenable to genetic transformation.

2. GREEN FLUORESCENT PROTEIN (GFP) FUSION PROTEINS AS IN VIVO MARKERS FOR CELLULAR STRUCTURES AND F-ACTIN

The green fluorescent protein (GFP) of the jellyfish, *Aequorea victoria*, is strongly fluorescent when expressed in cells of numerous organisms including plants (Leffel et al., 1997; Misteli & Spector, 1997). The GFP fluorophore appears to form spontaneously by a post-translational cyclation of the tripeptide, Ser65, Tyr66 and Gly67, which requires no exogenous co-factors other than oxygen (Prendergast, 1999).

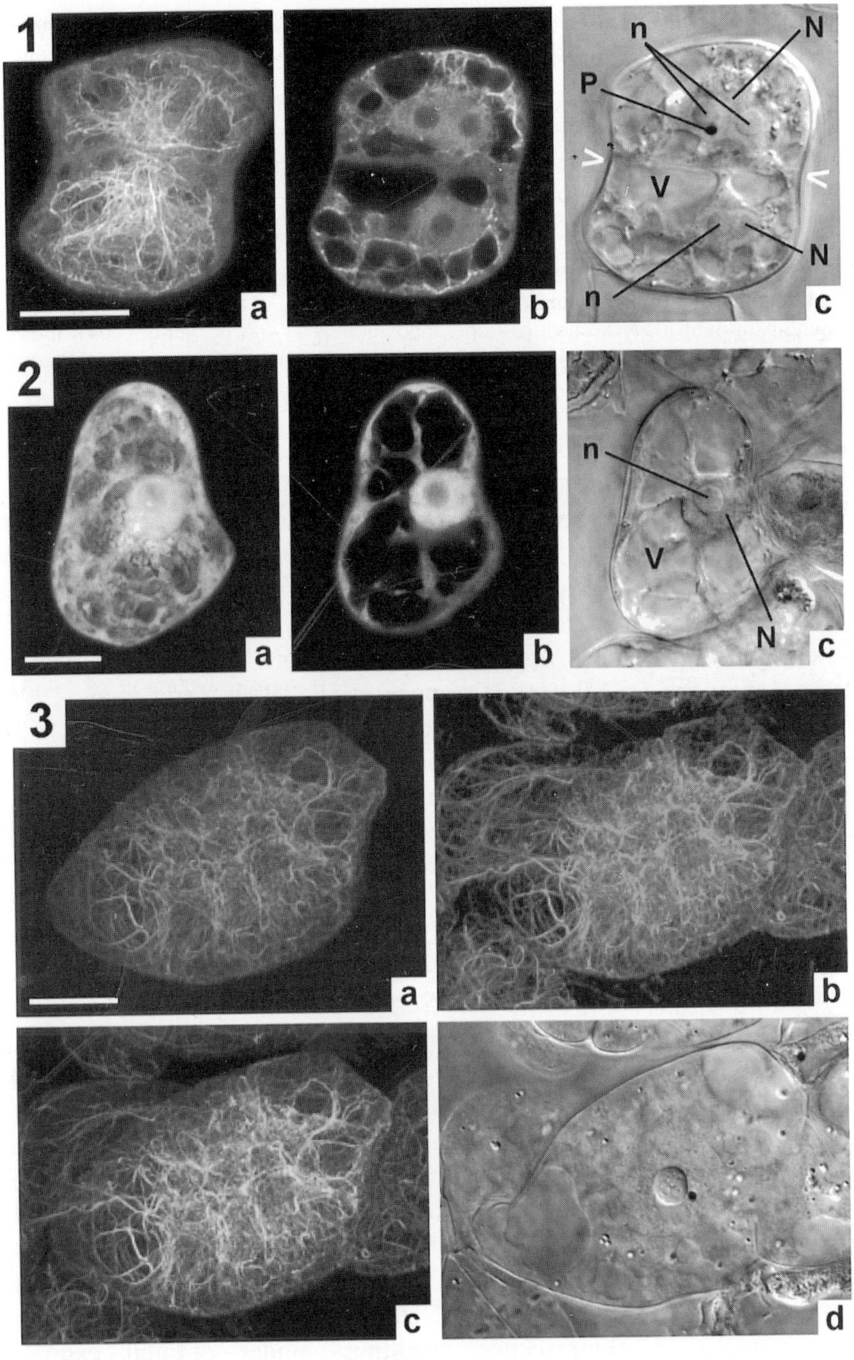

Figures 1-3. Figure 1. Actin labeling by GFP-mTn in a pair of living tobacco BY-2 suspension cells derived from a recent cell division. The *GFP-mTn* sequence was transiently

expressed in these cells under the control of the CaMV 35S promoter following gene transfer by particle bombardment. a) Projection of serial confocal optical sections. b) Single confocal optical section through the nuclei. c) Nomarski optics transmitted light reference image. N: nuclei, n: nucleoli, p: gold particle, V: vacuole, arrow heads: newly formed cell wall. Scale bar: 20 µm. *Figure 2.* Distribution of untagged GFP in a living tobacco BY-2 suspension cell. The *GFP* sequence was transiently expressed in this cell under the control of the CaMV 35S promoter following gene transfer by particle bombardment. a) Projection of serial confocal optical sections. b) Single confocal optical section through the nucleus. c) Nomarski optics transmitted light reference image. N: nucleus, n: nucleolus, V: vacuole. Scale bar: 20 µm. *Figure 3.* Co-labeling of F-actin in a permeabilized tobacco BY-2 suspension cell by transiently expressed *GFP-mTn* and by rhodamine-phalloidin. a) GFP-mTn fluorescence. b) Rhodamine-phalloidin fluorescence. c) Superimposition of a) and b): yellow color indicates overlapping staining patterns. d) Nomarski optics transmitted light reference image. Scale bar: 20 µm.

Expression of GFP at easily detectable, high levels is generally tolerated very well and does not affect cell viability or physiology. Since the cloning of a *GFP* cDNA by Prasher et al. (1992), GFP has been widely used as a non-invasive marker for gene expression, protein localization and intracellular structures (Tsien, 1998). The wild-type *GFP* cDNA has been mutagenized extensively over the past few years to enhance its usefulness as a marker gene. The most important improvements achieved to date are implemented in the *EGFP* gene (Cormack et al., 1996; Yang et al., 1996b) which is commercially available (Clontech, Palo Alto, CA, USA). The *EGFP* sequence contains a large number of silent mutations which increase the efficiency of mRNA translation in mammalian and in plants cells by adjusting codon usage (Yang et al., 1996b) and by eliminating cryptic splice sites which interfere with the expression of the wild-type *GFP* cDNA in certain types of plant cells (Haseloff et al., 1997). In addition, non-silent point mutations have been introduced, which affect the formation and the fluorescence characteristics of the EGFP fluorophore. The two most relevant of these mutations are Phe64Leu and Ser65Thr. The Phe64Leu substitution facilitates fluorophore formation at higher temperatures (37°C) and results in elevated levels of functional protein in cells that are kept under these conditions (Cormack et al., 1996; Palm et al., 1997; Patterson et al., 1997). Replacing Ser at position 65 by Thr shifts the main absorption peak of GFP from 395 nm to 490 nm (Heim et al., 1995). In contrast to wild-type GFP, which absorbs maximally in the UV range, EGFP can therefore be excited optimally using visible (blue) light. This greatly reduces problems with phototoxicity when GFP expression is analyzed in living cells. Other advantages that arise from the Ser65Thr mutation include: 1) an about sixfold increase in fluorescence brightness under optimal excitation conditions, 2) dramatically reduced photobleaching, 3) faster fluorophore formation and 4) the possibility of using common FITC (fluorescein isothiocyanate) filter sets to observe GFP expression (Cubitt et al., 1995). In many different types of cells, including plant cells, the presence and exact localization of EGFP can be conveniently and non-destructively monitored *in vivo* using standard fluorescence microscopy equipment.

GFP consists of 238 amino acids, which form a barrel shaped, compact protein that encloses the fluorophore in its center (Ormo et al., 1996; Yang et al., 1996a). Fluorophore formation and fluorescence characteristics are not affected when GFP is expressed as a fusion protein attached to the N- or the C-terminus of various different polypeptides. In many cases, proteins tagged with GFP have been shown to remain functional (Cubitt et al., 1995; Tsien, 1998). Expression of sequences that encode translational fusions of GFP to full length proteins, or protein domains, that are known to be targeted to certain cellular structures has become a standard approach to non-invasively visualize such structures and their dynamic behavior in living cells (Tsien, 1998). GFP-based markers have been developed that allow in vivo imaging in plant cells of organelles such as nuclei (Grebenok et al., 1997), mitochondria (Köhler et al., 1997b), plastids (Köhler et al., 1997a), endoplasmic reticulum (Haseloff et al., 1997), Golgi stacks (Boevink et al., 1998; Kost & Chua, unpublished observations), vacuoles (Di Sansebastiano et al., 1998) and peroxisomes (Jedd & Chua, unpublished observations). In addition, such markers were also established for microtubules (Marc et al., 1998) and actin filaments (Kost et al., 1998), which are the two key elements of the plant cytoskeleton.

Expression of GFP fused to the N-terminus or, in a few cases, to the C-terminus of different actin isoforms, has been successfully employed to visualize filamentous actin structures in living cells of *Dictyostelium discoideum* (Neujahr et al., 1997; Westphal et al., 1997), *Drosophila melanogaster* (Verkhusha et al., 1999), *Saccharomyces cerevisiae* (Doyle & Botstein, 1996), chicken (Small et al., 1999) and a number of mammalian organisms (Ballestrem et al., 1998; Fischer et al., 1998; Heinzen et al., 1999; Ludin & Matus, 1998; Small et al., 1999). However, actin carrying a directly attached fluorescent tag (GFP or conjugated fluorochromes) labels not only actin filaments but also pools of monomeric actin (Westphal et al., 1997; Yumura & Fukui, 1998) which generally contain a large proportion (around 50%) of the total cellular actin protein (Schafer & Cooper, 1995; Theriot, 1994). Doyle and Botstein (1996) have reported that only a subpopulation of the filamentous actin structures in yeast cells were stained by an actin-GFP fusion protein, which indicates that such fusion proteins may not be readily integrated into certain types of actin filaments. Furthermore, Aizawa et al. (1997) have found that GFP-actin fusion proteins can severely affect cell morphology, cell motility and cytokinesis when expressed in *D. discoideum*.

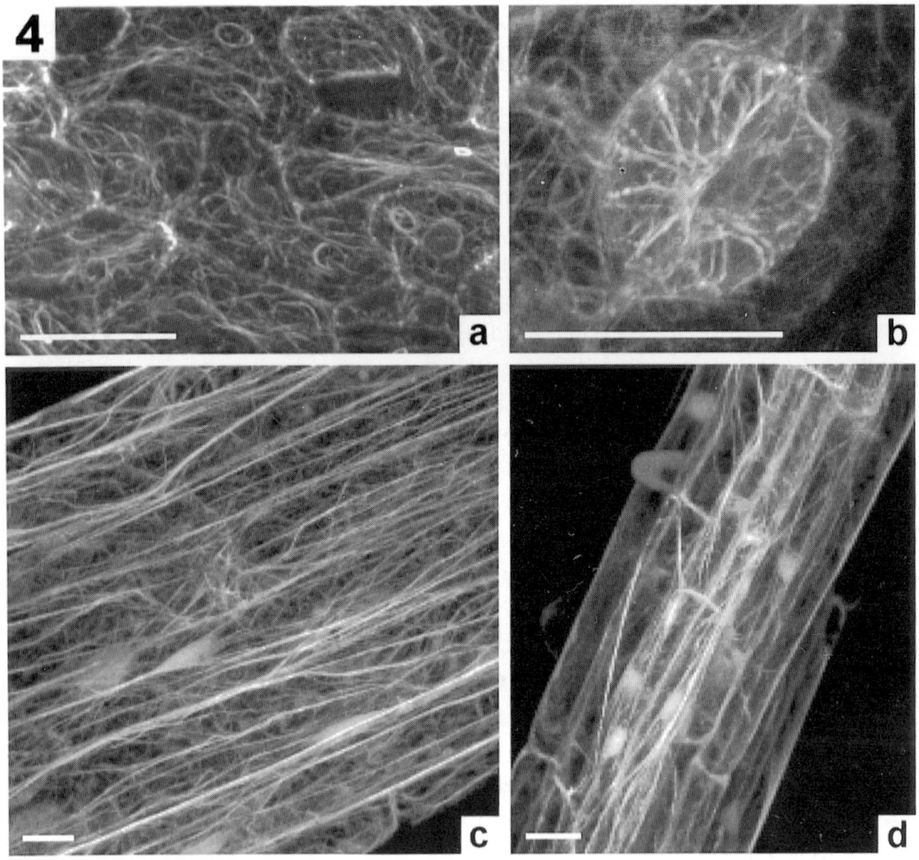

Figure 4. Actin labeling by GFP-mTn in living cells of different tissues of transgenic *A. thaliana* plants. The *GFP-mTn* sequence was expressed in these cells under the control of the CaMV 35S promoter following *Agrobacterium*-mediated stable transformation. Projections of serial confocal optical sections are shown. a) Epidermal cotyledon cells. b) Stomatal complex in the cotyledon epidermis. c) Cortical and epidermal hypocotyl cells. d) Cortical and epidermal root cells. Scale bars: 20 μm.

A few reports describe the use of GFP fused to full length or truncated actin binding proteins as markers for filamentous actin structures. GFP attached to full length Sac6p (a fimbrin homologue) or Abp1p (actin binding protein-1) was found to label the same structures as a GFP-actin fusion protein in yeast (Doyle & Botstein, 1996). Also, F-actin dynamics during the protrusion of lamellipodia of *D. discoideum* was investigated using cells that expressed GFP fused to full length coronin (Fukui *et al.*, 1999). In the same organism, the 240 amino acid actin binding domain of Abp-120 (actin binding protein 120/ gelation factor) tagged with GFP was found to colocalize with rhodamine-phalloidin (Pang et al., 1998). In another study, inducible expression of GFP attached to a 280-amino acid C-terminal fragment of moesin, which contains an actin binding domain, was employed to observe the accumulation of filamentous actin structures in embryos of *D. melanogaster* at different developmental stages (Edwards et al., 1997).

We have constructed the *GFP2* sequence, which is optimized for expression in plants and encodes a protein with essentially the same characteristics as EGFP (Kost et al., 1998). The 3' end of this sequence (after removal of the STOP codon) was fused in frame via a fragment that encodes a Gly-Ala-Gly-Ala-Gly-Ala-Gly-Ala-Gly-Ala-Gly (5x Gly-Ala) linker to an *Arabidopsis thaliana* actin cDNA (GenBank accession: U37281) and to a partial cDNA that codes for the C-terminal F-actin binding domain of mouse talin (amino acids 2345-2541; McCann & Craig, 1997).

The two resulting sequences were expressed in plant cells to test whether the fusion proteins they encode, GFP2-atAct and GFP2-mTn, respectively, can serve as *in vivo* markers for plant actin filaments. Stable expression of GFP2-atAct allowed the visualization of filamentous structures in tobacco BY-2 suspension cells. However, transgenic BY-2 lines expressing GFP2-atAct could not be maintained for extended periods of time, indicating that the fusion protein may interfere with cell viability or cell division (Spielhofer & Chua, unpublished observations). Also, transient expression of GFP2-atAct very severely inhibited tobacco pollen germination and pollen tube growth (Kost & Chua, unpublished observations). In contrast, we found that expression of the GFP2-mTn fusion protein can be employed to non-destructively label actin filaments in a range of different plant cell types, including tobacco BY-2 cells and pollen tubes (Figs. 1-8; Kost *et al.,* 1998).

3. THE MOUSE TALIN I/LWEQ F-ACTIN BINDING DOMAIN

Talin, a multi-domain 270 kDa protein consisting of 2541 amino acids, is a key component of focal adhesions (Burridge & Connell, 1983; DePasquale & Izzard, 1991; Nuckolls et al., 1990), which are membrane associated multi-protein complexes in animal and *D. discoideum* cells that link actin stress fibers to the extracellular matrix.

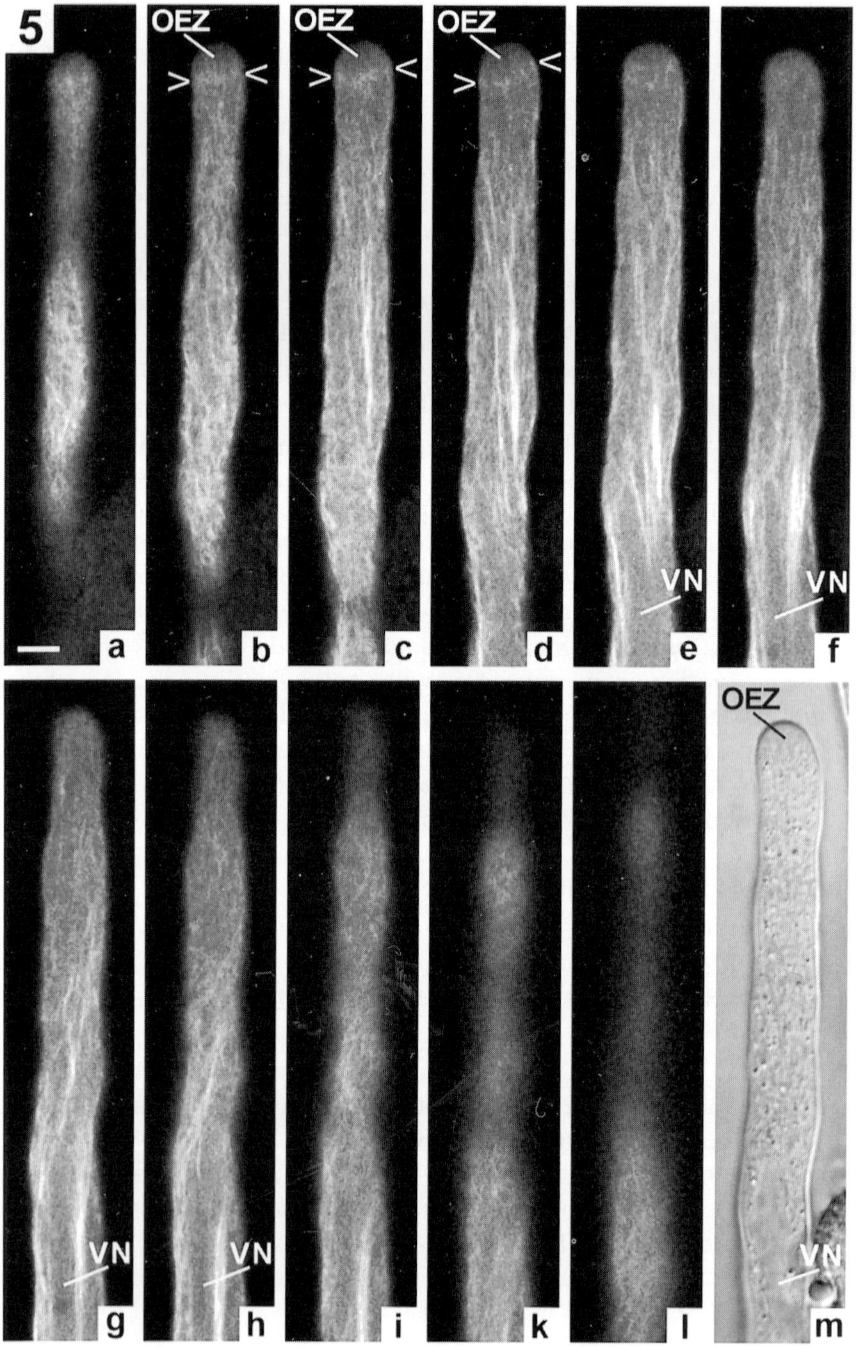

Figure 5. Actin labeling by GFP-mTn in a living, normally elongating cultured tobacco pollen tube. The *GFP-mTn* sequence was transiently expressed in this pollen tube under the control

of the LAT52 promoter following gene transfer by particle bombardment. a)-l) Serial confocal optical sections (1 μm apart from each other). m) Nomarski optics transmitted light reference image. OEZ: organelle exclusion zone; VN: vegetative nucleus (generative cell is not visible), arrow heads: actin ring marking the border of the OEZ. Scale bar: 5 μm.

Interfering with the activity of the talin protein or reducing talin gene expression compromises the ability of cells to adhere to and to spread on substrate surfaces (Albiges-Rizo et al., 1995; Niewoehner et al., 1997; Nuckolls et al., 1992; Priddle et al., 1998; Sydor et al., 1996). Numerous reports have shown that talin binds *in vitro* to G-actin (Kaufmann et al., 1991; Muguruma et al., 1990), to F-actin (Hemmings et al., 1996; McCann & Craig, 1997, 1999; Muguruma et al., 1990, Schmidt et al., 1993), to membrane phospholipids (Muguruma et al., 1995; Niggli et al., 1994) and to several proteins present in focal adhesions, including membrane-spanning extracellular matrix receptors of the integrin family (Calderwood et al., 1999; Horwitz et al., 1986; Patil et al., 1999), vinculin (Bass et al., 1999; Burridge & Mangeat, 1984) and FAK (Focal Adhesion Kinase; Chen et al., 1995). Furthermore, talin has been demonstrated to have the ability to nucleate actin polymerization (Kaufmann et al., 1991; Muguruma et al., 1995; Niggli et al., 1994) and to bundle and cross-link actin filaments (McCann & Craig, 1997; Muguruma et al., 1995; Zhang et al., 1996). All together, these findings indicate that talin may be a multi-functional protein that organizes the formation of actin stress fibers and links them to focal adhesions.

The binding sites in the talin protein that interact with each of its various ligands have not been mapped with great certainty to date. Talin can be proteolytically cleaved into a N-terminal 47 kDa head domain and a C-terminal 200 kDa tail fragment (Rees et al., 1990). Two reports have consistently found F-actin binding activity in the C-terminal 200 kDa talin fragment, but not in the N-terminal 47 kDa head domain (Muguruma et al., 1995; Niggli et al., 1994). In one study, Hemmings et al. (1996) have identified, by analyzing co-sedimentation with F-actin of a set of overlapping recombinant talin polypeptide fragments, three distinct potential talin F-actin binding sites, one in the N-terminal 47 kDa head domain and two in the C-terminal 200 kDa fragment. Whereas two of these sites do not show any sequence similarity with known F-actin interacting domains, the third one, which is located at the very C-terminal end of the talin protein, overlaps with the I/LWEQ domain that was demonstrated recently to bind actin filaments with high affinity, both *in vitro* and *in vivo* (McCann & Craig, 1997, 1999).

The I/LWEQ domain, which encompasses the last 197 amino acids at the C-terminus of mouse talin, was originally identified as a sequence module that is highly conserved among a group of actin binding proteins (McCann &

Craig, 1997) which includes different talin homologues, Hip1 and Sla2 (McCann & Craig, 1999).

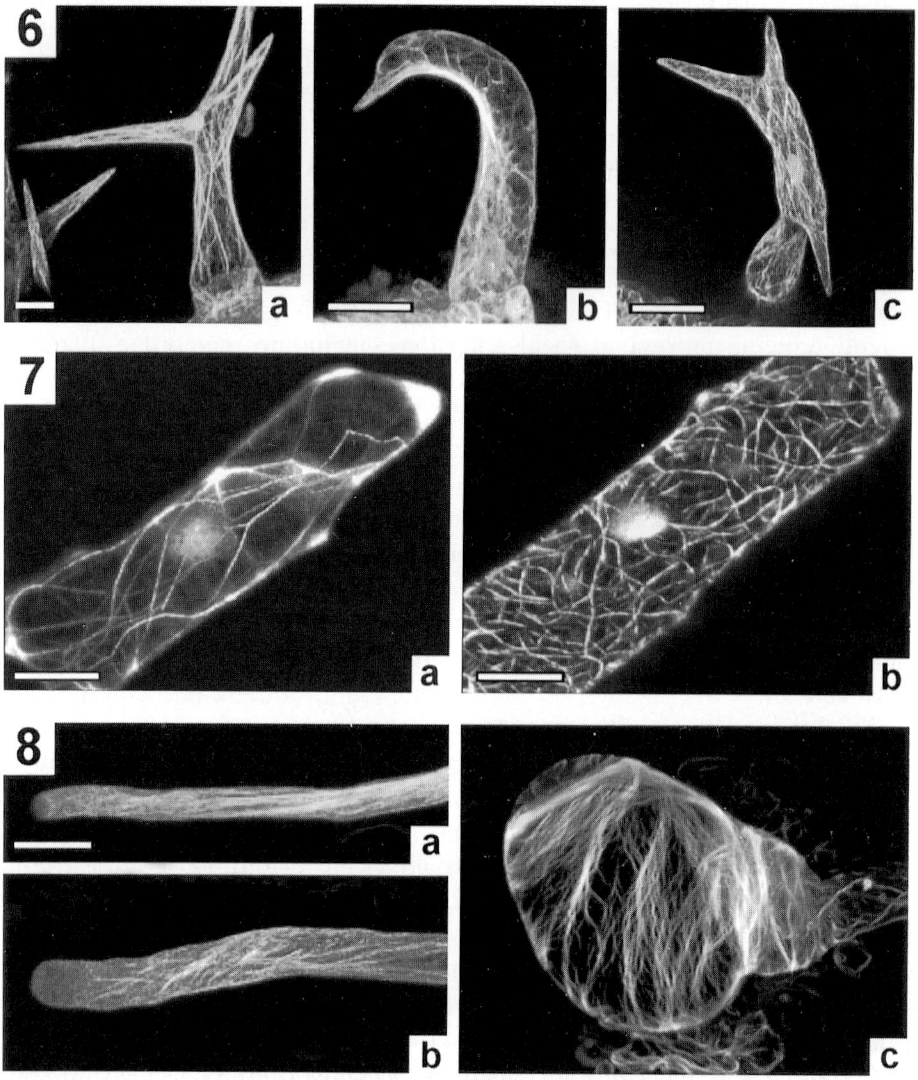

Figures 6-8. Figure 6. Actin labeling by GFP-mTn in living wild type and mutant *A. thaliana* trichomes (leaf hairs). The *GFP-mTn* sequence was expressed in the trichomes under the control of the CaMV 35S promoter. The *35S::GFP-mTn* expression cassette was introduced into wild-type plants by *Agrobacterium*-mediated stable transformation and transferred into mutant lines by crossing. Projections of serial confocal optical sections are shown. a) Wild-type trichomes. Scale bar: 100 μm. b) A trichome formed by a *wurm* mutant. Scale bar: 50 μm. c) A trichome formed by an *alien* mutant. Scale bar: 50 μm. *Figure 7.* Effects of co-

expression of an *A. thaliana* actin depolymerizing factor (ADF) on GFP-mTn labeled filamentous actin structures in living onion epidermal cells. The *GFP-mTn* and *ADF* sequences were transiently expressed in these cells under the control of the CaMV 35S promoter following gene transfer by particle bombardment. Projections of serial confocal optical sections are shown. a) A cell that expressed *GFP-mTn* only. b) A cell that co-expressed *GFP-mTn* and an *A. thaliana* ADF. Scale bars: 100 µm. *Figure 8.* Effects of co-expression of mutant forms of an *A. thaliana* Rac type small GTPase on GFP-mTn labeled filamentous actin structures in living tobacco pollen tubes. The *GFP-mTn* and *Rac* sequences were transiently expressed in these cells under the control of the LAT52 promoter following gene transfer by particle bombardment. Projections of serial confocal optical sections are shown. a) A pollen tube that expressed *GFP-mTn* only. b) A pollen tube that co-expressed *GFP-mTn* and a dominant negative mutant form of an *A. thaliana* Rac type small GTPase. c) A pollen tube that co-expressed *GFP-mTn* and a constitutive active mutant form of an *A. thaliana* Rac type small GTPase. Scale bar: 20 µm.

In co-sedimentation experiments, the isolated mouse talin I/LWEQ domain fused to GST (*GST-mTn*) was found to bind to vertebrate and yeast F-actin with a higher affinity than full length talin. At sufficiently high concentrations, *GST-mTn* was able to completely displace previously bound, full-length talin from actin filaments (McCann & Craig, 1997). This indicates that the I/LWEQ domain is responsible for all of the F-actin binding capacity of the native talin protein, and that the two additional F-actin binding sites identified by Hemmings et al. (1996) are cryptic and require some sort of exposure for function (McCann & Craig, 1997). Whereas the F-actin binding characteristics of full length talin and of *GST-mTn* have been well characterized, the interaction of these two polypeptides with G-actin requires further analysis.

McCann and Craig (1997) have clearly demonstrated that, in contrast to full length talin, *GST-mTn* does not nucleate actin polymerization. In the same study, *GST-mTn* was found to display some F-actin bundling activity, which was much weaker than the one exhibited by full length talin. Actin bundling and cross-linking are thought to require two actin binding domains that are held together in one complex. In accordance with this concept and with the hypothesis that the I/LWEQ domain constitutes the sole talin F-actin binding site, native talin has been demonstrated to interact with F-actin as a homodimer (Goldmann et al., 1994). The weak F-actin bundling activity of *GST-mTn* was therefore attributed to the known ability of GST to self-associate (McCann & Craig, 1997) and appears unlikely to be an intrinsic feature of the I/LWEQ domain itself. Supporting this assumption, a recombinant polypeptide that consisted of the C-terminal 272 talin residues and contained the I/LWEQ domain was observed to localize to filamentous actin structures, but had no discernable effects on actin organization, when microinjected into fibroblast cells (Hemmings et al., 1996).

In summary, the evidence summarized above indicates that the I/LWEQ domain can bind with a high affinity to F-actin without displaying the actin filament nucleating and cross-linking activities of the full length talin protein. With these properties, the mouse talin I/LWEQ domain appears to be an ideal tag to target attached GFP to filamentous actin structures in living cells.

4. GFP FUSED TO THE MOUSE TALIN I/LWEQ DOMAIN (GFP-MTN) LABELS PLANT ACTIN FILAMENTS *IN VIVO*

Transient or stable expression of constructs consisting of a strong promoter fused to an open reading frame that encodes GFP2 attached at the C-terminus via a 5x Gly-Ala linker to the C-terminal 197 amino acid F-actin binding domain of mouse talin (GFP-mTn) allows non-invasive visualization of the actin cytoskeleton in different types of plant cells by fluorescence microscopy (Kost et al., 1998).

Figure 1 shows the actin cytoskeleton, as observed by confocal imaging of GFP-mTn fluorescence, in a pair of tobacco BY-2 suspension cells derived from a recent cell division. These two cells transiently expressed the *GFP-mTn* sequence under control of the CaMV 35S promoter following gene transfer by particle bombardment. Unlike GFP-mTn, untagged GFP was found to be evenly dispersed throughout the cytoplasm and to slightly accumulate in the nucleus when expressed in BY-2 cells (Fig. 2), which is a typical distribution pattern for untagged GFP in plant cells (e.g. Haseloff et al., 1997). GFP-mTn labeling of filamentous actin structures in BY-2 cells resulted in similar pictures of F-actin organization as those obtained when conventional F-actin staining techniques were applied to BY-2 cells (Hasezawa et al., 1989; Katsuta et al., 1990) or to other types of plant cells (Seagull et al., 1987). When *GFP-mTn* expressing BY-2 cells were co-labeled with rhodamine-phalloidin after membrane permeabilization, perfectly overlapping staining patterns were observed with the two different F-actin probes (Fig. 3). Neither the presence of GFP-mTn at easily detectable, high levels, nor confocal imaging of GFP-mTn fluorescence, had apparent effects on the morphology or on the viability of BY-2 cells. Regularly, brightly fluorescent, *GFP-mTn* expressing BY-2 cells were found to undergo normal cell division (Fig. 1), indicating that the fusion protein did not interfere with this process, which depends strictly on a functional actin cytoskeleton (Kost et al., 1999b; Staehelin & Hepler, 1996).

Introduction of the *GFP-mTn* sequence under control of the constitutive active CaMV 35S promoter into *A. thaliana* plants by *Agrobacterium*-mediated stable transformation and allowed the visualization of F-actin organization in different cell types in a variety of tissues, including epidermal and cortical cell layers of cotyledons, hypocotyls and roots (Fig. 4). Transgenic *A. thaliana* lines that constitutively expressed *GFP-mTn* at readily detectable levels were maintained over several generations and did not show any morphological or developmental defects (Spielhofer, Jaideep Mathur & Chua, unpublished observations). This confirms that *GFP-mTn* expression can be employed to label the actin cytoskeleton in many different types of plant cells in an absolutely non-invasive manner.

Because pollen tubes are extremely sensitive to commonly used fixation and membrane permeabilization procedures (Doris & Steer, 1996; He & Wetzstein, 1995), it has a been very difficult to employ traditional F-actin visualization techniques to investigate the organization of the actin cytoskeleton in these cells. Recent results obtained using rhodamine-phalloidin microinjection and electron microscopy after physical fixation by rapid freeze-substitution, have indicated that pollen tubes do not contain a dense network of actin filaments in the apex (Miller et al., 1996), in contrast to what was described in earlier studies (reviewed in Derksen et al., 1995). We have found that transient expression of the *GFP-mTn* sequence under control of the pollen-specific LAT52 promoter (Twell et al., 1989) can be employed to visualize F-actin in living, elongating tobacco pollen tubes (Fig. 5). Supporting the observations made by Miller at al. (1996), our results confirmed that the pollen tube apex contains fine and elaborate F-actin structures, rather than a dense network of actin filaments. Very brightly fluorescent pollen tubes that expressed *GFP-mTn* at high levels often contained thick actin cables, displayed abnormal morphology and/or stopped growth prematurely (Kost & Chua, unpublished observations; Kost et al., 1998). However, more moderately fluorescent pollen tubes, with actin staining similar to that shown in Figure 5, were morphologically absolutely normal (Fig. 5m). Even after confocal imaging of GFP-mTn fluorescence, such pollen tubes elongated at average growth rates (Kost et al., 1998), demonstrating that *GFP-mTn* expression at sub-maximal levels allows non-invasive visualization of the actin cytoskeleton also in pollen tubes. Pollen tubes are not only particularly sensitive to high levels of *GFP-mTn* expression, but also to treatment with drugs that affect actin organization (Kost et al., 1998; Taylor & Hepler, 1997). This may illustrate the fundamental importance of the actin cytoskeleton for the maintenance of pollen tube morphology and growth.

We have employed expression of *GFP-mTn* not only to visualize filamentous actin structures in untreated cells that are kept under standard conditions, but also in mutant or experimentally manipulated cells. The *GFP-mTn* sequence under control of the CaMV 35S promoter was crossed into *A. thaliana* mutants that display defects in trichome (leaf hair) morphogenesis (Hülskamp et al., 1994). Visualization of GFP-mTn labeled actin filaments by confocal microscopy revealed that F-actin organization in trichomes formed by several of these mutants, including *wurm* and *alien*, was strikingly different from the one observed in wild-type trichomes (Fig. 6; Mathur et al., 1999). These observations have contributed to the identification of a crucial function of the actin cytoskeleton during the later stages of trichome morphogenesis (Mathur et al., 1999). Figures 7 and 8 show that F-actin organization in onion epidermal cells and in pollen tubes, respectively, as visualized by transient expression of *GFP-mTn*, can be dramatically altered by co-expression of actin regulatory proteins. Co-expression of an *A. thaliana* actin depolymerizing factor (ADF) transformed the long, fine and net-longitudinally orientated actin filaments, which are observed in onion epidermal cells that express only *GFP-mTn* (Fig. 7a), into a dense network of randomly arranged, thick and short actin cables (Fig. 7b; Dong & Chua, unpublished observations). Transient expression of a constitutive active mutant *A. thaliana* Rac-type small GTPase, along with GFP-mTn, was found to depolarize tobacco pollen tube growth and to induce the formation of huge balloons at the tip, which contained excessive numbers of thick actin cables that were arranged in a helical pattern (Fig. 8c; Kost et al., 1999a). In contrast, a dominant negative mutant version of the same protein, when transiently co-expressed in pollen tubes with *GFP-mTn*, reduced actin bundling (Fig. 8b) and resulted in the formation of F-actin structures that looked less organized as compared to those observed in control pollen tubes, which expressed *GFP-mTn* alone (Figs. 5 and 8a; Kost et al., 1999a).

5. CONCLUSIONS AND OUTLOOK

Visualization of filamentous actin structures by *GFP-mTn* expression and fluorescence microscopy can greatly facilitate the observation of the organization and the dynamic behavior of actin filaments in all types of plant cells that are amenable to genetic transformation. This technique appears to be non-invasive and allows imaging of the native structure of the F-actin network in living cells. Its application can be expected to contribute significant new information on the functions of the plant actin cytoskeleton during both cellular processes and plant development.

Whereas the involvement of actin filaments in cellular events such as cell division, cell elongation, tip growth, cellular morphogenesis, cell-to-cell communication, organelle movements and positioning, gravity response, pathogen response, and guard cell function has been well studied and documented (see other chapters of this volume), an integrated understanding of the role of the actin cytoskeleton in plant organogenesis and development is still missing (Kost et al., 1999b). Visualization of GFP-mTn labeled filamentous actin structures in cells within developing organs of both wild type plants and mutants that are defective in organogenesis will contribute to a better definition of this role. Two-photon microscopes, which allow the observation of fluorescent structures deep inside living tissues under conditions that minimize phototoxicity (Piston, 1999), are becoming more widely available and will help tremendously the visualization of the actin cytoskeleton in *GFP-mTn* expressing cells in developing plant organs.

Mutant versions of the GFP from *Aequorea victoria* (Ellenberg et al., 1999) and a red fluorescent protein whose gene has been cloned recently from *Discosoma* sp., an Indo-Pacific coral species (Matz et al., 1999), display clearly distinct excitation and emission characteristics, which allows simultaneous imaging of fluorescence emitted from these proteins in separate channels. It should be possible to perform double or triple labeling studies in living plant cells, which express two or three of these fluorescent proteins carrying tags that cause them to accumulate at different intracellular localizations. We are in the process of constructing sequences that encode marker proteins with different fluorescence characteristics attached to tags which target them to a variety of cellular structures including actin, microtubules and cell organelles. Using these constructs, we hope to be able to investigate interactions between actin filaments and microtubules, as well as between these cytoskeletal elements and cell organelles in living plant cells during processes such as cell division and cell elongation.

The application of tagged fluorescent proteins (e.g. GFP-mTn) and modern fluorescence microscopy to study the dynamic organization of individual intracellular structures, as well as interactions between different such structures, in living cells and tissues, is a fascinating new area of research that has started to transform common views on basic cellular mechanisms and that is expected to continue to contribute major insights into such mechanisms and their importance for the development of multicellular organisms.

6. REFERENCES

Aizawa H, Sameshima M and Yahara I (1997) A green fluorescent protein-actin fusion protein dominantly inhibits cytokinesis, cell spreading, and locomotion in *Dictyostelium*. Cell Struc Funct 22: 335-345

Albiges-Rizo C, Frachet P and Block MR (1995) Down regulation of talin alters cell adhesion and the processing of the $\alpha 5\beta 1$ integrin. J Cell Sci 108: 3317-3329

Andersland JM and Parthasarathy MV (1992) Phalloidin binds and stabilizes pea root actin. Cell Motil Cytoskeleton 22: 245-249

Ballestrem C, Wehrle-Haller B and Imhof BA (1998) Actin dynamics in living mammalian cells. J Cell Sci 111: 1649-1658

Baluška F, Vitha S, Barlow PW and Volkmann D (1997) Rearrangements of F-actin arrays in growing cells of intact maize root apex tissues: A major developmental switch occurs in the postmitotic transition region. Eur J Cell Biol 72: 113-121

Baskin TI, Miller DD, Vos JW, Wilson JE and Hepler PK (1996) Cryofixing single cells and multicellular specimens enhances structure and immunocytochemistry for light microscopy. J Microsc 182: 149-161

Bass MD, Smith BJ, Prigent SA and Critchley DR (1999) Talin contains three similar vinculin-binding sites predicted to form an amphipathic helix. Biochem J 341: 257-263

Blancaflor EB and Hasenstein KH (1997) The organization of the actin cytoskeleton in vertical and graviresponding primary roots of maize. Plant Physiol 113: 1447-1455

Boevink P, Oparka K, Santa Cruz S, Martin B, Betteridge A and Hawes C (1998) Stacks on tracks: the plant Golgi apparatus traffics on an actin/ER network. Plant J 15: 441-447

Burridge K and Connell L (1983) A new protein of adhesion plaques and ruffling membranes. J Cell Biol 97: 359-367

Burridge K and Mangeat P (1984) An interaction between vinculin and talin. Nature 308: 744-746

Calderwood DA, Zent R, Grant R, Rees DJG, Hynes RO and Ginsberg MH (1999) The talin head domain binds to integrin β subunit cytoplasmic tails and regulates integrin activation. J Biol Chem 274: 28071-28074

Cardenas L, Vidali L, Dominguez J, Perez H, Sanchez F, Hepler PK and Quinto C (1998) Rearrangement of actin microfilaments in plant root hairs responding to Rhizobium etli nodulation signals. Plant Physiol 116: 871-877

Chen H-C, Appeddu PA, Parsons JT, Hildebrand JD, Schaller MD and Guan J-L (1995) Interaction of focal adhesion kinase with cytoskeletal protein talin. J Biol Chem 270: 16995-16999

Cho S-O and Wick SM (1991) Actin in the developing stomatal complex of winter rye: A comparison of actin antibodies and rh-phalloidin labeling of control and CB-treated tissues. Cell Motil Cytoskelet 19: 25-36

Cleary AL (1995) F-actin redistributions at the division site in living Tradescantia stomatal complexes as revealed by microinjection of rhodamine-phalloidin. Protoplasma 185: 152-165

Cooper JA (1987) Effects of cytochalasin and phalloidin on actin. J Cell Biol 105: 1473-1478

Cormack BP, Valdivia RH and Falkow S (1996) FACS-optimized mutants of the green fluorescent protein (GFP). Gene 173: 33-38

Cubitt AB, Heim R, Adams SR, Boyd AE, Gross LA and Tsien RY (1995) Understanding, improving and using green fluorescent proteins. Trends Biochem Sci 20: 448-455

DePasquale JA and Izzard CS (1991) Accumulation of talin in nodes at the edge of the lamellipodium and separate incorporation into adhesion plaques at focal contacts in fibroblasts. J Cell Biol 113: 1351-1359

Derksen J, Rutten T, Van Amstel T, de Win A, Doris F and Steer M (1995) Regulation of pollen tube growth. Acta Bot Neerl 44: 93-119

Di Sansebastiano G-P, Paris N, Marc-Martin S and Neuhaus J-M (1998) Specific accumulation of GFP in a non-acidic vacuolar compartment via a C-terminal propeptide-mediated sorting pathway. Plant J 15: 449-457

Doris FP and Steer MW (1996) Effects of fixatives and permeabilisation buffers on pollen tubes: Implications for localisation of actin microfilaments using phalloidin staining. Protoplasma 195: 25-36

Doyle T and Botstein D (1996) Movement of yeast cortical actin cytoskeleton visualized *in vivo*. Proc Natl Acad Sci USA 93: 3886-3891

Edwards KA, Demsky M, Montague RA, Weymouth N and Kiehart DP (1997) GFP-moesin illuminates actin cytoskeleton dynamics in living tissue and demonstrates cell shape changes during morphogenesis in *Drosophila*. Dev Biol 191: 103-117

Ellenberg J, Lippincott-Schwartz J and Presley JF (1999) Dual-colour imaging with GFP variants. Trends Cell Biol 9: 52-56

Emons AMC and Derksen J (1986) Microfibrils, microtubules and microfilaments of the trichoblast of *Equisetum hyemale*. Acta Bot Neerl 35: 311-320

Endle M-C, Stoppin V, Lambert A-M and Schmit A-C (1998) The growing cell plate of higher plants is a site of both actin assembly and vinculin-like antigen recruitment. Eur J Cell Biol 77: 10-18

Fischer M, Kaech S, Knutti D and Matus A (1998) Rapid actin-based plasticity of dendric spines. Neuron 20: 847-854

Fukui Y, de Hostos E, Yumura S, Kitanishi-Yumura T and Inoue S (1999) Architectural dynamics of F-actin in eupodia suggests their role in invasive locomotion in *Dictyostelium*. Exp Cell Res 249: 33-45

Goldmann WH, Bremer A, Haener M, Aebi U and Isenberg G (1994) Native talin is a dumbbell-shaped homodimer when it interacts with actin. J Struct Biol 112: 3-10

Grebenok RJ, Lambert GM and Galbraith DW (1997) Characterization of the targeted nuclear accumulation of GFP within the cells of transgenic plants. Plant J 12: 685-696

Haseloff J, Siemering KR, Prasher DC and Hodge S (1997) Removal of a cryptic intron and subcellular localization of green fluorescent protein are required to mark transgenic Arabidopsis plants brightly. Proc Natl Acad Sci USA 94: 2122-2127

Hasezawa S, Hogetsu T and Syono K (1989) Changes of actin filaments and cellulose fibrils in elongating cells derived from tobacco protoplasts. J Plant Physiol 134: 115-119

He Y and Wetzstein HY (1995) Fixation induces differential tip morphology and immunolocalization of the cytoskeleton in pollen tubes. Physiol Plant 93: 757-763

Heim R, Cubitt AB and Tsien RY (1995) Improved GFP fluorescence. Nature 373: 663-664

Heinzen RA, Grieshaber SS, Van Kirk LS and Devin CJ (1999) Dynamics of actin-based movement by *Rickettsia rickettsii* in vero cells. Infect Immun 67: 4201-4207

Hemmings L, Rees DJG, Ohanian V, Bolton SJ, Gilmore AP, Patel B, Priddle H, Trevithick JE, Hynes RO and Critchley DR (1996) Talin contains three actin-binding sites each of which is adjacent to a vinculin-binding site. J Cell Sci 109: 2715-2726

Heslop-Harrison J and Heslop-Harrison Y (1991) The actin cytoskeleton in unfixed pollen tubes following microwave-accelerated DMSO-permeabilization and TRITC-phalloidin staining. Sex Plant Reprod 4: 6-11

Horwitz A, Duggan K, Buck C, Beckerle MC and Burridge K (1986) Interaction of plasma membrane fibronectin receptor with talin - a transmembrane linkage. Nature 320: 531-533

Huelskamp M, Misera S and Jürgens G (1994) Genetic dissection of trichome cell development in *Arabidopsis*. Cell 76: 555-566

Kakimoto T and Shibaoka H (1987a) Actin filaments and microtubules in the preprophase band and phragmoplast of tobacco cells. Protoplasma 140: 151-156

Kakimoto T and Shibaoka H (1987b) A new method for preservation of actin filaments in higher plant cells. Plant Cell Physiol 28: 1581-1585

Katsuta J, Hashiguchi Y and Shibaoka H (1990) The role of the cytoskeleton in positioning of the nucleus in premitotic tobacco BY-2 cells. J Cell Sci 95: 413-422

Kaufmann S, Piekenbrock T, Goldmann WH, Baermann M and Isenberg G (1991) Talin binds to actin and promotes filament nucleation. FEBS Lett 284: 187-191

Kim M, Hepler PK, Eun S-O, Ha KS and Lee Y (1995) Actin filaments in mature guard cells are radially distributed and involved in stomatal movement. Plant Physiol 109: 1077-1084

Köhler RH, Cao J, Zipfel WR, Webb WW and Hanson MR (1997a) Exchange of protein molecules through connections between higher plant plastids. Science 276: 2039-2042

Köhler RH, Zipfel WR, Webb WW and Hanson MR (1997b) The green fluorescent protein as a marker to visualize plant mitochondria *in vivo*. Plant J 11: 613-621

Kost B, Lemichez E, Spielhofer P, Hong Y, Tolias K, Carpenter C and Chua N-H (1999a) Rac homologues and compartmentalized phosphatidylinositol 4,5-bisphosphate act in a common pathway to regulate polar pollen tube growth. J Cell Biol 145: 317-330

Kost B, Mathur J and Chua N-H (1999b) Cytoskeleton in plant development. Curr Opin Plant Biol 2: 462-470

Kost B, Spielhofer P and Chua N-H (1998) A GFP-mouse talin fusion protein labels plant actin filaments *in vivo* and visualizes the actin cytoskeleton in growing pollen tubes. Plant J 16: 393-401

Lancelle SA, Callaham DA and Hepler PK (1986) A method for rapid freeze fixation of plant cells. Protoplasma 131: 153-165

Lancelle SA, Cresti M and Hepler PK (1987) Ultrastructure of the cytoskeleton in freeze-substituted pollen tubes of *Nicotiana alata*. Protoplasma 140: 141-150

Lancelle SA and Hepler PK (1992) Ultrastructure of freeze-substituted pollen tubes of *Lilium longiflorum*. Protoplasma 167: 215-230

Ledbetter MC and Porter KR (1963) A 'microtubule' in plant cell fine structure. J Cell Biol 19: 239-250

Leffel SM, Mabon SA and Stewart CN, Jr. (1997) Applications of green fluorescent protein in plants. Biotechniques 23: 912-918

Lehrer SS (1981) Damage to actin filaments by glutaraldehyde: protection by tropomyosin. J Cell Biol 90: 459-466

Lloyd CW (1988) Actin in plants. J Cell Sci 90: 185-188

Ludin B and Matus A (1998) GFP illuminates the cytoskeleton. Trends Cell Biol 8: 72-77

Marc J, Granger CL, Brincat J, Fisher DD, Kao T-H, McCubbin AG and Cyr RJ (1998) A GFP-MAP4 reporter gene for visualizing cortical microtubule rearrangements in living epidermal cells. Plant Cell 10: 1927-1939

Mathur J, Spielhofer P, Kost B and Chua N-H (1999) The actin cytoskeleton is required to elaborate and maintain spatial patterning during trichome cell morphogenesis in *Arabidopsis thaliana*. Development 126: 5559-5568

Matz MV, Fradkov AF, Labas YA, Savitsky AP, Zaraisky AG, Markelov ML and Lukyanov SA (1999) Fluorescent proteins from nonbioluminescent *Anthozoa* species. Nature Biotechnol 17: 969-973

Maupin-Szamier P and Pollard TD (1978) Actin filament destruction by osmium tetroxide. J Cell Biol 77: 837-852

McCann RO and Craig SW (1997) The I/LWEQ module: A conserved sequence that signifies F-actin binding in functionally diverse proteins from yeast to mammals. Proc Natl Acad Sci USA 94: 5679-5684

McCann RO and Craig SW (1999) Functional genomic analysis reveals the utility of the I/LWEQ module as a predictor of protein:actin interaction. Biochem Biophys Res Commun 266: 135-140

McCurdy DW, Sammut M and Gunning BES (1988) Immunofluorescent visualization of arrays of transverse cortical actin microfilaments in wheat root-tip cells. Protoplasma 147: 204-206

Meagher RB, McKinney EC and Kandasamy MK (1999) Isovariant dynamics expand and buffer the responses of complex systems: The diverse plant actin gene family. Plant Cell 11: 995-1005

Miller DD, de Ruijter NCA, Bisseling T and Emons AMC (1999) The role of actin in root hair morphogenesis: Studies with lipochito-oligosaccharide as a growth stimulator and cytochalasin as an actin perturbing drug. Plant J 17: 141-154

Miller DD, Lancelle SA and Hepler PK (1996) Actin microfilaments do not form a dense meshwork in Lilium longiflorum pollen tube tips. Protoplasma 195: 123-132

Misteli T and Spector DL (1997) Applications of the green fluorescent protein in cell biology and biotechnology. Nat Biotechnol 15: 961-964

Muguruma M, Matsumura S and Fukazawa T (1990) Direct interactions between talin and actin. Biochem Biophys Res Commun 171: 1217-1223

Muguruma M, Nishimuta S, Tomisaka Y, Ito T and Matsumura S (1995) Organization of the functional domains in membrane cytoskeletal protein talin. J Biochem 117: 1036-1042

Neujahr R, Heizer C, Albrecht R, Ecke M, Schwartz J-M, Weber I and Gerisch G (1997) Three-dimensional patterns and redistribution of myosin II and actin in mitotic Dictyostelium cells. J Cell Biol 139: 1793-1804

Niewoehner J, Weber I, Maniak M, Mueller-Taubenberger A and Gerisch G (1997) Talin-null cells of *Dictyostelium* are strongly defective in adhesion to particle and substrate surfaces and slightly impaired in cytokinesis. J Cell Biol 138: 349-361

Niggli V, Kaufmann S, Goldmann WH, Weber T and Isenberg G (1994) Identification of functional domains in the cytoskeletal protein talin. Eur J Biochem 224: 951-957

Nuckolls GH, Romer LH and Burridge K (1992) Microinjection of antibodies against talin inhibits the spreading and migration of fibroblasts. J Cell Sci 102: 753-762

Nuckolls GH, Turner CE and Burridge K (1990) Functional studies of the domains of talin. J Cell Biol 110: 1635-1644

Ormo M, Cubitt AB, Kallio K, Gross LA, Tsien RY and Remington SJ (1996) Crystal structure of the Aequorea victoria green fluorescent protein. Science 273: 1392-1395

Palevitz BA (1987) Actin in the preprophase of *Allium cepa.* J Cell Biol 104: 1515-1519

Palevitz BA, Ash JF and Hepler PK (1974) Actin in the green alga, *Nitella.* Proc Natl Acad Sci USA 71: 363-366

Palm GJ, Zdanov A, Gaitanaris GA, Stauber R, Pavlakis GN and Wlodawer A (1997) The structural basis for spectral variations in green fluorescent protein. Nature Struct Biol 4: 361-365

Pang KM, Lee E and Knecht DA (1998) Use of a fusion protein between GFP and an actin-binding domain to visualize transient filamentous-actin structures. Curr Biol 8: 405-408

Parthasarathy MV, Perdue TD, Witztum A and Alvernaz J (1985) Actin network as a normal component of the cytoskeleton in many vascular plant cells. Amer J Bot 72: 1318-1323

Patil S, Jedsadayanmata A, Wencel-Drake JD, Wang W, Knezevic I and Lam SC-T (1999) Identification of a talin-binding site in the integrin β_3 subunit distinct from the NPLY regulatory motif of post-ligand binding functions. J Biol Chem 274: 28575-28583

Patterson GH, Knobel SM, Sharif WD, Kain SR and Piston DW (1997) Use of the green fluorescent protein and its mutants in quantitative fluorescence microscopy. Biophys J 73: 2782-2790

Pickett-Heaps JD and Northcote DH (1966) Organization of microtubules and endoplasmatic reticulum during mitosis and cytokinesis in wheat meristems. J Cell Sci 1: 109-120

Pierson ES (1988) Rhodamine-phalloidin staining of F-actin in pollen after dimethylsulphoxide permeabilization. Sex Plant Reprod 1: 83-87

Piston DW (1999) Imaging living cells and tissues by two-photon excitation microscopy. Trends Cell Biol 9: 66-69

Prasher DC, Eckenrode VK, Ward WW, Prendergast FG and Cormier MJ (1992) Primary structure of the *Aequorea victoria* green-fluorescent protein. Gene 111: 229-233

Prendergast FG (1999) Biophysics of the green fluorescent protein. Methods Cell Biol 58: 1-18

Priddle H, Hemmings L, Monkley S, Woods A, Patel B, Sutton D, Dunn GA, Zicha D and Critchley DR (1998) Disruption of the talin gene compromises focal adhesion assembly in undifferentiated but not differentiated embryonic stem cells. J Cell Biol 142: 1121-1133

Rees DJG, Ades SE, Singer SJ and Hynes RO (1990) Sequence and domain structure of talin. Nature 347: 685-689

Schafer DA and Cooper JA (1995) Control of actin assembly at filament ends. Annu Rev Cell Dev Biol 11: 497-518

Schmidt JM, Robson RM, Zhang J and Stromer MH (1993) The marked pH dependence of the talin-actin interaction. Biochem Biophys Res Commun 197: 660-666

Schmit A-C and Lambert A-M (1990) Microinjected fluorescent phalloidin *in vivo* reveals the F-actin dynamics and assembly in higher plant mitotic cells. Plant Cell 2: 129-138

Seagull RW, Falconer MM and Weerdenburg CA (1987) Microfilaments: Dynamic arrays in higher plant cells. J Cell Biol 104: 995-1004

Small J-V, Rottner K, Hahne P and Anderson KI (1999) Visualising the actin cytoskeleton. Microsc Res Tech 47: 3-17

Sonobe S and Shibaoka H (1989) Cortical fine actin filaments in higher plant cells visualized by rhodamine-phalloidin after pretreatment with m-maleimidobenzoyl N-hydroxy-succinimide ester. Protoplasma 148: 80-86

Staehelin LA and Hepler PK (1996) Cytokinesis in higher plants. Cell 84: 821-824

Staiger CJ and Schliwa M (1987) Actin localization and function in higher plants. Protoplasma 141: 1-12

Sydor AM, Su AL, Wang F-S, Xu A and Jay DG (1996) Talin and vinculin play distinct roles in filopodial motility in the neuronal growth cone. J Cell Biol 134: 1197-1207

Tang XJ, Lancelle SA and Hepler PK (1989) Fluorescence microscopic localization of actin in pollen tubes: comparison of actin antibody and phalloidin staining. Cell Motil Cytoskel 12: 216-224

Taylor LP and Hepler PK (1997) Pollen germination and tube growth. Annu Rev Plant Physiol Plant Mol Biol 48: 461-491

Tewinkel M, Kruse S, Quader H, Volkmann D and Sievers A (1989) Visualization of actin filament pattern in plant cells without prefixation: A comparison of differently modified phallotoxins. Protoplasma 149: 178-182

Theriot JA (1994) Regulation of the actin cytoskeleton in living cells. Sem Cell Biol 5: 193-199

Tiwari SC and Polito VS (1988) Organization of the cytoskeleton in pollen tubes of *Pyrus communaris*: a study employing conventional and freeze-substitution electron microscopy, immunofluorescence and rhodamine-phalloidin. Protoplasma 147: 100-112

Traas JA, Doonan JH, Rawlins DJ, Shaw PJ, Watts J and Lloyd CW (1987) An actin network is present in the cytoplasm throughout the cell cycle of carrot cells and associates with the dividing nucleus. J Cell Biol 105: 387-395

Tsien RY (1998) The green fluorescent protein. Annu Rev Biochem 67: 509-544

Twell D, Wing R, Yamaguchi J and McCormick S (1989) Isolation and expression of an anther-specific gene from tomato. Mol Gen Genet 217: 240-245

Valster AH and Hepler PK (1997) Caffeine inhibition of cytokinesis: effect on the phragmoplast cytoskeleton in living Tradescantia stamen hair cells. Protoplasma 196: 155-166

Verkhusha VV, Tsukita S and Oda H (1999) Actin dynamics in lamellipodia of migrating border cells in the Drosophila ovary revealed by a GFP-actin fusion protein. FEBS Lett 445: 395-401

Vitha S, Baluška F, Mews M and Volkmann D (1997) Immunofluorescence detection of F-actin on low melting point wax sections from plant tissues. J Histochem Cytochem 45: 89-95

Wasteneys GO, Willingale-Theune J and Menzel D (1997) Freeze shattering: A simple and effective method for permeabilizing higher plant cell walls. J Microsc 188: 51-61

Westphal M, Jungbluth A, Heidecker M, Muehlbauer B, Heizer C, Schwartz J-M, Marriott G and Gerisch G (1997) Microfilament dynamics during cell movement and chemotaxis monitored using a GFP-actin fusion protein. Curr Biol 7: 176-183

White RG, Badelt K, Overall RL and Vesk M (1994) Actin associated with plasmodesmata. Protoplasma 180: 169-184

Wulf E, Deboben A, Bautz FA, Faulstich H and Wieland T (1979) Fluorescent phallotoxin, a tool for the visualization of cellular actin. Proc Natl Acad Sci USA 76: 4498-4502

Yang F, Moss LG and Phillips GN, Jr. (1996a) The molecular structure of green fluorescent protein. Nature Biotechnol 14: 1246-1251

Yang TT, Cheng L and Kain SR (1996b) Optimized codon usage and chromophore mutations provide enhanced sensitivity with the green fluorescent protein. Nucleic Acids Res 24: 4592-4593

Yumura S and Fukui Y (1998) Spatiotemporal dynamics of actin concentration during cytokinesis and locomotion in *Dictyostelium*. J Cell Sci 111: 2097-2108

Zhang D, Wadsworth P and Hepler PK (1993) Dynamics of microfilaments are similar, but distinct from microtubules during cytokinesis in living, dividing plant cells. Cell Motil Cytoskel 24: 151-155

Zhang J, Robson RM, Schmidt JM and Stromer MH (1996) Talin can crosslink actin filaments into both networks and bundles. Biochem Biophys Res Commun 218: 530-537

Contributors

Name	*E-mail*	*Page*
Abe S	abe@mcb.agr.ehime-u.ac.jp	129
Allen NS	nina_allen@ncsu.edu	145
Baluška F	baluska@uni-bonn.de	191, 457, 557, 619
Barlow PW	peter.barlow@bbsrc.ac.uk	457, 587, 619
Blackman LM	leila@bio.usyd.edu.au	497
Blancaflor EB	eblancaflor@noble.org	601
Braun M	mbraun@uni-bonn.de	237
Chaffey N	nigel.chaffey@bbsrc.ac.uk	587
Chua N-H	chua@rockvax.rockefeller.edu	637
Cleary A	cleary@rsbsgpo.anu.edu.au	411
Clore A	amy_clore@ncsu.edu	129
Collings DA	collings@bilbo.bio.purdue.edu	145
Davies E	edavies@unity.ncsu.edu	129
Dong C-H		637
Emons AMC	annemie.emons@algem.pcm.wau.nl	347, 373
Eun S-O	seun@pop.postech.ac.kr	427
Foissner I	ilse.foissner@sbg.ac.at	259
Franklin-Tong V	v.e.franklin-tong@bham.ac.uk	347

Name	*E-mail*	*Page*
Geitmann A	anja.geitmann@guest.pcm.wau.nl	347
Gibbon BC	bgibbon@ag.arizona.edu	45
Grolig F	grolig@mailer.uni-marburg.de	165
Hasenstein KH	hasenstein@louisiana.edu	601
Heath B	brent@yorku.ca	275
Hepler P	hepler@marlin.bio.umass.edu	323
Huber SC	steve_huber@ncsu.edu	119
Hwang J-U	thecute@postech.ac.kr	427
Jasik J	jasik@mpiz-koeln.mpg.de	619
Kandasamy M	kandu@arches.uga.edu	3
Kendrick-Jones J	jkj@mrc-lmb.cam.ac.uk	29
Kobayashi I	issei@bio.mie-u.ac.jp	573
Kobayashi Y		573
Kost B	benedikt@ima.org.sg	637
Kovar DR	dkovar@bilbo.bio.purdue.edu	67
Larkins B	larkins@ag.arizona.edu	129
Lee Y	ylee@pop.postech.ac.kr	427
Li H		301
Lichtscheidl IK	irene@pflaphy.pph.univie.ac.at	191
Lütz-Meindl U	ursula.meindl@sbg.ac.at	213
Mathur J	jaideep@ima.org.sg	637
McCurdy D	bidwm@cc.newcastle.edu.au	87
McKinney E		3
McLean G	gmclean@intranet.reeusda.gov	517
Meagher RB	meagher@arches.uga.edu	3
Menzel D	dmenzel@uni-bonn.de	213
Muday G	muday@pop.wfu.edu	541

Name	***E-mail***	***Page***
Nick P	pnick@uni-freiburg.de	477
Overall RL	roverall@postbox.usyd.edu.au	497
Pierson ES	epierson@sci.kun.nl	165
Radford JS	janine.radford@sci.monash.edu.au	497
Reichelt S	s_reichelt@hotmail.com	29
de Ruijter N	norbert.deruijter@algem.pcm.wag-ur.nl	373
Schmit A-C	anne-catherine.schmit@ibmp-ulp.u-strasbg.fr	437
Shimmen T	shimmen@sci.himeji-tech.ac.jp	103
Snowman BN	bns838@bcssrv3.bham.ac.uk	347
Spielhofer P		637
Staiger CJ	cstaiger@bilbo.bio.purdue.edu	45, 67, 87
Stanković B	bstankovic@facstaff.wisc.edu	129
Szymanski DB	dszyman@purdue.edu	391
Takagi S	shingot@bio.sci.osaka-u.ac.jp	203
Vidali L	lvidali@bio.umass.edu	323
Volkmann D	unb110@uni-bonn.de	457, 557, 619
Vitha S	vitha@scs.unr.edu	619
Waller F	@uni-freiburg.de	477
Wang Q-Y	yan@aktivnet.de	477
Wasteneys GO	geoffw@rsbs.anu.edu.au	237, 259
White RG	rosemary.white@pi.csiro.au	497
Winter H	winter@biologie.uni-osnabrueck.de	119
Yang Z	yang@mail.ucr.edu	301
Yokota E	yokota@sci.himeji-tech.ac.jp	103
Zambryski P	zambrysk@nature.berkeley.edu	517
Zonia L	zonia@ueb.cas.cz	361

Developments in Plant and Soil Sciences

1. J. Monteith and C. Webb (eds.): *Soil Water and Nitrogen in Mediterranean-type Environments.* 1981 ISBN 90-247-2406-6
2. J.C. Brogan (ed.): *Nitrogen Losses and Surface Run-off from Landspreading of Manures.* 1981 ISBN 90-247-2471-6
3. J.D. Bewley (ed.): *Nitrogen and Carbon Metabolism.* 1981 ISBN 90-247-2472-4
4. R. Brouwer, I. Gašparíková, J. Kolek and B.C. Loughman (eds.): *Structure and Function of Plant Roots.* 1981 ISBN 90-247-2510-0
5. Y.R. Dommergues and H.G. Diem (eds.): *Microbiology of Tropical Soils and Plant Productivity.* 1982 ISBN 90-247-2624-7
6. G.P. Robertson, R. Herrara and T. Rosswall (eds.): *Nitrogen Cycling in Ecosystems of Latin America and the Caribbean.* 1982 ISBN 90-247-2719-7
7. D. Atkinson, K.K.S. Bhat, M.P. Coutts, P.A. Mason and D.J. Read (eds.): *Tree Root Systems and Their Mycorrhizas.* 1983 ISBN 90-247-2821-5
8. M.R. Sarić and B.C. Loughman (eds.): *Genetic Aspects of Plant Nutrition.* 1983 ISBN 90-247-2822-3
9. J.R. Freney and J.R. Simpson (eds.): *Gaseous Loss of Nitrogen from Plant-Soil Systems.* 1983 ISBN 90-247-2820-7
10. United Nations Economic Commission for Europe (ed.): *Efficient Use of Fertilizers in Agriculture.* 1983 ISBN 90-247-2866-5
11. J. Tinsley and J.F. Darbyshire (eds.): *Biological Processes and Soil Fertility.* 1984 ISBN 90-247-2902-5
12. A.D.L. Akkermans, D. Baker, K. Huss-Danell and J.D. Tjepkema (eds.): Frankia *Symbioses.* 1984 ISBN 90-247-2967-X
13. W.S. Silver and E.C. Schröder (eds.): *Practical Application of* Azolla *for Rice Production.* 1984 ISBN 90-247-3068-6
14. P.G.L. Vlek (ed.): *Micronutrients in Tropical Food Crop Production.* 1985 ISBN 90-247-3085-6
15. T.P. Hignett (ed.): *Fertilizer Manual.* 1985 ISBN 90-247-3122-4
16. D. Vaughan and R.E. Malcolm (eds.): *Soil Organic Matter and Biological Activity.* 1985 ISBN 90-247-3154-2
17. D. Pasternak and A. San Pietro (eds.): *Biosalinity in Action.* Bioproduction with Saline Water. 1985 ISBN 90-247-3159-3
18. M. Lalonde, C. Camiré and J.O. Dawson (eds.): Frankia *and Actinorhizal Plants.* 1985 ISBN 90-247-3214-X
19. H. Lambers, J.J. Neeteson and I. Stulen (eds.): *Fundamental, Ecological and Agricultural Aspects of Nitrogen Metabolism in Higher Plants.* 1986 ISBN 90-247-3258-1
20. M.B. Jackson (ed.): *New Root Formation in Plants and Cuttings.* 1986 ISBN 90-247-3260-3
21. F.A. Skinner and P. Uomala (eds.): *Nitrogen Fixation with Non-Legumes* (Proceedings of the 3rd Symposium, Helsinki, 1984). 1986 ISBN 90-247-3283-2
22. A. Alexander (ed.): *Foliar Fertilization.* 1986 ISBN 90-247-3288-3
23. H.G. v.d. Meer, J.C. Ryden and G.C. Ennik (eds.): *Nitrogen Fluxes in Intensive Grassland Systems.* 1986 ISBN 90-247-3309-X
24. A.U. Mokwunye and P.L.G. Vlek (eds.): *Management of Nitrogen and Phosphorus Fertilizers in Sub-Saharan Africa.* 1986 ISBN 90-247-3312-X

Developments in Plant and Soil Sciences

25. Y. Chen and Y. Avnimelech (eds.): *The Role of Organic Matter in Modern Agriculture.* 1986 ISBN 90-247-3360-X
26. S.K. De Datta and W.H. Patrick Jr. (eds.): *Nitrogen Economy of Flooded Rice Soils.* 1986 ISBN 90-247-3361-8
27. W.H. Gabelman and B.C. Loughman (eds.): *Genetic Aspects of Plant Mineral Nutrition.* 1987 ISBN 90-247-3494-0
28. A. van Diest (ed.): *Plant and Soil: Interfaces and Interactions.* 1987 ISBN 90-247-3535-1
29. United Nations Economic Commission for Europe and FAO (eds.): *The Utilization of Secondary and Trace Elements in Agriculture.* 1987 ISBN 90-247-3546-7
30. H.G. v.d. Meer, R.J. Unwin, T.A. van Dijk and G.C. Ennik (eds.): *Animal Manure on Grassland and Fodder Crops.* Fertilizer or Waste? 1987 ISBN 90-247-3568-8
31. N.J. Barrow: *Reactions with Variable-Charge Soils.* 1987 ISBN 90-247-3589-0
32. D.P. Beck and L.A. Materon (eds.): *Nitrogen Fixation by Legumes in Mediterranean Agriculture.* 1988 ISBN 90-247-3624-2
33. R.D. Graham, R.J. Hannam and N.C. Uren (eds.): *Manganese in Soils and Plants.* 1988 ISBN 90-247-3758-3
34. J.G. Torrey and J.L. Winship (eds.): *Applications of Continuous and Steady-State Methods to Root Biology.* 1989 ISBN 0-7923-0024-6
35. F.A. Skinner, R.M. Boddey and I. Fendrik (eds.): *Nitrogen Fixation with Non-Legumes* (Proceedings of the 4th Symposium, Rio de Janeiro, 1987). 1989 ISBN 0-7923-0059-9
36. B.C. Loughman, O. Gašparíková and J. Kolek (eds.): *Structural and Functional Aspects of Transport in Roots.* 1989 ISBN 0-7923-0060-2; Pb 0-7923-0061-0
37. P. Plancquaert and R. Haggar (eds.): *Legumes in Farming Systems.* 1990 ISBN 0-7923-0134-X
38. A.E. Osman, M.M. Ibrahim and M.A. Jones (eds.): *The Role of Legumes in the Farming Systems of the Mediterranean Areas.* 1990 ISBN 0-7923-0419-5
39. M. Clarholm and L. Bergström (eds.): *Ecology of Arable Land – Perspectives and Challenges.* 1989 ISBN 0-7923-0424-1
40. J. Vos, C.D. van Loon and G.J. Bollen (eds.): *Effects of Crop Rotation on Potato Production in the Temperate Zones.* 1989 ISBN 0-7923-0495-0
41. M.L. van Beusichem (ed.): *Plant Nutrition – Physiology and Applications.* 1990 ISBN 0-7923-0740-2
42. N. El Bassam, M. Dambroth and B.C. Loughman (eds.): *Genetic Aspects of Plant Mineral Nutrition.* 1990 ISBN 0-7923-0785-2
43. Y. Chen and Y. Hadar (eds.): *Iron Nutrition and Interactions in Plants.* 1991 ISBN 0-7923-1095-0
44. J.J.R. Groot, P. de Willigen and E.L.J. Verberne (eds.): *Nitrogen Turnover in the Soil-Crop System.* 1991 ISBN 0-7923-1107-8
45. R.J. Wright, V.C. Baligar and R.P. Murrmann (eds.): *Plant-Soil Interactions at Low pH.* 1991 ISBN 0-7923-1105-1
46. J. Kolek and V. Kozinka (eds.): *Physiology of the Plant Root System.* 1992 ISBN 0-7923-1205-8
47. A.U. Mokwunye (ed.): *Alleviating Soil Fertility Constraints to Increased Crop Production in West Africa.* 1991 ISBN 0-7923-1221-X; Pb 0-7923-1222-8

Developments in Plant and Soil Sciences

48. M. Polsinelli, R. Materassi and M. Vincenzini (eds.): *Nitrogen Fixation* (Proceedings of the 5th Symposium, Florence, 1990). 1991 ISBN 0-7923-1410-7
49. J.K. Ladha, T. George and B.B. Bohlool (eds.): *Biological Nitrogen Fixation for Sustainable Agriculture.* 1992 ISBN 0-7923-1774-2
50. P.J. Randall, E. Delhaze, R.A. Richards and R. Munns (eds.): *Genetic Aspects of Plant Mineral Nutrition.* 1993 ISBN 0-7923-2118-9
51. K.S. Kumarasinghe and D.L. Eskew (eds.): *Isotopic Studies of Azolla and Nitrogen Fertilization of Rice.* 1993 ISBN 0-7923-2274-6
52. F.A. Bliss and G. Hardarson (eds.): *Enhancement of Biological Nitrogen Fixation of Common Baen in Latin America.* 1993 ISBN 0-7923-2451-X
53. M.A.C. Fragoso and M.L. van Beusichem (eds.): *Optimization of Plant Nutrition.* 1993 ISBN 0-7923-2519-2
54. N.J. Barrow (ed.) *Plant Nutrition – From Genetic Engineering to Field Practice.* 1993 ISBN 0-7923-2540-0
55. A.D. Robson (ed.): *Zinc in Soils and Plants.* 1993 ISBN 0-7923-2631-8
56. A.D. Robson, L.K. Abbott and N. Malajczuk (eds.): *Management of Mycorrhizas in Agriculture, Horticulture and Forestry.* 1994 ISBN 0-7923-2700-4
57. P.H. Graham, M.J. Sadowsky and C.P. Vance (eds.): *Symbiotic Nitrogen Fixation.* 1994 ISBN 0-7923-2781-0
58. F. Baluška, M. Čiamporová, O. Gašparíkova and P.W. Barlow (eds.): *Structure and Function of Roots.* 1995 ISBN 0-7923-2832-9
59. J. Abadía (ed.): *Iron Nutrition in Soils and Plants.* 1995 ISBN 0-7923-2900-7
60. P.S. Curtis, E.G. O'Neill, J.A. Teeri, D.R. Zak and K.S. Pregitzer (eds.): *Belowground Responses to Rising Atmospheric* CO_2. Implications for Plants, Soil Biota, and Ecosystem Processes. 1994 ISBN 0-7923-2901-5
61. P.C. Struik, W.J. Vredenberg, J.A. Renkema and J.E. Parlevliet (eds.): *Plant Production on the Threshold of a New Century.* 1994 ISBN 0-7923-2903-1
62. L.O. Nilsson, R.F. Hüttl and U.T. Johansson (eds.): *Nutrient Uptake and Cycling in Forest Ecosystems* (CEC/IUFRO Symposium, Halmstad, Sweden, 1993). 1995 ISBN 0-7923-3030-7
63. H.P. Collins, G.P. Robertson and M.J. Klug (eds.): *The Significance and Regulation of Soil Biodiversity* (International Symposium, East Lansing, 1993). 1995 ISBN 0-7923-3138-9
64. R.A. Date, N.J. Grundon, G.E. Rayment and M.E. Probert (eds.): *Plant-Soil Interactions at Low pH: Principles and Management* (3rd International Symposium, Brisbane, 1993). 1995 ISBN 0-7923-3198-2
65. J.K. Ladha and M.B. Peoples (eds.): *Management of Biological Nitrogen Fixation for the Development of More Productive and Sustainable Agricultural Systems.* (15th Congress of Soil Science, Acapulco, Mexico, 1994) 1995 ISBN 0-7923-3413-2; Pb 0-7923-3414-0
66. C. Rodriguez-Barrueco (ed.): *Fertilizers and Environment.* 1996 ISBN 0-7923-3729-8
67. B. Buerkert, B.E. Allison and M. von Oppen (eds.): *Wind Erosion in Niger.* Implications and Control Measures in a Millet-based Farming System. 1996 ISBN 0-7923-3885-5

Developments in Plant and Soil Sciences

68. O. van Cleemput, G. Hofman and A. Vermoesen (eds.): *Progress in Nitrogen Cycling.* 1996 ISBN 0-7923-3962-2
69. N. Ahmad (ed.): *Nitrogen Economy in Tropical Soils.* 1996 ISBN 0-7923-4094-9
70. M. Rahman, A. Kumar Podder, C. van Hove, Z.N. Tahmida Begum, T. Heulin and A. Hartmann (eds.): *Biological Nitrogen Fixation Associated with Rice Production.* 1996 ISBN 0-7923-4197-X
71. D. Rosen, E. Tel-Or, Y. Hadar and Y. Chen (eds.): *Modern Agriculture and the Environment.* 1997 ISBN 0-7923-4295-X
72. G.H. Elkan and R.G. Upchurch (eds.): *Current Issues in Symbiotic Nitrogen Fixation.* 1997 ISBN 0-7923-4367-0
73. H.M. Anderson, P.W. Barlow, D.T. Clarkson, M.B. Jackson and P.R. Shewry (eds.): *Plant Roots – From Cells to Systems.* 1997 ISBN 0-7923-4369-7
74. Zhu Zhao-liang, Wen Qi-xiao and J.R. Freney (eds.): *Nitrogen in Soils of China.* 1997 ISBN 0-7923-4372-7
75. J.K. Ladha, F.J. de Bruijn and K.A. Malik (eds.): *Opportunities for Biological Nitrogen Fixation in Rice and Other Non-Legumes* (2nd Working Group Meeting, Faisalabad, Pakistan, 1996). 1997 ISBN 0-7923-4514-2
76. R.W. Bell and B. Rerkasem (eds.): *Boron in Soils and Plants* (International Symposium, Chiang Mai, Thailand, 1997). 1997 ISBN 0-7923-4705-6
77. B. Dell, P.H. Brown and R. Bell (eds.): *Boron in Soils and Plants: Reviews.* 1997 ISBN 0-7923-4729-3
78. T. Ando, K. Fujita, T. Mae, H. Matsumoto, S. Mori and J. Sekiya (eds.): *Plant Nutrition for Sustainable Food Production and Environment.* 1997 ISBN 0-7923-4796-X
79. K.A. Malik, M.S. Mirza and J.K. Ladha (eds.): *Nitrogen Fixation with Non-Legumes.* 1998 ISBN 0-7923-4873-7
80. P.A. Finke, J. Bouma and M.R. Hoosbeek (eds.): *Soil and Water Quality at Different Scales.* 1998 ISBN 0-7923-4994-6
81. V. Balasubramanian, J.K. Ladha and G.L. Denning (eds.): *Resource Management in Rice Systems: Nutrients.* 1999 ISBN 0-7923-5199-1
82. J.E. Box, Jr. (ed.): *Root Demographics and Their Efficiencies in Sustainable Agriculture, Grasslands and Forest Ecosystems.* 1998 ISBN 0-7923-5230-0
83. G. Hardarson and W. Broughton (eds.): *Molecular Microbial Ecology of the Soil.* Results from an FAO/IAEA Co-ordinated Research Programme, 1992–1996. 1998 ISBN 0-7923-5252-1
84. C. Kirda, P. Moutonnet, C. Hera and D.R. Nielsen (eds.): *Crop Yield Response to Deficit Irrigation.* Report of an FAO/IAEA Co-ordinated Research Program by using Nuclear Techniques. 1999 ISBN 0-7923-5299-8
85. M.J. McLaughlin and B.R. Singh (eds.): *Cadmium in Soils and Plants.* 1999 ISBN 0-7923-5843-0
86. D. Anac and P. Martin-Prével (eds.): *Improved Crop Quality by Nutrient Management.* 2000 ISBN 0-7923-5850-3
87. A. Stokes (ed.): *The Supporting Roots of Trees and Woody Plants: Form, Function and Physiology.* 2000 ISBN 0-7923-5964-X
88. P.A. Finke, M.F.P. Bierkens and P. De Willigen: *Upscaling and Downscaling Methods for Environmental Research.* 2000 ISBN 0-7923-6339-6